LASER-INDUCED BREAKDOWN SPECTROSCOPY

LASER-INDUCED BREAKDOWN SPECTROSCOPY

Edited by

JAGDISH P. SINGH

Institute for Clean Energy Technology and Department of Physics and Astronomy, Mississippi State University, Starkville, MS, United States

SURYA N. THAKUR

Department of Physics, Banaras Hindu University, Varanasi, India

ELSEVIER

Elsevier
Radarweg 29, PO Box 211, 1000 AE Amsterdam, Netherlands
The Boulevard, Langford Lane, Kidlington, Oxford OX5 1GB, United Kingdom
50 Hampshire Street, 5th Floor, Cambridge, MA 02139, United States

Notices
Knowledge and best practice in this field are constantly changing. As new research and experience broaden our understanding, changes in research methods, professional practices, or medical treatment may become necessary.

Practitioners and researchers must always rely on their own experience and knowledge in evaluating and using any information, methods, compounds, or experiments described herein. In using such information or methods they should be mindful of their own safety and the safety of others, including parties for whom they have a professional responsibility.

To the fullest extent of the law, neither the Publisher nor the authors, contributors, or editors, assume any liability for any injury and/or damage to persons or property as a matter of products liability, negligence or otherwise, or from any use or operation of any methods, products, instructions, or ideas contained in the material herein.

Library of Congress Cataloging-in-Publication Data
A catalog record for this book is available from the Library of Congress

British Library Cataloguing-in-Publication Data
A catalogue record for this book is available from the British Library

ISBN: 978-0-12-818829-3

For information on all Elsevier publications
visit our website at https://www.elsevier.com/books-and-journals

Publisher: Susan Dennis
Acquisitions Editor: Kathryn Eryilmaz
Editorial Project Manager: Ruby Smith
Production Project Manager: Niranjan Bhaskaran
Cover Designer: Mark Rogers

Typeset by SPi Global, India

In memory of Vaidehi Thakur.

Contents

II

New LIBS techniques

6. Ultrafast and filament-LIBS

S.S. HARILAL, P.K. DIWAKAR, AND G. MILOSHEVSKY

7. Molecular laser-induced breakdown spectroscopy

CHRISTIAN G. PARIGGER, DAVID M. SURMICK, CHRISTOPHER M. HELSTERN, GHANESHWAR GAUTAM, ALEXANDER A. BOL'SHAKOV, AND RICHARD E. RUSSO

III

LIBS applications

8. Combustion applications of laser-induced breakdown spectroscopy

FANG Y. YUEH, MARKANDEY M. TRIPATHI, JAGDISH P. SINGH, BANU SEZER, AND KEMAL EFE ESELLER

9. LIBS application to liquid samples

CHET R. BHATT, CHRISTIAN L. GOUEGUEL, JINESH C. JAIN, DUSTIN L. MCINTYRE, AND JAGDISH P. SINGH

10. LIBS application to powder samples

CHET R. BHATT, HERVE K. SANGHAPI, FANG Y. YUEH, AND JAGDISH P. SINGH

11. LIBS of slurry samples

K.K. AYYALASOMAYAJULA, FANG Y. YUEH, AND JAGDISH P. SINGH

12. Defense applications

JENNIFER L. GOTTFRIED

13. Applications of LIBS in drug analysis

PRAVIN KUMAR TIWARI, PRADEEP KUMAR RAI, AND AWADHESH KUMAR RAI

14. LIBS imaging applications

V. MOTTO-ROS, S. MONCAYO, C. FABRE, AND B. BUSSER

15. Food science application

KEMAL EFE ESELLER, ISMAIL HAKKI BOYACI, BANU SEZER, GONCA BILGE, AND HALIL BERBEROGLU

16. LIBS study of geological samples coupled with chemometric methods

ABHISHEK KUMAR RAI, JAYANTA KUMAR PATI, AND AWADHESH KUMAR RAI

17. Study of the different parts of a tokamak using laser-induced breakdown spectroscopy

GULAB SINGH MAURYA, PRAVIN KUMAR TIWARI, ROHIT KUMAR, RAJESH KUMAR SINGH, AND AWADHESH KUMAR RAI

18. High-energy materials application

MANOJ KUMAR GUNDAWAR

19. Industrial applications

REINHARD NOLL

20. Laser-induced breakdown spectroscopy in planetary science

ROGER C. WIENS, XIONG WAN, JEREMIE LASUE, AND SYLVESTRE MAURICE

21. Coal analysis

ZHE WANG, ZONGYU HOU, LEI ZHANG, SHUNCHUN YAO, SAHAR SHETA, AND MUHAMMAD SHER AFGAN

22. LIBS for aerosol analysis

DANIEL DIAZ, DAVID W. HAHN, AND ULRICH PANNE

23. Field-portable and handheld LIBS

GIORGIO S. SENESI, RUSSELL S. HARMON, AND RICHARD R. HARK

24. Self-calibrated methods for LIBS quantitative analysis

STEFANO LEGNAIOLI, BEATRICE CAMPANELLA, STEFANO PAGNOTTA, FRANCESCO POGGIALINI, AND VINCENZO PALLESCHI

25. Scope of future development in LIBS

JAGDISH P. SINGH, FANG Y. YUEH, AND CHET R. BHATT

Contributors

Muhammad Sher Afgan State Key Lab of Power Systems, International Joint Laboratory on Low Carbon Clean Energy Innovation, Department of Energy and Power Engineering, Tsinghua University, Beijing, China

K.K. Ayyalasomayajula Center for Informatics and Analytics, University of Mississippi Medical Center, Jackson, MS, United States

Halil Berberoglu Physics Department, Ankara Hacı Bayram Veli University, Ankara, Turkey

Chet R. Bhatt National Energy Technology Laboratory, U.S. Department of Energy, Morgantown, WV, United States

Gonca Bilge Department of Food Engineering, Konya Food and Agriculture University, Konya, Turkey

Alexander A. Bol'shakov Applied Spectra, Inc., West Sacramento, CA, United States

Ismail Hakkı Boyaci Department of Food Engineering, Hacettepe University, Ankara, Turkey

B. Busser Grenoble Alpes University Hospital, CHUGA; Institute for Advanced Biosciences (Inserm U 1209 / CNRS 5309), Grenoble Alpes University, Grenoble, France

Beatrice Campanella Applied and Laser Spectroscopy Laboratory, Institute of Chemistry of Organometallic Compounds, Pisa, Italy

Daniel Diaz Mechanical and Aerospace Engineering Department, University of Florida, Gainesville, FL; College of Engineering, University of Arizona, Tucson, AZ, United States

P.K. Diwakar South Dakota School of Mines, Rapid City, SD, United States

Kemal Efe Eseller Electrical and Electronics Engineering Department, Atilim University, Ankara, Turkey

C. Fabre GeoRessources Laboratory (UMR 7359), University of Lorraine—CNRS, Vandoeuvre-les-Nancy, France

Ghaneshwar Gautam Fort Peck Community College, Poplar, MT, United States

J.J. Gonzalez Applied Spectra, Inc., West Sacramento; Lawrence Berkeley National Laboratory, Energy Technologies Area, Energy Storage and Distributed Resources Division, Laser Technologies Group, Berkeley, CA, United States

Jennifer L. Gottfried CCDC Army Research Laboratory, Weapons and Materials Research Directorate, Aberdeen Proving Ground, MD, United States

Christian L. Goueguel Logiag Inc., Chateauguay, QC, Canada

Manoj Kumar Gundawar Advanced Centre of Research in High Energy Materials (ACRHEM), University of Hyderabad, Gachibowli, Hyderabad, Telangana, India

David W. Hahn Mechanical and Aerospace Engineering Department, University of Florida, Gainesville, FL; College of Engineering, University of Arizona, Tucson, AZ, United States

S.S. Harilal Pacific Northwest National Laboratory, Richland, WA, United States

Richard R. Hark Institute for the Preservation of Cultural Heritage, Yale University, West Haven, CT, United States

Russell S. Harmon Department of Marine, Earth, and Atmospheric Sciences, North Carolina State University, Raleigh, NC, United States

Christopher M. Helstern University of Tennessee Space Institute, Center for Laser Applications, Tullahoma, TN, United States

Zongyu Hou State Key Lab of Power Systems, International Joint Laboratory on Low Carbon Clean Energy Innovation, Department of Energy and Power Engineering, Tsinghua University, Beijing, China

Jinesh C. Jain National Energy Technology Laboratory, U.S. Department of Energy, Morgantown, WV, United States

Rohit Kumar Department of Physics, Chaudhary Mahadeo Prasad Degree College, Prayagraj, Uttar Pradesh, India

Jeremie Lasue IRAP, Univ. Paul Sabatier –CNRS –Obs. Midi-Pyrenees, Toulouse, France

Stefano Legnaioli Applied and Laser Spectroscopy Laboratory, Institute of Chemistry of Organometallic Compounds, Pisa, Italy

X.L. Mao Lawrence Berkeley National Laboratory, Energy Technologies Area, Energy Storage and Distributed Resources Division, Laser Technologies Group, Berkeley, CA, United States

Sylvestre Maurice IRAP, Univ. Paul Sabatier –CNRS –Obs. Midi-Pyrenees, Toulouse, France

Gulab Singh Maurya Ultrafast Spectroscopy Lab, Department of Chemistry, IIT Kanpur, Kanpur, Uttar Pradesh, India

Dustin L. McIntyre National Energy Technology Laboratory, U.S. Department of Energy, Morgantown, WV, United States

G. Miloshevsky Virginia Commonwealth University, Richmond, VA, United States

S. Moncayo Ramón y Cajal Health Research Institute (IRYCIS), Madrid, Spain

V. Motto-Ros Light and Matter Institute (UMR 5306), Lyon 1 University—CNRS, Villeurbanne, France

Reinhard Noll Fraunhofer-Institut für Lasertechnik (ILT); RWTH Aachen University, Lehrstuhl für Lasertechnik (LLT), Aachen, Germany

Stefano Pagnotta Applied and Laser Spectroscopy Laboratory, Institute of Chemistry of Organometallic Compounds, Pisa, Italy

Vincenzo Palleschi Applied and Laser Spectroscopy Laboratory, Institute of Chemistry of Organometallic Compounds, Pisa, Italy

Ulrich Panne Federal Institute for Materials Research and Testing (BAM); Department of Chemistry, Humboldt University of Berlin, Berlin, Germany

Christian G. Parigger University of Tennessee Space Institute, Center for Laser Applications, Tullahoma, TN, United States

Jayanta Kumar Pati Department of Earth and Planetary Sciences, Nehru Science Centre; National Center of Experimental Mineralogy and Petrology, University of Allahabad, Allahabad, India

Francesco Poggialini Applied and Laser Spectroscopy Laboratory, Institute of Chemistry of Organometallic Compounds, Pisa, Italy

Abhishek Kumar Rai Department of Earth and Planetary Sciences, Nehru Science Centre, University of Allahabad, Allahabad, India

Awadhesh Kumar Rai Laser Spectroscopy Research Lab, Department of Physics, University of Allahabad, Prayagraj, Uttar Pradesh, India

Pradeep Kumar Rai Department of Nephrology, Opal Hospital, Varanasi, India

V.N. Rai Laser Plasma Division, Raja Ramanna Centre for Advanced Technology; School of Physics, Devi Ahilya University, Indore, India

Richard E. Russo Applied Spectra, Inc., West Sacramento; Lawrence Berkeley National Laboratory, Energy Technologies Area, Energy Storage and Distributed Resources Division, Laser Technologies Group, Berkeley, CA, United States

Herve K. Sanghapi Alcorn State University, Lorman, MS, United States

Giorgio S. Senesi CNR—Istituto per la Scienza e Tecnologia dei Plasmi (ISTP)—Sede di Bari, Bari, Italy

Banu Sezer Department of Food Engineering, Hacettepe University, Ankara, Turkey

Sahar Sheta State Key Lab of Power Systems, International Joint Laboratory on Low Carbon Clean Energy Innovation, Department of Energy and Power Engineering, Tsinghua University, Beijing, China

Jagdish P. Singh Institute for Clean Energy Technology (ICET); Department of Physics, Mississippi State University; JPS Advanced Technology R & D LLC, Starkville, MS, United States

Rajesh Kumar Singh Institute for Plasma Research, Gandhinagar, Gujarat, India

David M. Surmick University of Massachusetts, Lowell, MA, United States

Surya N. Thakur Laser and Spectroscopy Laboratory, Department of Physics, Banaras Hindu University, Varanasi, India

Pravin Kumar Tiwari Laser Spectroscopy Research Laboratory, Department of Physics, University of Allahabad, Prayagraj, Uttar Pradesh; Institute for Plasma Research, Gandhinagar, Gujarat, India

Markandey M. Tripathi Institute for Clean Energy Technology, Mississippi State University, Starkville, MS, United States

Xiong Wan Shanghai Institute of Technical Physics (SITP), Chinese Academy of Sciences (CAS), Shanghai, China

Zhe Wang State Key Lab of Power Systems, International Joint Laboratory on Low Carbon Clean Energy Innovation, Department of Energy and Power Engineering, Tsinghua University, Beijing, China

Roger C. Wiens Los Alamos National Laboratory, Los Alamos, NM, United States

Shunchun Yao School of Electric Power, South China University of Technology, Guangzhou, China

J. Yoo Applied Spectra, Inc., West Sacramento, CA, United States

Fang Y. Yueh Institute for Clean Energy Technology (ICET); Department of Physics, Mississippi State University; JPS Advanced Technology R & D LLC, Starkville, MS, United States

Lei Zhang State Key Laboratory of Quantum Optics and Quantum Optics Devices, Institute of Laser Spectroscopy, Shanxi University, Taiyuan, China

Preface

The history of laser-induced breakdown spectroscopy (LIBS) is similar to the history of "spectroscopy" itself. Spectroscopy formally began in 1814, when Fraunhofer recorded and measured the dark spectral lines in the solar spectrum; this was about 150 years after the first observation of a "spectrum" by Sir Isaac Newton. It took another four and a half decades for Kirchhoff and Bunsen to establish the fact that these "dark lines" belonged to elements that produced brightly colored emission lines when fed into the flame of a Bunsen burner. A variety of sources emitting line spectra, such as electric arcs, sparks, and discharge tubes, were developed to accumulate and analyze the spectral data for a large number of elements. The success of the Bohr model in theoretically explaining the observed spectral lines of hydrogen led to the concept of the electronic structure of atoms, and a new understanding of the periodic table emerged informed by spectral data. The invention of the laser in 1960 may be compared to the discovery of the spectrum because it provided a tool for investigating the dynamics of excited atoms and molecules by measurement of their temporal decay. Soon after its invention, a pulsed ruby laser beam was focused onto a metallic surface, producing a spark similar to that produced between two metal rods by an electric field. However, for two decades after the advent of laser-induced plasma emission spectroscopy, scientific interest was more focused on the basic physics of plasma formation and efforts to reduce the background continuum associated with the spectral lines. The acronym LIBS was coined by Radziemski and coworkers in the early 1980s; they played a major role in the experimental development of this versatile analytical technique. The plasma plume produced by a single laser pulse from a sample can reveal the presence of all of its constituent atoms by spectral analysis. LIBS is a unique technology that is capable of providing distinct spectral signatures characteristic of all chemical species in all environments.

Since the publication of the first edition of this book, there have been many developments in the components used in LIBS leading to a significant reduction in the size of the equipment; this and new statistical methods of analyzing the spectral data have led to many industrial applications. All these applications depend on the detection of atoms in a variety of complex samples. One is reminded of the famous suggestion of Richard Feynman that the most valuable information on scientific knowledge in a single sentence using the fewest words is: "All things are made of atoms." In a different context, Feynman said, "The most remarkable discovery in all of astronomy is that the stars are made of atoms of the same kind as those on the earth." This came much before August 2012 when ChemCam, with the LIBS instrument, started exploring the surface of Mars on the Curiosity rover and is still gathering spectral data on the rocks and soil in search of evidence of life in the remote past. It is the first time that the LIBS technique has been used on another planet and this historic event has generated a great deal of enthusiasm among students and young researchers in dealing with the compositional analysis of materials, including biological materials. In addition to the fundamental processes and underlying physics of LIBS, the present edition of the book has 15 new chapters containing recent advances in the field, such as molecular LIBS, self-calibrated methods, high-energy materials, LIBS imaging, food science, coal, flame, defense, matrix effects, LIBS for planetary science, analysis of medical drugs, and industrial applications. The chapters have been arranged such that there is an easy and smooth progression from fundamentals to experimental techniques and applications.

The purpose of this book is to give a simple and comprehensive review of LIBS, including recent advances in the field. It is well structured and organized in three broad sections dealing with fundamentals, new techniques, and applications of LIBS. Part I of the book comprises five chapters. The first chapter deals with the temporal and focusing properties of laser, laser-produced plasmas, and the spectral recording of the emission using spectrographs equipped with charged-coupled device and intensified charged-coupled device detectors. Chapter 2 contains a brief account of the electronic structure of atoms, the periodic table, and selection rules for radiative transitions. Intensities of spectral lines and mechanisms of spectral broadening are described along with their roles in the determination of temperature and electron density in laser-induced plasma. Chapter 3 deals with "ablation," the most significant process of plasma plume formation in LIBS. It describes the explosive vaporization of material and its rate, which depends on the type of material, the laser wavelength, the ambient atmosphere, and the geometry of the laser beam. A significant fraction of the incident laser pulse energy is absorbed in the expanding plasma plume, causing the atoms and ions to reach different states of excitation and hence subsequent optical emission. The physics of these optical processes forms the subject matter of Chapter 4, where the origins of continuum and line emission are discussed to emphasize the temporal characteristics of the spectral emission. The geometrical shape of the plasma plume and the time window for LIBS detection have been discussed. Recent developments in theoretical models for the dynamics of plasma plume have been described. The experimental setup consisting of a laser, the plasma plume, and a spectrometer form the contents of Chapter 5. The type of sample (solid, liquid, or gas) and its location in the laboratory or in the field demand a specific arrangement of the experimental components for LIBS detection. Instruments using optical fiber for delivery of laser pulses as well as for the collection of emission from the plasma plume are discussed along with handheld LIBS systems.

Part II of the book contains a variety of recent LIBS techniques. The use of ultrafast lasers for generation of the plasma plume and the subsequent detection of its emission forms the subject matter of Chapter 6. It also provides a comparison of this technique with nanosecond LIBS. The usefulness of femtosecond laser and laser filamentation for standoff analysis has also been discussed. Chapter 7 discusses the technique of laser ablation molecular isotopic spectrometry for the identification of atomic isotopes. Diatomic molecules are formed due to recombination in a LIBS experiment and analysis of their emission spectra reveals excitation temperatures up to 10^4 K following laser-induced optical breakdown.

Part III of the book embodies recent developments in the applications of LIBS. Chapter 8 deals with flame and offgas measurements. Chapter 9 is about the application of LIBS to liquid samples where the signals are generally weaker than in solids and require enhancement techniques. The difficulties associated with underwater measurements and their possible solutions have been discussed along with matrix and pressure effects. Chapter 10 deals with LIBS application to powder materials widely used in industry. The detection of impurities and trace elements in powder samples used in the glass and pharmaceutical industries is discussed. The possibilities of using LIBS for source characterization and abundance assessment for critical elements have been demonstrated. Chapter 11 contains the results of using LIBS in the challenging problem of quantitative analysis of slurry samples. Spectral data have been analyzed using three models based on univariate calibration, multiple linear regression, and partial least square regression in conjunction with a new sampling method, "spin-on-glass." Portable LIBS systems and standoff detection systems for defense applications form the subject matter of Chapter 12. Detection of buried landmines and explosive residue on personnel, energetic material characterization, and discrimination of chemical as well as biological materials are discussed in detail. Chapter 13 contains applications of LIBS to the analysis of medical drugs. Chemometric analysis of the LIBS data set has shown to be useful in the discrimination of drug samples with similar elemental compositions. The scope of this technique in the identification of counterfeit medicines in the market is discussed. Applications of LIBS imaging in a number of diverse fields are discussed in Chapter 14. It has unique specificities, such as micrometric-scale resolution with limits of detection up to ppm in material samples, including biological materials. Chapter 15 deals with the applications of LIBS to different food matrices for the purpose of detecting both adulteration and geographical as well as biological origin. LIBS has been shown to have great potential as a process control tool in the food industry. The application of LIBS for analysis of geological samples is discussed in Chapter 16. The analysis of moldavite spectra is expected to aid in studies of planetary analogs, including the study of impactites littered on the Martian surface. The characterization of select topsoil samples suggests that this method is capable of establishing the compositional heterogeneity of soil samples. Chapter 17 contains the development of an experimental setup to record online LIBS spectra of plasma-facing components (PFCs) of the Aditya tokamak in India. Calibration-free laser-induced breakdown spectroscopy (CF-LIBS) has been found to be very suitable for quantitative analysis of the PFCs. Chapter 18 deals with the scope of LIBS analysis of high-energy materials (HEMs). A comparative investigation of five HEMs and five non-HEMs has been carried out to assess the importance of statistical methods for their identification. The principal component analysis score plots show well-separated clusters for HEMs and non-HEMs for the full spectrum as well as for subspectra. It is concluded that it may not be necessary to record the full spectrum for the detection of the two types of materials. Industrial applications of LIBS are discussed in Chapter 19, with emphasis on the research publications from the years 2014 to 2019. In view of investigating objects at distances ranging from a few centimeters up to several meters, LIBS has a distinct advantage for monitoring samples in a process line even under harsh industrial conditions. Studies involving identification of steel blooms in a rolling mill, sorting of alloyed metal scrap, and recovery of valuable materials from end-of-life electronic equipment are discussed. The role of LIBS in exploring planetary environments using rovers is discussed in Chapter 20. The first extraterrestrial LIBS instrument, ChemCam, returned 700,000 spectra and provided major insights into the geochemistry and past history of Mars. The designs of SuperCam, developed by NASA to roam the surface of the red planet in 2121, and the Venus Elemental and Mineralogy Camera are described in addition to several similar systems. LIBS also has many advantages for outer solar system studies where lighter elements are likely to be observed on icy, volatile-rich moons of the outer planets. Chapter 21 emphasizes the importance of providing fast qualitative and quantitative information by LIBS required for coal washing, coal blending, as well as coal combustion optimization. The challenges on the way to full commercialization and industrialization of LIBS coal analysis are discussed in detail. LIBS is well suited for the analysis of aerosol particles due to the unique point sampling nature of laser-induced plasmas as described in Chapter 22. A detailed treatment of the breakdown process is discussed in terms of gas-phase processes associated with plasma initiation and propagation and the future course of investigation in this challenging area is indicated. Chapter 23 is concerned with the functioning and use of handheld equipment for LIBS. The fundamental concepts and applications of self-calibrated methods for quantitative analysis by LIBS forms the subject matter of Chapter 24. The hypotheses that form the basis of classical CF-LIBS are critically reviewed and its evolution in the last two decades is summarized. Monte Carlo LIBS and extended C-sigma methods appear to be promising alternatives to standard CF-LIBS. A short review of the main applications of CF-LIBS is given for such samples where the availability of matrix-matched standards is needed for uni- or multivariate calibration curves. Chapter 25 looks at the future developments in LIBS technology, new methods of spectral data analysis, and many of its applications in unexplored areas.

We gratefully acknowledge the most valuable and imaginative contributions by the authors who spared time from their busy schedule of teaching and research for this book. We are thankful to the members of the Institute for Clean Energy Technology at the Mississippi State University and those of the Laser and Spectroscopy Laboratory at the Banaras Hindu University for their enthusiastic help during the preparation of the manuscript. Special thanks are due to Professor David L. Monts and Professor S. B. Rai for their many valuable suggestions. SNT is beholden to Dr. Punam and Dr. Vineeta for taking care of his health, to Sudheer and Sangeeta for help with the computer, and to little Leo and Mia for keeping him entertained with their innocent inquiries. JPS greatly appreciates his wife Shila Singh for her patience and encouragement and his grandchildren Devansh, Shivansh, Arya, and Aniya for amusing him during the preparation of the book.

Jagdish P. Singh
Surya N. Thakur

PART I

Basic physics and instrumentation

CHAPTER

1

Fundamentals of LIBS and recent developments

Surya N. Thakur[a], *Jagdish P. Singh*[b]

[a]Laser and Spectroscopy Laboratory, Department of Physics, Banaras Hindu University, Varanasi, India
[b]Institute for Clean Energy Technology, Mississippi State University, Starkville, MS, United States

1 Introduction

The devastating power of the laser was demonstrated soon after its invention when a focused laser beam produced a bright flash in air similar to the spark produced by lightning discharge between two clouds [1]. Another spectacular effect involved the production of luminous clouds of vaporized material blasted from a metallic surface and often accompanied by a shower of sparks when the laser was focused on a metal surface [2,3]. These laser effects have found many technological applications in the fields of metalworking, plasma production, and semiconductors. When a pulsed laser beam of high intensity is focused, it generates a plasma from the material. This phenomenon has opened up applications in many fields of science from thin-film deposition to elemental analysis of samples. The possibility of using a high-power, short-duration laser pulse to produce a high-temperature, high-density plasma was pointed out by Basov and Krokhin [4] to study inertial confinement fusion. Laser ablation of solids into background gases is now a proven method of cluster assembly [5,6]. In this method, a solid target is vaporized by a powerful laser pulse to form a partially ionized plasma that contains atoms and small molecules. Not much is known about the formation and transport of particles in laser ablation plumes. In recent years there has been notable interest both in an increased understanding of laser-induced plasmas (LIPs) and in the development of their applications. Emission spectroscopy is used for elemental analysis of targets from which the luminous plasma is generated and it can also be applied to determine the temperature, electron density, and atom density in the LIP [7].

The history of laser spark spectroscopy runs parallel to the development of high-power lasers, starting with the early use of a ruby laser for producing sparks in gases [8]. In subsequent years the spectral analysis of LIP became an area of study that has significantly matured. The current developments of this technique for chemical analysis can be traced to the work of Radziemski and Cremers [9] and their coworkers at Los Alamos National Laboratory in the 1980s. It was this research group that first coined the acronym

Laser-Induced Breakdown Spectroscopy
https://doi.org/10.1016/B978-0-12-818829-3.00001-0

LIBS for laser-induced breakdown spectroscopy. During the last two decades, LIBS has undergone a dramatic transformation in terms of hardware, software, and application areas. It has become a powerful sensor technology for both laboratory and field use. To obtain a reliable quantitative elemental analysis of a sample using LIBS, one needs to control several parameters that can strongly affect the measurements. Some of these parameters are the laser wavelength, its irradiance, the morphology of the sample surface, the amount of ablated and vaporized sample, and the ability of the resulting plasma to absorb the optical energy. If these and related parameters are properly optimized, the spectral line intensities will be proportional to the elemental concentration. During the last decade some very significant progress has been made in the three basic components of the LIBS setup. There has been much progress in instrumentation, including small, low-cost lasers with high beam quality, improved spectrometers, and charge-coupled device (CCD) detectors. A deeper understanding of the mechanism of LIP, its emission, as well as developments in data processing algorithms has led to a variety of new applications of LIBS. The classic example of these developments is represented by the use of LIBS equipment on board the *Curiosity* rover that has been exploring the surface of planet Mars since August 2012. It has provided the first ever use of spectroscopy on a planet other than the Earth. It has caught the imagination of people in general and has prompted a large number of scientists to design and build LIBS systems dedicated to operate in very harsh conditions. In the following sections we briefly describe the basic components and the underlying physical processes that are essential to appreciate the range of applications and power of LIBS.

2 Lasers for LIBS

The main properties of laser light that distinguish it from conventional light sources are intensity, directionality, monochromaticity, and coherence. In addition, the laser may operate to emit radiation continuously or it may generate radiation in short pulses. Some lasers can generate radiation with the aforementioned properties, which is tunable over a wide range of wavelengths. Generally, pulsed lasers are used in the production of plasmas and also in LIBS. We consider only those properties of lasers relevant to plasma production in gaseous, liquid, and solid samples so that the role of various types of laser systems used in LIBS experiments is clearly understood in the later chapters of this book. It is possible to generate short-duration laser pulses with wavelengths ranging from the infrared (IR) to the ultraviolet (UV), with powers of the order of millions of watts. Several billions to trillions of watts and more have been obtained in a pulse from more sophisticated lasers. Such high-power pulses of laser radiation can vaporize metallic and refractory surfaces in a fraction of a second. It is to be noted that not only the peak power of the laser, but also the ability to deliver the energy to a specific location is of great importance. For LIBS, the power per unit area that can be delivered to the target is more important than the absolute value of the laser power. The power per unit area in the laser beam is termed "irradiance" and is also called "flux" or "flux density." Conventional light sources with kilowatt power cannot be focused as well as laser radiation, and therefore are not capable of producing the effects that lasers can.

The next property of laser radiation that is of interest is the directionality of the beam. Laser radiation is confined to a narrow cone of angles, which is of the order of a few tenths of a milliradian for gas lasers to a few milliradians for solid-state lasers. Because of the narrow divergence angle of laser radiation, it is easy to collect all the radiation with a simple lens. The narrow beam angle also allows focusing of the laser light to a small spot. Therefore the directionality of the beam is an important factor in the ability of lasers to deliver high irradiance to a target. Coherence of the laser is also related to the narrowness of the beam divergence angle and it is indirectly related to the ability of the laser to produce high irradiance. However, coherence is not of primary concern in LIBS. Provided that a certain number of watts per square centimeter are delivered to a surface, the effect will be much the same whether the radiation is coherent or not. The monochromaticity of the laser as such plays a small role as far as plasma production is concerned because it is the power per unit area on the target that matters irrespective of the fact that the radiation is monochromatic or covers a broad band. In special cases, one may require highly monochromatic laser radiation to probe the plasma using resonance excitation of atomic species. The frequency spread of gas lasers is of the order of one part in 10^{10} or even better, and for solid lasers it is of the order of several megahertz. In specifying these frequency spreads, we have taken the width of a single cavity mode of the laser, although most lasers operate in more than one cavity mode so that the total frequency spread may cover the entire line width of the laser transition. The frequency spread of each of the cavity modes is much narrower than the line width of the laser transition and the former is used to characterize the frequency stability of the laser.

2.1 Mode properties of lasers

The optical cavity of a laser is determined by the configuration of the two end mirrors. The stationary patterns of the electromagnetic waves formed in the cavity are called modes. For a cavity formed by two confocal spherical mirrors separated by a distance L, the frequency ν of a mode (mnq) is given by:

$$\nu_{mnq} = (c/2L)[q + ½(m + n + 1)] \tag{1}$$

where c is the velocity of light, q is a large integer, and m and n are small integers. The axial modes correspond to $m = n = 0$ and involve a standing wave pattern with an integral number q of half wavelengths with $q\lambda/2 = L$, between the two mirrors with a node at each mirror. The separation between the frequencies of two consecutive axial modes ($c/2L$) is of the order of gigahertz for typical solidstate lasers. The transverse modes of the laser are designated by TEM_{mn}. They affect the focusing properties of the laser beam. The smallest focal spots and highest irradiance are obtained with beams containing the lowest transverse modes with the smallest pair of values m, n. The higher transverse modes have radial intensity distributions that are less and less concentrated along the resonator axis with increasing values of m or n. These modes are also known as off-axis modes and their diffraction losses are much higher than that of the fundamental modes TEM_{00q}. Some of the patterns for transverse modes are shown in Fig. 1.

The presence of higher transverse modes (large m, n) increases the divergence angle and affects the focusing of the laser beam. If there is no control over the mode properties, the modes present in the laser pulse can change from one shot to the next and different pulses from a high-power laser would be focused differently. High brightness is essential for delivering high irradiance. The brightness of a source is the power emitted per unit area per unit solid angle. As laser power increases, the number of transverse modes increases with little increase in brightness. The technology to produce high irradiance in a laser beam thus involves decreasing beam divergence as much as increasing power.

2.2 Spatial intensity distribution and focusing of the laser beam

To determine the irradiance produced by a laser, it is necessary to know the spot size to which the beam can be focused. It is impossible

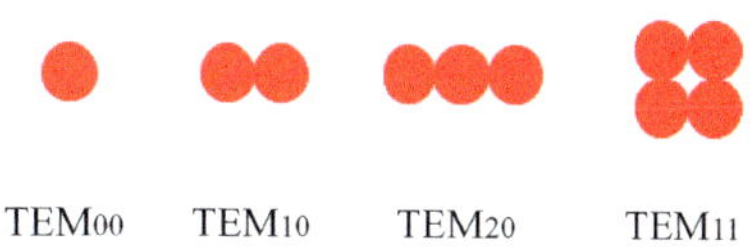

FIG. 1 Some low-order transverse modes showing off-axis spatial intensity distributions with increasing order of the mode.

to focus the beam to a geometrical point and the minimum spot size is dependent on diffraction. Since optical systems are not perfect, the actual spot size is larger than the limit set by diffraction. Maximum irradiance is obtained with minimum focal area of the laser spot.

The spatial distribution of the output of a continuous gas laser follows the mode patterns as shown in Fig. 1 and for the lowest transverse mode, the intensity distribution is given by

$$I_{00}(r) = \exp(-2r/w) \quad (2)$$

where w is called the Gaussian radius of the TEM_{00} mode. The output of a high-power solid-state laser has a complicated spatial intensity distribution and does not exhibit the recognizable mode patterns as shown in Fig. 1. The output is a superposition of many modes along with distortions caused by inhomogeneities of the crystal. This irregular spatial distribution leads to problems in focusing the laser beam to the minimum size. A schematic spatial profile of a solid laser is shown in Fig. 2.

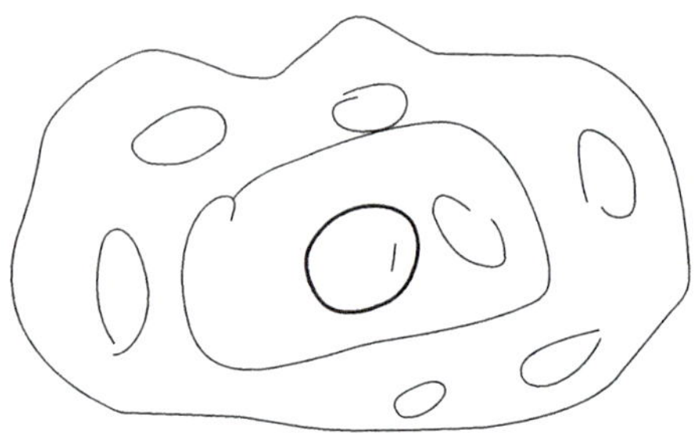

FIG. 2 Schematic contour of irradiance in an unfocused high-power ruby laser.

The spatial profile of the laser beam can change during the course of the laser pulse [10]. Many methods have been employed for improvement of the mode properties of high-power solid-state lasers. One is shown in Fig. 3 where an aperture was introduced at the focus of a lens system contained within the laser cavity. This arrangement can reduce the off-axis mode content significantly, because the high-order modes have large diffraction losses at the aperture. The output from a ruby laser can be made more spatially uniform than that shown in Fig. 2 and can have a divergence angle close to the diffraction limit [11].

The number of axial modes in a laser output can be reduced with an optical cavity in which one mirror is made up of a number of uncoated interferometric flats. Laser oscillation occurs at those wavelengths which are simultaneously modes of the total cavity and of the individual interferometers formed by each pair of flat parallel surfaces. Since the gain of the laser is nonlinear, the output power is funneled into a single or a few axial modes. In an optical cavity, introduction of a dye mode selector has been used to produce a single TEM_{00} axial mode output from a ruby laser [12].

The design and fabrication of a single-mode laser oscillator followed by amplifiers has led to diffraction-limited lasers of high brightness [14]. An important concept in the context of lasers is the distinction between near- and far-field spatial patterns. In the near field, the intensity pattern is the same as at the output mirror of the laser and it follows the mode patterns shown in Fig. 1. If a is the aperture

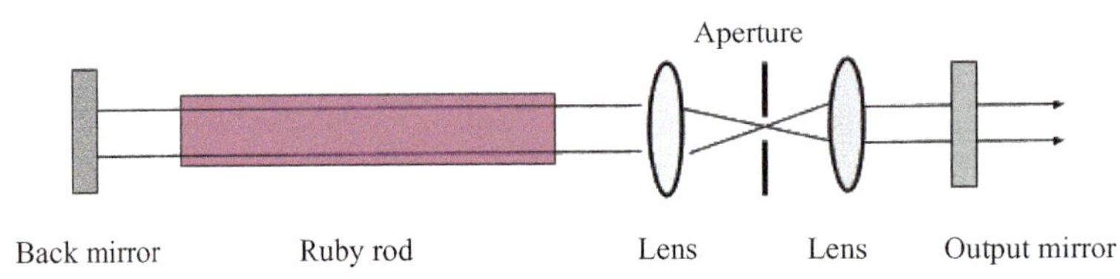

FIG. 3 Schematic diagram of a laser cavity with an aperture to remove higher-order transverse modes.

diameter of the output mirror and the laser beam is approximated by a Gaussian beam, then the near-field pattern persists for a distance of the order of a^2/λ where λ is the wavelength of laser light. However, for larger distances from the output mirror, the well-defined mode pattern in the near field would be washed out by diffraction effects; the spreading angle is of the order of λ/a. Gaussian beams have the same phase across the entire wave front, and they are capable of being focused to the minimum possible size [13].

Ruby lasers with an ordinary optical cavity can be focused with a simple lens to produce spots with diameters of the order of 300 μm, whereas those with an apertured optical cavity have focal spot sizes of the order of a few microns.

A focal area of 10^{-3} cm^2 is typical for a ruby laser focused with a simple lens and the following peak power and irradiance can be obtained by different types of ruby lasers:

Laser type	Peak power	Irradiance
Normal pulse laser	10^5 W	10^8 W cm^{-2}
Q-switched laser	10^8 W	10^{11} W cm^{-2}
Picosecond pulse laser	10^{11} W	10^{14} W cm^{-2}

2.3 Time behavior of laser pulses

Solid-state lasers, such as ruby lasers, Nd:glass, and Nd:YAG lasers that produce high power, are generally operated with widely different pulse durations and with different methods of pulsing. If the laser is pumped by a flashlamp, pulse widths in the range of 100–1000 μs are typical. In many cases, the laser emission is not uniform, but consists of many microsecond duration spikes called relaxation oscillations whose amplitudes and spacings are not uniform. The presence of these spikes in the laser pulse causes irregular heating of the target surface and is not suitable for producing a uniform plasma plume.

Laser pulse durations in the range of 10–1000 ns can be produced by Q-switching techniques, where laser operation is suppressed and population inversion in the solid rod increases greatly over the normal threshold condition [14]. If the Q-switching component in the laser cavity is changed to a transparent condition, the laser rod, now in a highly inverted state, is coupled to the two mirrors of the cavity and the stored energy is emitted in a pulse of much higher power and much shorter duration than without Q-switching.

It is possible to produce laser pulses of picosecond duration by the phenomenon of mode locking. If there are N resonant axial modes of the cavity simultaneously present in the line width of the lasing transition, then these can be coupled by using a Q-switching dye with nonlinear transparency. This coupling of modes leads to a locking of the phases of different axial modes. In the time domain, a single ultrashort pulse circulates in the cavity with a time period equal to the round-trip transit time ($2L/c$). The laser output is in the form of identical pulses whose spacing is equal to $2L/c$. The width τ of each pulse is approximately the inverse of the frequency spread of the laser output. Thus for N axial modes in the lasing transition linewidth, the pulse duration is given by:

$$\tau = 1/[Nc/2L] = 2L/Nc \quad (3)$$

Femtosecond laser pulses are produced by the technique of colliding pulse mode locking, which utilizes the collision of two counter-propagating pulse trains in a thin saturable dye jet. The interaction of the counter-propagating pulses creates a transient grating of the population of dye molecules, which synchronizes, stabilizes, and shortens the pulse. The operation of femtosecond pulses is very sensitive to mirror coatings. The short duration and high electric field intensities encountered in amplifying femtosecond pulses introduce new problems in amplifier design. Improvement in amplification techniques has permitted generation of femtosecond laser pulses of gigawatt intensities [15].

2.4 Measurement of laser power and energy

To study the physics of LIPs, reliable measurements of laser beam power, beam energy, beam divergence angle, and spatial intensity distribution of the beam cross-section are needed.

The most common detectors used in the measurement of laser power are referred to as square law detectors because they respond to the square of the electric field. Photomultiplier tubes (PMTs) and single-stage vacuum photoemissive detectors are sensitive in the UV, visible, and near IR, whereas photoconductive detectors are used for lasers emitting at wavelengths longer than 1 μm. For pulsed lasers, the phototube output is displayed on a fast oscilloscope to determine the pulse shape. The response speed of the photodetector must be fast and circuitry must be carefully designed to preserve the pulse shape. Intense laser output tends to saturate the output of the detectors so absorbing filters are used to keep the detector in the linear portion of their operating range and to make them blind to background radiation.

Another widely used detector is the semiconductor photodiode, which is a photovoltaic device. Laser radiation incident on the detector produces a voltage across the p-n junction even in the absence of an external bias. When light falls on a back-biased diode, the reverse current increases sharply. Room temperature devices are used in the visible region and up to 3.6 μm, whereas liquid nitrogen-cooled devices operate up to 5.7 μm.

The total energy in a laser pulse is measured by calorimetric methods using blackbody absorbers of low thermal mass in contact with thermocouples or other temperature-measuring devices. In one common form, the absorber is a small hollow cone of carbon such that radiation entering the base of the cone cannot be reflected out of the cone. Thermistor beads, forming an element of a balanced bridge circuit, are placed intimately in contact with the cone. As the cone is heated by a pulse of energy, the resistance of the thermistors changes, resulting in an imbalance of the bridge and a voltage pulse that decays as the cone cools to ambient temperature. The magnitude of the voltage pulse gives a measure of the energy in the laser pulse.

2.5 Varieties of lasers

A laser is not one single device, but there are a wide variety of different lasers with many different characteristics. Each type has its own properties of wavelength and operating parameters. Even within one type, there are many varieties of construction. Now, several thousand laser lines are known that span a whole spectral range from extreme UV to the far-IR region. Developments in LIBS have taken place by using the laser wavelengths provided by existing technology. In 1962 a ruby laser at 694 nm was used by Breach and Cross [16], but its pulse-to-pulse stability was very poor and LIBS was not considered to be a very reliable technique for spectrochemical analysis. The next phase of LIBS development was marked by the sophisticated pulsed-laser technology of the 1980s that led to very reliable Nd:YAG lasers in the near-IR, visible, and UV regions and to excimer lasers in the UV region. At present many more laser wavelengths have become available to study their effects on LIBS measurements [17–19]. Lasers commonly employed in LIBS are listed in Table 1 along with properties associated with these lasers. These are representative values, not necessarily the highest or the best ever achieved.

Femtosecond lasers produce a much better-quality plasma than nanosecond lasers. This is primarily due to a more controlled process of laser ablation. The ultrashort pulses produce mostly atomic emission from the clean craters created by ablation. The atomic plume is thus produced close to stoichiometric sample evaporation [21]. However, at present the use of femtosecond laser sources is confined to the laboratory because of their lack of portability and ruggedness and above all their high cost.

TABLE 1 Characteristic properties of some lasers for laser-induced breakdown spectroscopy [20].

Laser type	Wavelength	Pulse duration	Energy/ pulse
CO_2 repetitive	10.6 μm	10–100 μs	0.1–5 J
CO_2 Q-switched	10.6 μm	200 ns	0.1 J
Er:YAG Q-switched	2.94 μm	170 ns	25 mJ
Nd:YAG	1.06 μm	5–10 ns	1–3 J
Ruby normal pulse	694.3 nm	0.2–10 ms	1–500 J
Ruby Q-switched	694.3 nm	5–30 ns	1–50 J
Ruby picosecond pulse	694.3 nm	10 ps	0.01–0.5 J
Nd:YAG second harmonic	532.0 nm	4–8 ns	0.5–2 J
Nd:YAG third harmonic	354.7 nm	4–8 ns	0.2–0.7 J
N_2 laser	337.1 nm	3–6 ns	0.1–0.6 mJ
XeCl excimer	308 nm	20–30 ns	0.5–1 J
Nd:YAG fourth harmonic	266 nm	3–5 ns	0.1–0.3 J
KrF excimer	248 nm	25–35 ns	0.5–1 J
ArF excimer	193 nm	8–15 ns	8–15 mJ

In recent years, compact diode pumped Q-switched Nd:YAG lasers have been developed for use in the laboratory as well as for industry. These lasers have high beam quality, pulse energies in the range of 100 μJ at 1064 nm, with pulse widths of a few nanoseconds. They can generate stable laser emission with repetition rate in the multikilohertz range, compared to traditional Nd:YAG lasers whose repetition rate is typically 10–30 Hz with energy per pulse of a few 100 mJ [22]. Pulsed fiber lasers have also been developed, capable of multikilohertz repetition rate with pulse duration of about 40 ns, but these are large in size and consume more power [23].

3 Laser-induced plasmas

To produce a spark in air or a gas requires laser intensities of the order of 10^{11} W cm^{-2}. Sparks are caused by the breakdown of the gas due to the electric field associated with the light wave. Breakdown thresholds are of the order of 10^6–10^7 V cm^{-1}. The spark is accompanied by production of charged particles, absorption of laser light, and reradiation of light from the spark. If the temperature of the plasma at the position of the gas breakdown becomes high enough, X-ray emission is also observed, in addition to visible and UV radiation. This phenomenon was termed laser-induced breakdown in analogy with the electrical breakdown of gases [24]. The breakdown results from strong ionization and absorption by gases that are usually transparent to light. The breakdown is marked by a threshold irradiance below which virtually no effects are observed. The onset of the breakdown is a sudden, dramatic phenomenon occurring at an easily determined threshold. Its spatial as well as temporal profiles make interesting study [25].

The breakdown in the focal volume of the lens in which the peak laser irradiance occurs can be understood as occurring in two steps. First, the production of the initial ionization and the subsequent cascade by which the ionization grows resulting in the breakdown. Multiphoton ionization, where simultaneous absorption of many quanta by an atom produces an ion-electron pair, is considered to be a plausible mechanism for the initial ionization. An alternative possibility is multiphoton excitation of an atom to an excited state with many other excited states between it and the free electron continuum. Single-photon absorption processes may rapidly ionize the atom from this excited level. A free electron in the focal volume absorbs photons and gains enough energy to ionize additional atoms by collisions. In each such ionization process, the colliding electron is replaced by two electrons with lower energy in the free electron continuum. These in turn absorb photons so that an avalanche or cascade of ionization will occur. The absorption of a photon by an electron may be visualized in two equivalent ways: (1) it can be considered as an inverse Bremsstrahlung process in which a single light quantum is absorbed by an electron in the field of an atom or ion, and (2) it can be considered as analogous to microwave-generated breakdown, in which the electron oscillates in the electric field of the incident radiation.

3.1 Laser-induced breakdown in gases

One of the most striking observations is the extinction of laser light by a plasma plume produced by breakdown in gases. When the laser irradiance is less than the threshold, no significant attenuation is observed, but with laser irradiance exceeding the threshold, the absorption is so strong that it is often used as a critical test of whether breakdown has actually occurred. Fig. 4 shows the shape of the original laser pulse and the pulse transmitted through the plasma when breakdown occurs. It is evident that early in the blue profile, there is little attenuation, but at later times, after breakdown occurs, the plasma becomes very opaque. The abrupt shutoff of the transmitted light occurs simultaneously with initiation of the spark. When light transmission is studied for a series of laser pulses with increasing energy, breakdown occurs earlier in the pulse as laser irradiance increases. The time to breakdown as a function of intensity depends on the focal area [26]. For a small focal volume, a higher laser intensity is required to produce breakdown within the same time.

Although the laser-induced blue-white spark appears spatially uniform to the naked eye, it is indeed elongated along the direction of the incoming laser beam. For laser powers of the order of 100 MW, the spark may be 1 cm long and a few millimeters in diameter. A schematic shape of the spark is shown in Fig. 5 where its expansion back toward the laser essentially fills the converging cone of laser radiation. The growth of the spark in a direction opposite to the light flux has led to the model of a radiation-supported detonation wave. A detonation wave is a shockwave that is fed by release of energy behind the shock wave front. In this case the energy is supplied by the absorption of the incoming laser beam. This is analogous to the detonation of reacting gases, with the reaction energy of the gases replaced by the absorbed laser energy. A shockwave propagates from the focal region into the undisturbed gas and absorption of energy from the laser beam drives the shockwave, causing it to spread. The motion of the luminous front has been measured as a function of time and two time regions have been identified. The plasma front has been found

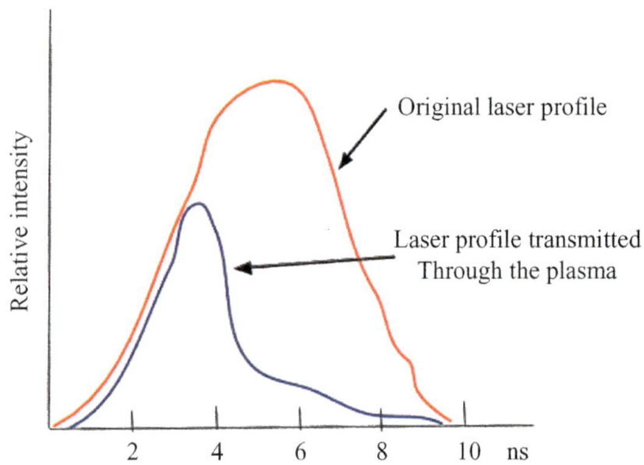

FIG. 4 Schematic temporal profile of laser pulse in the absence and presence of gas breakdown showing attenuation of the laser beam.

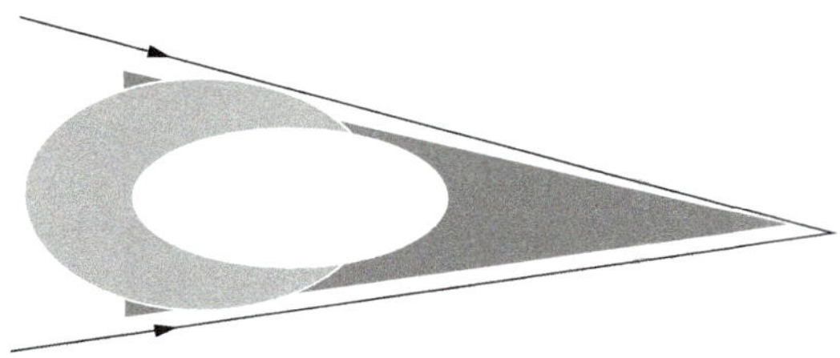

FIG. 5 Schematic shape of laser-produced spark in air. The intense core is indicated by the *white contour*. *Arrows* indicate the propagation of the focused laser beam.

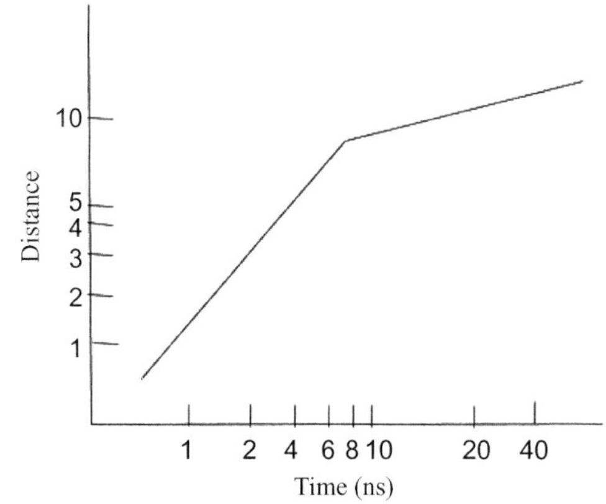

FIG. 6 Relative displacement of expanding luminous front as a function of time.

to move faster before the end of the laser pulse but its expansion is slowed down after the end of the pulse. This is schematically shown in Fig. 6.

Spectroscopic investigations of laser-induced spark in air show that its emission consists of spectral lines of N and O atoms as well as a strong continuum [27]. It has been found that in the early part of the development of the spark, the continuum is the dominant component of emission in addition to broad lines of ions and neutral atoms. When the spark has expanded and cooled, fewer broadened lines from neutral atoms are observed.

3.2 Plasma production from solid targets

When a high-power laser beam strikes a solid surface, it produces a plasma plume due to rapid melting and/or vaporization of the sample surface. The vaporization of a tungsten surface by a Nd:glass laser pulse was found to be accompanied by a shower of sparks characteristic of molten material expelled along with vaporization, whereas a plume of glowing material was emitted by a pulse from a ruby laser beam on a carbon target [28]. The plasma is produced by vaporization of the opaque target surface and subsequent absorption of laser light in this vaporized material. The phenomena observed in this interaction are in many ways similar to the phenomena accompanying gas breakdown, but the initial density of the material is much lower in the latter case. Plasma production studies are carried out at laser irradiances of the order of 10^9 W/cm^2 or greater, which produce a denser, more absorbing blow-off material. There is a great difference in the behavior of surfaces struck by laser pulses with millisecond duration as compared to those with pulse durations in the nanosecond region. The short pulses of very high power do not produce much vaporization, but instead remove only a small amount of material from the surface, whereas longer, low-power pulses produce deep, narrow holes in the target.

The interaction of the laser with a target surface is considerably modified by the presence of material emitted from the surface by short-pulsed high-power laser irradiation [29]. It exerts a high pressure on the surface and changes the vaporization characteristics of the surface. Since the laser flux density is very high, the ejected material can be heated further by absorption of incoming laser radiation. It becomes thermally ionized and opaque to the incident radiation. The absorbing plasma prevents light from reaching the target surface, which is effectively cut off from the incoming radiation for a large fraction of the laser pulse. At the end of the laser pulse, the blow-off material becomes so hot that it begins to radiate thermally and some of this radiation may reach the surface, causing further vaporization. The temporal evolution of the depth vaporized by the high-power laser pulse is schematically shown in Fig. 7.

The processes involved in vaporization by a Q-switched laser can be understood in terms of a simple model [30]. It takes into account the pressure produced by a small amount of the blow-off material early in the laser pulse. This recoil pressure raises the boiling point of the target above its usual vaporization temperature. If the increase in vaporization temperature

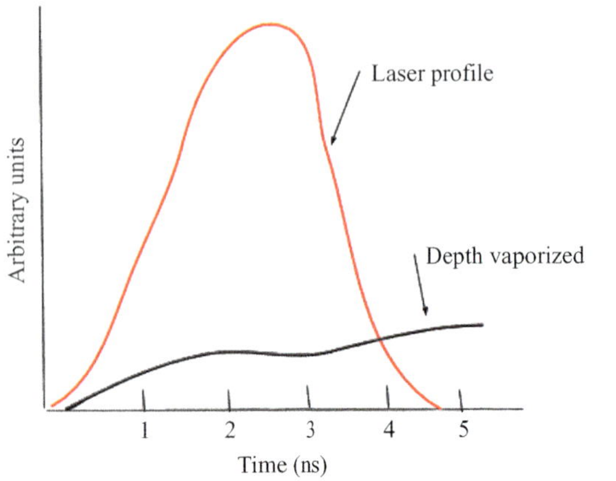

FIG. 7 Schematic representation of depth vaporized in a metal target as a function of time showing the effect of shielding by the blow-off material.

is sufficiently high, the surface will be prevented from vaporizing further and the material will continue to heat to a high temperature (above the normal vaporization temperature) as more and more laser light from the pulse is absorbed by the target surface. Eventually, the target surface will reach the critical point and at that point vaporization can occur. This model has been used to estimate the maximum depth at which the critical temperature is exceeded. At depths greater than this, removal of the material that is heated above the critical point will continue to exert a sufficiently high pressure so that no vaporization will occur. The heat will eventually be conducted into the interior of the target. This model does not take into account the shielding of the target surface from the incoming laser light as the blow-off material becomes hot, ionized, and opaque.

The absorption of laser radiation at a surface can produce large pressure waves in the target material. One mechanism is evaporation of material from the surface, with recoil of the heated material against the surface leading to motion of the target as a whole. There is another mechanism that does not necessarily involve removal of any material from the surface. In this case, as laser radiation is absorbed in a thin layer near the surface, the internal energy of that layer increases and it will expand by thermal expansion. The thermal energy is deposited very rapidly by a short pulse laser and the expanding layer of material exerts pressure on the adjacent layer, thus sending a compressive shockwave into the target.

The LIP is a very complex system affected by many factors, including the nature of the target and the surrounding environment. The unstable plasma and surrounding gas interaction, the inhomogeneous spatial distribution, and uncontrolled temporal evolution make it very difficult for LIBS to provide a stable spectral signal. The understanding of the factors that affect the physics and chemistry in the plasma greatly enhances our ability of LIBS for qualitative and quantitative analysis. In recent times, the effect of pressure of the atmosphere surrounding a target, by a single gas or a mixture of many gases, has been experimentally explored to model the characteristics of the plasma [31,32]. Many groups are involved in developing numerical approaches to adequately describe the plasma expansion, and to link it with experimental research to improve the knowledge of this complex phenomenon. Much work has already been done to comprehend the complex processes that occur during the formation and analysis of plasma [33–36]. The main difficulty in developing such models is that the plume dynamics and plasma chemical reactions need to be investigated in a coupled fashion.

3.3 Radiation from laser-induced plasmas

The plasmas produced from solid targets also exhibit strong anisotropy in their expansion. The flow of the plasma has maximum velocity perpendicular to the surface, and it is independent of the angle at which the laser beam is incident on the surface. Photographic measurements determine the motion of the

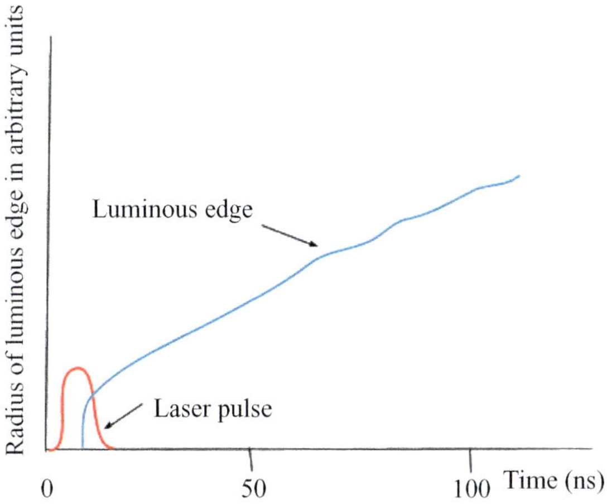

FIG. 8 Size of the luminous edge of expanding plasma produced by a Q-switched laser as a function of time.

plasma that emits light by recombination or deexcitation of atoms [9]:

$$X^+ + e^- \rightarrow X + h\nu \,(\text{recombination}) \tag{4}$$

$$X^{+*} \rightarrow X^+ + h\nu \,(\text{deexcitation of ions}) \tag{5}$$

$$X^* \rightarrow X + h\nu \,(\text{deexcitation of atoms}) \tag{6}$$

A schematic diagram of expanding plasma is shown in Fig. 8 where the radius of its outer luminous edge is plotted as a function of time. The results of one of the earliest studies of plasma production by a Q-switched ruby laser from a carbon target indicated that a bright plume of emission began somewhat after the peak of the 45-ns laser pulse, reaching its maximum intensity about 120 ns after the start of the laser pulse [37].

Optical spectroscopic studies of laser-produced plasmas reveal both continuum and line radiation. The continuum radiation originates near the target surface and covers the spectral range from about 2 to 600 nm. The line spectrum shows the presence of highly ionized atoms as well as neutral atoms. The most highly ionized species are present near the plasma center, while lines of lower ionization and neutral species are observed near the outer regions of the plasma plume. The spectra of neutral atoms are found to originate in a larger spatial region, indicating that neutral atom emission dominates after the plasma has expanded and cooled. The time variation of spectral line intensities indicates that the highest ionized states are present fairly early and lower ionized states appear later.

4 Progress in detection of LIBS

The early measurements of spectral emission from LIPs employed photographic detection using prism or grating spectrographs. The system was far from satisfactory because emission spectra consist of lines as well as continuum. Photometric detection with provision for time resolving the emission signals was not widely available and spectrally resolved light could be detected using a PMT only since the early 1980s [38]. The availability of a gated integrator made it possible to integrate the PMT current only during a time period selected by a gate pulse. The gate pulse is synchronized in time to the arrival of the laser pulse at the target. To suppress the detection of continuum from LIPs present in the early part of the pulsed emission, the high voltage to the dynodes is gated so that full gain of the PMT is not realized until several microseconds after plasma formation. The disadvantage of gated PMT detection is that its gain does not remain constant. The other detector for spectrally resolved emission is a photodiode array (PDA) consisting of a series of photosensitive silicon detector elements known as pixels, lined up in a row. Time resolution in this case is achieved by time gating the voltage applied to the microchannel plate image intensifier in front of the PDA. Time resolution of a few nanoseconds could be obtained with a PDA and its gain could be controlled by a factor of 10^6. Due to the nature of the photodiode detectors, the PDA can only be cooled to temperatures reached by thermoelectric devices. At such temperatures, the dark current is on the order of 500 counts/pixel/s, which is relatively high so that PDAs are best suited for medium- and high-intensity signals.

The selection of the spectral range of plasma emission to be recorded in an experiment could be made by using (1) a narrow bandpass filter, (2) a monochromator, or (3) a spectrograph between the detector and the plasma plume. If only a single emission line is to be recorded at a time, the filter-detector combination or the monochromator-detector combination would be used depending on the presence of isolated or closely spaced emission lines. The simultaneous recording of several lines with high resolution would require a spectrograph-detector combination. The PMT can be used in all of these cases, and it is positioned behind the filter or the exit slit of the monochromator. In the latter case, the wavelength of the monochromator can be scanned to record several spectral lines in emission from the LIP provided the nature of plasma plume does not change from one pulse to another during the period of the scan. In the case of a sample containing many elements, PDAs can be used with the spectrograph to record several spectral lines simultaneously. A slit-PMT combination located in the focal plane of the spectrograph has to be used for each spectral line and the number of lines to be recorded is limited by the length of the focal plane. Another disadvantage of PMT detection is that.

continuous wavelength coverage of the spectrum is not possible. PDA detection is more versatile because it has continuous wavelength coverage over the array length and it can record a spectrum from a single laser shot. A typical detection system is shown in Fig. 9.

There has been tremendous growth in the range and sophistication of photodetectors since 1990 due to progressive research and improvements in optical technology. The advent of high-quality solid-state detectors has led to a quantum leap in applications of LIBS [39–43].

4.1 CCD and ICCD detectors

A CCD is a microelectronic device that is used in memory, signal processing, and imaging applications. CCDs were initially conceived as an electronic analog of the magnetic bubble device. To function as memory, there must be a physical quantity that represents a bit of information, a means of recognizing the presence or

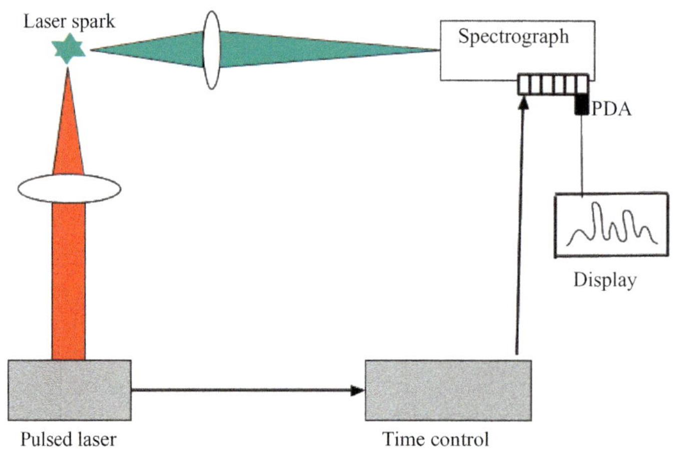

FIG. 9 Schematic diagram for spectral analysis of plasma plume with a time-gated photodiode array.

absence of the bit, and a means of creating and destroying the information. In the CCD, a bit of information is represented by a packet of electrons. These charges are stored in the depletion region of a metal insulator semiconductor (MIS) capacitor and moved about in the CCD circuit by placing the MIS capacitors to allow the charge to spill from one capacitor to the next and hence the name charge-coupled device.

The CCD must perform four tasks in generating an image, namely charge generation, charge collection, charge transfer, and charge detection. The first step occurs when free electrons are liberated due to incident photons. In the second step, the photoelectrons are collected in the nearest collecting site, referred to as pixels. Pixels are defined by electrodes called gates formed on the surface of the CCD. The third operation is accomplished by manipulating the voltage on the gates in a systematic way so that signal electrons move down vertically from one pixel to the next. At the end of the columns is a horizontal register of pixels. This register collects a line at a time and then transports the charge packets in a serial fashion to an output amplifier. The final operating step is performed by the CCD when the charge packet from the horizontal register is converted to an output voltage by the on-chip amplifier. This voltage is amplified, processed, and digitally encoded off-chip and stored in a computer to reconstruct the image on a television monitor.

CCDs provide the multichannel advantage of array detectors and since it is a two-dimensional array, it can record multiple spectra simultaneously. The large format, two-dimensional nature of CCDs is ideal for high-resolution or echelle spectroscopy. High-resolution spectra with overlapping orders are produced by a grating; each order contains information in successive spectral regions. The different order spectra are separated in the orthogonal direction by a cross-dispersing element. The resulting two-dimensional spectrum is imaged onto the CCD. In this way, it is possible to obtain spectra covering the UV to the near-IR range with 0.01 nm resolution.

The CCD is the most sensitive multichannel detector. It can be cooled with liquid nitrogen to 140 K where the dark current is less than 1 electron/pixel/h. At this temperature the detector can be exposed to a signal for hours without any significant contribution from the dark current. CCDs have a large dynamic range, which is defined as the ratio of the smallest distinguishable measurable charge to the largest before saturation. A 16-bit converter used with the CCD will allow the measurement of signals that are 1/65536th of the full-scale signal. CCDs also offer variable gain, which is important in the measurement of weak signals. By increasing the gain to measure signal levels that are very close to the noise, the signal-to-noise ratio (SNR) can be improved while maintaining the same integration time. In other words, one can achieve the same SNR in less time.

Experiments involving rapid kinetic measurements require an intensified CCD (ICCD). The ICCD is a CCD with a multichannel plate intensifier attached. Light hits the photocathode on the front of the multichannel plate and is converted to electrons that are multiplied and hit the phosphor coating to produce photons that are detected by the CCD.

Since the intensifier adds noise to the signal, causes blurring of the image, and has a nonuniform photocathode response, ICCDs are used for time-domain measurements. The intensifier is gated and the time between the pulsing of the laser and opening of the multichannel plate can be set to within better than 5 ns accuracy.

4.2 The spectrograph-detector combination

LIBS makes use of the atomic emission from plasma plumes generated by a laser from solid, liquid, or gaseous samples to identify the constituent elements present in the sample. An ideal experimental system should be capable of simultaneous multielemental monitoring of both high- and low-Z elements. In many applications, rapid, near real-time standoff detection capabilities are required. Typically, a lens or a fiber optic collects the radiation from the plasma and couples it to a spectrograph. Emission from different atomic species may occur at different times during the pulsed laser spark and time-resolved detection is necessary to obtain a spectral fingerprint of the atomic species that are present in the sample. The wavelengths of the atomic emission lines most commonly analyzed with LIBS range from 190 to 850 nm. Detection below 190 nm is limited by atmospheric absorption but some elements with nonmetallic character have their strongest lines in the near-vacuum UV (110–190 nm). Special efforts are required to minimize attenuation due to ambient air in the visible UV region [44–47]. An ideal spectrograph-detector combination to detect all possible elements in a sample should have the following features:

1. Wide wavelength coverage (130–950 nm) to record simultaneously several elements.
2. High resolution (0.003–0.01 nm) to resolve closely spaced spectral lines and to avoid interference.
3. A large dynamic range (6–7 orders of magnitude) for the detector to provide the optimum SNR for a large range of elemental concentrations.
4. High quantum efficiency of the detector particularly in the near IR and UV.
5. Short readout and data-acquisition time (less than the time lap between laser pulses) for rapid analysis.

The ICCD array detector coupled to a grating spectrograph or integrated into a compact high-resolution Czerny-Turner spectrometer has been widely used as the detector platform for a great variety of LIBS applications. In some applications ensemble-averaged spectra are used to smooth pulse-to-pulse variations frequently seen in LIBS [48,49]. In applications that require rapid sorting or emission from single particles, single-shot spectral measurements have to be made [50,51]. The use of non-ICCD arrays in LIBS is not common although correlation analysis in the identification of stainless-steel standards has been carried out using this detector [52,53]. The much lower costs of non-ICCD detectors are an important factor in their increasing use in research laboratories. The performance and sensitivity of a non-ICCD array and an ICCD array detector system have been compared in a recent publication [54].

Many applications of LIBS require remote and rapid multichannel analysis in hostile environments, which implies large spectral coverage with high resolution [55]. Conventional Czerny-Turner spectrometers provide high resolution only in a limited spectral range and it takes many laser shots to make sequential measurements for the analysis of many elements. In contrast an echelle spectrometer coupled with an ICCD detector can cover a large spectral range. Bauer and coworkers [56] were the first to couple an ICCD camera to an echelle spectrometer, but they had to use a mobile mirror to obtain large spectral coverage. The efforts of several workers have shown that a large ICCD camera is necessary for wide spectral coverage without any moving parts [57–60].

There has been tremendous progress in the development of miniature spectrometers coupled to compact lasers and high-quality CCD, ICCD, and multisensor modalities for fast LIBS. These compact laser and spectrometer designs are also capable of high sensitivity, spectral resolution, and wavelength coverage comparable to bench-top LIBS systems. Wormhoudt et al. described a miniature spectrometer coupled to a microchip laser for determination of carbon in steel [61]. Eseller et al. used a miniature spectrometer with spectral coverage 620–800 nm for online simultaneous multispecies impurity monitoring [62]. Connors et al. used a handheld device for the investigation of geochemical samples, including igneous rocks and soils [63]. According to these authors, it is only in the last two years that the technology has progressed to the point of enabling handheld systems for use outside the laboratory environment. Afgan et al. [64] performed quantitative analysis with a handheld micro-LIBS device to detect Si, Cr, Mn, and Ni. The results are found to be better than handheld XRF instruments and comparable to those obtained using bench-top LIBS. This device shows feasibility of the technology for real industrial applications. Barnett et al. [65] used a spatial heterodyne spectrometer for standoff LIBS. This grating-based interferometer has no moving parts, offers a large field of view, and has high light throughput and high spectral resolution in a small package. It is found good up to a distance of 20 m.

5 Applications of LIBS

The technological developments leading to the emergence of broadband high-resolution spectrometers has led LIBS into the 21st century with unprecedented capabilities to extract spectral information from microplasmas. It is now possible to detect almost all chemical elements in the periodic table by analyzing the UV, visible, and IR emissions prevalent in laser-generated sparks. Broadband high-resolution detection enables simultaneous analysis of multiple component elements of targeted samples. For the first time in the history of LIBS [66], there is hope to obtain qualitative as well as quantitative information on complex biological molecules in a sample. LIBS-based technologies are developing rapidly. It is not inconceivable that it would be possible to develop LIBS sensors capable of the detection and identification of almost all forms of matter. In such a case, it is difficult to make future predictions about the course of LIBS applications. In the following section, we will attempt to briefly summarize some novel features of this rapidly expanding field.

The use of femtosecond laser pulses in LIBS experiments has led to better precision and better reproducibility in emission measurement as compared to nanosecond pulses. This improvement is attributed to high peak powers in the range of 10^{14} W/cm^2. Femtosecond lasers consistently create well-defined craters and lead to better ablative reproducibility than nanosecond lasers [67]. Extremely short femtosecond laser pulses account for some remarkable features as atomizers. In contrast to nanosecond lasers, the impact of femtosecond laser energy on the sample has ceased before the plasma is formed. There is no shielding by the plasma and hence no dissipation of laser energy by it. The ablation threshold is lower than for nanosecond lasers and the energy is more localized in the sample leading to better spatial resolution [68]. Femtosecond LIBS is being used for enhancement of signal and measurement of atom density distributions in the LIP [69,70].

The analysis of single microscopic particles, aerosols, and cells has received great interest in recent years. A novel feature of LIBS for single particle analysis is its ability to provide elemental mass composition and size data for individual particles [71]. The presence of aerosols in ambient air has been cause of great concern because of their hazardous effects on human health, visibility, and climate change [72]. LIBS has found increasing application in studies on aerosols, including effluent waste and real-time monitoring [73,74]. Bioaerosols, which include pollen, fungi, bacteria, and viruses, are found nearly everywhere; although their concentration is not high, they can cause disease or allergic reaction when inhaled even in very minute amounts [75–77].

The use of LIBS technology in field-portable instruments has given rise to a spurt of research activity to deal with social problems arising from criminal and terrorist activities [78–81]. LIBS is the preferred detection and identification technique because of its many characteristic features, including flexibility of point detection or operation in a standoff mode, and fast, real-time response.

The performance of LIBS can be enhanced with the use of an array of Geiger photodiodes as the detector in echelle spectrometers. Single-photon detection in room temperature conditions is possible without complex gating-timing circuitry [82]. A compact design and high sensitivity would make this instrument very handy for standoff detection when low levels of plasma emission are to be collected. This development is very attractive in view of earlier work, which shows that LIBS would provide an extremely useful tool for space and planetary exploration [83].

The application of LIBS technology on the surface of planet Mars since 2012 has proved to be a platform to educate the public and especially students, and enhance its capabilities for use outside the laboratory environment. Some of the leading LIBS experts are of the opinion that ChemCam, aboard the *Curiosity* rover, has a strong educational role for undergraduates in universities [84]. With a view to train undergraduate students an attempt has already been made to briefly introduce LIBS in a textbook on atomic spectroscopy [85].

Russo and coworkers [86–88] developed a technique for recording molecular spectra from the expanding plasma plume following laser ablation of targets at atmospheric as well as at reduced pressures. This technique is important because of its ability to measure isotope abundance at relatively low resolutions and it can be used in the laboratory as well as in the field with equal ease. It is called laser ablation molecular isotope spectroscopy, which measures the emission spectra of molecules and molecular ions. It is a direct and rapid technique that measures emission from LIP and exploits relatively large isotope shifts in transient molecular isotopologues. Semiquantitative isotope analysis at a distance of more than 7 m, without using calibration standards, has also been demonstrated. This group has also proposed installing equipment on a future *Curiosity* rover-like vehicle to conduct absorption measurements. LAOCIS stands for laser ablation optical cavity isotopic spectrometer and is used for high-resolution absorption measurements. The module would be installed on a front arm of the rover and the ablating laser would be in its mast. As the laser-ablated plasma extends upward and perpendicular to the target surface, light from an array of diode lasers would be used for simultaneous multiisotope detection.

In principle, LIBS can directly and simultaneously detect all neutral and ion spectral features of all atomic species present in any type of sample using a single laser shot. However, several aspects of laser-material interaction remain unknown. The signal from a specific analyte atom can depend on the sample matrix. This so-called "matrix effect" requires the preparation of a high-quality standard sample, which is extremely difficult for highly heterogeneous samples like complex minerals, soil, or dirt. The true cause of the "matrix effect" is difficult to identify unless one uses theoretical modeling and accurate experimental measurements. Calibration-free LIBS (CF-LIBS) refers to a procedure that can provide analytical results without using calibration standards. According to Tognoni et al. [89] the CF approach analyzes the matrix together with the analyte rather than considering it as an external interference. These authors emphasize the analytical considerations and figure of merit. Gornushkin and Panne [90] provided a detailed discussion on the modeling aspects of CF-LIBS. Recent advances in CF-LIBS indicate an encouraging step toward eliminating the need of calibration standards. Multivariate analysis is a valuable part of LIBS data analysis because all of the information in the spectrum is used for quantitative analysis and matrix effects are taken into account in calibrations. Thus chemometric analysis will continue to be a significant step in data interpretation and classification. The unique spectra can be used for identification purposes, and would allow a user to quickly analyze samples and group them into specific types without knowing much about the samples.

In a recent review by Harmon et al. [91], application of LIBS for the analysis of geological and environmental materials has been discussed in great detail. It includes the potential of LIBS for on-site monitoring of natural waters affected by industrial contamination and also in situ analysis of fluid inclusions in rocks and minerals. Identification of minerals and their chemical composition is critical to the genesis of any particular rock or soil unit and the role played by LIBS is discussed at great length. Chemometric analysis of LIBS data in the context of the Mars Science Laboratory project has been summarized. The chemical composition of rock type for simulated Martian samples could be determined using multivariate analysis techniques. Senesi et al. [92] developed a portable handheld LIBS analyzer for real-time chemical analysis under simulated field conditions to identify elements and minerals. It is believed to be an impressive impulse for the scientific investigation of cultural heritage materials.

LIBS is expected to play a great role in chemical imaging with high resolution required for a variety of applications related to medicine and energy research [84]. High-definition hyperspectral imaging for LIBS looks like a key area, provided it is possible to rapidly image a tissue section to provide information similar to histological analysis. Lasers are currently being used to "drill" cavities in teeth before filling. It has been found that the LIBS emission from such a tooth can be used to differentiate between healthy tooth tissue and the carious tissue of the cavity. However, this technology has not been adopted and the dentist uses training and experience to decide how much tissue to remove. Similarly, LIBS emission could be used in transcranial surgery to monitor the depth into the skull the surgeon is required to cut [93]. An extremely thin and almost transparent "window" of skull bone over the brain has to be left during transcranial surgery. The optical procedures in the brain are performed through this "window" without exposing it to the environment. These applications, however, need to be preceded by a lot of research on biological samples. It is hoped that the trends in 21st century medicine to perform less invasive, faster, and cheaper procedures would be greatly helped by the relevant research in LIBS applications.

References

[1] P.D. Maker, R.W. Terhune, C.M. Savage, Optical third harmonic generation, in: P. Grivet, N. Bloembergen (Eds.), Quantum Electronics, Columbia University Press, New York, 1964, p. 1559.

[2] S. Namba, P.H. Kim, Electron and laser beam processing, Jpn. J. Appl. Phys. 3 (1964) 536.

[3] R.H. Fairbanks, C.M. Adams, Effects of high-power laser radiation, Weld. J. 43 (1964) 97.

[4] N.G. Basov, O.N. Krokhin, Conditions for heating up of a plasma by the radiation from an optical generator, Sov. Phys. - JETP 19 (1964) 123.

[5] H.W. Kroto, J.R. Heath, S.C. O'Brien, R.F. Curl, R. E. Smalley, C_{60}: buckminsterfullerene, Nature 318 (1985) 162.

[6] F. Kokai, K. Takahashi, K. Shimizu, M. Yudakasa, S. Iijima, Growth dynamics of carbon-metal particles and nanotubes synthesized by CO2 laser vaporization, Appl. Phys. A69 (1999) S691.

[7] D.A. Rusak, B.C. Castle, B.W. Smith, J.D. Winefordner, Fundamentals and applications of laser-induced breakdown spectroscopy, Crit. Rev. Anal. Chem. 27 (1997) 257.

[8] R.G. Meyerand, A.F. Haught, Gas breakdown at optical frequencies, Phys. Rev. Lett. (9) (1963) 401.

[9] L.J. Radziemski, D.A. Cremers (Eds.), Laser-Induced Plasma and Applications, Marcel Dekker, New York, 1989.

[10] E.S. Dayhoff, B. Kessler, High-speed sequence photography of a ruby laser, Appl. Opt. 1 (1962) 339.

[11] J.A. Baker, C.W. Peters, Mode selection and enhancement with a ruby laser, Appl. Opt. 1 (1962) 674.

[12] F.J. McClung, D. Weiner, Longitudinal mode control in giant pulse lasers, IEEE J. Quantum Electron. QE-1 (1965) 94.

[13] A.L. Bloom, Gas Lasers, Chapter 4, Wiley, New York, 1968.

[14] F.J. McClung, R.W. Hellwarth, Characteristics of giant optical pulsations from ruby, Proc. IEEE 51 (1963) 46.

[15] C.V. Shank, R.L. Fork, R.T. Yen, Moving from the picosecond to the femtosecond time regime, in: K.B. Eisenthal, R. M. Hochstrasser, W. Kaiser, A. Laubereau (Eds.), Picosecond Phenomena III, Springer Verlag, New York, 1982, p. 1.

[16] F. Brech, L. Cross, Optical microemission stimulated by a ruby maser, Appl. Spectrosc. 16 (1962) 59.

[17] W. Sdorra, J. Brust, K. Niemax, Basic investigations for laser microanalysis, Mikrochim. Acta 108 (1992) 1.

[18] C.W. Ng, W.F. Ho, N.H. Cheung, Spectrochemical analysis of liquids using laser-induced plasma emissions, Appl. Spectrosc. 51 (1997) 976.

[19] M. Martin, M.D. Chang, Detection of chromium aerosol using time-resolved laser-induced plasma spectroscopy, Appl. Spectrosc. 54 (2000) 1279.

[20] R. Beck, W. Englisch, K. Gürs, Table of Laser Lines in Gases and Vapors, third Revised and Enlarged Edition, Springer-Verlag, Berlin, 1980.

[21] D.W. Hahn, N. Omenetto, LIBS Part II: Review of instrumental and methodological approaches to material analysis and applications to different fields, Appl. Spectrosc. 66 (2012) 347.

[22] B. Noharet, T. Irebo, H. Karlsson, Compact industrial LIBS systems can assist aluminum recycling, Laser Focus World (2014) 50.

[23] M. Scharun, C. Fricke-Begemann, R. Noll, LIBS with multi-kHz fiber laser for mobile metal analysis tasks, Spectrochim. Acta B87 (2013) 198.

[24] E. Damon, R. Thomlinson, Observation of ionization of gases by a ruby laser, Appl. Opt. 2 (1963) 546.

[25] Y.-L. Chen, J.W.L. Lewis, C. Parigger, Spatial and temporal profiles of pulsed laser-induced air plasma emissions, J. Quant. Spectrosc. Radiat. Transf. 67 (2000) 91.

[26] R.W. Waynant, J.H. Ramsey, Laser-induced ionization of gases, J. Opt. Soc. Am. 55 (1965) 602.

[27] S.L. Manel'shtam, Investigation of the spark discharge produced in air by focusing laser radiation, Sov. Phys. - JETP 20 (1965) 1344.

[28] J.F. Ready, Effects of High-Power Laser Radiation, Academic Press, New York, 1971. p. 95.

[29] G.A. Askar'yan, E.M. Moroz, Pressure on evaporation of matter in a radiation beam, Sov. Phys. JETP 16 (1963) 1638.

[30] J.F. Ready, Effects due to absorption of laser radiation, J. Appl. Phys. 36 (1965) 462.

[31] S. Amoruso, J. Schou, J.G. Lunney, Influence of the atomic mass of the background gas on laser ablation plume propagation, Appl. Phys. A92 (2008) 907.

[32] A.J. Effenberger Jr., R. Scott, Effect of atmospheric conditions on LIBS spectra, Sensors 10 (2010) 4907.

[33] S. Merk, A. Demidov, D. Shelly, I.B. Gornushkin, U. Panne, B.W. Smith, N. Omenetto, Investigation of the photoionization properties of pharmaceutically relevant substances, Appl. Spectrosc. 67 (2013) 851.

[34] A.R. Casavola, G. Colonna, M. Capitelh, Modeling laser-induced plasma expansion under equilibrium conditions, J. Thermophys. Heat Transf. 22 (2008) 407.

[35] J. van Dijk, G.M.W. Kroesen, A. Bogaerts, Plasma modeling and numerical solution, J. Phys. D. Appl. Phys. 42 (2009) 190301.

[36] V. Morel, A. Bultel, B.G. Cheron, Modeling of thermal and chemical non-equilibrium in a laser-induced aluminum plasma, Spectrochim. Acta B 65 (2010) 830.

[37] J.F. Ready, Development of plume of material vaporized by giant-pulse laser, Appl. Phys. Lett. 3 (1963) 11.

[38] Y. Talmi (Ed.), Multichannel image detectors, in: ACS Symp., ACS, Washington, DC, 1983. Series No. 102.

[39] M.J. Pilon, M.B. Denton, R.G. Schleicher, P.M. Moran, S. B. Smith, Evaluation of a new array detector atomic emission spectrometer, Appl. Spectrosc. 44 (1990) 1613.

[40] S. Vasile, P. Gothoskar, R. Farrell, D. Sdrulla, Photon detection with high gain avalanche photodiode arrays, IEEE Trans. Nucl. Sci. 45 (1997) 720.

[41] Q.S. Hanely, C.W. Earle, F.M. Pennebaker, S.P. Madden, M.B. Denton, Charge-transfer devices in analytical chemistry, Anal. Chem. 68 (1996) 661A.

[42] J.M. Harnly, R.E. Fields, Solid-state array detectors for analytical spectrometry, Appl. Spectrosc. 51 (1997) 334A.

[43] F.M. Pennebaker, D.A. Jones, C.A. Gresham, R. W. Williams, R.E. Simon, M.F. Schappert, M.B. Denton, Spectroscopic instrumentation in the 21st century, J. Anal. At. Spectrom. 13 (1998) 821.

[44] C.J. Lorenzen, C. Carlhoff, U. Hahn, M. Jogwich, Application of laser-induced emission spectral analysis for industrial process and quality control, J. Anal. At. Spectrom. 7 (1992) 1029.

[45] V. Strum, L. Peter, R. Noll, Steel analysis with LIBS in vacuum ultraviolet, Appl. Spectrosc. 54 (2000) 1275.

[46] M.A. Khater, J.T. Costello, E.T. Kennedy, Optimization of the emission characteristics of laser-produced steel plasma, Appl. Spectrosc. 56 (2002) 970.

[47] S. Kaski, H. Hakkanen, J. Korppi-Tommola, Laser-induced plasma spectroscopy to as low as 130 nm when a gas-purged spectrograph and ICCD detection are used, Appl. Opt. 42 (2003) 6036.

[48] F. Colao, R. Fantoni, V. Lazic, V. Spizzichino, LIBS for semi-quantitative and quantitative analyses of artworks, Spectrochim. Acta B 57 (2002) 1219.

[49] A.K. Rai, F.-Y. Yueh, J.P. Singh, H. Zhang, High temperature fiber optic LIBS sensor for elemental analysis of molten alloy, Rev. Sci. Instrum. 73 (2002) 3599.

[50] C. Aragon, V. Madurga, A.J. Aguilera, Application of LIBS to the analysis of composition of thin films produced by pulsed laser deposition, Appl. Surf. Sci. 197–198 (2002) 217.

[51] J.E. Carranza, B.T. Fisher, G.D. Yoder, D.W. Hahn, On-line analysis of ambient air aerosols using LIBS, Spectrochim. Acta B 56 (2001) 851.

[52] I.B. Gornushkin, B.W. Smith, H.J. Nasajpour, J. W. Winefordner, Identification of solid materials by correlation analysis using a microscopic LIPS, Anal. Chem. 71 (1999) 5157.

[53] S.I. Gornushkin, I.B. Gornushkin, J.M. Anzano, B. W. Smith, J.D. Winefordner, Effective normalization technique for correction of matrix effects in LIBS detection of magnesium in powdered samples, Appl. Spectrosc. 56 (2002) 433.

[54] J.E. Carranza, E. Gibb, B.W. Smith, D.W. Hahn, J. D. Winefordner, Comparison of nonintensified and intensified CCD detectors for LIBS, Appl. Opt. 42 (2003) 6016.

[55] P. Fichet, P. Mauchien, C. Moulin, Determination of impurities in uranium and plutonium dioxides by LIBS, Appl. Spectrosc. 53 (1999) 1111.

[56] H.E. Bauer, F. Leis, K. Niemax, LINS with an echelle spectrometer and ICCD detection, Spectrochim. Acta B 53 (1998) 1815.

[57] P. Lindblom, New compact echelle spectrographs with multichannel time-resolved recording capabilities, Anal. Chim. Acta 380 (1998) 353.

[58] S.R. Goode, S.L. Morgan, R. Hoskins, A. Oxsher, Identifying alloys by LIBS, J. Anal. At. Spectrom. 15 (2000) 1133.

[59] S. Florek, C. Haisch, M. Okruss, H. Becker-Ross, A new, versatile echelle spectrometer relevant to LIP applications, Spectrochim. Acta B 56 (2001) 1027.

[60] P. Fichet, D. Menut, R. Brennetot, E. Vors, A. Rivoallan, Analysis by LIBS of complex solids, liquids and powders, with an echelle spectrometer, Appl. Opt. 42 (2003) 6029.

[61] J. Wormhoudt, E.J. Iannarill, S. Jone, K.D. Annen, A. Freedman, Determination of carbon in steel by LIBS using a microchip laser and miniature spectrometer, Appl. Spectrosc. 59 (2005) 1098.

[62] K.E. Eseller, F.-Y. Yueh, J.P. Singh, Non-intrusive, on-line, simultaneous multi-species impurity monitoring in hydrogen using LIBS, Appl. Phys. B Lasers Opt. 102 (2011) 963.

[63] B. Connors, A. Somers, D. Day, Application of handheld LIBS to geochemical analysis, Appl. Spectrosc. 70 (2016) 810.

[64] M.S. Afgan, Z. Hou, Z. Wang, Quantitative analysis of common elements in steel using a handheld micro-LIBS instrument, J. Anal. At. Spectrom. 32 (2017) 1905.

[65] P.D. Barnett, N. Lamsal, M. Angel, Standoff LIBS using a miniature wide field of view heterodyne spectrometer with spectrometer, Appl. Spectrosc. 71 (2017) 583.

[66] L. Radziemski, From LASER to LIBS the path of technology development, Spectrochim. Acta B 57 (2002) 1109.

[67] K.L. Eland, D.N. Stratis, D.M. Gold, S.R. Goode, S. M. Angel, Energy dependence of emission intensity and temperature in a LIBS plasma using femtosecond excitation, Appl. Spectrosc. 55 (2001) 286.

[68] V. Margetic, T. Ban, F. Leis, K. Niemax, R. Hergenroder, Hydrodynamic expansion of a femtosecond laser produced plasma, Spectrochim. Acta B 58 (2003) 415.

[69] J. Scaffidi, J. Pender, W. Pearman, S.C. Goode, B. W. Colson Jr., J.C. Carter, S.M. Angel, Dual-pulse LIBS with combinations of femtosecond and nanosecond laser pulses, Appl. Opt. 42 (2003) 6099.

[70] O. Samek, F. Leis, V. Margetic, R. Malina, K. Niemax, R. Hergenroder, Imaging of the expansion of femtosecond-laser-produced silicon plasma atoms by off-resonant planar laser-induced fluorescence, Appl. Opt. 42 (2003) 6001.

[71] D.W. Hahn, M.M. Lunden, Detection and analysis of aerosol particles by LIBS, Aerosol Sci. Technol. 33 (2000) 30.

[72] J.H. Seinfeld, S.N. Pandis, Atmospheric Chemistry and Physics: From Pollution to Climate Change, Wiley, New York, 1998.

[73] J.P. Singh, F.-Y. Yueh, H. Zhang, R.L. Cook, Study of LIBS as a process monitor and control tool for hazardous waste remediation, Process. Control. Qual. 10 (1997) 247.

[74] J.E. Carranza, B.T. Fisher, G.D. Yoder, D.W. Hahn, On-line analysis of ambient air aerosols using laser-induced breakdown spectroscopy, Spectrochim. Acta B56 (2001) 851.

[75] W.C. Hinds, Aerosol Technology: Properties, Behavior and Measurement of Airborne Particles, second ed., Wiley, New York, 1999.

[76] A.R. Boyain-Goitia, D.C.S. Beddows, B.C. Griffiths, H. H. Telle, Single-pollen analysis by LIBS and Raman spectroscopy, Appl. Opt. 42 (2003) 6119.

[77] A.C. Samuels, F.C. DeLucia Jr., K.L. McNesby, A. W. Miziolek, LIBS of bacterial spores, molds, pollens, and protein, Appl. Opt. 42 (2003) 6205.

[78] J.M. Anzano, I.B. Gornushkin, B.W. Smith, J. D. Winefordner, LIBS for plastic identification, Polym. Eng. Sci. 40 (2000) 2423.

[79] R.T. Wainner, R.S. Harmon, A.W. Miziolek, K. L. McNesby, P.D. French, Analysis of environmental lead contamination: comparison of LIBS field and laboratory instruments, Spectrochim. Acta B 56 (2001) 777.

[80] C.R. Dockery, S.R. Goode, LIBS for the detection of gunshot residues on the hands of a shooter, Appl. Opt. 42 (2003) 6153.

[81] F.C. DeLucia Jr., R.S. Harmon, K.L. McNesby, R. J. Winkel Jr., A.W. Miziolek, LIBS analysis of energetic materials, Appl. Opt. 42 (2003) 6148.

[82] R.A. Myers, A.M. Karger, D.W. Hahn, Geiger photodiode array for compact, lightweight LIBS instrumentation, Appl. Opt. 42 (2003) 6072.

[83] A.K. Knight, N.L. Scherbarth, D.L. Cremers, M.J. Ferris, Characterization of LIBS for application to space exploration, Appl. Spectrosc. 54 (2000) 331.

[84] M. L'Heureux, Analysis of the state of the art: LIBS, Spectroscopy 30 (2015) 74.

[85] S.N. Thakur, D.K. Rai, 'LIBS on Planet Mars' in Atom, Laser and Spectroscopy, second ed., PHI Learning, Delhi, 2013. p. 286.
[86] R.E. Russo, A.A. Bol'shakov, X. Mao, C.P. McKay, D. L. Perry, O. Sorkhabi, Laser ablation molecular isotopic spectrometry, Spectrochim. Acta B 66 (2011) 99.
[87] X. Mao, A.A. Bol'shakov, D.L. Perry, O. Sorkhabi, R. E. Russo, LAMIS: parameter influence on boron isotope measurements, Spectrochim. Acta B 66 (2011) 604.
[88] A.A. Bolshakov, X. Mao, J.J. Gonzalez, R.E. Russo, LAMIS: current state of the art, J. Anal. At. Spectrom. 31 (2016) 119.
[89] E. Tognoni, G. Cristoforetti, S. Legnaioli, V. Pailegchi, Calibration-free LIBS: state of the art, Spectrochim. Acta B 65 (2010) 1.
[90] I.B. Gornushkin, U. Panne, Radiative models of laser-induced plasma and pump-probe diagnostics relevant to LIBS, Spectrochim. Acta B 65 (2010) 345.
[91] R.S. Harmon, R.E. Russo, R.R. Hark, Applications of LIBS for geochemical and environmental analysis: a comprehensive review, Spectrochim. Acta B87 (2013) 11.
[92] G.S. Senesi, D. Manzini, O. De Pascale, Handheld LIBS instrument for the diagnosis of the conservation state of stone monuments, in: IMEKO Conf. Metrolo. Archeo. & Cultural Heritage, Lecce, Italy, October 23–25, 2017, 2017.
[93] D.C. Jeong, P.S. Tsai, D. Kleinfeld, Prospect for feedback guided surgery with ultra-short pulsed laser light, Curr. Opin. Neurobiol. 22 (2012) 24.

CHAPTER

2

Atomic emission spectroscopy

Surya N. Thakur

Laser and Spectroscopy Laboratory, Department of Physics, Banaras Hindu University, Varanasi, India

1 Introduction

Light from the Sun sustains all forms of life on Earth. It carries the visual information that makes us aware of our surroundings and the materials around us. Light belongs to the family of electromagnetic waves and this discovery by Maxwell in the mid-19th century brought one of the most profound changes to the history of science. Spectroscopy deals with the science of interaction between electromagnetic radiation and microscopic constituents (atoms and molecules) of materials of all kinds. Electromagnetic waves are generated by changes of internal energy of atoms and molecules, and these waves (including light) are very specific to the atomic species that generate them. It would not be out of place to say that atoms communicate with the space around them by means of light emitted by them, which travels in a vacuum with speed c that cannot be exceeded by any other moving object. Our world view and conception of vast heavens is based on our ability to see with the naked eye and also with the aid of instruments like telescopes and microscopes. Spectroscopy is one of the most important tools that have helped our understanding of the structure and dynamics of atoms and molecules.

The natural phenomenon of the rainbow, exhibiting concentric arcs of different colors, has attracted human attention from the dawn of civilization. However, the cause of its appearance due to refraction of sunlight from tiny water droplets in the atmosphere was not known until the second half of the 17th century. It was not until 1666 that Newton passed a beam of sunlight through a glass prism and found that it was dispersed into a continuous series of rainbow colors. Newton coined the word "spectrum" to describe the dispersion of white light into its constituent colors and this marked the beginning of the science of spectroscopy. The word "VIBGYOR" was coined to represent violet, indigo, blue, green, yellow, orange, and red in the decreasing order deviation from the direction of the incident ray of white light. It was believed, for nearly 135 years after Newton's discovery, that sunlight has only visible constituents. The fact that the Sun's radiation has components beyond the visible portion of the spectrum was first demonstrated by W. Herschel in 1800 by discovering the "infrared," and then by J.W. Ritter who discovered "ultraviolet" radiation. Until this time, spectroscopy was limited to visual observations only, but then thermometric followed by photographic tools started to be used to analyze the spectrum.

Laser-Induced Breakdown Spectroscopy
https://doi.org/10.1016/B978-0-12-818829-3.00002-2

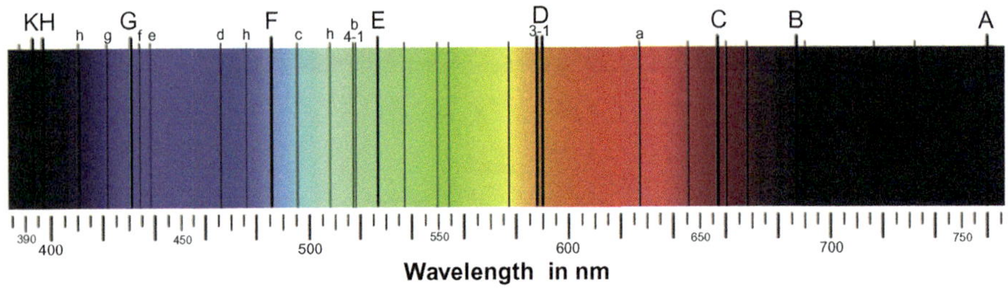

FIG. 1 Fraunhofer lines in the solar spectrum. A and B are spectral lines of oxygen, C is the doublet line of hydrogen, and D is Hα line of the sodium atom.

Joseph Fraunhofer, in 1814, extended the discovery of Newton by observing that the solar spectrum exhibited a large number of dark lines when the dispersion was increased by using a much larger prism (Fig. 1). These dark lines in the solar spectrum were the first spectral lines ever observed, now known as the Fraunhofer lines after their discoverer. Fraunhofer measured the wavelength of these lines, providing the first quantitative analysis of the spectrum and identifying them with letters as shown in Fig. 1. He also studied the spectra of stars and planets by using a telescope objective and prisms of different types of glasses, thus laying the foundation for the science of astrophysics. He made a transmission diffraction grating, which dispersed light in much the same way as a prism except that the order of deviation of colors was reversed. His achievements are all the more impressive considering that he died at an early age of 39.

Kirchhoff and Bunsen in 1859, more than three decades after the death of Fraunhofer, recorded the spectra of many substances by burning them in the flame from a Bunsen burner. From the spectra thus recorded, they established that each chemical element has its unique line spectrum. These emission lines were bright and the wavelengths of some of them were identical to some of the dark Fraunhofer lines. It was shown that a vapor at high temperature, which radiates line spectrum, must at the same temperature absorb the spectral lines emitted by it. On the basis of these investigations, Kirchhoff and Bunsen explained Fraunhofer lines as a result of selective absorption of the continuous spectrum emitted from the hot interior of the Sun by elements present in its relatively cooler atmosphere (Fig. 1). It is to be noted that helium was first discovered in the solar spectrum during the total solar eclipse of 1868, in Guntur town in India, by French astronomer Jules Janssen as a yellow line at 587.49 nm. Thus Kirchhoff and Bunsen were the first to establish spectroscopy as a tool for spectrochemical analysis to determine the elemental composition of materials. In the following section we describe the internal energy states of atoms and the characteristic spectral lines in their emission spectra.

2 Atomic states and spectral lines

Quantum theoretical interpretation of the spectrum of a hydrogen atom leads to its electronic structure in which the tiny nucleus (diameter 10^{-13} cm) is surrounded by a spherical electron charge cloud (diameter 10^{-8} cm). The ground state energy of a hydrogen atom is $-R$ in wavenumber units (cm^{-1}), where R is the Rydberg constant, and atom can be visualized as a sphere of about 1 Å diameter. The excited states with their energy in terms of principal quantum number n are shown in Fig. 2, where the energy separation between consecutive excited levels goes on decreasing with

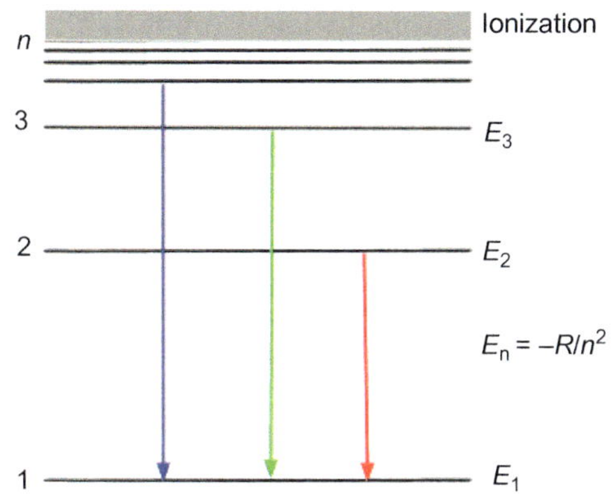

FIG. 2 Energy-level diagram of a hydrogen-like atom showing the ionization limit and emission transitions to the ground state of the atom.

increase in the degree of excitation until it merges into a continuum of energy at the limit of ionization. The size of the atom as visualized in terms of the diameter of the electron cloud goes on increasing with the degree of excitation until the electron is no longer bound to the nucleus at the limit of ionization. The frequency of transition between a pair of energy levels E' and E'' ($<E'$), in wavenumber units (cm^{-1}), is given by:

$$\tilde{\upsilon} = (\nu/c) = E' - E'' \tag{1}$$

where ν is the frequency in hertz (Hz). Some of the transitions from excited states to the ground state of a hydrogen-like atom are also shown in Fig. 2.

In thermal equilibrium at temperature T, the number density of atoms in an atomic state of energy E_n is given by the Boltzmann equation:

$$N_n = N_1 \exp(-E_n/kT) \tag{2}$$

So, most of the atoms are in the lowest (or ground) state E_1 and their number density in the excited states exponentially decreases with the degree of excitation. An atom from the ground state can reach the excited state by impact of a photon or by impact of a particle (such as electrons). The condition of photon impact excitation requires Eq. (1) to be satisfied so that the energy of the photon has to exactly match the difference of excited and lower atomic states. The condition of particle impact excitation is that the kinetic energy of the colliding particle must be greater than or equal to the energy difference between the two states, i.e.:

$$\tfrac{1}{2}mv^2 \geqq E' - E'' \tag{3}$$

where m is the mass of the particle and v is its velocity.

Once in the excited state, a free atom emits a photon spontaneously after a time of the order of a few nanoseconds and this process is termed "spontaneous emission." In the presence of an external photon, which satisfies Eq. (1), the atom emits a photon that is similar in all respects to the incident photon and the process is termed "stimulated or induced emission." The process of induced emission involving a pair of atomic states is the exact reverse of the process of absorption from the lower to the upper state. The Einstein coefficient of spontaneous emission A and the coefficient of induced emission (or absorption) B are related by the following equation:

$$A = \left(8\pi h\upsilon^3/c^3\right)B \tag{4}$$

where h is Planck's constant, c is the speed of light, and ν is the frequency of transition in Hertz.

2.1 Many-electron atoms

In a hydrogen atom the quantum state is described with the same set of four quantum numbers n, l, m, and s that describe the single electron around the nucleus. The model for the ground state of a neutral atom of nuclear charge Z is constructed by assigning the Z electrons to hydrogenic orbitals (defined by quantum numbers n, l, and m) in such a way that electronic configuration of the lowest potential energy is obtained. The orbital capacities and order of assigning electrons to atomic orbitals are governed, respectively, by Pauli's exclusion principle and Hund's rule of maximum multiplicity. Thus an orbital specified by n, l, and m can accommodate a maximum of two electrons: one with $s = ½$ and the other with $s = -½$. Hund's rule has its basis in Coulomb's law and it requires that the configuration of minimum potential energy would be obtained by allowing the electrons to stay as far apart as possible.

The application of Pauli's principle together with Hund's rule has been quite successful in predicting the ground state electronic configurations of lighter atoms. This is, however, not so for heavier atoms. In the case of iron (Fe) with $Z = 26$ the ground state electronic configuration would be $1s^22s^22p^63s^23p^63d^8$ with the two unpaired electrons in the d orbitals. However, this configuration is not consistent with the chemical and magnetic properties of iron, which requires four unpaired electrons in the d orbitals with the ground-state configuration $1s^22s^22p^63s^23p^63d^64s^2$. This discrepancy is due to repulsion produced on $n = 3$ electrons by the inner shell electrons. The occupied d orbitals are repelled to energies greater than that of the unoccupied 4s orbital and as a result the 4s orbital becomes occupied with two paired electrons in preference to the 3d orbitals. A modern form of periodic table due to Longuet-Higgins [1] is shown as Table 1 where each group contains elements with similar electronic configuration.

TABLE 1 The periodic table.

Orbital	Elements
7p	113,114,115,116,117,118
6d	Rf,Db,Sg,Bh,Hs,Mt,Ds,Rg,Cn
7s	Fr, Ra
7s	Ac
5f	Th, Pa,U,Np,Pu, Am, Cm,Bk, Cf, Es, Fm, Md, No, Lr
6p	Tl, Pb, Bi, Po, At, Rn
5d	Hf, Ta, W, Re, Os, Ir,Pt, Au, Hg
6s	Cs, Ba
6s	La
4f	Ce, Pr, Nd, Pm, Sm, Eu, Gd, Tb, Dy, Ho, Er, Tm, Yb, Lu
5p	In, Sn, Sg, Tc, I, X
5s	Rb, Sr
4d	Y,Zr, Nb, Mo, Tc,Ru, Rh, Pd, Ag,Cd
4p	Ga, Ge, As, Se, Br, Kr,
4s	K, Ca
3d	Sc, Ti, V, Cr, Mn, Fe, Co, Ni,Cu, Zn
3p	Al, Si, P, S, Cl, Ar
3s	Na, Mg
2p	B, C, N, O, F, Ne
2s	Li, Be
1s	H, He

Atomic weight increases as the table is read from left to right and then upward.

2.2 Classification of electronic states

There are two ways in which individual electronic l and s values may be added vectorially to give the resultant atomic angular momentum quantum number J. In the first scheme the l and s of each electron may be added vectorially to give a resultant one-electron angular momentum quantum number j. The j values of all atomic electrons may then be added vectorially to give J. This is known as the j–j coupling scheme. In the second scheme, the individual l values of each electron may be added vectorially to give a total orbital angular momentum quantum number L. The s values of each electron may similarly be added vectorially to give a total spin angular momentum quantum number S. The values of L and S may then be added vectorially to give J for the atom. This is known as the L–S (or Russell–Saunders) coupling scheme. The two addition methods for electronic angular momenta correspond to different physical situations. In L–S coupling, L, S, as well as J can be used to describe electronic states but in j–j coupling, L and S have no physical meaning and J is the only good angular momentum quantum number.

The orbital angular momentum vector of an electron is located at the nucleus, whereas the spin angular momentum vector is located, roughly, on the electronic orbit; the coupling of l and s is very weak unless the electron spends a considerable portion of its time near the nucleus. The probability density of an electron is large near the nucleus in atoms with high nuclear charge (heavy atoms), which exhibit appreciable spin-orbit coupling and their atomic angular momentum is given by j–j coupling. On the other hand, interelectronic repulsion is strong if several electrons have high probability densities in the same region of space. If the atom has a nucleus of low nuclear charge, all the individual electronic orbital angular momenta, located at the nucleus, will couple strongly and similarly with the individual spin momenta, located on the electronic charge distribution. The coupling of l and s will be secondary to the coupling of l–l and s–s and hence L–S coupling will adequately describe the situation. For most atoms, even when appreciable spin-orbit interaction occurs, L–S coupling is retained, with an appropriate perturbation treatment to account for the interaction. In the remainder of this discussion we will describe the stationary states of atoms in the framework of L–S coupling.

The general nomenclature of an atomic state is based on the representation ${}^{2S+1}L_J$, where the state is labeled as S, P, D, F, G, etc. corresponding to $L = 0, 1, 2, 3, 4$, etc., respectively.

Thus with $L = 2$ and $S = 1$, the values of J will be 3, 2, and 1 and the resulting states will be represented as ${}^3D_3$, ${}^3D_2$, and ${}^3D_1$, respectively. It is to be noted that the total angular momentum quantum number J and the corresponding magnetic quantum number M are always good quantum numbers, irrespective of the coupling scheme, and the atomic wavefunction is represented as ψ_{JM}. The corresponding atomic state is $(2J + 1)$-fold degenerate in the absence of an external electric or magnetic field. For further information on atomic structure the reader is advised to read some of the excellent books [2–7].

3 Radiation from atoms

The intensity of a spectral line depends on the atomic population of the initial level and also on the intrinsic probability to the final level. If we assume an external electromagnetic radiation as a time-dependent perturbation to the atom, it is easy to visualize that this perturbation should be as effective for the induced emission $E_2 \rightarrow E_1$ as for the absorption transition $E_2 \leftarrow E_1$ between a pair of atomic states E_2 and E_1 ($<E_2$). Thus in the light of Eq. (4), $B_{21} = B_{12}$ if the two states are nondegenerate. The B coefficient can be calculated from quantum

mechanics by treating the oscillating electric and magnetic fields of incident light as a time-dependent perturbation leading to the following expression:

$$B_{21} = (2\pi^2/3\varepsilon_0 h^2)\left[Mx_{21}{}^2 + My_{21}{}^2 + Mz_{21}{}^2\right] \quad (5a)$$

where ε_0 is the permittivity of free space and Mx, My, and Mz are the components of the transition moment vector $\mathbf{M_{12}}$ given by

$$\mathbf{M_{12}} = \int \Psi^*{}_{J2M2} e\mathbf{r}\Psi_{J1M1} d\tau \quad (5b)$$

In Eq. (5b), Ψ_{J2M2} and Ψ_{J1M1} are the wavefunctions of the upper and lower states, respectively, e is the electronic charge, and $\mathbf{r}$ is the operator corresponding to the displacement of electronic charge in the atom as a result of the transition ($e\mathbf{r}$ is the dipole moment vector) and the transition between the two states is said to be allowed if the integral is different from zero.

3.1 Electric dipole selection rules

The quantum numbers that define the wavefunctions of the two states satisfy certain relations for a nonzero transition moment and these relations are called selection rules. The selection rules for the total angular momentum quantum number J for dipole-allowed transitions are given by:

$$\Delta J = J_2 - J_1 = 0, +1, -1 \quad (6a)$$

If the spin-orbit coupling is so weak that the orbital motion is practically the same as in the absence of electron spin, then the orbital quantum number L retains its significance and the corresponding selection rules are:

$$\Delta L = L_2 - L_1 = +1, -1 \quad (6b)$$

The selection rules for M are:

$$\Delta M = M_2 - M_1 = 0, +1, -1 \quad (6c)$$

The selection rule $\Delta M = 0$ corresponds to emission of light from the atom with its electric vector oscillating along the z-axis. This radiation cannot be seen along the z-direction and it gives rise to linearly polarized light when viewed along a direction in the x–y plane. The selection rules $\Delta M = +1$ and $\Delta M = -1$ correspond to right and left circular polarization when viewed in the z-direction and linearly polarized when viewed in a direction in the x–y plane.

3.2 Parity selection rules

The energy states of free atoms can be divided into two classes according to their parities. If the orbital angular momentum quantum number l_i of individual electrons is taken into consideration, the parity of the sum Σl_i is the parity of the resulting energy state. The energy state is said to have.

odd or even parity depending on the odd or even value of Σl_i. Odd parity of a state is indicated by a superscript o such as $^2P^o_{3/2}$. The parity property is well defined for an atom with any number of electrons with any kind of coupling of their orbital and spin angular momenta. The rigorous selection rule based on parity is known as the Laporte rule. Only transitions between even and odd states are allowed for dipole radiation. It can be shown that all the selection rules in Eq. (6) obey the parity selection rule.

3.3 Forbidden transitions

Transitions that are forbidden in the dipole approximation may appear very weakly as quadrupole radiation or magnetic dipole radiation. The matrix elements for quadrupole radiation take the following form:

$$Q_{12} = \int \Psi^*{}_{J2M2}\, exy\Psi_{J1M1} d\tau \quad (7a)$$

and it does not vanish when Ψ_{J2M2} and Ψ_{J1M1} have the same parity because the parity of the operator xy is clearly even. The selection rules for quadrupole-allowed transitions are:

$$\Delta L = 0, +2, -2;\ \Delta J = 0, +1, -1, +2, -2 \quad (7b)$$

The Laporte rule in this case states that only transitions from even to even or odd to odd terms (states) give rise to quadrupole radiation.

Magnetic dipole transitions appear with the same strength (with intensity 10^{-6} times the intensity of electric dipole-allowed transitions) as electric quadrupole transitions. They result from the fact that the radiation field produced by a system of moving charges in the atom cannot be adequately described in terms of electric dipole, quadrupole, or octopole radiation but also need terms containing magnetic dipole, quadrupole, etc. The selection rules for magnetic dipole-allowed transitions are:

$$\Delta L = 0; \Delta J = +1, -1 \tag{7c}$$

The forbidden transitions that are forbidden as dipole radiation only to a first approximation may become allowed under special conditions. Intercombination lines resulting from transitions between states of different multiplicities fall in this category. The electric dipole selection rules hold strictly only for free atoms and external fields due to ions in a crystal or those in a discharge; even the fields resulting due to neutral atoms can give rise to forced electric dipole transitions of observable intensity.

3.4 Line strength

If we adopt the following notation for the x-component of the electric dipole transition moment: $\int \Psi^*{}_2\ ex\Psi_1 d\tau = e|x_{12}|$, Eq. (5a) takes the form:

$$B_{21} = (2\pi^2 e^2/3\varepsilon_0 h^2)\left[|x_{12}|^2 + |y_{12}|^2 + |z_{12}|^2\right] \tag{8a}$$

The electric dipole line strength S of the transition is defined as:

$$S = S_{21} = S_{12} = e^2\left(|x_{12}|^2 + |y_{12}|^2 + |z_{12}|^2\right) \tag{8b}$$

From Eqs. (4), (5), and (8b), we get the following relations for nondegenerate states:

$$B_{21} = (2\pi^2/3\varepsilon_0 h^2)S_{12} \tag{8c}$$

$$A_{21} = (16\pi^3\nu^3/3\varepsilon_0 h^2)S_{12} \tag{8d}$$

If the two states involved in the transition are degenerate and the degeneracies of E_1 and E_2 are g_1 and g_2, respectively, then the line strength is given by:

$$S = S_{21} = S_{12} = e^2\Sigma^{g_1}{}_{M1}\Sigma^{g_2}{}_{M2}\left(|x_{12}|^2 + |y_{12}|^2 + |z_{12}|^2\right) \tag{9a}$$

The degeneracies will cause the atomic populations to be divided among g_1 and g_2 number of sublevels for E_1 and E_2, respectively, and the number of atoms in E_1 will reduce by a factor $1/g_1$ and those in E_2 by a factor $1/g_2$. The expressions for Einstein's coefficients will take the following form:

$$B_{12} = (2\pi^2/3\varepsilon_0 h^2)S/g_1 \tag{9b}$$

$$B_{21} = (2\pi^2/3\varepsilon_0 h^2)S/g_2 \tag{9c}$$

$$A_{21} = (16\pi^3\nu^3/3\varepsilon_0 h^2)S/g_2 \tag{9d}$$

3.5 Oscillator strength

The classical model of emission from an atom has the electron performing simple harmonic motion at a characteristic frequency ν but with amplitude decreasing with time because of the energy radiated away. The emission of *radiation* acts as the damping agent for the electron's motion and the damping constant is given by:

$$\gamma = (2\pi e^2\nu^2)/(3\varepsilon_0 mc^3) \tag{10}$$

where m is the electron mass. The lifetime of a classical oscillator is the inverse of the damping constant ($\tau = 1/\gamma$) and it is of the order of 10^{-8} s for emission in the visible region.

When light passes through an absorbing medium, the energy absorbed by atoms is proportional to the thickness of the medium and also to the incident flux of light. The transmitted light intensity at frequency ν across a thickness l of the absorbing medium is given by:

$$I(l) = I(0)\exp(-k_\nu l) \tag{11}$$

where k_ν is in units of cm^{-1} and is called the absorption coefficient. Absorption is also expressed in terms of atomic absorption coefficient (or atomic cross-section),

$$\alpha_\nu(k_\nu/N_1),$$

where N_1 is the population density of the lower energy state E_1 and α_ν has the dimension of area. It is to be noted that k_ν is dependent on the profile of the spectral transition (line shape) and its relation with B_{12} is given by:

$$\int k_\nu d\nu = (h\nu_0/c)N_1 B_{12} \quad (12a)$$

where the integration is to be carried over the line profile and ν_0 is the frequency corresponding to the peak of the line profile.

If the number density N_2 of atoms in the upper energy state E_2 is appreciable, there would be stimulated emission, putting photons back into the incident beam and the relation between k_ν and B_{12} would become:

$$\int k_\nu d\nu = (h\nu_0/c)N_1 B_{12}(1 - g_1 N_2/g_2 N_1) \quad (12b)$$

The quantum mechanical and classical models of light emission are related through a quantity called oscillator strength. This is achieved by equating the light absorbed by N_1 atoms in the transition $E_2 \leftarrow E_1$ to that absorbed at the same frequency by N classical oscillators so that $N = f_{12}N_1$. The oscillator strength f is related to Einstein's coefficient B by:

$$f_{12} = \left(4mh\varepsilon_0/e^2\right)\nu_{12}B_{12} \quad (13a)$$

If we replace ε_0 by $1/4\pi$ the relation between f and B (in cgs units) becomes:

$$f_{12} = \left(\mathrm{mh}/\pi e^2\right)\nu_{12}B_{12} \quad (13b)$$

The normal meaning of f is the oscillator strength f_{12} but the emission f value, f_{21} is a negative quantity given by:

$$f_{21} = -(g_1/g_2)f_{12} \quad (13c)$$

The oscillator strength f is also interpreted as the effective number of electrons per atom for a particular transition. Thus f should be approximately 1 for one valence electron atom (hydrogen and alkalis) and 2 for the alkaline earths. Since each electron can participate in several different transitions, the total oscillator strength is accordingly split between several spectral lines. Thus f represents the fraction of the available electrons participating in a particular transition and has values in the range 0.01–1 for strong spectral lines.

3.6 Intensities of spectral lines

The observed intensity of a spectral line depends on two factors. The first factor is an intrinsic property of an atom and can be described in terms of transition probability or line strength or f value. The second factor depends on the conditions of excitation. When the emitting medium is such that a photon produced by spontaneous emission has no appreciable chance of being reabsorbed, the medium is termed optically thin. This condition holds in a gas discharge at low pressure where kinetic energy of the electrons is distributed approximately according to Maxwell's distribution law. The concept of an electron temperature is valid in such conditions, which has a very much higher value than the temperature defined by the kinetic energy of atoms or ions. Since the probability of electron impact excitation of atoms is very high in these conditions, the population of excited states tends to be in accordance with the electron temperature.

Equilibrium conditions exist in a discharge source whenever the exchange of energy between atoms, and between atoms and electrons, is rapid compared to the rate of excitation. For gas discharges at or above atmospheric pressure, the gas temperature may approach the electron temperature. If two spectral lines originating from upper energy levels E_1 and E_2 in the condition of thermal equilibrium are observed to have intensities I_1 and I_2, respectively, then these are related to the transition probabilities A_1, A_2, and line strengths S_1, S_2 as follows:

$$\begin{aligned} I_1/I_2 &= A_1N_1/A_2N_2 \\ &= (A_1g_1/A_2g_2)\exp[(E_2-E_1)/kT] \\ &= \left(S_1\nu^3{}_1/S_2\nu^3{}_2\right)\exp[(E_2-E_1)/kT] \end{aligned} \quad (14)$$

where k is the Boltzmann constant and T is the gas temperature.

3.7 Self-absorption

Spectral lines with lowest excited level of the atom as their upper state and ground state of the atom as their lower state are called resonance lines. The light emitted in such lines has a very large probability of reabsorption before leaving the plasma discharge and this phenomenon is called self-absorption. Self-absorption tends to broaden the profile of the spectral line and in extreme cases makes the peak appear greatly flattened. If the excitation temperature drops in the outer regions of the discharge, the passage of light from the central region not only broadens the profile but also shows an intensity dip at the line center. This phenomenon is called self-reversal and such lines give the wrong impression of being doublets.

The deviations from thermal equilibrium can occur at very low gas densities and emission of spectral lines may exhibit peculiar intensity features if metastable levels are involved. A level is called metastable if all transitions to lower levels occur with very small probabilities due to selection rules. The concentration of atoms in a metastable level can be abnormally high and in extreme conditions they lose their excitation energy by collision and not by radiation. Metastability of a level is, however, a matter of degree depending on the values of transition probability and gas density. In the case of interstellar space or in stellar nebula, the gas density is extremely low and a metastable level with a lifetime of 0.1 s, which may emit only quadrupole radiation, will lose all its excitation energy by radiation. The intensity of such a line will only depend on the rate at which the metastable level is excited and the line may appear quite strong while it is not observed in laboratory sources.

3.8 Continuous emission and bremsstrahlung

There exist free (unbound) states of an electron-ion system for which the energy is positive. The transitions between a free state and the stationary state E_j of the atom give rise to a continuum emission whose frequencies are:

$$h\nu = \xi - E_j + \tfrac{1}{2}mv^2 \quad (15)$$

where $\tfrac{1}{2}mv^2$ is the kinetic energy of the free electron and ξ is the ionization energy of the atom.

This emission process is also called radiative recombination, which occurs when an ion captures an electron and makes a transition to the bound energy state E_j. This continuum is characterized by discontinuities whenever $h\nu$ becomes large enough to reach the next bound level. The rate at which the binary reaction involving electron and ion occurs per unit volume is proportional to the densities of both electrons and ions. It can be calculated directly from quantum mechanics in analogy to spontaneous emission of line radiation. However, it is more conveniently done by use of Kirchhoff's law, which says that in thermal equilibrium the emission can be obtained from the net absorption (difference of absorption and induced emission) by multiplying the Planck's function.

Free-free emission transitions correspond to loss of kinetic energy by an electron in the field of an ion. This emission results from deceleration of the electron and is known as bremsstrahlung, meaning "braking radiation." The time between emissions by the electron is much longer than the time between its collisions with other electrons. The bremsstrahlung emission does not lead to significant distortion of the electron velocity distribution, which may always be assumed to be near-Maxwellian. It is difficult to separate emission by radiative recombination from that due to bremsstrahlung except to say that the former is dominant at higher frequencies and the latter at lower frequencies.

4 Broadening of spectral lines

The intensities of spectral lines are greatly dependent on the environment of the atom that emits the radiation. In the ideal case of a free atom the radiated intensity of a line is spread over a frequency-dependent Lorentzian profile having the form:

$$I(\nu) = I_0(\gamma/4\pi)^2 / \left[(\nu - \nu_0)^2 + (\gamma/4\pi)^2\right] \qquad (16a)$$

where I_0 is the intensity at the center of the line profile ν_0 and γ is the radiation damping constant of Eq. (10). This spread of intensity over a range of frequencies is called natural broadening of the spectral line and $(\gamma/2\pi)$ is called full width at half maximum (FWHM). Except at very low atomic densities the ideal condition is never realized in practice and natural broadening is always accompanied by Doppler broadening, which dominates the line shape near its center. Doppler broadening arises due to random thermal motions of the emitting atoms and the resulting profile has a Gaussian shape with FWHM given by:

$$\Delta\nu_D = (2\nu_0/c)R\left(T\left(\log 2/M\right)\right)^{1/2}$$
$$= 7.16 \times 10^{-7}\nu_0(T/M)^{1/2} \qquad (16b)$$

where ν_0 is the frequency of the line center, M is the atomic mass, and T is the equilibrium temperature.

When the radiating atom is surrounded by dense plasma both of the foregoing broadening mechanisms are completely negligible in comparison to the broadening caused by the charged particles. Since the atom interacts with the charged particles through the electric fields produced by them, this type of broadening is called Stark broadening. There are two major reasons for determining the line shapes of spectral lines originating in plasma. The first reason is the use of measured line shapes to determine physical properties of the emitting plasma such as the charged particle densities and temperature. The second reason is to determine the absorption and induced emission coefficients that depend on the oscillator strength and densities of emitting atoms in addition to the line shape.

4.1 Stark broadening

The emitting species (atoms or ions) in plasma are under the influence of electric fields by fast-moving electrons and relatively slow-moving ions. The perturbing electric field acting on an atom at a distance r from an ion or electron is $F = (e/4\pi\varepsilon_0 r^2)$. The interaction between the atom and this electric field is described by the Stark effect, which splits and shifts the energy levels of the atom. Perturbation to the energy levels caused by the electric field is proportional to F only in the case of hydrogen atoms leading to the linear Stark effect. For all other atoms the perturbation to the energy levels is proportional to F^2 and the resulting shift and splitting is called the quadratic Stark effect. It is obvious that the extent of the Stark effect will be negligibly small for large values of r. If we assume that the shift in the energy level of the emitting atom depends on its excitation energy as well as on its separation from the ion (or the electron), then the center of the spectral line ν_0 shifts to $\nu(r)$ such that (see energy level diagram in Fig. 2):

$$\Delta\nu(r) = \nu_0 - \nu(r) = (1/h)[\Delta E_2(r) - \Delta E_1(r)] \qquad (17a)$$

It can be seen from this relation that a + ve value of ΔE amounts to a downward shift of the corresponding energy level. A + ve value of $\Delta\nu(r)$ corresponds to a shift of the line center towards red and a –ve value indicates a shift to the blue.

An interaction causing a frequency shift $\Delta\nu$, which depends on r and tends to zero for large r, can be written as:

$$\Delta\nu(r) = C_n/r^n \tag{17b}$$

where the value of n and the interaction constant C_n depend on the type of interaction.

Thus in the case of hydrogen and hydrogen-like atoms exhibiting the linear Stark effect (proportional to $F = (e/4\pi\varepsilon_0 r^2)$), $n = 2$ and the interaction constant is C_2. The linear Stark effect splits the energy levels symmetrically resulting in a symmetrically broadened but unshifted line. In the case of all other atoms the quadratic Stark effect is proportional to F^2 so that $n = 4$ and the corresponding interaction constant is C_4 in Eq. (17b). It is to be noted that because of the F^2 dependence the Stark shift is the same for electrons and ions. The quadratic Stark effect splits the energy levels asymmetrically and also shifts their center of gravity downward (to lower energies). Since the shift of energy levels is larger for higher excitation energies the frequencies of the transitions are reduced. This implies that $\Delta\nu < 0$ in Eq. (17b) and hence $C_4 < 0$. A spectral line broadened by the quadratic Stark effect becomes asymmetric and its center shifts to longer wavelengths.

Stark broadening may be qualitatively described in terms of impact approximation where the emitting atom is assumed to collide with fast-moving electrons for a duration extremely small in comparison to the time between two successive collisions. This impact on the atom is assumed to cause a phase change in the optical wave train emitted by the atom before and after the collision and not to stop it altogether by knocking out the atom from its excited state. The interaction between the electron and the emitting atom is thus regarded as an optical collision, which results in a sudden phase change of the light wave emanating from the atom and there is no effect on the state of the atom [8–11].

The interaction between slowly moving ions and radiating atoms can be described by quasi-static approximation in which the perturbation remains nearly constant over the whole time that the atom is radiating. Following Holtsmark [12] the motions of ions are neglected and their perturbing action is included in an electric field F that produces a static Stark effect. In the next step the statistical average of the Stark effect over various values of the ion field F is taken. In the final step of the calculation, each element of the spectral line is considered to be broadened and shifted by the electron impacts. The quadratic Stark effect produces asymmetrically broadened lines, whereas the linear Stark effect gives rise to symmetrically broadened line shapes. A large amount of theoretical and experimental work has been carried out in the case of spectral line broadening by charged particles. With the help of existing theoretical models, it is possible to calculate line profiles that fit the experimental ones [13–15].

5 Applications

In the preceding sections we have seen that the intensities, shapes, and widths of atomic lines depend on the atomic structure as well as on the temperature, pressure, and electron density of the discharged plasma. Analysis of the spectral lines can give information about the physical state of the emitting gas without interfering with the plasma. The use of laser-based optical techniques in recent years has replaced spectroscopy as a diagnostic tool to some extent. Nevertheless, spectroscopy still plays a major part in determining the physical processes occurring in the plasma. It is the most reliable method of detecting and identifying trace elements in a source. In the following sections we present a brief account of these applications.

5.1 Determination of electron temperature

When the temperature of a molecular vapor is increased, molecules tend to dissociate into atoms and atoms into ions plus electrons; some of the molecules, atoms, and ions are excited to higher energy states, and the kinetic energy of all these particles and of the free electrons increases. The spectroscopic determination of electron temperature of a source of radiation is based on the assumption that local equilibrium conditions must exist in each small volume that contributes to emission. Complete thermodynamic equilibrium exists when all forms of energy distribution are described by the same temperature. In the following sections we first discuss the approximate conditions that may prevail in plasma and then describe methods of determining the temperature of the source from the measurements of spectral lines emitted by the plasma.

5.1.1 *Temperature and equilibrium*

Maxwell's distribution for velocities of particles of mass M gives the number of particles dN with velocity between v and $v + dv$ in terms of their number density N and temperature T as follows:

$$dN(v) = N(M/2\pi kT)^{3/2} \exp\left(-mv^2/2kT\right)4\pi v^2 dv \tag{18a}$$

The Boltzmann distribution of particles N_j having excitation energy E_j is:

$$N_j = N\left[g_j/U(T)\right] \exp\left(-E_j/kT\right) \tag{18b}$$

where $U(T) = \Sigma g_j \exp.(-E_j/kT)$ is called the state sum or partition function.

The condition of equilibrium for ions, electrons, and neutral atoms is given by Saha's equation:

$$N_e N_i/N_0 = \left[2(2\pi mkT)^{3/2}\right]/h^3[U_i(T)/U_0(T)] \exp\left(-\xi/kT\right) \tag{18c}$$

where N_e is the number density of electrons and N_i and N_0 that of the ions and neutral atoms, respectively, regardless of the energy levels they occupy, m is the mass of the electron, and ξ is the ionization energy of the atom. The factor of 2 represents the state sum for the two possible spin states.

The equilibrium distributions for kinetic energy, excitation energy, and ionization energy are represented by the Maxwell, Boltzmann, and Saha equations, respectively, and it may happen that there is equilibrium distribution of one of these forms of energy but not for the others. In that case the temperature parameters for Eqs. (18a), (18b), and (18c) are all different. Complete thermodynamic equilibrium exists when all forms of energy distribution are described by the same temperature parameter. Thermodynamic arguments require that for equilibrium to hold, for *every* photon emitted by the system, a photon of the same energy is absorbed and for every excitation by electron collision there must be a deexcitation by electron collision. In practice, however, photons do leak out from the plasma, no matter how large or dense the plasma is, otherwise we would be unable to observe the plasma. Thus the condition of near thermodynamic equilibrium requires that such losses be small compared to the total energy. Many plasmas can be described by a state known as local thermodynamic equilibrium (LTE), in which it is possible to find a temperature parameter for every point in a region of space that fits the Boltzmann and Saha relations for the population density of excited and ionic states and the Maxwell distribution of velocities among the electrons. The criterion for LTE is that collisional processes must be much more important than radiative, so that the deficit of radiative energy is extremely small. In other words, the probability of deexcitation by inelastic collision for an excited state must be very large compared to that of spontaneous emission. This is possible at very high electron densities in the plasma such that:

$$N_e \gg A_{21}/(v\sigma_{21}) \tag{19a}$$

If S is the line strength, the excitation cross-section corresponding to electron velocity v at threshold is given by $\sigma_{12}(v) \sim \pi(e/4\pi\varepsilon_0 h)^2 S/v^2$. Putting this value of σ_{12} in Eq. (19a) makes the right-hand side proportional to vA_{21}/S. Since A_{21} is proportional to $S\nu^3$ and v is proportional to $T^{1/2}$, the value of

N_e is proportional to $\nu^3 T^{1/2}$ and the numerical relationship for LTE is given by:

$$N_e \gg 1.6 \times 10^{12} T^{1/2} (E_2 - E_1)^3 \ \text{cm}^{-3} \qquad (19b)$$

where T is the electron temperature in K and $E_2 - E_1$ is the energy difference in electron volts between the two neighboring states with an allowed transition.

It is possible to determine the relative population of atoms in various excited states even in the absence of LTE provided the collisional cross-sections and radiative transition probabilities are known. Another approximation is to assume that level E_2 in Fig. 2 is populated entirely by electron collision and depopulated entirely by spontaneous radiation. This is called coronal equilibrium (CE) since it is applicable to the Sun's corona where temperature is high (10^6 K), electron density is low (10^8 cm^{-3}), and radiation density is also low. The populations of higher states are much lower in CE than in LTE. CE can hold only if N_e is below a critical value and it holds good only for the lower excited states. It is quite possible that the atomic populations of higher states follow LTE, while those of the lower states follow CE.

5.1.2 *Temperature from relative intensities of lines*

The method for determination of temperature in LTE plasma is based on the fact that the number densities in various excited states follow Boltzmann distribution. The temperature in terms of relative intensities of lines from the same element and the same state of ionization is given by:

$$kT = (E_2 - E_1)/\log_e\left(I_1 \lambda_1{}^3 g_2 f_2 / I_2 \lambda_2{}^3 g_1 f_1\right) \qquad (19c)$$

where I_1 is the total intensity integrated over the line profile, λ_1 is the wavelength, f_1 is the oscillator strength of the spectral line with excitation energy E_1, and I_2, λ_2, and f_2 are the corresponding quantities for the line with excitation energy E_2. The statistical weights for energy states E_1 and E_2 are g_1 and g_2, respectively.

Relative intensities can be measured with an accuracy of better than 10% but errors in the oscillator strengths are more. Since kT is of the order of the largest energy separation $E_2 - E_1$ between excitation energies of nonresonance lines, the uncertainty of oscillator strengths is reflected in the errors associated with the temperature.

The method just described, based on the relative intensities of lines from the same atom and ionization stage, generally leads to inaccurate temperature. The main reason being the small separation between E_1 and E_2, which is typically smaller than or equal to kT and it renders the line intensity ratio somewhat insensitive to temperature variations. This shortcoming is removed if spectral lines from successive ionization stages of the same atom are compared. The effective energy difference is enhanced by the ionization energy leading to increased sensitivity to temperature changes. In LTE the relation between relative intensities and the source temperature is given by:

$$(I'/I) = (f(g'\lambda^3/fg\lambda'^3)\left(4\pi^{3/2}a_0{}^3 N_e\right)^{-1}(kT/E_H)^{3/2}\exp\left[(E + \Delta E_\infty - E' - E_\infty)/kT\right] \qquad (19d)$$

where primed quantities correspond to the spectral line from the higher ionization stage, ΔE_∞ is the reduction in the ionization energy E_∞ of the lower ionization stage due to plasma effect, a_0 is the Bohr radius, and E_H is the ionization energy of hydrogen.

5.1.3 *Temperature from Doppler profile*

The most reliable spectroscopic technique for measuring the kinetic temperature of atoms and ions is based on the measurement of the widths of Doppler-broadened spectral lines. In the case of Maxwellian velocity distribution of emitting species, such lines have Gaussian profiles with FWHM given by Eq. (16b). One must make sure that the thermal Doppler effect is the major cause of line broadening in the source before using this method of temperature measurement. It has been found that even gross motions in the source could simulate a Gaussian profile. To avoid such misinterpretations it is advisable to observe the plasma from different angles to watch for shifts of the intensity maxima.

5.2 Determination of the electron density

The most powerful spectroscopic technique of determining the electron density N_e of a discharge plasma comes from the measurement of the Stark broadening of spectral lines. In this method, absolute intensities of spectral lines are not required, merely line shapes and FWHM are sufficient. Since broadening is quite appreciable for electron density $N \geq 10^{15}\ \mathrm{cm}^{-3}$, standard spectrometers often suffice to record the spectra for measurements of line shape. The electron density N_e is extracted by matching the line width (or the entire line shape) with the calculated one. Details of line shape calculations can be found in a book by Griem [16]. In this section we will summarize only the salient points.

5.2.1 Hydrogen and hydrogen-like ions

Hydrogen and hydrogen-like ions exhibit linear Stark effect. The broadening of spectral lines is found to be dependent on the optical transition and a judicious choice is important. At low plasma densities a spectral line with large Stark broadening is desirable but at high densities a line with relatively smaller broadening is useful so that its outer wings do not overlap with neighboring lines. The FWHM (in Ångstrom units) of the spectral line in the quasistatic approximation is given by:

$$\Delta\lambda = 8.16 \times 10^{-19}\left(1 - 0.7N_D{}^{-1/3}\right)\lambda_0{}^2\left(n_2{}^2 - n_1{}^2\right)\left(Z_p{}^{1/3}/Z_e\right)N^{2/3} \quad (20a)$$

where $N_D = (4\pi/3)N\rho_D^3$ is the number of particles in the Debye sphere, N is in cm^{-3}, λ_0 is the line center, n_2 and n_1 are the principal quantum numbers of the upper and lower states, respectively, Z_p is the nuclear charge on the perturbing ion, and Z_e is that on the emitting particle (atom or ion).

In the foregoing discussion we have neglected contributions to FWHM from the plasma electrons. Although the line shapes depend on the electron contribution, the FWHM are generally insensitive. Eq. (20a) represents a very good estimate of FWHM in those hydrogenic lines that do not have a strong undisplaced Stark component as, for example, the Lyman β, Lyman δ, Balmer β, and Balmer δ transitions. On the other hand, the FWHM of hydrogenic lines with strong central Stark components are dominated by the interaction of the electron with the emitting atom such as Lyman α and Balmer α transitions. Such lines have a Lorentzian line shape and FWHM for Lyman α transition in the impact approximation is given by:

$$\Delta\lambda \approx 1.62 \times 10^{-17}\left(\mathrm{N}/\mathrm{T}^{1/2}\right)\left[13.76 - \log\left(\mathrm{N}^{1/2}/\mathrm{T}\right)\right] \quad (20b)$$

where $\Delta\lambda$ is in Ångstrom, T is in K, and N is in cm^{-3}.

It is seen from Eqs. (20a) and (20b) that the ion broadening in the quasistatic approximation varies as $N^{2/3}$ and is virtually independent of temperature, whereas the collisional broadening, in the impact approximation, varies approximately as N and it is very much temperature dependent. In case T is not reliably known, it is advisable to determine electron density N from the Balmer β line whose FWHM is ion dominated. It is to be noted that electron densities determined from Eq. (20) are only crude estimates of N and one must compute the entire line profile to extract the total line width for an accurate value of N. Some very precise hydrogen line shape measurements have been reported by Wiese and coworkers [17, 18].

5.2.2 *Many-electron atoms and ions*

Plasma generated by high-power lasers focused on gas or solid targets have electron energies in the range of 50–10 keV and hydrogen as well as other low Z atoms are fully stripped and do not radiate. It is found that carbon is completely stripped when $T \geq 100$ eV and copper is stripped when $T \approx 1$ keV [19], which suggests that sufficiently high-Z atoms should be involved in spectroscopic measurements on hot plasma. Spectral lines of neutral atoms and nonhydrogenic ions have Stark broadening mostly due to electron impacts. For such high-Z ions the Stark widths are too small to be resolved and they may be dominated by Doppler broadening. Nonhydrogenic high-Z ions are therefore not very suitable for determining electron densities in very hot plasma.

5.3 Qualitative atomic emission analysis

One of the major applications of atomic emission spectra is the identification of elements present in the source of light. The use of this technique for qualitative analysis of samples, which could be fed into flames to emit characteristic light (yellow for Na, red for Ca), dates back to Bunsen and Kirchhoff and other spectroscopists of the 19th century. Major developments in astrophysics have resulted from the studies of the spectra of radiating stellar bodies, which provided information about their chemical composition, temperature, distance, mass, etc. Some elements are more easily excited than others for their emission spectra to be observed and recorded. Thus nonmetal atoms are more difficult to excite than metal atoms because of their high ionization potential. It is found that the presence of easy-to-excite atoms in a sample suppresses the emission from atoms that are relatively difficult to excite. Thus emission from helium is suppressed in the presence of nitrogen, that of nitrogen in the presence of mercury, and the emission from mercury is suppressed if the sample also contains potassium.

When an element is excited in an arc or discharge source, a number of lines of varying intensities are observed over a wide range of the spectrum. As one dilutes the amount of the element in the arc, the number of lines observable is reduced and ultimately only a few lines of that element remain observable. These lines are known as persistent lines. It has been found that the persistent lines are also the lines of largest intensity. From this description of the persistent lines it is apparent that in the qualitative analysis for a particular element we need to look only for the persistent lines of that element. If they are absent it may be safely assumed that the element is not present in the sample. Tables of persistent lines of many elements are to be found in a book by Brode [20] and in a publication by Meggers [21]. Persistent lines for some elements are given in Table 2. In most cases of qualitative analysis, the major constituents of the discharge are readily determined by inspection of the strongest lines and other lines belonging to these constituents can then be identified by the use of the tabulation of spectral lines of the elements [22, 23]. After this procedure, the lines left unidentified are generally weak and they may be hitherto unobserved lines of the major constituents or lines of the unknown minor constituents.

In the latter case, they must be among the stronger lines of these elements. The unknown element can then be identified by comparing these lines with the persistent lines of various elements.

The qualitative analysis of a sample may be divided into two parts:

1. Search for a definite element
2. Identification of unknown elements

In the search for a definite element, the identification of the persistent lines should be sufficient to confirm the presence of the element. The identification of unknown elements requires a more complete identification of all spectral lines in the emission spectrum of the sample.

TABLE 2 Wavelengths (in Ångstrom) of persistent lines of some elements.

Element	Wavelength	Element	Wavelength	Element	Wavelength	Element	Wavelength
Ag I	3280.68	Cd I	2288.02	Be I	2348.61	Cu II	2192.26
	3382.90		2312.84		2650.78		2246.99
	5209.07		2573.09		3321.01	Fe I	3581.20
	5465.49		2748.58		3321.09		3719.94
Ag II	2246.41		3261.06		3321.34		3737.13
	2437.80		3403.65	Be II	3130.42		3245.56
Al I	3032.16		3466.20		3131.07		3745.90
	3092.71		3610.51	Bi I	2061.70		3748.26
	3944.03		6438.47		2276.58	Fe II	2382.04
	3961.53	Cd II	2144.38		2780.52		2395.63
Al II	1671.0		2265.02		2809.63		2404.88
	1856.0	Cl II	4794.54		2897.98		2410.52
	1858.1		4810.06		2938.30		2413.31
	1862.5		4819.46		2989.03	Hg I	1849.68
	2631.55	Co I	3453.51		3067.72		2536.52
	2669.17		3465.80	Bi II	1909.41		3650.15
	2816.18		3529.81		4722.55		3654.83
	6231.76	Co II	2286.16	Br II	4704.86		3663.28
	6243.36		2307.86		4785.50		4046.56
Au I	2427.95		2363.79		4816.71		4358.35
	2675.95		2378.62	C I	2296.89		5460.74
	2802.19		2388.92		2478.57	K I	4044.14
B I	2496.78		2519.82	C II	2836.71		4047.20
	2497.73		3405.12		2837.60		7664.91
B II	3451.41	Cr I	4254.35		4267.02		7698.98
Ba I	2304.23		4274.80		4267.27	Mg I	2852.13
	2335.27		4289.02	Ca I	4226.73		3829.35
	3071.59		5204.52		4425.44		3832.31
	5424.62		5206.04		4434.96		3838.20
	5519.12		5208.44		4454.78		5167.34
	5535.55	Cr II	2835.61	Ca II	3158.87		5172.70
	5777.67		2843.25		3179.33		5183.62
Ba II	3891.79		2849.84		3933.67	Mn I	4030.76
	4130.66		2855.68		3968.47		4033.07
	4554.04		2860.92	Cu I	5218.20		4034.49
	4934.09			Cu II	2135.95	Mn II	2576.10

At least three lines of which there is no possible doubt as to their origin should be identified so as to positively identify the element.

5.4 Quantitative atomic emission analysis

Determination of the elemental composition of a gaseous or condensed phase sample by means of laser-induced plasma requires the measurement of the intensity of those spectral lines that are characteristic of the individual elements present in the sample. The intensities must then be related to the number density of atoms or ions present in the plasma. The first step in the quantitative analysis is the preparation of standard samples in which the concentration of the element of interest is varied in a precisely known manner. It is obvious that the intensities of spectral lines of this element in the emission spectra recorded for different standard samples will be proportional to its concentration. The next step in the analysis is the measurement of intensities of one or more spectral lines of the said element from spectra of all the standard samples as well as the samples to be analyzed. The last step of the analysis is to plot a working curve of line intensity against known concentrations of the element for the standard samples. The concentrations of the element in the unknown samples are determined from the working curve from the measured intensities of its spectral lines.

The procedure outlined in the previous paragraph for quantitative analysis is based on the assumption that the excitation conditions for the unknown as well as the standard samples are exactly identical. This is never realized in practice and fluctuations in emission from the plasma may cause the relative intensities of a line for samples of different elemental concentrations not to reflect their relative concentrations. This shortcoming, due to uncontrolled random fluctuations of emission intensity arising from the difference in excitation conditions, leads to a less accurate working curve and consequent error in quantitative estimates of the element in unknown samples. To avoid this error it is necessary to compare the intensities of lines in the unknown and standard samples with some common line of another element that remains unchanged in both spectra. Such a spectral line is called the internal standard because its intensity is also affected by the same random fluctuations as that of the element under investigation. Two types of internal standard lines may be used:

1. A weak line of the element that is the major constituent of the sample whose intensity remains constant for all standard samples.
2. A persistent line of an added small amount of an element that is known not to be present in either the unknown or the standard samples.

Since the lines from the element to be analyzed as well as the internal standard originate in the plasma from the same sample, it is obvious that their relative intensities are less sensitive to slight variation in the excitation conditions. The working curve plotted by using this relative intensity against the concentration of the element in the standard samples can reduce the errors in quantitative analysis by a very large factor. For further details of quantitative analysis the reader is referred to the books by Brode [20], Sawyer [24], and Radziemski and Cremers [25].

References

[1] H.C. Longuet-Higgins, A periodic table: the Aufbau-prinzip as a basis for classification of the elements, J. Chem. Educ. 34 (1957) 30.

[2] G. Herzberg, Atomic Spectra and Atomic Structure, Dover, New York, 1944.

[3] H.G. Kuhn, Atomic Spectra, Longman, London, 1964.

[4] V. Kondratyev, The Structure of Atoms and Molecules, Foreign Language Publishing House, Moscow, 1960.

[5] B. Cagnac, J.C. Pebay-Peyroula, Modern Atomic Physics: Fundamental Principles, The Macmillan Press, London, 1975.

[6] P. Bousquet, Instrumental Spectroscopy, Dunod, Paris, 1969.
[7] A.P. Thorne, Spectrophysics, Chapman & Hall and Science Paperbacks, London, 1974.
[8] V. Weisskopf, Die breite der spektrallinien in gasen, Phys. Z. 34 (1933) 1.
[9] E. Lindholm, Binary collision theory of the spectral line shape, Arkiv Mat. Astron. Fysik. 28B (1941) 3.
[10] H.M. Foley, The pressure broadening of spectral lines, Phys. Rev. 69 (1946) 616.
[11] P.W. Anderson, A method of synthesis of the statistical and impact theories of pressure broadening, Phys. Rev. 86 (1952) 809.
[12] J. Holtsmark, Uber die verbreiterung von spektrallinien, Ann. Phys. 58 (1919) 577.
[13] W.R. Hindmarsh, Collisional broadening of spectral lines in neutral atoms, Prog. Quantum Electron. 2 (1972) 143.
[14] D.D. Burgess, Spectroscopy of laboratory plasmas, Space Sci. Rev. 13 (1972) 493.
[15] H.R. Griem, Plasma Spectroscopy, McGraw-Hill, New York, 1964.
[16] H.R. Griem, Spectral Line Broadening by Plasmas, Academic Press, New York, 1974.
[17] W.L. Wiese, D.E. Kelleher, D.R. Paquette, Detailed study of the stark broadening of Balmer lines in a high-density plasma, Phys. Rev. A6 (1972) 1132.
[18] W.L. Wiese, D.E. Kelleher, V. Helping, Variations in Balmer-line stark profiles with atom-ion reduced mass, Phys. Rev. A11 (1975) 1854.
[19] D. Mosher, Coronal equilibrium of high-atomic-number plasmas, Phys. Rev. A10 (1974) 2330.
[20] W.R. Brode, Chemical Spectroscopy, second ed., John Wiley, New York, 1952.
[21] W.F. Meggers, The strongest lines of singly ionized atoms, J. Opt. Soc. Am. 31 (1941) 44, 605.
[22] C.E. Moore, Selected Tables of Atomic Spectra, Washington, N.S.R.D.S.-N.B.S. 3, Section 1 (1965) and Section 2 (1967).
[23] C.E. Moore, Bibliography on the Analysis Of Optical Atomic Spectra, Section 1, H to V, N.B.S. Special Publ. 306, Washington, (1968).
[24] R.A. Sawyer, Experimental Spectroscopy, third ed., Dover, New York, 1963.
[25] L.J. Radziemski, D.A. Cremers (Eds.), Laser Induced Plasma and Applications, Marcel Dekker, New York, 1989.

CHAPTER

3

Laser ablation

Richard E. Russo[a,b], X.L. Mao[b], J. Yoo[a], J.J. Gonzalez[a,b]

[a]Applied Spectra, Inc., West Sacramento, CA, United States

[b]Lawrence Berkeley National Laboratory, Energy Technologies Area, Energy Storage and Distributed Resources Division, Laser Technologies Group, Berkeley, CA, United States

1 Introduction

The definition of ablation from the Merriam-Webster Dictionary is "loss of a part by melting or vaporization." For ablation to occur, energy absorption is needed. The energy can be provided in the form of electrical discharges (e.g., an arc and spark) or in the form of light (e.g., as a laser). Laser ablation means using laser light energy to remove a portion of a sample by melting, fusion, sublimation, ionization, erosion, and/or explosion. Laser ablation results in the formation of a gaseous vapor, luminous plasma, and in the production of fine particles. By measuring the emission spectrum from the laser-induced plasma, qualitative and quantitative information about the sample's chemical composition can be obtained. This measurement technology is known as laser-induced breakdown spectroscopy (LIBS). LIBS is an exciting field of study, both theoretically and experimentally, due to the wealth of diverse mechanisms underlying the physical processes and its significant potential for spectrochemical analysis. This chapter discusses the fundamental mechanisms of laser ablation processes and their relation to LIBS.

"The history of the interaction of high-power lasers with solid matter is as old as the laser itself" [1]. In 1917 Albert Einstein [2] first proposed that stimulated emission of light (the process that makes lasing possible) should occur in addition to absorption and spontaneous emission. It then took over 40 years until the development of the theoretical principles of lasers was established by Arthur Schawlow and Charles Townes [3] in 1958 and 2 more years to develop the first laser source by Theodore Maiman [4, 5]. The laser was built using a rod of synthetic ruby as the active medium. In 1962 the first account of laser ablation was presented by Breech and Cross [6] at the International Conference on Spectroscopy held at the University of Maryland. A ruby laser was used to vaporize and excite atoms from solid surfaces, and the plasma spectrum was used to characterize the elemental composition of the sample. This paper began the field of laser microprobe emission spectroscopy, which was one of the first real applications of laser ablation. Throughout 1963 and 1964, about a dozen publications detailed early laser ablation experiments [1]. Many phenomena, which were first observed in those years, are still the subject of study today. During the 1970s and

https://doi.org/10.1016/B978-0-12-818829-3.00003-4

early 1980s, the development of lasers and the understanding of laser ablation were incremental and steady. The fastest-growing applications during the 1980s were driven by the needs of materials science. Several laser ablation-based methods reached maturity in the late 1980s, including pulsed laser deposition [7] for making high-T_c superconductor thin films, micromachining [8], and laser-based medical applications such as laser-based ophthalmology (laser-assisted in situ keratomileusis) [9], removal of birthmarks tattoos, and smoothing of wrinkled skin in dermatology [10, 11]; laser surgery for internal arthroscopic cutting and for arterial angioplasty [12]; and for dental applications [13].

For analytical purposes, laser ablation became widely used for a range of microanalysis applications. Matrix-assisted laser desorption/ionization [14] revolutionized the identification and study of large molecular weight biomolecules and polymers. Laser ablation as a sampling technique coupled to established analytical techniques such as inductively coupled plasma atomic emission spectrometry [15] and inductively coupled plasma mass spectrometry (ICP-MS) [16], improved analytical capability for direct solid analysis by minimizing the time of analysis, reducing hazardous chemical exposure and waste, and providing a reliable alternative to acid dissolution for chemical analysis. Laser ablation has been recognized for its powerful advantages: rapid, in situ, multi-element analysis of any kind of sample with no sample preparation. Diagnostics and theoretical studies have advanced laser ablation research to a very active field supported by a wide range of applications. The basis of LIBS is rooted in laser ablation. Laser ablation is the first step in the LIBS process, and its influence will be reflected in the "figures of merit," temporal and spatial resolution, sensitivity, precision, and accuracy. The influence of laser ablation on LIBS is addressed in this chapter.

2 Fundamental ablation processes

Laser ablation is governed by a variety of distinct nonlinear mechanisms. Once the laser beam illuminates the sample, mass leaves the surface of a sample in the form of electrons, ions, atoms, molecules, clusters, and particles, with each of the processes separated in time and space. An understanding of the fundamental mechanisms involved in each of these processes is critical for efficiently coupling the laser beam to the sample and removing mass in the appropriate form for analysis. Understanding laser-material interaction will allow ablation of stoichiometric vapor and control of the laser-induced plasma properties for optimum LIBS performance.

Laser ablation will be divided into three main processes for discussion in this chapter: bond breaking and plasma ignition, plasma expansion and cooling, and particle ejection and condensation. These laser ablation processes occur over several orders of magnitude in time, starting with electronic absorption of laser optical energy (10^{-15} s) to particle condensation (10^{-3} s) after the laser pulse is completed.

Fig. 1 shows a summary of these three processes and various mechanisms occurring during each. During the plasma ignition process, the mechanisms and plasma properties strongly depend on laser irradiance and pulse duration. For a nanosecond laser pulse with irradiances less than 10^8 W/cm^2, the dominant mechanism is thermal vaporization: the temperature of the solid surface increases, and a well-defined phase transition occurs from solid to liquid, liquid to vapor, and vapor to plasma. For a picosecond laser pulse with irradiance between 10^{10} and 10^{13} W/cm^2, both thermal and nonthermal mechanisms such as Coulomb explosion exist. For irradiances higher than 10^{13} W/cm^2 with femtosecond laser pulse, Coulomb explosion is the main bond-breaking mechanism.

When laser pulse duration is in the nanosecond time region, the later part of the laser pulse

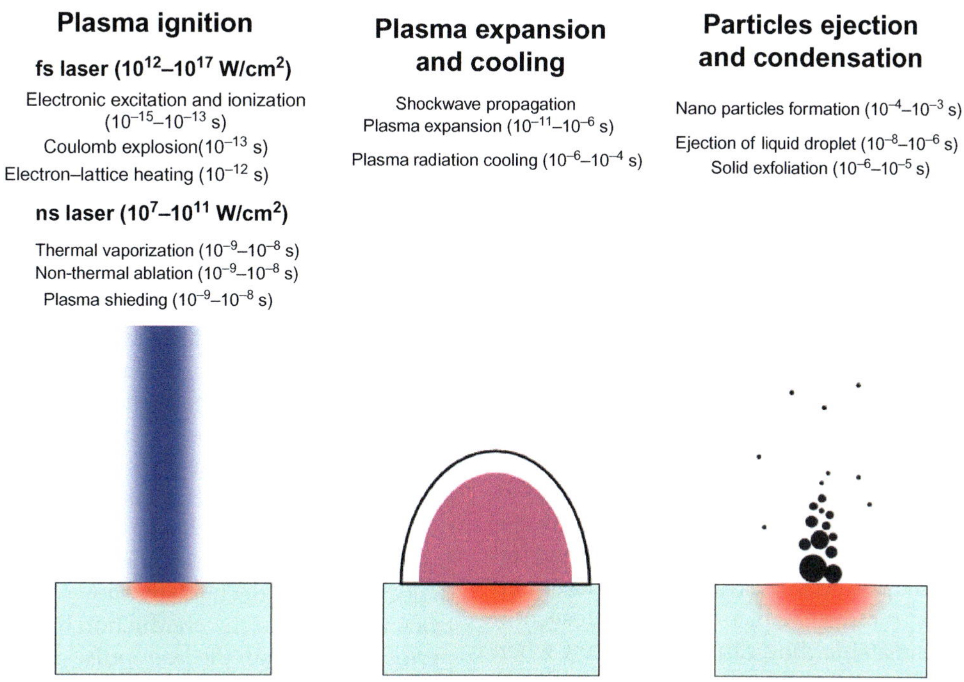

FIG. 1 A summary of laser ablation processes and various mechanisms occurring during each process.

can be absorbed by the laser-induced plasma, which is called plasma shielding. For picosecond pulsed laser ablation, the laser pulse is too short to be absorbed by the plasma. Plasma shielding will influence how much of the solid mass is converted into vapor and the properties of vapor. However, an air plasma can form during picosecond laser pulse duration due to seed electrons from the target surface; this air plasma can absorb part of the picosecond pulse. With femtosecond laser pulses, plasma shielding can be neglected because, to the best of our knowledge, no mass can be ejected from the surface during the short pulse duration.

Plasma expansion begins after the plasma ignition process. The plasma expansion process will be governed by the initial plasma properties (at the end of the laser pulse) and the expansion medium. The properties (electron number density, temperature, and expansion speed) of the plasma are initially strongly dependent on the laser properties. Plasma expansion will be related to the initial mass and energy in the vapor plume and the gas environment. Plasma expansion will be adiabatic until approximately 1 μs after the laser pulse. After that time, line radiation will be a dominant energy loss influencing the temperature decrease.

Particle formation will be influenced by these primary processes. Nano-sized particles will be formed from condensation of the vapor. Condensation starts when the vapor plume temperature reaches the boiling temperature of the material (~3000 K) and stops at the condensation temperature of the material (<2000 K). Liquid ejection of particles can occur by high-pressure gradient forces within the highly expanding vapor plume acting as the molten surface. Solid sample exfoliation can occur from the large thermal stress gradients of the fast heating process; thermal stresses can break the sample into irregular-shaped particles, ejecting them from the surface. More detailed explanations for each of the processes diagramed in Fig. 1 are presented in detail in the following sections.

2.1 Plasma ignition processes

Understanding plasma ignition processes will help to determine optimum conditions for LIBS measurements. The plasma ignition processes include bond breaking and plasma shielding during the laser pulse. The plasma conditions, after the laser pulse terminates, will determine expansion and cooling.

Bond-breaking mechanisms influence the amount and forms of energy (kinetic, ionization, and excitation) that atoms and ions acquire. Plasma shielding can increase the energy by additional heating before the laser pulse is finished. These mechanisms strongly depend on laser irradiance and pulse duration, as described previously for nano-, pico-, and femtosecond lasers. Plasma shielding can be dominant when laser irradiance reaches certain thresholds. In the following sections, these different mechanisms will be discussed in detail for the three (nanasecond, picosecond, and femtosecond) laser pulse durations.

2.1.1 *Femtosecond laser ablation*

When a femtosecond laser pulse interacts with a solid sample, different electronic mechanisms are excited, depending on the sample material. For conducting samples, free electrons inside the solid can directly absorb laser energy and form hot electron-hole plasma. For semiconductor and wide bandgap dielectrics, the electron-hole plasma is created through nonlinear processes such as multiphoton absorption and ionization, tunneling, and avalanche ionization. At high energy, the electron-hole plasma created on the surface of the solid will induce emission of X-rays, hot electrons, photoemission, and produce highly charged ions through a phenomenon called Coulomb explosion or nonthermal melting.

For wide bandgap dielectrics, the simultaneous absorption of multiple photons results in a photoionization rate that is strongly dependent on the laser intensity [17]. The rate of multiphoton absorption can be expressed as σI^n, where I is the laser intensity and σ is the n-photon absorption cross-section for a valence band electron to be excited to the conduction band. The number of photons required is determined by the smallest n that satisfies the relation $nh\omega > E_g$, where E_g is the bandgap energy of the dielectric material and $h\omega$ is the photon energy. A second photoionization process, tunneling ionization, may come into play under an extremely strong laser electromagnetic field interaction with dielectrics. In the strong-field regime, the superposition of the nuclear Coulomb field and the laser electric field results in an oscillating finite potential barrier through which bound electrons can tunnel, thus escaping the atom. In dielectrics, this mechanism allows valence electrons to tunnel to the conduction band in a time period shorter than the laser pulse duration. Both multiphoton and tunneling ionization can be treated under the same theoretical framework developed by Keldysh. The transition from multiphoton to tunneling ionization is characterized by the Keldysh parameter γ [18]:

$$\gamma = \frac{\omega\left(2m_e E_g\right)^{1/2}}{eE_A} \tag{1}$$

where m_e and e are the effective mass and charge of the electron and E_A is the amplitude of the laser electric field oscillating at frequency ω. When γ is much larger than one, which is the case for high-intensity laser interactions with dielectrics, multiphoton ionization dominates the excitation process.

For semiconductor samples, where the photon energy is larger than the bandgap, single photon absorption is the dominant mechanism for exciting valence electrons to the conduction band [19]. In the case of semiconductors with an indirect bandgap, such as silicon, single photon absorption can still occur with photons of energy greater than the gap, but phonon assistance is necessary to conserve momentum.

Once an electron-hole plasma is formed inside the solid, the carriers can absorb additional laser photons, sequentially moving to higher energy states. The absorption coefficient α_0 depends on the imaginary part of the refractive index κ, which is related to the dielectric function ε. According to the Drude model [20, 21], ε can be expressed as:

$$\varepsilon = 1 - \omega_p^2 \left[\frac{\tau^2}{1+\omega^2\tau^2} + i\frac{\tau^2}{\omega\tau(1+\omega^2\tau^2)} \right] \quad (2)$$

where τ, the scattering time, is typically a fraction of a femtosecond and depends on the conduction electron energy. ω_p is the plasma frequency defined by:

$$\omega_p = \sqrt{\frac{e^2N}{\varepsilon_0 m_e}} \quad (3)$$

where N is the carrier density and ε_0 is the electric permittivity.

Photon absorption increases the carrier energy of the electron-hole plasma; when the energy of carriers is well above the bandgap (or Fermi level in a metal), collisional ionization generates additional excited carriers. A high-energy electron can ionize another electron from the valence band, resulting in two excited electrons with lower energy at the conduction band [22, 23]. These electrons can be heated by the laser through free carrier absorption and impact additional valence band electrons. This process can repeat itself as long as the laser electromagnetic field is present and intense, leading to the so-called electronic avalanche. Avalanche ionization requires seed electrons to be present in the conduction band, which can be excited by photoionization. The following rate equation can describe the injection of electrons into the conduction band of dielectrics using femtosecond-to-picosecond laser pulses under the combined action of multiphoton excitation and avalanche ionization [24]:

$$\frac{dN}{dt} = aIN + \sigma NI^n \quad (4)$$

where a is a constant. Within the femtosecond timescale, a large number of excited electrons can leave the solid, the lattice modes remain vibrationally cold, and the irradiated solid consists of charged ions and an electron-hole plasma. After about 10% of the valence electrons are removed, the lattice is weakened and begins to melt. This is called nonthermal melting (Coulomb explosion) caused by high-energy electrons and ions [25–29].

Coulomb explosion processes were studied by time-resolved reflectivity and X-ray diffraction using pump-probe experiments [30]. Optical reflectivity spectra for femtosecond laser-excited silicon are plotted in Fig. 2. Negative times mean that the probe beam arrived at the surface before the pump ablation laser pulse. The dashed-dotted and dashed lines represent the well-known reflectivity spectra of crystalline and molten silicon, respectively. When the probe pulse occurred 120 fs before the ablation pulse, the reflectivity spectrum was similar to that of crystalline Si. An abrupt increase in the optical reflectivity occurred within 300 fs after the ablation pulse, and the reflectivity spectrum changes from crystalline to liquid character within this short time.

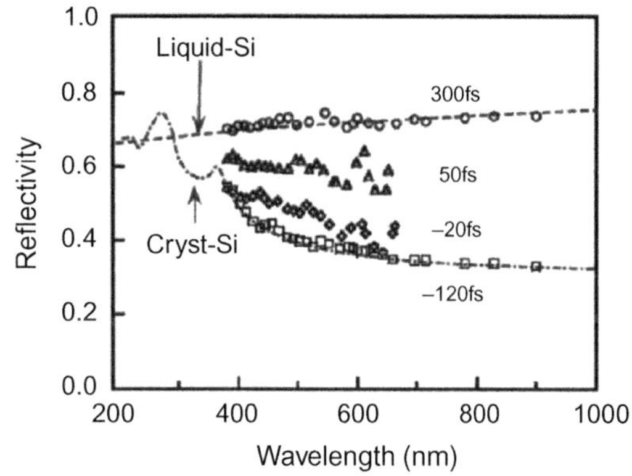

FIG. 2 Spectra of the optical reflectivity of silicon. *Dash-dotted curve*: crystalline silicon. *Dashed line*: molten silicon. *Data points*: spectra measured at various time delays between the pump pulse and the optical probe pulse (optical pump/optical probe measurements) [30].

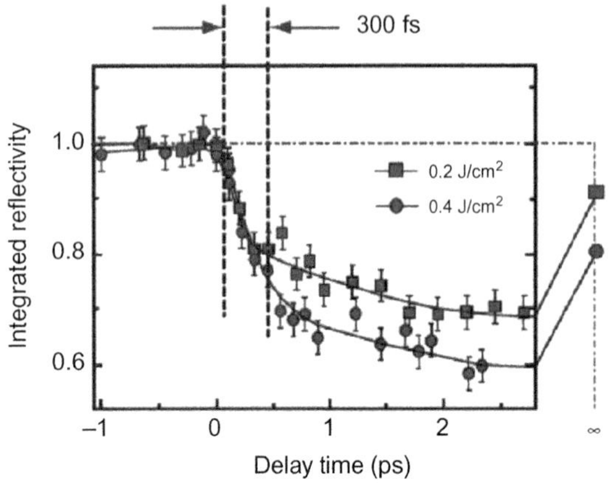

FIG. 3 X-ray diffraction from the (111) lattice planes of Ge versus time delay for two different energy fluences of the pump pulse [30].

High-energy femtosecond laser-material interactions can produce X-ray radiation with duration comparable to that of the laser pulse. This X-ray pulse can be used to detect lattice changes by X-ray diffraction using a pump-probe method with femtosecond time resolution. Fig. 3 shows time-resolved X-ray diffraction [30] measured from the (111) lattice planes of Ge as a function of the delay time between the X-ray probe and the laser excitation pulse for two different laser fluences. Negative times indicate that the X-ray probe pulse arrived before the laser excitation pulse. The diffraction intensity remained unchanged when the X-ray pulse arrived before the excitation (ablation) pulse. A sharp decrease in diffraction was observed after the arrival of the ablation pulse. The initial decrease in reflectivity takes approximately 300 fs and indicates lattice melting. Conventional electron-phonon interactions occur on the picosecond timescale. Therefore the fast decrease in diffraction (within a few hundred femtoseconds) and optical reflectivity demonstrate that a portion of the Ge crystal underwent nonthermal melting.

2.1.2 *Picosecond laser ablation*

For picosecond laser ablation, the lattice could be melted through thermal and/or nonthermal processes, depending on the laser irradiance. Electrons are ejected from the target surface during the laser pulse. The free electrons can interact with the air and absorb laser energy to initiate an air plasma during the picosecond laser pulse duration. The plasma forms long before the plume forms [31–33]. Fig. 4 shows the measured air plasma electron number density N_e as a function of distance z from the target surface at a 150 ps delay time between the pump and probe beams. The electron density at z was measured from the interference pattern (insert of Fig. 4) using the expression:

$$N_e(z) = \frac{2\varepsilon_0 m_e \omega^2 \lambda q(z)}{e^2 l(z)} \tag{5}$$

where $q(z)$ and $l(z)$ are, respectively, the average phase shift and width of the plasma at location z. ω and λ are circular frequency and wavelength of the probe beam. The electron number density of this air plasma was on the order of 10^{20} cm^{-3}, which is higher than the air density. The air plasma was observed immediately and expanded longitudinally during the laser pulse. Its longitudinal extent remained approximately constant after about 100 ps, after which the plasma expanded principally in the lateral (radial) direction. The evolution of the

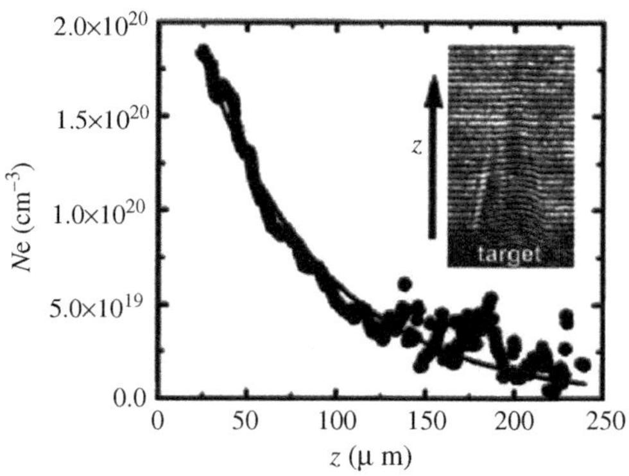

FIG. 4 Electron number density profile along the incident laser axis. The *solid curve* is a least square fit of the experimental data showing exponential decay. The *inset* is an interferogram of the picosecond laser ablation plasma [33].

laser-ablated air plasma was simulated with a two-fluid plasma model [31, 34].

The air plasma above the sample would absorb a part of the incoming laser beam radiation. Unlike nanosecond laser ablation, plasma shielding is not caused by absorption from the vapor plume; on the picosecond timescale, plasma shielding is caused by the air plasma. To confirm this plasma shielding mechanism in picosecond laser ablation, the lateral expansion of early-stage ablation plasma induced by a 1064 nm, 35 ps laser pulse on a copper target was measured. A relation of $t^{1/2}$ was found for the lateral expansion of the air plasma. Measurements of energy absorption by the air plasma (~10% of incoming laser energy) confirmed plasma shielding for picosecond laser ablation.

2.1.3 Nanosecond ablation

When the pulse duration is on the order of a few nanoseconds, and laser irradiance is on the order of 10^7–10^{11} W/cm^2, some of the mechanisms involved in ablation are: melting, fusion, sublimation, vaporization, ionization, etc. If the laser irradiance is high enough, nonthermal ablation is also important and can coexist with these thermal mechanisms.

When the laser irradiance is less than 10^8 W/cm^2, thermal processes are dominant. The temperature at the target surface will rise during the laser pulse, and eventually the target will melt and vaporize. The temperature distribution in the target can be calculated with the heat conduction equation [35]:

$$\frac{\partial T(x,t)}{\partial t} = \frac{\partial}{\partial x}\left[\left(\frac{\kappa}{C_p\rho_s}\right)\frac{\partial T(x,t)}{\partial x}\right] + \frac{\alpha}{C_p\rho_s}I(x,t) \quad (6)$$

where T represents the temperature inside the target, x is the position from the surface, and κ, C_p, ρ_s, and α denote the thermal conductivity, heat capacity, mass density, and absorption coefficient of the solid target material, respectively.

The thermal evaporation rate J_v is the function of surface temperature. Assuming thermal equilibrium:

$$J_v = 1.06\times 10^6 \exp\left(-\frac{L_v}{k_B}\left(\frac{1}{T_s}-\frac{1}{T_B}\right)\right)\sqrt{\frac{M}{2\pi k_B T_s}} \quad (7)$$

where L_v is the heat of vaporization, M is the mass of vapor, k_B is the Boltzmann constant, and T_b and T_s are the boiling point temperature and surface temperature of the sample, respectively.

Vaporized mass can be ionized by absorbing the incoming laser beam, forming a plasma. Laser radiation is absorbed primarily by inverse Bremsstrahlung, which involves the absorption of a photon by free electrons during collision with heavy particles (ions and atoms). The inverse Bremsstrahlung absorption coefficient is given by:

$$\alpha_{IB} = \left[QN_eN_0 + \frac{4e^6\lambda^3N_eZ^2N_i}{3hc^4m_e}\times\left(\frac{2\pi}{3m_ek_BT_e}\right)^{1/2}\right] \times\left[1-\exp\left(-\frac{hc}{\lambda k_BT_e}\right)\right] \quad (8)$$

where Q is the cross-section for photon absorption by an electron during the collision with atoms, c is the speed of light, h is Planck's constant, Z is the charge on ions, N_e, N_0, and N_i are number density of electron, atoms, and ions, respectively, and T_e is the electron temperature. The first term on the right side of Eq. (8) is the electron-atom interaction and the second term is related to electron-ion interaction. Multiphoton ionization in the vapor can also contribute to this process if the laser intensity is high and laser wavelength is short.

When the plasma plume is near the critical density, the later part of the laser beam pulse energy would be partially absorbed before it reaches the target. Plasma shielding was observed by the transmitted laser pulse

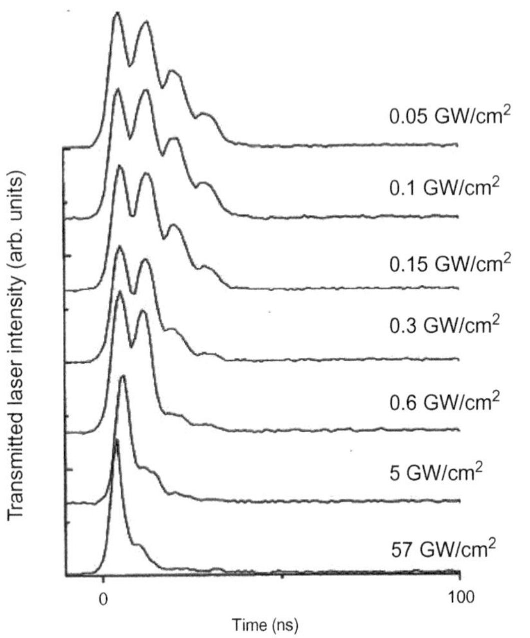

FIG. 5 Transmitted laser beam temporal profiles through a glass sample at various power densities [36].

temporal profile through a glass sample (Fig. 5). The temporal profiles of the transmitted laser pulse were similar to the original laser pulse at low laser irradiance. When the laser irradiance was >0.3 GW/cm^2, the later part of laser pulse became truncated [36].

2.2 Plasma expansion processes

After the laser pulse ends, the induced plasma plume will continue to expand into the ambient. The electron number density and temperature of the plasma change as the plasma expands. Plasma expansion depends on the amount and properties of the ablated mass, how much energy was coupled into the mass, the spot size of the laser beam, and the environment (gases, liquid, and pressure). Most LIBS spectra are recorded from several hundreds of nanoseconds to several microseconds after the laser pulse. Understanding plasma expansion during this time period is critical for optimization of LIBS measurements.

2.2.1 Expansion of the evaporated material plume and shockwaves

After the laser pulse ends, hot electrons and atomic and ionic mass leave the sample surface. The expansion of the evaporated material into vacuum can be described by the Euler equations of hydrodynamics, expressing the conservation of mass density, momentum, and energy [35]:

$$\frac{\partial \rho}{\partial t} = -\frac{\partial(\rho v)}{\partial x} \tag{9}$$

$$\frac{\partial(\rho v)}{\partial t} = -\frac{\partial}{\partial x}\left[p + \rho v^2\right] \tag{10}$$

$$\frac{\partial}{\partial t}\left[\rho\left(E_d + \frac{v^2}{2}\right)\right] = -\frac{\partial}{\partial x}\left[\rho v\left(E_d + \frac{p}{\rho} + \frac{v^2}{2}\right)\right] + \alpha_{IB} I \tag{11}$$

where ρ is the mass density, v is the velocity, E_d is the internal energy density, and p is the local pressure. This theory governing plasma expansion can be used for both nanosecond and femtosecond laser ablation.

In a vacuum, the laser-induced plasma-plume expands adiabatically. The expansion speed can be expressed by [37]:

$$v_p = \sqrt{\frac{4\gamma + 10}{3}\frac{E}{M_v}} \tag{12}$$

where v_p is the velocity, γ is the specific heat ratio, E is the energy supporting the expansion, and M_v is the total vaporized sample mass within the vapor plume. Most of the plume energy is kinetic energy.

When ablation occurs in a gas or liquid environment, the ejected mass compresses the surrounding media and produces shockwaves. The plume is the ablated mass from the sample target. The plasma is a mixture of atoms and ions and mass from both the ablated target material and the ambient gas. The interaction between the plume and surrounding media slows the expansion of the plasma. At the same time, the ambient media perform work on the vapor. The vapor temperature will be higher than that for free expansion; temperature and number density of ablated mass depend on the properties of the surrounding media.

Once the external shockwave is formed, its expansion distance can be described by Sedov's theory. The expansion distance H, representing the location of the shockwave front, can be calculated as a function of time [38]:

$$H = \lambda_0 \left(\frac{E_0}{\rho_1}\right)^{1/(2+d)} t^{2/(2+d)} \tag{13}$$

where the parameter d is the dimensionality of the propagation (for spherical propagation $d = 3$, for cylindrical propagation $d = 2$, and for planar propagation $d = 1$), λ_0 is a dimensionless constant, E_0 has the unit of "energy per area" in the case of one-dimensional expansion (planar propagation), "energy per length" for two-dimensional expansion (cylindrical propagation), and "energy" for three-dimensional expansion (spherical propagation), and ρ_1 is air density. By fitting the experimental data using Eq. (13), the dimensionality of expansion can be determined.

Early-stage plasma expansion from femtosecond laser ablation of stainless-steel targets was investigated by time-resolved shadowgraph imaging (Fig. 6). At early times, the femtosecond laser-induced plasma expanded primarily in the direction perpendicular to the sample surface; the expansion distance was approximately 10 μm after 130 ps. There was

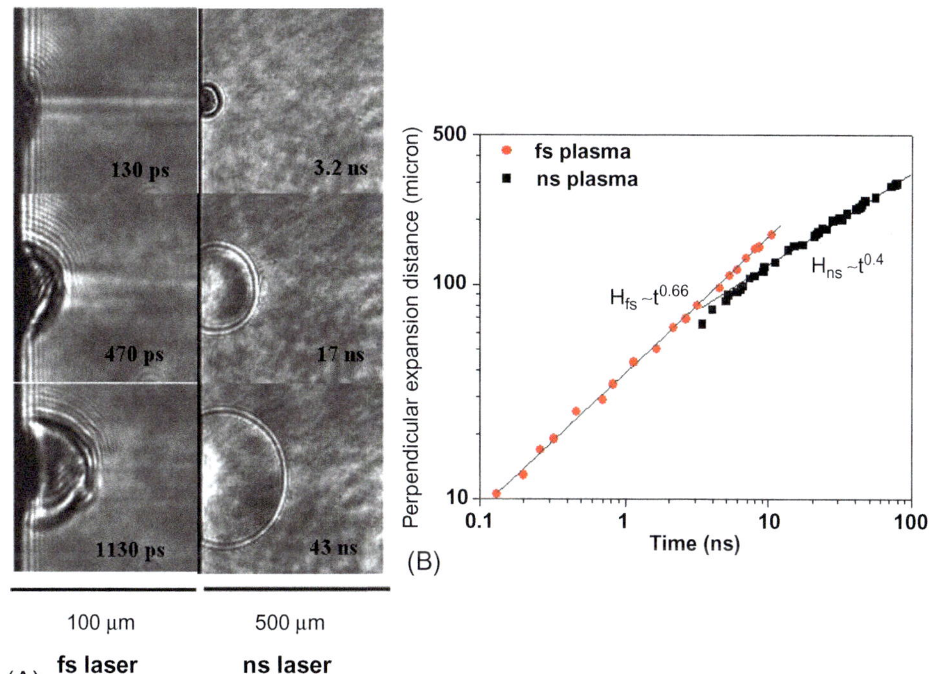

FIG. 6 (A) Sequence of shock wave images obtained by laser shadowgraph for femtosecond and nanosecond laser ablation. Please note that the scale for femtosecond laser and nanosecond laser is different. (B) Perpendicular expansion distance of shockwave as a function of time for femtosecond and nanosecond laser ablation [39].

no lateral expansion until nanoseconds after the ablation laser pulse, after which the lateral expansion slowly increased. If the laser has a nanosecond pulse duration, the perpendicular expansion distance of the plasma is proportional to $t^{2/5}$, and can be predicted by Sedov's blast wave theory for spherical propagation (three-dimensional expansion). The perpendicular expansion of the plasma generated by the femtosecond laser-pulsed ablation was proportional to $t^{2/3}$, corresponding to one-dimensional expansion (measured in the shadowgraphs of Fig. 6, which showed no lateral expansion at early times [39]).

Once the plume pressure is equal to the pressure of the surrounding media the expansion stops. The stopping time and distance of the vapor plume can be expressed as [40]:

$$t_s = \xi_{st}\left(\frac{E}{p_g}\right)^{1/3}\frac{1}{c_g} \tag{14}$$

$$R_s = \zeta_{st}\left(\frac{E}{p_g}\right)^{1/3} \tag{15}$$

where ξ_{st} and ζ_{st} are constants, p_g is pressure, and c_g is the sound velocity of the gas. The stopping time is in the range of microseconds. LIBS emission signals are usually measured microseconds after the laser pulse, a time at which vapor plume expansion has stopped. The final distance determines the volume of the vapor plume. LIBS performance depends on the electron number density and temperature of the plasma, which strongly depend on the plume volume.

2.3 Plasma emission spectra

2.3.1 Femtosecond pulsed laser plasma emission

When a femtosecond laser pulse is focused in air, optical emission of nitrogen molecular lines will exist in the air plasma several picoseconds after the laser pulse; molecular structure is preserved for this time period [41]. A spectrum, measured using gated integration with a delay slightly after the laser pulse (1 ns), is presented in Fig. 7. The laser energy was 200 mJ. The two electronic systems observed are the 2^+ N_2 ($C^3\Pi_u \rightarrow B^3\Pi_g$) and the 1^- N_2^+ ($B^2\Sigma_u^+ \rightarrow X^2\Sigma_g^+$) bands. No oxygen lines were observed in this early time period. The bands are identified in Fig. 7 with notation indicating the upper (v') and lower (v'') vibrational levels. The estimated vibrational temperature (T_v) of the 2^+ system was between 7000 and 8000 K with accuracy ~1000 K. A streak camera was used to obtain a fast time-resolution measurement of the total emission, which is shown in Fig. 8. A narrow 60 ps duration peak was observed just after the laser pulse followed by a slower decrease in amplitude lasting 0.5 ns. The peak is due to molecular line emission, because there were no other spectral features measured at this short delay.

Conventional LIBS measurements are made using nanosecond-to-microsecond delays after the laser pulse. Emission spectra at these times

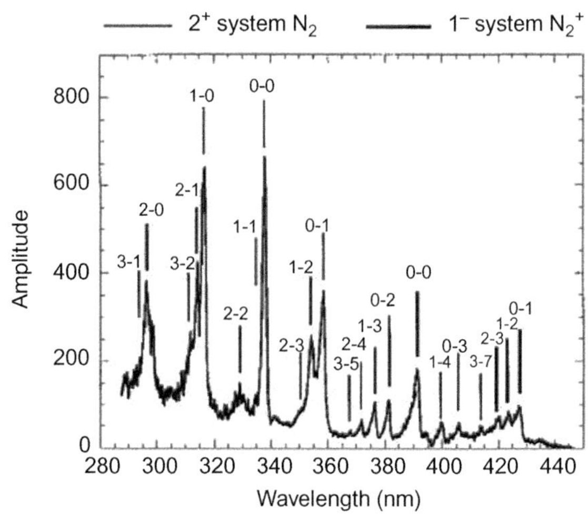

FIG. 7 Molecular line emission spectrum obtained at a laser energy of 200 mJ with the maximum of the gate pulse ~1 ns before the laser pulse. The upper-lower vibrational levels of the transitions are indicated [41].

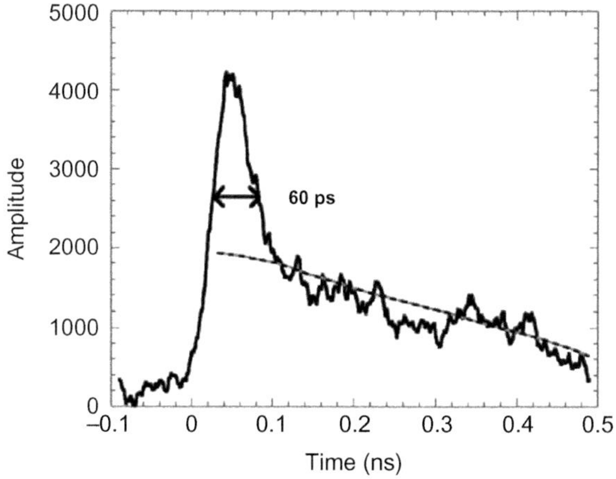

FIG. 8 Streak camera traces lineouts of the light emitted on a fast timescale. The laser pulse is at $t = 0$. Streak speeds are 15 mm/ns [41].

depend on the laser-induced plasma properties; when the plasma is hot and dense, the spectrum is mostly composed of continuum emission. During plasma expansion, the temperature and number density decrease; ionic lines followed by atomic emission lines appear. Continuum emission was observed within 1 ns after the femtosecond laser ablation pulse; Fig. 10 shows a continuum emission spectrum measured at increasing time delays [41]. Within the time measurement resolution (4.5 ns), the amplitude of continuum emission increased ~10 times after the laser pulse, and decreased by about two orders of magnitude in 80 ns. Ionic lines began to appear about 10 ns after the laser pulse. The femtosecond pulsed laser-induced plasma has a shorter overall lifetime compared to those plasmas initiated using longer laser pulses.

2.3.2 Nanosecond pulsed laser plasma emission

For the nanosecond pulsed laser-induced plasma, continuum emission appears during the laser pulse and lasts for several hundred nanoseconds. Ion emission also dominates on the nanosecond timescale. Atomic and molecular line emission occurs after ~1 μs. Molecular line emission measured at later times is from the recombination of species in the plasma. C—N and C—C swan bands are often observed on the microsecond timescale. To the best of our knowledge, there are no reports of the original molecular structure being preserved for nanosecond pulsed plasma emission.

2.4 Electron density and plasma temperature

Plasma temperature and electron number density can be estimated from the continuum emission and peak width of atomic and ionic emission lines. A Lorentz function can be used to fit the line spectra (Fig. 9). Stark line broadening from collisions of charged species is the primary mechanism influencing the emission spectra in conventional LIBS experiments [42].

The full width half maximum of Stark broadened lines is related [42, 43] to the electron number density N_e by Eq. (10):

$$\Delta\lambda_{1/2} = 2W\left(\frac{N_e}{10^{16}}\left[1 + 1.75A\left(\frac{N_e}{10^{16}}\right)^{\frac{1}{4}}\left(1 - \frac{3}{4}N_D^{-\frac{1}{3}}\right)\right]\right) \quad (16)$$

where N_D is the number of particles in the Debye sphere and is estimated from:

$$N_D = 1.72 \times 10^9 \frac{T_e^{\frac{3}{2}}}{N_e^{\frac{1}{2}}} \quad (17)$$

where W is the electron impact parameter in nm and A is the ion impact parameter; W and A are functions of temperature and can be obtained approximated by second-order polynomials from Ref. [43].

Under the assumption of local thermal equilibrium, the plasma temperature T can be determined by the line-to-continuum intensity ratio $\varepsilon_c/\varepsilon_l$, where ε_c is the continuum emission coefficient and ε_l is the integrated emission coefficient over the line spectral profile. The line emission

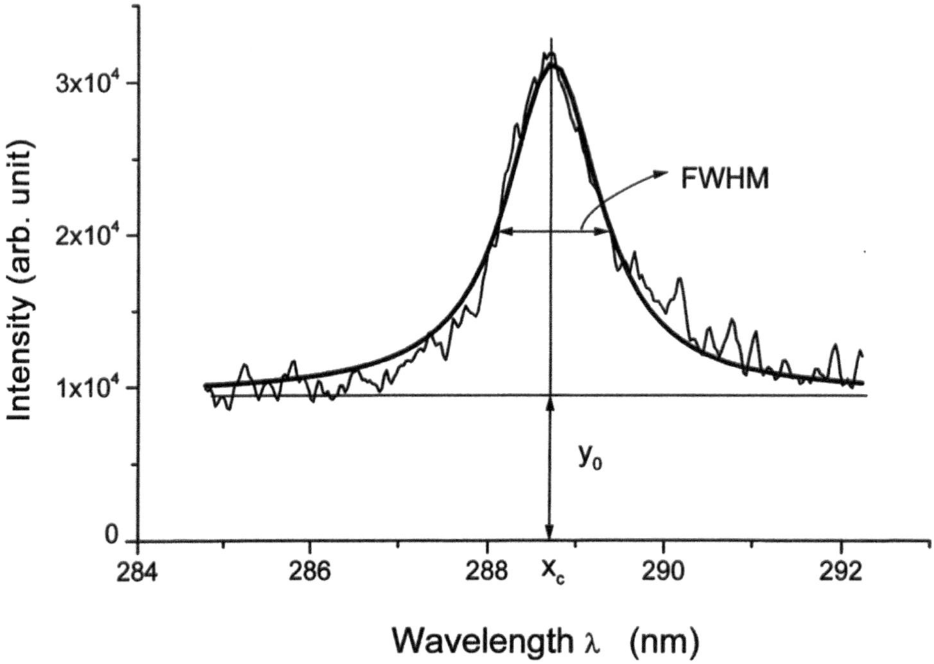

FIG. 9 Lorentzian fitting of the Stark broadened profile for Si(I) 288.16 nm at *d* 1.8 mm. The full width half maximum (FWHM) was used for the calculation of the electron number density. The ratio of the integrated spectral line intensity (*A*) and continuum intensity (y_0) at the wavelength (x_c) were used for the calculation of plasma temperature [42].

coefficient ε_l can be expressed in terms of the electron temperature and density [44]:

$$\varepsilon_l = \left(\frac{h\nu_l}{4\pi}\right) A_{21} \frac{g_2}{2Z_{ion}(T)} \frac{h^3}{(2\pi m_e k_B)^{3/2}} N_e N_i T_e^{-3/2} \exp\left(\frac{E_{ion} - E_2}{kT_e}\right) \tag{18}$$

where A_{21} is the Einstein transition probability of spontaneous emission, E_{ion} is the ionization potential, E_2 and g_2 are upper-level energy and degeneracy, respectively, v_l is the frequency of the emission line, and $Z_{ion}(T)$ is the partition function for ions, which is given by [45]:

$$Z_{ion}(T) = \sum_i g_i \exp\left(-\frac{E_i}{kT_e}\right) \tag{19}$$

with g_i the degeneracy or statistical weight of the ith energy level E_i.

At early times (10–100 ns), the plasma temperature is relatively high and the second ionization state can be important at this time. By including the second ionization contribution, the expression for continuum radiation can be rewritten as [44, 45]:

$$\varepsilon_c = \left(\frac{16\pi e^6}{3c^3(6\pi m_e{}^3 k_B)^{1/2}}\right) N_e\left(N_i^+ + 4N_i^{++}\right) T_e^{-1/2} \left[\xi\left(1 - \exp\frac{-h v_c}{k_B T_e}\right) + G\left(\exp\frac{-h v_c}{k_B T_e}\right)\right] \tag{20}$$

where N_i^+ and N_i^{++} are the number density of single and double charged ions, G is the free-free Gaunt factor, which is assumed to be unity by Kramer's rule [29], v_c is the frequency of the continuum emission, and ξ is the free-bound continuum correction factor.

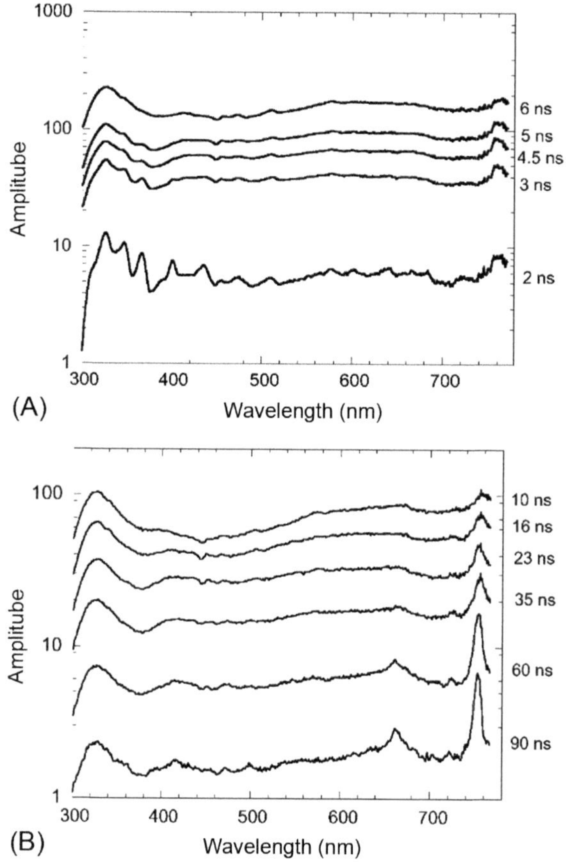

FIG. 10 Continuum emission spectra at successive delays after the laser pulse. The spectra in both (A) and (B) are all plotted to the same scale. The time values indicate the delay of the sampling pulse with respect to the laser pulse. (A) 2–6ns and (B) 10–90ns [41].

From Eqs. (18) and (20), the line-to-continuum ratio is:

$$\frac{\varepsilon_l}{\varepsilon_c}=\frac{\frac{1}{1+4\left(\frac{N_i^{++}}{N_i^{+}}\right)}C_r A_{21}\frac{g_2}{Z_{ion}}\frac{\lambda_c^2}{\lambda_l T_e}\exp\left(\frac{E_{ion}-E_2}{k_B T_e}\right)}{\left[\xi\left(1-\exp\left(\frac{-hv_c}{k_B T_e}\right)\right)+G\left(\exp\frac{-hv_c}{k_B T_e}\right)\right]} \quad (21)$$

where λ_c, λ_l are the continuum and center wavelength of the spectral line, respectively, and C_r is a constant. The ion ratio $\frac{N_i^{++}}{N_i^{+}}$ can be calculated using the Saha equation [46]:

$$\frac{N_i^{++}}{N_i^{+}}=\frac{(2\pi m_e kT_e)^{3/2}}{h^3}\frac{2Z^{++}(T)}{Z^{+}(T)}\frac{1}{N_e}\exp\left(-\frac{E_{ion}^{++}}{k_B T_e}\right) \quad (22)$$

$$\frac{N_i^{+}}{N_a}=\frac{(2\pi m_e k_B T)^{3/2}}{h^3}\frac{2Z^{+}(T)}{Z^{0}(T)}\frac{1}{N_e}\exp\left(-\frac{E_{ion}^{+}}{k_B T_e}\right) \quad (23)$$

where E_{ion}^{+} and E_{ion}^{++} are the first and second ionization potentials, respectively, and $Z^{++}(T)$, $Z^{+}(T)$, and $Z_0(T)$ are the partition function for second ionized, first ionized, ions and atoms, respectively.

The evolution of the plasma temperature and electron number density were evaluated from the Si I line emission versus time (Fig. 11). Since continuum emission dominates initially, the electron number density and temperature could not be obtained for a delay time less than 30 ns. After 30 ns, but before 300 ns, the line and continuum intensities were comparable; good measurement precision for the temperature calculation could be obtained. For later times ($t > 300$ ns) the continuum was very weak, and the line-to-continuum ratio would be sensitive to the errors of continuum determination. The data in Fig. 11 show that in the early stage of plasma evolution (30–300 ns), temperature and electron number density decreased rapidly with time. Within 300 ns, the plasma temperature and electron number density decreased following a power-law dependence proportional to t^n (insert of Fig. 11). The exponent n, obtained by experimental data regression, was -0.70 and -0.99 for the temperature and electron number density, respectively. The analytical performance of LIBS is determined by the temperature and electron number density of laser-induced plasma.

When the continuum emission becomes weak, the plasma temperature calculated from the line-to-continuum ratio is not accurate. The

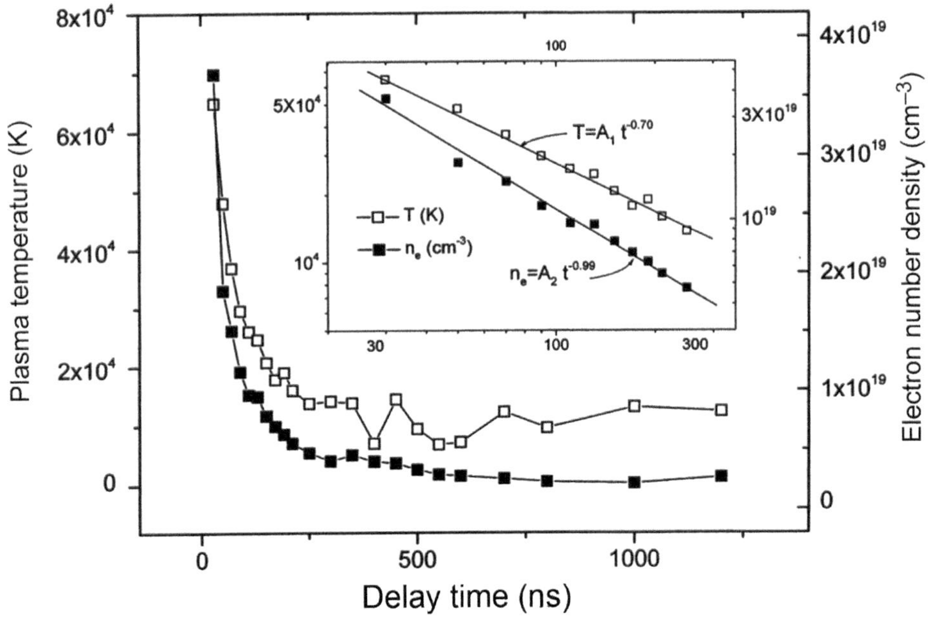

FIG. 11 Temporal evolution of plasma temperature (T) and electron number density (n_e). The *inset* shows in log/log scale in the early phase (30–300 ns) of the plasma.

plasma temperature can be estimated by using a Boltzmann plot based on measuring the relative intensities of lines with known transition probabilities and degeneracies. Temperature can be determined from the slope of a Boltzmann plot using $\ln(I\lambda/gA)$ versus the upper energy level of the transition, where I is the line intensity, g is the statistical weight, and A is the transition probability.

3 Particle formation processes

A significant quantity of the ablated mass is not excited vapor, but is in the form of particles. Particle formation occurs from condensed vapor, liquid sample ejection, and solid sample exfoliation. The mass ablated as particles does not contribute to a LIBS measurement unless these particles can be reevaporated and excited by the plume itself. Particles are important for laser ablation ICP analysis. For LIBS, laser parameters must be established to eliminate particle formation.

3.1 Particle ejection

In most cases, laser-ablated mass consists primarily of particles. Kelly et al. reported that homogeneous boiling within the molten layer was a significant mechanism responsible for particle removal during high-power nanosecond pulsed laser ablation [47, 48]. Time-resolved shadowgraph images show that the violent ejection of particles occurs on the microsecond time scale (Fig. 12). After the laser pulse, there is a time period in which no particle ejection is observed; approximately 400 ns after the laser pulse, mass leaving the silicon surface begins to appear. The ejection of these particles lasts for about 30 μs. The largest particles were estimated to be on the order of 10 μm in size [49].

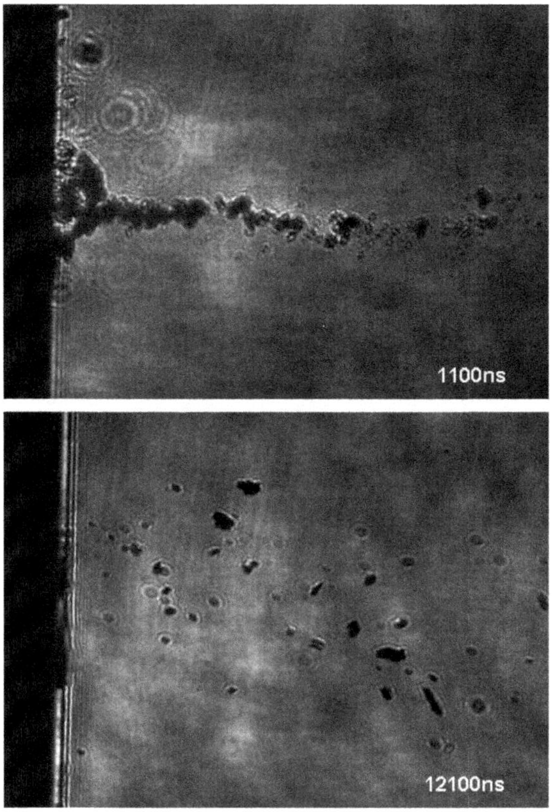

FIG. 12 Liquid ejection from Si single crystal sample.

Much of the theoretical foundation on explosive boiling was established by Martynyuk [50, 51]. A rapid heating rate is required to induce explosive boiling. For a 3 ns laser pulse duration, the heating rate can exceed 10^{12} °C/s. Since thermal diffusion takes place on the order of 10^{-11} s, a melt layer readily forms and propagates into the bulk sample (silicon in this case) during the laser pulse. The liquid silicon is heated above its boiling temperature and becomes metastable. Near the critical point, density fluctuations can generate vapor bubbles in the superheated liquid silicon. Vapor bubbles greater than a critical radius, r_c, will grow; bubbles of size less than r_c will collapse [52]. Once vapor bubbles of size r_c are generated in the superheated liquid, they undergo a rapid transition into a mixture of vapor and liquid droplets. Rapid expansion of the high-pressure bubbles in the liquid leads to a violent ejection of molten droplets from the target surface [53]. This phase explosion process is detrimental to LIBS in that a significant portion of ablated mass is not utilized for analysis.

3.2 Nanoparticle formation

When the laser-induced vapor plume cools to below the boiling point of the sample material, atoms begin to condense and form nanoparticles. The size of particles is determined by the cooling time and the density of the vapor plume. Currently, numerous groups study particle formation mechanisms because of their influence for ICP-MS and nanotechnology applications [54–61]. For LIBS, the particles represent a loss of signal. However, by understanding particle formation mechanisms, laser parameters can be established to minimize this loss and ultimately increase LIBS sensitivity.

4 Laser ablation parameters

As discussed earlier, laser ablation involves complex nonlinear phenomena that depend on the laser and sample material properties. The experimental isolation and effect of each laser and material parameter is very difficult without considering the interaction with other parameters. A general description of each of these parameters and their influence on ablation behavior is presented in this section.

The primary mechanisms operative during ablation depend on laser irradiance. In general, for low-intensity nanosecond laser pulses, the dominant mechanism is thermal vaporization. For picosecond laser pulses and high-intensity nanosecond laser pulses, both thermal and nonthermal mechanisms exist. For high-intensity femtosecond laser pulses, Coulomb explosion is the primary ablation mechanism. Therefore laser pulse duration and irradiance are the most important factors for defining experimental conditions. In addition to pulse duration, some of the other parameters considered during the selection of the laser are wavelength, energy, beam profile, repetition rate, fluence, etc. The influence of environmental ambient gas and pressure and the sample's properties on laser ablation processes will also be discussed.

4.1 Nanosecond pulsed lasers

Nanosecond pulsed lasers are most commonly used for analytical applications, especially LIBS. Analytical performance using nanosecond lasers for LIBS has been studied in many papers, including the influence of the wavelength, energy, repetition rate, dual and multipulse regime, and ambient gas [62–64].

4.1.1 Laser wavelength

The laser wavelength effect on LIBS could be addressed from two points of view: (1) the laser-material interaction (energy absorption) and (2) plasma development and properties (plasma-material interaction).

(a) *Laser-sample interaction*: Shorter wavelengths offer higher photon energies for bond breaking and ionization processes. For example, the ultraviolet (UV) wavelength 193 nm has a photon energy of 6.4 eV compared to 1064 nm that provides 1.16 eV. For most materials, bonding energy is a few eV. When the photon energy is higher than the bond energy, photon ionization occurs and nonthermal mechanisms will play an important role in the ablation process. In addition, shorter optical penetration depth exits with UV wavelengths, providing more laser energy per unit volume for ablation. In general, the shorter the laser wavelength, the higher the *ablation rate* and the lower the *elemental fractionation*.

Ablation rate is a parameter used to describe the amount of ablated mass per laser pulse per unit area. *Ablation rate* also is an indirect indicator of the coupling efficiency between the laser energy and the target material, and a measurement of the spatial resolution (lateral and depth resolution). An example of different ablation rates was presented by Gunther et al. [65]. They studied the ablation rate in metals using a 266 nm Nd:YAG laser and 193 nm excimer laser (Table 1). The ablation rate depended on the laser wavelength for samples with low optical absorbance; Fig. 13 shows the rate behavior for NIST Standard Reference Materials 610–614 (series of glasses). Different ablation rates using

TABLE 1 Ablation rates for metals and CaF_2 (infrared grade) and NIST silicate glasses at 23 J/cm^2.

Metal or compound	Ablation rate (mm/pulse) Nd:YAG	Ablation rate (mm/pulse) excimer	Reflectance at 260 nm
Al	1.04	0.7	92
Si	0.39		71
Ti	0.20		
Cr	0.12	0.13	45.7
Ni	0.13	0.30	47.5
Cu	0.13	0.29	35.5
Zn	0.55	0.65	35
Mo	0.14		61
Pt	0.18	0.25	44.5
Au	0.54	0.75	35.6
ZnSe	0.2		
CaF_2	2.0		
SRM NIST 610	0.34	0.28	
SRM NIST 612	0.60	0.29	
SRM NIST 614	0.67	0.29	

266 nm wavelength (Fig. 13A) imply different laser-material coupling efficiency for each of these samples. By using a 193 nm wavelength, the three glass samples exhibited the same ablation rate (Fig. 13B); optical penetration depth was very shallow for these three samples at this wavelength.

During the interaction of nanosecond laser pulses with materials, there is enough time for a thermal wave to propagate into the sample and create a relatively large molten layer. Evaporation occurs from the molten liquid, which can cause *preferential evaporation* or *elemental fractionation*. *Elemental fractionation* is due to many factors, including wavelength, laser energy, pulse duration, sample properties, etc.; wavelength is not the most critical parameter influencing fractionation [66]. Fractionation can be minimized or enhanced in any sample, depending on the laser beam irradiance.

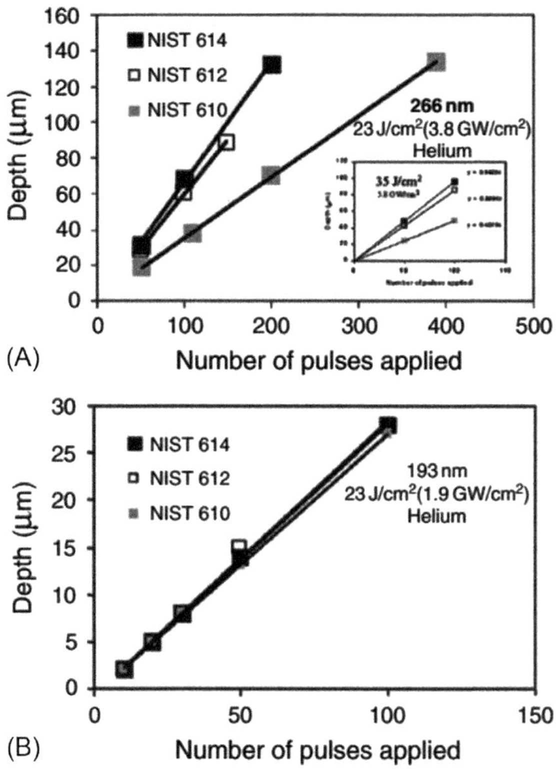

FIG. 13 (A) Depth versus applied number of pulses for the widely used Standard Reference Materials NIST 610, 612, and 614 (266 nm Nd:YAG). Despite their similarity in their major element composition, largely different ablation rates have been found ranging from 0.34 to 0.67 μm/pulse using an energy of 23 J/cm^2 and from 0.49 to 0.96 μm/pulse at 35 J/cm^2 (inlet—5.4 GW/cm^2). (B) Depth versus number of laser pulses for the Standard Reference Materials NIST 610, 612, and 614 using the 193 nm excimer LA system with an energy density of 23 J/cm^2 (3.8 GW/cm^2) [65].

(b) *Plasma development and properties*: The initiation of plasma and its properties also depends on the laser wavelength. Plasma formation requires vaporization of the sample surface as a first step. The initiation of nanosecond laser-induced plasma over the target surface could be promoted by two different photon absorption processes. One is inverse Bremsstrahlung absorption by which free electrons gain kinetic energy from the laser beam and during collisions among sample atoms and ions, electrons, and gas species. The second mechanism is photoionization of excited species and, at sufficiently high laser intensity, multiphoton ionization of excited or ground state atoms.

Laserna et al. [67] studied the influence of wavelength (1064, 532, and 266 nm) on the plasma formation threshold for metal samples. Although laser energy coupling is more effective at shorter wavelengths, the fluence threshold for plasma formation was greater for 266 nm compared to 532 and 1064 nm. These studies agree with the assumption of plasma ignition by inverse Bremsstrahlung, which is approximately proportional to λ^3, (see Eq. 8), considerably more favorable for infrared (IR) than UV wavelengths.

Russo et al. [66] studied the influence of laser wavelength on fractionation in laser ablation by comparing three different UV wavelengths (266,

213, and 157 nm). It was found that the shorter the wavelength, the more controlled and reproducible was the ablation rate. Also, the shorter the wavelength, the lower was the fluence required to initiate ablation. These data support the proposed mechanism of photoionization and/or multiphoton ionization due to the higher photonic energy provided by UV wavelengths (7.9, 5.8, and 4.7 eV for 157, 213, and 266 nm, respectively). For these wavelengths, the inverse Bremsstrahlung process was less important.

When the inverse Bremsstrahlung process occurs, part of the laser beam heats the plasma. Reheating the plasma can increase the lifetime and intensity of the emission lines, which would be beneficial to LIBS. However, an increase in the broadband background emission can also occur. The overall effect of inverse Bremsstrahlung plasma reheating needs to be better investigated and understood.

At high fluence, the efficiency of inverse Bremsstrahlung can be such that the plasma acts to shield the laser pulse energy (plasma shielding) from reaching the sample surface. Longer wavelengths favor the inverse Bremsstrahlung plasma shielding processes, but lower the ablation rate and increase the chances of elemental fractionation [36, 68–70]. In general, most LIBS studies are based on the 1064 nm (IR) Nd:YAG wavelength, whereas most laser ablation ICP-MS studies are based on the fourth or fifth harmonic (266 or 213 nm) of the Nd:YAG laser.

4.1.2 Laser energy

The primary energy-related parameters influencing laser-material interaction are fluence (energy per unit area, J/cm^2) and irradiance (energy per unit area and time, W/cm^2). Laser ablation processes (i.e., melting, fusion, sublimation, erosion, explosion, etc.) are dependent on laser energy and pulse duration, and these different processes have different fluence (or irradiance) thresholds [71–75]. These processes define the characteristics of laser-induced plasma (temperature and electron number density) and the characteristics of ablated mass. The effect of laser energy alone is difficult to quantify. In general, ablated mass quantity and *ablation rate* increase with increase in laser energy (when compared to the same pulse duration and spot size).

Russo et al. [36, 76–80] investigated the plasma shielding effect on mass ablation rate. By using different lasers, increased laser irradiance led to increased ablated mass, as shown in Fig. 14 (~0.3 GW/cm^2 for a copper target). However, plasma shielding caused saturation in mass removal and constant ablation rate with further increase in irradiance. Studies of mass ablation rate dependence on experimental parameters (laser irradiance, spot size, pulse width, surface condition, etc.) at atmospheric pressure showed that, in general, the mass ablation rate increased with increasing irradiance and decreasing pulse duration. Shorter pulses (on the order of picoseconds or femtoseconds) produce higher mass ablation rates, likely because they are not affected by plasma

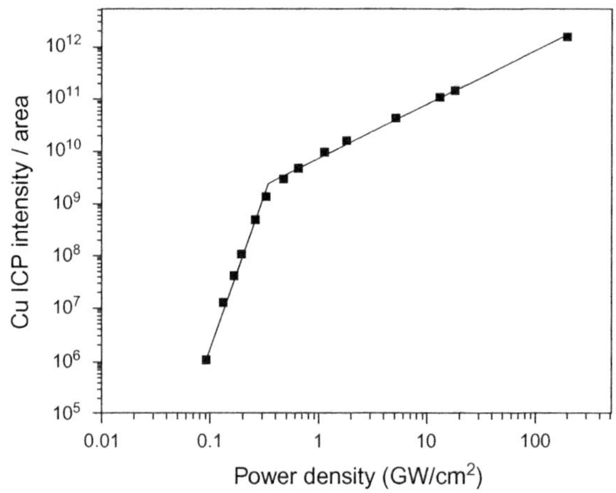

FIG. 14 Inductively coupled plasma emission intensity/area versus laser power density showing the two distinct mass ablation rates and roll-off at 0.3 GW/cm^2 [36].

shielding. Also, the fraction of the pulse energy lost by thermal diffusion (heat-affected zone) in the sample is lower for shorter pulses [39, 80, 81].

Theoretical studies have been conducted to model the processes underlying nanosecond pulsed laser ablation. Bogaerts et al. [82] presented a model to describe nanosecond pulsed laser ablation with "typical" experimental conditions: wavelength of 266 nm, a Gaussian-shaped laser pulse, and pulse duration of 10 ns. Copper (Cu) was used as the sample. The model supported that laser-material interactions during nanosecond pulsed laser ablation include heating, melting, and vaporization of the material. For a laser irradiance of 10^7 W/cm^2, target heating was moderate, and neither melting nor vaporization took place. When the laser irradiance increased to 10^8 W/cm^2 or 2×10^8 W/cm^2, target heating and melting became more pronounced, but evaporation was still limited; the vapor plume was short and cool and no plasma was formed. At higher laser irradiance (5×10^8 W/cm^2), target evaporation became much more significant, the plume became longer and hotter, and a plasma was established. Plasma shielding was predicted at this irradiance. For laser irradiances between 5×10^8 and 10^{10} W/cm^2, target heating, melting, and vaporization increased, and the plume lifetime became longer. The vapor density, temperature, and degree of ionization increased with irradiance. Although this model did not include all mechanisms underlying nanosecond pulsed laser ablation, it qualitatively supported measured behavior.

Dumitru et al. [83] developed a numerical model based on enthalpy to describe nanosecond laser ablation. The authors reported that vaporization stopped before the end of the laser pulse. The effect was related to shielding by the plume, which is continuously growing and whose adsorption is increasing. Differences in ablated volumes were reported when the calculation was performed with different pulse durations (1–100 ns); the lower the pulse duration, the higher the ablated volume.

4.1.3 Ambient

The simplicity of LIBS derives from the fact that sample preparation is not required and in situ analysis at atmospheric pressure can be performed. However, the interest of seeking new applications for LIBS, for example, designing an analytical instrument that can operate in Venus or Mars atmospheres [84, 85], has led to studies of LIBS performance under different ambient conditions.

The laser-induced plasma size, propagation speed, stability, energy, and emission properties depend strongly on the ambient gas into which the plasma expands. Plasma properties have been studied in different gases [82, 86–90], as a function of pressure [86, 91], and even in liquids [92–94]. The ambient gas either helps or prevents the coupling of laser energy into the plasma. For example, the ambient can shield the sample from the laser beam (plasma shielding) if gas breakdown occurs before sample vaporization occurs [86]. The ambient can also quench the luminous plume by collisional cooling; a shorter plasma lifetime with lower temperatures was found in air atmosphere compared to argon [95]. This observation was explained by the lower conductivity and specific heat of argon with respect to air. Wisbrun et al. [96] found that the collisional translation energy was also dependent on the atomic mass of the ambient gas, being less effective when the atomic mass increased, thereby causing a plasma with longer lifetime.

These opposite effects may be less significant for gaseous or aerosol samples, but will be more important for solid samples in the form of reduced ablation rates (plasma shielding), higher continuum background, and shorter lifetime (fast dissipation). The ambient gas can also be helpful in confining the expanding plume and minimizing background emission by atmospheric elements. In a particular case where the UV spectrum ranged from 100 to 200 nm, the use of inert gas avoided absorption by oxygen molecules [97].

Ambient pressure will influence plasma expansion and LIBS emission intensities. For low ambient pressure (<1 mbar), the ablated vapor expands almost freely and the outer part of the plasma becomes colder due to energy loss. Confining the plasma to pressures higher than 1 mbar causes a reduction in energy loss and more uniform distribution of the energy within the plasma. Different gases show different behaviors at different pressures. For example, comparison of iron emission line intensities (Fe I 374.979 nm) at 760 Torr was shown to be greater in helium than in argon or air (Fig. 15). However, at a reduced pressure of 100 Torr, an intensity increase of 10-fold in argon and 2–3 in air was observed. These

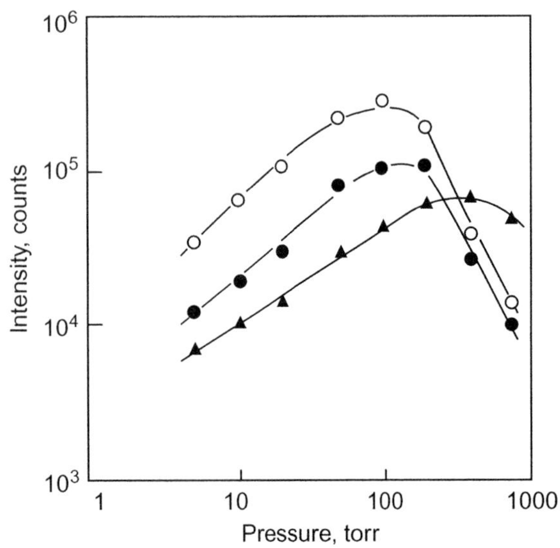

FIG. 15 Dependence of the emission intensities of Fe I 374.949 nm on the pressure of the ambient gases: Ar (*open circle*), air (*filled circle*), and He (*triangle*). The sample was a standard Al alloy containing 0.97% Fe [86].

observations were attributed to the fact that denser plasmas were formed in Ar compared to He [86].

An increase in plasma size with decreased pressure is commonly measured [98]. However, further reduction of pressure down to 10^{-3} Torr resulted in a decrease in LIBS signal intensity as a result of a reduction of collisional excitation of the emission lines. Free expansion (short lifetime) of the plasma in vacuum is not recommended for LIBS, although many other laser ablation applications find the use of vacuum suitable [7, 99, 100].

Several studies of LIBS at high pressures have been pursued. Deguchi et al. [101] used LIBS to detect carbon in fly ash, char, and pulverized coal under high-pressure and high-temperature conditions typical of gasification thermal power plants. High-pressure environments are also interesting for performing in situ geochemical analysis in hostile environments such as the deep sea, volcanoes, or even on other planets [85, 102].

One of the drawbacks of high-pressure environments is broadening of spectral line emission as well as self-absorption, therefore diminishing LIBS performance. This behavior was reported by Nyga et al. [92] during the ablation of samples submerged in water (Fig. 16). However, minimization of line broadening and quenching was reported by using a double pulse approach. The first pulse creates a gas environment "bubble" above the sample surface. The second pulse fired into the "bubble" allows LIBS analysis in a gas environment provided by the first pulse. Arp et al. [85] studied the performance of LIBS under pressures as high as 90 atm in nitrogen and 56 atm in CO_2 (simulating Venus planet conditions). Although some elements showed strong emission line broadening and self-absorption, other lines appeared unaffected by the high pressure.

There is no single optimal ambient gas environment for all laser ablation applications.

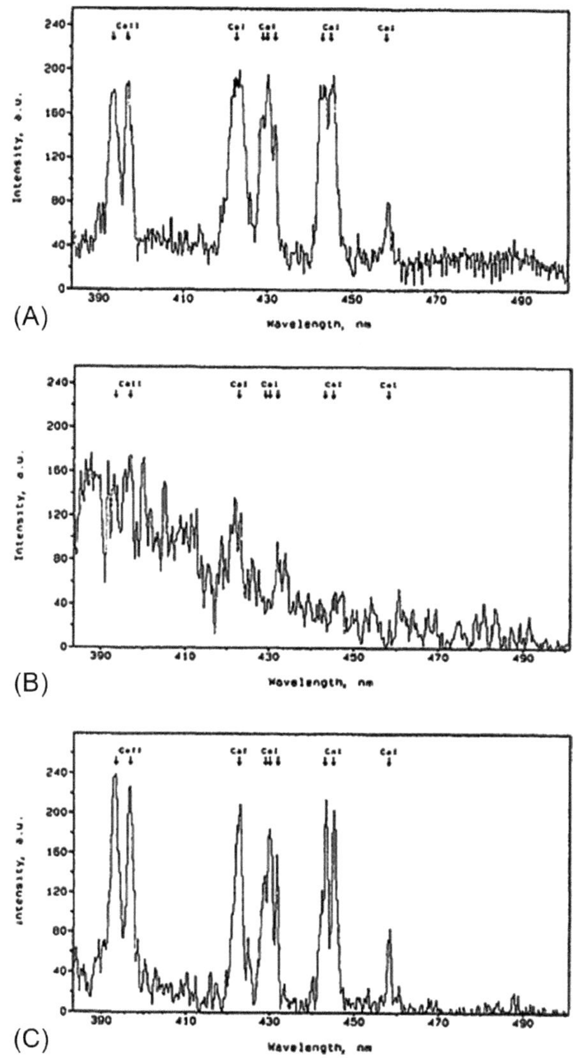

FIG. 16 Optical emission spectra from calcite (A) in air, (B) in water using the single pulse technique, and (C) in water using the double pulse technique. The *arrows* mark the wavelengths of the identified calcium lines. The pulse energy was 10 mJ for all spectra shown [92].

Characterization of pathways for energy transfer between the plasma, ambient gas, and analyte needs to be discerned and then exploited to optimize sensitivity and detection limits for a particular application.

5 Picosecond pulsed lasers

For picosecond pulsed laser ablation, the pulse duration is similar to the lattice heating time ($\tau_l \sim \tau_p$). In this case, laser ablation is accompanied by electron heat conduction and formation of a melted zone in the target. Evaporation occurs as a direct solid-vapor (or solid-plasma) transition. However, the presence of a liquid phase (as in the nanosecond case) can still cause preferential evaporation or elemental fractionation. Free electron emission is strong for picosecond laser ablation [103]. A comparison by Angel et al. [81] of LIBS measurements using nanosecond versus picosecond laser pulses showed that mass removal from the sample was more reproducible when using picosecond compared to nanosecond laser pulses. Less heat, stress damage, and redeposition of mass were observed in scanning electron microscopy images of the craters generated by picosecond laser pulses compared to those by nanosecond laser pulses (Fig. 17). The emission line intensity-to-background ratio was many times higher when using picosecond laser pulses (Fig. 18). The volume of the plasma was about an order of magnitude smaller in the picosecond case, because the plasma was not reheated by the laser pulse. In the case of nanosecond excitation, plasma reheating tends

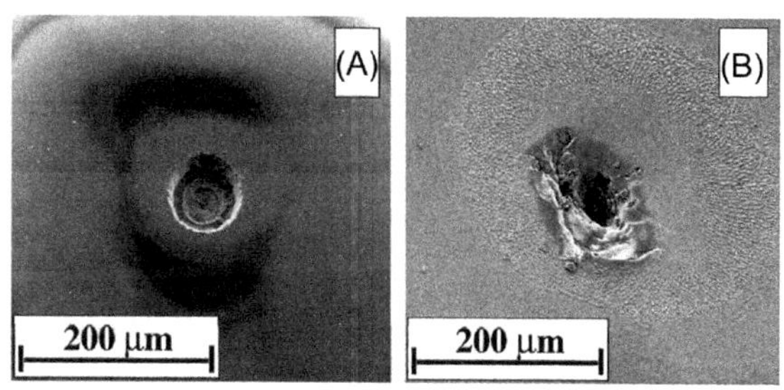

FIG. 17 Scanning electron microscopy images of holes produced from 50 consecutive laser shots of the glass sample by using picosecond (A) and nanosecond (B) excitation [81].

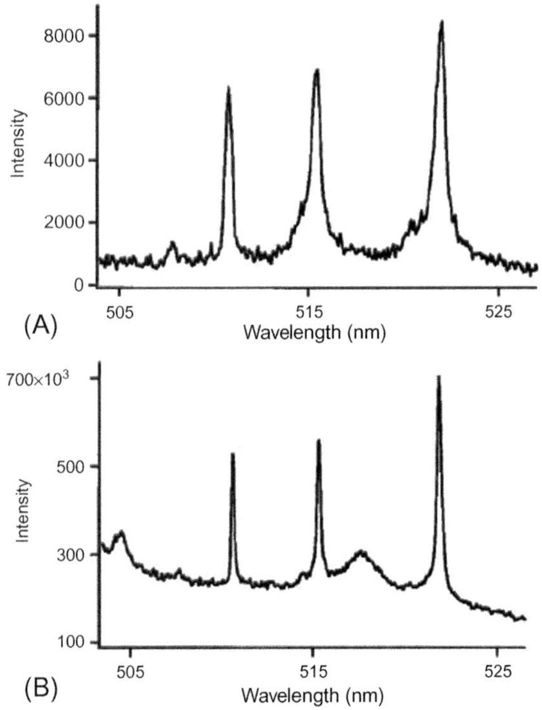

FIG. 18 Laser-induced breakdown spectroscopy spectra of a Cu metal sample with the use of one laser pulse and a nongated detector for picosecond (A) and nanosecond (B) excitation. The exposure time was 0.1 s for each measurement [81].

to elongate the plasma in the direction of the laser beam.

In spite of the differences between nanosecond and picosecond laser pulses, a mix of mechanisms is expected depending on the fluence. Rieger et al. [104] found similar emission behavior from the laser-induced plasma generated by picosecond and nanosecond laser pulses using only microjoules of laser energy. The plasma emission properties with pulse durations of 50 ps and 10 ns at energies above approximately 3 μJ were found to be similar. This behavior was attributed to the fact that even though the initial plasma conditions during the time of irradiation of the laser pulses are quite different, and the plasma formation threshold for picosecond laser pulse is lower, after 10 ns the energy absorbed by the plasma for both laser pulses will create a similar blast wave expanding into the background air. Most of the observed emission occurs in this expanding blast wave, which primarily depends on the absorbed energy. However, as one approaches the energy threshold for plasma formation for 10 ns pulses (~1 mJ) the emission rapidly diminishes. For the case of 50 ps pulses, significant plasma emission is still observed for 1 mJ pulses since the much higher intensities are still significantly above the plasma breakdown threshold and significant plasma heating occurs for these ultrashort pulses (Fig. 19).

6 Femtosecond pulsed lasers

For femtosecond laser pulse durations, the pulse length is similar to the electron cooling time ($\tau_e \sim \tau_p$) in a solid sample. Ablation with femtosecond laser pulses is considered as a direct solid-vapor (or solid-plasma) transition.

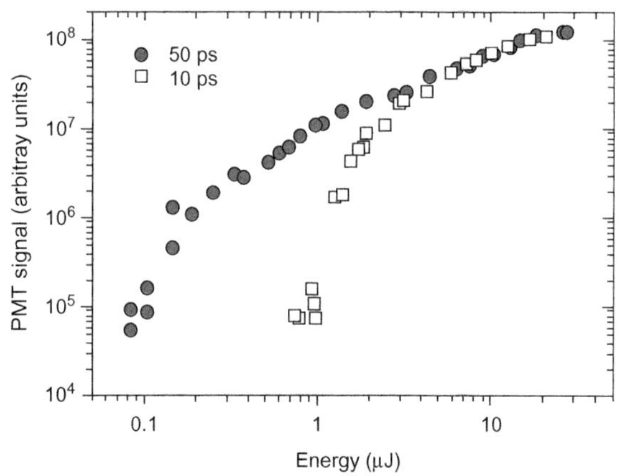

FIG. 19 Time-integrated emission of Si I 288 nm as measured with a photomultiplier at an average angle of 77 degrees from the target normal as a function of the laser pulse energy for single shots on silicon wafer. For signals below approximately 3×10^7 on the vertical axis the arbitrary units correspond to total photons emitted per steradian. Above approximately 3×10^7 arbitrary units the signals start to become slightly saturated and the response is no longer linearly related to the emission [104].

The lattice is heated on a subpicosecond timescale by nonthermal melting, followed by the creation of vapor and plasma with rapid expansion rates.

During the pulse, thermal conduction into the sample can be neglected. Some of the benefits of this fast interaction are a more reproducible ablation process and very precise laser machining of solids. Chichkov et al. [105] showed a qualitative comparison of holes drilled in 100 μm thick steel foils (in vacuum) with 10^4 laser pulses in the three regimes: femtosecond, picosecond, and nanosecond (Fig. 20).

Russo et al. [39] compared the crater depth produced by 266 nm nanosecond and femtosecond ablation in silicon (Fig. 21). For nanosecond and femtosecond laser ablation, the crater depth increased linearly with the number of pulses. However, for the same number of pulses, the femtosecond craters were about two times deeper than the nanosecond crater using the same fluence.

Within the framework of LIBS, significant differences in the temporal evolution of emission lines were also reported for various laser pulse durations. Le Drogoff et al. [106, 107] reported that the plasma temperature increased with the laser pulse duration, while the electron density remained relatively constant for the typical delays used in LIBS experiments. For pulses longer that 5 ps, the laser-plasma interaction results in plasma heating. Therefore as the laser pulse

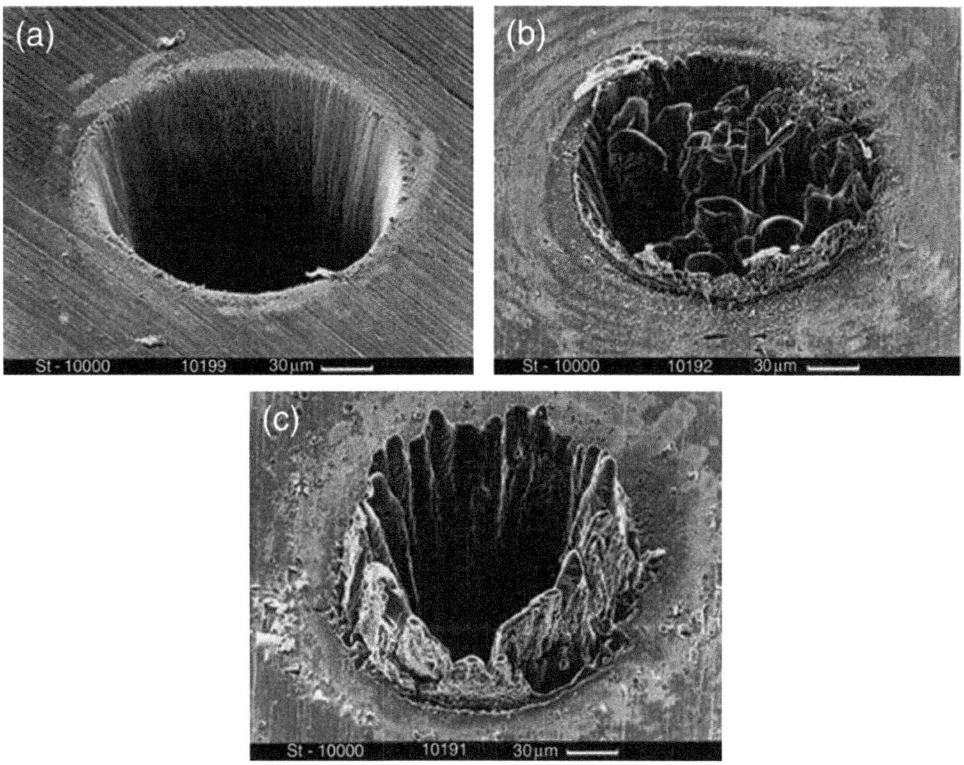

FIG. 20 Scanning electron microscopy photographs of pulsed laser ablation holes drilled (at 780 nm) in a 100 μm thick steel foil with (A) a femtosecond pulse with 200 fs, 120 μJ, $F = 0.5\,J/cm^2$, (B) a picosecond pulse with 80 ps, 900 μJ, $F = 3.7\,J/cm^2$, and (C) a picosecond pulse with 3.3 ns, 1 mJ, $F = 4.2\,J/cm^2$ [105].

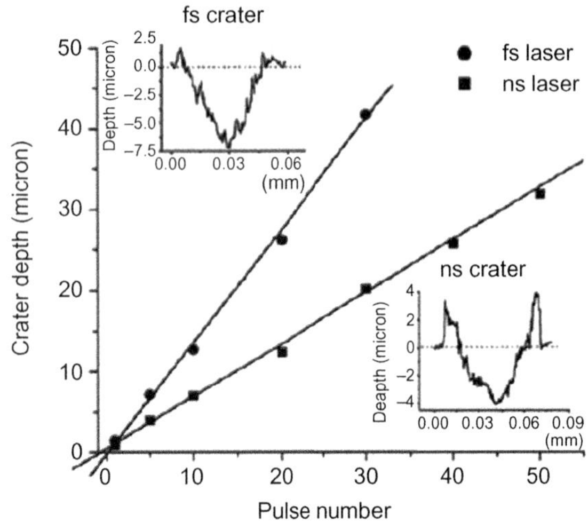

FIG. 21 Ablation depth versus pulse number for femtosecond and nanosecond laser ablation. The *inserts* show the femtosecond and nanosecond crater profile after five laser pulses [39].

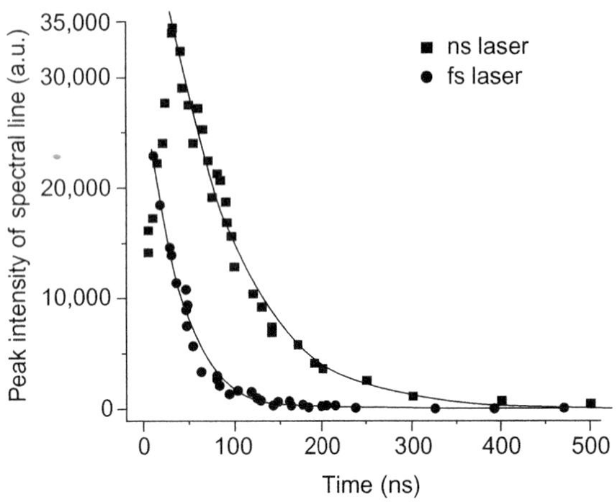

FIG. 22 Peak intensity of the Si I 288.16 nm emission line as a function of time for femtosecond and nanosecond laser ablation at a distance of 0.6 mm above the sample surface [39].

duration increases (from femtosecond to nanosecond), the plasma takes longer to decay. This effect was also shown by Russo et al. [39]. For early times (<30 ns) (Fig. 22) the emission intensity of the femtosecond plasma decreased, while the nanosecond-induced plasma became hotter due to absorption of the trailing part of the laser pulse at the beginning of the plasma process. Similar results were reported by Sirven et al. [108] where the characteristics of the temporal dynamics of the emission were similar in both cases (nanosecond and femtosecond), except at early moments after the excitation. Although the results obtained in this study with nongated detection for both regimes were satisfactory (in terms of signal-to-noise ratio) the continuum background in the nanosecond regime was about three times higher than in the femtosecond pulsed case (Fig. 23).

Angel et al. [81] also reported an increase in mass ablated with the use of a hybrid femtosecond laser, 140 fs pulse width, and energies between 0.26 and 0.94 mJ. They showed that at high laser irradiances, higher plasma temperatures and emission intensities were reached. Although the intensity increase appeared to be mostly related to the amount of ablated material, the increased temperature allowed a higher degree of dissociation and excitation, thus improving analytical sensitivity (Fig. 24).

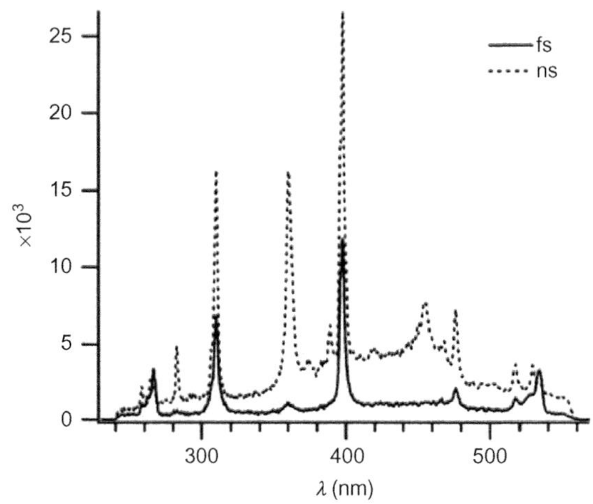

FIG. 23 Time-integrated spectra of a sample of metallic aluminum performed at 67 J/cm^2 in two laser-induced breakdown spectroscopy experiments after a 100 fs excitation (*plain line*) and a 7 ns excitation (*dotted line*) [108].

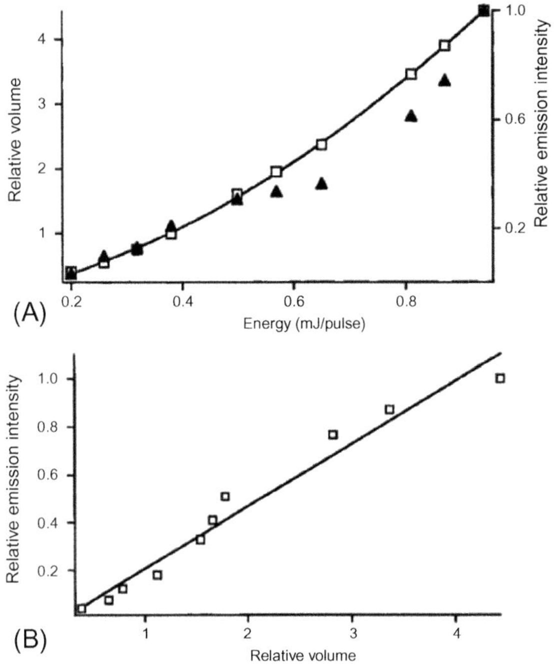

FIG. 24 (A) The estimated relative crater volume (*filled triangles*) and relative emission intensity of the 404.6 nm Fe line (*open squares*) versus laser energy with 140 fs excitation at 10 Hz. The relative volume was estimated by assuming a radius-cubed dependence. (B) A plot of relative emission intensity for the 404.6 nm Fe line versus the relative volume of the ablation crater [81].

As was mentioned earlier, plasma decays faster when it is induced by femtosecond laser pulses compared to nanosecond pulses. The main reason for this observation was that the plasma induced by nanosecond laser pulses absorbs part of the laser energy and is reheated, causing elongation of its lifetime and intensifying line emissions. However, an increase in the continuum background is also observed, which decreases technique sensitivity for nanosecond laser pulses.

Further improvement in sensitivity during femtosecond laser ablation-based analysis may be possible by choosing suitable ambient gases and pressures. Yalcin et al. [109] with the use of a femtosecond laser in air reported that the characteristic emission lines from Al, Mg, Si, and Cu elements exhibited significant enhancement in signal intensity at a few torr background air pressure as compared to atmospheric air pressure. These observations were attributed to a longer lifetime of the plasma expanding to a larger size at lower background pressures. It was also observed that signal enhancement at low pressure was dependent on the measurement delay time and on the transition being observed. With a delay time of 200 ns, the integrated intensity of the neutral Al I line at 396 nm exhibited 67 times enhancement in signal intensity at 4 Torr of pressure with respect to atmospheric pressure. A signal enhancement was only four times when no measurement delay time was used (Fig. 25). The best signal-to-noise ratio of 850 was observed at 4 Torr pressure for an 85 ns delay time.

Another way of improving the performance of femtosecond laser-induced plasma for LIBS could be by the use of a combination of pulses, for example, femtosecond pulses to generate a more efficient ablation and nanosecond or picosecond pulses to heat or reheat the plasma. Optimal conditions for plasma reheating with ultrafast pulses was proposed by Semerok et al. [110]. The optimal plasma reheating regime with the highest plasma intensity and reproducibility was determined to correspond to the double pulse delay of 100–200 ps. The pulses in the 50 fs–2 ps range were observed to give approximately the same ablation efficiency with a metal target. However, in these cases the double pulse approach is not being used to improve the ablation process (laser-material interaction) but to improve the plasma emission characteristics. A more detailed perspective of these approaches will be given in subsequent chapters.

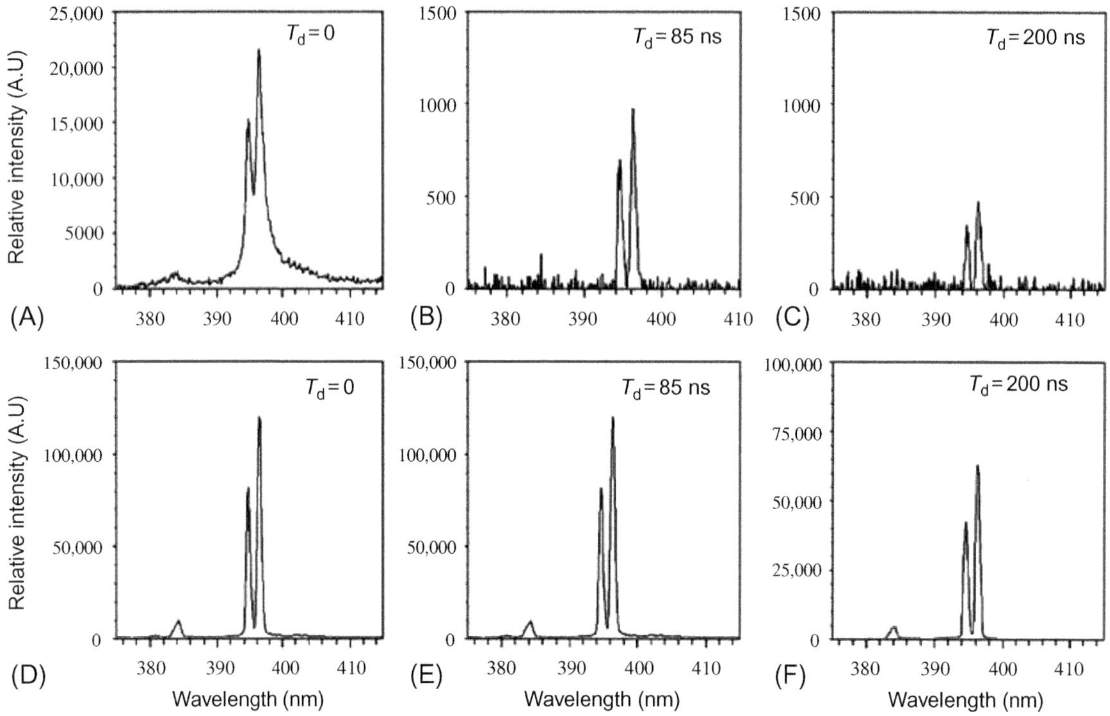

FIG. 25 Al I 394.6 and 396.15 nm lines at three different delay times (0.85 and 200 ns) and two different pressures: (A–C) for atmospheric pressure, (D–F) for 4 Torr pressure. Laser pulse energy 10 μJ [109].

7 Perspectives, future, and trends

Laser ablation is the underlying process to LIBS. In this respect, the laser parameters will have a dramatic influence on LIBS analytical performance metrics. A better understanding of fundamental laser ablation processes will guaranty better LIBS performance by being able to choose the parameters that couple laser energy with plume temperature and electron number density. With the varied available laser parameter space (energy, wavelength, pulse duration, fluence, etc.), and sample chemical systems, an empirical approach to LIBS optimization for each application would be onerous. A balanced approach consisting of iterative theory, modeling, and experimentation will be necessary to drive LIBS into new regimes.

Acknowledgment

The US Department of Energy, Office of Basic Energy Sciences, Chemical Sciences, and the Office of Nonproliferation and National Security (NA22) supported this research at the Lawrence Berkeley National Laboratory under Contract DE-AC02-05CH11231.

References

[1] J.C. Miller, Laser Ablation: Principles and Applications, Springer Series in Materials Science. Springer-Verlag Berlin Heidelberg, 2013.
[2] A. Einstein, Zur quantentheorie der strahlung, Phys. Z. 18 (1917) 121.
[3] A.L. Schawlow, C.H. Townes, Infrared and optical masers, Phys. Rev. 112 (1940) 1940–1949.
[4] T.H. Maiman, Stimulated optical emission in fluorescent solids. I. Theoretical considerations, Phys. Rev. 123 (1961) 1145–1150.

[5] T. Maiman, R. Hoskins, I. D'haenens, C. Asawa, V. Evtuhov, Stimulated optical emission in fluorescent solids. II. Spectroscopy and stimulated emission in ruby, Phys. Rev. 123 (2006) 1145–1150.

[6] F. Breech, L. Cross, Optical micro emission stimulated by a ruby maser, Appl. Spectrosc. 16 (2) (1962) 59.

[7] M.N.R. Ashfold, F. Claeyssens, G.M. Fuge, S.J. Henley, Pulsed laser ablation and deposition of thin films, Chem. Soc. Rev. 33 (2004) 23–31.

[8] T. Otani, L. Herbst, M. Heglin, S. Govorkov, A. Wiessner, Microdrilling and micromachining with diode-pumped solid-state lasers, Appl. Phys. A 79 (2004) 1335–1339.

[9] S. Taneri, J. Zieske, D. Azar, Evolution, techniques, clinical outcomes, and pathophysiology of LASEK: review of the literature, Surv. Ophthalmol. 49 (2004) 576–602.

[10] A. Troilius, Effective treatment of traumatic tattoos with a Q-switched Nd:YAG laser, Lasers Surg. Med. 22 (1998) 103–108.

[11] B. Drnovšek-Olup, M. Beltram, J. Pižem, Repetitive Er:YAG laser irradiation of human skin: a histological evaluation, Lasers Surg. Med. 35 (2004) 146–151.

[12] O. Topaz, E. Rozenbaum, A. Schumacher, M. Luxenberg, Solid-state mid-infrared laser facilitated coronary angioplasty: clinical and quantitative coronary angiographic results in 112 patients, Lasers Surg. Med. 19 (1998) 260–272.

[13] A. Moritz, Treatment of periodontal pockets with a diode laser, Lasers Surg. Med. 22 (1998) 302–311.

[14] R. Kaufmann, Matrix-assisted laser desorption ionization (MALDI) mass spectrometry: a novel analytical tool in molecular biology and biotechnology, J. Biotechnol. 41 (1995) 155–175.

[15] L. Moenke-Blankenburg, T. Schumann, J. Nölte, Direct solid soil analysis by laser ablation inductively coupled plasma atomic emission spectrometry, J. Anal. At. Spectrom 9 (1994) 1059–1062.

[16] R.E. Russo, X. Mao, H. Liu, J. Gonzalez, S.S. Mao, Laser ablation in analytical chemistry—a review, Talanta 57 (2002) 425–451.

[17] S.S. Mao, F. Quere, S. Guizard, X.L. Mao, R.E. Russo, G. Petite, et al., Dynamics of femtosecond laser interactions with dielectrics, Appl. Phys. A 79 (2004) 1695–1709.

[18] L.V. Keldysh, Ionization in the field of a strong electromagnetic wave, Sov. Phys. JETP 20 (1965) 1307.

[19] S.K. Sundaram, E. Mazur, Inducing and probing nonthermal transitions in semiconductors using femtosecond laser pulses, Nat. Mater. 1 (2002) 217–224.

[20] P.Y. Yu, M. Cardona, Fundamentals of Semiconductors, Springer Berlin Heidelberg, Berlin Heidelberg, 2010.

[21] N.W. Ashcroft, N.D. Mermin, Solid State Physics, Saunders, Philadelphia, 1976.

[22] N. Bloembergen, Laser-induced electric breakdown in solids, IEEE J. Quantum Electron. 10 (1974) 375.

[23] E. Yablonovitch, N. Bloembergen, Avalanche ionization and the limiting diameter of filaments induced by light pulses in transparent media, Phys. Rev. Lett. 29 (1972) 907–910.

[24] B.C. Stuart, M.D. Feit, S. Herman, A.M. Rubenchik, B. W. Shore, M.D. Perry, Nanosecond-to-femtosecond laser-induced breakdown in dielectrics, Phys. Rev. B 53 (1996) 1749–1761.

[25] P. Saeta, J. Wang, Y. Siegal, N. Bloembergen, E. Mazur, Ultrafast electronic disordering during femtosecond laser melting of GaAs, Phys. Rev. Lett. 67 (1991) 1023–1026.

[26] J.S. Preston, H.M. Van Driel, J.E. Sipe, Order-disorder transitions in the melt morphology of laser-irradiated silicon, Phys. Rev. Lett. 58 (1987) 69–72.

[27] D.H. Lowndes, G.E. Jellison Jr., Chapter 6 Time-resolved measurements during pulsed laser irradiation of silicon, Semiconduct. Semimet. 23 (1984) 313, https://doi.org/10.1016/S0080-8784(08)62439-4.

[28] Y. Siegal, E.N. Glezer, E. Mazur, Dielectric constant of GaAs during a subpicosecond laser-induced phase transition, Phys. Rev. B 49 (1994) 16403–16406.

[29] L. Spitzer, Physics of Fully Ionized Gases, 2nd rev., Dover, Mineola, NY, 2006, 170.

[30] K. Sokolowski-Tinten, D. von der Linde, Ultrafast phase transitions and lattice dynamics probed using laser-produced X-ray pulses, J. Phys. Condens. Matter 16 (2004) R1517–R1536.

[31] R.E. Russo, G. Klunder, P. Grant, B. Andresen, Laser ablation ion-storage time-of-flight mass spectrometry, Appl. Phys. A 69 (1999) 895–897.

[32] S.S. Mao, X.L. Mao, R. Greif, R.E. Russo, Dynamics of an air breakdown plasma on a solid surface during picosecond laser ablation, Appl. Phys. Lett. 76 (2000) 31.

[33] S.S. Mao, X. Mao, R. Greif, R.E. Russo, Initiation of an early-stage plasma during picosecond laser ablation of solids, Appl. Phys. Lett. 77 (2000) 2464–2466.

[34] S.S. Mao, X.L. Mao, R. Greif, R.E. Russo, Simulation of a picosecond laser ablation plasma, Appl. Phys. Lett. 76 (2000) 3370.

[35] A. Bogaerts, Z. Chen, R. Gijbels, A. Vertes, Laser ablation for analytical sampling: what can we learn from modeling? Spectrochim. Acta Part B 58 (2003) 1867–1893. http://linkinghub.elsevier.com/retrieve/pii/S0584854703001666.

[36] X.L. Mao, R.E. Russo, Observation of plasma shielding by measuring transmitted and reflected laser pulse temporal profiles, Appl. Phys. A 64 (1997) 1–6.

[37] Y.B. Zel'dovich, Y.P. Raizer, In: W. Hayes (Ed.), Physics of Shock Waves and High-Temperature Hydrodynamic Phenomena, vol. I, 1966. and Vol. II (1967), 2002, Academic Press, ISBN: 9780323147859.
[38] L.I. Sedov, Similarity and Dimensional Methods in Mechanics, tenth ed., CRC Press, 2018.
[39] X. Zeng, X.L. Mao, R. Greif, R.E. Russo, Experimental investigation of ablation efficiency and plasma expansion during femtosecond and nanosecond laser ablation of silicon, Appl. Phys. A 80 (2005) 237–241. http://www.springerlink.com/index/10.1007/s00339-004-2963-9.
[40] N. Arnold, J. Gruber, J. Heitz, Spherical expansion of the vapor plume into ambient gas: an analytical model, Appl. Phys. A 69 (1999) S87–S93.
[41] D. Comtois, F. Martin, F. Vidal, H. Pépin, H. P. Mercure, I. Gallimberti, et al., Spectroscopic study of ultrashort pulse laser-breakdown plasmas in air, Appl. Spectrosc. 56 (2002) 1444–1452.
[42] H. Liu, X.L. Mao, J.H. Yoo, R.E. Russo, Early phase laser induced plasma diagnostics and mass removal during single-pulse laser ablation of silicon, Spectrochim. Acta Part B 54 (1999) 1607–1624.
[43] X. Zeng, X.L. Mao, S.-B. Wen, R. Greif, R.E. Russo, Energy deposition and shock wave propagation, J. Phys. D Appl. Phys. 37 (2004) 1132–1136.
[44] G.J. Bastiaans, R.A. Mangold, The calculation of electron density and temperature in Ar spectroscopic plasmas from continuum and line spectra, Spectrochim. Acta Part B 40 (1985) 885–892.
[45] H.R. Griem, Principles of Plasma Spectroscopy, Cambridge University Press, Cambridge, 2009.
[46] G. Befeki, Principles of Laser Plasmas, Wiley-Interscience, New York, 1976.
[47] R. Kelly, A. Miotello, Comments on explosive mechanisms of laser sputtering, Appl. Surf. Sci. 96–98 (1996) 205–215.
[48] A. Miotello, R. Kelly, Critical assessment of thermal models for laser sputtering at high fluences, Appl. Phys. Lett. 67 (1995) 3535–3537.
[49] J.H. Yoo, S. Jeong, R. Greif, R.E. Russo, Explosive change in crater properties during high power nanosecond laser ablation of silicon, J. Appl. Phys. 88 (2000) 1638.
[50] M.M. Martynyuk, The role of evaporation and boiling of liquid metal in the process of electric explosion, Zh. Tekh. Fiz. 44 (1974) 1262–1270.
[51] M.M. Martynyuk, Explosive mechanism of destruction of metals by a powerful stream of electromagnetic radiation, Zh. Tekh. Fiz. 46 (1976) 741–746.
[52] S.S. Mao, F. Quéré, S. Guizard, X. Mao, R.E. Russo, G. Petite, et al., Dynamics of femtosecond laser interactions with dielectrics, Appl. Phys. A 79 (2004) 1695–1709.
[53] V.P. Carey, Liquid-Vapor Phase-Change Phenomena (1992 edition), Open Library, 1992.
[54] B. Luk'yanchuk, W. Marine, S. Anisimov, Condensation of vapor and nanoclusters formation within the vapor plume, produced by ns-laser ablation of Si, Laser Phys. 8 (1998) 291.
[55] J. Koch, H. Lindner, A. von Bohlen, R. Hergenröder, K. Niemax, Elemental fractionation of dielectric aerosols produced by near-infrared femtosecond laser ablation of silicate glasses, J. Anal. At. Spectrom 20 (2005) 901–906.
[56] J. Kosler, R. Pedersen, C. Kruber, P. Sylvester, Analysis of Fe isotopes in sulfides and iron meteorites by laser ablation high-mass resolution multi-collector ICP mass spectrometry, J. Anal. At. Spectrom 20 (2005) 192–199.
[57] C.Y. Liu, X.L. Mao, J.J. Gonzalez, R.E. Russo, Study of particle size influence on laser ablation inductively coupled plasma mass spectrometry using an in-line cascade impactor, J. Anal. At. Spectrom 20 (2005) 200–203.
[58] H. Kuhn, D. Gunther, Laser ablation-ICP-MS: particle size dependent elemental composition studies on filter-collected and online measured aerosols from glass, J. Anal. At. Spectrom 19 (2004) 1158–1164.
[59] D. Geohegan, A. Puretzky, G. Duscher, S. Pennycook, Time-resolved imaging of gas phase nanoparticle synthesis by laser ablation, Appl. Phys. Lett. 72 (1998) 2987.
[60] M. Kuwata, B. Luk'yanchuk, T. Yabe, Nanocluster formation within the vapor plume, produced by nanosecond-laser ablation: effect of initial density and pressure distributions, Jpn. J. Appl. Phys. 40 (2001) 4262–4268.
[61] A. Pereira, A. Cros, P. Delaporte, S. Georgiou, A. Manousaki, W. Marine, et al., Surface nanostructuring of metals by laser irradiation: effects of pulse duration, wavelength and gas atmosphere, Appl. Phys. A 79 (2004) 1433–1437.
[62] L. St-Onge, M. Sabsabi, P. Cielo, Analysis of solids using laser-induced plasma spectroscopy in double-pulse mode, Spectrochim. Acta Part B 53 (1998) 407–415.
[63] F. Colao, V. Lazic, R. Fantoni, S. Pershin, A comparison of single and double pulse laser-induced breakdown spectroscopy of aluminum samples, Spectrochim. Acta Part B 57 (2002) 1167–1179, https://rdm.pure.elsevier.com/en/publications/a-comparison-of-single-and-double-pulse-laser-induced-breakdown-s.
[64] G. Abdellatif, H. Imam, A study of the laser plasma parameters at different laser wavelengths, Spectrochim. Acta Part B 57 (2002) 1155–1165.

[65] I. Horn, M. Guillong, D. Gunther, Wavelength dependant ablation rates for metals and silicate glasses using homogenized laser beam profiles—implications for LA-ICP-MS, Appl. Surf. Sci. 182 (2001) 91–102.
[66] R.E. Russo, X.L. Mao, O.V. Borisov, H. Liu, Influence of wavelength on fractionation in laser ablation ICP-MS, J. Anal. At. Spectrom 15 (2000) 1115–1120.
[67] L.M. Cabalín, J.J. Laserna, Experimental determination of laser induced breakdown thresholds of metals under nanosecond Q-switched laser operation, Spectrochim. Acta Part B 53 (1998) 723–730.
[68] X.L. Mao, W. Chan, M. Shannon, R.E. Russo, Plasma shielding during picosecond laser sampling of solid materials by ablation in He versus Ar atmosphere, J. Appl. Phys. 74 (1993) 4915.
[69] X. Mao, W.-T. Chan, M. Caetano, M.A. Shannon, R. E. Russo, Preferential vaporization and plasma shielding during nano-second laser ablation, Appl. Surf. Sci. 96–98 (1996) 126–130.
[70] J.A. Aguilera, C. Aragón, F. Peñalba, Plasma shielding effect in laser ablation of metallic samples and its influence on LIBS analysis, Appl. Surf. Sci. 127–129 (1998) 309–314.
[71] X.L. Mao, A.C. Ciocan, O.V. Borisov, R.E. Russo, Laser ablation processes investigated using inductively coupled plasma–atomic emission spectroscopy (ICP–AES), Appl. Surf. Sci. 127–129 (1998) 262–268.
[72] H.C. Liu, X.L. Mao, J.H. Yoo, R.E. Russo, Nonlinear changes in plasma and crater properties during laser ablation of Si, Appl. Phys. Lett. 75 (1999) 1216–1218.
[73] D. Von der Linde, K. Sokolowski-Tinten, The physical mechanisms of short-pulse laser ablation, Appl. Surf. Sci. 154–155 (2000) 1–10.
[74] W. Marine, N.M. Bulgakova, L. Patrone, I. Ozerov, Electronic mechanism of ion expulsion under UV nanosecond laser excitation of silicon: experiment and modeling, Appl. Phys. A 79 (2004) 771–774.
[75] J. Reif, F. Costache, S. Eckert, M. Henyk, Mechanisms of ultra-short laser pulse ablation from ionic crystals, Appl. Phys. A 79 (2004) 1229–1231.
[76] W.-T. Chan, R.E. Russo, Study of laser-material interactions using inductively coupled plasma-atomic emission spectrometry, Spectrochim. Acta Part B 46 (1991) 1471–1486.
[77] A. Fernandez, X.L. Mao, W. Chan, M. Shannon, R. E. Russo, Correlation of spectral emission intensity in the inductively coupled plasma and laser-induced plasma during laser ablation of solid samples, At. Spectrosc. 67 (1995) 2444–2450.
[78] R.E. Russo, X.L. Mao, W. Chan, M. Bryant, W. Kinard, Laser ablation sampling with inductively coupled plasma atomic emission spectrometry for the analysis of prototypical glasses, J. Anal. At. Spectrom 10 (1995) 295–301.
[79] X.L. Mao, W. Chan, R.E. Russo, Influence of sample surface condition on chemical analysis using laser ablation inductively coupled plasma atomic emission spectroscopy, Appl. Spectrosc. 51 (1997) 1047–1054.
[80] O.V. Borisov, X.L. Mao, A.C. Ciocan, R.E. Russo, Time-resolved parametric studies of laser ablation using inductively coupled plasma atomic emission spectroscopy, Appl. Surf. Sci. 127–129 (1998) 315–320.
[81] K.L. Eland, D.N. Stratis, D.M. Gold, S.R. Goode, S. M. Angel, Energy dependence of emission intensity and temperature in a LIBS plasma using femtosecond excitation, Appl. Spectrosc. 55 (2001) 286–291.
[82] A. Bogaerts, Z. Chen, Nanosecond laser ablation of Cu: modeling of the expansion in He background gas, and comparison with expansion in vacuum, J. Anal. At. Spectrom 19 (2004) 1169–1176.
[83] G. Dumitru, V. Romano, H.P. Weber, Model and computer simulation of nanosecond laser material ablation, Appl. Phys. A 79 (2004) 1225–1228.
[84] F. Colao, R. Fantoni, V. Lazic, A. Paolini, LIBS application for analyses of martian crust analogues: search for the optimal experimental parameters in air and CO_2 atmosphere, Appl. Phys. A 79 (2004) 143–152.
[85] Z.A. Arp, D.A. Cremers, R.D. Harris, D.M. Oschwald, G. R. Parker Jr., D.M. Wayne, Feasibility of generating a useful laser-induced breakdown spectroscopy plasma on rocks at high pressure: preliminary study for a Venus mission, Spectrochim. Acta Part B 59 (2004) 987–999.
[86] Y. Iida, Effects of atmosphere on laser vaporization and excitation processes of solid samples, Spectrochim. Acta Part B 45 (1990) 1353–1367.
[87] S.S. Harilal, C.V. Bindhu, V.P.N. Nampoori, C.P. G. Vallabhan, Influence of ambient gas on the temperature and density of laser produced carbon plasma, Appl. Phys. Lett. 72 (1998) 167–169.
[88] V. Detalle, M. Sabsabi, L. St-Onge, A. Hamel, R. Héon, Influence of Er:YAG and Nd:YAG wavelengths on laser-induced breakdown spectroscopy measurements under air or helium atmosphere, Appl. Optics 42 (2003) 5971–5977.
[89] S. Amoruso, B. Toftmann, J. Schou, Expansion of a laser-produced silver plume in light background gases, Appl. Phys. A 79 (2004) 1311–1314.
[90] R.E. Russo, X.L. Mao, M. Caetano, M.A. Shannon, Fundamental characteristics of laser-material interactions (ablation) in noble gases at atmospheric pressure using inductively coupled plasma-atomic emission spectroscopy, Appl. Surf. Sci. 96–98 (1996) 144–148.
[91] G. Cristoforetti, S. Legnaioli, V. Palleschi, A. Salvetti, E. Tognoni, Influence of ambient gas pressure on laser-induced breakdown spectroscopy technique in the parallel double-pulse configuration, Spectrochim. Acta Part B 59 (2004) 1907–1917.

[92] R. Nyga, W. Neu, Double-pulse technique for optical emission spectroscopy of ablation plasmas of samples in liquids, Opt. Lett. 18 (1993) 747–749.
[93] A. Pichahchy, D. Cremers, M. Ferris, Elemental analysis of metals under water using laser-induced breakdown spectroscopy, Spectrochim. Acta Part B 52 (1997) 25–39.
[94] D. Beddows, O. Samek, M. Lika, H. Telle, Single-pulse laser-induced breakdown spectroscopy of samples submerged in water using a single-fibre light delivery system, Spectrochim. Acta Part B 57 (2002) 1461–1471.
[95] D.E. Kim, D.W. Kim, H.K. Park, K.J. Oh, K.J. Yoo, Quantitative analysis of aluminum impurities in zinc alloy by laser-induced breakdown spectroscopy, Appl. Spectrosc. 51 (1997) 22–29.
[96] R. Wisbrun, I. Schechter, R. Niessner, H. Schroeder, K. L. Kompa, Detector for trace elemental analysis of solid environmental samples by laser plasma spectroscopy, At. Spectrosc. 66 (1994) 2964–2975.
[97] L. Radziemski, D. Cremers, K. Benelli, C. Khoo, R. Harris, Use of the vacuum ultraviolet spectral region for laser-induced breakdown spectroscopy-based Martian geology and exploration, Spectrochim. Acta Part B 60 (2005) 237–248.
[98] Y.-I. Lee, T.L. Thiem, G.-H. Kim, Y.-Y. Teng, J. Sneddon, Interaction of an excimer-laser beam with metals. Part III: the effect of a controlled atmosphere in laser-ablated plasma emission, Appl. Spectrosc. 46 (2016) 1597–1604.
[99] R.J. Lade, F. Claeyssens, K.N. Rosser, M.N.R. Ashfold, 193-nm laser ablation of CVD diamond and graphite in vacuum: plume analysis and film properties, Appl. Phys. A 69 (1999) S935–S939.
[100] K. Tanaka, N. Shirai, T. Chen, Pulsed laser ablation of a single crystal cadmium tungstate (010) surface studied by time-of-flight mass spectrometry, J. Appl. Phys. 83 (1998) 7975.
[101] M. Noda, Y. Deguchi, S. Iwasaki, N. Yoshikawa, Detection of carbon content in a high-temperature and high-pressure environment using laser-induced breakdown spectroscopy, Spectrochim. Acta Part B 57 (2002) 701–709.
[102] B. Sallé, J.-L. Lacour, E. Vors, P. Fichet, S. Maurice, D. A. Cremers, et al., Laser-Induced Breakdown Spectroscopy for Mars surface analysis: capabilities at stand-off distances and detection of chlorine and sulfur elements, Spectrochim. Acta Part B 59 (2004) 1413–1422. http://linkinghub.elsevier.com/retrieve/pii/S0584854704001569.
[103] S.S. Mao, X. Mao, J.H. Yoo, R. Greif, R.E. Russo, Thermionic electron emission from narrow band-gap semiconductors under picosecond laser excitation, J. Appl. Phys. 83 (1998) 4462–4465.
[104] G.W. Rieger, M. Taschuk, Y.Y. Tsui, R. Fedosejevs, Comparative study of laser-induced plasma emission from microjoule picosecond and nanosecond KrF-laser pulses, Spectrochim. Acta Part B 58 (2003) 497–510.
[105] B.N. Chichkov, C. Momma, S. Nolte, F. von Alvensleben, A. Tünnermann, Femtosecond, picosecond and nanosecond laser ablation of solids, Appl. Phys. A 63 (1996) 109–115.
[106] B. Le Drogoff, J. Margot, M. Chaker, M. Sabsabi, O. Barthelemy, T.W. Johnston, et al., Temporal characterization of femtosecond laser pulses induced plasma for spectrochemical analysis of aluminum alloys, Spectrochim. Acta Part B 56 (2001) 987–1002.
[107] B. Le Drogoff, J. Margot, F. Vidal, S. Laville, M. Chaker, M. Sabsabi, et al., Influence of the laser pulse duration on laser-produced plasma properties, Plasma Sources Sci. Technol. 13 (2004) 223–230.
[108] J.-B. Sirven, B. Bousquet, L. Canioni, L. Sarger, Time-resolved and time-integrated single-shot laser-induced plasma experiments using nanosecond and femtosecond laser pulses, Spectrochim. Acta Part B 59 (2004) 1033–1039.
[109] Ş. Yalçın, Y.Y. Tsui, R. Fedosejevs, Pressure dependence of emission intensity in femtosecond laser-induced breakdown spectroscopy, J. Anal. At. Spectrom 19 (2004) 1295–1301.
[110] A. Semerok, C. Dutouquet, Ultrashort double pulse laser ablation of metals, Thin Solid Films 453–454 (2004) 501–505.

CHAPTER

4

Physics and dynamics of plasma in laser-induced breakdown spectroscopy

V.N. Rai[a,b], Surya N. Thakur[c]

[a]Laser Plasma Division, Raja Ramanna Centre for Advanced Technology, Indore, India [b]School of Physics, Devi Ahilya University, Indore, India [c]Laser and Spectroscopy Laboratory, Department of Physics, Banaras Hindu University, Varanasi, India

1 Introduction

The interaction of high-power laser light with a target material has been an active topic of research not only in plasma physics but also in the field of material science, chemical physics, and particularly analytical chemistry [1]. The high-intensity laser beam impinging on a target (solid, liquid, or gas) may dissociate, excite, and/or ionize the constituent atomic species of the solid and produce plasma, which expands either in the vacuum or in the ambient gas depending on the experimental requirements. Various processes may occur as a result of laser-matter interaction such as ablation of material, target acceleration, high-energy particle emission, and generation of various parametric instabilities as well as the emission of radiation ranging from the visible to hard X-rays depending on the intensity of the laser. These processes have many applications but we are mainly interested here in the study of optical emission from plasma. For other applications of laser-produced plasma, readers can find detailed information in a series of review articles and books [1–6a]. The spectroscopic study of optical emission from laser-produced plasma is known as laser-induced breakdown spectroscopy (LIBS), which began in 1962 soon after the invention of the laser [6b]. In brief, LIBS is an atomic emission technique suitable for quick and online elemental analysis of any phase of material (solid, liquid, gas, and aerosol). It has several advantages over conventional laboratory-based analytical techniques such as a high spatial resolution (due to small focal spot), absence of sample preparation, studies of hostile environments like melting or burning samples, and remote detection. Several review articles are available in Refs. [1, 7–16] on LIBS. The physics and dynamics of plasma relevant to LIBS is discussed in the following sections, which are such that a beginner can understand the basic processes occurring during LIBS and access up-to-date references in one place.

Laser-Induced Breakdown Spectroscopy
https://doi.org/10.1016/B978-0-12-818829-3.00004-6

2 Basics of laser-matter interaction

When a high-power laser pulse is focused onto a material target (solid, liquid, gas, and aerosol), the intensity in the focal spot produces rapid local heating and intense evaporation followed by plasma formation. The interaction between a laser beam and a solid is a complicated process dependent on many characteristics of both the laser and the solid material. Various factors affect the ablation of material, which include the laser pulse width, its spatial and temporal fluctuations, as well as its power fluctuations. The mechanical, physical, and chemical properties of the target material also play an important role in laser-induced ablation. The phenomena of laser-target interactions have been reviewed by several authors [17,18], while the description of melting and evaporation at metal surfaces has been reported by Ready [18]. It has been found that thermal conductivity is an important parameter for free vaporization of a solid into a vacuum. The plasma expands normally to the target surface at a supersonic speed in a vacuum or ambient gas. The hot expanding plasma interacts with the surrounding gas mainly by two mechanisms: (1) the expansion of high-pressure plasma compresses the surrounding gas and drives a shockwave, and (2) during this expansion, energy is transferred to ambient gas by the combination of thermal conduction, radiative transfer, and heating by the shockwave. The evolution of plasma depends on the intensity of the laser, its wavelength, size of focal spot, target vapor composition, ambient gas composition, and pressure. It has been found that plasma parameters such as radiative transfer, surface pressure, plasma velocity, and plasma temperature are strongly influenced by the nature of the plasma. Since vaporization and ionization take place during the initial fraction of laser pulse duration, the rest of the laser pulse energy is absorbed in the vapor and expanding plasma plume. This laser absorption in the expanding vapor/plasma generates three different types of waves as a result of different mechanisms of propagation of the absorbing front into the cool transparent gas atmosphere. These waves are (1) laser-supported combustion (LSC) waves, (2) laser-supported detonation (LSD) waves, and (3) laser-supported radiation (LSR) waves [18]. Each wave is distinguished on the basis of its velocity, pressure, and the effect of its radial expansion during the subsequent plasma evolution, which is strongly dependent on the intensity of irradiation.

At low irradiation, laser-supported combustion waves are produced, which comprise a precursor shock that is separated from the absorption zone and the plasma. The shockwave results in an increase in ambient gas density, pressure, and temperature, whereas the shock edges remains transparent to the laser radiation. The front edge of the expanding plasma and the laser absorption zone propagate into the shocked gas and give rise to a laser-supported combustion wave (Fig. 1). In the early days, thermal conduction was assumed to be the primary propagation mechanism. However, many investigators [19–21] studied one-dimensional propagation using a variety of transport models and found that the major mechanism causing LSC wave propagation was radiative transfer from the hot plasma to the cool high-pressure gas in the shockwave. The plasma radiation lies primarily in the extreme ultraviolet (UV) and is generated by the photorecombination of electrons and ions into the ground-state atom.

At intermediate irradiance, the precursor shock is sufficiently strong and the shocked gas is hot enough to begin absorbing the laser radiation without requiring additional heating by energy transport from the plasma. The laser absorption zone follows directly behind the shockwave and moves at the same velocity. This is similar to the chemical detonation wave and has been modeled by Ramsden and Savic [22] and Raiser [23]. The propagation of the laser-supported detonation wave is entirely

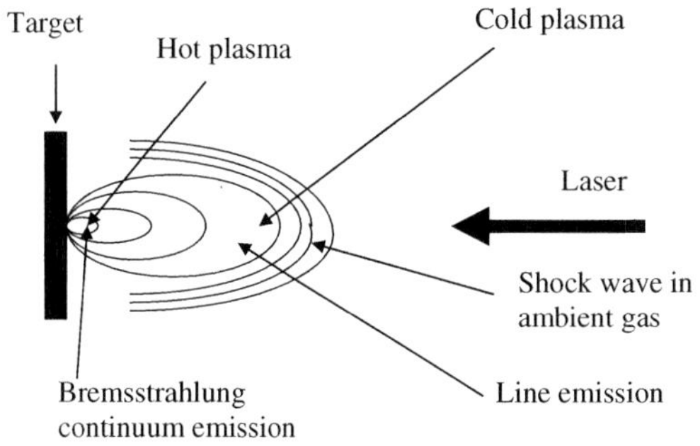

FIG. 1 Schematic diagram of expanding laser-produced plasma in ambient gas. The plasma plume is divided into many zones having high-density hot plasma and low-density cold plasma. The furthest zone from the target has minimum plasma density and temperature. Laser is absorbed in low-density corona.

controlled by the absorption of the laser energy. Several workers have theoretically and experimentally studied the ignition and propagation of the LSD wave away from metal surfaces [24–27].

At high irradiance, plasma radiation is so hot that, prior to the arrival of the shockwave, the ambient gas is heated to temperatures at which laser absorption begins. In the ideal condition, laser absorption is initiated without any density change, and the pressure profile results mainly from strong local heating of the gas rather than a propagating shockwave. This configuration is nothing but an overdriven absorption wave [23]. These supersonic waves have been modeled numerically by Bergel'son et al. [24] and it has been found that once the transient plasma initiation and formation processes are completed, then quasi-steady approximation is suitable. The laser-supported radiative wave velocity increases much more rapidly with irradiance than those of the LSC and LSD waves. Temperature and pressure increases are quite low, which indicates that the LSR wave is effective in channeling the absorbed energy into heating a large amount of gas rather than increasing the local enthalpy.

3 Processes in laser-produced plasma

As discussed in the previous section, the interaction of a high-intensity laser light with a solid target initially increases the surface temperature of the sample such that material transfer across the surface becomes significant (vaporization). As a result of material vaporization and plasma formation, target erosion appears in the form of craters on the sample surface. Theoretical considerations on plasma production and heating by means of laser beams have been proposed by several workers [25–28]. The initiation of plasma formation over a target surface begins in the hot target vapor. First of all, absorption of laser radiation takes place via electron-neutral inverse Bremsstrahlung (IB), but when sufficient electrons are generated, the dominant laser absorption mechanism makes a transition to electron-ion IB. Photoionization of excited states can also contribute in the case of interactions with short wavelength radiations. The same absorption processes are also responsible for absorption by the ambient gas. The laser-produced plasma expands into the vacuum or into the surrounding gas atmosphere, where the free electrons present in the plasma [1–6] modify the propagation of laser light. The plasma formed by a high-intensity or short-pulse duration laser has a very steep density and temperature gradient in comparison to the plasma formed by the low-intensity or longer-pulse duration laser. The density gradient in the plasma plays a very important role in the mechanism of light absorption and in the partition of absorbed energy between thermal and nonthermal particle distribution. There are three basic mechanisms through which intense laser light may interact with plasma [5]. The first mechanism is an IB, where the electric field of the incident light rattles electrons, which then lose this energy in collision with ions. This mechanism is important with shallow density gradients in the plasma. Parametric processes also take part most efficiently when the density gradient is shallow. There are three-wave parametric interaction processes in which an intense laser field drives one or more longitudinal plasma waves out of the noise and also starts parametric decay processes, where laser light decays into a high-frequency electron acoustic wave and a low-frequency ion acoustic wave conserving energy and momentum. Another important short-pulse laser absorption mechanism is resonance absorption. With a *p*-polarized light obliquely incident on the plasma surface, the radial component of the electric field resonates with plasma frequency and causes a large transfer of energy to electrons near the critical density (N_c) surface. Critical density for a given laser wavelength is:

$$N_c = 10^{21}/\lambda^2\,\text{cm}^{-3} \tag{1}$$

where λ is in microns. Energy absorbed at or below the critical density in plasma is then conducted toward the target surface by various transport processes. The study of energy coupling to the target has many subareas such as laser light absorption, nonlinear interaction, electron energy transport, and ablation of material from the target surface. One of the important processes during laser-plasma interaction is the emission of radiation from the plasma ranging from visible to hard X-ray wavelengths [1] and it is very relevant for the understanding of LIBS. It has been found that X-rays are emitted from all parts of the absorption, interaction, and transport regime. At densities near and slightly above the critical, almost 70% of the incident laser energy may be reemitted as X-rays with energy ranging from 50 eV to 1 keV or above depending on the temperature of the plasma. However, as the plasma expands away from the target surface, its density as well as the temperature decrease. As the plasma temperature decreases the wavelength of emission from the plasma increases, that is, emission shifts from X-rays to the visible wavelength region.

4 Spectral emission from plasma

The spectral composition of emission from the plasma has both line as well as continuum components. A study of the characteristic line emissions from plasma can give information about the composition of target material, that is, the elements present in the target. LIBS has been proved as a versatile technique of elemental analysis for any type of samples [1, 8–16].

4.1 Continuum emission

Continuum radiation is emitted by the plasmas as a result of free-free and free-bound transitions. Free-free transitions (Bremsstrahlung emission) are due to the interaction of electrons with positively charged ions of charge Z and density N_Z. Neglecting the correction factor (the Gaunt factor), the free-free spectral power density for a Maxwellian electron distribution at a temperature T_e is given by [29,30]:

$$I_\nu \approx Z_{\text{eff}} \frac{N_e N_Z}{T_e^{1/2}} \exp\left(-\frac{h\nu}{T_e}\right) \tag{2}$$

where Z_{eff} is the effective charge for Bremsstrahlung emission, which is generally different from Z due to the partial screening of the nuclear charge Z_N by the $(Z_N - Z)$ electrons [31]. Eq. (2) shows that the logarithmic slope of the spectrum can provide information about the electron temperature T_e of plasma. The inverse process is called IB and is responsible for the absorption of laser radiation at or below the critical density region. When the electron distribution function is not Maxwellian, it is in principle possible to deconvolute it from the high photon energy nonthermal spectrum (since free-free radiation is very sensitive to high-energy tails). In the case of laser-produced plasmas the high-energy electrons (usually called suprathermal electrons) have long mean free paths and can therefore penetrate into the cold solid target. In this case the Bremsstrahlung spectrum is modified, because the electrons (of all energies) are slowed down by their interaction with the cold solid matter [32].

In free-bound (recombination radiation) transition, a free electron (kinetic energy ξ_e) is captured by an ion of charge Z in a bound-level n of the ion of charge (Z − 1) (ionization energy χ^n) and as a result a photon of energy $h\nu = \xi_e + \chi^n$ is emitted. The photon energy is a function of ξ_e, but only photons with $h\nu > \chi^n$ are emitted. The spectra therefore show discontinuities, generally called recombination edges. Above the edges, the spectral dependence is the same as for free-free transition, and T_e can therefore again be deduced from the logarithmic slope. Nonthermal electrons do not contribute to the free-bound spectrum, because the recombination cross-section decreases rapidly for $\xi_e \gg \chi^n$. De Michelis and Mattioli [31] and references therein have discussed this subject in detail.

4.2 Line emission

The emissivity of a spectral line (i.e., number of photons emitted per unit time and per unit volume) is equal to the product of its radiative transition probability times the emitting excited-level density. On the other hand, the absorption coefficient for line radiation is obtained from the radiative transition probability by using the Einstein relations. For the interpretation of line emission, the excited-level population should be known.

4.3 Emission from laser-produced plasma

Emission from laser-produced plasma in general and optical emission in particular has been studied extensively using temporally and spatially resolved spectroscopy [1, 9–16]. It has been observed that immediately after plasma formation an intense continuum is emitted, which remains close to the target surface. This emission from the dense plasma is like a black body continuum radiation. However, as the plasma expands away from the surface of the target, it cools and the emission is dominated by the optical line emission. During this expansion the spectral lines from highly ionized atoms are observed close to the target surface, whereas those from neutral atoms are observed in the plume away from the target. Thus line emissions are superimposed on the continuum emission. The line emission from the multiply ionized species occurs at the time of plasma formation, whereas emissions from singly ionized and neutral species are observed nearly 500 ns after plasma formation. The rates of decay of the continuum and line emissions are different. The continuum emission due to Bremsstrahlung from hot plasma decays faster in comparison to the line emission. Thus it is necessary to record the emission after a certain time delay to obtain clear information about the line emission from the cold plasma for the purpose of elemental analysis.

5 Models for the distribution of charged and excited states in plasma

Interpretation of the radiation emitted by the plasmas requires knowledge of both the charge-state distribution and the excited-level populations of different ions. This is possible by obtaining the solution of a complex system of rate equations, describing the population and depopulation of all the levels by processes such as ionization, recombination, collisional excitation and deexcitation, radiative decay, and absorption, as well as stimulated emission. Any given charge state is connected with its two neighboring states by the processes of ionization and recombination. Considering the difficulties associated with solving these equations, the approximations used in order of increasing electron density are the corona model (CM), collisional-radiative model (CRM), local thermodynamic equilibrium (LTE), and models suitable for ultrahigh density ($N_e \geq 10^{24}$ cm^{-3}) plasmas [29,31,33].

5.1 Corona model

In this approximation there is a balance between collisional ionization (and excitation) and recombination (and spontaneous decay). This model therefore depends critically on the knowledge of atomic cross-sections. Assuming that free electrons have a Maxwellian velocity distribution, it is required that only a negligible number of ions be in excited levels (as compared to the ground level). Two neighboring ionization states, of charge Z and $(Z + 1)$, are then connected by:

$$N_Z N_e S_Z(T_e) = N_{Z+1} N_e \alpha_{Z+1}(T_e, N_e)$$

or

$$\frac{N_Z}{N_{Z+1}} = \frac{\alpha_{Z+1}(T_e, N_e)}{S_Z(T_e)} \tag{3}$$

where S_z and α_{Z+1} are the ionization and recombination rate coefficients, respectively. Eq. (3) is, to a first approximation, independent of N_e. In this approximation a balance between the rate of collisional excitation from the ground level and the rate of spontaneous radiative decay determines the population densities of excited levels. This model requires that the electron density be sufficiently low for collisions not to interfere with the radiative emission.

5.2 Local thermodynamic equilibrium model

In the LTE model it is assumed that mainly particle collision processes determine the distribution of population densities of the electrons, which are so fast that the latter take place with sufficient rapidity and the distribution responds instantaneously to any change in the plasma condition. In such circumstances each process is accompanied by its inverse and these pairs of processes occur at equal rate by the principle of detailed balance. Thus the distribution of population densities of the energy levels of the electrons is the same as it would be in a system in complete equilibrium. The population distribution is determined by the statistical mechanical law of equipartition among the energy levels and does not require knowledge of atomic cross-sections. Actually, the plasma temperature and density vary in space and time, but the distribution of population densities at any instant and point in space depends entirely on the local values of temperature, density, and chemical composition of the plasma. Uncertainty in prediction of spectral line intensities from LTE model plasma depends mainly on the uncertainty in the values of these plasma parameters and atomic transition probabilities. For analytical plasmas the condition of LTE is considered very much vital for obtaining any reliable quantitative information. In the case of thermal equilibrium, all the processes in the plasma are collision dominated as already discussed and the plasma can be considered as having a single temperature $T_i = T_e$. However, in the expanding plasma this is possible only locally and for specific time segments during the evolution. The following criterion must be satisfied by the plasma to be in LTE [33,34]:

$$N_e \geq 1.6 \times 10^{12} \Delta E^3 T_e^{1/2} \tag{4}$$

where ΔE (eV) is the largest observed transition energy for which the condition holds and T_e is the excitation temperature (K). It should be noted that the choice of time delay is crucial for obtaining the best operating conditions in the LIBS plasma to ensure that LTE prevails during the measurements for obtaining quantitative results. However, it has also been found that LTE is not an indispensable condition for qualitative analysis, once the measurement conditions are kept constant and are exactly reproducible.

5.3 Collisional radiative model

In the case of high-density plasma, both of the foregoing models cannot be used safely (although LTE is sometimes marginally satisfied). Salzmann [35] addressed the applicability of the CM and LTE approximations to laser-produced plasmas. In the intermediate density range the rate equations have to be solved up to the thermal limit. The total ion densities of two successive ionization states of charge Z and $(Z + 1)$ are connected by:

$$N_Z S_Z^{\text{eff}} = N_{Z+1}\left(\alpha_{Z+1}^{\text{eff}} + N_e \beta_{Z+1}^{\text{eff}}\right) \tag{5}$$

where S_Z^{eff}, $\alpha_{Z+1}^{\text{eff}}$ and β_{Z+1}^{eff} are the effective (or net) ionization, recombination, and three-body recombination rate coefficients, respectively. The last term dominates only at high densities, when $\frac{N_Z}{N_{Z+1}}$ tends to the Saha equation value. At low electron densities the three-body recombination is negligible, and Eq. (5) tends to Eq. (3) (excited-state populations become negligible with respect to the ground-state population). The CRM system can be solved locally for given N_e and T_e values. It can, however, also be coupled to a hydrodynamic code describing the plasma evolution. Capitelli et al. [36] studied nonequilibrium and equilibrium problems in laser-induced breakdown plasma. Particularly, they focused on the problem associated with the fluid dynamics of the expanding plume with time-dependent CRMs for describing the population densities of excited states and with the time-dependent Boltzmann equation for characterizing the electron energy distribution function in the LIBS plasmas. It was found that the violation of equilibrium conditions in laser-produced plasma near the surface could be caused by the decrease in plasma temperature due to expansion. This can occur if the characteristic time of such temperature decrease is less than or comparable with that corresponding to the ionization balance, which is estimated as $\tau_{\text{ion}} \approx (N_a k_{\text{ion}})^{-1}$, where N_a (cm^{-3}) and k_{ion} (cm^3 s^{-1}) are the number density of heavy particles and ionization rate coefficients, respectively. For typical laser plasma conditions ($N_a \approx 3 \times 10^{19}$ cm^{-3}; $T_e \approx 2$ eV), τ_{ion} is in the range $\approx 10^{-6}$–10^{-5} s. This quantity has to be compared with the characteristic time of laser plasma expansion, i.e., $\tau_{\text{exp}} \approx d/v$, where d is the laser spot diameter and $v \approx 10^5$–10^6 cm/s is the plasma expansion velocity. It can be concluded that the equilibrium conditions are violated whenever:

$$\tau_{\text{ion}} \geq \tau_{\text{exp}} \tag{6}$$

which takes place at laser spot sizes $d < 1$ cm. In this condition, deviation from the equilibrium state has a recombination character. LIBS plasmas, characterized by large electron densities and electron temperatures, apparently seem to satisfy the LTE conditions. However, the characteristic equilibrium times for the different phenomena can occur in the same temporal scale during which LIBS measurements are performed. Dedicated experiments and the development of a unitary theory taking into account the fluid dynamics and kinetic aspects are necessary to completely master the experimental conditions for the development of calibration-free LIBS [37].

6 Measurement of spectroscopic and plasma parameters

During the LIBS experiments, some important parameters such as emission line shape functions, electron density, and plasma temperature are required to derive reliable quantitative data. Some relevant information about the measurement of these parameters is provided in the following sections.

6.1 Line broadening in plasma emission

In the spectroscopic study of line emission from the plasma, intensity of the spectral lines as well as their profiles plays an important role. The emission line profile is important because it contains information related to the emitter and the surrounding plasma environment. An introduction to the theory of line broadening can be found in the books by Griem [29,38] and Sobelman et al. [39]. Various types of line broadening have been observed in plasma emission. Natural broadening occurs due to the finite lifetime of excited states and results in a Lorentzian profile. This must be considered in particular cases of highly ionized ions like Fe^{25+}. Doppler broadening occurs due to thermal or directed motion of the emitting ions and has a Gaussian line shape with a width proportional to the square root of the emitter temperature. This is the dominant broadening mechanism at low electron densities. Stark broadening, which is also known as pressure broadening, is observed because the emitting ions experience an electric field due to the presence of plasma electrons and ions around them. This field varies statistically for individual emitters and fluctuates in time. The net result is an ensemble-averaged line shape with an overall width related to the average strength of the perturbation in the bound states.

During the early stage of plasma formation in the LIBS experiment the electron density remains very high ($n_e \sim 10^{15}$–10^{18} cm^{-3}). As a consequence, the line profiles are dominated by Stark/pressure broadening for a considerable period of time [14]. Doppler as well as natural broadening are generally negligible during this period. When the expanded plasma cools and the electron density decreases, the dominance of Stark broadening is reduced and finally Doppler broadening starts playing a leading role. It is well known that the measured line profiles normally also contain contributions from instrument resolution width if spectrometers are used for wavelength analysis. This indicates that the measured profiles must be deconvoluted prior to their analysis for the extraction of plasma parameters. The instrumental contribution to line width needs to be measured experimentally for a given spectrometer, which is dependent on parameters such as slit width, grating dispersion, and dynamic behavior of the photon detector. The central peak of the spectrometer slit function can be approximated to a good degree of accuracy by a Lorentzian profile. Since both (the Stark-broadened) spectral lines and the spectrometer exhibit Lorentz shape functions, convolution and deconvolution become easy. In this case, line widths can simply be added or subtracted in the convolution/deconvolution process as given here:

$$\Delta\lambda_{\text{Total}} = \Delta\lambda_{\text{line}} + \Delta\lambda_{\text{spectrometer}} \quad (7)$$

This shows that actual line width can easily be extracted from the measured line width by simply subtracting the instrumental width.

Samek et al. [14] fitted the line shape for Ca I and Ca II for the highest concentration in a liquid sample, and nearly perfect Lorentzian profiles were obtained even for a delay time of <1 μs, which shows that the problem of self-absorption is less probable in liquid plasma than in solid plasma due to comparatively less concentration.

6.2 Electron density

The width of Stark-broadened spectral lines in plasmas depends mainly on the electron density N_e as discussed in Chapter 2. Both the linear and the quadratic Stark effect are encountered in spectroscopy. However, only the hydrogen atom and H-like ions exhibit the linear Stark effect, whereas all other atoms exhibit the quadratic Stark effect. This is the reason why ideally information on electron density is extracted from the lines of H or H-like ions, where half width of the line profile can be calculated easily with greater accuracy. In the case of the linear Stark effect, the relation between electron density and line width is given by a simple relation [29]:

$$N_e = C(N_e, T_e)\Delta\lambda_{\text{FWHM}}^{3/2} \quad (8)$$

Here $\Delta\lambda$ is the "true" full width at half maximum (FWHM) and the parameter C depends (only weakly) on N_e and T_e, which can normally be treated as being constant. The constant C for the H Balmer lines is available in Ref. [29]. The first choice for electron density determination in LIBS plasmas containing hydrogen is the H_β line (with an error of 5%) [29]. This is because of its large intensity and sufficiently large line broadening, which can be measured precisely using a spectrometer of moderate resolution. The possibility of self-absorption in this case is relatively small. The second best choice among the Balmer series is the H_γ line. The H_α line is suitable in cases where electron density is not too high ($N_e \sim 10^{17}$ cm^{-3}), because at higher electron densities this strong line is quite susceptible to self-absorption, which severely distorts the line profile. In the case of non-H-like atoms, where the quadratic Stark effect is dominant, the relation between electron density and line width [29] is:

$$\Delta\lambda_{\text{FWHM}} \approx 2\left[2 + 1.75 \times 10^{-4} N_e^{1/4} \alpha \left(1 - 0.068 N_e^{1/6} T^{-1/2}\right)\right] \times 10^{-16} w N_e \quad (9)$$

The first term in the brackets gives the contribution from electron broadening, and the second term stems from ion broadening. Here, w is the electron impact parameter at $N_e = 10^{16}$ cm^{-3}, and α is the ion-broadening parameter. The parameters w and α can be found easily from Ref. [29]. Since the second term in Eq. (9) is normally small, the expression reduces to:

$$\Delta\lambda_{\text{FWHM}} \approx 2 \times 10^{-16} w N_e \quad (10)$$

which is normally used for calculations in the case of plasmas generated from solid targets.

It has been reported by Samek et al. [14] that the electron densities derived from the H_β and H_γ line data were in good agreement with each other, while the electron density estimated from the H_α line was up to a factor of 10 larger. This study suggested that the H_α line is not purely Stark broadened. Additional broadening may be due to the onset of self-absorption. The electron densities extracted from the H_δ line data fluctuate significantly because the H_δ line overlaps with strong emission from a nitrogen line. It is also found that in addition to the hydrogen lines, resonance lines of aluminum and calcium at 396.15 and 422.67 nm, respectively, can be used to estimate electron densities considering that these lines exhibit quadratic Stark broadening.

6.3 Plasma temperature

Plasma temperatures are often determined from the measurement of ratios of the intensities of (1) ion-to-neutral lines and (2) neutral-to-neutral lines, usually for the same element. In case (1) the line intensities are combined with the Saha equation and the electron density measurements to determine the ionization temperature of the plasma:

$$\frac{I_{\text{ion}}}{I_{\text{atom}}} = 2 \times \frac{(2\pi m_e kT)^{3/2}}{N_e h^3} \left(\frac{gA}{\lambda}\right)_{\text{ion}} \left(\frac{\lambda}{gA}\right)_{\text{atom}} \times \exp\left[\frac{-(V^+ + E_{\text{ion}} + E_{\text{atom}})}{kT_{\text{ion}}}\right] \quad (11)$$

Here, I is the integrated emission intensity of the ion or atom, N_e is the electron density (cm^{-3}), gA is the product of the statistical weight and the Einstein coefficient for spontaneous emission of the upper level (s^{-1}), λ is the wavelength (nm), T_{ion} is the ionization temperature (K), V^+ is the ion potential of the atom (J), E_{ion} is the excitation energy of the ionic line (J), E_{atom} is the excitation energy of the atomic line (J), k is the Boltzmann constant (J/K), and h is the Planck's constant (J s).

In case (2) the line intensities are combined with the Boltzmann equation to determine the excitation temperature of the plasma and the relation is given by:

$$\frac{I_1}{I_2} = \frac{g_1 A_1}{g_2 A_2} \cdot \frac{\lambda_2}{\lambda_1} \exp\left(-\frac{|E_1 - E_2|}{kT_e}\right) \quad (12)$$

where 1 and 2 refer to the individual lines in the pair. The accuracy in the temperature determination increases with an increase in energy difference $E_1 - E_2$ in Eq. (12). The accuracy may be improved by measuring a number of different line pairs and taking the average. However, it should be noted that the accuracy of the measurement largely depends on the values of A coefficients and their error ranging from 5% to 50%. The electron densities obtained from the hydrogen Balmer lines were used, along with line intensities of Ca I at $\lambda = 422.6$ nm and Ca II at $\lambda = 393.3$ nm, to calculate the ionization temperature from Eq. (11) [14].

Normally in the plasmas generated from liquids, very few (resonance) lines are observed. Consequently, line pairs available for calculating the excitation temperature for such cases using the two-line Boltzmann method, as given in Eq. (12), are rather sparse. Samek et al. [14] attempted to determine the excitation temperature using line pairs from Ca II, Cu I, and the H_β and H_γ lines. They relied on the temperature extracted from H_β/H_γ line-pair data because of the large errors in estimating the excitation temperature from the Ca II or Cu I line pairs (a consequence of the small energy gap $E_1 - E_2$ and a small FWHM). From the evaluation of the hydrogen line shapes and intensities they found that the excitation temperature agrees reasonably well with the ionization temperature in the LIBS plasma generated from the laminar liquid jet stream in the time window 1–5 μs. The errors for excitation and ionization temperature (10% and 8%, respectively) were estimated according to the procedure described by Simeonsson and Miziolek [40]. They noted during this study that the temperatures observed for the plasma generated from an aqueous solution are generally lower than those obtained from solid samples.

Abdellatif and Imam [41] used Eq. (12) to evaluate the spatial profile of plasma temperature from aluminum plasma along the axial direction normal to the target surface and found a peak at 500 μm from the surface.

6.4 Optical thickness and self-absorption

In the case of very high plasma density conditions, the plasma itself absorbs its own emission. This is mainly true for the resonance lines connected to the ground state, but other lines may also be affected. The absorption causes a distortion in the spectral line profile resulting in a broadened line. This effect is known as self-absorption. During the LIBS experiments the plasma temperature tends to drop toward the outer parts of the plasma plume, and when light passes through these colder parts an effect known as self-reversal can occur. Such lines show a (severe) dip at the center of the ordinarily well-behaved line shape function, giving doubt whether there are two lines.

The foregoing discussion shows that for evaluating the plasma parameters and extracting quantitative data from the line intensities, it is important to verify that the plasma is not optically thick for the lines being used for this purpose. The ratio of emission intensities of resonant and nonresonant lines should be verified according to a procedure for the "optically thin" limit described by Cremers and Radziemski [42], Simeonsson and Miziolek [40], and Sabsabi and Cielo [43]. It is necessary that the observed intensity ratios are consistent with those predicted by the statistical weights of the upper levels indicating that the plasma is optically thin. Normally during the measurement of major elements in composite solid samples, or in pure samples, severe self-reversal and self-absorption are observed for many lines during a few microseconds after plasma formation. The resonance line of calcium at 422.67 nm has shown huge self-reversal and self-absorption over a long period of plasma evolution [14]. This phenomenon has less effect in the case of liquid and gaseous samples in comparison to solid samples but it is observable at higher analyte concentrations or at higher laser intensities.

7 Characteristics of LIBS plasma

After discussing various processes for laser-produced plasma and its emission, it is time to discuss generation, its characteristics, as well as the parameters affecting LIBS plasma. A schematic diagram of the LIBS setup is shown in Fig. 2. It consists of a laser, a focusing system, the target sample to be analyzed, and a telescope for imaging the plasma plume on the entrance slit of the spectrometer, where a gated intensified charge-coupled device camera is used as a detector. Small differences in LIBS instrumentation has been noted in terms of laser wavelengths used for plasma formation. Ciucci et al. [44], Cremers [45], and Wisbrun et al. [46] used an Nd:YAG laser operating in its fundamental wavelength ($\lambda = 1064$ nm). Singh et al. [47] and Rai et al. [48] used the second harmonic of Nd:YAG, whereas Barbini et al. [49] used an Nd:YAG laser operating in the third harmonic ($\lambda = 355$ nm). In most cases, laser duration is between 5 and 10 ns and the fluence ranges from 1 to 50 J/cm^2. When the laser is focused on solid samples of a different nature, irradiances of the order of a few GW/cm^2 can be achieved. Much higher laser irradiances are necessary when the laser is focused on liquid, gas, or aerosol mixtures [47–49, 51, 52]. Yalcin et al. [50] used a modified Nd:YAG laser ($\lambda = 532$ nm) with pulse energies of 40–150 mJ (pulse duration between 10 and 13 ns) focused on a 21-μm spot size, thus reaching irradiance values ranging from 500 to 1500 GW/cm^2. Normally, the line emission intensities of selected elements are recorded during a temporal window (gate width) t_b after a given time delay (t_d) from plasma formation [45–51].

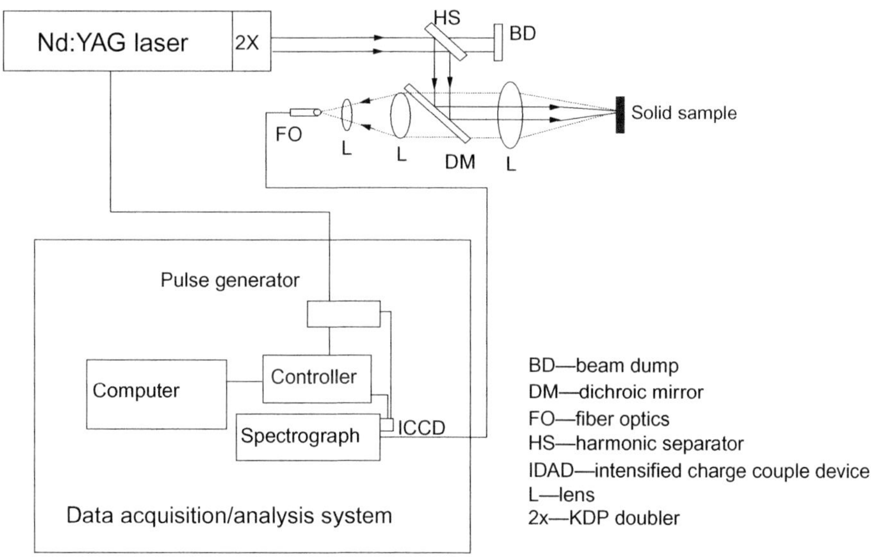

FIG. 2 Schematic diagram of an experimental system for recording the laser-induced breakdown spectroscopy (LIBS) spectra from solid samples. In fact, any type of sample can be placed at laser focus to obtain LIBS spectra of that sample [48].

Detection of the strong background continuum at early times can introduce noise in the measurements, because in the beginning the plasma is hot enough and emits Bremsstrahlung radiation (Fig. 3) [52]. Thus a judicious selection of t_d and t_b is necessary to optimize the signal-to-noise ratio as well as signal-to-background ratio. Experimental investigations show that LIBS measurements are generally made for t_d and t_b values ranging in the microsecond regime. These characteristic times can be compared with the characteristic times indicating the different kinetics in the plasma under consideration.

As discussed earlier the plasma is formed by nanosecond laser pulses as a result of photon absorption processes such as IB and photoionization. Theory and experiment [53] have shown that electron densities as large as 10^{20} cm^{-3} and electron temperatures of the order of 2 eV are typical of the laser ablation process in the nanosecond regime. However, after nearly 5 ns, electron (T_e) and ion temperature (T_i) become almost equal as a result of the thermalization process occurring due to collisions between electrons, ions, and neutrals. This is one of the requirements for plasma to be in LTE and is the reason why LIBS measurements are made a few microseconds after plasma formation when the LIBS plasma is under typical recombination conditions. Lower electron density N_e and temperatures T_e exist during LIBS sampling in the microsecond regime. Typical density and temperatures lie in the range $10^{15} < N_e < 10^{18}$ cm^{-3} and $0.5 < T_e < 2$ eV, respectively, which are strongly dependent on the time delay t_d. Higher electron densities are found for plasmas produced at atmospheric conditions, because of the confinement of free expansion of plasma by the air. However, these values also depend on the type of laser and the sample used for the experiment. Normally, lower electron densities and lower temperatures are observed in the plasma plume away from the target sample surface. The reported N_e and T_e value ranges are sufficient to satisfy the LTE plasma condition, where the Saha and Boltzmann equations hold for the number densities of plasma constituents, including excited-state number densities.

These equations also imply the validity of the Maxwell distribution function for describing the electron energy distribution function. Moreover, these distributions (Saha, Boltzmann, and Maxwell) are thought to adapt themselves to the local values of N_e and T_e during the process of plasma expansion. The choice of experimental arrangement should be aimed at optimizing the reproducibility of the measurements, while at the same time achieving plasma conditions where the hypotheses of LTE and thin plasma are fulfilled. It has been noted that LIBS plasma as well as emission from it depend on the experimental configuration as well as on the plasma

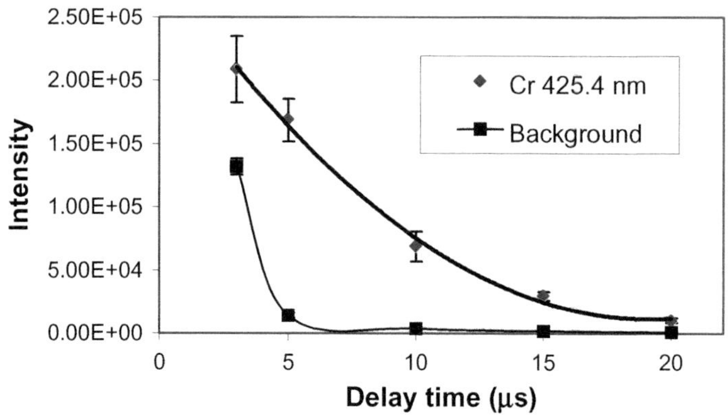

FIG. 3 The variation in the atomic line and background emission intensity from chromium (Cr) in aqueous solution with gate delay (t_d). Gate width (t_b) = 10 μm [52].

parameters. This dependence will be discussed in the following sections. A variety of different experimental configurations have been reported in the references, where different arrangements are adopted to solve different analytical problems.

8 Parameters affecting LIBS plasma

Various types of lasers are used in LIBS, typically ranging from UV excimer lasers to infrared (IR) solid-state lasers. Each laser has a different absorption characteristic during plasma production, which affects the behavior of the resulting plasma. However, the shape and size of the laser-induced plasma plume is dependent on many other experimental parameters, including the thermal and mechanical properties of the target material. Effects of these parameters on LIBS plasma will be discussed in the following sections.

8.1 Effect of laser characteristics

There are two main mechanisms for electron generation and its growth before plasma formation. The first mechanism involves absorption of laser radiation by free electrons present in the target vapor when they collide with neutrals. If the electrons gain sufficient energy, they can impact and ionize atoms or molecules present in the sample vapor. Electron concentration will increase exponentially with time due to cascade breakdown. The second mechanism, called multiphoton ionization, involves the simultaneous absorption of a sufficient number of photons by an atom or molecule to cause its ionization. Multiphoton ionization plays an important role only for short-wavelength lasers (<1 μm). Both cascade and multiphoton ionization require high laser irradiances, usually $\geq 10^8\ \mathrm{W/cm^2}$. Finally, plasma formation takes place when the laser intensity exceeds this threshold value, which is easily attained by proper focusing of the laser beam. In the diffraction limit, the diameter w_2 of the laser beam at the focus of the lens can be calculated as:

$$w_2 \approx 2.44\lambda \frac{f}{w_1} \tag{13}$$

where w_1 is the diameter of the laser beam before focusing, f is the focal length of the lens, and λ is the laser wavelength.

Many authors have studied laser ablation in various materials as well as its relation with threshold intensity, which is also dependent on the properties of the target [10]. Hwang et al. [54] reported that the mass removed from the target depends not only on the intensity of the laser but also on several other factors. A relation has been found for the ablated mass $m(t)$ from the target surface on the basis of a heat conduction mechanism as:

$$m(t) = A(aIt) + B(AI)^2 t^{3/2} \tag{14}$$

where A, B are proportional to the target thermal properties, a is the energy coupling factor, I is the laser irradiance, and t is the laser pulse duration. Cabalin and Laserna [55] made a systematic study for determining the threshold intensity for plasma formation in a group of nine metals largely spread in physical and thermal properties, like melting and boiling temperatures, thermal conductivity, etc. They found that the threshold intensity correlated fairly well with thermal properties such as melting and boiling temperature, suggesting that thermal effects play a significant role during laser ablation with nanosecond pulses. They also investigated the behavior of line emission intensity versus incident intensity and found an initial linear correlation, probably due to an increase in the amount of ablated material. For higher laser intensity, the emission signal reaches a saturation regime, attributed to the self-absorption of the emission by the plasma formed in front of the sample, or due to poor coupling of the laser because of plasma shielding. The starting point of the saturation regime is element dependent and correlated with the value of the threshold intensity. Similar behavior with respect of laser coupling efficiency was found for nanosecond and picosecond lasers by Russo [56].

It has been found that spectral emission intensity is also dependent on the plasma temperature, which is significantly increased by higher laser power density. Yueh et al. [52] reported an increase in the sensitivity of the calibration curve of Re when the laser energy was increased from 200 to 250 mJ (Fig. 4). Higher power density has been found beneficial for improving the analytical sensitivity [57], but in the limit of self-absorption. Several techniques have been used for manipulating the laser intensities. Chaleard et al. [58] spatially filtered the laser beam before focusing it on the sample surface to obtain a more homogeneous energy distribution and consequently a more regular crater. Ideally, a "flat-top" distribution of the laser beam energy is supposed to be better for target ablation because each point of the target surface would receive the same amount of energy, which is generally absent in the lasers used for this purpose.

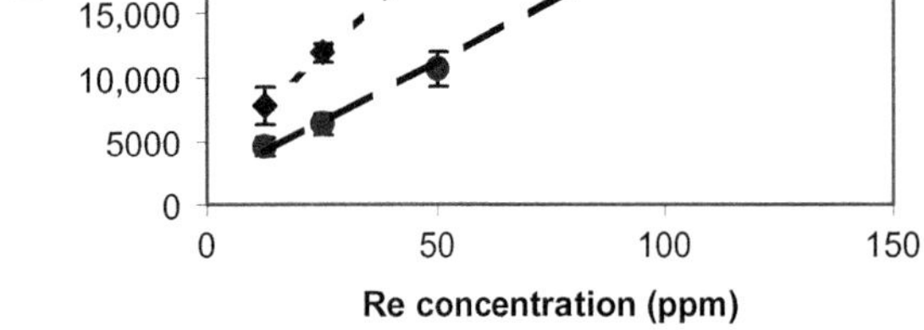

FIG. 4 The laser-induced breakdown spectroscopy calibration curve for Re obtained from liquid jet measurement with a delay time of 8 μs and gate delay of 15 μs [52].

The effect of pulse-to-pulse laser stability on emission intensity has been investigated by Castle et al. [59] by simultaneously monitoring the analyte (Cu) signal and the laser pulse energy by means of a photodiode. The absence of significant correlation indicated that the laser pulse variance played a small role in the overall variance of LIBS measurement. It seems that along with the fluctuation in laser intensity, instabilities in the plasma also add up in the standard deviation of the LIBS signal. Wisbrun et al. [46] measured the behavior of the signal-to-noise ratio in sand and soil samples depending on the laser pulse energy, which was varied between 0 and 320 mJ. In their experimental conditions, the signal-to-noise ratio increased with laser pulse energy until approximately 100 mJ, and then remained almost constant.

8.2 Effects of wavelength and pulse duration of a laser

As stated earlier, various types of lasers have been used for the study of LIBS and the resulting plasma having different properties [60,61]. In a large number of studies, Nd:YAG lasers ($\lambda = 1064$ nm) have been used, and in many cases the fundamental radiation has been converted into second ($\lambda = 532$ nm), third ($\lambda = 355$ nm), or fourth ($\lambda = 266$ nm) harmonics depending on the application. Fabbro et al. [62] used an Nd: YAG laser that was frequency doubled and quadrupled to study the effect of wavelengths and found a relation for the mass ablation rate $\dot{m}$ ($\mathrm{kg/s\,cm^2}$) in terms of wavelength (λ) and the absorbed laser flux I_L ($\mathrm{W/cm^2}$):

$$\dot{m} = 110\frac{\left(I_L{}^{1/3}\right)}{\left(10^{14}\right)}\lambda^{-4/3} \tag{15}$$

Another scaling for mass ablation rate has been reported by Dahmain [63] for a laser energy lower than 10^{13} W/cm^2 and $Z \leq 13$ as:

$$\dot{m}\,(\text{kg/s cm}^2) = 65\left[\frac{I_L(\text{W/cm}^2)}{10^{13}}\right]^{5/9} \times \lambda^{-4/9}(\mu m) \times z^{1/4} \tag{16}$$

These relations show that mass ablation rate will increase strongly with shorter wavelengths. Eq. (1) also shows that the critical density of the plasma will increase for a laser of lower wavelength, which is an indication that laser energy will be absorbed in the plasma more efficiently (up to higher plasma density). In a nanosecond laser ablation process, target evaporation begins just after the impact of the leading edge of the laser pulse on the surface. The interaction of the following part of the laser pulse with the vapor in the vicinity of the target surface leads to strong heating and ionization of the vapor resulting in plasma formation. Although some species can be directly vaporized as ionized particles, plasma formation can be mainly ascribed to the processes involved in the laser-vapor interaction. Amoruso et al. [64] evaluated the efficiency of the mechanisms of energy absorption in the plasma for a visible and UV laser during ablation of an aluminum sample. They observed that the primary mechanism of laser absorption and ionization of the relatively cold neutral vapor formed by the leading edge of the ablating laser pulse is very strong for the radiation wavelength $\lambda = 532$ nm. This ionization can be mainly ascribed to the electron-neutral IB process. However, for $\lambda = 355$ nm it was found that direct photoionization of excited states in the vapor was the most effective process (the IB process is less efficient in the UV than in the visible part of the spectrum, because of the λ^3 dependence of IB).

Berman and Wolf [65] compared the analytical results obtained in the detection of Ni in water by using alternatively the fundamental Nd:YAG wavelength and the third-harmonic UV radiation. They observed lower continuum intensity in the UV-generated spectrum, leading to a better signal-to-noise ratio. The calibration curves for the same spectral lines obtained with UV irradiation revealed a higher slope and allowed a better limit of detection value. These results are in close agreement with the results obtained on liquids by Ng et al. [66], who found that the main difference in the plasma plume generated by the visible or UV laser of the same fluence is in the T_e value, which is significantly lower in the second case, while the N_e value is approximately the same.

Shorter pulses (picosecond) are reported to produce higher mass ablation rates, probably because the fraction of the pulse energy loss by thermal diffusion in the sample is much lower than in the case of nanosecond pulses [67]. A comparison of emission from plasma produced by nanosecond, picosecond, and femtosecond lasers has been made by many authors [68,69]. They found that the line emission and continuum background emission intensity as well as plasma temperature decay very rapidly after excitation using short time duration laser pulses (picosecond and femtosecond) compared to the nanosecond pulse excitation. It was also noted that gated detection is not necessary with short laser pulses, because background emission is much lower in the case of LIBS using a shorter-pulse laser than LIBS using a nanosecond laser.

8.3 Effect of target material

The physical and mechanical properties of the target materials have an important influence on the shape and size of the crater on its surface. The reflectivity of the target material determines the fraction of laser energy coupled to the target. It has been found that laser energy can be absorbed effectively by a reflecting surface due to the phase change of the material at high temperature, which is possible only at a very high intensity of the laser [70]. Allemand [71] indicated that the reflectivity of the sample surface, density, specific heat, and boiling point of the pure metal target have an important influence on the shape and size of the craters that can be given by the following relation:

$$D = \frac{A(1-R)}{\rho c T_b} \tag{17}$$

where D is the diameter of the crater, A is the proportionality constant, R is the reflectivity of the surface, ρ is the density of the material, C is the specific heat, and T_b is the boiling temperature.

A comparison of the crater size of the homogeneous material revealed that thermal conductivity is an important parameter. The volume heated by laser pulse is found to depend on the thermal conductivity of the material [72]. The heating of material around the crater increased with increase in incident laser intensity, because evaporation depends mainly on the boiling point of the material at fixed pressure.

Dimitrov et al. [73,74] investigated the dependence of evaporation processes and the dynamics of the plasma plume on the orientation of the target surface with respect to the laser beam. When a metal target is irradiated by laser, the ablated products expand almost perpendicular to the target surface. When the target surface is inclined with respect to the direction of the laser beam, the path length of the radiation in the plasma is shortened and results in decreased absorption by the laser-produced plasma.

Mao et al. [67,75] reported that the absorbed energy in a copper target is much higher when irradiated by UV laser for producing a plasma plume. They experimentally demonstrated that the transition between thermal and explosive regimes in laser ablation occurs, for the different wavelengths, at power density values scaling inversely as the $(1-R)$ factor, where R is the reflectivity of copper. Russo et al. [76] analyzed the influence of laser wavelength on the specific problem of fractionation in laser ablation and it was found that the shorter the wavelength, the more controlled and reproducible is the ablation rate. The intensity requirement to initiate the ablation process is also found to be lower for shorter-wavelength lasers.

Cabalin and Laserna [55] found that the thresholds for plasma formation and the onset of saturation regime were shifted toward lower fluence for IR wavelength, while the energy threshold values were higher for IR radiation. They found that reflectivity does not seem to be a relevant parameter at high fluences (i.e., above the threshold) because the plasma formation changes the properties of the target surface.

8.4 Effect of observation window

Initially, laser-induced plasma has very high electron temperature. However, plasma temperature decreases as the plasma expands away from the target surface. Early stages of plasma evolution are dominated by the Bremsstrahlung continuum emission. The evolution of line emission has been observed after the expansion of plasma, which is followed by the decrease in plasma temperature. These line emissions are observed superimposed on the continuum emission. Only the line emission from the plume is important for the compositional analysis of the sample target. Normally, the continuum emission from the plasma decays very fast in comparison to the line emission. Recording of line emission for the purpose of elemental analysis would depend on choosing a proper delay time, when the ratio of line emission to background (continuum) emission is very high. At large delay times the absolute intensity of the line emission can be too low for efficient detection. The best compromise between high line intensity and low background is determined on a case-by-case basis. Depending on the density of the plasma, ambient gas and other factors, plasma lifetime ranges from approximately 300 ns to >40 μs [77].

Ciucci et al. [44] compared the time evolution of the plasma emission obtained by irradiating the same sample in air with the fundamental wavelength of an Nd:YAG laser and the UV radiation of an excimer laser. They observed that in the case of UV excitation the plasma emission was initially dominated by the background up to approximately 400 ns, when the atomic line transitions started to appear. The continuum emission generated by an Nd:YAG laser showed comparatively longer times ranging in several microseconds. The timescales of the plasma induced by the two different sources differ remarkably due to a more rapid decay of continuum emission in the experiment performed with UV excitation as compared to IR excitation. It has been observed that the delay time (gate delay) and integration time (gate window) both play an important role in the optimization of signal-to-background and signal-to-noise ratio. The optimum value of gate delay and gate width changes due to the composition of the target as well as the other plasma parameters that affect the plasma emission. Wisbrun et al. [46] reported a systematic analysis of two elements (Zn and Cd) in a sand sample. They found an optimal value for the delay time of approximately 0.5 μs with an integration time of approximately 1.5 μs for these elements.

During the plasma evolution, ion density remains higher near the target surface, whereas neutral density dominates as a result of recombination as the plasma expands away from the target surface. This is the reason why the ratio between the population of neutral and ionized species changes with time. Yueh et al. [52] reported that the intensities of line emission from magnesium ion and background emission were found to be high in comparison to that of line emission from neutral Mg when the LIBS spectra were recorded at 2-μs gate delay. However, ion emission and background emission intensity decayed fast as the gate delay was increased to 4–5 μs (Fig. 5). Leis et al. [78] reported the study of evolution of the emission intensity of two iron lines (Fe I 285.2 nm and

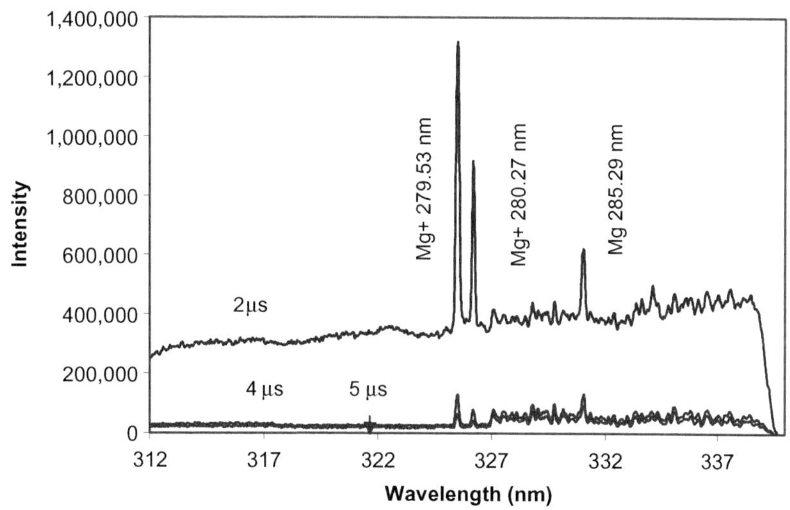

FIG. 5 A typical laser-induced breakdown spectroscopy spectra of Mg recorded at different delay times from a liquid jet [52].

Fe II 288.4 nm) in a pure iron sample, recorded during the first 15 μs of plasma lifetime. During the first 3 μs, the intensity of the ionic line exceeded the neutral by 50 times, whereas at a delay of approximately 10 μs, the neutral line intensity was eight times that of the ionic. Aragon et al. [79] studied the evolution of line intensity and the line-to-continuum ratio and found its dependence on the specific line and on the pulse energy. They concluded that the line emission intensities and the line-to-continuum emission ratios cannot be maximized simultaneously for all the elements investigated by using a single detection time window. They suggested using a large integration time (2–15 μs), which allows the collection of most of the LIBS signal for all the elements, while eliminating the undesirable initial part of the emission.

Sabsabi and Cielo [43] analyzed the LIBS spectrum of aluminum alloys recorded after a delay time of 10 μs and a gate width of 10 μs. They also reported the temperature evolution in a nanosecond time duration Nd:YAG laser-induced plasma from aluminum and copper targets in air [80]. A quick decrease in plasma temperature was found during the first few microseconds (from approximately 1 eV to approximately 0.5 eV) in both the cases, whereas a small change was noted at later times in the microsecond scale (two decay rates). Leis et al. [78] determined the time-resolved temperature for a series of binary Fe—Cr alloys with iron content ranging from 10% to 100%. The time evolution in plasma temperature was found similar for all the samples, whereas the absolute value differed by approximately 50% among the samples, with the pure iron showing the higher temperature.

8.5 Effect of geometric setup

The geometrical shape of the plasma and spatial emission intensity profile are strongly dependent on the laser power density and other parameters such as optical alignment for focusing the laser and collecting the emission from a plasma plume for recording the spectrum. Therefore it is necessary to understand the dependence of the LIBS signal on optical alignment and the collection of emitted light for recording the spectrum.

Eppler et al. [81] compared the precision of the results obtained using a spherical and cylindrical lens for focusing the laser pulse on the surface of soil samples, where more precise results were found in the case of the cylindrical lens. In the case of LIBS of a solid sample, normally the laser is focused perpendicular to the target surface. The high intensity of laser may produce a breakdown in air before the focus at the surface, particularly when dust particles are occasionally irradiated. This problem is avoided by setting the distance between the focusing lens and the target a little shorter than the focal length to produce a stable breakdown, while maximizing the interaction area [46,79]. This perpendicular irradiation configuration may cause some problems in the case of liquid samples due to the splashing of liquid on the lenses, which strongly reduces the sensitivity of LIBS as the plasma plume (ejected droplets) expands in the direction perpendicular to the sample surface (liquid surface) [77,82]. Fichet et al. [83] suggested a configuration with the laser beam directed on the surface with a tilt angle of 15 degrees (75 degrees with respect to the perpendicular), which keeps the optics protected from splashing.

The collection of plasma emission and its detection is normally performed axially, that is, along the direction perpendicular to the target surface because of its simplicity and reproducibility. This configuration was found less sensitive to changes in the surface-to-lens distance, which occurs when several shots are fired at the same place on the sample surface and a crater is formed. With on-axis collection, the change in lens-to-surface distance causes minimum perturbations on the LIBS signal, because of the depth of focus of the detection optics, which is typically longer than crater depth [58]. This is why lenses having focal lengths of a few centimeters are typically used for LIBS measurements. There are certain other types of changes that may occur in the plasma emission, i.e., those produced by time evolution of the plasma plume as the plume expansion is governed by the strong explosion law. If we assume a one-dimensional model, where the plasma expands along the x-direction and the target surface is at $x = 0$, the distance $x(t)$, reached at time t, can be expressed as:

$$x(t) = k\left(\frac{E_L}{\rho_0}\right)^{m/2} t^m \tag{18}$$

where E_L is the energy deposited by the laser during the ablation process and ρ_0 is the ambient gas density. The coefficient m depends on the plasma expansion geometry and is equal to 2/3 for planar propagation and 2/5 for spherical symmetry [84,85]. Eq. (18) shows that the plasma is moving fast in the initial stages, but slows down at later times. This indicates that the time gates and the region of the plasma to be inspected must therefore be matched to intercept the emission signal at a certain spatial location [57].

Another arrangement used for collecting the plasma emission is in the direction perpendicular to the laser axis. Several studies have investigated the dependence of the emission intensity on the distance of the collection axis from the target surface. Kim et al. [86] studied the spatial distribution of the emission intensity by scanning the direction perpendicular to the target surface and found highest intensity at a distance of 3 mm from the target surface for a time delay of 30 μs in air. Liu et al. [87] and Ciucci et al. [88] studied the early phase of the plasma and reported a double-peaked intensity distribution along the normal to the irradiated surface. They collected the emission signal from the region where the maximum intensity was found, avoiding at the same time the region close to the target surface and at the plasma-air interface, where self-absorption was dominant even for nonresonance lines.

Lee et al. [89] studied the spatial features of laser-induced plasmas from different metallic targets by measuring the emission intensity along the axial direction perpendicular to the target using a time-integrated system. It was noted that for copper plasma the spatial extension was limited to approximately 2 mm in the axial direction, with a peak intensity lying at a distance 1 mm from the target surface, whereas lead plasma extended for approximately 5 mm from the target, with peak intensity at approximately 2 mm from it. This indicates that the spatial location sampled for analytical measurements should be identified carefully, which is dependent on the target material. In this experiment, lead plasma expanded more in spite of being heavier than copper. It was found that thermal conductivity of the element played an important role. Copper being more conducting showed less plasma temperature and resulted in less expansion in comparison to the lead plasma.

Normally, LIBS measurements are performed by integrating the emitted intensity over the line of sight in the plasma, which takes care of the inhomogeneity of the plume arising due to different temperatures and electronic densities present in the path. During this measurement the whole of the plasma plume is imaged on the entrance slit of the spectrometer for obtaining better reproducibility. Aragon et al. [79] used a demagnification factor of five in the optical collection optics to form an image of the whole plume on the entrance slit of the spectrometer. Similarly, Chaleard et al. [58] used a demagnification factor of 30 for the collecting optics to image the whole volume of the plasma on the 200-μm entrance slit of the spectrometer to improve the reproducibility of the measurements. It is to be noted that shot-to-shot variations in the atomic densities and temperature distribution inside the plasma affect the reproducibility of the measurement, particularly when the central part of the plasma is probed.

One of the important techniques of collecting emissions from the plume is by placing a fiber optic near the plume, at a distance of a few millimeters, to avoid damage by excessive heating. The collection angle of the fiber end allows gathering of light from a broad volume of the plasma plume. The use of optical fiber bundles has also been proposed for the simultaneous acquisition of spectral emission from different regions inside the plasma plume [90]. Various configurations have been proposed, including the use of two optical fibers [91–93]: one for delivering the laser beam and one for collecting the plume emission, or one single fiber for both purposes [94,95]. A comparison has been carried out between the performance of the fiber-coupled LIBS and lens-coupled LIBS [95]. After optimization of different experimental parameters, almost similar detection limits were achieved for spectral emission of some of the elements. The use of a single optical fiber for delivering the laser pulse and collecting the emission from the plume makes the LIBS system suitable for the study of a sample present at a remote distance, which presents a unique advantage for its industrial applications.

8.6 Effect of ambient gas

The laser-produced hot plasma expands away from the target. The presence of ambient gas around the plasma affects the dynamics of the plasma plume as well as the emission characteristics of the plasma. Piepmeier and Olsten [96] reported the effects of the surrounding atmosphere on the emission spectra, on the crater size, and on the amount of the sample ablated by changing the ambient air pressure. Grant and Paul [97] measured the spatially resolved emission intensity from an excimer (XeCl) laser-produced plasma on a steel target along the axial direction in the presence of different gas atmospheres (air, helium, and argon) at pressures of 0.5, 50, and 760 Torr. The behavior of the emission intensity with change in the three variables (gas, pressure, and axial distance) was too complex to reach a unique interpretation. However, the best signal-to-noise ratio was observed for an argon atmosphere at ~50 Torr pressure, when the LIBS signal was recorded approximately 6 mm from the target surface. Leis et al. [78] found an argon ambient pressure of ~140 Torr for yielding a higher intensity value for the Si I 288.2 nm line in a steel sample. Sdorra and Niemax [98] compared the effects of different ambient gases (argon, neon, helium, nitrogen, and air) in the generation of plasma by focusing a nanosecond Nd:YAG laser on a copper target. In a similar experimental condition (pressure lower than atmospheric), argon was found to produce a comparatively higher plasma temperature, higher electron density, and higher emission intensity, but a lower mass ablation rate for some of the elements. The decay rate of the temperature during the first 40 μs after the plasma formation was found slower for argon than for the other gases due to its low thermal conductivity. The same behavior was also found by Iida [99]. However, a helium atmosphere produced a comparatively lower temperature, electron density, and emission intensity. This indicates that argon is most efficiently heated by IB and produces buffer plasma, which optically shields the target surface and as a result reduces the amount of ablated mass. The shielding effect takes place in the presence of other gases also but for a pulse energy >20 mJ, whereas for argon it was observed for 10 mJ of energy per pulse. It was noted that argon yields the highest analyte emission intensity, except at high pressures, where neon offers the best performance.

Kim et al. [86] also observed an increase in emission intensity along with longer plasma lifetime in an argon atmosphere and attributed this phenomenon to smaller conductivity (0.0387 cal/cm s degree in STP) and specific heat

(0.0763 cal/g degree in STP) of argon gas in comparison to those of air. Such differences in the thermal properties of ambient gas result in higher-temperature plasma leading to stronger and longer emission and slower cooling. Another important effect produced by the ambient argon is protection of the excited atoms from forming stable compounds such as oxides, which might reduce the LIBS emission from the analyte. Wisbrun et al. [46] found that the argon atmosphere was most favorable both in terms of higher analyte emission intensity (1.8 times the intensity obtained in air for the Zn line at 481.1 nm) and better reproducibility (relative standard deviation = 12% over 18 measurements, compared to 18% in air). It is important to note that as the atomic mass of the ambient gas increases, the collisional translational energy transfer is less effective and plasma life becomes longer.

Normally at low ambient pressure (<1 Torr), the ablated vapor (plasma) expands almost freely, and the outer part of the plasma becomes colder in comparison to its core because of higher energy loss. An increase in the pressure to approximately 1 Torr confines the plasma, causes a reduction in energy loss, and produces a more uniform distribution. Hermann et al. [100] studied the time and space evolution of electron density and temperature by changing the laser power density and ambient pressure. In their experimental conditions, the electron density was found to be more sensitive to laser power density changes than electron temperature. Measurements of line emission intensity, plasma density, and temperature decay rate indicated an increase in plasma lifetime with an increase in laser power density. These observations can be explained by the assumption that the plasma density increases with laser power density due to increased ablation, and self-absorption of the spectral lines becomes more important leading to reduced radiation loss and enhanced plasma lifetime. In the case of higher pressure of ambient gas, elastic and inelastic collisions occur between target vapor and ambient gas species that become more frequent in comparison to those in free expansion in a vacuum. Confinement of plasma by ambient gas makes the decrease in plasma density slower and at the same time the kinetic energy of the plasma particles is partially transformed into excitation energy as a result of inelastic collisions.

Thiem et al. [101] noted that analysis of experimental data at atmospheric pressure is difficult due to the presence of a broadband background emission spectrum generated by atmospheric elements. They preferred the use of a vacuum chamber to obtain a better signal-to-noise ratio. Special ambient gases are necessary in particular cases, such as for observing UV spectrum ranging down to 180 nm where the experimental housing must be filled with nitrogen or some other inert gas to avoid the absorption of the LIBS signal by oxygen molecules. In the previous section we saw the effects of external conditions on plasma emission. The next section discusses the effects of plasma dynamics on its emission.

9 Dynamics of plasma and its emission

Investigation of dynamics and emission from laser-produced plasma under different experimental conditions [45, 104–109] is an important subject that has its technological application in various fields of research such as material science, chemical physics, and plasma physics as well as in inertial and magnetic confinement fusion [110–121]. As discussed earlier, laser-produced plasma expands perpendicular to the surface of the target and is independent of the direction of incident laser beam. Various types of phenomena take place in the plasma depending on its density and temperature such as emission of radiation ranging from visible to X-ray wavelength, generation of high-energy electrons and ions, as well as different waves and instabilities [103]. Any change in the free expansion of plasma affects all of its associated properties. The optical emission from laser-produced plasma also depends on the dynamics of the plasma. Fundamental processes such as the distribution of excited-state species in the plasma depend mainly on the fluid dynamics of ablated particles as well as on the balancing of other elementary processes [104,122]. These aspects have been described extensively both theoretically and experimentally, and related models have been applied successfully to the interpretation of data obtained in a wide range of conditions [107, 114, 116, 122–124].

Various theoretical models have been developed to improve basic knowledge as well as to determine the best experimental conditions of laser-induced plasma processes. The main difficulty in modeling laser-induced plasma is that both the dynamics of the plume and the plasma chemical reactions occurring in it should be taken into account. To reproduce the temporal and spatial evolution of the experimental plasma a lot of theoretical investigations have been reported on the simulation of plume expansion [125]. Different approaches have been proposed: an analytical approach [126,127], a fluid dynamic model [128–130], and Monte Carlo methods [131,132]. Experimentally, plasma dynamics has been studied using different types of lasers, including the combination of lasers with different time durations and wavelengths for different applications [133–136]. Optical emission from the plasma has been used as a diagnostic to better understand the dynamics of the plasma [137–140]. Out of all the complicated theories, Rai et al. [114–116] analyzed the dynamics of plasma excited by the dual-pulse laser as well as emission from it in a very simple way to explain the experimental observations, particularly the enhancement in plasma emission obtained in similar conditions. This analysis has also been simplified to better understand the dynamics of single-laser-produced plasma. For this purpose, a single laser pulse has been assumed as the combination of two pulses as the pedestal (initial part of the laser) and the main pulse. Normally, a sufficient amount of plasma is formed by the pedestal of the laser pulse, whereas the rest of the laser pulse is absorbed by the plasma.

9.1 Model of plasma emission during dual-pulse LIBS

The plasma dynamics and emission from plasma under dual laser pulse excitation have been analyzed and reported by Rai et al. [114,116]. They considered that the first laser creates the plasma expanding normal to the target surface. This plasma has cylindrical shape. A second laser pulse delayed by a few microseconds is absorbed in the expanding plasma, followed by further ablation of target material by the residual second laser pulse. Absorption of the second laser in the plasma has been assumed to be dominated by IB absorption, which occurs due to electron-ion collisions in the plasma. Finally, they obtained and reported an expression for the enhancement in emission (E) under double-pulse excitation in comparison to single-pulse excitation that is given as:

$$E = 1 + \left(\frac{\dot{m}_2}{\dot{m}_1}\right)^2 \frac{\left[1 - \exp\left\{-\dfrac{32 C_{s1} \Delta t \nu_{ei}}{15 \quad c}\right\}\right]}{\left(\dfrac{\Delta t}{T_1} + \dfrac{C_{s2} T_2}{C_{s1} T_1}\right)} \tag{19}$$

where $\dot{m}_1$ and $\dot{m}_2$ are the mass ablation rate after the first and second laser pulse. The parameters of both lasers are considered to be the same (time duration τ_L). C_{s1} and C_{s2} are the ion acoustic speed of plasma after the first and second laser pulse. T_1 and T_2 are the times of recording the emission after the first and second laser pulse. Δt is the time delay between two laser pulses. ν_{ei} ($=3.62 \times 10^{-6} n_i T_e^{-3/2} \ln \Lambda \ \mathrm{s}^{-1}$) is the electron-ion collision frequency [116] and c is the speed of light. For the calculation of enhancement in emission (E), one can take the value of T_1 as the time between the first laser pulse and the time of peak emission observed during a double-pulse LIBS experiment. Eq. (19) has been used to study the effect of different parameters on the enhancement in plasma emission under dual-pulse excitation. It explained several experimental observations of dual-pulse LIBS as reported earlier [115]. According to Eq. (19), zero delay between lasers ($\Delta t = 0$) provides an enhancement of unity ($E = 1$), just like a single-pulse LIBS emission from added laser intensity.

9.2 Model of plasma emission during single-pulse LIBS

Eq. (19), obtained for dual-pulse excitation of plasma, was simplified for the single-pulse case in light of the foregoing discussion that the plasma is formed and starts expanding when the initial portion of the laser pulse interacts with the matter [116]. So, a single laser pulse was assumed operating as a dual pulse with certain considerations. One can consider that after plasma formation by the initial portion of the laser pulse, it expands for the period τ_L (duration of laser pulse) during which the laser light is absorbed in the expanding plasma. This indicates that a Δt delay between two lasers in Eq. (19) can be replaced by τ_L for the single-pulse case. In light of these considerations, Eq. (19) was simplified in two ways as described next to better understand the dependence of important plasma parameters on the emission characteristics of the plasma as well as on its temporal behavior.

9.2.1 Case 1

In this case, assumption was made that $\Delta t = \tau_L$ (laser pulse duration) and $T_1 = \tau_L/2$ (time of peak laser intensity). T_2 is the time at which the measurement of the plasma emission takes place. Another consideration was that $C_{s2} = 2C_{s1}$ (plasma expansion speed was more after full laser pulse). In this situation, Eq. (19) can be written as:

$$E = 1 + \frac{\left(\frac{\dot{m}_2}{\dot{m}_1}\right)^2 \left[1 - \exp\left\{-\frac{32 C_{s1}\tau_L \nu_{ei}}{15\quad c}\right\}\right]}{\left(2 + 4\frac{T_2}{\tau_L}\right)} \tag{20}$$

This equation provides information on the emission from the plasma at different time durations, to be compared with the variation of emission with the gate delay (T_2). In other words, the temporal evolution of the plasma emission during single-pulse LIBS can be obtained using this expression.

9.2.2 Case 2

In this case, the consideration was that $\Delta t = \tau_L$ and $T_1 = T_2 = T$ is the gate delay or the time of measurement of the emission after the laser peak. The ratio of the laser pulse time duration τ_L and gate delay T (time of measurement) will be negligibly small for a nanosecond laser pulse ($\tau_L/T \approx 0$). After these assumptions, Eq. (19) can be written as:

$$E = 1 + \left(\frac{\dot{m}_2}{\dot{m}_1}\right)^2 \left(\frac{C_{s1}}{C_{s2}}\right) \left[1 - \exp\left\{-\frac{32 C_{s1}\tau_L \nu_{ei}}{15\quad c}\right\}\right] \tag{21}$$

This shows that plasma emission under single-pulse laser excitation is mainly dependent on the square of the ratio of the mass ablation rate after full pulse and the initial mass ablation rate, the ratio of plasma sound velocity, the plasma collision frequency, and the pulse duration of the laser.

10 Effect of plasma parameters on LIBS emission

To achieve a better understanding of the physical processes arising during LIBS, variation in the intensity of emission for different plasma parameters during the laser-plasma interaction was studied. The important parameters are mass ablation rate, plasma density, plasma temperature, electron-ion collision frequency, as well as laser pulse duration. In fact, the effect of laser intensity on the plasma emission was taken care of by the increase in plasma density and the temperature since both are directly proportional to the laser intensity.

10.1 Effect of delay between lasers

Enhancement in plasma emission for different values of interpulse delay between lasers (Δt) in the range from 0 to 30 μs has been studied using Eq. (19). It is assumed here that mass ablation rate and plasma expansion speed in single- and double-pulse LIBS are nearly the same, so that $\dot{m}_1 = \dot{m}_2$ and $C_{s1} = C_{s2}$. Also, T_1 and T_2 are taken as 4 and 2 μs, respectively, which implies that in the case of single-pulse LIBS, emission is recorded after 4 μs (for normalization), whereas in the double-pulse LIBS, emission is measured 2 μs after firing the second laser. The variation in emission enhancement was calculated for three values of electron-ion collision frequency as $\nu_{ei} = 3.62 \times 10^9$, 1×10^{10}, and $3.62 \times 10^{10}\,s^{-1}$. Fig. 6 shows the variation in the enhancement of emission E with change in delay between the two lasers (Δt) for different electron-ion collision frequencies. The enhancement is unity for $\Delta t = 0$ but it exhibits a sharp increase followed by a slow decrease with increasing Δt. The peak of the emission shifts to lower delay as the electron-ion collision frequency increases from 3.62×10^9 to 3.62×10^{10}. Enhancement is nearly ~2 times in the case of $\nu_{ei} \sim 3.62 \times 10^{10}\,s^{-1}$ for $\Delta t \sim 2$ μs. For lower values of electron-ion collision frequencies, enhancement is less and emission peak is broadened. Higher electron-ion collision frequency corresponds to high plasma density and low plasma temperature, which is possible when a moderate intensity laser is used for creating low-temperature plasma, which interacts with the second laser before its adequate expansion. The lower electron-ion collision frequency corresponds to the lower density and high plasma temperature. In this case, plasma expands fast and the absorption of the second laser pulse is possible in the expanded plasma even for higher interpulse separation. In this case, laser absorption will extend to a longer path length in plasma resulting in a spread in the emission profile peaking at a comparatively higher interpulse delay with decreased emission intensity. Such behavior is possible when a high-intensity laser is used as the first laser for producing preplasma. In this case, plasma will expand faster as a result of high temperature providing a broad density profile. Finally, less enhancement in emission will be obtained with a broadened profile peaking at comparatively higher interpulse delay. The peak values of interpulse delay Δt were found to be 1, 3, and 5 μs for electron-ion collision frequencies of ~3.62×10^{10}, 1×10^{10}, and $3.62 \times 10^9\,s^{-1}$, with the corresponding enhancements of 1.96, 1.52, and 1.27, respectively (Fig. 6). These theoretically obtained results were compared with experimental observations reported earlier. It was found that the emission profile of magnesium (Mg) ion (279.55 nm) peaks at an interpulse delay of 2–3 μs with an enhancement of ~6 when the first laser energy was ~100 mJ (moderate intensity). However, as the energy of the first laser increased from 100 to 180 mJ, the emission profile became much broader and the peak occurred at an interpulse delay of larger than 5 μs. The theoretically obtained emission profile peaks at an ~3-μs interpulse delay for an electron-ion collision frequency of ~$1 \times 10^{10}\,s^{-1}$ with an enhancement $E \sim 1.52$, whereas the peak at 5 μs occurs for an ion collision frequency of ~$3.62 \times 10^9\,s^{-1}$ with an enhancement of $E \sim 1.27$. The large difference in the amplitudes of emission enhancement in theoretical and experimental results is due to consideration of $\dot{m}_1 = \dot{m}_2$. In fact, $\dot{m}_2$ will be many times more after the second laser in

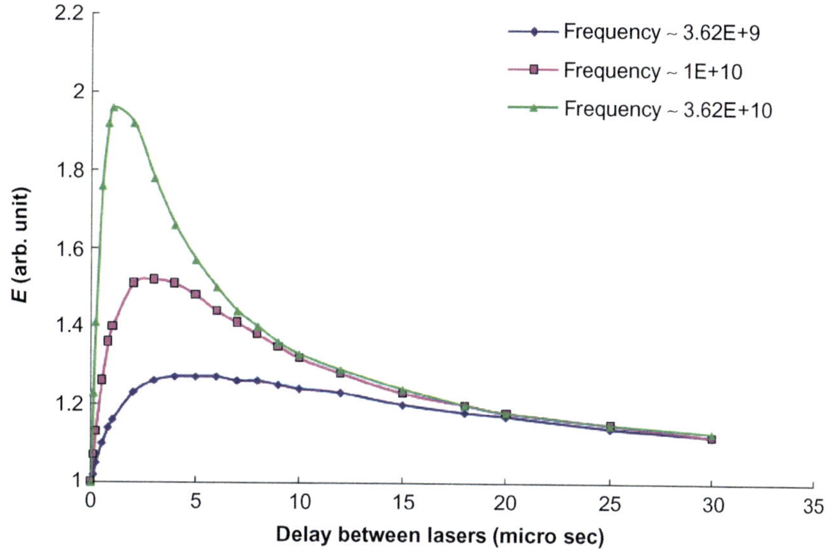

FIG. 6 Variation in enhancement in emission with change in delay between lasers for different electron-ion collision frequencies ν_{ei} ($\dot{m}_1 = \dot{m}_2$ and $C_{s1} = C_{s2}$) [114].

dual-pulse LIBS as reported earlier [114]. This comparison clearly shows that this theory successfully explained the experimental findings of enhancement in emission during dual-pulse LIBS.

10.2 Effect of plasma density (laser energy)

It is well known that an increase in laser intensity increases the plasma density as well as plasma temperature during laser-matter interaction. Plasma density increases due to increased material ablation, whereas plasma temperature increases due to absorption of laser energy into preformed plasma due to the pedestal (initial part) of the laser.

The effect of plasma density and temperature on the plasma emission was studied using Eq. (21) for different values of electron-ion collision frequencies that also depend on the plasma density and temperature. Ultimately, this collision frequency plays an important role in the absorption of laser energy into plasma. The calculation for emission was made after considering $\frac{\dot{m}_2}{\dot{m}_1} = 4$ and $\frac{C_{s2}}{C_{s1}} = 2$. Fig. 7 shows the variation in emission intensity with an increase in plasma density (10^{13}–10^{20} cm^{-3}) for different values of plasma temperature (1, 5, and 10 eV). It shows that initially, plasma emission increases slowly followed by a fast nearly linear increase in emission between 10^{15} and 10^{19} cm^{-3}. However, a further increase in plasma density beyond 10^{19} cm^{-3} shows saturation in emission. It has also been noted that the emission intensity is high for lower values of plasma temperature, which is expected as the value of electron-ion collision frequency remains high for low plasma temperature. This increases the absorption of the laser intensity by the plasma, which results in a stronger emission. Another observation is that emission intensity starts increasing and saturating at comparatively lower values of plasma density for low-temperature plasma. Again the role of plasma frequency is found to be more important. Finally, one can say that an increase in laser intensity increases the plasma density as well as the temperature, whereas the balance between the two effects leads toward saturation in emission.

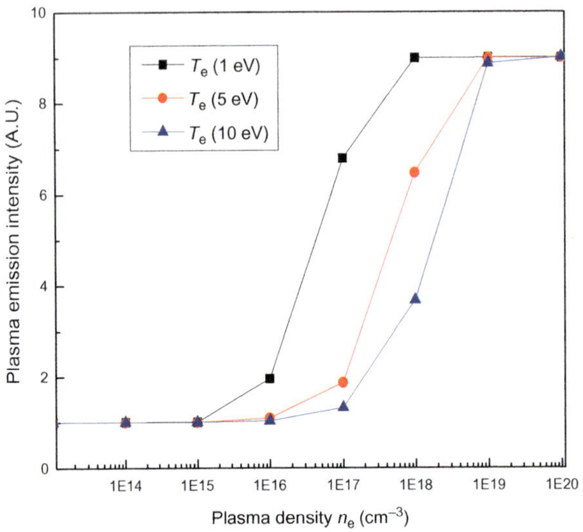

FIG. 7 Variation in emission intensity with an increase in plasma density for plasma temperatures of 1, 5, and 10 eV [116].

Saturation in the plasma emission is also possible due to shielding by the laser-produced plasma near a critical density (10^{21} cm^{-3}) surface formed due to high laser intensity, where electron-ion collision frequency is $>10^{13}$ s^{-1}. In this case excess laser energy is being reflected from the critical density surface of the plasma, where the plasma frequency becomes equal to the laser frequency. In this case, the laser cannot propagate toward the higher plasma density side. At higher plasma density, self-absorption of emission is also possible leading to saturation in the emission when the experiment is performed at higher laser intensity. The presence of self-absorption can be observed from the measurement of FWHM of the Lorentzian profile of line emission. Both processes are possible in the high-density plasma, which can contribute toward saturation in the emission. The breakdown in gases surrounding the target also shields the laser energy reaching the target surface, which results in a decrease and/or saturation in material ablation from the target surface and consequently in the plasma emission. Similar saturation effects have been reported earlier [113] in different experiments.

10.3 Effect of pulse duration of a laser

The pulse duration of a laser also plays an important role in the emission from laser-induced plasma. The study of variations in plasma emission with electron density and temperature for lasers of different pulse durations can provide a better understanding of the factors affecting plasma emission and its saturation. For this purpose the calculation is performed using Eq. (21) with a consideration that $\frac{\dot{m}_2}{\dot{m}_1} = 4$, $\frac{C_{s2}}{C_{s1}} = 2$, and $T_e = 1$ eV and the results are shown in Fig. 8 [116]. It shows the variation in plasma emission with change in plasma density produced by lasers of different pulse durations τ_L (1, 10, and 20 ns). In this case also the plasma emission starts increasing at $\sim 10^{16}$ cm^{-3} for a laser pulse of 1 ns time duration. This increase becomes fast with an increase in plasma density, which saturates at $\sim 10^{19}$ cm^{-3}. Similar variations are also noted for laser pulses of 10 and 20 ns, but with minor change in the threshold for initiation and saturation of emission, which occur at lower density for lasers of larger pulse duration. A longer-pulse laser produces comparatively more plasma emission as a result of more

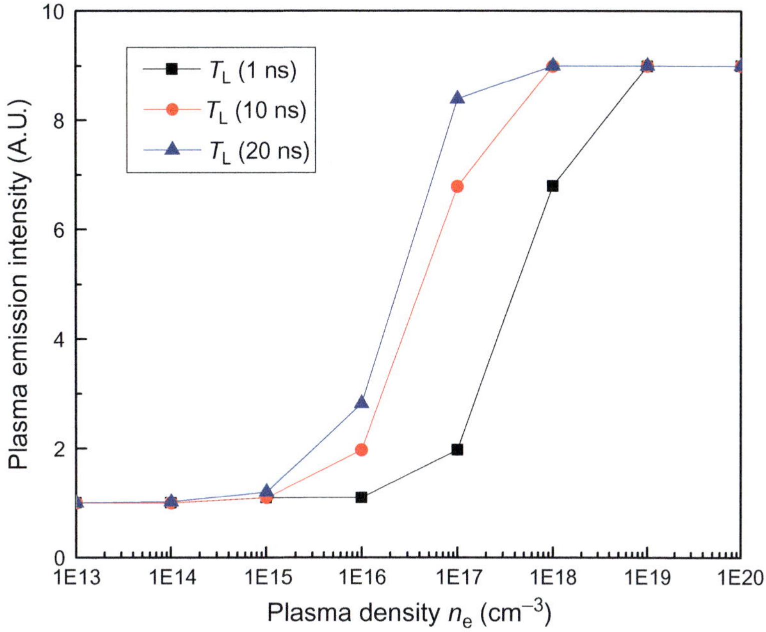

FIG. 8 Variation in emission intensity with an increase in plasma density for plasma produced by 1-, 10-, and 20-ns pulsed lasers [116].

material ablation. Here again the electron-ion collision frequency and the laser-plasma interaction time play an important role. In this case, the plasma collision frequency increases with an increase in plasma density, which helps in increasing the absorption of laser light in plasma through IB and consequently the plasma emission. The longer-pulse laser provides a longer time for plasma to expand, to absorb the laser, and to efficiently ablate the materials from the target.

The effect of plasma temperature on its emission has also been obtained using Eq. (21), where the plasma density was taken constant at $\sim 10^{17}$ cm^{-3} keeping all the other parameters the same. In this case emission from the plasma is calculated by varying the plasma temperature for lasers of different pulse durations [116]. It is noted that plasma emission is very high for low-temperature plasma, which decreases first very fast and then slowly with an increase in plasma temperature. The reason for the decrease in plasma emission with an increase in plasma temperature is due to the decrease in electron-ion collision frequency as $\nu_{ei} \propto (T_e)^{-3/2}$. Here again plasma emission is high for longer time duration lasers as seen in the earlier case. In this case, saturation is seen at lower plasma temperatures near $\sim$1 eV. These results indicate that saturation in the plasma emission is possible only when the plasma density is high and plasma temperature is low. In other words, one can say that instead of plasma density and plasma temperature, electron-ion collision frequency plays an important role in deciding the saturation in plasma emission. The condition for saturation is found as $\nu_{ei} > 10^{13}$ s^{-1}, which was discussed earlier [116].

10.4 Effect of plasma confinement

It is well known that laser-produced plasma expands away very fast from the target surface and emits various types of emission ranging from X-ray to visible regions. Both types of plasma emission increase after plasma confinement [111–113]. Two types of plasma confinement have been reported, namely, magnetic and spatial confinement for enhancing the emission from laser-produced plasma [113,141]. In both cases, plasma expansion is decelerated using different techniques. The effect of plasma expansion velocity on plasma emission has been studied using Eq. (21) to better understand the confinement process. This calculation was performed for $\nu_{ei} \sim 3.62 \times 10^{12}\ s^{-1}$ corresponding to a plasma density of $\sim 10^{17}\ cm^{-3}$ and a plasma temperature of 1 eV. The value of C_{s2}/C_{s1} was changed from 0.1 to 10. Fig. 9 shows that the emission intensity is very high when the value of C_{s2}/C_{s1} is low toward 0.1 (a case of plasma deceleration), whereas it decreases very fast as the value of C_{s2} increases. This observation shows that deceleration in plasma expansion enhances the plasma emission. It seems that the plasma volume increases with an increase in C_{s2} and the corresponding energy density of absorbed laser decreases, which is reflected as decreased plasma emission for higher values of C_{s2}. This indicates that enhancement in emission is mainly dependent on the order of plasma confinement. This supports the results of enhancement in emission by magnetic and/or spatial confinement [113,141]. Increased enhancements in spatial confinement are due to comparatively better confinement/deceleration of the plasma. In the case of magnetic confinement, the plasma diffuses across the magnetic field as a result of plasma instability and fluctuations in the plasma density, which degrade the plasma confinement, whereas the spatial confinement is free from these problems. However, the combination of both confinement techniques is beneficial as has been reported by Guo et al. [141]. Enhancement in plasma emission is independent of the technique used for plasma confinement. Only one factor is important and that is the efficiency of the plasma confinement. This result further indicates that the combination of spatial confinement with dual-pulse LIBS [142] or other techniques of enhancing the LIBS signal [143–145] may provide even more enhancement in emission than the combination of magnetic and spatial confinement [141]. Here again the emission is more for lasers of longer pulse duration.

10.5 Temporal evolution of emission from confined plasma

The temporal evolution of plasma emission has been studied using Eq. (20). For this purpose, consideration was made that $T_1 = \tau_L/2$ (peak time of laser) and T_2 is varied, which is equivalent to the time delay at which the emission intensity is recorded during LIBS. To see the effect of plasma confinement and the laser pulse duration on the temporal evolution, calculations have been performed first for different values of C_{s2}/C_{s1} as 2, 0.5, and 0.1, keeping laser pulse duration constant at 10 ns, and then for 1-, 10-, and 20-ns laser pulses, keeping C_{s2}/C_{s1} constant at ~ 2 [116]. Fig. 10 shows the variation of plasma emission with time delay for decreasing values of C_{s2}/C_{s1}, which also shows the effect of plasma confinement. For $C_{s2}/C_{s1} \sim 2$ the plasma emission first decreases very fast and then slowly with an increase in time delay. After including the confinement effect by decreasing the values of C_{s2}/C_{s1} to 0.5 and 0.1, plasma emission increases for all the values of delay time. Even the rate of decay in emission intensity changes, which indicates that plasma emission lasts for a longer time in the case of plasma confinement. It further shows that maximum enhancement in the emission after confinement occurs around the time delay of 20 ns. This happens because of the slowing down of decay in plasma emission as a result of plasma confinement. The variation in temporal evolution of plasma emission obtained from the laser pulses of 1, 10, and 20 ns has also been reported [116]. The plasma emission is found to

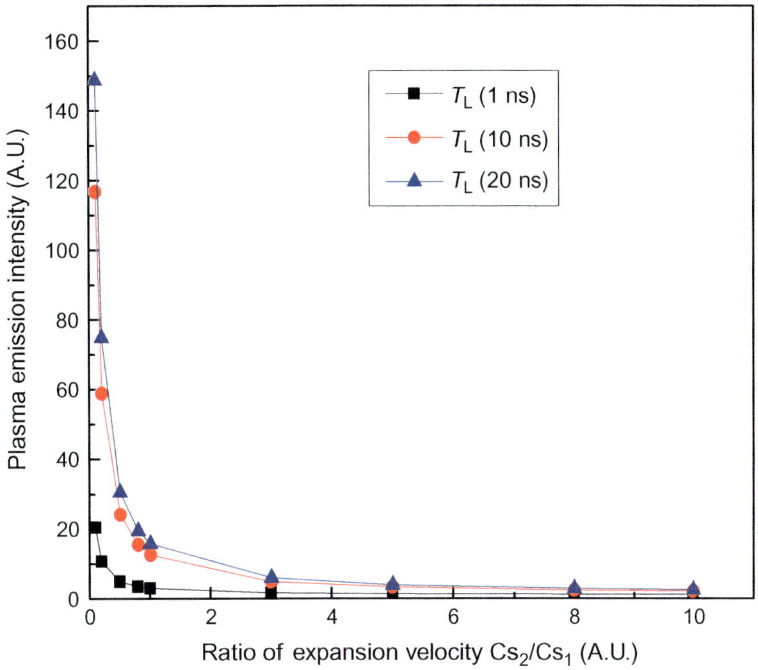

FIG. 9 Variation in emission intensity with an increase in ratio of expansion velocity (C_{s2}/C_{s1}) for plasma produced by 1-, 10-, and 20-ns pulsed lasers [116].

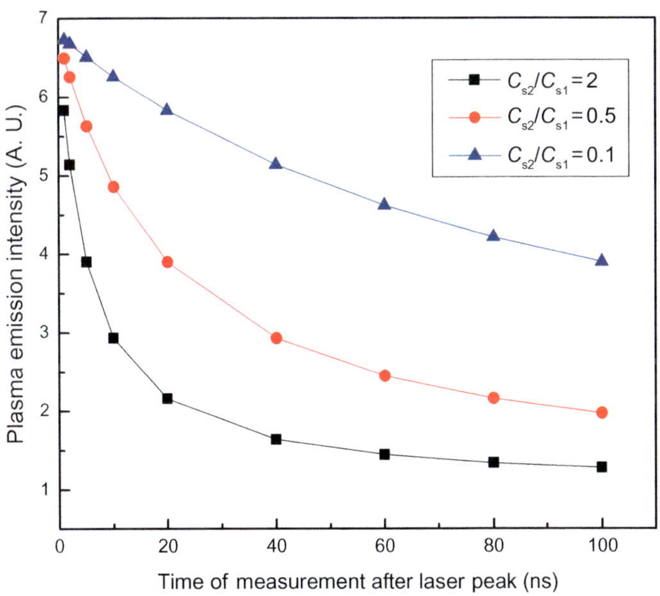

FIG. 10 Variation in emission intensity with time of measurement after a laser peak for plasma having a ratio of expansion velocity (C_{s2}/C_{s1}) as 0.1, 0.5, and 2 [116].

be less for 1-ns laser-produced plasma, which initially decays fast and then slowly. The rate of plasma decay decreases with an increase in the pulse duration of the laser. The emission is higher for the plasma produced by a longer pulsed laser, which is due to efficient heating of the plasma for a longer time. It also shows that the plasma produced by a longer pulsed laser lasts for a longer time. These observations are in agreement with the experimental observations.

A similar observation has been reported experimentally by Rai et al. [113]. They studied temporal evolution of chromium line emission from aluminum alloy in the absence and presence of a magnetic field (magnetic confinement) and reported that plasma emission was maximum at ~2-μs time delay after which it started decreasing. The emission intensity lasted for nearly 20 μs. An overall increase in intensity was noted for all the delay in the presence of a magnetic field (magnetic confinement), whereas the shape of the curve remained the same.

These results are similar, as shown in Fig. 10. The decay in emission with time indicates that initially the plasma is hot and the emission from it is dominated by the background emission, which contains mainly Bremsstrahlung emission. As the gate delay increased, the plasma expanded away from the target and cooled resulting in a decrease in emission. The presence of the magnetic field increases the effective density of the plasma in the emission volume as a result of confinement of laser-produced plasma. It ultimately increases the probability of radiative recombination, electron-ion collision frequency, as well as the absorbed energy per unit volume. All these factors play important roles in increasing plasma emission.

During magnetic confinement, an enhancement in plasma emission and deceleration in plasma expansion are found correlated with plasma β $(=8\pi n_e kT_e/B^2)$ [48,113]. It shows that plasma β depends mainly on the density and temperature of the plasma, which decreases as the plasma expands away from the target. It also depends on the intensity of the magnetic field, which remains constant during this experiment. It is reported that initially for a higher value of β, plasma deceleration and enhancement in emission are negligible [113]. However, as the β decreases below 10, the ratio of plasma expansion velocities v_2/v_1 decreases (plasma decelerates in a magnetic field) and the ratio of emission intensity I_2/I_1 (value of enhancement) increases in the presence of a magnetic field. This indicates that Bremsstrahlung emission, which occurs at higher plasma temperatures and density ($\beta > 10$) and comparatively at early times (smaller gate delay) after the plasma formation, cannot contribute to the enhancement in the emission. Mainly, the line emissions are being affected due to increase in recombination radiations or heating of the plasma as a result of its confinement after a certain time of expansion. These reported results [113] are also in qualitative agreement with the theoretical findings discussed in this chapter. Finally, further investigations are needed for better plasma confinement techniques, including improvement in the reported magnetic and spatial confinements to increase the plasma emission even more than the already reported values [141]. Other combinations of emission-enhancing processes [142–145] must be investigated. Efficient theoretical models for plasma dynamics and emission have to be worked out with less assumption and considering all the affecting parameters to better explain the experimental results quantitatively.

11 Conclusion

LIBS is now a very active field in analytical science. Considerable progress in the area of basic and applied research of LIBS has been made during the last two decades. A number of portable and handheld LIBS systems are now available from various companies for field applications where sensitivity is not much of a problem. Various new techniques have been investigated for enhancing the sensitivity of LIBS. But still, LIBS is being used in a limited number of applications for which the analytical requirements are low and the advantages of LIBS as a rapid, noncontact, and in situ method are essential. However, to fully realize the potential of LIBS, a better understanding of the physics and dynamics of LIBS plasma is needed. Many experimental and theoretical investigations have been performed in this direction to improve the sensitivity of LIBS and serious efforts are still required in this direction. There are some areas that require much attention such as the development of theoretical models that provide improved understanding of LIBS events for attaining better sensitivity as well as incorporation of the latest advances in lasers, spectrometers, and detection technologies to facilitate acquisition and analysis of data in the field.

From a theoretical point of view, there is a need to simulate and understand the complete LIBS phenomenon, including the initial laser-material interaction, ignition and growth of the microplasma, fluid dynamics, plasma physics, chemistry of the plasma, as well as detailed chemical kinetics involving electrons, atoms, ions, and emission from plasma that ultimately lead to the analytical spectra that are detected. LIBS plasmas characterized by large electron densities and electron temperatures normally satisfy the LTE condition, but it is quite likely that nonequilibrium plasma is also present during the LIBS experiment. Such experimental conditions cannot be handled theoretically in the absence of a suitable model. This indicates the requirement of a multidisciplinary team of scientists to be engaged in experimental and theoretical research in the related fields. This level of understanding of the LIBS process would prove valuable in advancing LIBS sensing that will move the technology forward. Further study will be beneficial in advanced chemometrics to improve LIBS for the identification of elements in materials and quantification of plasma parameters such as electron density and temperature to improve pulse-to-pulse reproducibility of the LIBS signal. An improvement in this direction would lead to a better interpretation of LIBS information. Advances in basic LIBS methodology can also enhance the applications of this technique.

For the use of LIBS in a broader field of applications, it is necessary that the sensitivity of this system be raised considerably. Out of some of the individual techniques for enhancing the sensitivity of LIBS as discussed earlier, either a combination of these techniques or other new techniques have to be investigated for this purpose. Accuracy of the measurements can be improved by reducing pulse-to-pulse fluctuations. A number of statistical methods can be obtained to reduce the standard deviation in the data points. Some of the applications of LIBS require advances in specific areas. Use of LIBS at large distances requires a detailed understanding of atmospheric propagation as well as innovative approaches for implementing LIBS hardware.

Finally, it seems that LIBS has lot of scope for its improvement and optimization. It requires extensive studies and investigations in the relevant multidisciplinary fields to further enrich this technique for its many more novel applications.

References

[1] L.J. Radziemski, D.A. Cremers, Laser-Induced Plasma and Applications, Marcel Dekker, New York, 1989.

[2] G. Bekefi, Principle of Laser Plasmas, John Wiley & Sons, New York, 1976.

[3] R.E. Cairn, J.J. Sanderson, Laser Plasma Interaction, Institute of Physics Publishing, Edinberg, 1980.

[4] C.E. Max, in: R. Balian, J.C. Adams (Eds.), Laser Plasma Interaction, North Holland, Amsterdam, 1982.

[5] W.L. Kruer, The Physics of Laser Plasma Interaction, Addison-Wesley, New York, 1988.

[6] [(a)]S.F. Jacob, M.O. Scully, I.I.I. Sergent, C. D. Cantrel III, Laser-Induced Fusion and X-Ray Laser

Studies, Addison-Wesley, New York, 1976; [(b)] F. Brech, L. Cross, Optical microemission stimulated by a ruby maser, Appl. Spectrosc. 16 (1962) 59.
[7] L.J. Radziemski, R.W. Solarz, J.A. Paisner, Laser Spectroscopy and Its Applications, Marcel Dekker Inc, New York & Basel, 1987.
[8] R.S. Adrain, J. Watson, Laser microspectral analysis: a review of principles and applications, J. Phys. D. Appl. Phys. 17 (1984) 1915.
[9] V. Majidi, M.R. Joseph, Spectroscopic applications of laser-induced plasmas, Crit. Rev. Anal. Chem. 23 (1992) 143.
[10] T.L. Thiem, Y. Lee, J. Sneddon, Laser in atomic spectroscopy: selected applications, Microchem. J. 45 (1992) 1.
[11] F.Y. Yueh, J.P. Singh, H. Zhang, Laser-induced breakdown spectroscopy: elemental analysis, in: Encyclopedia of Analytical Chemistry, vol. 3, Wiley, New York, 2000, pp. 2065–2087.
[12] K. Song, Y. Lee, J. Sneddon, Applications of laser-induced breakdown spectrometry, Appl. Spectrosc. Rev. 32 (1997) 183.
[13] D.A. Rusak, B.C. Castle, B.W. Smith, J.D. Winefordner, Fundamentals and applications of laser-induced breakdown spectroscopy, Crit. Rev. Anal. Chem. 27 (1997) 257.
[14] O. Samek, D.C.S. Beddows, J. Kaiser, S.V. Kukhlevsky, M. Liska, H.H. Telle, J. Young, Applications of laser-induced breakdown spectroscopy to in situ analysis of liquid samples, Opt. Eng. 39 (2000) 2248.
[15] A.K. Rai, V.N. Rai, F.Y. Yueh, J.P. Singh, Trends in Applied Spectroscopy, vol. 4, Research Trends, Trivendrum, India, 2002, pp. 165–214.
[16] E. Togoni, V. Palleschi, M. Corsi, G. Cristoforetti, Quantitative micro-analysis by laser-induced breakdown spectroscopy: a review of the experimental approaches, Spectrochim. Acta B 57 (2002) 1115.
[17] C. Gray-Morgan, Laser-induced breakdown of gases, Rep. Prog. Phys. 38 (1975) 621.
[18] J.F. Ready, Effect of High Power Laser Radiation, Academic Press, New York, 1971.
[19] R.G. Root, A.N. Pirri, Paper # AIAA 79-1489-1979, 12th AIAA Fluid and Plasma Dynamics Conference, Williamsburg, Virginia, July 23–25, 1979.
[20] N.H. Kemp, R.G. Root, Analytical study of laser-supported combustion waves in hydrogen, J. Energy 3 (1979) 40.
[21] D.R. Keefer, H.L. Crowder, R.E. Elkins, Paper # AIAA 82-0404-1981, (1981).
[22] S.A. Ramsden, P. Savic, A radiative detonation model for the development of a laser-induced spark in air, Nature 203 (1964) 1217.
[23] Y.P. Raiser, Heating of a gas by a powerful light pulse, Soviet Physics-JEPT 21 (1965) 1009.
[24] V.I. Bergel'son, T.V. Loseva, I.V. Nemechinov, T.I. Orlova, Propagation of plane supersonic radiation waves, Sov. J. Plasma Phys. 1 (1975) 498.
[25] G. Weyl, A. Pirri, R. Root, Laser ignition of plasma off aluminum surfaces, AIAA J. 19 (1981) 460.
[26] W.E. Maher, R.B. Hall, R.R. Johnson, Experimental study of ignition and propagation of laser-supported detonation waves, J. Appl. Phys. 45 (1974) 2138.
[27] J.B. Steverding, Ignition of laser detonation waves, J. Appl. Phys. 45 (1974) 3507.
[28] A.N. Pirri, R.G. Root, P.K.S. Wu, Plasma energy transfer to metal surfaces irradiated by pulsed lasers, AIAA J. 16 (1978) 1296.
[29] H.R. Griem, Plasma Spectroscopy, McGraw Hill, New York, 1964.
[30] Y.B. Zeldovich, Y.P. Raizer, Physics of Shock Waves and High Temperature Hydrodynamic Phenomenon, vol. I, Academic Press, New York, 1966.
[31] C. De Michelis, M. Mattiali, Spectroscopy and impurity behaviour in fusion plasmas, Rep. Prog. Phys. 47 (1984) 1233.
[32] M.H. Key, R.J. Hutcheon, D.R. Bates, B. Bederson (Eds.), Advances in Atomic and Molecular Physics, vol. 16, Academic Press, New York, 1980.
[33] R.H. Huddlestone, S.L. Leonard, Plasma Diagnostics Techniques, Academic Press, New York, 1965.
[34] A.P. Thorne, Spectrophysics, Chapman and Hall, London, 1988.
[35] D. Salzmann, Non-LTE population probabilities of the excited ionic levels in a steady state plasma, J. Quant. Spectrosc. Radiat. Transf. 27 (1982) 359.
[36] M. Capittelli, F. Capittelli, A. Eletskii, Non-equilibrium and equilibrium problems in laser-induced plasmas, Spectrochim. Acta B 55 (2000) 559.
[37] A. Ciucci, M. Corsi, V. Palleschi, S. Rastelli, A. Salvetti, E. Togoni, New procedure for quantitative elemental analysis by laser-induced plasma, Appl. Spectrosc. 53 (1999) 960.
[38] H.R. Griem, Spectral Line Broadening by Plasmas, Academic Press, New York, 1974.
[39] I.I. Sobelman, L.A. Vainshtein, E.A. Yukov, Excitation of Atoms and Broadening of Spectral Lines, Springer Verlag, Berlin, 1981.
[40] J.B. Simeonsson, A.W. Miziolek, Time-resolved emission studies of ArF-laser-produced microplasmas, Appl. Opt. 32 (1993) 939.
[41] G. Abdellatif, H. Imam, A study of the laser plasma parameters at different laser wavelengths, Spectrochim. Acta B 57 (2002) 1155.
[42] D.A. Cremers, L.J. Radziemski, Detection of chlorine and fluorine in air by laser-induced breakdown spectrometry, Anal. Chem. 55 (1983) 1252.
[43] M. Sabsabi, P. Cielo, Quantitative analysis of aluminum alloys by laser-induced breakdown spectroscopy and plasma characterization, Appl. Spectrosc. 49 (1995) 499.
[44] A. Ciucci, V. Palleschi, S. Ratelli, R. Barbini, F. Colao, R. Fantoni, A. Palucci, S. Ribezzo, H.J.L. Van der

Steen, Trace pollutants analysis in soil by a time-resolved laser-induced breakdown spectroscopy technique, Appl. Phys. B Lasers Opt. 63 (1996) 185.

[45] D.A. Cremers, The analysis of metals at a distance using laser-induced breakdown spectroscopy, Appl. Spectrosc. 41 (1987) 572.

[46] R. Wisbrun, I. Schechter, R. Niesner, H. Schroder, K. L. Kompa, Detector for trace elemental analysis of solid environmental samples by laser plasma spectroscopy, Anal. Chem. 66 (1994) 2964.

[47] J.P. Singh, F.Y. Yueh, H. Zhang, R.L. Cook, Study of laser induced breakdown spectroscopy as a process monitor and control tool for hazardous waste remediation, Process. Control. Qual. 10 (1997) 247.

[48] [(a)]V.N. Rai, H. Zhang, F.Y. Yueh, J.P. Singh, A. Kumar, Effect of steady magnetic field on laser-induced breakdown spectroscopy, Appl. Opt. 42 (2003) 3662. [(b)] V.N. Rai, J.P. Singh, F.Y. Yueh, R.L. Cook, Study of optical emission from laser-produced plasma expanding across an external magnetic field, Laser Part. Beams 21 (2003) 65.

[49] R. Barbini, F. Colao, R. Fantoni, A. Palucci, F. Capitelli, Application of laser-induced breakdown spectroscopy to the analysis of metals in soils, Appl. Phys. A Mater. Sci. Process. 69 (1999) 175.

[50] S. Yalsin, D.R. Crosley, G.P. Smith, G.W. Faris, Influence of ambient conditions on the laser air spark, Appl. Phys. B Lasers Opt. 68 (1999) 121.

[51] V.N. Rai, J.P. Singh, C. Winstead, F.Y. Yueh, R.L. Cook, Laser-induced breakdown spectroscopy of hydrocarbon flame and rocket engine simulator plume, AIAA J. 41 (2003) 2192.

[52] F. Y. Yueh, V. N. Rai, J. P. Singh and H. Zhang, Paper # AIAA-2001-2933, 32nd AIAA Plasmadynamics and Laser Conference, 11–14 June 2001, Anaheim, CA, USA (2001).

[53] S. Amoruso, Modeling of UV pulsed-laser ablation of metallic targets, Appl. Phys. A Mater. Sci. Process. 69 (1999) 323.

[54] Z.W. Huang, Y.Y. Teng, K.P. Li, J. Sneddon, Interaction of a laser beam with metals. Part I: quantitative studies of plasma emission, Appl. Spectrosc. 45 (1991) 435.

[55] L.M. Cabalin, J.J. Laserna, Experimental determination of laser induced breakdown thresholds of metals under nanosecond Q-switched laser operation, Spectrochim. Acta B 53 (1998) 723.

[56] R.E. Russo, Laser ablation, Appl. Spectrosc. 49 (1995) 14A.

[57] R.E. Russo, X.L. Mao, Chemical analysis by laser ablation, in: M.C. Millar (Ed.), Laser Ablation and Desorption, Academic Press, San Diego, 1998, pp. 375–412.

[58] C. Chaleard, P. Mauchien, N. Andre, J. Uebbing, J. L. Lacour, C. Geertsen, Correction of matrix effects in quantitative elemental analysis with laser ablation optical emission spectrometry, J. Anal. At. Spectrom. 12 (1997) 183.

[59] B.C. Castle, K. Talabardon, B.W. Smith, J. D. Winefordner, Variables influencing the precision of laser-induced breakdown spectroscopy measurements, Appl. Spectrosc. 52 (1998) 649.

[60] H.T. Buscher, R.G. Tomlinson, E.K. Damon, Frequency dependence of optically induced gas breakdown, Phys. Rev. Lett. 15 (1965) 847.

[61] R.A. Bingham, P.L. Salter, Analysis of solid materials by laser probe mass spectrometry, Anal. Chem. 48 (1976) 1735.

[62] F. Fabbro, E. Fabre, F. Amiranoff, C. Garbeau-Labaune, J. Virmont, M. Weinfield, C.E. Max, Laser-wavelength dependence of mass-ablation rate and heat-flux inhibition in laser-produced plasmas, Phys. Rev. A 26 (1980) 2289.

[63] F. Dahmain, Ablation scaling in laser-produced plasma with laser intensity, laser wavelength, and target atomic number, Phys. Fluids B 4 (1992) 1585.

[64] S. Amoruso, M. Armenante, V. Berardi, R. Bruzzesse, N. Spinelli, Absorption and saturation mechanisms in aluminium laser ablated plasmas, Appl. Phys. A Mater. Sci. Process. 65 (1997) 265.

[65] L.M. Berman, P.J. Wolf, Laser-induced breakdown spectroscopy of liquids: aqueous solutions of nickel and chlorinated hydrocarbons, Appl. Spectrosc. 52 (1998) 438.

[66] C.W. Ng, W.F. Ho, N.H. Cheung, Spectrochemical analysis of liquids using laser-induced plasma emissions: effects of laser wavelength on plasma properties, Appl. Spectrosc. 51 (1997) 976.

[67] X.L. Mao, A.C. Ciocan, O.V. Borisov, R.E. Russo, Laser ablation processes investigated using inductively coupled plasma–atomic emission spectroscopy (ICP–AES), Appl. Surf. Sci. 127-129 (1998) 262.

[68] K.L. Eland, D.N. Stratis, T. Lai, M.A. Berg, S.R. Goode, S. M. Angel, Some comparisons of LIBS measurements using nanosecond and picosecond laser pulses, Appl. Spectrosc. 55 (2001) 279.

[69] K.L. Eland, D.N. Stratis, D.M. Gold, S.R. Goode, S. M. Angel, Energy dependence of emission intensity and temperature in a LIBS plasma using femtosecond excitation, Appl. Spectrosc. 55 (2001) 286.

[70] E.H. Piepmeier, Laser ablation for atomic spectroscopy, in: E.H. Piepmeier (Ed.), Analytical Applications of Lasers, Wiley, New York, 1986.

[71] C.D. Allemand, Spectroscopy of single-spike laser-generated plasmas, Spectrochim. Acta B 27 (1972) 185.

[72] A.M. Prochorov, V.A. Batanov, F.V. Bunkin, V.B. Fedorov, Metal evaporation under powerful optical radiation, IEEE J. Quantum Electron. QE-9 (1973) 503.

[73] G. Dimitrov, T. Maximova, Improvement of the reproducibility and sensitivity of laser microspectral analysis, Spectrosc. Lett. 14 (1981) 737.

[74] G. Dimitrov, T. Zheleva, Decreasing the laser radiation absorption in the plasma plume and improving the sensitivity in quantitative laser microspectral analysis, Spectrochim. Acta B 39 (1984) 1209.
[75] X.L. Mao, A.C. Ciocan, R.E. Russo, Preferential vaporization during laser ablation inductively coupled plasma atomic emission spectroscopy, Appl. Spectrosc. 52 (1998) 913.
[76] R.E. Russo, X.L. Mao, O.V. Borisov, H.C. Liu, Influence of wavelength on fractionation in laser ablation ICP-MS, J. Anal. At. Spectrom. 15 (2000) 1115.
[77] Y.W. Kim, in: L.J. Radziemski, D.A. Cremers (Eds.), Laser-Induced Plasmas and Applications, Marcel Dekker Inc., New York, 1989.
[78] F. Leis, W. Sdorra, J.B. Ko, K. Niemax, Basic investigations for laser microanalysis: I. Optical emission spectrometry of laser-produced sample plumes, Mikrochim. Acta II (1989) 185.
[79] C. Aragon, J.A. Aguilera, F. Penalba, Improvements in quantitative analysis of steel composition by laser-induced breakdown spectroscopy at atmospheric pressure using an infrared Nd: YAG laser, Appl. Spectrosc. 53 (1999) 1259.
[80] M. Sabsabi, P. Cielo, Quantitative analysis of copper alloys by laser-produced plasma spectrometry, J. Anal. At. Spectrom. 10 (1995) 643.
[81] A.S. Eppler, D.A. Cremers, D.D. Hickmott, M.J. Ferris, A.C. Koskelo, Matrix effects in the detection of Pb and Ba in soils using laser-induced breakdown spectroscopy, Appl. Spectrosc. 50 (1996) 1175.
[82] R.A. Multari, L.E. Foster, D.A. Cremers, M.J. Ferris, Effect of sampling geometry on elemental emissions in laser-induced breakdown spectroscopy, Appl. Spectrosc. 50 (1996) 1483.
[83] P. Fichet, P. Mauchien, J.F. Wagner, C. Maulin, Quantitative elemental determination in water and oil by laser induced breakdown spectroscopy, Anal. Chim. Acta 429 (2001) 269.
[84] L.I. Sedov, Similarity and Dimensional Methods in Mechanics, MIR Publishers, Moscow, 1981.
[85] M. Gatti, V. Palleschi, A. Salvetti, D.P. Singh, M. Vaselli, Spherical shock waves in laser produced plasmas in gas, Opt. Commun. 69 (1988) 141.
[86] D.E. Kim, K.J. Yoo, H.K. Park, K.J. Oh, D.W. Kim, Quantitative analysis of aluminum impurities in zinc alloy by laser-induced breakdown spectroscopy, Appl. Spectrosc. 51 (1997) 22.
[87] H.C. Liu, X.L. Mao, J.H. Yoo, R.E. Russo, Early phase laser induced plasma diagnostics and mass removal during single-pulse laser ablation of silicon, Spectrochim. Acta B 54 (1999) 1607.
[88] A. Ciucci, V. Palleschi, S. Rastelli, A. Salvetti, D.P. Singh, E. Tognoni, Effect of imperfect focusing in laser-induced plasma spectroscopy measurements, Nuovo Cimento-D 20 (1998) 1469.
[89] Y.I. Lee, S.P. Sawan, T.L. Thiem, Y.Y. Teng, J. Sneddon, Interaction of a laser beam with metals. Part II: Space-resolved studies of laser-ablated plasma emission, Appl. Spectrosc. 46 (1992) 436.
[90] R. Krasniker, V. Bulatov, I. Schechter, Study of matrix effects in laser plasma spectroscopy by shock wave propagation, Spectrochim. Acta B 56 (2001) 609.
[91] R.E. Neuhauser, U. Panne, R. Niessner, Laser-induced plasma spectroscopy (LIPS): a versatile tool for monitoring heavy metal aerosols, Anal. Chim. Acta 392 (1999) 47.
[92] B.J. Marquardt, P.N. Stratis, D.A. Cremers, S.M. Angel, Novel probe for laser-induced breakdown spectroscopy and Raman measurements using an imaging optical fiber, Appl. Spectrosc. 52 (1998) 1148.
[93] R.E. Neuhauser, U. Panne, R. Niessner, Utilization of fiber optics for remote sensing by laser-induced plasma spectroscopy (LIPS), Appl. Spectrosc. 54 (2000) 923.
[94] A.I. Whitehouse, J. Young, I.M. Botheroyd, S. Lawson, C.P. Evans, J. Wright, Remote material analysis of nuclear power station steam generator tubes by laser-induced breakdown spectroscopy, Spectrochim. Acta B 56 (2001) 821.
[95] A.K. Rai, H. Zhang, F.Y. Yueh, J.P. Singh, A. Weisburg, Parametric study of a fiber-optic laser-induced breakdown spectroscopy probe for analysis of aluminum alloys, Spectrochim. Acta 56 (2001) 2371.
[96] E.H. Piepmeier, D.E. Osten, Atmospheric influences on Q-switched laser sampling and resulting plumes, Appl. Spectrosc. 25 (1971) 462.
[97] K.J. Grant, G.L. Paul, Electron temperature and density profiles of excimer laser-induced plasmas, Appl. Spectrosc. 44 (1990) 1349.
[98] W. Sdorra, K. Niemax, Basic investigations for laser microanalysis: III. Application of different buffer gases for laser-produced sample planes, Mikrochim. Acta 107 (1992) 319.
[99] Y. Iida, Effects of atmosphere on laser vaporization and excitation processes of solid samples, Spectrochim. Acta 45 (1990) 1353.
[100] J. Hermann, C. Baulmer Leborgne, D. Hong, Diagnostics of the early phase of an ultraviolet laser induced plasma by spectral line analysis considering self-absorption, J. Appl. Phys. 83 (1998) 691.
[101] T.L. Thiem, R.H. Salter, J.A. Gardner, Y.I. Lee, J. Sneddon, Quantitative simultaneous elemental determinations in alloys using laser-induced breakdown spectroscopy (LIBS) in an ultra-high vacuum, Appl. Spectrosc. 48 (1994) 58.
[102] L.J. Dawson, On the production of plasma by giant pulse lasers, Phys. Fluids 7 (1964) 981.
[103] N. Smijesh, K.H. Rao, D. Chetty, I.V. Litvinyuk, R. T. Sang, Plasma plumes produced by laser ablation of Al with single and double pulse schemes, Opt. Lett. 43 (2018) 6081.

[104] D. Batani, Short-pulse laser ablation of materials at high intensities: influence of plasma effects, Laser Part. Beams 28 (2010) 235.

[105] E. Schwarz, S. Gross, B. Fischer, I. Muri, J. Tauer, H. Kofler, E. Winter, Laser-induced optical breakdown applied for laser spark ignition, Laser Part. Beams 28 (2010) 109.

[106] A. Kumar, S. George, R.K. Singh, V.P.N. Nampoori, Influence of laser beam intensity profile on propagation dynamics of laser-blow-off plasma plume, Laser Part. Beams 28 (2010) 387.

[107] J. Krasa, A. Lorusso, V. Nassisi, L. Velardi, A. Belyhan, Revealing of hydrodynamic and electrostatic factors in the center-of-mass velocity of an expanding plasma generated by pulsed laser ablation, Laser Part. Beams 29 (2011) 113.

[108] R. Hai, X. Mao, G.C.-Y. Chan, R.E. Russo, H. Ding, V. Zorba, Internal mixing dynamics of Cu/Sn-Pb plasmas produced by femtosecond laser ablation, Spectrochim. Acta 148 (2018) 92.

[109] A. Rastegan, M. Lenzner, K.P. Jean-Claude Diels, L. Arissian, High-resolution remote spectroscopy and plasma dynamics induced with UV filaments, Opt. Lett. 44 (2019) 147.

[110] J. Thomas, H. Joshi, A. Kumar, R. Philip, Pulse width dependent dynamics of laser-induced plasma from a Ni thin film, J. Phys. D: Appl. Phys. 52 (2019), https://doi.org/10.1088/1361-6463/aaff43.

[111] V.N. Rai, M. Shukla, H.C. Pant, Some studies on picosecond laser produced plasma expanding across a uniform external magnetic field, Laser Part. Beams 16 (1998) 431.

[112] V.N. Rai, M. Shukla, H.C. Pant, Pramana, density oscillations in laser produced plasma decelerated by external magnetic field, J. Phys. 55 (2000) 773.

[113] V.N. Rai, A.K. Rai, F.Y. Yueh, J.P. Singh, Optical emission from laser-induced breakdown plasma of solid and liquid samples in the presence of a magnetic field, Appl. Opt. 42 (2003) 2085.

[114] V.N. Rai, F.Y. Yueh, J.P. Singh, Theoretical model for double pulse laser-induced breakdown spectroscopy, Appl. Opt. 47 (2008) G30.

[115] V.N. Rai, F.Y. Yueh, J.P. Singh, Time-dependent single and double pulse laser-induced breakdown spectroscopy of chromium in liquid, Appl. Opt. 47 (2008) G21.

[116] V.N. Rai, Theoretical aspect of enhancement and saturation in emission from laser produced plasma, Laser Part. Beams 30 (2012) 621.

[117] D.H.H. Hoffman, Ion and laser beams as tools for high energy density physics, Laser Part. Beams 27 (1) (2009).

[118] Y.L. Wang, W. Xu, Y. Zhou, L.-Z. Chu, G.-S. Fu, Influence of pulse repetition rate on the average size of silicon nanoparticles deposited by laser ablation, Laser Part. Beams 25 (2007) 9.

[119] A. Kumar, A.L. Verma, Nonlinear absorption of intense short pulse laser over a metal surface embedded with nanoparticles, Laser Part. Beams 29 (2011) 333.

[120] A. Huber, B. Schweer, V. Philipps, R. Leyte-Gonzales, N. Gierse, M. Zlobinski, S. Brezinsek, V. Katov, P. Mertens, U. Samm, G. Sergienko, Study of the feasibility of applying laser-induced breakdown spectroscopy for the in-situ characterization of deposited layer in fusion devices, Phys. Scr. T145 (2011).

[121] R. Fazeli, M.H. Mahdieh, G.J. Tallents, Enhancement of line X-ray emission from iron plasma created by laser irradiation of porous targets, Laser Part. Beams 29 (2011) 193.

[122] M. Capitelli, A. Casavola, G. Colonna, A. De Giacomo, Laser-induced plasma expansion: theoretical and experimental aspects, Spectrochim. Acta B 59 (2004) 271.

[123] A. De Giacomo, M. Dell' Aglio, O. De Pascale, Single pulse-laser induced breakdown spectroscopy in aqueous solution, Appl. Phys. A Mater. Sci. Process. 79 (2004) 1035.

[124] S.S. Mao, X. Mao, R. Grief, R.E. Russo, Initiation of an early-stage plasma during picosecond laser ablation of solids, Appl. Phys. Lett. 76 (2000) 3370.

[125] A. Bogaerts, Z. Chen, R. Gijbels, A. Vertes, Laser ablation for analytical sampling: what can we learn from modeling, Spectrochim. Acta B 58 (2003) 1867.

[126] S.I. Anisimov, D. Bauerle, B.S. Luk'yanchuk, Gas dynamics and film profiles in pulsed-laser deposition of materials, Phys. Rev. B 48 (1993) 12076.

[127] N. Arnold, J. Gruber, J. Heitz, Spherical expansion of the vapor plume into ambient gas: an analytical model, Appl. Phys. A Mater. Sci. Process. 69 (1999) S87.

[128] A. Casavola, G. Colonna, A. De Giacomo, O. De Pascale, M. Capitelli, Experimental and theoretical investigation of laser-induced plasma of a titanium target, Appl. Opt. 42 (2003) 5963.

[129] H.C. Le, D.E. Zeitoun, J.D. Parisse, M. Snetis, W. Marine, Modeling of gas dynamics for a laser-generated plasma: propagation into low-pressure gases, Phys. Rev. E 62 (2000) 4152.

[130] G. Colonna, A. Casavola, M. Capitelli, Modelling of LIBS plasma expansion, Spectrochim. Acta B 56 (2001) 567.

[131] T.E. Itina, W. Marine, M. Autric, Monte Carlo simulation of pulsed laser ablation from two-component target into diluted ambient gas, J. Appl. Phys. 82 (1997) 3536.

[132] I. Noor Batcha, R.R. Lucchese, Monte Carlo simulations of gas-phase collisions in rapid desorption of molecules from surfaces, J. Chem. Phys. 86 (1987) 5816.

[133] P.K. Diwakar, D.W. Hahn, Study of early laser-induced plasma dynamics: transient electron density gradients via Thomson scattering and Stark Broadening, and the implications on laser-induced breakdown

spectroscopy measurements, Spectrochim. Acta B 63 (2008) 1038.

[134] S. Eschlbock-Fuchs, M.J. Haslinger, A. Hinterreiter, P. Kolmhofer, N. Huber, R. Rossler, J. Heitz, J.D. Pedarning, Influence of sample temperature on the expansion dynamics and the optical emission of laser-induced plasma, Spectrochim. Acta B 87 (2013) 36.

[135] J.R. Freeman, S.S. Harilal, P.K. Diwakar, B. Berhof, A. Hassanein, Comparison of optical emission from nanosecond and femtosecond laser produced plasma in atmosphere and vacuum conditions, Spectrochim. Acta B 87 (2013) 43.

[136] E. Jobiliong, H. Suyanto, A.M. Marpaung, S.N. Abdul Madgid, N. Edris, R. Hedurg, M. Ramli, M. Pardede, M.M. Suliyanti, K. Kagawa, K.-I. Fukamoto, M.O. Tjia, T.J. Lie, Z.S. Lie, K.H. Kurniawan, Spectral and dynamic characteristics of helium plasma emission and its effect on a laser-ablated target emission in a double-pulse laser-induced breakdown spectroscopy (LIBS) experiment, Appl. Spectrosc. 69 (2015) 115.

[137] D. Zhang, A. Chen, X. Wang, Y. Wang, L. Sui, D. Ke, S. Li, Y. Jiang, M. Jin, Influence of the distance between target surface and focal point on the expansion dynamics of a laser-induced silicon plasma with spatial confinement, Spectrochim. Acta B 143 (2018) 71.

[138] H. Hou, B. Yang, X. Mao, V. Zorba, P. Ran, R. E. Russo, Characteristics of plasma plume in ultrafast laser ablation with a weakly ionized air channel, Opt. Express 26 (2018) 13425.

[139] V.N. Lednev, P.A. Sdvizhenskii, R.D. Asyutin, M. Y. Grishin, R.S. Tretyakov, S.M. Parsin, Surface plasma influence on nanosecond laser ablation, Appl. Opt. 58 (2019) 1496.

[140] C. Du, C. Yang, M. Zhang, Investigation on the dynamic characteristics of LIBS for heavy metal Mn in liquid matrix, Optik 180 (2019) 602.

[141] L.B. Guo, W. Hu, B.Y. Zhang, X.N. He, C.M. Li, Y.S. Zhou, Z.X. Cai, X.Y. Zeng, Y.F. Lu, Enhancement of optical emission from laser-induced plasmas by combined spatial and magnetic confinement, Opt. Express 19 (2011) 14067.

[142] V.N. Rai, F.Y. Yueh, J.P. Singh, Study of laser-induced breakdown emission from liquid under double-pulse excitation, Appl. Opt. 42 (2003) 2094.

[143] Y. Li, D. Tian, Y. Ding, G. Yang, K. Liu, C. Wang, X. Han, A review of laser-induced breakdown spectroscopy signal enhancement, Appl. Spectrosc. Rev. 53 (2018) 1.

[144] M. Dell'Aglio, R. Alrifai, A. De Giacomo, Nanoparticle enhanced laser induced breakdown spectroscopy (NELIBS), a first review, Spectrochim. Acta B 148 (2018) 105.

[145] Z. Yu, S. Yao, L. Zhang, Z. Lu, Z.S. Lie, J. Lu, Surface-enhanced laser-induced breakdown spectroscopy utilizing metallic target for direct analysis of particle flow, J. Anal. At. Spectrom. 34 (2019) 172.

CHAPTER

5

Instrumentation for LIBS and recent advances

V.N. Rai[a,b], Surya N. Thakur[c]

[a]Laser Plasma Division, Raja Ramanna Centre for Advanced Technology, Indore, India
[b]School of Physics, Devi Ahilya University, Indore, India
[c]Laser and Spectroscopy Laboratory, Department of Physics, Banaras Hindu University, Varanasi, India

1 Introduction

Laser-induced breakdown spectroscopy (LIBS) is a laser diagnostics system, where a laser beam is focused onto a material that generates transient high-density plasma as the laser intensity exceeds the breakdown threshold of the material (~1–10 MW/cm^2). The ultraviolet (UV) and visible emission from the plasma are spectrally resolved and recorded for qualitative and quantitative analysis of the sample. LIBS was first used for the determination of the elemental composition of materials in the form of gases, liquids, and solids during the 1960s [1, 2]. Research on LIBS is progressing very fast with inclusion of new technologies and it has now achieved a new height. Field-portable and handheld instruments capable of in situ and real-time analysis of samples have been developed in recent years with the availability of reliable, smaller, and less costly laser systems along with sensitive optical detectors, such as the intensified charge-coupled device (ICCD). All types of LIBS systems are now commercially available. Several review articles and books have been published on this topic [3–14].

A short duration laser pulse of sufficient energy focused onto the surface of a material sample instantly increases its temperature above the vaporization temperature, regardless of the type of material. Compared with the rate of energy delivery from the laser pulse, energy dissipation through vaporization is relatively slow and the underlying layer of material reaches critical temperatures and pressures before the surface layer vaporizes, which forces the surface to explode. Generally, material ablation and plasma formation take place during the initial period of the laser pulse, whereas the rest of the laser energy is absorbed by the ablated material to form luminous plasma. The temperature of the plasma-emitting UV and visible radiation is in the range of 10^4–10^5 K, whereas the electron number density ranges from 10^{15} to 10^{19} cm^{-3} and the plasma plume may last from microseconds to milliseconds. The laser-induced

Laser-Induced Breakdown Spectroscopy
https://doi.org/10.1016/B978-0-12-818829-3.00005-8

plasma may be coupled with various detection methods, such as mass spectrometry (MS), atomic emission spectrometry (AES), atomic absorption spectrometry, and laser-excited atomic fluorescence spectrometry. In some experiments, laser-induced plasma (LIP) is used as an atom or ion reservoir directly, or the ablated material is transported to a second source for further excitation or ionization. The direct detection of atomic emission from LIP is referred to as LIBS, which has proved its potential in elemental analysis and feasibility for miniaturization.

LIBS has many advantages as an analytical technique. There is no need for sample preparation, which avoids further contamination of the material to be analyzed [10, 11]. The analysis process is fast and can be used for both nonconducting and conducting samples, regardless of their physical states, i.e., aerosols, gases, liquids, or solids. LIBS is applicable to the analysis of extremely hard materials that are difficult to digest or dissolve, such as ceramics and semi-/superconductors as well as biological samples. Its capability for simultaneous multielement determination, localized microanalysis, and surface analysis is also of great importance. It has been used successfully in hazardous and difficult environmental conditions to study remotely located samples for online and real-time information about their spectra. LIBS has been found useful in elemental process monitoring and field-portable analyzers for in situ trace metal analysis of real samples where accuracy and precision are not the main requirements [11].

2 Typical LIBS setup

Various types of LIBS experimental setups have been used, which differ mainly in the form of collection optics for the radiation emitted by the plasma plume. In one of the arrangements the emission from plasma is collected in the direction perpendicular to the direction of the incident laser. In addition to the difficulties of alignment and reduced sensitivity the collected emission exhibits spatial dependence leading to loss of spectral information about emission from the whole plasma plume. These shortcomings are removed in another arrangement, where the focusing lens itself acts as the collecting lens for plasma emission. In this case, collected emission corresponds to the integrated value of radiation from all the spatial locations of the plasma. The typical schematic diagrams of the experimental setup [14–16] for recording the laser-induced breakdown emission from the solid, liquid, and gaseous samples are shown in Figs. 1, 2, and 3, respectively.

The LIBS experimental setup for studying solid samples consists of a Q-switched, frequency-doubled Nd:YAG laser (Continuum Surelite III) that delivers an energy of ~300 mJ at 532 nm with a 5-ns pulse. This laser was operated at 10 Hz and was focused on the target with the help of a dichroic mirror and quartz-focusing lens of 20 cm focal length. The combination of dichroic mirror and the same focusing lens (Fig. 1) was used to collect the optical emission from the LIP. Two UV-grade quartz lenses of focal lengths 100 and 50 mm were used to couple the plasma emission to an optical fiber bundle. The fiber bundle consists of 80 single fibers of 0.01 mm core diameter. The rectangular exit end of the optical fiber was coupled to the spectrograph (Model HR 460, Instrument SA, Inc., Edison, NJ) and used as an entrance slit. The spectrograph was equipped with 1200- and 2400-L/mm diffraction gratings of dimension 75 × 75 mm. A 1024 × 256-element ICCD (Princeton Instrument Corporation, Princeton, NJ), with a pixel width of 0.022 mm, was attached to the exit focal plane of the spectrograph and used to detect the dispersed light from the LIP. The detector was operated in gated mode with the control of a high-voltage pulse generator (PG-10, Princeton Instruments Corporation, Princeton, NJ) and was

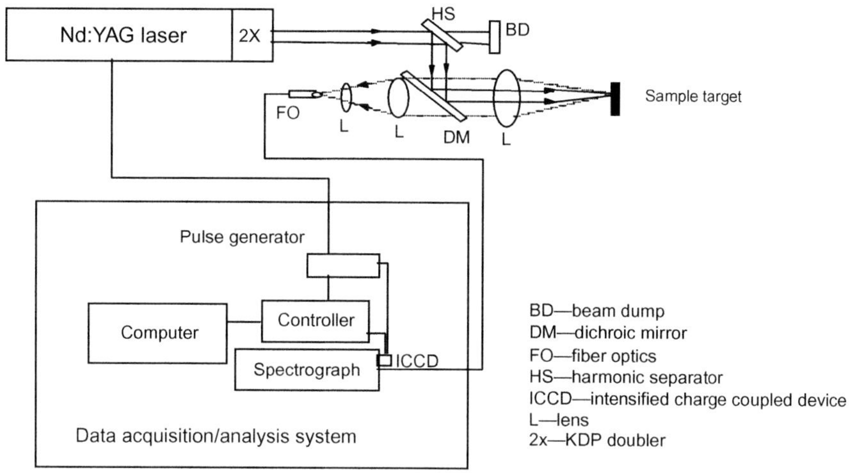

FIG. 1 Schematic diagram of experimental system for recording the laser-induced breakdown spectroscopy of solid samples [14A].

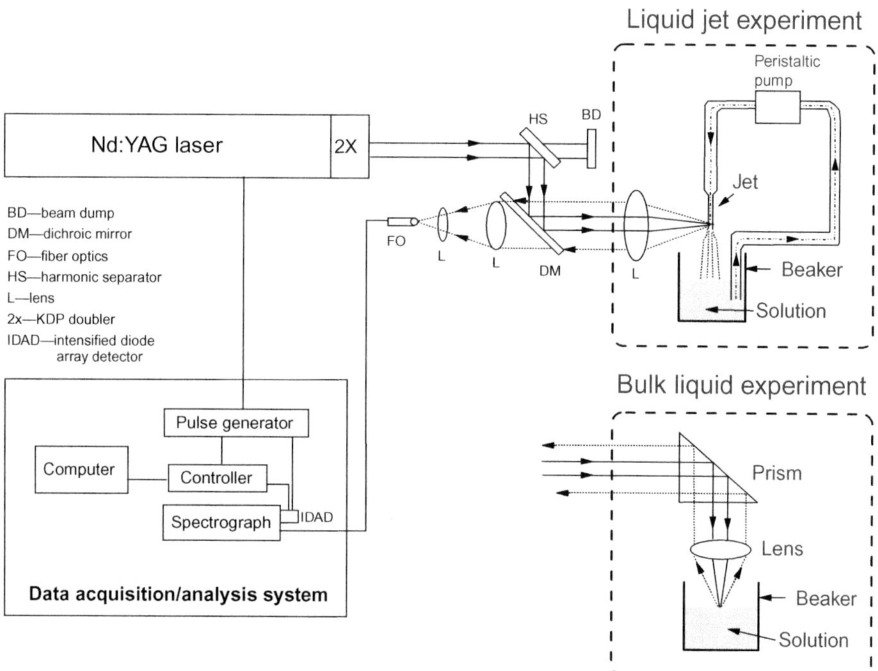

FIG. 2 Schematic diagram of experimental system for recording laser-induced breakdown spectroscopy of liquid samples. In these experiments, plasma is produced on the surface of bulk liquid or liquid jet [14B].

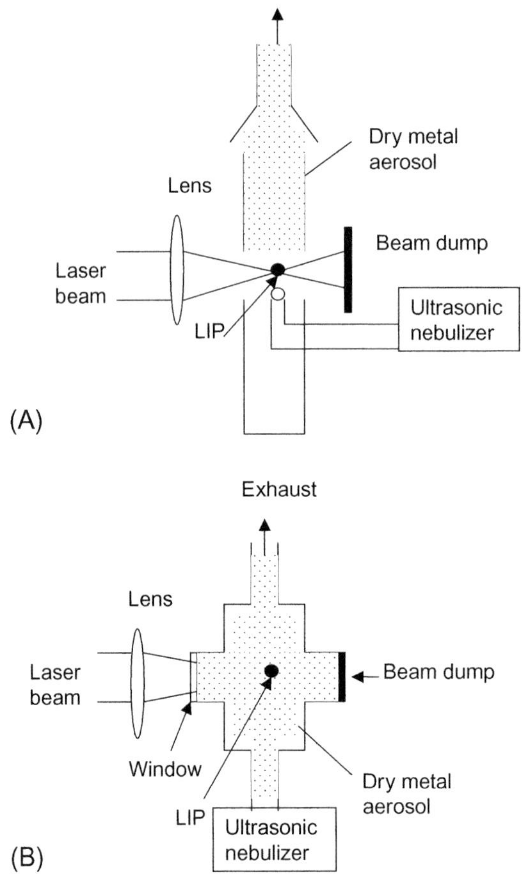

FIG. 3 Laser-induced breakdown spectroscopy calibration system for gaseous samples: (A) open system and (B) closed system. The rest of the instrumentation is similar to that shown in Figs. 1 and 2 [15, 16].

synchronized with the output of the laser pulse. Data acquisition and analysis were performed using a personal computer. The gate delay time and gate width were adjusted to maximize the signal-to-background (S/B) and signal-to-noise (S/N) ratios. Emission spectra were recorded mainly using a 2400-L/mm grating for better spectral resolution. Around 100 pulses were accumulated to obtain one spectrum and 30 such spectra were recorded for each experimental condition to increase the sensitivity of the system and to reduce the standard deviation in the recorded data.

3 LIBS instrumentation

The principles of LIBS are similar to those of conventional plasma AES, such as inductively coupled atomic emission spectrometry (ICP-AES), microwave-induced plasma-AES, direct current plasma-AES, arc-AES, and spark-AES. The main difference between LIBS and the conventional AES is that there is no need to transport the sample to the plasma in LIBS. As already discussed, plasma is formed in or on the sample in situ by the use of a focused laser beam. LIBS instrumentation consists of three major parts: (1) a laser to generate the LIP; (2) a sample container (ablation chamber) to house the samples in an inert gaseous environment under a vacuum or simply in air; and (3) a detection system to collect, resolve, and measure the atomic emission lines from the LIP. The detection system usually consists of a dispersing element (a monochromator or a grating), an optical detector, detection electronics, and a computer.

The basic purpose of a laser for LIBS is to produce sufficient and stable pulsed energy to generate the plasma. Various lasers with wavelengths ranging from the infrared (IR) to UV regions of the spectrum have been used in LIBS and have been summarized by Lee et al. [13]. These include solid-state lasers such as the Nd:YAG laser (1064, 532, and a pulse duration of 5–10 ns); the ruby laser (693 nm, and a pulse duration of 20 ns); gas lasers such as the CO_2 laser (10.6 μm, and a pulse duration of 100 ns) and the N_2 laser (337 nm, and a pulse duration of 30 ps–10 ns); and excimer lasers (193 nm [ArF], 248 nm [KrF], 308 nm [XeCI], and a pulse duration of 10–20 ns). Among all these, Nd:YAG lasers are the most widely used. The typical output energies for these lasers are tens of millijoules to hundreds of millijoules per pulse, and peak power is in the range of megawatts. These laser beams are focused to the spots a few tens of micrometers in diameter producing 10^{10}–10^{12} W/cm^2 irradiance. The characteristics of a laser, such as energy, energy stability, wavelength, pulse duration, beam quality, and mode quality, together with the properties of the target material, affect the production and characteristics of the plasma. Typically, 100 mJ/pulse energy is sufficient to generate plasma for the analysis of most of the materials. With the use of laser wavelengths associated with gaseous atomic transitions of analyte elements, much lower laser energy at the sub-millijoule level is adequate for the generation of the plasma, where enhanced signals at resonant wavelengths can be observed when the laser is scanned across them [17]. This technique is known as resonant laser-induced breakdown spectrometry (RLIBS). RLIBS requires less laser energy and may result in less spectral interference in real sample analysis, but it requires a tunable laser, which makes the whole instrumental system more complex. Laser wavelength is a major influential factor for LIBS along with laser energy. UV laser radiation for LIBS has an advantage because of its low UV reflectivity from most of the metal surfaces, which usually leads to more efficient energy coupling and high optical resolution [3].

LIBS experiments can be performed in air, low-pressure inert gas, and vacuum. Some considerations concerning the construction of vacuum and gas chambers mainly include two aspects: (1) to extend the analytical spectral range to the deep-UV region for elements such as carbon, phosphorus, sulfur, chlorine, bromine, iodine, oxygen, and nitrogen, and (2) to improve the detectability by inert gas purging. Research on laboratory applications of LIBS is frequently carried out on a sample kept in a vacuum chamber, but field applicable LIBS is usually performed in the atmospheric environment. Detailed discussions about sample chambers can be found in the literature [5, 12]. The detection system consists of focusing optics (used with an optical fiber cable for remote sensing), a dispersing element, a detector, signal processing electronics, and a computer for data processing and storage. Many types of detectors have been used for recording LIBS. In the early days, a photographic plate was used as a detector, which had the advantage of a wide wavelength range with relatively low cost but had the disadvantage of being time consuming with low reproducibility. Photographic detection has been gradually replaced by detectors for spectrally resolved emission such as a photomultiplier tube (PMT), a photodiode array (PDA), or a charge-coupled device (CCD), which provide fast and accurate measurements. The PMT is placed behind the exit slit of the dispersive unit that produces a photocurrent proportional to the intensity of the incident radiation. The problem with this system is that many PMTs have to be used for simultaneous multielement analysis. Optical multichannel analyzers can be used for multielement determinations, but at the cost of more complicated instrumentation with a limited coverage of number of elements. Photodiode array detectors are good for simultaneous multielement analysis. In early studies of LIBS, Radziemski et al. [18] and Cremers et al. [19] used various types of PMTs sensitive over the range of 200–900 nm. They also used a time-gated linear diode array coupled to a multichannel analyzer. The array consisted of 1024 diodes 2.54 cm in length, where each channel recorded the signal seen by one diode during each shot. The array was time gated by switching on the high voltage of the intensifier during the time of interest, typically 200 ns to several microseconds. Time gating was essential because of the strong continuum emission from the plasma during early stages (500 ns) after the plasma formation. The system was sensitive for the wavelength range of 350–800 nm. In fact, PDAs [20–23], gated and/or intensified, and multichannel analyzers [24, 25] are now frequently used as detectors in LIBS. However, sensitive and imaging scientific CCD cameras as detectors have been gaining more popularity in recent years and can be time gated to isolate temporal intervals, during the evolution of the plasma, for optimum signal measurements in different applications. It has been reported that CCD is nearly three orders of magnitude more sensitive than PDA [26], but the dynamic range is slightly less than that obtained with PDA. However, the spectral range is more limited because the light is detected through a transparent electrode. Castle et al. [27] and Zhang et al. [16] used an ICCD as a detector for LIBS where a programmable pulse delay generator was used to gate the ICCD to obtain an optimum S/B noise ratio and the firing of the laser and data collection were under the control of a computer. In another case, Castle et al. [28] employed a linear CCD as the detector, which was nongateable but had the provision of external triggering and a timing circuit that was designed to control laser firing and data collection. This linear CCD had 2046 pixels, and the system covered the spectral range of 339–462 nm.

3.1 Echelle spectrometer

Czerny-Turner spectrographs are usually employed to disperse the emission collected from LIBS plasma, where suitable detectors coupled to it offer the possibility of time-resolved measurements. These detection systems are intrinsically limited in resolution as well as in spectral coverage. The multielemental detection capability offered by the LIBS technique demands a spectrograph with a wider spectral coverage. In multielemental analysis, sequential measurements of parts of the spectrum of interest are performed, inspecting each time a different sample of the material ablated from the target surface. In principle this procedure limits the LIBS application to homogeneous samples but most of the samples are inhomogeneous, which is why the spectra vary from shot to shot as a result of changes in the sample composition as well as due to stochastic fluctuations in the plasma [29]. Therefore simultaneous measurement of the complete optical spectrum is necessary for getting optimum information for analytical purposes. Instruments, which allow simultaneous measurements, are Paschen-Runge spectrometers or the more compact Echelle spectrometers as discussed in a review article by Detalle et al. [30]. Echelle spectrometers offer excellent spectral resolving power ($/\Delta\lambda \geq$ 10,000 and more) in combination with a spectral coverage of several hundred nanometers. In combination with ICCDs, Echelle spectrometers represent a very powerful tool for elemental analysis as demonstrated by Haisch et al. [31] who found substantial improvement in the detection limits obtained for several elements with an Echelle system as compared with those obtained with a conventional Czerny-Turner system.

The principle of the Echelle spectrometer has been described by Detalle et al. [30]. It has a focal length of 25 cm with a numerical aperture 1:10 and a quartz prism positioned in front of the grating separates the different orders of spectra and produces a two-dimensional pattern. The flat image plane is $24.85 \times 24.5\ mm^2$. This system provides maximum resolution in the wavelength range between 200 and 780 nm. The linear dispersion per pixel ranges from 0.005 nm (at 200 nm) to 0.019 nm (at 780 nm), which is based on the spectral resolution $\lambda/\Delta\lambda = 40{,}000$. The detector in this system is an ICCD camera, having a CCD array of 1024×1024 pixels ($24 \times 24\ \mu m^2$) and a microchannel plate. A fast pulse generator delivers a 5-ns pulse to the intensifier to ensure synchronization of the measurements with the laser pulse. The spectral response in a particular order of diffraction of the Echelle spectrometer is nonlinear when measured using the black-body radiation from a deuterium lamp and maximum sensitivity is found in the center of the given order. Each diffraction order has similar shape but a different sensitivity, which requires a correction factor when the measurement is made in different spectral ranges with different sensitivities. Normally, a black-body radiation calibration spectrum is recorded to obtain the intrinsic response of the Echelle/ICCD system, which is then used to normalize the acquired spectrum.

3.2 Specialty of the Echelle spectrometer

In recent years the Echelle spectrometer has been found very successful in the acquisition of spectral data from which relevant physical or chemical information can be extracted. Calibration curves of various elements have been obtained, limits of their detection determined, and the excitation and ionic temperatures of the LIP as well as the electron density have been measured. The reproducibility of experimental results shows that the dynamic range of the detector makes it possible to measure simultaneously the intensities of spectral lines of the major and trace elements in the sample. The detectability of elements at low concentration is facilitated by the excellent resolution of this system. One has to be aware of the advantages as well as shortcomings of this system for its judicious application.

3.2.1 *Advantages*

The main advantage of the very compact Echelle system is its excellent resolution, which is comparable to that of a Czerny-Turner spectrometer of 1 m focal length having a grating of 3600 grooves/mm. This resolution is useful in avoiding spectral interference. The second important feature of this system is its coverage of a very broad spectral range (200–780 nm), which makes it possible to record several lines of the same element. Thus, a large dynamic range of concentrations can be measured, because a saturated line can be replaced by another one of the same element in the spectrum. The system has been found appropriate for the analysis of complex matrices, such as alloys containing a large number of elements. The broad spectral range, combined with the good spectral resolution of the system, is also found suitable for multielemental analysis. An Echelle spectrometer is an excellent tool for instantaneous recognition of the elements present in an unknown sample as long as the element emits in the plume of LIP. Multielemental analysis in various types of samples, solids (conductor or not), liquids, or gases can be easily carried out with an Echelle spectrometer and it has been used to classify very different types of alloys [32].

The main difficulty faced by the LIBS technique is the variation in the intensities of spectral lines, which depend on the complex process of laser-matter interaction. The solution often used to solve this problem is internal standardization. Generally, the ratio of the line intensity of the impurity or trace element to that of the major element of the matrix is used in determining the concentration of the trace element.

Some research groups showed the possibility of normalizing the analyte signals using the variation in the excitation temperature of the plasma [33], which is difficult to realize with a conventional spectrometer, but becomes easy with an Echelle system. One can also follow the evolution of emission spectra of all the elements present in the matrix and carry out a compositional assessment considering that the sum of concentrations is always 100%.

3.2.2 Limitations

The mode of dispersion in an Echelle spectrometer involves a change in resolution with wavelength and the response of the intensifier is not equal throughout the spectral range (maximum at ~400 nm). Thus spectral sensitivity is low for wavelengths above 600 nm resulting in a loss of luminosity and resolution of the system for the longest wavelengths. Echelle dispersion also involves loss of zones of the spectrum located between the different orders, called dead zones. Although this has little consequence for the UV region, the dimensions of these dead zones increase with wavelength to reach several nanometers beyond 700 nm. Thus spectral lines falling in the dead zone may not be accessible. As discussed earlier, Echelle dispersion is nonlinear even in the same order of dispersion, which makes it necessary to correct the signal with the help of a reference spectrum corresponding to the radiation of a black body, whose emission is linear in the spectral range covered by the spectrometer.

A CCD detector can receive only a limited number of electrons on each of its pixels. With an increase in the gain or the gate width, the emission lines with very large intensities may saturate some pixels of the CCD. The electrons accumulated on the saturated pixels jump, mainly onto the closest pixels located below, leading to the phenomenon of blooming as discussed by Detalle et al. in the case of the Al emission line at 308.22 nm [30]. The pixels located below the saturated one correspond to a different order and different wavelengths are affected by the signal of the saturating emission line. Detalle et al. [30] observed the appearance of a line corresponding to the theoretical position of Nb (316.37 nm) on the rebuilt linear spectrum, whereas the sample was completely free of this element. The spectral lines observed as a consequence of blooming are called "ghost lines" and to avoid them the experiments must be performed under conditions of nonsaturation. The dynamic range of the system thus presents a limit, not due to the lack of gain of the intensifier, but due to the impossibility of eliminating the signal from the major lines.

Another limitation of the Echelle system is related to its low rate of data acquisition. The CCD detector has a dimension of 1024 × 1024 pixels, whereas the rate of transfer of the PC board is 500 kHz for 16-bit resolution. The transfer time for all data from the CCD is thus higher than 2 s and one acquisition is actually possible every 3 s. Therefore the present configuration of the system does not make it possible to carry out rapid sampling. Of course, the speed of analysis is a relative concept, but it is one of the main factors in the choice of one technique or method over another.

4 Fiber optic LIBS

LIBS is most suitable for field-based industrial applications, which include real-time, online analysis of material for process control and monitoring. Most of the experimental techniques discussed so far are laboratory based, where plasma is generated by focusing the high-intensity laser beam on the sample surface with an assembly of lenses. The light emitted from the plasma is collected by either the same assembly of lenses or a separate assembly of lenses and is focused on the entrance slit of the spectrometer for further analysis [10]. Such an experimental setup is not well suited for field measurements, which requires flexible optical access to the test facility and minimal on-site alignment.

Recent advances in fiber optic materials have opened up many new areas of applications for the LIBS technique. Using optical fiber, one can send the laser beam to the desired location and perform remote measurements. The low breakdown threshold of the optical fiber material did not permit delivery of laser radiation on the target and the applications of optical fiber in LIBS were initially limited to delivering the plasma emission to the detection system [16, 23, 34]. The next development in LIBS studies was the use of two optical fibers: one of the optical fibers was delivering the laser beam for creating spark by focusing the laser radiation on the surface of sample, and the other one was collecting radiation from the spark emission. Adjustment of two optical fibers is a very delicate and difficult task, especially when LIBS is being used in harsh and hazardous environmental conditions, such as those found in the aluminum, glass, and steel industries. Therefore it is more desirable to use only one optical fiber to transmit the laser beam as well as to collect the emission from LIP [35, 36].

A simple and robust fiber optic probe that uses one optical fiber both for delivering the laser power to produce a spark and for collecting the resulting radiation from the spark for quantitative elemental analysis with greater accuracy and a lower detection limit has been developed [33]. To obtain maximum emission intensity and a better S/B ratio, several parameters have to be optimized such as detector gain, damage threshold of fiber optics, focal lengths of different lenses, gate delay and gate width, sample surface, etc.

4.1 Fiber optic LIBS probe

A schematic diagram of the fiber optic LIBS probe is shown in Fig. 4 [11–36]. The second harmonic (532 nm) of a pulsed Nd:YAG laser (Big Sky, Model CFR 400) operating at 10 Hz, with pulse duration 8 ns, beam diameter 7 mm, and the full angle divergence 1.0 m rad, was directed into the optical fiber by a 532/1064-nm beam splitter and a 532-nm dichroic mirror. A specially coated 45-degree dichroic mirror, which reflects at 532 nm and transmits in wavelength ranges 180–510 and 550–1000 nm, was used to reflect the laser beam and to transmit the LIBS signal to the detection system. This simple design avoids any damage to the detector from the reflected laser light. To transmit sufficient laser energy below the damage threshold of the fiber optic cable, the laser beam was focused 5 mm away from the fiber tip via a 10-cm focal length lens. A cap with a 0.8-mm pinhole was placed at the fiber input end to avoid the possibility of damage to the core and cladding of the fiber. The laser beam transmitted through the optical fiber was collimated with a 10-cm focal length lens and then focused on the sample with a 5-cm focal length lens. The same lenses and optical fiber assembly were used to collect the emission from the LIP and the collimated emission was passed through the dichroic mirror and focused onto an optical fiber bundle with a 20-cm focal length lens. The fiber bundle, a round-to-slit type, consists of 78 fibers, each having a 100 μm diameter and a 0.16 numerical aperture. The slit-type end of the fiber bundle delivers the emission to the entrance slit of a 0.5-m focal length spectrometer (Model HR 460, JOBIN YVON-SPEX) equipped with a 2400-lines/mm grating blazed at 300 nm. An ICCD (Model ITF/CCD, Princeton Instruments) was used as the detector with its controller (Model ST 133, Princeton Instruments). A programmable pulse delay generator (MODEL PG-200, Princeton Instruments) was used to gate the ICCD and the data acquisition was under the control of a computer (Dell Dimension M 200a) running WinSpec/32 software (Princeton Instruments). Multiple (100) laser-shot spectra were stored in one file, where 50 such spectra were recorded for analysis to get an average area/intensity value for the spectral lines under investigation.

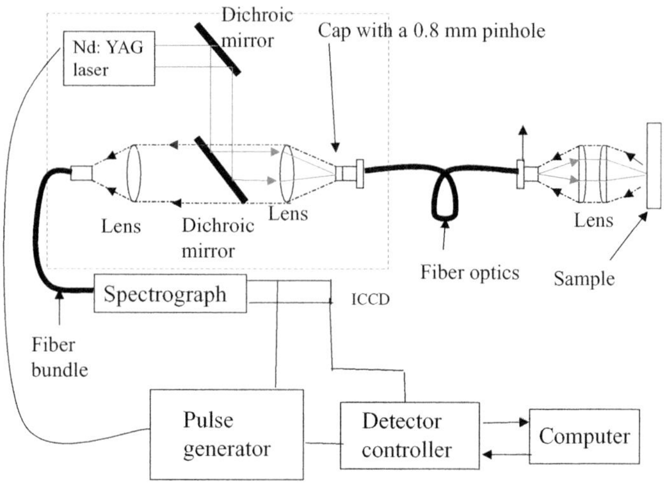

FIG. 4 Schematic diagram of fiber optic probe for remote analysis of a laser-induced breakdown spectroscopy (LIBS) signal from a sample. A similar system is used for recording LIBS from samples submerged under water or molten liquid samples in industries. In the molten metal case, optical fiber is passed through a ceramic tube and dry air, N_2, or argon gas is passed through it to create a bubble at the place of plasma formation. In some cases, fiber is kept close enough to the sample for creating plasma without a focusing lens [11].

LIBS spectra of different Al alloys recorded with an optical fiber probe were corrected for baseline spectral intensity by integrating the peak area under each line [11, 36]. The quantitative spectral analysis involves relating the spectral line intensity of an element in the plasma to the concentration of that element in the target. The most important minor elements in Al samples were analyzed, which included copper, magnesium, manganese, nickel, chromium, and iron. To optimize the signal for the quantitative analysis of these elements in the aluminum alloys, the LIBS spectra were recorded by changing the various experimental parameters (laser energy, sample surface, detector gain, gate delay, width, etc.).

4.2 Transmission property of optical fiber

The LIBS system used by Rai et al. [36] had a silica core/silica cladding multimode fiber (FG-I.0-UAT from ThorLabs Inc.). The stability of silica cladding allows the fiber to transmit high laser energy and the low OH silica core design provides superior UV transmission, required to transfer the LIBS signal. The length of the fiber was ~3 m, and it was fitted with SMA 905 stainless-steel fiber connectors (ThorLabs Inc.) at both ends. The fiber was polished with a 0.3-mm grain size aluminum oxide powder in the final step. The core and cladding diameters were kept as 1 and 1.25 mm, respectively, and its maximum power capability was ~5 GW/cm^2. The low numerical aperture of 0.16 enables the fiber to produce low-beam divergence and uniform spot size that facilitates focusing the beam after transmission through the fiber. A spherical planoconvex fused silica lens of 10 cm focal length was used to couple the laser beam into the fiber. With this lens, a 30-mJ laser beam causes breakdown in air, which is the maximum laser energy that can be transferred through the fiber. A metal cover with a 0.8-mm pinhole at the center was placed just in front of the fiber to avoid any damage to the boundary of the core cladding during alignment. The fiber was placed about 5 mm behind the focal point in the diverging beam, where only about 0.6–0.7 mm of the core diameter was illuminated. A simple calculation indicates that 30 mJ pulse energy with a spot size of 0.5 mm diameter would produce an energy density of ~2 GW/cm^2. This is lower than the damage threshold of the fiber. However, at this laser energy level, damage to the input surface of the fiber is still possible by randomly occurring hot spots in the laser profile. This system had an energy transmission efficiency of about 88%, which is fairly high. To improve the S/B ratio, various experimental parameters were tested and the optical fiber was damaged several times during these tests. Most of the damage occurred inside the fiber when the laser energy input was more than 20 mJ. Later, it was found that as long as the laser energy remains below ~20 mJ at the fiber input end, the fiber is not damaged and most of the experiments were performed by using laser energy below this threshold. The core cladding was also damaged due to breakdown several times and it occurred between 2 and 5 cm away from the fiber input end that is very likely at the location of the first reflection of the laser beam inside the fiber. Precautions are therefore necessary to avoid any damage in the optical fiber during the experiments. It was noted that while the laser power at the input end remained approximately constant with time (30–40 min), the laser power decreased slightly at the output end of the fiber. This suggests that the recording time for one set of data should be kept as short as possible if the laser energy is near threshold.

5 Portable LIBS devices

In recent years, miniaturizing LIBS instrumentation has become a necessity because of efforts to move LIBS systems for field applications and for at-site analysis, particularly of environmental samples [37]. This has been possible by using optical fiber to deliver a laser beam from a miniaturized laser system providing the radiation necessary for plasma production. LIBS instruments using fiber optic cables have gained popularity for local and remote sensing in hostile environments, such as urban and industrial dumps or heavily polluted zones [24, 38]. The fiber optics system is used both to deliver the laser radiation to the sample and to collect the emission signal from the LIP as discussed earlier, which allow in situ analysis of relatively inaccessible samples, such as painted surfaces and nuclear reactors. Barbini et al. [34] and Ciucci et al. [25] recently developed a mobile instrument to apply the LIBS technique in conjunction with light detection and ranging apparatus. The single-laser shot mobile device, equipped with the relevant data analysis software, was able to provide a real-time response regarding the presence of hazardous species (such as antimony, barium, copper, chromium, lead, and mercury) in a variety of polluted environmental solids (such as rock, soil, sand, and ash). An S/N ratio of 2 was obtained for mercury with a concentration as low as 80 ppb. This type of LIBS instrument is suitable for remote sensing and on-site analysis in some cases.

The fiber optic probes mentioned in previous sections are not truly field-portable LIBS instruments, due to the nonportability of the lasers employed, and the length of optical fibers is also not unlimited (several meters in the aforementioned cases) due to signal and laser radiation attenuation in the optical fibers. A real field-portable LIBS instrument can only be realized by using a battery power supply, optical fibers, and miniature laser. Such a device was first developed in the research group of Cremers at Los Alamos National Laboratory [23]. It had a weight of 14.6 kg and a compact size of $46 \times 33 \times 24\ cm^3$ to fit into a small suitcase. The handheld probe employed a compact, low-cost, passively Q-switched Nd:YAG laser to make it portable. The laser had a low pulse energy (15–20 mJ pulse at 1064 nm, 4–8 ns duration) and repetition rate (<1 Hz), but it had the ability to operate from 12 V DC batteries. A spark was produced on the sample by focusing the laser with a 50-mm focal length lens of 12 mm diameter. A fused silica fiber optic bundle of 2 m length was used to collect the emission from the LIP and transmit it to a 1/8-m spectrograph. The end of the fiber optic bundle was positioned 5 cm from the LIP. It was not necessary to focus the LIP light onto the fiber with a lens due to the already sufficient emission collected in this configuration. The spectrally resolved emission was recorded with a compact CCD system. A compact computer was used for data processing and storage. The performance of the portable LIBS device was compared with that of a laboratory-based system using lead-containing paint samples and soil samples containing barium, beryllium, lead, and strontium. The results were identical in all aspects, indicating that downsizing the instrument did not affect its analytical performance. The same samples were also analyzed with ICP-AES and portable X-ray fluorescence (XRF) and it was found that results were in agreement with each other.

A more compact portable LIBS instrument has been built in Winefordner's research group at the University of Florida [28] using rechargeable batteries. This device is useful for field application, where regular power supplies are not available. It consists of a Kigre Nd:YAG laser (1064 nm, 21 mJ, 3.6 ns duration, 1/3 Hz), spectrometer coupled with detector, computer, electronics, optics, and rechargeable battery, all arranged in a suitcase of $48.3 \times 33 \times 17.8\ cm^3$, with a total weight of 13.8 kg. An average incidence laser power density of $0.92\ GW/cm^2$ was achieved for the production of the LIP. A linear CCD (2046 pixels) was used as the detector having a spectral range of 339–462 nm. This system was optimized using the battery power supply, and after parametric studies of spatial development of plasma, lens-to-sample distance, and spatial filtering its performance was evaluated on paint, steel, and biological samples for the determination of lead, manganese, and calcium.

Portable LIBS instrumentation has shown advantages [9, 23] over a portable XRF system, which is currently the choice for many types of field screening measurements. In the case of LIBS, one may choose different analytical lines to avoid spectral interferences. A portable LIBS device provides analytical results 3–30 times faster than a portable XRF unit for homogeneous samples. The microprobe capability of LIBS makes it possible to analyze very small samples (e.g., lead in solder joint) and uneven surfaces (e.g., irregular rock surfaces and uneven ground surfaces). Finally, a LIBS probe can provide fiber optic delivery of the laser pulses to subsurface soil for remote, in situ monitoring, which is not possible with XRF. Nowadays, various kinds of portable and handheld LIBS systems are available commercially.

5.1 Handheld LIBS

Fast technological progress within the last decade has enabled the development of sophisticated self-contained handheld LIBS instruments that are much better than the portable LIBS systems discussed earlier [39–41]. Several commercial handheld LIBS instruments have become available that have capabilities and features comparable to the traditional bench-top instruments, including resolution to 0.1 nm, spectral ranges from vacuum UV to near-IR, argon purging, and rastering. Such handheld LIBS instruments are a boon in respect of their portability, easy operation, and application in the field. The SciAps Z500 handheld LIBS analyzer [42] has multiple features that make it particularly useful for various applications.

This handheld LIBS analyzer (Fig. 5) provides a spectral range from 185 to 865 nm, argon purging, and elemental mapping capability. LIBS measurements using this system are carried out by placing the nose of the instrument against a sample and then pulling the trigger. It has various user-selectable settings that can be used to determine the way in which the data are collected, analyzed, and displayed. This system consists of a 5-mJ diode-pumped solid-state laser (either 1534 nm, 6 ns full width at half maximum [FWHM] pulse duration, or 1064 nm, 2 ns FWHM pulse duration) operating at 10 Hz. The internal spectrometer uses time-gated CCD

FIG. 5 Front and back view of the SciAps Z-500 handheld laser-induced breakdown spectroscopy analyzer [42].

detectors. This unit is equipped with an XYZ stage to enable beam rastering for better averaging of data in nonhomogeneous samples. The raster pattern and spacing can be decided by the user as per their requirement. The instrument is purged by argon using small replaceable 21-g cartridges that fit in the handle of the system. Argon pressure is monitored before every measurement where a microsolenoid valve controls the flow. The gas is delivered directly to the area of the sample surface where the LIBS plasma is formed. A miniature camera in the nose of the instrument enables sample viewing, aiming, and documentation-archiving of the samples [42]. This handheld LIBS system is powered by one or two hot-swappable batteries, which provides 8 h of operation. Additional useful features of the analyzer include the Android operating system, Wi-Fi, Bluetooth, USB, and GPS. Resulting data can be stored locally on the unit or can be emailed to remote locations via the Android share function using one of the available wireless connections. These facilities in the system make this instrument very useful and user friendly. This system has been successfully used for the analysis of geochemical samples [42].

6 Signal enhancement in LIBS

It has been realized that poor detection limit is the most serious limitation of the LIBS technique in comparison to other analytical techniques. The main reason for this phenomenon includes the fluctuations in laser energy, discrepancy in the resolution of spectrometers, uneven surfaces of samples, matrix effect, inhomogeneity, and instability of LIP along with influence of the external environment [43, 44]. It has been realized that accuracy and precision in LIBS quantitative and qualitative analysis can be improved by a certain modification in the conventional LIBS instruments and the advanced data processing methods [45, 46]. Several research groups have made modifications to the experimental setup to improve the LIBS detection limits with the help of different techniques such as double-pulse (DP) excitation [47–49], magnetic and/or spatial confinement [50, 51], spark discharge [52, 53], nanoparticle (NP)-enhanced laser-induced breakdown spectroscopy (NELIBS) [54] as well as surface-enhanced laser-induced breakdown spectroscopy (SENLIBS) [55] and other methods.

6.1 Double-pulse LIBS

DP-LIBS uses a pair of laser pulses to ablate the material and further excite the resulting plasma to enhance the sensitivity of LIBS and has been found most promising. This technique is known as dual-pulse LIBS, or repetitive spark pair (RSP), or DP excitation (Fig. 6). In some cases the dual pulse is delivered by a unique laser [56, 57], while in other experiments, use of two different lasers has been reported [58–62].

The second technique is more flexible with spatial arrangement of the two laser beams, their pulse energies, and the time delay between two pulses. The use of a single laser, however, makes the system more compact and avoids the problems of alignment between the two laser pulses, ensuring better reproducibility. In 1997 Pichahchy et al. [56] applied the RSP technique to analyze the metals under water. The use of two laser pulses separated in time by tens of microseconds produced hotter plasmas (9000 K compared to 3000 K for a single spark) and resulted in better detection limits for the elements. The reason for the strong excitation produced by the second pulse was attributed to the formation of a gas bubble in the water at the solid surface. The presence of a solid-gas interface would allow the interaction of the second pulse with the plasma in a way similar to the gaseous environment.

In 1998 St-Onge et al. [61] reported the study of parameters affecting the performance of dual-pulse LIBS on metal samples in air. They found that the volume of the emitting plasma increases under the effect of the second laser pulse resulting in signal enhancement. This is due to more uniform absorption of the second laser pulse, whose energy is then distributed over a larger volume. St-Onge et al. [62] also used a UV laser pulse to increase the ablation of the sample and an IR laser pulse to maximize the heating efficiency. In this configuration a significant signal enhancement was noted, the extent of which varied depending on the ionization state and energy levels giving rise to the spectral line of

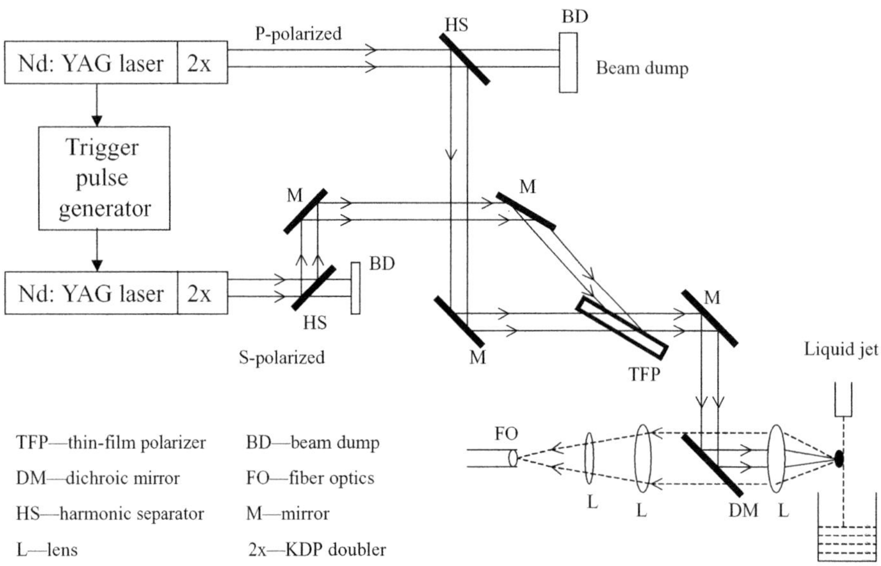

FIG. 6 Schematic diagram of an optical system for recording the laser-induced breakdown spectroscopy spectrum of a liquid sample (jet) in double laser pulse excitation mode. The rest of the system such as dispersing device, detectors, and data acquisition components is the same as shown in Fig. 1 [63].

interest. A correlation has been established between the observed increase in intensity and the theoretical increase expected as a result of the higher plasma temperature generated by a combination of the UV-IR pulses. Enhancement in the signal was found greater than predicted by the increased temperature. This shows that an increase in plasma volume also contributes to the enhancement in intensity of the emission. In contrast to the aforementioned studies, where the plasma obtained by the first ablating pulse is reheated by a second pulse, Stratis et al. [60] used a prepulse parallel to the sample surface and focused it to form air plasma, followed by a second ablating pulse perpendicular to the sample surface and delayed in time by a few microseconds. In this case, increase in intensity of the spectral line was correlated with an increased mass ablation. They simultaneously measured the time-resolved, spatially integrated emission intensity from two directions—perpendicular to the target surface and parallel to the target surface—resulting in a slight difference, which indicates the importance of the collection geometry in the LIBS measurements. Rai et al. [63] used two lasers (Nd:YAG) for LIBS experiments and the spatially integrated emission in the direction opposite to the direction of the laser beams was collected (Fig. 6). LIBS signal was enhanced by more than six times when the time separation between two laser pulses was ~2–3 μs. Various DP-LIBS experiments have been performed using a combination of laser parameters such as its time duration, wavelengths, and energy ratio [64]. Finally, a large number of experimental results have proved that DP-LIBS is an effective technique to achieve better analytical performance. The collinear DP-LIBS provides limit of detection for Mg, Cr, B, and Li as 0.06, 0.12, 0.8, and 0.8 ppm, respectively [63, 65]. The combination of femtosecond (fs) UV laser and nanosecond UV laser in DP-LIBS in reheating configuration on single crystalline silicon provided a 360-fold increase in the emission enhancement and a 50-fold enhancement in limit of detection (LOD) [66].

6.1.1 *Effect of delay between laser pulses*

The delay between laser pulses plays an important role in the performance of DP-LIBS. The physical processes in the plasma, related material ablation from the target, and the emission from plasma are all found to be dependent on the delay between the laser pulses. A brief discussion about the effect of various time delays ranging from picoseconds (ps) to microseconds (μs) is provided next.

The effect of a 1–1000-ps delay time between fs laser pulses on the ion yield, crater volume, and ablation depth has been investigated by Donnelly et al. [67]. More than a two times increase in the ablation depth and volume was noted for the short time delay in comparison to single-pulse (SP) LIBS. The volume and depth start decreasing after a 10-ps delay and reach the level of SP-LIBS for a ≥100-ps delay. However, the ion yield reaches maximum near an ~100-ps delay. A three to five times signal enhancement was noted by Somekawa et al. [68] in a collinear DP-LIBS configuration having two polarizers and a 50–150-ps delay on a copper target.

Weidman et al. [69] used a 1.064/10.6-μm crossed beam DP-LIBS scheme to enhance the signal intensity and S/R for the organic material. They reported a 230, 20, and 7 times enhancement in the line intensity of atomic carbon, molecular CN, and C_2, respectively, at 100-ns interpulse delay between lasers for polystyrene. Coons et al. [70] used an Nd:YAG and CO_2 laser in the crossed beam configuration for the analysis of an Al target having Mg impurity. They reported an ~100 times enhancement in the emission from Al II (281.6 nm) and a relatively lower enhancement of 5–15 times for neutral Al lines (309.3) at a 750-ns interpulse delay. Another experiment [71] reported that enhancement factor was higher for ionized species at 50–300 times, whereas for neutral atomic emission it was 25–60 times.

Rai et al. [63] studied the signal enhancement in collinear DP-LIBS and reported a $\geq$six times enhancement in the Mg ion emission at an interpulse delay of ~2–3 μs. Enhancement in the emission intensity was slightly less, whereas interpulse delay was slightly more in the case of neutral Mg emission. Hai et al. [72] reported an ~6.5 times enhancement in emission from an Mo line at 550.65 nm compared to SP-LIBS when the interpulse separation was ~1.5 μs. Ahmed and Baig [73] used the combination of fundamental and second harmonics of an Nd: YAG laser in collinear DP-LIBS configuration for the study of emission from magnesium plasma. They observed maximum emission intensity for an interpulse delay of ~5 μs. It was demonstrated that optimum interpulse delays between two lasers are independent of the wavelength of lasers.

Finally, it seems that ps and ns interpulse separation provided more enhancements in comparison to μs delay in most cases.

6.2 Other DP-LIBS

Smith et al. [74] used a different LIBS technique to improve the detection limits by applying selective elemental excitation using a tunable diode laser to the LIP. Tunable diode laser-induced atomic fluorescence was used for selective isotope detection of uranium-containing samples. To detect the fluorescence signal, two techniques were employed: (1) fast wavelength scanning of the diode laser during the lifetime of the plasma produced by each shot of the ablating laser, and (2) time-integrated measurement with the diode laser wavelength fixed at the isotope line center. The optimal experimental conditions were found by means of a systematic scanning of the pressure of argon in the experimental chamber. The LOD in the optimal conditions was of the order of 0.6 ppm. Many other combinations of LIBS and laser-induced fluorescence have been reported [75–77] to enhance the performance of LIBS as an extremely sensitive technique.

In some DP-LIBS systems a second laser is replaced by other types of excitation sources. Such DP-LIBS systems are also important for different applications and are described in brief next.

6.2.1 Spark discharge LIBS

Another technique known as spark discharge LIBS (SD-LIBS) uses conventional LIBS instruments with a fast high-voltage pulse discharge circuit. Spark-assisted LIBS can be operated at ambient pressure and requires a capacitor and high-voltage power supply. In this case a capacitor charged to high voltage is discharged through the LIP. It creates less damage on the surface of the sample [78]. The effect of capacitance and voltage on the optical emission of soil plasma with a nanosecond pulse discharge circuit was investigated in detail [79]. The large values of capacitance led to an increase in both peak intensity and noise level. A similar observation was noted with an increase in discharge voltage. Different types of circuits were used to create the spark and were used for the analysis of soil and Si [52, 80–83]. The spark created by a 6-nF capacitor charged at 11 kV high voltage excites all the weak lines in LIBS (both metal and nonmetal in soil samples) with an enhancement factor in the signal by ~223 times. The LOD for Pb, Mg, and Sn in the soil was found to be 1.5, 34, and 0.16 μg/g, respectively. The enhancement of signal in laser ablation-spark-induced breakdown spectroscopy (same as SD-LIBS) makes possible the detection of As and Co, which is not visible in SP-LIBS [52].

6.2.2 Glow discharge and microwave LIBS

Some other excitation sources such as glow discharge [84] and microwave were used as additional excitation sources with LIBS [81]. Glow discharge (steady state and pulsed) was used to excite the material ablated by a laser system [85, 86]. An enhancement factor of ~75 was reported for the pure sample of Cu in comparison with SP-LIBS experiments in identical conditions. This was due to reexcitation of plasma by synchronous microsecond-pulsed glow discharge. The only disadvantage of this technique is the requirement of a vacuum chamber to produce the glow discharge.

Microwave LIBS was also used successfully to enhance emission intensity by elongating the lifetime of laser-ablated plasma. Around 15 times enhancement in the peak intensity of Pb was obtained with a 2-kW microwave power injection [87]. An enhancement of ~23 times was noted in the LIBS signal by the use of microwave radiation for copper in soil samples [88].

6.3 Magnetic and spatially confined LIBS

A cost-effective method of enhancing the sensitivity of LIBS is the combination of magnetic and spatial confinement of LIBS plasma. Normally, plasma is confined by the magnetic field when it is propagating across the magnetic field. This is because no charged particle can cross the lines of the magnetic field. This process affects the emission characteristics of ions and neutrals in the plasma. The enhancement in emission intensity in magnetically confined LIBS may be caused either by increased electron impact excitation or by radiative recombination [14, 89]. Rai et al. [14] investigated the effect of a steady magnetic field of ~0.5 T in the air. They reported an enhancement factor of ~2 and 1.5 for solid and liquid samples, respectively. Many authors have studied the emission characteristics of plasma having different elements in various configurations of magnetic fields [90, 91].

Spatially confined LIBS is a simple and cost-effective method used for improving the detection sensitivity when the laser plasma is confined in a small hemispherical or cylindrical cavity. Normally, a shockwave is produced by expanding plasma when a sample is ablated by a laser pulse. The expanding shockwave at very high speed compresses the plasma after reflection from the wall. This effect compresses the plasma at the center resulting in a highly increased collision among the particles (electrons, ions, and neutrals) within the plasma [92–94]. Finally, it increases the number of atoms in excited states that enhances the emission intensity from the plasma. Various types of cavity have been studied to optimize its dimension for better performance of LIBS [95–98]. Spatial confinement of plasma in a hemispherical aluminum cavity (diameter ~11.1 mm) provided an enhancement factor of ~11–12. The combined effect of spatial and magnetic confinement provided an enhancement of ~22 to ~24 in the emission intensity of Co and Cr lines, respectively [98, 99]. Similarly, a combination of spatial confinement with DP-LIBS [100, 101] and spark discharge [102] was also investigated.

6.4 Nanoparticle-enhanced LIBS

Metal nanostructures, particularly the noble metals (Au and Ag) and their plasmonic properties, have been widely used in surface-enhanced spectroscopy [103, 104]. During the interaction of an electromagnetic field with a metal nanostructure, the free electrons of NPs oscillate under the effect of the electric field. When the frequency of the electromagnetic radiation coincides with the plasma frequency of free electrons in the NPs, a resonance takes place that is popularly known as localized surface plasmon resonance, which enhances the absorption of radiation by the NPs [105]. Recently, NPs have been used for enhancing the performance of material ablation by lasers, particularly LIBS. The use of NPs for enhancing the LIBS signal and its analytical performance is called NELIBS [106]. This technique is found beneficial particularly for the study of transparent samples such as glasses and gems. In this case, NPs are deposited on the transparent sample's surface. To avoid direct laser interaction with the glass sample, a laser beam in resonance with the surface plasmon of the NPs was used. In such a situation, most of the laser energy is absorbed by the NPs. The strong absorption of laser by the NPs induces breakdown on the NP layer deposited on the sample surface. In this case, direct interaction of laser with sample is avoided.

As a consequence, NP explosion takes place and a few superficial layers of the sample are transported in the plasma. Finally, the plasma emission has contribution from the NPs and the elements present in the samples. Most of the emissions in NELIBS come from a zone of the plasma some tens of microns closer to the target in comparison to normal LIBS that has a large plasma plume [107].

The analysis of metallic alloys by NELIBS was reported for the first time in 2013 [108]. Analytical performance of NELIBS showed an enhancement in the signal by two orders of magnitude on conductive samples that allows detection of test elements up to the sub-ppm level. The use of different plasmonic NPs was also reported. Only 0.004% contribution from NPs was observed in the total ablated material plasma. NELIBS has been used for the detection of silver in a liquid solution of silver nitrate with LOD below the ppb level [109]. This technique was applied to the elemental analysis of glasses and gemstones successfully without any damage to the samples, which is not possible with direct laser interaction. The strongest point of the NELIBS technique is the possibility of performing single-shot elemental analysis with sensitivity better than that obtained with traditional LIBS. This technique can be useful for forensic and biomedical applications. The detailed physical process, sample preparation, and recent experimental results related to NELIBS have been reviewed recently [107].

6.5 Surface-enhanced LIBS

Another technique known as surfaceenhanced laser-induced breakdown spectroscopy (SENLIBS) enhances the LIBS signal with the help of a metallic target [110]. SENLIBS has been used for liquid and powder samples on the surface of a metallic substrate. In this case, plasma produced on the substrate surface also indirectly ablates the sample material. Finally, plasma is composed of a mixture of species from the sample and the metallic substrate. SENLIBS is able to improve the analytical performance of LIBS, particularly the LOD and the matrix effect. SENLIBS has been found to be a simple and effective enhancement technique that has been applied successfully for the analysis of many types of samples, including oil [111, 112], sunscreen [113], heavy metals in liquids [114, 115], as well as powders [116]. The LOD of heavy elements such as Ni, Cr, and Cd in solution deposited on a polished aluminum substrate was reported as 22, 19, and 184 μg/L, respectively. However, the laser-pretreated metallic microstructures provided further improvement in LOD to sub-μg/L level for Cr and Ni and μg/L for Cd [117]. Recently, the SENLIBS technique was used to enhance emission intensity and stabilize plasma for direct analysis of silica particle flow with the help of a metallic target [118]. In this case, a metallic target was placed in the path of a silica particle flow to provide a target surface to fix the laser ablation position and to enhance the plasma emission intensity. LIBS signal was recorded in the absence (LIBS) and presence (SENLIBS) of the metallic target. It was noted that sampling rate was improved in the case of SENLIBS and the metallic target showed a lower breakdown threshold compared to silica particle flow. Signal intensity was enhanced by more than three times for the silicon emission line at 288.16 nm. Relative standard deviation (RSD) for pulse-to-pulse signal decreased from 116% to 48%. An increase in S/N ratio was found from 9.89 to 18.63. Finally, these results indicate the applicability of SENLIBS for elemental analysis in different types of samples.

6.6 Filament-induced breakdown spectroscopy

Fs laser filamentation in air occurs due to the balance between the self-focusing of laser, plasma defocusing, and diffraction that allows the delivery of light intensity (10^{13}–10^{14} W/cm^2) [119] to the target kept at a remote distance. Filaments have been found propagating over a distance on the order of kilometers. It can propagate through the turbulent environment. Ns lasers cannot propagate more than ~30 m in air due to diffraction, which limits the use of LIBS to standoff distances of ~30 m. However, filament-induced breakdown spectroscopy (FIBS) or remote FIBS have been demonstrated up to ~90 m [120, 121]. It is beneficial in many circumstances, where the target samples are located in difficult and hazardous environments such as biochemical agents in the atmosphere, nuclear-polluted industrial or urban sites, and molten alloys in an industrial site. Recently, single-shot FIBS has been demonstrated to detect the uranium from uranium oxides [122].

7 Applications of LIBS

LIBS has been proved to be a versatile technique for the detection and identification of elements in a variety of samples that cannot be easily analyzed by other spectroscopic methods. This capability of LIBS has been utilized in various fields ranging from laboratory to field applications. Some of the applications of LIBS are described in brief next to show their importance.

7.1 Environmental monitoring

7.1.1 Off-gas emission

The detection of hazardous and toxic trace elements in the off-gas from waste processing systems is very important for public health. LIBS has been used for in situ off-gas monitoring by focusing the laser beam on the gas stream through a window and collecting the optical emission through an optical fiber. Neuhauser et al. [123] tested an online lead (Pb) aerosol detection system with aerosol diameters ranging between 10 and 800 nm where a detection limit of 155 μg m^{-3} has been achieved. LIBS has also been demonstrated as a process monitor and control tool for waste remediation [15]. The toxic metals from three plasma torch test facilities were monitored and it was found that LIBS can be integrated with a torch-control system to minimize toxic metal emission during plasma torch waste remediation. The possibility of using metal hydride to calibrate metals in off-gas emission was also investigated [124] by using a static sample cell to perform LIBS measurements where the signal was found to be affected by gas composition, gas pressure, and laser intensity.

The use of LIBS as a continuous emission monitor (CEM) requires the quantitative trace-level determination of toxic metals. A system has been developed to monitor the concentration of selected toxic metals in near real time [16]. The concentrations of Be and Cr were measured at all the tested metal levels, while that of Cd was measured during medium- as well as high-metal feed tests and the concentration of Pb was measured only at high concentrations. It was concluded that the LIBS system can be used as a CEM to monitor mainly the concentrations of Be, Cr, and Cd, but further improvements in the sensitivity of this system are important for the monitoring of Pb, Hg, As, and Sb.

7.1.2 Study of soil, concrete, and paint

Detection of contaminated soil and concrete is an important area of environmental applications of LIBS. Yamamoto et al. [23] used a portable LIBS system to detect toxic metals in soil and detection limits of Ba, Be, Pb, and Sr were found to be 265, 9.3, 298, and 42 ppm, respectively. Cremers et al. [38] detected Ba and Cr in soil using an optical fiber probe for remote operation with LODs of 26 and 50 ppm for Ba and Cr, respectively. The effect of a matrix was also studied in a soil sample. The LODs for Pb and Ba in a sand matrix were found to be 17 and 76 ppm (by weight), respectively, with a precision of 7% RSD, whereas those in the soil were 112 and 63 ppm, respectively, with 10% RSD. The LIBS signal was found to be affected by chemical speciation as well as matrix composition and its accuracy could be degraded if calibrations were not matrix specific.

Pakhomov et al. [22] have applied LIBS for the detection of Pb in contaminated concrete. A time-resolved LIBS spectrum was recorded for the quantitative measurement of the Pb content in concrete. Pb calibrations were obtained by using the ratio of the integrated emission of a lead line (405.78 nm) and that of an oxygen line (407.59 nm). It was found that the absolute Pb signal was independent of the laser pulse energy for laser energy between 250 and 400 mJ. The presence of Pb in the paint is a potential health threat, especially to children. Yamamoto et al. [23] successfully demonstrated the feasibility of using LIBS to determine Pb in the paint surface.

7.1.3 *Study of radioactive elements*

LIBS has also been used to monitor the level of radioactive elements in a process stream. Watcher and Cremers [125] found a detection limit of 100 ppm for uranium in solution. LIBS is preferable to other radiological measurements because a nuclear detector may not be able to differentiate the radionuclides U, Pu, and Np. The LIBS spectra of U, Pu, and Np were recorded in a globe box and the emission lines suitable for the detection of these radioactive elements were identified by Singh et al. [126]. The preliminary studies show that LIBS is suitable for the measurement of radioactive elements in waste streams. LIBS has also been used as a tool for the detection of radiation embrittlement [9] in a nuclear power plant by determining the copper concentration in A533b steel. As copper is a key impurity contributing to radiation embrittlement, the Cu concentration in the steel may be an indicator of radiation embrittlement and expected material lifetime.

7.2 Industrial applications

Nowadays, all the metal-producing industries are facing a major challenge of increasing productivity at reduced cost and maximizing the benefits from existing equipment. During refining, it is critical that operating parameters be adjusted and controlled so that the chemistry of the molten metal remains within predetermined limits. LIBS has been successfully used to obtain the composition of alloys along with their quantitative analysis in solid as well as in molten states [36, 127, 128]. The analyte lines of Cu, Cr, Mn, Fe, and Zn were used to obtain the calibration curve for their quantitative analysis. This type of analysis was performed by recording the spectra of alloy samples in the laboratory, but LIBS has been found most suitable for field-based industrial applications, which include real-time and online analysis of molten material for process control and monitoring. Many groups [129–133] have used a LIBS probe that uses one optical fiber for delivering the laser pulses to the target at a remote place for producing a microplasma as well as for collecting the resulting radiation from the LIP for quantitative elemental analysis.

7.3 LIBS in nuclear fusion research

The tokamak is a toroidal plasma machine used for nuclear fusion research. It produces pulsed hot plasma for comparatively longer times. During its operation it produces high-temperature fluxes and high-energy particle loads that are transported to the wall of the tokamak/plasma-facing components (PFCs) in the diverter area inside the tokamak [134, 135]. The interaction between plasma and wall/PFC plays an important role in nuclear fusion reactors that results in material erosion, redeposition of material, and the retention of fuel gas atoms as deuterium and tritium between the deposited layers of materials. The control of tritium in the tokamak vessel is crucial, which is important for the design and operation of the International Thermonuclear Experimental Reactor (ITER) and other nuclear fusion devices [136]. For safety reasons, the maximum quantity of tritium in the ITER vessel is limited to 700 g [137], which requires an in situ monitoring of tritium retention in the PFCs. Therefore it is important to better understand the interaction of high flux of deuterium/tritium plasma with the wall/PFC. Tungsten or lithiated tungsten is being considered as a suitable PFC material. LIBS was proposed as an in situ technique for measuring the fuel retention, local growth, and composition of the deposited layers on PFCs [138]. Most fusion-related LIBS measurements are carried out by using a single-pulse laser with ns pulse duration. Ns DP-LIBS improved the S/N ratio during the study of hydrogen isotope detection. Some studies show that shorter pulse duration lasers are advantageous for determining the surface characteristics and retained fuel depth profiles of the analyzed samples. The comparison of LIBS results obtained using ns and ps lasers in ITER-relevant conditions have been studied and reported recently [139]. It was found that in the case of the ps laser, an acceptable accuracy of the detection of deuterium was possible at considerably lower values of laser fluence. LIBS has also been applied for the analysis of ITER-like deposits consisting of a mixture of W/Al/C and D (Al as substitute for Be) for elemental composition and depth profile analysis under different laser energy density conditions. They also show that LIBS can provide a simple method to discriminate in situ the codeposited layers from the substrate. Finally, more experiments and analyses are needed in this direction to optimize the method, particularly to deliver the concentration of the retained fuel. Even the combination of other laser-based diagnostics such as laser-induced ablation spectroscopy and laser-induced desorption spectroscopy could provide a powerful tool for elemental composition and depth profile as well as for fuel retention analysis in the codeposited layers on substrate [140].

7.4 LIBS in space research

7.4.1 Rocket engine health monitor

Detection and characterization of metallic species in the exhaust plume of hydrocarbon-fueled rocket engines can indicate the onset of wear and/or corrosion of metal in the rocket engine. This information on engine wear obtained during engine operation is very useful, allowing the possibility of engine shutdown before any catastrophic failure. It has been observed that catastrophic engine failure is generally preceded by a bright optical emission, which results from the erosion of metal from the engine parts. This is because of high temperature in the rocket plume (~2000 K), which partially vaporizes and atomizes the metal species, leading to atomic emission in the near UV and visible region (300–760 nm). The performance of LIBS was evaluated by Rai et al. [141] in detecting the trace of elements in the fuel plume of a hybrid rocket engine simulator at Stennis Space Center, USA. Copper wire was inserted in the ignition chamber of the engine and its vaporized trace was recorded in the rocket plume outside the exit nozzle. The trace of copper was recorded near the nozzle exit during an initial fraction of a second when the burnt fuel plume started building up. However, it decreased when the plume attained its full length, high temperature, and high speed. It was interesting to note that copper emission was observed throughout the plume away from the exit nozzle as well as in the luminous zone. This observation was attributed to better mixing of the metal vapor (away from the exit nozzle) along with a decrease in background emission (due to luminous zone). It was found that the measurements made away from the luminous part of the plume could provide more meaningful information about the health of the rocket engine.

7.4.2 Probe for Mars expedition

Cremers et al. [142] evaluated LIBS for its use on the lander and rover for the Mars expedition. The main interest was the use of LIBS for standoff measurements of geological samples up to 20 m from the instrument. The objective was to develop a very compact instrument operating at a remote distance from the target to detect at least 10 species in the rocks with detection limits <100 ppm, including Ba, Li, Rb, and Sr with detection limits <20 ppm. The ability to measure these separately in dust and pristine rocks was required. Minor and trace element compositions are also important in determining the provenance of rocks and dusts. In an experiment under simulated Martian atmospheric conditions (5–12 mbar CO_2), it was noted that bulk matrix affected the calibration for Sr. Accuracy and precision were obtained in the detection of various other elements. Another project called MALIS (Mars elemental analysis by laser-induced breakdown spectroscopy) was developed to demonstrate LIBS capability in Martian atmospheric conditions [143]. The Mars 2020 mission will launch a rover that will land on and explore the surface of Mars between 30°N and 30°S latitude and below 1.0 km elevation as part of the NASA Mars exploration program. The SuperCam instrument suite containing a LIBS system with various other systems such as UV, visible, and IR imaging systems will investigate the soil and rocks on the Mars surface [144]. This SuperCam also has a Mars microphone system to record the shockwave (acoustic signal) generated by laser rock interaction on Mars. This system has also been tested successfully in a controlled Martian environment [145].

8 Analytical performance of LIBS

The observed analytical figures of merit (precision, accuracy, and LOD) in LIBS experiments are inferior to those for other atomic spectrometric techniques such as ICP-AES and ICP-MS. The main reason for this deficiency is the multitude of experimental parameters that influence the analytical signal. These factors include laser wavelength, laser power, incidence angle, pulse-to-pulse variation, beam profile, beam shape, pulse duration, sample matrix, freshness of surfaces for solid samples, purging gas, ambient pressure, etc. Even the experimental arrangement and sampling geometry affect the LIBS measurements significantly [146]. These factors have to be identified and optimized for better analytical performance. Spatial and temporal dependence of emission signals from the LIP have been studied by several research groups [147, 148]. It has been realized that the investigation of methods for measuring relative mass removal in the LIP would probably continue for a long time and will be valuable, not only for LIBS experiments, but also for LIP-ICP-AES and LIP-ICP-MS experiments. It is expected that the fundamental studies of ablation and the excitation processes in the LIP using different wavelengths and ultrashort pulses would enhance the analytical capabilities of LIBS. A good review of these aspects has been published by Rusak et al. [7].

The major limitations of LIBS for practical applications result from self-absorption, line broadening, and the high intensity of the background continuum along with strong matrix effects [149]. Some of these limitations can be minimized or avoided by working in a controlled atmosphere and using time-resolved spectroscopic measurements or time-integrated and spatially resolved measurement techniques. In time-resolved spectroscopy, the temporal evolution of the plasma is obtained by recording the plasma emission spectra at various delay times. The LIBS spectra of magnesium in a liquid matrix were reported by Rai et al. [150] who found that 500 ns after the laser irradiation, the observed spectrum consisted of continuum and ion emission lines, as the plasma temperature was high, but after 10 μs the intensity of the continuum and ion lines decreased, and that of lines due to neutral atoms increased as a result of electron–ion recombination.

The purpose of time-integrated and spatially resolved spectroscopy is to measure the emission from the LIP at different positions of the plasma. Lee et al. [12, 21, 151] carried out experiments on copper and lead using ArF, XeCI, and Nd:YAG lasers. They found that the plasma consisted of two distinct regions when the ambient pressure was reduced below 50 Torr in air or argon atmospheres. The region near the target surface referred to as the inner sphere plasma emitted a strong signal of copper ions and continuum background. The other region referred to as the outer sphere plasma surrounded the inner sphere plasma and emitted blue–green copper atomic lines with a relatively low background continuum and without ion lines. They reported LOD values in the range of several ppm to hundreds of ppb in solid samples, while the RSD values varied from a few percent to 80%. Wachter and Cremers [53] examined the effects of laser pulse repetition rate, detector gating, and number of averaged laser shots on precision. It was found that precision increased with repetition rate and total number of laser pulses averaged, but was independent of the gating parameters. It was also found that the RSD decreased from 13.3% for 50 laser shots to 1.8% for 1600 laser shots. The freshness of the sample surface also affects precision and the lowest RSD has been obtained when each shot samples a totally new portion of the material [152]. Eppler et al. [153] found an increase in precision using a cylindrical lens instead of a spherical lens, but the RSD was independent of the choice of lens. The reduction in RSD was attributed to the greater amount of material sampled by the cylindrical lens. Castle et al. [27] systematically studied variables that influence the precision of LIBS measurements with special emphasis on the effect of temporal development of the emission, sample translational velocity, number of spectra accumulated, laser pulse stability, detector gate delay, surface roughness, and use of background correction.

9 Conclusion

LIBS has been found to be an attractive candidate for elemental analysis in samples present in any form: solid, liquid, gas, and aerosol. It has shown its peculiar capability such as either no or little sample preparation, compact instrumentation, as well as remote detection. Developments of LIBS techniques have been very rapid in recent years. Nowadays, commercial portable as well as handheld LIBS systems are available in the market suitable for field application. But still SP-LIBS suffers from disadvantages of low sensitivity and high limit of detection. Various new approaches have been applied to improve the sensitivity and accuracy of conventional LIBS such as DP-LIBS, spatial and/or magnetically confined LIBS, spark discharge, glow and microwave discharge, NELIBS, as well as SEN-LIBS, etc. Out of these techniques DP-LIBS is a widely used method with significant enhancement in its sensitivity. With the development of new techniques, even DP-LIBS can be considered as a portable LIBS system for future applications. Other methods of signal enhancement in LIBS have been investigated by many researchers, but the mechanisms of signal enhancement are still not very clear and in some cases experimental setup becomes complicated and costly. To further improve the sensitivity of LIBS systems, more work is needed in the field of laser, material, and plasma interaction. New technological developments can help to make more sensitive, compact, and cost-effective LIBS instruments that will be more useful for in situ analysis in field applications instead of laboratory-based experiments. All the applications discussed in this chapter need a sensitive LIBS system for better performance in elemental analysis. Another important factor in LIBS instrumentation is chemometrics, which is a data analysis technique. Much work is needed in this direction to reduce the effect of laser and plasma instability observed in the form of standard deviation in data points. The vast applications of LIBS in various fields as discussed here and the progress in improving the performance of LIBS indicate that LIBS can be a promising tool for quantitative and qualitative analysis of trace elements in the near future.

References

[1] F. Brech, L. Cross, Optical microemission stimulated by a ruby maser, Appl. Spectrosc. 16 (1962) 59.

[2] E.R. Runge, R.W. Minck, F.R. Bryan, Spectrochemical analysis using a pulsed laser source, Spectrochim. Acta B 20 (1964) 733.

[3] L.J. Radziemski, D.A. Cremers, Laser-Induced Plasma and Applications, Marcel Dekker, New York, 1989, pp. 295–325.

[4] L. Moenke-Blankenburg, in: J.D. Winefordner, I.M. Kolthoff (Eds.), Laser Microanalysis, John Wiley & Sons, New York, 1989.

[5] L. Moenke-Blankenburg, in: J. Sneddon, T.L. Thiem, Y.I. Lee (Eds.), Laser in Analytical Atomic Spectroscopy, Wiley-VCH, New York, 1997, , pp. 125–195.

[6] S.A. Drake, J.F. Tyson, Interaction of laser radiation with solid materials and its significance to analytical spectrometry: a review, J. Anal. At. Spectrom 8 (1993) 145.

[7] D.A. Rusak, B.C. Castle, B.W. Smith, J.D. Winefordner, Fundamentals and applications of laser-induced breakdown spectroscopy, Crit. Rev. Anal. Chem. 27 (1997) 257.

[8] K. Song, Y.I. Lee, J. Sneddon, Applications of laser-induced breakdown spectrometry, Appl. Spectrosc. 32 (1997) 183.

[9] F.Y. Yueh, J.P. Singh, H. Zhang, in: R.A. Meyers (Ed.), Encyclopedia of Analytical Chemistry, Vol. 3, Wiley, New York, 2000, p. 2065.

[10] O. Samek, D.C.S. Beddows, J. Kaiser, S.V. Kukhlevsky, M. Liska, H.H. Telle, J. Young, Application of laser-induced breakdown spectroscopy to in situ analysis of liquid samples, Optim. Eng. 38 (2000) 2248.

[11] A.K. Rai, V.N. Rai, F.Y. Yueh, J.P. Singh, Trends in Applied Spectroscopy, Vol. 4, Research Trends, Trivandrum, India, 2002, p. 165.

[12] Y.I. Lee, K. Song, J. Sneddon, in: J. Sneddon, T.L. Thiem, Y.I. Lee (Eds.), Laser in Analytical Atomic Spectroscopy, Wiley-VCH, New York, 1977, p. 197.

[13] Y.I. Lee, Y.J. Yoo, J. Sneddon, Recent advances in laser-induced breakdown spectroscopy, Spectroscopy 13 (1998) 14.

[14] (a) V.N. Rai, A.K. Rai, F.Y. Yueh, J.P. Singh, Optical emission from laser-induced breakdown plasma of solid and liquid samples in the presence of a magnetic field, Appl. Optics 42 (2003) 2085. (b) F.Y. Yueh, V.N. Rai, J.P. Singh, H. Zhang, Paper # AIAA-2001-2933, in: 32nd AIAA Plasma dynamics and Laser Conference, 11–14 June (2001), Anaheim, CA, USA, 2001.

[15] J.P. Singh, F.Y. Yueh, H. Zhang, R.L. Cook, Study of laser-induced breakdown spectroscopy as a process monitor and control tool for hazardous waste remediation, Process. Control. Qual. 10 (1997) 247.

[16] H. Zhang, F.Y. Yueh, J.P. Singh, Laser-induced breakdown spectrometry as a multimetal continuous-emission monitor, Appl. Optics 38 (1999) 1459.

[17] X. Hou, P. Stchur, T. Sun, K.X. Yang, R.G. Michel, 37th Eastern analytical symposium and exposition, in: Paper # 37, Nov. 15-20, Somerset, NJ, 1998.

[18] L.J. Radziemski, T.R. Loree, D.A. Cremers, N.M. Hoffman, Time-resolved laser-induced breakdown spectrometry of aerosols, Anal. Chem. 55 (1983) 1246.

[19] D.A. Cremers, L.J. Radziemski, Detection of chlorine and fluorine in air by laser-induced breakdown spectrometry, Anal. Chem. 55 (1983) 1252.

[20] C.M. Davies, H.H. Telle, J.D. Montgomery, R.E. Corbett, Quantitative analysis using remote laser-induced breakdown spectroscopy (LIBS), Spectrochim. Acta B 50 (1995) 1059.

[21] Y.I. Lee, S.P. Sawan, T.L. Thiem, Y.Y. Teng, J. Sneddon, Interaction of a laser beam with metals. Part II: Spece-resolved studies of laser-ablated plasma emission, Appl. Spectrosc. 46 (1992) 436.

[22] A.V. Pakhomov, W. Nichols, J. Borysow, Laser-induced breakdown spectroscopy for detection of lead in concrete, Appl. Spectrosc. 50 (1996) 880.

[23] K.Y. Yamamoto, D.A. Cremers, M.J. Ferris, L.E. Foster, Detection of metals in the environment using a portable laser-induced breakdown spectroscopy instrument, Appl. Spectrosc. 50 (1996) 222.

[24] B.J. Marquardt, S.R. Goode, S.M. Angel, In situ determination of lead in paint by laser-induced breakdown

spectroscopy using a fiber-optic probe, Anal. Chem. 68 (1996) 977.

[25] A. Ciucci, V. Palleschi, S. Rastelli, R. Barbini, F. Colao, R. Fantoni, A. Palucci, S. Ribezzo, H.J.L. Van der Steen, Trace pollutants analysis in soil by a time-resolved laser-induced breakdown spectroscopy technique, Appl. Phys. B63 (1996) 185.

[26] R.E. Russo, X.L. Mao, W.T. Chan, M.F. Bryant, W. F. Kinard, Laser ablation sampling with inductively coupled plasma atomic emission spectrometry for the analysis of prototypical glasses, J. Anal. At. Spectrom 10 (1995) 295.

[27] B.C. Castle, K. Talabardon, B.W. Smith, J.D. Winefordner, Variables Influencing the Precision of Laser-Induced Breakdown Spectroscopy Measurements, Appl. Spectrosc. 52 (1998) 649.

[28] B.C. Castle, A.K. Knight, K. Visser, B.W. Smith, J.D. Winefordner, Battery powered laser-induced plasma spectrometer for elemental determinations, J. Anal. At. Spectrom 13 (1998) 589.

[29] H.E. Bauer, F. Leis, K. Niemax, Laser-induced breakdown spectrometry with an echelle spectrometer and intensified charge coupled device detection, Spectrochim. Acta B 53 (1998) 1815.

[30] V. Detalle, R. Heon, M. Sabsabi, L. St. Onge, An evaluation of a commercial Echelle spectrometer with intensified charge-coupled device detector for materials analysis by laser-induced plasma spectroscopy, Spectrochim. Acta B 56 (2001) 1011.

[31] C. Haisch, U. Panne, R. Niessner, Combination of an intensified charge coupled device with an echelle spectrograph for analysis of colloidal material by laser–induced plasma spectroscopy, Spectrochim. Acta B 53 (1998) 1657.

[32] S.R. Goode, S.L. Morgan, R. Hoskins, A. Oxsher, Identifying alloys by laser-induced breakdown spectroscopy with a time-resolved high resolution echelle spectrometer, J. Anal. At. Spectrom 15 (2000) 1133.

[33] U. Panne, C. Haisch, M. Clara, R. Niessner, Analysis of glass and glass melts during the vitrification process of fly and bottom ashes by laser-induced plasma spectroscopy. Part I: Normalization and plasma diagnostics, Spectrochim. Acta B 53 (1998) 1957.

[34] R. Barbini, F. Colao, R. Fantoni, A. Palussi, S. Ribezzo, H.J.L. Van Der Steen, M. Angelone, Semi-quantitative time resolved LIBS measurements, Appl. Phys. B 65 (1997) 101.

[35] A.I. Whitehouse, J. Young, I.M. Botheroyd, S. Lawson, C.P. Evans, J. Wright, Remote material analysis of nuclear power station steam generator tubes by laser-induced breakdown spectroscopy, Spectrochim. Acta B 56 (2001) 821.

[36] A.K. Rai, H. Zhang, F.Y. Yueh, J.P. Singh, A. Weisburg, Parametric study of a fiber-optic laser-induced breakdown spectroscopy probe for analysis of aluminum alloys, Spectrochim. Acta B 56 (2001) 2371.

[37] X. Hou, B.T. Jones, Field instrumentation in atomic spectroscopy, Microchem. J. 66 (2000) 115.

[38] D.A. Cremers, J.E. Barefield II, A.C. Koskelo, Remote elemental analysis by laser-induced breakdown spectroscopy using a fiber-optic cable, Appl. Spectrosc. 49 (1995) 857.

[39] R. Harmon, F. DeLucia, A. Miziolek, K. McNesby, R. Walters, P. French, Laser-induced breakdown spectroscopy (libs)—an emerging field-portable sensor technology for real-time, in-situ geochemical and environmental analysis, Geochem.: Explor., Environ., Anal. 5 (2005) 21.

[40] J. Rakovsky, O. Musset, J. Buoncristiani, V. Bichet, F. Monna, P. Neige, P. Veis, Testing a portable laser-induced breakdown spectroscopy system on geological samples, Spectrochim. Acta B 74–75 (2012) 57.

[41] C. Pasquini, J. Cortej, L. Silva, F. Gonzaga, Laser induced breakdown spectroscopy, J. Braz. Chem. Soc. 18 (2007) 463.

[42] B. Connors, A. Somers, D. Day, Application of handheld laser-induced breakdown spectroscopy (LIBS) to geochemical analysis, Appl. Spectrosc. 70 (2016) 810.

[43] I.B. Gornushkin, S.V. Shabanov, S. Merk, E. Tognoni, U. Panne, Effects of non-uniformity of laser induced plasma on plasma temperature and concentrations determined by the Boltzmann plot method: implications from plasma modeling, J. Anal. At. Spectrom 25 (2010) 1643.

[44] S. Eschlbock-Fuchs, A. Demidov, I.B. Gornushkin, T. Schmid, R. Rossler, N. Huber, U. Panne, J.D. Pedarnig, Tomography of homogenized laser-induced plasma by Radon transform technique, Spectrochim. Acta B 123 (2016) 59.

[45] Y. Yu, L.B. Guo, Z.Q. Hao, X.Y. Li, M. Shen, Q.D. Zeng, K.H. Li, X.Y. Zeng, Y.F. Lu, Z. Ren, Accuracy improvement on polymer identification using laser-induced breakdown spectroscopy with adjusting spectral weightings, Opt. Express 22 (2014) 3895.

[46] V.K. Unnikrishnan, K.S. Choudhari, S.D. Kulkarni, R. Nayak, V.B. Kartha, C. Santosh, Analytical predictive capabilities of laser induced breakdown spectroscopy (LIBS) with principal component analysis (PCA) for plastic classification, RSC Adv. 3 (2013) 25872.

[47] C. Gautier, P. Fichet, D. Menut, J. Dubessy, Applications of the double-pulse laser-induced breakdown spectroscopy (LIBS) in the collinear beam geometry to the elemental analysis of different materials, Spectrochim. Acta B 61 (2006) 210.

[48] G. Cristoforetti, S. Legnaioli, L. Pardini, V. Palleschi, A. Salvetti, E. Togoni, Spectroscopic and shadowgraphic analysis of laser induced plasmas in the

orthogonal double pulse pre-ablation configuration, Spectrochim. Acta B 61 (2006) 340.

[49] M.E. Asgill, M.S. Brown, K. Frische, W.M. Roquemore, D.W. Hahn, Double-pulse and single-pulse laser-induced breakdown spectroscopy for distinguishing between gaseous and particulate phase analytes, Appl. Optics 49 (2010) C110.

[50] A.M. Popov, F. Colao, R. Fantoni, Enhancement of LIBS signal by spatially confining the laser-induced plasma, J. Anal. At. Spectrom 24 (2009) 602.

[51] A. Hussain, Q. Li, Z. Hao, X. Gao, J. Lin, The effect of an external magnetic field on the plume expansion dynamics of laser-induced aluminum plasma, Plasma Sci. Technol. 17 (2015) 693.

[52] K.X. Li, W.D. Zhou, Q.M. Shen, Z.J. Ren, B.J. Peng, Laser ablation assisted spark induced breakdown spectroscopy on soil samples, J. Anal. At. Spectrom 25 (2010) 1475.

[53] M.V. Belkov, V.S. Burakov, A. De Giocomo, V.V. Kiris, S.N. Raikov, N.V. Tarasenko, Comparison of two laser-induced breakdown spectroscopy techniques for total carbon measurement in soils, Spectrochim. Acta B 64 (2009) 899.

[54] M. Dell'Aglio, R. Alrifai, A. De Giocomo, Nanoparticle enhanced laser induced breakdown spectroscopy (nelibs), a first review, Spectrochim. Acta B 148 (2018) 105.

[55] Z. Yu, S. Yao, L. Zhang, Z. Lu, Z.S. Lie, J. Lu, Surface-enhanced laser-induced breakdown spectroscopy utilizing metallic target for direct analysis of particle flow, J. Anal. At. Spectrom 34 (2019) 172.

[56] A.E. Pichahchy, D.A. Cremers, M.J. Ferris, Elemental analysis of metals under water using laser-induced breakdown spectroscopy, Spectrochim. Acta B 52 (1997) 25.

[57] R. Sattmann, V. Sturn, R. Noll, Laser-induced breakdown spectroscopy of steel samples using multiple Q-switch Nd: YAG laser pulses, J. Phys. D28 (1995) 2181.

[58] D.N. Stratis, K.L. Eland, S.M. Angel, Effect of pulse delay time on a pre-ablation dual-pulse LIBS plasma, Appl. Spectrosc. 55 (2001) 1297.

[59] D.N. Stratis, K.L. Eland, S.M. Angel, Enhancement of aluminum, titanium, and iron in glass using pre-ablation spark dual-pulse LIBS, Appl. Spectrosc. 54 (2000) 1719.

[60] D.N. Stratis, E.L. Eland, S.M. Angel, Dual-pulse LIBS using a pre-ablation spark for enhanced ablation and emission, Appl. Spectrosc. 54 (2000) 1270.

[61] L. St-Onge, M. Sabsabi, P. Cielo, Analysis of solids using laser-induced plasma spectroscopy in double-pulse mode, Spectrochim. Acta B 53 (1998) 407.

[62] L. St-Onge, V. Detalle, M. Sabsabi, Enhanced laser-induced breakdown spectroscopy using the combination of fourth-harmonic and fundamental Nd: YAG laser pulses, Spectrochim. Acta B 57 (2002) 121.

[63] V.N. Rai, F.Y. Yueh, J.P. Singh, Study of laser-induced breakdown emission from liquid under double-pulse excitation, Appl. Optics 42 (2003) 2094.

[64] Y. Li, D. Tian, Y. Ding, G. Yang, K. Liu, C. Wang, X. Han, A review of laser-induced breakdown spectroscopy signal enhancement, Appl. Spectrosc. Rev. 53 (2018) 1.

[65] D.H. Lee, S.C. Han, T.H. Kim, J.I. Yun, Highly sensitive analysis of boron and lithium in aqueous solution using dual-pulse laser-induced breakdown spectroscopy, Anal. Chem. 83 (2011) 9456.

[66] Y. Lu, V. Zorba, X. Mao, R. Zheng, R.E. Russo, UV fs–ns double-pulse laser induced breakdown spectroscopy for high spatial resolution chemical analysis, J. Anal. At. Spectrom 28 (2013) 743.

[67] T. Donnelly, J.G. Lunney, S. Amoruso, R. Bruzzese, X. Wang, X. Ni, Double pulse ultrafast laser ablation of nickel in vacuum, J. Appl. Phys. 106 (2009) 013304.

[68] T. Somekawa, M. Otsuka, Y. Maeda, M. Fujita, Signal enhancement in femtosecond laser induced breakdown spectroscopy with a double-pulse configuration composed of two polarizers, Jpn. J. Appl. Phys. 55 (2016) 058002.

[69] M. Weidman, M. Baudelet, S. Palanko, M. Sigman, P.J. Dagdigian, M. Richardson, Nd:YAG-CO2 double-pulse laser induced breakdown spectroscopy of organic films, Opt. Express 18 (2010) 259.

[70] R.W. Coons, S.S. Harilal, S.M. Hassan, A. Hassanein, The importance of longer wavelength reheating in dual-pulse laser-induced breakdown spectroscopy, Appl. Phys. B 107 (2012) 873.

[71] D.K. Killinger, S.D. Allen, R.D. Waterbury, C. Stefano, E.L. Dottery, Enhancement of Nd:YAG LIBS emission of a remote target using a simultaneous CO2 laser pulse, Opt. Express 15 (2007) 12905.

[72] R. Hai, P. Liu, D. Wu, H. Ding, J. Wu, G.N. Luo, Collinear double-pulse laser-induced breakdown spectroscopy as an in-situ diagnostic tool for wall composition in fusion devices, Fusion Eng, Des. 89 (2014) 2435.

[73] R. Ahmed, M.A. Baig, On the optimization for enhanced dual-pulse laser-induced breakdown spectroscopy, IEEE Trans. Plasma Sci. 38 (2010) 2052.

[74] B.W. Smith, A. Quentmeir, M. Bolshov, K. Niemax, Measurement of uranium isotope ratios in solid samples using laser ablation and diode laser-excited atomic fluorescence spectrometry, Spectrochim. Acta B 54 (1999) 943.

[75] I.B. Gornushkin, S.A. Baker, B.W. Smith, J.D. Winefordner, Determination of lead in metallic reference materials by laser ablation combined with laser excited atomic fluorescence, Spectrochim. Acta B 52 (1997) 1653.

[76] F. Hilbk Kortenbruck, R. Noll, P. Wintjens, H. Falk, C. Becker, Analysis of heavy metals in soils using laser-induced breakdown spectrometry combined with laser-induced fluorescence, Spectrochim. Acta B 56 (2001) 933.

[77] H.H. Telle, D.C.S. Beddows, G.W. Morris, O. Samek, Sensitive and selective spectrochemical analysis of metallic samples: the combination of laser-induced breakdown spectroscopy and laser-induced fluorescence spectroscopy, Spectrochim. Acta B 56 (2001) 947.

[78] H. Sobral, A. Robledo-Martinez, Signal enhancement in laser-induced breakdown spectroscopy using fast square-pulse discharges, Spectrochim. Acta B 124 (2016) 67.

[79] W. Zhou, K. Li, H. Quian, Z. Ren, Y. Yu, Effect of voltage and capacitance in nanosecond pulse discharge enhanced laser-induced breakdown spectroscopy, Appl. Optics 51 (2012) B42.

[80] L.I. Kexue, W. Zhou, Q. Shen, J. Shao, H. Qian, Signal enhancement of lead and arsenic in soil using laser ablation combined with fast electric discharge, Spectrochim. Acta B 65 (2010) 420.

[81] W. Zhou, K. Li, Q. Shen, Q. Chen, J. Long, Optical emission enhancement using laser ablation combined with fast pulse discharge, Opt. Express 18 (2010) 2573.

[82] W. Zhou, K. Li, X. Li, H. Quian, J. Shao, X. Fang, P. Xie, W. Liu, Development of a nanosecond discharge-enhanced laser plasma spectroscopy, Opt. Lett. 36 (2011) 2961.

[83] X. Li, W. Zhou, K. Li, H. Quian, Z. Ren, Laser ablation fast pulse discharge plasma spectroscopy analysis of Pb, Mg and Sn in soil, Opt. Commun. 285 (2012) 54.

[84] K.A. Tereszchuk, J.M. Vadillo, J.J. Laserna, Glow-discharge-assisted laser-induced breakdown spectroscopy: Increased sensitivity in solid analysis, Appl. Spectrosc. 62 (2008) 1262.

[85] Y. Liu, M. Baudelet, M. Richardson, Elemental analysis by microwave-assisted laser-induced breakdown spectroscopy: evaluation on ceramics, J. Anal. At. Spectrom. 25 (2010) 1316.

[86] K.A. Tereszchuk, J.M. Vadillo, J.J. Laserna, Energy assistance in laser induced plasma spectrometry (LIPS) by a synchronized microsecond-pulsed glow discharge secondary excitation, J. Anal. At. Spectrom 22 (2007) 183.

[87] Y. Ikeda, R. Tsuruoka, Characteristics of microwave plasma induced by lasers and sparks, Appl. Optics 51 (2012) 183.

[88] Y. Liu, B. Bousquet, M. Baudeletnas, M. Richardson, Improvement of the sensitivity for the measurement of copper concentrations in soil by microwave-assisted laser-induced breakdown spectroscopy, Spectrochim. Acta B 73 (2012) 89.

[89] Q. Xiao, R. Hai, H. Ding, A. Huber, V. Philipps, N. Gierse, G. Sergienko, In-situ analysis of the first wall by laser-induced breakdown spectroscopy in the TEXTOR tokamak: Dependence on the magnetic field strength, J. Nucl. Mater. 463 (2015) 911.

[90] L. Cheng, G. Xun, L. Qi, S. Chao, L. Jingquan, Spectral enhancement of laser-induced breakdown spectroscopy in external magnetic field, Plasma Sci. Technol. 17 (2015) 919.

[91] S.S. Harilal, M.S. Tillack, B. O'Shay, C.V. Bindu, F. Najmabadi, Confinement and dynamics of laser-produced plasma expanding across a transverse magnetic field, Phys. Rev. E 69 (2004) 026413.

[92] X.K. Shen, J. Sun, H. Ling, Y.F. Lu, Spectroscopic study of laser-induced Al plasmas with cylindrical confinement, J. Appl. Phys. 102 (2007) 30.

[93] L.B. Guo, C.M. Li, W. Hu, Y.S. Zhou, B.Y. Zhang, Z. X. Cai, X.Y. Zeng, Y.F. Lu, Plasma confinement by hemispherical cavity in laser-induced breakdown spectroscopy, Appl. Phys. Lett. 98 (2011) 1459.

[94] X. Gao, L. Liu, C. Song, J. Lin, The role of spatial confinement on nanosecond YAG laser-induced Cu plasma, J. Phys. D Appl. Phys. 48 (2015) 175205.

[95] L.B. Guo, Z.Q. Hao, M. Shen, W. Xiang, X.N. He, Z.Q. Xie, M. Gao, X.Y. Li, X.Y. Zeng, Y.F. Lu, Accuracy improvement of quantitative analysis by spatial confinement in laser-induced breakdown spectroscopy, Opt. Express 21 (2013) 18188.

[96] S.S. Mao, X. Zeng, X. Mao, R.E. Russo, Laser-induced breakdown spectroscopy: flat surface vs. cavity structures, A. Anal. At. Spectrom. 19 (2004) 495.

[97] X. Zeng, X. Mao, S.S. Mao, S.B. Wen, R. Greif, R.E. Russo, Laser-induced shockwave propagation from ablation in a cavity, Appl. Phys. Lett. 88 (2006) 061502.

[98] L.B. Guo, W. Hu, Y.B. Zhang, X.N. He, C.M. Li, Y.S. Zhou, Z.X. Cai, X.Y. Zeng, Y.F. Lu, Enhancement of optical emission from laser-induced plasmas by combined spatial and magnetic confinement, Opt. Express 19 (2011) 14067.

[99] L.B. Guo, B.Y. Zhang, X.N. He, C.M. Li, Y.S. Zhou, T. Wu, J.B. Park, X.Y. Zeng, Y.F. Lu, Optimally enhanced optical emission in laser-induced breakdown spectroscopy by combining spatial confinement and dual-pulse irradiation, Opt. Express 20 (2012) 1436.

[100] X. Su, W. Zhou, H. Quian, Optical emission character of collinear dual pulse laser plasma with cylindrical cavity confinement, J. Anal. At. Spectrom 29 (2014) 2356.

[101] L.B. Guo, Y.B. Zhang, X.N. He, C.M. Li, Y.S. Zhou, T. Wu, J.B. Park, X.Y. Zeng, Y.F. Lu, Optimally enhanced optical emission in laser-induced breakdown spectroscopy by combining spatial confinement and dual-pulse irradiation, Opt. Express 20 (2012) 1436.

[102] Z. Hou, Z. Wang, J. Liu, W. Ni, Z. Li, Combination of cylindrical confinement and spark discharge for signal improvement using laser induced breakdown spectroscopy, Opt. Express 22 (2014) 12909.

[103] (a) V.N. Rai, Optical properties of silver-island films having an overlayer of RhB dye, Appl. Optics 26 (1987) 2395 (b) V.N. Rai, A.K. Srivastava, C. Mukherjee, S.K. Deb, Surface enhanced absorption and transmission from dye coated gold nanoparticles in thin films, Appl. Optics 51 (2012) 2606.

[104] V.N. Rai, A.K. Srivastava, C. Mukherjee, S.K. Deb, Localized surface plasmon resonance and refractive index sensitivity of vacuum-evaporated nanostructured gold thin films, Ind. J. Phys. 19 (2016) 107.

[105] S.A. Maier, Plasmonics: Fundamentals and Applications, Springer Science, New York, NY, 2007.

[106] A. De Giacomo, M. Dell'Aglio, R. Gaudiuso, C. Coral, G. Valenza, Perspective on the use of nanoparticles to improve LIBS analytical performance: nanoparticle enhanced laser induced breakdown spectroscopy (NELIBS), J. Anal. At. Spectrom 31 (2016) 1566.

[107] M. Dell'Aglio, R. Alrifai, A. De Giacomo, Nanoparticle enhanced laser induced breakdown spectroscopy (nelibs), a first review, Spectrochim. Acta B 148 (2018) 105.

[108] A. De Giacomo, R. Gaudiuso, C. Coral, M. Dell'Aglio, O. De Pascale, Nanoparticle-enhanced laser-induced breakdown spectroscopy of metallic samples, Anal. Chem. 85 (2013) 10180.

[109] A. De Giacomo, C. Coral, G. Valenza, R. Gaudiuso, M. Dell'Aglio, Nanoparticle enhanced laser-induced breakdown spectroscopy for microdrop analysis at subppm level, Anal. Chem. 88 (2016) 5251.

[110] M.A. Aguirre, S. Legnaioli, F. Almodovar, M. Hidalgo, V. Palleschi, A. Canals, Elemental analysis by surface-enhanced Laser-Induced Breakdown Spectroscopy combined with liquid–liquid microextraction, Spectrochim. Acta B 79–80 (2013) 88.

[111] J. Xiu, X. Bai, V. Motto-Ros, J. Yu, Characteristics of indirect laser-induced plasma from a thin film of oil on a metallic substrate, Front. Phys. 10 (2015) 231.

[112] J. Xiu, X. Bai, E. Negre, V. Motto-Ros, J. Yu, Indirect laser-induced breakdown of transparent thin gel layer for sensitive trace element detection, Appl. Phys. Lett. 102 (2013) 244101.

[113] L. Zhen, F. Cao, J. Xiu, X. Bai, V. Motto-Ros, N. Gilon, H. Zeng, J. Yu, On the performance of laser-induced breakdown spectroscopy for direct determination of trace metals in lubricating oils, Spectrochim. Acta B 99 (2014) 1.

[114] X. Yang, Z. Hao, C.M. Li, J.M. Li, R. Yi, M. Shen, K. Li, L. Guo, X.Y. Li, Y.F. Lu, X.Y. Zeng, Sensitive determinations of Cu, Pb, Cd, and Cr elements in aqueous solutions using chemical replacement combined with surface-enhanced laser-induced breakdown spectroscopy, Opt. Express 24 (2016) 13410.

[115] X. Yang, Z. Hao, M. Shen, R. Yi, J. Li, H. Yu, L. Guo, X. Li, X. Zeng, Y. Lu, Simultaneous determination of La, Ce, Pr, and Nd elements in aqueous solution using surface-enhanced laser-induced breakdown spectroscopy, Talanta 163 (2017) 127.

[116] Y. Tian, H.C. Cheung, R. Zheng, Q. Ma, Y. Chen, N. Delepine-Gilon, J. Yu, Elemental analysis of powders with surface-assisted thin film laser-induced breakdown spectroscopy, Spectrochim. Acta B 124 (2016) 16.

[117] S. Niu, L. Zheng, A. Qayyum Khan, G. Feng, H. Zeng, Laser-induced breakdown spectroscopic detection of trace level heavy metal in solutions on a laser-pretreated metallic target, Talanta 179 (2018) 312.

[118] Z. Yu, S. Yao, L. Zhang, Z. Lu, Z.S. Lie, J. Lu, Surface-enhanced laser-induced breakdown spectroscopy utilizing metallic target for direct analysis of particle flow, J. Anal. At. Spectrom 34 (2019) 172.

[119] S.L. Chin, T.J. Wang, C. Marceau, J. Wu, J.S. Liu, O. Kosareva, N. Panov, Y.P. Chen, J.F. Diagle, S. Yuan, A. Azam, W.W. Liu, T. Sidemann, H.P. Zeng, M. Richardson, R. Li, Z.Z. Xu, Advances in intense femtosecond laser filamentation in air, Laser Phys. 22 (2012) 1.

[120] W. Liu, H.L. Xu, G. Mejean, Y. Kamali, J.F. Diagle, A. Azam, P.T. Simard, P. Mathieu, G. Roy, S.L. Chin, Efficient non-gated remote filament-induced breakdown spectroscopy of metallic sample, Spectrochim. Acta B 62 (2007) 76.

[121] M. Durand, A. Howard, B. Prade, A. Mysyrowicz, A. Durecu, B. Moreau, D. Fleury, O. Vasseur, H. Borchert, K. Diener, R. Schmitt, F. Theberge, M. Chateauneuf, J.F. Diagle, J. Dubois, Kilometer range filamentation, Opt. Express 21 (2013) 26836.

[122] L.A. Finney, P.J. Skrodzki, M. Burger, J. Nees, S. S. Harilal, I. Jovanovic, Single-shot, multi-signature remote detection of uranium by filament-induced breakdown spectroscopy, Opt. Lett. 44 (2019) 2783.

[123] R.E. Neuhauser, U. Panne, R. Neissner, G.A. Petrucci, P. Cavalli, N. Omenetto, On-line and in-situ detection of lead aerosols by plasma-spectroscopy and laser-excited atomic fluorescence spectroscopy, Anal. Chim. Acta 346 (1997) 37.

[124] J.P. Singh, H. Zhang, F.Y. Yueh, K.P. Karney, Investigation of the effects of atmospheric conditions on the

quantification of metal hydrides using laser-induced breakdown spectroscopy, Appl. Spectrosc. 12 (1996) 764.
[125] J.R. Watcher, D.A. Cremers, Determination of uranium in solution using laser-induced breakdown spectroscopy, Appl. Spectrosc. 41 (1987) 1042.
[126] J.P. Singh, F.Y. Yueh, H. Zhang, K.P. Karney, A preliminary study of the determination of uranium, plutonium and neptunium by laser-induced breakdown spectroscopy, Rec, Res. Dev. Appl. Spectrosc. 2 (1999) 59.
[127] A.K. Rai, F.Y. Yueh, J.P. Singh, High temperature fiber optic laser-induced breakdown spectroscopy sensor for analysis of molten alloy constituents, Rev. Sci. Instrum. 73 (2002) 3589.
[128] A.K. Rai, F.Y. Yueh, J.P. Singh, Laser-induced breakdown spectroscopy of molten aluminum alloy, Appl. Optics 42 (2003) 2078.
[129] C. Aragon, J.A. Aguilera, J. Campos, Determination of carbon content in molten steel using laser-induced breakdown spectroscopy, Appl. Spectrosc. 47 (1993) 606.
[130] L. Paksy, B. Nemet, A. Lengyel, L. Kozma, Production control of metal alloys by laser spectroscopy of the molten metals. Part I. Preliminary investigations, Spectrochim. Acta B 51 (1996) 279.
[131] J. Gruber, J. Heitz, H. Strasser, D. Bauerle, N. Ramasedar, Rapid in-situ analysis of liquid steel by laser-induced breakdown spectroscopy, Spectrochim. Acta B 56 (2001) 685.
[132] J. Gruber, J. Heitz, N. Arnold, N. Ramsedar, W. Meyer, F. Koch, In situ analysis of metal melts in metallurgic vacuum devices by laser-induced breakdown spectroscopy, Appl. Spectrosc. 58 (2004) 457.
[133] L. Peter, V. Sturm, R. Noll, Liquid steel analysis with laser-induced breakdown spectrometry in the vacuum ultraviolet, Appl. Optics 42 (2003) 6199.
[134] C. Li, X. Wu, C. Zhang, H. Ding, G. De Temmerman, H.J. Van der Meiden, Study of deuterium retention on lithiated tungsten exposed to high-flux deuterium plasma using laser-induced breakdown spectroscopy, Fusion Eng. Des. 89 (2014) 949.
[135] Z. Hu, N. Gierse, C. Li, J. Oelmann, D. Zhao, M. Tokar, X. Jiang, D. Nikolai, J. Wu, F. Ding, S. Brezinsek, H. Ding, G.N. Luo, C. Linsmeier, Laser induced ablation spectroscopy for in situ characterization of the first wall on EAST tokamak, Fusion Eng. Des. 135 (2018) 95.
[136] G. Federici, C.H. Skinner, J.N. Brooks, J.P. Coad, C. Grisolia, A.A. Haasz, et al., Plasma-material interactions in current tokamaks and their implications for next step fusion reactors, Nucl, Fusion 41 (2001) 1967.
[137] J. Roth, E. Tsitrone, A. Loarte, T. Loarte, G. Counsell, R. Neu, et al., Recent analysis of key plasma wall interactions issues for ITER, J. Nucl. Mater. 390–391 (2009) 1.
[138] N. Farid, C. Li, H. Wang, H. Ding, Laser-induced breakdown spectroscopic characterization of tungsten plasma using the first, second, and third harmonics of an Nd: YAG laser, J. Nucl. Mater. 433 (2013) 80.
[139] P. Paris, J. Butikova, M. Laan, A. Hakola, I. Jogi, J. Likonen, E. Grigore, C. Ruset, Comparison of LIBS results on ITER-relevant samples obtained by nanosecond and picosecond lasers, Nucl. Mater. Energy 18 (2019) 1.
[140] Q. Xiao, A. Huber, G. Sergienko, B. Schweer, P. Martens, A. Kubina, V. Philipps, H. Ding, Application of laser-induced breakdown spectroscopy for characterization of material deposits and tritium retention in fusion devices, Fusion Eng. Des. 88 (2013) 1813.
[141] V.N. Rai, J.P. Singh, C. Winstead, F.Y. Yueh, R.L. Cook, Laser-induced breakdown spectroscopy of hydrocarbon flame and rocket engine simulator plume, AIAA J. 41 (2003) 2192.
[142] D.A. Cremers, R.C. Wiens, M.J. Ferris, R. Brennetot, S. Maurice, Trends in Optics and Photonics, in: Laser-Induced Plasma Spectroscopy and Applications, OSA Technical Digest, 81, Optical Society of America, Washington, 2002, p. 5.
[143] R. Brennetot, J.L. Lacour, E. Vors, P. Fichet, D. Vailhen, S. Maurice, A. Rivoallan, Trends in optics ans photonics, in: Laser-Induced Plasma Spectroscopy and Applications, OSA Technical Digest, Vol. 81, Optical Society of America, Washington, 2002, p. 9.
[144] R. Wiens, S. Maurice, F. Perez, The Super Cam remote sensing instrument suite for the Mars 2020 rover mission: a preview, Spectroscopy 32 (2017) 50.
[145] N. Murdoch, B. Chide, J. Lasue, A. Cadu, A. Sournac, M. Bassas-Portus, X. Jacob, J. Merrison, J.J. Iversen, C. Moretto, C. Velasco, L.P.A. Hynes, V. Godiver, R.D. Lorenz, P. Cais, P. Bernadi, S. Maurice, R.C. Weins, D. Mimoun, Laser-induced breakdown spectroscopy acoustic testing of the Mars 2020 microphone, Planet. Space Sci. 165 (2019) 260.
[146] R.A. Multari, L.E. Foster, D.A. Cremers, M.J. Ferris, Effect of sampling geometry on elemental emissions in laser-induced breakdown spectroscopy, Appl. Spectrosc. 50 (1996) 1483.
[147] B.C. Castle, K. Visser, B.W. Smith, J.D. Winefordner, Spatial and temporal dependence of lead emission in laser-induced breakdown spectroscopy, Appl. Spectrosc. 51 (1997) 1017.
[148] K. Song, H. Cha, J. Lee, Y.I. Lee, Investigation of the line-broadening mechanism for laser-induced copper plasma by time-resolved laser-induced breakdown spectroscopy, Microchem. J. 63 (1999) 53.
[149] M. Austin, A. Briand, P. Mauchien, J.M. Mermet, Characterization by emission spectrometry of a laser-produced plasma from a copper target in air at atmospheric pressure, Spectrochim. Acta B 48 (1993) 851.

[150] V.N. Rai, H. Zhang, F.Y. Yueh, J.P. Singh, A. Kumar, Effect of steady magnetic field on laser-induced breakdown spectroscopy, Appl. Optics 42 (2003) 3662.
[151] Y.I. Lee, T.L. Thiem, G.H. Kim, Y.Y. Teng, J. Sneddon, Interaction of an excimer-laser beam with metals. Part III: The effect of a controlled atmosphere in laser-ablated plasma emission, Appl. Spectrosc. 46 (1992) 1597.
[152] R. Wisbrun, I. Schechter, R. Niessner, H. Schroder, K.L. Kompa, Detector for trace elemental analysis of solid environmental samples by laser plasma spectroscopy, Annal. Chem. 66 (1994) 2964.
[153] A.S. Eppler, D.A. Cremers, D.D. Hickmott, M.J. Ferris, A.C. Koskelo, Matrix effects in the detection of Pb and Ba in soils using laser-induced breakdown spectroscopy, Appl. Spectrosc. 50 (1996) 1175.

PART II

New LIBS techniques

CHAPTER

6

Ultrafast and filament-LIBS

S.S. Harilal[a], P.K. Diwakar[b], G. Miloshevsky[c]

[a]Pacific Northwest National Laboratory, Richland, WA, United States [b]South Dakota School of Mines, Rapid City, SD, United States [c]Virginia Commonwealth University, Richmond, VA, United States

1 Introduction

The history of spectroscopy can be traced back to Newton's era (16th century). Among the various spectroscopic tools, visible emission spectroscopy is one of the oldest plasma/flame diagnostic tools routinely used for identifying spectral features originated from excited atoms and molecules. Significant advances made in the spectroscopy field in the last several decades can be associated with the invention of lasers in the 1960s and their technological evolution. Shortly after the invention of the laser, Brech and Cross [1] reported microemissions from plasma produced by a ruby laser. Although the first report on emission analysis of laser-produced plasmas (LPPs) appeared in 1962, research and applications of laser-induced breakdown spectroscopy (LIBS) were limited until the 1990s. Since then significant advancement has been made and this could be related to several factors, including the availability of relatively inexpensive, compact, and stable pulsed Nd:YAG lasers, improvement in the performance of spectrographs (e.g., miniaturized spectrographs, Echelle, etc.), and sensitive light detection systems (e.g., intensified charge-coupled device cameras) along with newer applications, methods for improving analyte sensitivity (double-pulse, resonance-enhanced LIBS, nanoparticle-enhanced LIBS, hybrid methods such as spark-LIBS, etc.), and availability of short-pulse lasers [2, 3]. Extensive fundamental research, both experimental and modeling, also played a crucial role in advancing LIBS technology.

Although the field of LPPs and its applications has been growing since the development of the laser in the 1960s, the peak intensities achievable from early laser sources were stagnated for more than three decades. This scenario changed dramatically with the invention of chirped pulse amplification (CPA) and the development of high-fluence solid-state materials like Ti:sapphire, which led to revolutionary developments in the peak power of table-top lasers. CPA for lasers was introduced by Strickland and Mourou in 1985 for which they won the Nobel Prize in 2018 [4]. In CPA, short laser pulses are temporally and spatially stretched so that the optical elements in the amplifier never encounter high peak powers and the

Laser-Induced Breakdown Spectroscopy
https://doi.org/10.1016/B978-0-12-818829-3.00006-X

process is reversed after amplification. Hence, a typical ultrafast laser system contains a mode-locked femtosecond (fs) oscillator and its pump, stretcher, regenerative amplifier, and pump and compressor. The evolution of fs- or ultrafast laser technology in the last one-and-a-half decades is so amazing that currently very reliable, single-box turnkey systems are available in the commercial market. In addition to these, the cost of ownership of an fs-laser has reduced significantly. These factors led to the wide use of ultrafast lasers for LIBS and other laser ablation (LA) applications and evolved a new field called ultrafast LIBS or fs-LIBS. Researchers also noticed potential advantages for using fs-lasers instead of traditional long-pulse nanosecond (ns) lasers for performing LIBS, which include reduced optical continuum emission and damage to the sample, generation of colder plasma conditions, less susceptibility to sampling matrix and elemental fractionation effects, nongated detection, and better spatial and depth resolution [5, 6]. Ultrafast LA is also found to be capable of providing cleaner craters because of the reduced heat-affected zone (HAZ), high-precision micromachining, better ablation efficiency, and lower ablation threshold compared to ns-LA [5].

Although the procedure of generating LIBS plumes by LA is simple, the physics involved in such a process is notoriously complex and contains numerous interrelated processes like heating, melting, vaporization, ejection of particles, plasma creation and its expansion, and shockwave generation and confinement, to list a few [7]. It is well known that the fundamentals of LPP depend strongly on laser parameters such as deposited energy, wavelength, and pulse width. For example, the laser-target interaction will be stronger for ultraviolet ns-laser pulse ablation, while laser-plasma heating will be predominant with ns-near infrared (NIR) LPPs and these can be related to the critical density of the generated plasma [7]. The pulse width of the laser will also affect laser-target and laser-plasma coupling. For example, the ablation physics changes significantly when the duration of the plasma generation pulse is shorter than the characteristic timescales such as electron and phonon thermalization times, which influences the dynamics of the plasma plume generation, its expansion, and emission features. There exist several articles in the literature comparing the differences between ns- and fs-LIBS [5–8].

Compared to ns-LIBS, ultrafast LIBS warrants more diagnostic needs on the laser side. The fs-laser pulses are accompanied by weak, longer-duration pre- and postpulses. The origin of prepulses in fs-laser systems is due to incorrect recompression of the stretched pulses, leakage from the Pockels cells in the regenerative amplifier, or an amplified spontaneous emission background. If the contrast of the fs-laser system (ratio of the peak pulse to the prepulse or background) is poor, the prepulses can induce deleterious effects during laser-target interaction. Because of this, the intensity contrast ratio should be $\geq 10^6$ for fs-pulses with respect to the ns-prepulse to avoid prepulse damage to the material. However, the contrast issue may not be detrimental for most LIBS studies considering that very low-energy pulses are used for ablation. If a fs-laser pulse is propagated through a refractive medium, an attenuator, or a window of a vacuum chamber, some properties of the pulse (e.g., duration, chirp, wavefront) will be modified because of the nonlinear effects. For example, group velocity dispersion (GVD) in a transparent medium, which introduces a frequency-dependent delay of the spectral components, may lead to a dispersive temporal broadening of the pulse [9]. GVD and associated group delay dispersion are relevant considerations for pulses with widths ≤ 50 fs, but less significant for pulses ≥ 100 fs. Reflective optics is useful for avoiding/reducing any nonlinear effects during beam propagation.

The propagation of a long-pulse ns-laser beam through a medium is negligibly influenced by the media properties. However, for lasers with fs-duration, nonlinear absorption becomes significant in all materials, including air. If the fs-laser intensity is higher than the critical power for self-focusing, filamentation can occur in the beam during its propagation. So, even if a short focal length lens is used for generating LIBS plume on a target in the air, the filaments may be generated in the beam provided the localized beam intensity is higher than the critical power density. Therefore for ultrafast LIBS it is essential to use short focal length lenses as well as keeping the focal point just behind the target to avoid filamentation in air. On the other hand, the fs-laser filaments are useful for generating plasmas from a target material at large standoff distances. In recent times there has been significant interest in using ultrashort laser filaments for standoff LIBS applications, which are called filament-LIBS or remote LIBS [7, 10].

This chapter attempts to provide the reader with a brief discussion of the physics behind ultrafast LIBS generation, ablation threshold, crater properties, plasma emission, fundamental properties, etc. The differences between traditional ns-LIBS and fs-LIBS are discussed. A description of fs-laser filament generation conditions and its usefulness for generating a LIBS plume at standoff distances is also given. Also included in this chapter is the current status of modeling efforts on ultrafast LIBS plume generation, its expansion, and emission features. The chapter ends with a brief summary of future outlooks.

2 Nanosecond versus femtosecond LIBS

The characteristics of any LPP change significantly with numerous parameters and one of them is the pulse width (τ_p). At the early stages of laser-matter interaction, regardless of the τ_p of the laser, the photons will be coupled to the electrons in the system. For ns-lasers, the absorption is linear and obeys the Beer-Lambert law driven by a thermal process. However, absorption becomes nonlinear during fs-laser-material interaction. The differences in the physics involved during ns- and fs-laser-matter interaction can be understood by comparing the duration of the laser pulse to characteristic times such as electron-to-ion energy transfer time (τ_{ei}) and electron-heat conduction time (τ_{heat}), both of which typically occur on the order of several picoseconds (ps) after the onset of laser absorption. Hence, for ns-LIBS, processes like ionization, sample heating, vaporization, etc. occur during the laser pulse ($\tau_p \gg \tau_{ei} \sim \tau_{heat}$), while in the case of fs-LIBS, these phenomena do not occur until after the end of the laser pulse ($\tau_p < \tau_{ei} \sim \tau_{heat}$). Approximate timescales of various processes happening in ns- and fs-LIBS plume generation are given in Fig. 1 [7, 11, 12]. For ns-LIBS, a significant portion of the laser pulse will be used for plasma heating instead of target ablation and such a process is absent in fs-LIBS where the absorption of fs-laser photons leads to the presence of very hot electrons while keeping the lattice and ions at colder temperatures. Such strong electronic excitation during fs-LA may lead to nonthermally induced phase transitions. Because of these changes in ablation mechanisms, fs-LIBS offers potential benefits over ns-LIBS, which include greatly reduced thermal damage and HAZ on the target [13], reduced continuum emission [14], lower-temperature plasma generation with predominantly neutral atomic emission [5, 14], and nanoparticle emission [15]. The lower-temperature conditions present for fs-LA also lead to the generation of molecules at earlier times of its evolution in comparison with ns-LA [5, 6, 16].

In ns-LIBS, thermal processes lead to ablation and production of the plasma plume. For fs-LIBS, there are two competing mechanisms resulting in material removal and ablation: Coulomb explosion and thermal vaporization [17]. The ablation threshold (F_{th}) for fs-LIBS is found to be lower than that of ns-LIBS. The laser fluence threshold for ablation for ns-LIBS [18]

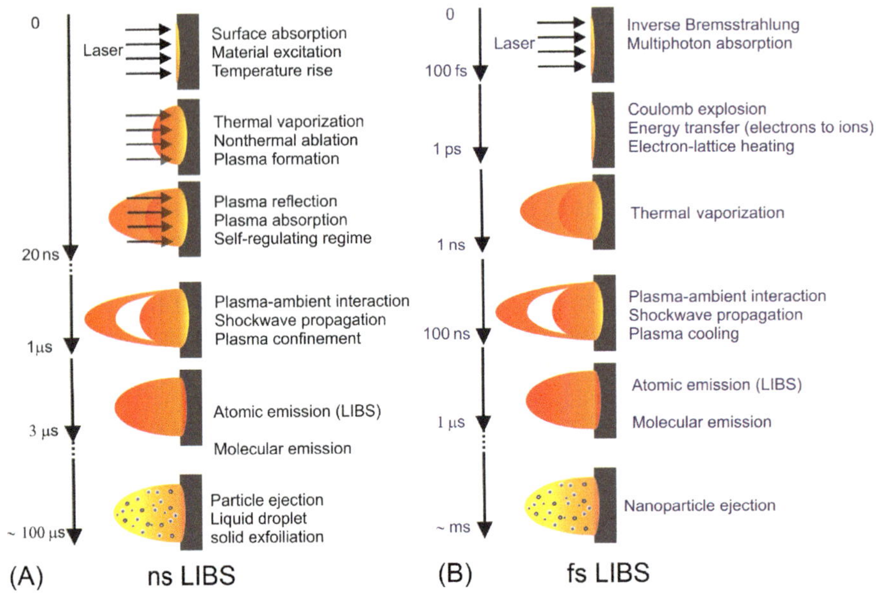

FIG. 1 Approximate timescales of various processes involved in (A) ns-laser-induced breakdown spectroscopy (LIBS) and (B) fs-LIBS [7, 11, 12].

approximately follows $F_{\mathrm{th}} \sim \tau_{\mathrm{p}}^{1/2}$, while a strong deviation from such a scaling was reported for pulses below ~10 ps [18]. Shaheen et al. [19] investigated the influence of target material on the ablation behavior of fs-laser pulses and two different ablation regimes were distinguished (nonthermal and thermal) depending on the fluence. The first regime is characterized by a relatively low ablation rate and is determined by the optical penetration depth, while the second regime is characterized by relatively high ablation rate and is determined by the thermal diffusion length.

There are distinct differences in ablation efficiency and crater shape between ns- and fs-LIBS [13]. Since laser-target coupling is very efficient in fs-LIBS, the ablation efficiency will be significantly higher compared to ns-LIBS for a fixed laser fluence, especially for NIR wavelengths [20]. However, the use of shorter wavelengths provides higher ablation efficiency for ns-LIBS where plasma heating through inverse Bremsstrahlung is negligible. Since the timescales of ns-LIBS are longer than the heat diffusion times, the shockwaves propagating through the melt layer and ablation-induced reactive forces will initiate splashing of the melt layer providing significant irregularities in the crater rims [21]. For fs-LIBS, the heat load to the surrounding material is minimized providing cleaner craters with well-defined edges. Cleaner ablation zones with less HAZ is beneficial, especially for performing LIBS with very delicate targets. Rudenko et al. [22] performed comprehensive modeling on the pathways of surface topology evolution during fs-laser interaction with the flat metal surface with respect to the laser fluence and the number of laser pulses. Their results showed formation of microvoids below the surface with near-threshold single pulse irradiation to volcano-like crater formation by spallation and material expansion with increasing number of pulses. Additional increase in the number of pulses and at near-threshold irradiation leads to ripple formation. Fig. 2 gives a schematic of the various processes and predicted surface

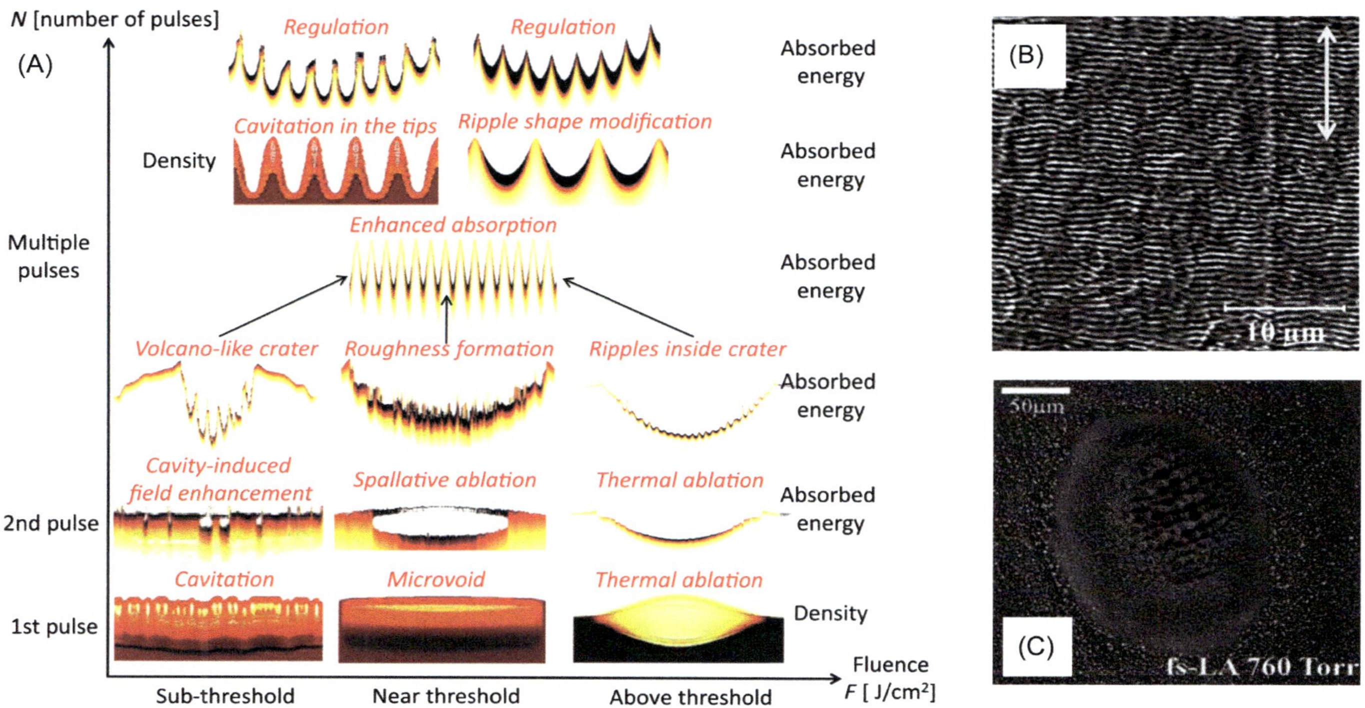

FIG. 2 (A) Schematics of coupled electromagnetic and hydrodynamic processes upon multipulse fs-laser irradiation as a function of laser fluence and the number of applied pulses (after Rudenko et al. [22]). (B) Morphology of the steel surface irradiated with 100-fs, 1-kHz pulses at 0.52 J/cm^2 and with a scan speed of $t = 6$ mm/s [23]. (C) Crater morphology of Cu after exposing to 40-fs, 100 shots at 14 J/cm^2 [21].

morphology along with images of craters recorded at near-threshold and well-above-threshold laser fluence [21–23].

The hydrodynamic expansion features of the plasma are also governed by initial plume conditions, which are different for ns- and fs-LIBS, as well as environmental conditions. In ns-LIBS, larger spot sizes lead to plume-sharpening effects caused by ion acceleration from enhanced laser–plasma coupling [24]. There exist several reports describing collimation of fs-LIBS in comparison with ns-LIBS plumes at atmospheric conditions [5, 14] and this can be related to the narrower angular distribution of emitted particles [20] that could affect the confinement properties of the plume. The position of the geometrical focal point of the focusing lens (in front, on, or behind the target) may also have an influence on laser–target coupling conditions regardless of the pulse width of the laser [25, 26]. The use of longer focal length lenses for generating fs-LIBS plumes may lead to the generation of filamentation, which can reduce the effective laser–target coupling. A detailed account of filament-LIBS is given in the latter part of this chapter.

There exist several articles comparing ns- and fs-LA emission. However, most of the reported experiments used widely different parameters (laser energy, fluence, etc.), so a direct comparison is always challenging. Several researchers compared the emission and noticed that fs-LIBS provides short-lived and relatively cool plasmas in comparison to the longer ns-pulses, where continuum radiation is predominant at early times [14, 27]. Emission from ions, neutral atoms, and molecules can be detected using LIBS. However, their emission delay with respect to the onset of plasma generation as well as duration varies significantly between ns- and fs-LIBS. Typically hotter ($\gtrsim 5$ eV) and denser ($\gtrsim 10^{18}$ cm^{-3}) conditions exist in ns-LIBS, which generate continuum radiation, larger ion fraction, and significant broadening of the lines due to the Stark effect at early times of plasma evolution [28, 29]. The continuum radiation caused by free–free and free–bound transitions at the early times of ns-LIBS emit broadband radiation, which persists up to ~1 μs. For fs-LIBS, the continuum radiation is short-lived, $\lesssim 100$ ns. When the plasma cools with time, emission from atoms predominates with a reduction of continuum emission and electron density giving rise to narrower line profiles with improved signal-to-noise and signal-to-background ratios. Hence, knowledge of the acquisition time window is one of the most important considerations for LIBS-based studies.

The LIBS plume typically radiates molecular emission when temperatures are ≤ 6500 K [30, 31]. Several studies reported strong molecular emissions from fs-LIBS at early times, whereas atomic and ionic emission lines are dominant in ns-LIBS [32, 33]. But, molecular emission will appear in ns-LIBS too when the plasma cools down to lower temperatures at later times of its evolution. For example, the emission features from ns- and fs-LIBS from an Al target in air generated using 17 mJ/pulse energy and recorded at 1 and 10 μs are given in Fig. 3. The molecular emission is found to be predominant in fs-LIBS even at early times (~1 μs). However, such emission features can be seen with ns-LIBS at later times (~10 μs). Since hotter conditions exist in ns-LIBS at early times, the persistence of the plasma will be significantly higher compared to fs-LIBS. So, in principle, similar signal-to-noise ratios can be obtainable for atomic and molecular emission although at different times, provided appropriate detection gate delay and widths are employed. The suggested temporal windows for atomic and molecular emission detection from ns- and fs-LIBS are given in Fig. 4 [33]. It should be mentioned that apart from the method of plasma generation, the optimum temporal gating parameters can vary considerably with other factors such as laser fluence, laser wavelength (especially for ns-LA), sample composition (trace or bulk), strength of the observed emission lines, nature and pressure of the ambient gas, etc.

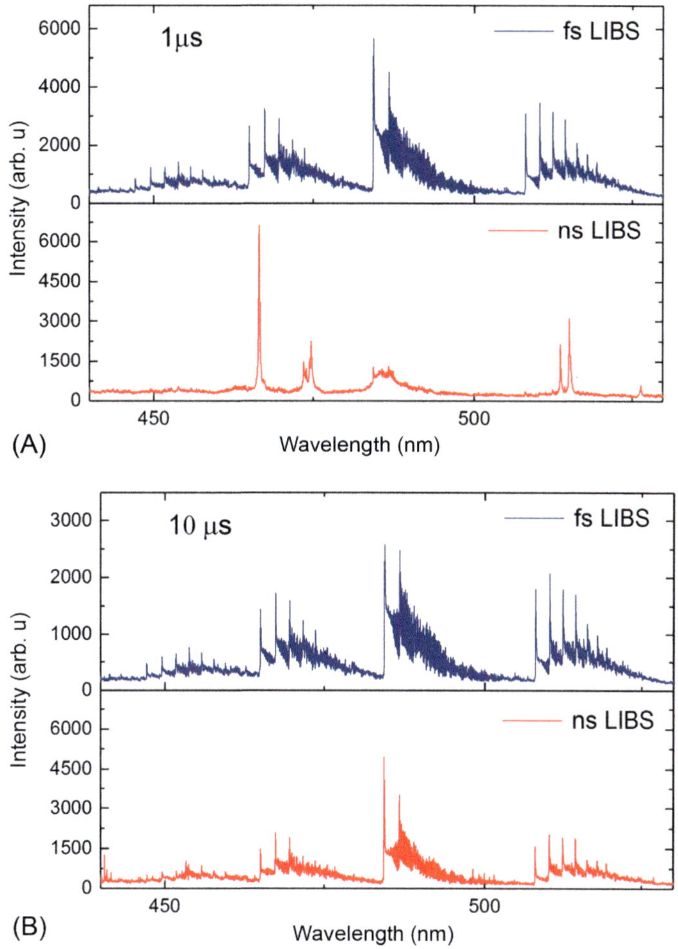

FIG. 3 The emission from Al laser-induced breakdown spectroscopy (LIBS) in air generated using 6-ns- and 30-fs-laser pulses recorded at (A) 1 μs and (B) 10 μs. The laser energy used for plasma generation was 17 mJ for both ns- and fs-LIBS [33].

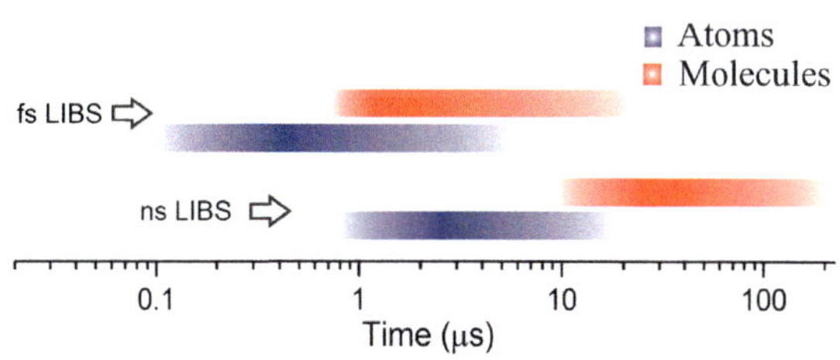

FIG. 4 Approximate temporal windows for atomic and molecular emission from ns- and fs-laser-induced breakdown spectroscopy.

When the fs-laser pulses were propagated in air, because of the nonlinear interaction between the laser pulses and air medium, filamentation would happen. Filamentation generation during fs-laser propagation is capable of overcoming the beam diffraction effects seen in ns-laser propagation and hence it is a very useful tool for remote analysis. Currently, significant efforts are under way for using fs-filamentation for LIBS [7]. Considering the timescales of interaction with the target material will be similar for fs- and filament-LIBS, significant differences in physical conditions of the generated plasma are not expected. However, energy coupling between the laser and target will be different for fs- and filament-LIBS considering that filaments carry a fraction of the energy of the propagating fs-laser beam. A detailed description of filament generation and the differences between fs-LIBS and filament-LIBS are given in the following section.

3 Filamentation and filament-LIBS

Self-focusing of the intense laser beam was problematic and generated deleterious effects in the past with the use of high-energy ns-lasers. However, such phenomena became more popular with the advent of ultrafast laser-generated filamentation because of its potential standoff applications. The filaments generated during fs-laser propagation can travel over a very large distance without suffering from diffraction effects seen during ns-laser propagation and can potentially be used to generate LIBS plumes at large standoff distances. It has been demonstrated that such self-guided filaments, on the order of 100 μm in diameter, can be generated over hundreds of meters to kilometers [34, 35]. These filaments can be used for detecting chemical and biological agents in the air using the LIDAR technique. Although the fs-laser filaments can be generated at large standoff distances, one of the major questions concerns its capability for solid target ablation and properties of subsequent plasma generation. In this section, a brief account of filament generation physics, methods for filament generation in the laboratory and field, fundamental differences in plumes generated by fs- and filament-LIBS, and approximate timescales of emission from various plumes are given.

When fs-pulses are sufficiently intense, self-focusing of the beam cannot be avoided when it is propagating in any transparent medium, including air because of the nonlinear Kerr effect. Self-focusing can be understood by considering the Gaussian intensity distribution of an fs-pulse where the intensity distribution across the wavefront is stronger at the center of the beam compared to the wings. During beam propagation, the high-intensity central part of the wavefront will experience a higher refractive index of the medium due to nonlinear effects than the low-intensity edges resulting in a curvature of the wavefront toward the axis of propagation. Such self-focusing of an ultrashort pulse leads to beam collapse and generation of a plasma. Because of the negative refractive index of the plasma medium, the beam becomes defocused. The repeated processes of Kerr self-focusing and plasma defocusing result in a continuous series of hot spots along the beam propagation axis. In other words, the balance or equilibrium between the focusing and defocusing of the beam leads to the formation of filaments or self-guided laser channels [36–38]. A pictorial view of filament generation during fs-laser propagation is given in Fig. 5A [7, 36]. Some of the nonlinear processes involved in the filamentation of an fs-laser are nonlinear phase changes of the laser pulse, multiphoton and tunnel ionization, and optical breakdown or plasma generation. A detailed account of these processes can be found elsewhere [39]. After filamentation, the laser beam generates a white spot surrounded by colored conical emission rings. The white light or supercontinuum is generated due to the broadening of the laser pulse due to self-phase modulation

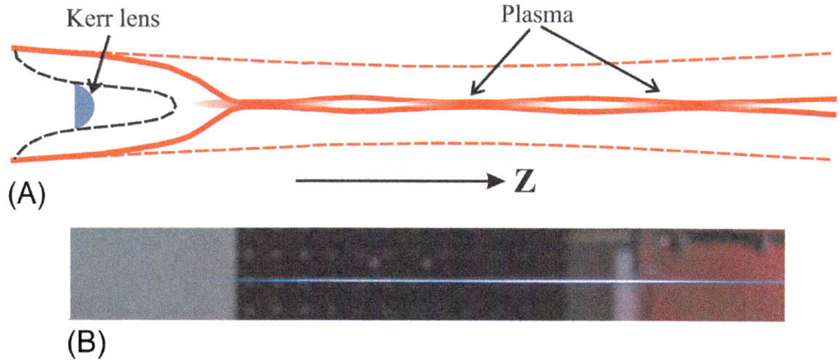

FIG. 5 (A) A schematic of the vertical slice of a propagating filament. The Kerr effect generates a positive refractive index profile that overcomes the diffraction of the beam, leading to focusing of the beam and plasma generation. The negative refractive index generated by the electrons in the plasma defocuses the beam. (B) A photograph of the filament generated in the laboratory.

and self-steepening. A photograph of the filament generated during the propagation of a 35-fs Ti:sapphire laser pulse in air is given in Fig. 5B. The photograph shows white light emission in the beam propagation path, which is due to the fluorescence of nitrogen molecules in air.

The critical power (P_{cr}) required for self-focusing for a Gaussian-shaped laser is $P_{cr} \approx 3.72\lambda^2/8\pi n_0 n_2$ where λ is the wavelength and n_0 and n_2 are the linear and the nonlinear refractive index coefficients [36]. It has to be mentioned that the nonlinear refractive index n_2 is not a constant in air and it changes with pulse duration. The reported critical power for filamentation during the propagation of a 800-nm Ti:sapphire ultrafast laser beam in air is $P_{cr} \sim 3$–10 GW [40]. When the power level of the laser used for filament generation is $\sim$1–10P_{cr}, a single filament will be formed that typically carries 10%–15% of the total pulse energy. The rest of the laser energy will be stored in the energy reservoir surrounding the filament. The critical power for self-focusing and subsequent plasma generation leads to intensity clamping, which sets the upper limit for the intensity at the self-focus. The plasma density in the filament is thus limited due to this balancing act. Such an intensity clamping phenomenon in ultrafast laser filamentation may have consequences, especially in the field of material ablation. For example, if we generate a filament at a standoff distance for material ablation, it is not possible to increase the intensity inside the self-focus by significantly increasing the laser energy of the pulse or critical power. Instead, when the energy of the pulse becomes higher, multiple filaments will occur in the beam [34].

For generating filaments in the laboratory, long focal length optics is routinely used to force self-focusing within the limited length of the air medium. When a positive lens is used for focusing an ultrafast laser, the beam will converge toward the geometrical focus until the laser power approaches critical power for self-focusing. So, the selection of the focal length of the lens is important, which controls the competition between the breakdown caused by geometrical focus and filamentation due to self-focusing. Liu et al. [41] used three different focal length lenses, namely, $f = 100$, 30, and 5 cm for generating optical breakdown plasmas and found that for a longer focal length lens ($f = 100$ cm), filamentation occurs well before the geometrical focus. For $f = 30$ cm, the filament is found to end at the geometrical focus with the formation of a stronger plasma than generated by the filament, while the self-focusing and geometrical focusing effects were indistinguishable when $f = 5$ cm (Fig. 6). Intensity clamping can still occur with geometrical

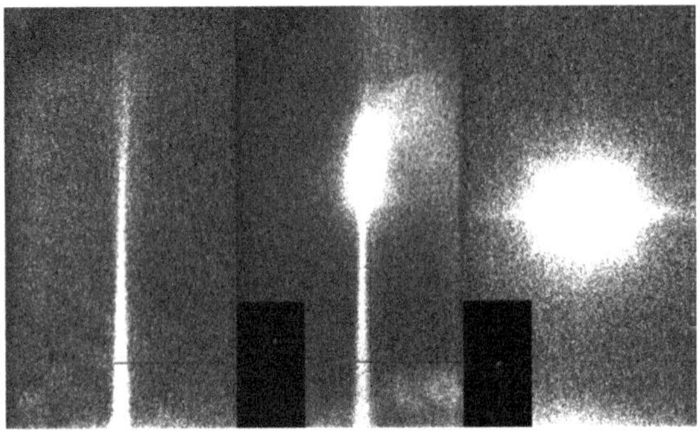

FIG. 6 Images taken via grating zeroth order when a 42-fs, 5-mJ/pulse was propagated through various focal length lenses (*left*: $f = 100$ cm; *middle*: $f = 30$ cm, and *right*: $f = 5$ cm). *Inset* shows low-contrast pictures. *Reprinted with permission from Chin et al. [41].*

focusing although the intensity will be slightly higher because of the increase in plasma density. The measured plasma density of the filaments generated through free propagation of the beam in 1 atm air was $\sim 1 \times 10^{14}$ cm^{-3}, and this value increased to $\sim 1 \times 10^{18}$ cm^{-3} when an $f = 10$ cm lens was used for optical breakdown [42].

The filaments carry only a small fraction of the laser energy ($\sim$10%) and the remaining energy will be stored in the energy reservoir surrounding it. The existence of a wide background or energy reservoir has been noticed even in the early demonstrations of ultrafast laser filamentation in air [43]. Previous studies also highlighted that for the maintenance of a high-intensity filament core it is important to have the presence of a good energy reservoir with 50% or more energy. Hence, the robustness of the filament depends critically on the diffraction of the energy reservoir, which has important implications in long-range filament generation. Although the energy reservoir is also useful for exciting gas molecules in LIDAR applications [44], its role for solid target ablation is questionable considering its intensities are reduced compared to the filament. For laboratory filament generation using various long focal length lenses ($f = 0.3$–10 m), the energy reservoir may contribute ablation along with filaments. Therefore filament generation conditions, such as self-collapsing or loosely focused, may influence the characteristics of the energy reservoir, which in turn may influence the characteristics of the produced LIBS plume from a solid target at a standoff distance. At larger standoff distances, filament generation should be solely due to self-collapse and hence ablation properties are expected to be similar assuming that there is no laser energy loss due to atmospheric absorption/scattering.

Traditionally, laser energy is the parameter used to increase the ablation efficiency and temperature of the ns-LIBS plume, and hence its emission intensity and persistence. However, intensity clamping in the filament core limits the amount of energy it can carry. So, an increase in laser energy will not generate stronger filaments; instead, it will lead to the stochastic formation of multiple filaments, which happens when the laser energy is $\geq 10P_{cr}$. Then, the filament beam will propagate nonlinearly with a turbulent exchange between nucleating and dying filaments and the photon bath itself. This presents a great challenge for ultrafast laser filament-LIBS since the increase in fs-laser energy will not be delivered to a certain point in the target for ablation; instead, it will generate multiple ablation zones, which may lead to large shot-to-shot fluctuations in LIBS signal. There exist a few studies on evaluating multiple filament formations and their role in ablation and LIBS plume emission [45].

3.1 Standoff filament-LIBS

The factors influencing standoff LIBS experiments are the energy coupling of the laser/filament with the target, the radiation properties of the LIBS plume, and the collection solid angle. When using filaments for standoff LIBS, the fs-laser beam should be propagated through an air medium and generate a filament at the standoff distance. The filament subsequently interacts with the sample of interest and generates a LIBS plume. The plasma generated should emit copious amounts of radiation in the spectral region of interest. For field experiments, unlike laboratory-based systems, one has to account for atmospheric turbulence, which leads to spreading of the laser (irradiance fluctuations) and pointing stability [46].

The majority of standoff plasma generation research presented in the literature has used ns-lasers and there exist a number of review articles on this topic [47–50]. Most of the reported ns-laser-based standoff systems used fundamental radiation from an Nd:YAG laser (1064 nm), and operations up to 120 m have been demonstrated using ~1 J of laser energy and a Galilean telescope for focusing and light collection [46, 50]. Similar to ns-LIBS, focused fs-lasers can also be used to generate plasmas at standoff distances. But, considering the complexity of the fs-laser system, it would be beneficial to use ns-lasers instead of fs-lasers if focusing optics are used. However, filaments generated by the fs-laser beam can be used for performing standoff LIBS. Several research groups started using filament-LIBS for standoff detection and this technique has been named remote filament-induced breakdown spectroscopy (R-FIBS). The studies explored the remote capability of R-FIBS, including explosive detection [51], plants [52], discrimination of trees [53], and cultural heritage monitoring [54]. Previous studies highlighted that filaments can be generated at distances over hundreds of meters to kilometers [34, 35, 55, 56]. Xu et al. [57] performed detailed studies of R-FIBS on metallic targets, showing several advantages over traditional ns-LIBS and projecting that performance out to kilometer-scale distances should be possible. So far, filaments have been used for the remote LA of solid samples up to ~500 m [35]. Rohwetter et al. [58] using the Teramobile laser facility demonstrated filament plasma generation at 180 m. R-FIBS on copper and steel plates has been demonstrated at 90-m standoff distances [59] and aluminum in a simulated polar environment at a distance of 60 m [60]. With the current advent of shorter-pulse (<30 fs) and high-power (10–100 TW) lasers, larger standoff distances should be feasible though further research in this direction is necessary.

4 Emission and plume characteristics: Ultrafast LIBS versus filament-LIBS

It is well known that laser irradiation conditions, particularly the spot size and the laser energy, affect the properties of the LIBS plume. The irradiation conditions in focused fs-LIBS and filament-LIBS are very different and will influence LIBS plasma properties such as temperature, density, and emission persistence. For example, if the filaments are generated in the laboratory using a loosely focused beam, the energy reservoir surrounding the filament may also be utilized for ablation. This may not be true for a filament-LIBS plume generated at larger standoff distances. In addition to these, multiple-filament generation at higher laser energies may lead to nonuniform ablation zones. Femtosecond filaments generated in the air are elongated and ablated plasma properties generated along the filament channel may have different properties. Several recent studies showed significant changes in radiation intensity of the loosely focused fs-filament ablation plumes along the filament channel although the temperature of the plasma is clamped [61]. The change in signal intensity at various locations along the filament channel was correlated to atomic number density, which is directly connected to ablation efficiency [61]. The ablation efficiency during filament-assisted plasma generation depends on both clamped filamentation intensity and the rest of the energy distributed in the energy reservoir [7]. Instead, Weidman et al. [62] noticed that the crater's diameter and volume remained constant for filament ablation at standoff distances >35 m. They also noticed that the craters generated during filament-assisted ablation are larger in diameter than the filament core (typically ~100 μm) and explained that the laser energy contained in the energy reservoir may also be used for ablation. A crater morphology study showed that there was a substantial reduction in evaporated mass for filament-induced ablation compared to sharply focused fs-LA [62, 63]. These results indicate that the differences in focusing conditions and the laser energy levels may also affect both the properties of the filaments and subsequently generated plasma properties.

There exist several studies comparing the difference between ns- and fs-LIBS; however, limited studies exist comparing the properties of the LIBS plume generated by focused fs-laser pulses and filament ablation. Since the interaction time of fs-laser/filament with the target materials is similar, the ablation mechanisms leading to fs- and filament-LIBS will also be similar. However, the persistence of LIBS may be different and it depends on effective energy coupling to the target and hence ablation efficiency. A study comparing the plasma properties showed that loosely focused filaments generate plumes with higher temperature and density (Fig. 7) along with increased persistence compared to filament-LIBS plumes generated by the self-collapse of the beam at longer distances [64]. However, regardless of the filament generation conditions, the LIBS plume generated by tightly focused laser pulses provides the highest temperature and density. Since these filament beam properties determine the ablation and plasma conditions, they, in turn, have a strong effect on the observed optical emission properties [64]. These results highlighted that the temporal window for detection may vary significantly for ultrafast LIBS plumes depending on their generation conditions. The suggested temporal windows for atomic and molecular emission for loosely focused laboratory filament-LIBS and self-collapsing filament-LIBS are ~0–1 and 0–0.2 μs, respectively [7].

The emission from ions, neutral atoms, and molecules can be detected using LIBS. However, their emission times with respect to the onset of plasma formation vary significantly and strongly depend on the temperature of the plasma. Compared to ns-LIBS, where laser–plasma coupling leads to strong heating, the temperatures of fs-LIBS are found to be lower. Because of this, ultrafast LIBS plume emission is predominantly contributed by atomic species with a near absence of ion and continuum emission and these features can be related to

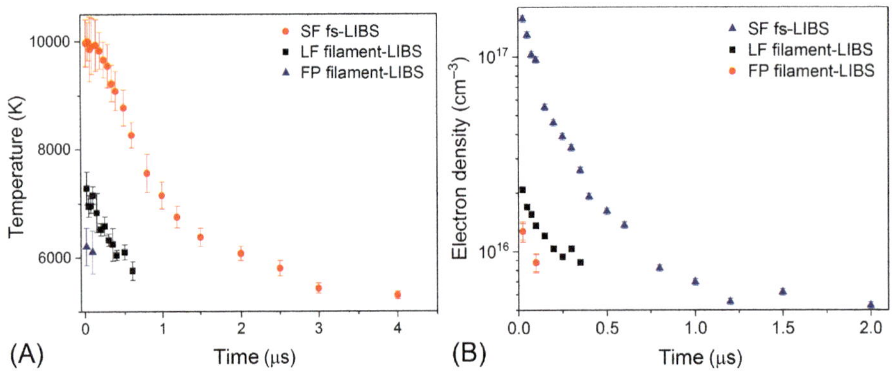

FIG. 7 The time evolution of (A) excitation temperature and (B) electron density of ablation plumes generated by sharply focused (SF, $f = 20$ cm) fs-pulses, loosely focused (LF, $f = 4$ m) laboratory filament ablation, and freely propagating (FP) filament ablation [64].

intensity clamping. An increase in laser energy during focused fs-LIBS experiments will lead to improved ablation rather than a significant increase in plasma temperature. Since lower-temperature conditions exist in ultrafast LIBS plumes even at early conditions, the molecular emission features can be detected immediately after ablation events. Since most of the laboratories perform ultrafast LIBS experiments with pulse energies <10 mJ, the strength of the shockwaves generated will be weaker in fs-LIBS compared to ns-LIBS where laser–plasma heating generates stronger shockwaves.

5 Modeling of ultrafast LIBS

Theoretical and computational models are essential for providing useful insights into laser–matter interaction, associated physical and chemical processes, and enhancing the analytical capabilities of LIBS [65]. However, a comprehensive model to study the complete LIBS phenomenon has not yet been developed. This is explained by the fact that the nonlinear interaction between a laser pulse and any material is extremely complex, involving many subareas such as laser light absorption, phase change, atomic and plasma physics, photon transfer, chemical kinetics, fluid dynamics, etc. Many computational studies have been focused on various aspects of the physical phenomena occurring during the absorption of laser energy, electron–lattice interactions, thermal effects and phase explosion, thermomechanical stresses and shocks, material ablation and ionization, plasma formation and expansion. The kinetic particle-in-cell (PIC), atomic molecular dynamics (MD) coupled with the two-temperature model (TTM), and computational fluid dynamics (CFD) approaches are used to study the evolution and dynamic changes in the fs-LPPs. A brief description of these models is provided in the following subsections along with simulation of LIBS spectral features.

5.1 PIC models

The coupled Maxwell–Lorentz–Newton equations can be used to accurately describe the behavior of LPPs in self-consistent electromagnetic fields on the kinetic level. The numerical implementation of these equations and kinetic simulations are carried out using the electromagnetic PIC models [66]. A spatial grid (cells) and large number of plasma particles (electrons, ions, and neutrals) are introduced into the computational domain. To speed up computations, they are lumped into the simulation macroparticles representing clusters on the order of 10^6–10^{10} real particles. Macroparticles have the same charge-to-mass ratio as the real plasma ions and electrons and are defined in spatial and velocity continuum phase space. The number and position of macroparticles in neighboring cells are used to evaluate the charge and current densities on each grid point. Maxwell's equations are solved on this grid, and the force acting on each macroparticle within a cell is calculated by interpolating values of the field from neighboring grid points. The positions and velocities of macroparticles are then advanced due to this force. The great advantage of a PIC model is the efficiency of force calculations [67]. The number of numerical operations for advancing the system of macroparticles on a single time step is proportional to the number of macroparticles N. By contrast, the number of operations on the order of N^2 is required in the MD simulations to evaluate the binary forces between all particle pairs. Moreover, PIC computation of the electric field on the grid points does not significantly increase by increasing the number of macroparticles, since the number of grid points is fixed.

PIC studies of LPPs have been performed using different levels of approximations, i.e., the dimensionality of spatial and velocity phase space (1D3V, 2D3V, and 3D3V). The basic principles of electromagnetic and quasistatic PIC methods and their applications were recently reviewed by Pukhov [68]. In some cases, the electrostatic PIC models, which eliminate the magnetic field and solve only the Poisson's equation for the electric field (the collisionless Vlasov–Poisson equations), were used to investigate laser–matter interactions [69, 70]. In particular, one of the early studies was focused on the initial stage of LPP plume expansion from a solid in vacuum using 1D-PIC modeling based on the solution of the collisionless Vlasov equations without accounting for the magnetic field [71]. The interaction between laser light and plasma was neglected, which is appropriate for ablation with ultrashort fs-pulses. This study was aimed at clarifying why the measured ion energy exceeds the thermal energy of ions in the expanding plume by several orders of magnitude. Enhancement of the kinetic energy of ions was attributed to Coulomb interaction between the charged particles in the plasma plume. Hybrid PIC models have also been developed, in which the electrons are treated as a background fluid. The expansion of a collisionless, hypersonic plasma plume into a vacuum was investigated using both fully kinetic and hybrid 2D-PIC models [72]. It is concluded that the isothermal electron fluid model is generally not valid for modeling the expansion of a collisionless plasma plume. The strong anisotropic thermodynamics for electrons should be taken into account in the fluid approximation.

The original PIC models were designed to describe collisionless, low-density, and high-temperature plasmas. However, the warm dense plasma of near-solid density and low temperatures is produced on the initial stage of laser–solid interactions. Over many years, PIC schemes were extended to account for coupled physical processes such as particle collisions, ionization, and strong laser field effects [73–76]. More realistic 2D- and 3D-PIC simulations have been carried out. The expansion of LPP has been studied in the presence of a magnetic field using a 3D-PIC model [77]. It is observed that the dynamics of plasma plumes was considerably altered with the development of flute-type instability in the presence of the magnetic field. 2D-PIC modeling of the propagation of laser wakes (plasma waves) in an inhomogeneous plasma with high-density walls was carried out to study the conditions for realizing the accelerating flying mirror [78]. However, the acceleration of plasma wakes was not observed in the PIC simulations. The experimental and 3D-PIC study of the coupling mechanisms between the ultraintense fs-laser pulse and high-density plasma in vacuum was conducted [79]. The dense plasma reflects a large fraction of the incident laser light. Therefore understanding coupling with laser light at the interface of the dense plasma and vacuum is essential for a wide range of applications. The transition from the vacuum heating (Brunel absorption) mechanism to the stochastic heating of electrons in the underdense plasma layer was revealed in this PIC study [79].

PIC simulations are computationally time consuming and limited to the micrometer-size domain and ps duration. This is because it is necessary to treat millions of macroparticles and resolve very small spatial and temporal scales to ensure numerical stability. The artificial, unphysical heating and macroparticle trapping that may arise in PIC modeling was reported [80]. These artifacts can be reduced by increasing resolution, number of macroparticles per cell, and using smoother grid interpolation functions. On the methodology side, there have been many attempts to develop new algorithms and improve PIC modeling. The Morse pair potential was implemented in the electromagnetic PIC models to investigate damage to a copper target from an fs-laser pulse [81]. Since the standard PIC models are designed to describe a gaseous plasma state, the motivation for this work was to include an interatomic attractive force in the PIC modeling and reproduce the low-temperature behavior of a solid plasma phase. The implicit time integration schemes are developed to increase the time step and speed up the PIC computations [82].

High-order PIC methods are implemented for reducing the number of degrees of freedom and improving the temporal and spatial resolution of the method [83]. The availability of massively parallel computers and improved algorithms for parallel PIC modeling will undoubtedly provide new insights into the fundamental mechanisms of fs-laser pulse–matter interactions and LIBS. 3D-PIC modeling of more realistic cases such as the fs-pulse-produced LIBS plasma expanding into a dense ambient would be possible.

5.2 TTM–MD approach

TTM–MD simulations of fs-laser–matter interactions on the atomic scale are very useful for understanding the nucleation kinetics, fast nonequilibrium transitions from solid to liquid and gas phases, and material removal [65]. The electrons play the main role in the initial stage of laser energy absorption and dissipation. The electronic subsystem of a material exposed to an ultrashort fs-laser pulse is excited into a highly nonequilibrium state. Due to the long electron–lattice relaxation time on the order of several ps, this state initially represents the hot electrons with a temperature of several thousands of kelvin and cold lattice atoms [65]. The electrons transfer their kinetic energy to a phonon subsystem heating up the lattice within a few ps. This process of electron–phonon coupling can be described by the TTM [84], in which the electrons and ions are treated as two separate subsystems with two distinct temperatures exchanging the energy. The continuous TTM solves two coupled heat conduction equations and predicts the temporal and spatial evolution of electron and lattice temperatures. However, the TTM has several limitations concerning the rapid exchange of energy between electrons and atoms after short-pulse laser heating. Nanoscale atomistic effects play a significant role and the thermophysical properties of the material are not well known under extreme conditions. The material undergoes a phase change with fragmentation, spallation, and development of cracks and voids. Therefore the continuum-based TTM approach has been coupled with the MD method, involving the solution of Newton's equations for each of the atoms [84]. The resulting TTM–MD model evaluates the heating of the electronic fluid by the laser, electron–phonon energy exchange, and electronic heat conduction.

In the combined TTM–MD model, the continuum TTM equation for the lattice temperature is substituted with the MD simulation of the movement of discrete atoms due to heating by the electronic fluid. Finite difference (FD) schemes are used to couple the electronic heat conduction with the MD simulation of atomic motions [85]. The nano-sized MD box represents a laterally confined small volume of material embedded within a large laser spot, so the effects of laser beam edges are neglected. The MD box is discretized into 1D cells with a sufficiently large number of atoms in each cell. The average kinetic energy of the thermal motion of atoms is used to evaluate the lattice temperature in the cells. Only the temperature change along the target depth is considered. The electrons as particles are not explicitly included in the TTM–MD model. The heat conduction equation for the electronic fluid is solved on the grid cells using an FD method. This adds an extra computational cost since hundreds of FD iterations are carried out during small MD time steps. A coupling term involving the electron temperature is added to the MD equations to account for the energy exchange between the electronic fluid and atoms. In classical MD, only the atoms interacting through interatomic potentials are considered with no explicit electrons taken into account. The challenge is the choice of proper potential functions describing the interaction between pairs of atoms. It is usually selected to reproduce known material properties such as crystal structure, cohesive energy, elastic constants, and melting temperature. For metals, a good choice is the embedded atom method potential combining both two-body and multibody interactions [86]. The momentum scaling model (MSM) coupled with the MD method (MSM–MD) was developed and applied to study fs-laser–matter interactions [87, 88]. The MSM–MD approach involves a direct rescaling of the kinetic energy of atoms by a factor involving the laser energy deposition on each atom, thus eliminating the need to solve the heat conduction equation for an electronic fluid. MD systems composed of tens to hundreds of millions of atoms and a finite-size laser spot on the surface of samples can be treated using the simplified MSM–MD approach.

TTM–MD studies of pulsed fs-LA have elucidated nonequilibrium electron–phonon coupling [89, 90], thermomechanical wave propagation [91], melting and disintegration mechanisms of metallic and nonmetallic materials [84, 92], laser-induced spallation [85], Coulomb explosion [65], and generation of nanoclusters (NCs) [88]. Several interesting features can be noted in these studies. Three distinct mechanisms of melting such as the nucleation and growth of the liquid phase, propagation of melting fronts, and ultrafast collapse of the overheated crystal structure have been revealed in early TTM–MD simulations [84]. Strong deviations of electron–phonon coupling and the electron heat capacity from the constant and linear temperature dependences have been found for several metals [89]. The hot electron blast force induces a thermomechanical shock near the irradiated surface, which travels into the material and converts to a tensile stress [91]. The growth and coalescence of voids is found to result in the formation of interconnected liquid regions and spallation of thin liquid layers from the surface [85].

A distinctive feature of fs-LA is the ejection of NCs comprising up to 80% of the ablated mass in an expanding plasma plume [88, 93]. The response of a sample composed of 20 million Al atoms to a Gaussian temporal and spatial distribution of laser pulse intensity within a finite-size spot was studied using the MSM–MD approach [88]. Snapshots of the plume are illustrated in Fig. 8 at 10, 20, and 50 ps for a laser fluence of 1500 J/m^2 [88]. The shape of the plume dynamically changes during expansion. At several tens of ps, it can be characterized as a long hollow ellipsoid with holes on the surface surrounded by Al atoms and small NCs. Later (50 ps), it disintegrates into large interconnected fragments with single Al atoms and many-atom NCs. The fractional distribution of ejected NCs is affected by the laser fluence. For a fluence of 500 J/m^2 (Fig. 9A), there are only single Al atoms during the first 5 ps and their portion drops to ~60% at 95 ps. The NCs created after 5 ps are primarily two-atom clusters varying around ~10% and increasing up to ~30% at 95 ps. The fraction of NCs with a larger number of

FIG. 8 3D snapshots of the Al plume for a laser fluence of 1500 J/m^2 at times of 10, 20, and 50 ps. *Reprinted with permission from Miloshevsky et al. [88].*

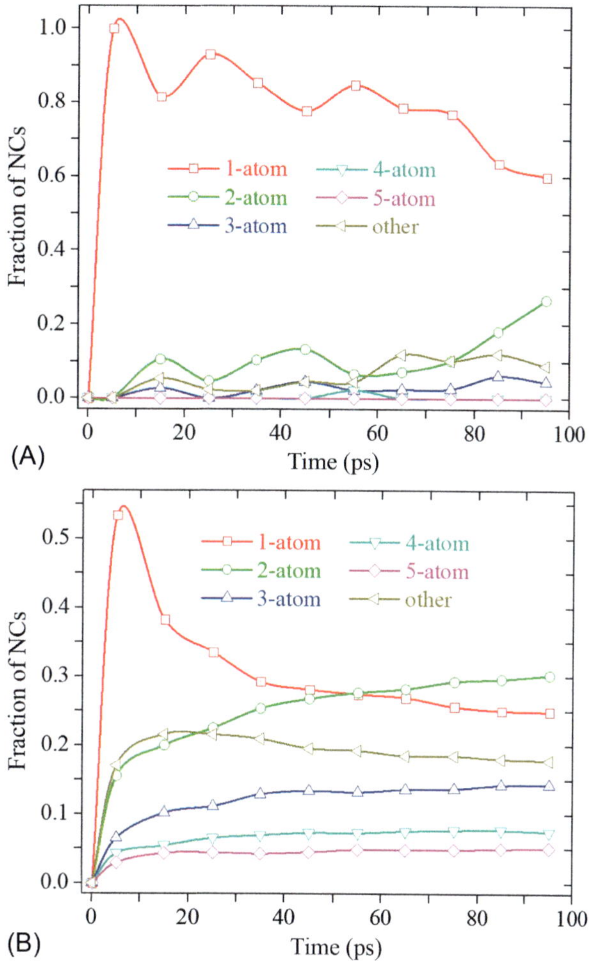

FIG. 9 Evolution of relative proportion (fraction) of different-size nanoclusters with one, two, three, etc. atoms in a plume for (A) 500 J/m^2 and (B) 1500 J/m^2 fluences. *Reprinted with permission from Miloshevsky et al. [88].*

atoms remains insignificant (≾10%). A higher fluence (1500 J/m^2) significantly alters the NC fractional distribution (Fig. 9B). The fraction of single Al atoms is reduced to ~55% on the initial stage of plume expansion and then drops to ~25% at 95 ps. However, the population of multiatom NCs is significantly increased. The two-atom NCs and "other" NCs with more than five atoms are most abundant, followed by three-atom, four-atom, and finally five-atom NCs. The fraction of two-atom NCs exceeds that of single Al atoms after ~55 ps. At higher laser fluences, single Al atoms and diatomic NCs remain the most prevalent components in the plasma plume [88].

It is crucial for LIBS applications to minimize the fraction of NCs in the plume to enhance plasma luminosity. Several TTM–MD studies have been focused on double-pulse LA with the goal of increasing the resolution and intensity of spectral lines. The melting and ablation efficiency of Al was investigated as a function of the relative intensity and time delay between two fs-laser pulses [94]. The Gaussian pulse shapes are found to be optimal and the ablation depth was decreased for delays longer than 10 ps when a second laser pulse was applied to a damaged sample. The other TTM–MD study shows that fragmentation occurs during the first pulse for delays longer than 50 ps [95]. However, the second pulse creates a high-pressure plasma and pushes the ablated layers back to the target. The crater depth becomes even smaller than that in the case of a single laser pulse. The mechanisms leading to the decrease of ablation depth were established because of (1) compensation of the rarefaction wave from the first pulse by the compression wave generated by the second pulse, and (2) absorption of the energy of the second pulse by the ablated material spalled by the first pulse [96]. The second laser pulse interacts with a hot, foamy structure increasing the decomposition of generated NCs and the production of single atoms. The plasma luminosity increases, which is essential for the increase in spectral line intensities in LIBS analysis. To provide further insights into the enhancement of LIBS performance, future TTM–MD simulations should be focused on double fs-laser pulses with a combination of different intensities, durations, shapes, and fluences.

5.3 CFD modeling

The CFD approach is based on the macroscopic continuum approximation. The fluid is treated as a structureless medium with density, pressure, temperature, and velocity fields varying in space and time in a continuous way. The physical length and timescales associated with the fluid flow are required to be much larger than the mean free path of atoms or molecules constituting a medium. This is the reason why the CFD model is well suited to treat the expansion of fluid plumes on much longer macroscopic scales compared to those accessible in the PIC and MD approaches. The Navier–Stokes hydrodynamic equations expressing the conservation of mass density, momentum, and energy are used to describe the expanding plasma plume as a viscous, heat-conducting fluid [97]. Most of the early research has focused on plasma expansion in a vacuum or low-pressure background gas using the 1D adiabatic models [65]. A particular feature of the later research was a combination of the experimental and 2D-CFD approaches for studies of both fs- and ns-LA.

The expansion of fs-laser-ablated Al plumes in Ar gas at 0.5 and 1 atm has been studied using shadowgraphy, fast-gated imaging, and CFD modeling [98]. The position of the shock front obtained from experimental and computational approaches is compared and found to be in good agreement. The computed spatial and temporal fields of pressure, temperature, density, and velocity have provided valuable insights into the dynamics of plasma expansion and shock-wave formation in fs-pulse-ablated Al plumes. The temperature is found to drop rapidly during the initial stage of expansion with the development of hot ($\sim$1400 K), cold ($\sim$300 K), and warm (>300 K) gas (plasma) later in the central, intermediate, and peripheral regions of the expanding plume. A map of the pressure field is shown in Fig. 10 for different time moments [98].

At $\sim$100 ns, expansion of the plume is slightly greater in the lateral direction (x-coordinate) than that in the vertical one (y-coordinate) due to the larger size of a plasma spot compared to the depth of a crater. However, the pressure at the developing shock front increases more in the vertical direction and it is significantly greater than the background Ar pressure of 0.1 MPa. At $\sim$500 ns, the shock front of the plume has traveled more than $\sim$0.4 mm in both directions with a slight advance in the direction perpendicular to the target surface. The pressure remains higher along the vertical direction and a region with pressure lower than background Ar pressure has developed at the center of the plume. At $\sim$1000 ns, the plume is more advanced in the vertical direction with the

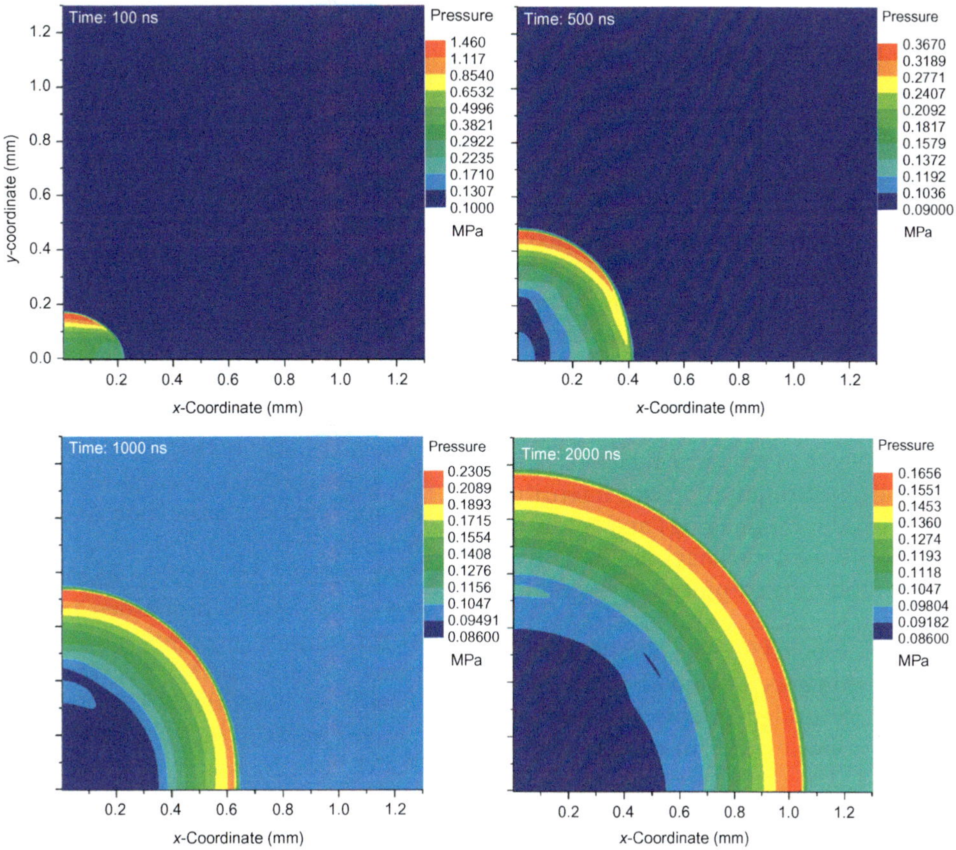

FIG. 10 Map of pressure field in the fs-pulse generated plume expanding in Ar gas at 1 atm for different time moments [98]. (A movie of this process is available at https://doi.org/10.1063/1.4873701.1.)

pressure remaining slightly higher than that in the lateral direction. The magnitude of pressure over the plume's front becomes more uniformly distributed. At ~2000 ns, the expansion has progressed >1 mm in both directions. A uniform pressure is developed over the entire shock front. At the plume's center, the region of pressure lower than background Ar pressure has increased up to ~0.6 mm. A large peak of mass density is observed in a very thin layer at the shock front with the rarefaction mass density much lower than background Ar density developed behind the shock during less than tens of ns. The shock initially propagates at a supersonic speed until its velocity becomes subsonic at late times and the vortical motion is developed in the rarefaction region. These results on the expansion of fs-LIBS are compared to the data on ns-LIBS of Al [97]. The similarities and differences in the expansion dynamics of ns-pulse- and fs-pulse-produced plumes are as follows: (1) magnitudes of pressure, temperature, mass density, and velocity fields are smaller in the fs-plume due to significantly lower fs-pulse laser energy (5 mJ compared to 100 mJ); (2) expansion of the fs-pulse-produced plume is three times less than that of the ns-plume; (3) fs-induced fields of gas dynamic parameters are much more evolved on the same timescale of ~2 μs; (4) oscillatory shock waves visible in the ns-plume are not observed in the internal region of the fs-plume; and (5) both fs- and ns-plumes acquire a nearly spherical shape at the end of expansion.

The expansion dynamics of shockwaves induced by an ns-pulsed laser on an unconfined flat surface and in a confined rectangular cavity has been investigated using both experimental and computational approaches [99]. In the first case, a hemispherical shape of the shockwave, especially at the latter stage of expansion, is observed. In the second case, the reflection of shockwaves from the channel sidewall and the compression of the plasma plume lead to its vertical stretch. Both experimental (spectrographic and intensified charge-coupled device imaging) and hydrodynamic (ANSYS Fluent package) methods have been used to study the ns-pulse LA of graphite in ambient argon under conditions of carbon nanotube synthesis [100]. The analyzed temperature, emission spectra, and electron densities are useful for understanding the synthesis of carbon nanotubes by the pulsed laser vaporization method. The early stages of expansion of the fs-laser-ablated uranium plasma plume in atmospheric oxygen have been examined using ANSYS Fluent CFD software [101]. This CFD model includes the kinetics of plasma chemistry. The formation of molecular uranium monoxide and dioxide (UO and UO_2) and trioxide (UO_3) was observed in the core and edge regions of the ablation plume, respectively. Future developments in the CFD modeling of LIBS-based applications will involve the implementation of particular models for describing a broad variety of processes and effects characteristic of ultrafast LA. The coupling of electromagnetic and hydrodynamic models involving Maxwell equations, electron-ion heat transfer equations, and compressible Navier–Stokes equations was used very recently to study the evolution of metal surfaces irradiated by multiple fs-laser pulses [22].

5.4 Modeling of LPP emission spectra

The modeling of LPP emission spectral features is an active area of research regardless of the different LA generation conditions. There exist several spectroscopic databases in the literature that cover information on various materials, spectral wavelength, energy levels, degeneracy, oscillator strength, etc. [102, 103]. Fundamental research on the modeling of LPP emission spectra focuses on the spectroscopic model development and accurate calculation of the atomic data such as the energy levels for ground and excited states of different ions, wave functions, oscillator strengths, ionization potentials, photoionization cross-sections, etc. For LIBS applications, these data can be calculated nowadays by using atomic structure codes such as FAC, HULLAC, CATS, RMBPT, etc. Databases of the atomic properties are usually created for various elements and their mixtures. For non-LTE LPP, both radiative and collisional processes are important. Therefore collisional–radiative models (e.g., ATOMIC, FLYCHK, SCRAM codes) should be used to determine the level of populations of different ions. A large system of rate equations is solved describing the population and depopulation of all the levels by collisional and radiative processes such as collisional excitation and deexcitation, collisional ionization, three-body recombination, spontaneous and stimulated emission, photorecombination, and dielectronic recombination. The atomic data and ionic populations are then utilized to determine the ion and electron concentrations and the average charge state of plasma, and to spectrally resolve the emissivities and opacities of LPP. This fundamental approach, involving the atomic structure and plasma emission codes, was applied to study the ionization potential depression in hot dense Al plasmas and its impact on the emission spectra using FLYCHK [104], opacities of light elements using ATOMIC [105], emission spectra of copper LPP induced by mid-infrared fs-pulses [106], and collisional ionization rates under warm and hot dense plasma conditions [107]. The theory of fundamental plasma processes and their connection to atomic and molecular emission spectra in LIBS were recently reviewed [108].

A number of rapid tools for LIBS analysis of experimental emission spectra are also being developed. The National Institute of Standards and Technology (NIST) LIBS web-based interface provides access to the NIST atomic spectra database for extracting and postprocessing spectroscopic information to evaluate synthetic spectra of chemical elements and mixtures [102]. This NIST LIBS analysis of LPP is limited by the assumption of LTE plasma conditions and availability of spectral line data for atoms and ions, especially for heavy elements. It was demonstrated that the LPP generated in a background gas by ns-laser pulses can be described by the LTE model utilizing the spectroscopic data taken from NIST [102] and Kurucz [103] databases [109]. However, the question of assuming the LTE condition remains open particularly for molecular emission bands [108]. A more advanced commercial PrismSPECT tool, part of the SPECT3D package [110], is based on a collisional–radiative code that can be applied for LIBS analysis of both LTE and non-LTE plasmas. PrismSPECT can solve time-dependent atomic rate equations and treat LPP mixtures with non-Maxwellian electron distributions. However, it is applicable to a limited set of chemical elements from hydrogen through argon and their mixtures.

As an example of spectral modeling, we present here our results on the strength of spectral lines of bromine (Br) traces in the emission spectra of a nitrogen–bromine (N—Br) LPP mixture in the visible wavelength range [111]. The electrical and electronic products, textiles, plastics, polymeric materials, paints, and gaseous HBr compounds contain Br from traces (~50 μg/g or 50 ppm) to several percents (up to ~3%) [112]. Detection, identification, and elimination of toxic Br compounds at the recycling process are needed due to their environmental contamination and health hazards. The REODP code [113] incorporating the atomic structure post-Hartree–Fock–Slater and collisional–radiative steady-state models [114, 115] was used to determine the detailed spectra of N—Br plasma mixtures with resolved emission lines and their intensities. The charge state of N and Br ions as a function of temperature is shown in Fig. 11A for a 99%N–1%Br mixture with a density of $10^{16}\,cm^{-3}$. At low temperatures ($\lesssim 1$ eV), the neutral N I and Br I atoms prevail. The number density of N II, Br II, and Br III ions dominates in a range of temperatures from ~1 to ~3 eV. In the temperature range from ~2.7 to ~5 eV, the plasma is predominantly composed of N III and Br IV ions. The emissivities of Br spectral lines in an N background for various Br fractions are shown in Fig. 11B for a temperature of 2 eV and density of $10^{16}\,cm^{-3}$. The intense emission lines of pure N plasma (100% N–0%Br) in the visible wavelength range are due to the 3d-4p, 3s-3p, 3p-3d, and 3d-4f transitions of N II ions at 679.445, 576.089, 548.266, and 458.12 nm. The first three photon wavelengths are found to be in very good agreement with those available in the NIST database, but the last spectral line is not reported in NIST atomic line data. There is also a number of weak N lines in the absence of Br. A plasma mixture of 99.99% N–0.01%Br (100 ppm of Br) starts to emit photons in two particular lines at 560.366 and 509.96 nm within the array of 5s-5p transitions of Br II ions. The increase in Br trace concentration to 1000 ppm (99.9%N–0.1%Br mixture) rises the emissivity of these two Br lines. Two additional Br II lines appear at 451.22 and 595.49 nm and the two other spectral lines at 521.97 and 528.326 nm correspond to the 4d-5p transitions of Br III ions (Fig. 11B). The calculated wavelengths were found to agree well with some of those listed in the NIST database for Br II and Br III ions. However, the line transitions of these ions are not provided by NIST and the

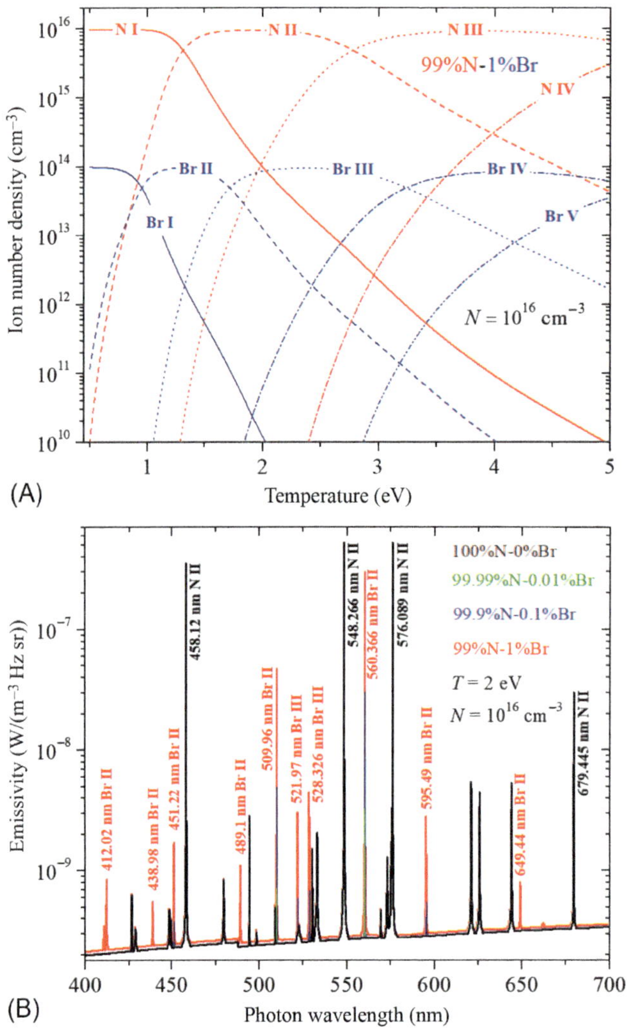

FIG. 11 (A) Ion composition of 99%N–1%Br laser-produced plasma (LPP) mixture as a function of temperature for a density of 10^{16} cm^{-3}, and (B) emissivities of N and Br spectral lines for 100%N–0%Br *(black)*, 99.99%N–0.01%Br *(green)*, 99.9%N–0.1%Br *(blue)*, and 99%N–1%Br *(red)* LPP mixtures. The intensity of emission in Br spectral lines increases with increasing Br percentage in the N—Br mixture.

identification of spectral lines is not possible. With a further increase in Br content to 10,000 ppm (99%N–1%Br mixture), the emissivity of all the Br lines increases and a set of new spectral lines originates from Br II ions. However, the intensity of major emission lines of Br in the visible range remains lower compared to that of N spectral lines even at 10,000 ppm. In contrast, the Br line emission in the near and middle ultraviolet wavelength range from the N—Br LPP mixture with 10,000 ppm of Br is found to be considerably higher than the emissivity of N lines [111]. It is observed in general that the emissivity of Br lines increases by several orders of magnitude with increasing Br additive trace concentration from 100 to 10,000 ppm.

The results can be useful for LIBS analysis of Br pollutions in air and technical polymers.

6 Summary

This chapter provided an overview of the current status of ultrafast and filamentation LIBS, including modeling efforts. It is well known that the physics behind LIBS plume generation, its emission and expansion are notoriously complex. The physical properties of LIBS plumes change significantly with time and space and have a direct correlation to numerous experimental parameters, including the pulse width of the laser. Currently, ns-lasers are routinely used for LIBS allowing for rapid, noncontact measurements, and standoff distances are only limited by the focusing difficulty raised due to diffraction of the laser beam. In the last decade, significant progress has been made in the field of ultrafast LIBS partly due to the availability of table-top, robust fs-laser systems. There are several potential advantages in using fs-lasers for ablation instead of using traditional ns-lasers and some of them have limited HAZ, precision micromachining, minimal elemental fractionation in analytical systems, lower ablation threshold, and better spatial and depth resolution. In addition to these, fs-LIBS provides cooler plasmas, which will lead to reduced continuum, emission predominantly from neutral atomic species, early molecular emission times, etc. Considering these potential benefits of fs-LA, ultrafast laser sources are superior for microanalysis.

When the fs-laser intensity is higher than the critical power for self-focusing, filamentation can occur in the beam during its propagation. These filaments can be generated at large standoff distances and capable of overcoming diffraction effects commonly seen in ns-laser propagation. There exist several demonstrations of filament-LIBS analyses at various standoff distances. Most of the previous efforts were focused on laboratory filament-LIBS where standoff distances were limited to $\sim$10 m. Higher-energy ultrafast laser pulses may be necessary to propagate and perform filament-LIBS at large standoff distances. An increase in laser energy will not be able to generate stronger filaments; instead, it will lead to the stochastic formation of multiple filaments, which happens when the laser energy is $\geq 10P_{cr}$. So, some of the existing challenges for increasing the standoff distances using filament-LIBS include enhancing the energy coupling of the filament with solid samples located at a standoff distance, multiple filament ablation and hence large pulse-to-pulse LIBS signal variation, etc. Both ns-long pulse and ultrafast LIBS analysis have matured over many years, but filament-LIBS is still in its infancy and hence significant research efforts are necessary before it can be used for large standoff distances.

This chapter also provided a brief overview of the modeling efforts of fs-LIBS. Although a large number of experimental articles are available in the field of fs-LIBS, very limited efforts have been made in the modeling of the physics behind ultrafast LA, plasma formation, and expansion. It has to be highlighted that computational models are crucial for providing useful insights into laser–matter interaction, associated physical and chemical processes, and enhancing the analytical capabilities of LIBS. Considering the complexity of the LIBS process, a comprehensive model to study the complete LIBS phenomenon has not yet been developed even for ns-LIBS. So, further research efforts are necessary in the modeling of ultrafast LIBS.

Acknowledgment

This work was supported by the DOE/NNSA Defense Nuclear Nonproliferation (DNN) R&D Office and the Defense Threat Reduction Agency under Grant No. HDTRA119-1-0019. Pacific Northwest National Laboratory is operated for the US Department of Energy by the Battelle under Contract No. DEAC05-76RLO1830.

References

[1] F. Brech, L. Cross, Optical microemission stimulated by a ruby laser, Appl. Spectrosc. 16 (1962) 59.

[2] J.P. Singh, S.N. Thakur, Laser-induced breakdown spectroscopy, Elsevier Science, 2007.

[3] D.W. Hahn, N. Omenetto, Laser-induced breakdown spectroscopy (LIBS), part I: review of basic diagnostics and plasma-particle interactions: still-challenging issues within the analytical plasma community, Appl. Spectrosc. 64 (2010) 335a–366a.

[4] D. Strickland, G. Mourou, Compression of amplified chirped optical pulses, Opt. Commun. 56 (1985) 219–221.

[5] E.L. Gurevich, R. Hergenroder, Femtosecond laser-induced breakdown spectroscopy: physics, applications, and perspectives, Appl. Spectrosc. 61 (2007) 233A–242A.

[6] T.A. Labutin, V.N. Lednev, A.A. Ilyin, A.M. Popov, Femtosecond laser-induced breakdown spectroscopy, J. Anal. At. Spectrom 31 (2016) 90–118.

[7] S.S. Harilal, B.E. Brumfield, N.L. LaHaye, K.C. Hartig, M.C. Phillips, Optical spectroscopy of laser-produced plasmas for standoff isotopic analysis, Appl. Phys. Rev. 5 (2018) 021301.

[8] R.E. Russo, X.L. Mao, J.L. Gonzalez, J. Yoo, Femtosecond versus nanosecond laser pulse duration for laser ablation chemical analysis, Spectroscopy 28 (2013) 24.

[9] A.M. Weiner, Ultrafast Optics, John Wiley & Sons, New York, 2009.

[10] S.L. Chin, Femtosecond Laser Filamentation, Springer-Verlag, New York, 2010.

[11] M. Sabsabi, Femtosecond LIBS, in: J.P. Singh, S. N. Thakur (Eds.), Laser-Induced Breakdown Spectroscopy, Elsevier, Amsterdam, 2007.

[12] B. Rethfeld, K. Sokolowski-Tinten, D. von der Linde, S. I. Anisimov, Timescales in the response of materials to femtosecond laser excitation, Appl. Phys. Mater. Sci. Process. 79 (2004) 767–769.

[13] B.N. Chichkov, C. Momma, S. Nolte, F. von Alvensleben, A. Tunnermann, Femtosecond, picosecond and nanosecond laser ablation of solids, Appl. Phys. A 63 (1996) 109–115.

[14] J.R. Freeman, S.S. Harilal, P.K. Diwakar, B. Verhoff, A. Hassanein, Comparison of optical emission from nanosecond and femtosecond laser ablated brass plasma in atmosphere and vacuum, Spectrochim. Acta, Part B 87 (2013) 65–73.

[15] K.K. Anoop, X.C. Ni, M. Bianco, D. Paparo, X. Wang, R. Bruzzese, S. Amoruso, Two-dimensional imaging of atomic and nanoparticle components in copper plasma plume produced by ultrafast laser ablation, Appl. Phys. A 117 (2014) 313–318.

[16] O. Albert, S. Roger, Y. Glinec, J.C. Loulergue, J. Etchepare, C. Boulmer-Leborgne, J. Perriere, E. Millon, Time-resolved spectroscopy measurements of a titanium plasma induced by nanosecond and femtosecond lasers, Appl. Phys. A 76 (2003) 319–323.

[17] R. Stoian, D. Ashkenasi, A. Rosenfeld, E.E.B. Campbell, Coulomb explosion in ultrashort pulsed laser ablation of Al2O3, Phys. Rev. B 62 (2000) 13167–13173.

[18] E.G. Gamaly, A.V. Rode, B. Luther-Davies, V. T. Tikhonchuk, Ablation of solids by femtosecond lasers: ablation mechanism and ablation thresholds for metals and dielectrics, Phys. Plasmas 9 (2002) 949–957.

[19] M.E. Shaheen, J.E. Gagnon, B.J. Fryer, Femtosecond laser ablation behavior of gold, crystalline silicon, and fused silica: a comparative study, Laser Phys. 24 (2014) 106102.

[20] B. Verhoff, S.S. Harilal, A. Hassanein, Angular emission of ions and mass deposition from femtosecond and nanosecond laser-produced plasmas, J. Appl. Phys. 111 (2012) 123304.

[21] S.S. Harilal, N. Farid, J.F. Freeman, P.K. Diwakar, N. L. LaHaye, A. Hassanein, Background gas collisional effects on expanding fs and ns laser ablation plumes, Appl. Phys. A 117 (2014) 319.

[22] A. Rudenko, C. Mauclair, F. Garrelie, R. Stoian, J.-P. Colombier, Amplification and regulation of periodic nanostructures in multipulse ultrashort laser-induced surface evolution by electromagnetic-hydrodynamic simulations, Phys. Rev. B 99 (2019) 235412.

[23] S. Razi, O. Varlamova, J. Reif, M. Bestehorn, S. Varlamov, M. Mollabashi, K. Madanipour, M. Ratzke, Birth of periodic Micro/Nano structures on 316L stainless steel surface following femtosecond laser irradiation; single and multi scanning study, Opt. Laser Technol. 104 (2018) 8–16.

[24] S.S. Harilal, Influence of spot size on propagation dynamics of laser-produced tin plasma, J. Appl. Phys. 102 (2007) 123306.

[25] J. Guo, J. Shao, T. Wang, C. Zheng, A. Chen, M. Jin, Optimization of distances between the target surface and focal point on spatially confined laser-induced breakdown spectroscopy with a cylindrical cavity, J. Anal. At. Spectrom 32 (2017) 367–372.

[26] S.S. Harilal, P.K. Diwakar, M.P. Polek, M.C. Phillips, Morphological changes in ultrafast laser ablation plumes with varying spot size, Opt. Express 23 (2015) 15608.

[27] L.A. Emmert, R.C. Chinni, D.A. Cremers, C.R. Jones, W. Rudolph, Comparative study of femtosecond and nanosecond laser-induced breakdown spectroscopy of depleted uranium, Appl. Optics 50 (2011) 313–317.

[28] S.S. Harilal, P. Skordzki, A. Miloshevsky, B. Brumfield, M.C. Phillips, G. Miloshevsky, On- and off-axis spectral emission features from laser-produced gas breakdown plasmas, Phys. Plasmas 24 (2017) 063304.

[29] S.D. Zhang, X.H. Wang, M.H. He, Y.B. Jiang, B. C. Zhang, W. Hang, B.L. Huang, Laser-induced

plasma temperature, Spectrochim. Acta B 97 (2014) 13–33.

[30] S.S. Harilal, E.J. Kautz, B.E. Bernacki, M.C. Phillips, P. Skrodzki, M. Burger, I. Jovanovic, The physical conditions for UO formation in laser-produced uranium plumes, Phys. Chem. Chem. Phys. 21 (2019) 16161.

[31] S.S. Harilal, B.E. Brumfield, B. Cannon, M.C. Phillips, Shock wave mediated plasma chemistry of AlO formation in laser ablation plumes, Anal. Chem. 88 (2016) 2296–2302.

[32] A.K. Shaik, N.R. Epuru, H. Syed, C. Byram, V.R. Soma, Femtosecond laser induced breakdown spectroscopy based standoff detection of explosives and discrimination using principal component analysis, Opt. Express 26 (2018) 8069–8083.

[33] S.S. Harilal, J. Yeak, B.E. Brumfield, M.C. Phillips, Dynamics of molecular emission features from nanosecond, femtosecond laser and filament ablation plasmas, J. Anal. At. Spectrom 31 (2016) 1192–1197.

[34] M. Rodriguez, R. Bourayou, G. Mejean, J. Kasparian, J. Yu, E. Salmon, A. Scholz, B. Stecklum, J. Eisloffel, U. Laux, A.P. Hatzes, R. Sauerbrey, L. Woste, J. P. Wolf, Kilometer-range nonlinear propagation of femtosecond laser pulses, Phys. Rev. E 69 (2004) 036607.

[35] M. Durand, A. Houard, B. Prade, A. Mysyrowicz, A. Durecu, B. Moreau, D. Fleury, O. Vasseur, H. Borchert, K. Diener, R. Schmitt, F. Theberge, M. Chateauneuf, J.F. Daigle, J. Dubois, Kilometer range filamentation, Opt. Express 21 (2013) 26836–26845.

[36] A. Couairon, A. Mysyrowicz, Femtosecond filamentation in transparent media, Phys. Rep. 441 (2007) 47–189.

[37] A. Braun, G. Korn, X. Liu, D. Du, J. Squier, G. Mourou, Self-channeling of high-peak-power femtosecond laser-pulses in air, Opt. Lett. 20 (1995) 73–75.

[38] L. Berge, S. Skupin, R. Nuter, J. Kasparian, J.P. Wolf, Ultrashort filaments of light in weakly ionized, optically transparent media, Rep. Prog. Phys. 70 (2007) 1633–1713.

[39] S.L. Chin, Femtosecond Laser Filamentation, Springer, New York, 2009, p. 162.

[40] W. Liu, S.L. Chin, Direct measurement of the critical power of femtosecond Ti: sapphire laser pulse in air, Opt. Express 13 (2005) 5750–5755.

[41] W. Liu, Q. Luo, S.L. Chin, Competition between multiphoton/tunnel ionization and filamentation induced by powerful fs laser pulses in air, Chin. Opt. Lett. 1 (2003) 56.

[42] F. Theberge, W.W. Liu, P.T. Simard, A. Becker, S. L. Chin, Plasma density inside a femtosecond laser filament in air: strong dependence on external focusing, Phys. Rev. E 74 (2006) 036406.

[43] E.T.J. Nibbering, P.F. Curley, G. Grillon, B.S. Prade, M. A. Franco, F. Salin, A. Mysyrowicz, Conical emission from self-guided femtosecond pulses in air, Opt. Lett. 21 (1996) 62–64.

[44] I. Dicaire, V. Jukna, C. Praz, C. Milian, L. Summerer, A. Couairon, Spaceborne laser filamentation for atmospheric remote sensing, Laser Photonics Rev. 10 (2016) 481–493.

[45] P.J. Skrodzki, M. Burger, I. Jovanovic, Transition of femtosecond-filament-solid interactions from single to multiple filament regime, Sci. Rep. 7 (2017) 12740.

[46] J.J. Laserna, R.F. Reyes, R. Gonzalez, L. Tobaria, P. Lucena, Study on the effect of beam propagation through atmospheric turbulence on standoff nanosecond laser induced breakdown spectroscopy measurements, Opt. Express 17 (2009) 10265–10276.

[47] D.A. Cremers, The analysis of metals at a distance using laser-induced breakdown spectroscopy, Appl. Spectrosc. 41 (1987) 572–579.

[48] B. Sallé, P. Mauchien, S. Maurice, Laser-Induced Breakdown Spectroscopy in open-path configuration for the analysis of distant objects, Spectrochim. Acta, Part B 62 (2007) 739–768.

[49] J.L. Gottfried, F.C. De Lucia, C.A. Munson, A. W. Miziolek, Standoff detection of chemical and biological threats using laser-induced breakdown spectroscopy, Appl. Spectrosc. 62 (2008) 353–363.

[50] F.J. Fortes, J.J. Laserna, The development of fieldable laser-induced breakdown spectrometer: no limits on the horizon, Spectrochim. Acta B 65 (2010) 975–990.

[51] D. Mirell, O. Chalus, K. Peterson, J.C. Diels, Remote sensing of explosives using infrared and ultraviolet filaments, J. Opt. Soc. Am. B 25 (2008) B108–B111.

[52] J.N. Kunz, D.V. Voronine, H.W.H. Lee, A.V. Sokolov, M.O. Scully, Rapid detection of drought stress in plants using femtosecond laser-induced breakdown spectroscopy, Opt. Express 25 (2017) 7251–7262.

[53] Y. Fu, M.Y. Hou, H.W. Zang, H.L. Li, S.L. Chin, H. L. Xu, Remote discrimination of willow, pine and poplar trees and their growing environments by femtosecond filament-induced breakdown spectroscopy, Spectrochim. Acta B At. Spectrosc. 155 (2019) 107–114.

[54] S. Tzortzakis, D. Anglos, D. Gray, Ultraviolet laser filaments for remote laser-induced breakdown spectroscopy (LIBS) analysis: applications in cultural heritage monitoring, Opt. Lett. 31 (2006) 1139–1141.

[55] B. La Fontaine, F. Vidal, Z. Jiang, C.Y. Chien, D. Comtois, A. Desparois, T.W. Johnston, J.C. Kieffer, H. Pepin, H.P. Mercure, Filamentation of ultrashort pulse laser beams resulting from their propagation over long distances in air, Phys. Plasmas 6 (1999) 1615–1621.

[56] G. Mechain, C. D'Amico, Y.B. Andre, S. Tzortzakis, M. Franco, B. Prade, A. Mysyrowicz, A. Couairon, E. Salmon, R. Sauerbrey, Range of plasma filaments created in air by a multi-terawatt femtosecond laser, Opt. Commun. 247 (2005) 171–180.

[57] H.L. Xu, J. Bernhardt, P. Mathieu, G. Roy, S.L. Chin, Understanding the advantage of remote femtosecond LIBS of metallic targets, J. Appl. Phys. 101 (2007) 033124.
[58] P. Rohwetter, K. Stelmaszczyk, L. Wöste, R. Ackermann, G. Méjean, E. Salmon, J. Kasparian, J.-P. Wolf, Filament-induced remote surface ablation for long range laser-induced breakdown spectroscopy operation, Spectrochim. Acta, Part B 60 (2005) 1025–1033.
[59] K. Stelmaszczyk, P. Rohwetter, G. Méjean, J. Yu, E. Salmon, J. Kasparian, R. Ackermann, J.-P. Wolf, L. Wöste, Long-distance remote laser-induced breakdown spectroscopy using filamentation in air, Appl. Phys. Lett. 85 (2004) 3977–3979.
[60] H.L. Xu, P.T. Simard, Y. Kamali, J.F. Daigle, C. Marceau, J. Bernhardt, J. Dubois, M. Chateauneuf, F. Theberge, G. Roy, S.L. Chin, Filament-induced breakdown remote spectroscopy in a polar environment, Laser Phys. 22 (2012) 1767–1770.
[61] S.S. Harilal, J. Yeak, M.C. Phillips, Plasma temperature clamping in fLIBS, Opt. Express 23 (2015) 27113.
[62] M. Weidman, K. Lim, M. Ramme, M. Durand, M. Baudelet, M. Richardson, Stand-off filament-induced ablation of gallium arsenide, Appl. Phys. Lett. 101 (2012) 034101.
[63] A. Valenzuela, C. Munson, A. Porwitzky, M. Weidman, M. Richardson, Comparison between geometrically focused pulses versus filaments in femtosecond laser ablation of steel and titanium alloys, Appl. Phys. B Lasers Opt. 116 (2014) 485–491.
[64] S.S. Harilal, J. Yeak, B.E. Brumfield, M.C. Phillips, Consequences of femtosecond laser filament generation conditions in standoff LIBS, Opt. Express 24 (2016) 17941.
[65] B. Rethfeld, D.S. Ivanov, M.E. Garcia, S.I. Anisimov, Modelling ultrafast laser ablation, J. Phys. D Appl. Phys. 50 (2017) 193001.
[66] J.P. Verboncoeur, Particle simulation of plasmas: review and advances, Plasma Phys. Controlled Fusion 47 (2005) A231–A260.
[67] P.L. Pritchett, Particle-in-cell simulation of plasmas—a tutorial, in: J. Büchner, M. Scholer, C.T. Dum (Eds.), Space Plasma Simulation, Springer Berlin Heidelberg, Berlin, Heidelberg, 2003, , pp. 1–24.
[68] A. Pukhov, Particle-in-cell codes for plasma-based particle acceleration, CERN Yellow Reports 1 (2016) 181–206.
[69] G.V. Miloshevsky, A. Hassanein, Self-confinement of finite dust clusters in isotropic plasmas, Phys. Rev. E 85 (2012) 056405.
[70] G.V. Miloshevsky, A. Hassanein, Dynamics of isotropic and drifting plasmas in the vicinity and within two-dimensional finite dust clusters, Plasma Phys. Controlled Fusion 54 (2012) 045006.
[71] O. Ellegaard, T. Nedelea, J. Schou, H.M. Urbassek, Plume expansion of a laser-induced plasma studied with the particle-in-cell method, Appl. Surf. Sci. 197–198 (2002) 229–238.
[72] Y. Hu, J. Wang, Expansion of a collisionless hypersonic plasma plume into a vacuum, Phys. Rev. E 98 (2018) 023204.
[73] A.J. Kemp, R.E.W. Pfund, J. Meyer-ter-Vehn, Modeling ultrafast laser-driven ionization dynamics with Monte Carlo collisional particle-in-cell simulations, Phys. Plasmas 11 (2004) 5648–5657.
[74] T. Taguchi, T.M. Antonsen, J. Palastro, H. Milchberg, K. Mima, Particle in cell analysis of a laser-cluster interaction including collision and ionization processes, Opt. Express 18 (2010) 2389–2405.
[75] A. Gonoskov, S. Bastrakov, E. Efimenko, A. Ilderton, M. Marklund, I. Meyerov, A. Muraviev, A. Sergeev, I. Surmin, E. Wallin, Extended particle-in-cell schemes for physics in ultrastrong laser fields: review and developments, Phys. Rev. E 92 (2015) 023305.
[76] D. Wu, X.T. He, W. Yu, S. Fritzsche, Particle-in-cell simulations of laser–plasma interactions at solid densities and relativistic intensities: the role of atomic processes, High Power Laser Sci. Eng. 6 (2018) e50.
[77] A. Roy, A. Endo, T. Mocek, 3-D particle-in-cell simulation of laser-produced plasma in axial magnetic field, IEEE Trans. Plasma Sci. 44 (2016) 574–581.
[78] Y. Hayashi, Y. Kuramitsu, Y.L. Liu, H. Habara, 2-D-particle-in-cell simulation of laser Wakefield in an inhomogeneous plasma, IEEE Trans. Plasma Sci. 47 (2019) 9–11.
[79] L. Chopineau, A. Leblanc, G. Blaclard, A. Denoeud, M. Thévenet, J.L. Vay, G. Bonnaud, P. Martin, H. Vincenti, F. Quéré, Identification of coupling mechanisms between ultraintense laser light and dense plasmas, Phys. Rev. X 9 (2019) 011050.
[80] E. Cormier-Michel, B.A. Shadwick, C.G.R. Geddes, E. Esarey, C.B. Schroeder, W.P. Leemans, Unphysical kinetic effects in particle-in-cell modeling of laser wakefield accelerators, Phys. Rev. E 78 (2008) 016404.
[81] R.A.R.A. Mitchell, D.W. Schumacher, E. Chowdhury, Modeling femtosecond pulse laser damage using particle-in-cell simulations, Optim. Eng. 53 (2014) 122507.
[82] M. Drouin, L. Gremillet, J.C. Adam, A. Héron, Particle-in-cell modeling of relativistic laser–plasma interaction with the adjustable-damping, direct implicit method, J. Comput. Phys. 229 (2010) 4781–4812.
[83] S.M. Copplestone, M. Pfeiffer, S. Fasoulas, C.-D. Munz, High-order particle-in-cell simulations of laser-plasma interaction, Eur. Phys. J. Spec. Top. 227 (2019) 1603–1614.
[84] D.S. Ivanov, L.V. Zhigilei, Combined atomistic-continuum modeling of short-pulse laser melting and disintegration of metal films, Phys. Rev. B 68 (2003) 064114.
[85] C. Wu, L.V. Zhigilei, Microscopic mechanisms of laser spallation and ablation of metal targets from large-scale molecular dynamics simulations, Appl. Phys. A 114 (2014) 11–32.

[86] M.S. Daw, M.I. Baskes, Embedded-atom method: derivation and application to impurities, surfaces, and other defects in metals, Phys. Rev. B 29 (1984) 6443–6453.

[87] S. Sonntag, C. Trichet Paredes, J. Roth, H.-R. Trebin, Molecular dynamics simulations of cluster distribution from femtosecond laser ablation in aluminum, Appl. Phys. A 104 (2011) 559–565.

[88] A. Miloshevsky, M.C. Phillips, S.S. Harilal, P. Dressman, G. Miloshevsky, Generation of nanoclusters by ultrafast laser ablation of Al: molecular dynamics study, Phys. Rev. Mater. 1 (2017) 063602.

[89] Z. Lin, L.V. Zhigilei, V. Celli, Electron-phonon coupling and electron heat capacity of metals under conditions of strong electron-phonon nonequilibrium, Phys. Rev. B 77 (2008) 075133.

[90] B.Y. Mueller, B. Rethfeld, Nonequilibrium electron–phonon coupling after ultrashort laser excitation of gold, Appl. Surf. Sci. 302 (2014) 24–28.

[91] Y. Gan, J.K. Chen, Thermomechanical wave propagation in gold films induced by ultrashort laser pulses, Mech. Mater. 42 (2010) 491–501.

[92] H. Tang, M. Bai, Y. Dou, Q. Ran, G.V. Lo, Computer simulations of laser-induced melting of aluminum, Nucl. Instrum. Methods Phys. Res., Sect. B 301 (2013) 36–40.

[93] S. Amoruso, R. Bruzzese, X. Wang, J. Xia, Propagation of a femtosecond pulsed laser ablation plume into a background atmosphere, Appl. Phys. Lett. 92 (2008) 041503.

[94] J. Roth, A. Krauß, J. Lotze, H.-R. Trebin, Simulation of laser ablation in aluminum: the effectivity of double pulses, Appl. Phys. A 117 (2014) 2207–2216.

[95] M.E. Povarnitsyn, V.B. Fokin, P.R. Levashov, T.E. Itina, Molecular dynamics simulation of subpicosecond double-pulse laser ablation of metals, Phys. Rev. B 92 (2015) 174104.

[96] G.D. Förster, L.J. Lewis, Numerical study of double-pulse laser ablation of Al, Phys. Rev. B 97 (2018) 224301.

[97] S.S. Harilal, G.V. Miloshevsky, P.K. Diwakar, N. L. LaHaye, A. Hassanein, Experimental and computational study of complex shockwave dynamics in laser ablation plumes in argon atmosphere, Phys. Plasmas 19 (2012) 083504.

[98] A. Miloshevsky, S.S. Harilal, G. Miloshevsky, A. Hassanein, Dynamics of plasma expansion and shockwave formation in femtosecond laser-ablated aluminum plumes in argon gas at atmospheric pressures, Phys. Plasmas 21 (2014) 043111.

[99] Y. Wang, H. Yuan, Y. Fu, Z. Wang, Experimental and computational investigation of confined laser-induced breakdown spectroscopy, Spectrochim. Acta B At. Spectrosc. 126 (2016) 44–52.

[100] T. Moscicki, J. Hoffman, Z. Szymanski, Laser ablation in an ambient gas: modelling and experiment, J. Appl. Phys. 123 (2018) 083305.

[101] M.S. Finko, D. Curreli, Simulation of uranium plasma plume dynamics in atmospheric oxygen produced via femtosecond laser ablation, Phys. Plasmas 25 (2018) 083112.

[102] Y. Ralchenko, A.E. Kramida, J. Reader, NIST ASD Team, NIST Atomic Spectra Database (version 3.1.5), National Institute of Standards and Technology, Gaithersburg, MD, 2008. http://physics.nist.gov/asd3.

[103] R.L. Kurucz, Atomic spectral line database, http://www.pmp.uni-hannover.de/cgi-bin/ssi/test/kurucz/sekur.html.

[104] T.R. Preston, S.M. Vinko, O. Ciricosta, H.-K. Chung, R. W. Lee, J.S. Wark, The effects of ionization potential depression on the spectra emitted by hot dense aluminium plasmas, High Energy Density Phys. 9 (2013) 258–263.

[105] J. Colgan, D.P. Kilcrease, N.H. Magee, G.S. J. Armstrong, J. Abdallah, M.E. Sherrill, C.J. Fontes, H.L. Zhang, P. Hakel, Light element opacities from ATOMIC, High Energy Density Phys. 9 (2013) 369–374.

[106] K.C. Hartig, J. Colgan, D.P. Kilcrease, J.E. Barefield II, I. Jovanovic, Laser-induced breakdown spectroscopy using mid-infrared femtosecond pulses, J. Appl. Phys. 118 (2015) 043107.

[107] S.M. Vinko, O. Ciricosta, T.R. Preston, D.S. Rackstraw, C.R.D. Brown, T. Burian, J. Chalupský, B.I. Cho, H. K. Chung, K. Engelhorn, R.W. Falcone, R. Fiokovinini, V. Hájková, P.A. Heimann, L. Juha, H.J. Lee, R.W. Lee, M. Messerschmidt, B. Nagler, W. Schlotter, J.J. Turner, L. Vysin, U. Zastrau, J. S. Wark, Investigation of femtosecond collisional ionization rates in a solid-density aluminium plasma, Nat. Commun. 6 (2015) 6397.

[108] A. De Giacomo, J. Hermann, Laser-induced plasma emission: from atomic to molecular spectra, J. Phys. D Appl. Phys. 50 (2017) 183002.

[109] J. Hermann, D. Grojo, E. Axente, C. Gerhard, M. Burger, V. Craciun, Ideal radiation source for plasma spectroscopy generated by laser ablation, Phys. Rev. E 96 (2017) 053210.

[110] J.J. MacFarlane, I.E. Golovkin, P. Wang, P.R. Woodruff, N.A. Pereyra, SPECT3D—a multi-dimensional collisional-radiative code for generating diagnostic signatures based on hydrodynamics and PIC simulation output, High Energy Density Phys. 3 (2007) 181–190.

[111] G. Miloshevsky, A. Hassanein, Emission spectra of binary plasma mixtures from visible to X-ray range, in: 2016 IEEE International Conference on Plasma Science (ICOPS), 2016, p. 1.

[112] I. Radivojevic, R. Niessner, C. Haisch, S. Florek, H. Becker-Ross, U. Panne, Detection of bromine in thermoplasts from consumer electronics by laser-induced plasma spectroscopy, Spectrochim. Acta B At. Spectrosc. 59 (2004) 335–343.

[113] G. Miloshevsky, A. Hassanein, Atomic and optical properties of warm dense copper, Phys. Rev. E 92 (2015) 033109.

[114] G.V. Miloshevsky, V.I. Tolkach, S. Rozin, G. Shani, Elastic scattering of electrons by gadolinium and barium atoms, Nucl. Instrum. Methods Phys. Res., Sect. B 168 (2000) 467–472.

[115] S.S. Harilal, G.V. Miloshevsky, T. Sizyuk, A. Hassanein, Effects of excitation laser wavelength on Ly-α and He-α line emission from nitrogen plasmas, Phys. Plasmas 20 (2013) 013105.

CHAPTER

7

Molecular laser-induced breakdown spectroscopy

Christian G. Parigger[a], *David M. Surmick*[b], *Christopher M. Helstern*[a], *Ghaneshwar Gautam*[c], *Alexander A. Bol'shakov*[d], *Richard E. Russo*[d]

[a]University of Tennessee Space Institute, Center for Laser Applications, Tullahoma, TN, United States [b]University of Massachusetts, Lowell, MA, United States [c]Fort Peck Community College, Poplar, MT, United States [d]Applied Spectra, Inc., West Sacramento, CA, United States

1 Introduction

Laser-induced breakdown spectroscopy (LIBS) serves the purpose of decoding light emitted from laser plasma. Analytical chemistry utilizes general plasma excitation that includes electric discharge sparks, or more recently, laser-induced sparks. Several extensive books on LIBS describe various aspects of laser-target interactions [1–7]. Moreover, data accumulation extends to extraterrestrial exploration, including LIBS on Mars [8]. This chapter communicates LIBS research on diatomic molecular emission spectra from cyanide (CN). Furthermore, the chapter makes various references to related diatomic molecular spectroscopy [9], to fundamentals of atomic, molecular, and optical physics [10, 11], and to detailed treatments of selected optical transitions, including carbon containing diatomic molecules [12]. Significant applications include studies of molecular isotopologues for enhancements of atomic isotope shifts and diagnostics with LIBS.

The CN molecule has been of interest in a variety of research areas, including spectroscopic characterization. CN is evident in many forms in nature and is instrumental in the production of amino acids, lipids, membranes, nucleotides, and proteins. Living organisms use cyanides as a source of nitrogen and plants can use cyanides to defend against disease [13]. Moreover, cyanides can be instrumental in human applications such as for gold extraction [14]. Cyanides also occur in industrial processes, such as chemical and drug production, electroplating, iron and steel production, and pesticide and plastic manufacturing [15]. Cyanides are connected to many living organisms and are created through synthetic processes, but they are highly toxic, can be lethal, and have been used as chemical weapons. To prevent health hazards to humans, identify CN poisoning occurrences,

Laser-Induced Breakdown Spectroscopy
https://doi.org/10.1016/B978-0-12-818829-3.00007-1

and protect the environment, adequate detection of cyanides is imperative. Current detection methods include optical methods, electrochemical methods, mass spectrometry, gas chromatography, and quartz crystal mass monitors [16]. Although these methods are sufficient in detecting cyanides, molecular emission spectroscopy can measure composition with a minimally invasive approach [17].

Recombination molecular spectra of CN occur subsequent to plasma generation with pulsed laser radiation [18]. The laser-induced dynamics are routinely measured using time-resolved spectroscopy [19]. Recorded LIBS studies show atomic and molecular signatures, including background contributions that depend on the laser pulse width in standard ambient temperature and pressure (SATP) environments [20]. Moreover, there are other LIBS implementations: distribution of molecular species is of interest in elemental mapping and in studies of explosive surrogates [21, 22]. In terms of measuring molecular spectra with relatively low background contributions from free electron radiation, ultrashort, femtosecond laser-induced breakdown spectroscopy (FLIBS) [23] can be advantageous for CN diagnostics [24–26]. FLIBS has also been applied in measurements of CN to interpret the molecular structure of explosives [26–29], or in femtosecond filamentation [30] remote sensing experiments for atmospheric sensing or filament-induced "fingerprinting" and identification of atmospheric molecules [31]. FLIBS [32–34] measurements and combustion applications of air-methane mixtures include spectroscopy of CN [35]. Bacteria detection with FLIBS reveals CN molecular bands following direct ablation of CN bonds [36], including comparisons of FLIBS to the classic nanosecond-LIBS [37].

This work communicates results from experiments using nanosecond laser spectroscopy [38, 39] that focus on molecular emission spectroscopy. In some LIBS applications, use of a second laser pulse appears advantageous, appropriately termed dual-pulse LIBS, but dual excitation can be traced back to emission spectroscopy with CO_2 laser radiation at 10.6 μm. CO_2 laser pulses indicate a nominal 100-ns initial pulse that is followed by a tail pulse of the order of 1 μs. Consequently, a relatively homogeneous plasma is generated. The CO_2 laser pulse shape can be interpreted as a double-pulse excitation [40] with 100 times longer excitation than for Nd:YAG radiation. Moreover, the focal volume is of the order of 1000 times larger for 10.6-μm than for 1.064-μm radiation due to the 10:1 wavelength ratio.

CN recombination radiation occurs within the first 100 ns [41] for laser-induced breakdown in 1:1 CO_2:N_2 gas mixtures when using 150-mJ, 6-ns, Q-switched Nd:YAG nanosecond laser pulses to create the laser plasma. Aspects of the analysis, including Abel inversions and computational modeling of the plasma, are discussed, and how one can make inferences of molecular excitation temperatures. For nominal 6-ns laser pulses and for pulse energies in the range of 100–800 mJ, expansion dynamics and turbulence due to shock phenomena are usually observed. Determination of spatial distributions serves the purpose of investigating the level of homogeneous distribution. Local thermodynamic equilibrium is frequently supposed to be valid for diagnostics with atomic and molecular species emission spectroscopy.

Furthermore, selected experimental results and analysis of AlO emission spectra elaborate the complexity of molecular LIBS [42]. Applications of laboratory measurements include analysis of astrophysical white dwarf C_2 spectra [43]. Additionally, of course, employing diatomic molecules for enhancement of isotope shifts is viewed as a major contribution to molecular LIBS, namely, laser ablation molecular isotopic spectrometry (LAMIS) [44, 45]. This technique is based on direct and rapid measurements of optical emissions from laser-induced plasma. LAMIS uses relatively large isotope shifts in the spectra of transient molecular isotopologues formed in laser ablation plasma. Larger isotope shifts (compared to those in atomic spectra) signify simpler measurement requirements. A spectrometer with modest spectral resolution can be suitable for both LIBS and LAMIS techniques, and thus elemental and isotopic measurements can be accomplished with the same instrument.

2 Experimental details of CN molecular emission spectroscopy

The experimental arrangement is best summarized as a set of components typical for time-resolved, laser-induced, optical emission spectroscopy, or nanosecond-LIBS [3]. Primary instrumentation includes the Q-switched Nd: YAG device (Quantel model Q-smart 850) operated at the fundamental wavelength of 1064 nm to produce full-width-at-half-maximum 6-ns laser radiation with an energy of 850 mJ per pulse, a laboratory-type Czerny-Turner spectrometer (Jobin-Yvon model HR 640) with a 0.64-m focal length, an intensified charge-coupled device (ICCD) (Andor Technology model iStar DH334T-25U-03) for recording of temporally and spatially resolved spectral data, a laboratory chamber or cell with inlet and outlet ports together with a vacuum system, electronic components for synchronization, and various optical elements for beam shaping, steering, and focusing. This particular arrangement has been used recently for optical breakdown shadowgraphy studies [46], or for emission spectroscopy of expanding gaseous hydrogen-nitrogen plasma [47] with laser pulses from a Quanta Ray DCR-2 (10) device.

For the generation of microplasma, a singlet lens (Thorlab model LA1509-C) is used close to the top entrance window of one arm of the chamber containing the 1:1 CO_2:N_2 atmospheric gas mixture (Airgas ultrahigh purity N_2 and research grade CO_2). For 1:2 imaging of the plasma onto the 100-μm spectrometer slit, a fused silica planoconvex lens (Thorlab model LA4545) is employed. Moreover, for the CN experiments reported here, the energy of the laser pulses is attenuated with beamsplitters and apertures to 150 mJ/pulse. The overall spatial distributions have been computed for the Thorlab LA4545 lens and for f/10 and f/5 focusing [48, 49]. Fig. 1A illustrates the computed radial distributions for f/5 focusing, with the singlet lens positioned at $x = 0$, and the radiation incident from the right. Fig. 1B displays the on-axis irradiance distributions. A shallower f/10 focusing reveals about one order of magnitude (or 2^3) larger focal volume than that obtained for f/5 focusing.

In previous experiments, captured shadowgraphs of the breakdown plasma [46] served the purpose of visualizing the plasma expansion when using 850-mJ, 6-ns radiation. One of the objectives of this work includes investigation of spatially resolved CN distribution using Abel inverse transforms along the lines reported for a 9:1 hydrogen-nitrogen atmospheric gas mixture [47]. However, visualization experiments and focal volume computations [48, 49] are helpful for justification of symmetric Abel inverse transform with extensions due to slight spatial asymmetries as for hydrogen and air plasmas [50–52]. Fig. 2 illustrates two shadowgraphs. These shadowgraphs were recorded in SATP air, revealing vertical stagnation layers that originate from multiple breakdown sites as indicated in the computed focal intensity distribution in Fig. 1.

The laser-supported plasma expansion is consistent with the Taylor-Sedov blast wave model [46, 53]. Recent investigations of optical breakdown in SATP air [46] apply the expansion formula [53] in the analysis of shadowgraphs of the type illustrated in Fig. 2A and B. Increased electron density and temperature occur in the outer region of the plasma kernel [47] as evidenced by the bright-to-dark boundaries that appear to cause multiple reflection inside the

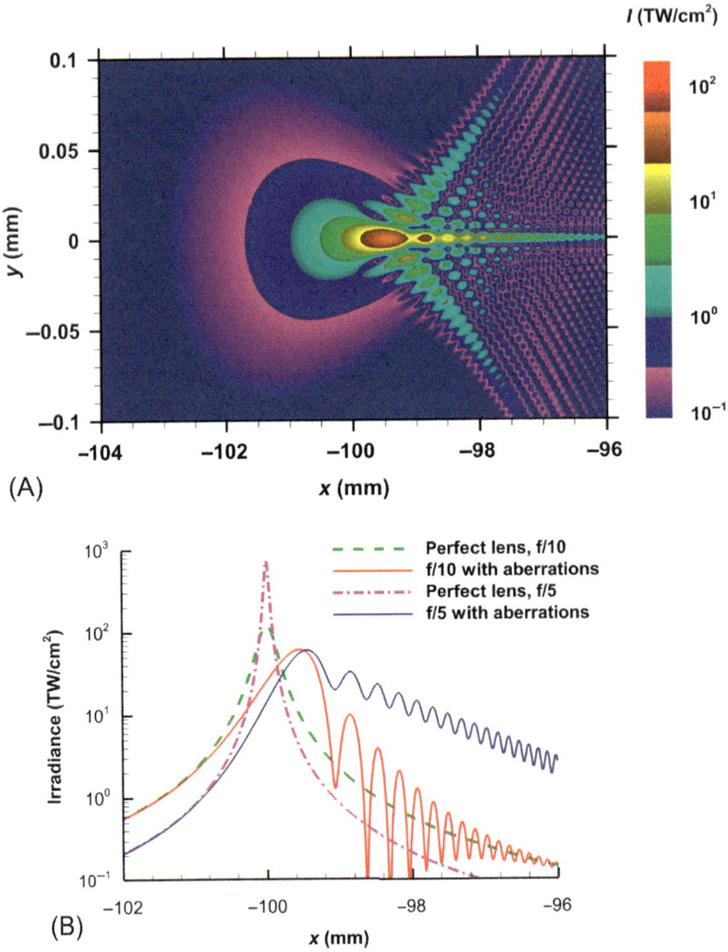

FIG. 1 (A) Spatial distribution of 1064-nm laser beam f/5 focusing with a 100-mm focal length lens. (B) On-axis irradiance distributions for f/5 and f/10 focusing with aberrations [41].

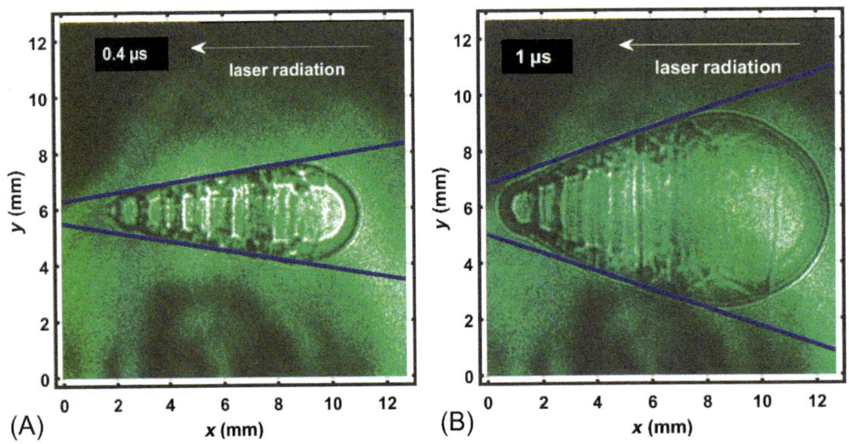

FIG. 2 (A) The shockwave maximum expands in excess of vertical Mach 6 (~2 km/s) and horizontal Mach 40 (~14 km/s) for a time delay of 0.4 μs. The *blue lines* indicate a cone and the significant, measured horizontal speed. (B) Vertical and horizontal expansion of ~Mach 3 (~1 km/s) and ~Mach 9 (~3 km/s), respectively, for a time delay of 1 μs from plasma generation [41].

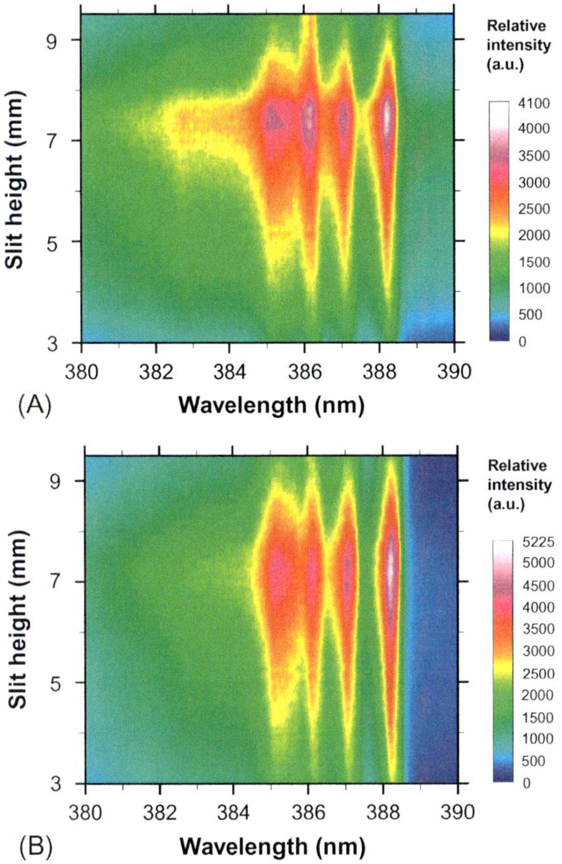

FIG. 3 Pseudocolored line-of-sight contour map of measured spectra. Time delay (A) 0.7 μs and (B) 1.95 μs, gate width 0.125 μs. The second-order carbon line indicates larger electron density at the top side (toward the incoming laser pulse), especially at the 0.7-μs time delay [41, 54].

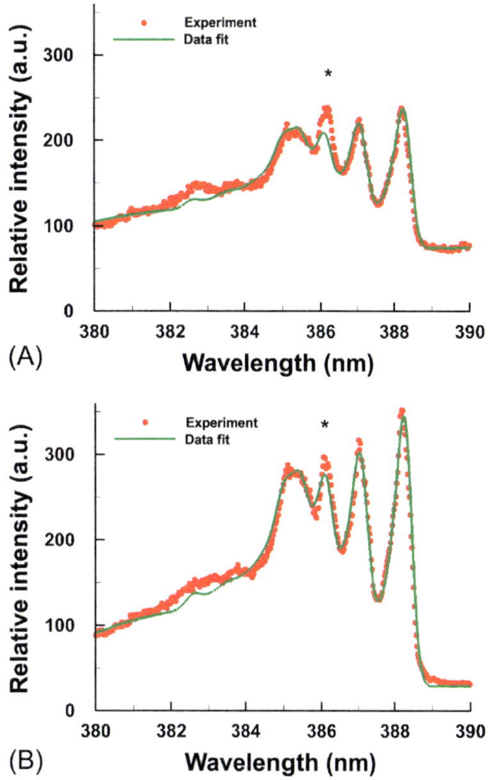

FIG. 4 Measured cyanide spectra and fitted temperature at a slit height of 7 mm (see Fig. 3). (A) $T = 9.5$ kK, time delay 0.7 μs. (B) $T = 8.6$ kK, time delay 1.95 μs. The second-order atomic carbon line at 193.09 nm. (*) indicates higher electron density for earlier time delays [41].

shockwave. For the measurement of CN recombination spectra, expansion dynamics are expected to be similar to the air results indicated in Figs. 3 and 4. However, for an energy/pulse of 150 mJ, a smaller number of breakdown spots are anticipated, leading to a smaller number of stagnation layers than recorded for the 850-mJ/pulse air breakdown events. The expanding "bubble" is also expected to grow to a nearly spherical distribution for shorter time delays with 150 mJ/pulse than for 850 mJ/pulse. Furthermore, the vertical extent is about a factor of 1.4 smaller for 150-mJ pulses than that for 850-mJ pulses, according to the Taylor-Sedov energy dependency [53] for the radius of the spherical expansion. It is worth noting that reported air shadowgraphs [46] for a 4-μs time delay and for a 850-mJ/pulse optical breakdown in air show a prolate spheroidal distribution that is nearly spherical. This behavior would be consistent with plasma emission measurements in investigations of phenomena near the focal region [50].

3 Results and discussion of CN experiments

The experimental series for the separate measurements of CN molecular distribution after optical breakdown includes evacuating the cell to a nominal mercury pump vacuum of the order of 10^{-4} Pa (10^{-6} Torr) and then introducing the N_2:CO_2 mixture [41]. Fig. 3 illustrates wavelength-calibrated, sensitivity-corrected, and background-subtracted maps of captured time-resolved data [41, 54, 55] following optical breakdown in the ultrahigh pure N_2 and research-grade CO_2 mixture.

In the experimental investigations, signatures of the 0-0, 1-1, 2-2, 3-3, 4-4 band heads begin to emanate for time delays of the order of 100 ns from optical breakdown. Moreover, the plasma expands and tends to propagate toward the laser side.

The detector pixels are binned in four tracks along the slit direction, resulting in obtaining 256 spectra for each time delay. The average of accumulated data from 100 consecutive optical breakdown events is recorded for 21 different time delays using a gate width of 125 ns. In Fig. 3, the vertical axis indicates the slit height, and the contour map represents the recorded detector counts versus wavelength and slit height. With 1:2 imaging and a pixel resolution of 13.6 μm, the plasma size amounts to ~3 mm. For the two time delays of 0.7 and 1.95 μs, Fig. 3A and B reveal that the CN band heads of the $\Delta v = 0$ sequence are well developed, and it also shows an atomic line near 386.2 nm, namely, the carbon C I 193.09-nm atomic line recorded in second order. The electron number density, n_e, can be determined from the Stark full width at half maximum, $\Delta\lambda_{\text{Stark}}$, or from the Stark shift, $\delta\lambda_{\text{Stark}}$, of that carbon line [51, 54]:

$$\begin{aligned} \triangle\lambda_{\text{Stark}} &= 2wn_e\left(10^{17}\,\text{cm}^{-3}\right), \text{or} \\ \delta\lambda_{\text{Stark}} &= dn_e\left(10^{17}\,\text{cm}^{-3}\right) \end{aligned} \quad (1)$$

where the width, w, and shift, d, parameters are $w = 0.00291$ nm and $d = 0.00294$ nm [54], respectively, at an electron temperature of 40 kK. At a 200-ns time delay, the full width of ~0.4 ± 0.1 nm in second order reveals an electron density of ~$34 \times 10^{17}\,\text{cm}^{-3}$. The shift of 0.2 ± 0.05 nm in second order yields a consistent result for the electron density. The accuracy of the Stark width parameter is listed at 30% [52]; consequently, the uncertainty of the determined electron density is estimated to be of the order of 50%.

Fig. 4 displays line-of-sight-measured and -fitted spectra near the very center of the detector at a slit height of 7 mm. Well-tested routines deliver comparisons of measured and computed spectra [56]. The fitted spectral resolution, $\delta\lambda$, amounts to $\delta\lambda = 0.43$ nm. At the center of the plasma, the analysis shows a CN excitation temperature of 9.5 and 8.6 kK for time delays of 0.7 and 1.95 μs, respectively. The temperature is higher near the edges than at the center. The C I 193.09-nm atomic line recorded in second order (Fig. 4A and B) indicates that the electron density is larger near the edges than at the center of the plasma.

The CN recombination radiation signals may be stronger for specfic ranges of temperature. Computation of the freely available chemical equilibrium with applications (CEA) code [57] elucidate CN mass fractions versus temperature. Several atoms and molecules, including ionic species, are part of the CEA computations, but the results for the CN mass fractions are of primary interest in this work. Fig. 5 shows the molar fraction distribution of CN, electron, and other species for the mixture as a function of temperature.

The recorded spectra are subjected to Abel inverse transformation to obtain the radial distribution of the plasma. Strictly speaking, radially symmetric profiles are required for Abel inverse transformation. In the analysis of the molecular CN spectra, the same symmetrization

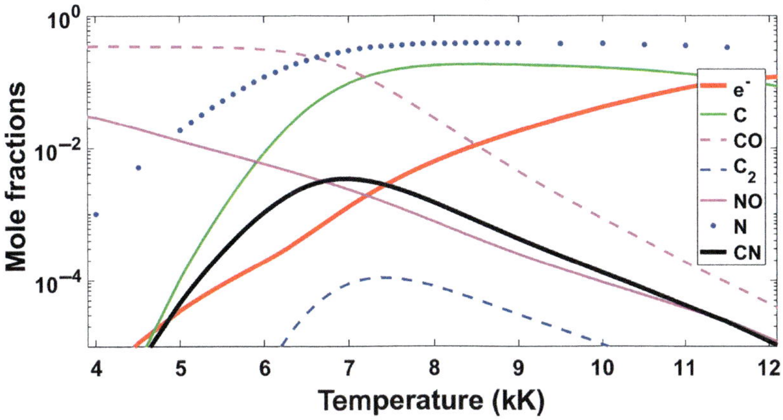

FIG. 5 Computed mole fractions versus temperature. Cyanide shows a maximum near 7 kK for a 1:1 atmospheric CO_2:N_2 mixture [41].

methods are applied as for atomic hydrogen spectra [58, 59]. The use of Chebyshev polynomials for the inversion of the integral equation:

$$I(z,\lambda) = 2\int_z^{\rho} \varepsilon(r,\lambda)\frac{r}{\sqrt{r^2 - z^2}}dr \qquad (2)$$

allows one to directly invert measured data. The choice of the number of typically 10 polynomials for the inversion [60, 61] is equivalent to the use of a digital filter resulting in the broadening of the computed radial spectra. A smaller number of polynomials would cause smaller spectral resolution. The measured line-of-sight data, $I(z,\lambda)$, along the slit dimension, z, are inverted for each wavelength, λ, to obtain the radial emission coefficient, $\varepsilon(r,\lambda)$, with the upper limit much larger, $\rho \gg R$, than the radius, R, of the plasma. Strictly speaking, line-of-sight LIBS records spectral radiance [62] in units of $\mathrm{Wcm^{-2}\,sr^{-1}}$-$\mathrm{nm^{-1}}$ versus wavelength. However, it is customary in the LIBS community to display intensity in arbitrary units corresponding to detector counts rather than absolutely calibrated spectral radiance. The recorded CN data are calibrated and corrected for system sensitivity using standard lamps.

Analysis of asymmetric plasma expansion would require Radon inverse transformations [63]. The radial spectra are computed from the two-dimensional data that were accumulated with an intensified camera attached to the spectrometer that allows one to resolve spectra along the slit. The radial intensity, $I(r, \lambda)$, corresponds to $\varepsilon(r, \lambda)dr$, with $dr \ll$ plasma extent. Fig. 6 illustrates $I(r, \lambda)$ of the Abel-inverted data in contour plots for time delays of (A) 0.7 μs and (B) 1.95 μs.

Near the edges of the expanding plasma, the CN signals are lower than near the center. Computations of equilibrium CN distributions versus temperature indicate about a factor of 3 smaller CN mass fractions at 9.5 kK than at 8.6 kK. Moreover, analysis of Abel inverse integral transforms reveals lower CN temperature in the plasma center, e.g., for a time delay of 1.2 μs. The signals of the atomic carbon line are clearly visible in Fig. 6A, and this line is spectrally wider with larger shifts near the plasma edges than at the center.

4 Carbon Swan C_2 astrophysics emission spectra

Laboratory molecular emission spectroscopy enjoys applications in astrophysics, namely, molecular sequences and progressions aid the

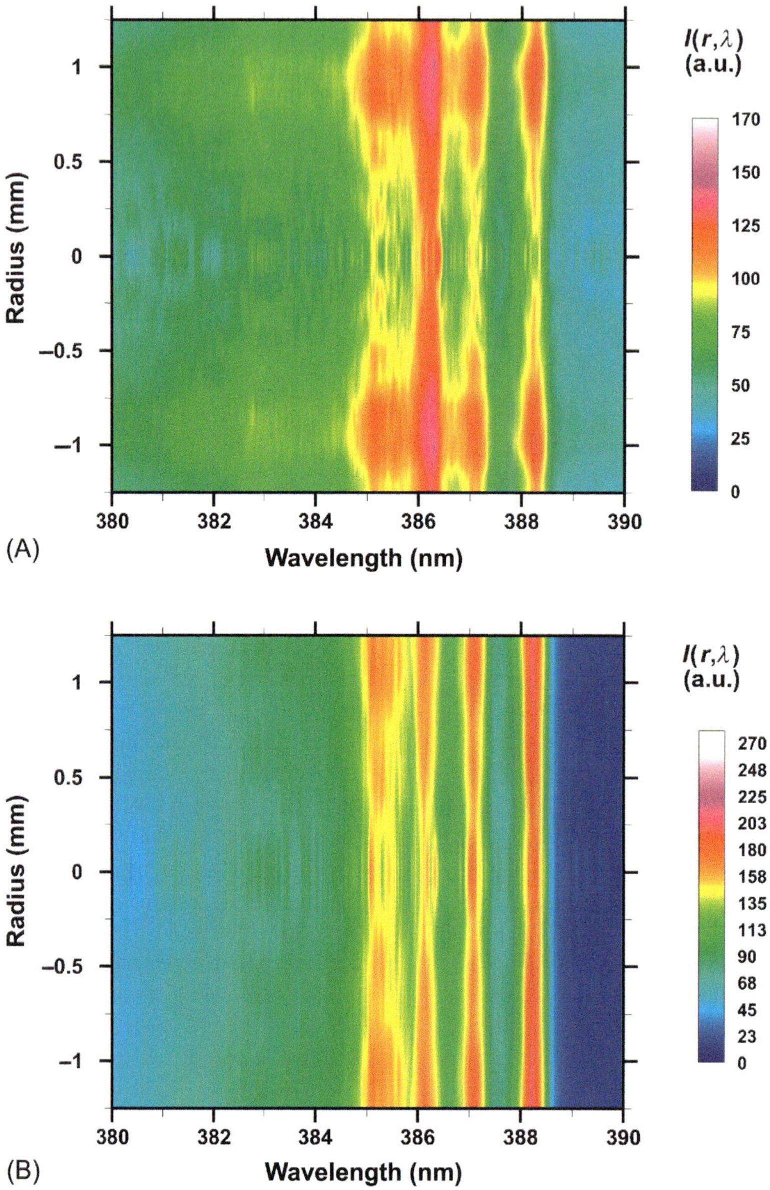

FIG. 6 (A) Abel inverted spectra versus radius for a time delay of 0.70 μs, gate width 0.125 μs. Near 1 mm, the cyanide (CN) $\Delta v = 0$ sequence shows 0-0, 1-1, 2-2, 3-3, and 4-4 band heads, and the second-order atomic carbon 193.09 nm that masks the 2-2 band. The CN signals are smaller near the center. The carbon line appears narrower at the center than at the 1-mm radial position, indicating lower electron density at the center. (B) Abel-inverted spectra appear nearly uniform for a 1.95-μs time delay [41].

interpretation of recorded spectra from white dwarf stars [43, 64], to name one example. In turn, various databases collected and maintained for optical region studies of the universe can be advantageous for LIBS. Of course, white dwarfs that show a temperature of 30 kK or of the order of that of Sirius B, the companion to the brightest star seen from earth, Sirius A, can serve as a gauge for atomic spectroscopy. As white dwarfs cool to a temperature of 8 kK or of the order of that of Procyon B, companion to the fifth brightest star seen from earth, Procyon A, molecular spectroscopy analysis becomes important.

Analysis of a particular white dwarf, GJ 841 B [43, 64, 65], is elaborated here. This particular white dwarf reveals well-developed C_2 Swan band absorption progressions, namely, absorptions from a common lower level, and sequences, namely, vibrational quantum number differences are constant. Fig. 7A illustrates the overall absorption spectrum, and Fig. 7B displays the inverted $\Delta v = 0$ sequence. Fig. 7B also shows the fitted spectrum, inferred temperature, and spectral resolution used for the fit.

Analysis of the C_2 absorption spectra suggests deviation from equilibrium due to significant differences in computed and recorded spectra. Fig. 8 conveys further results for the $\Delta v = -1$ and -2 bands. The determined temperature from fitting these selected bands is lower

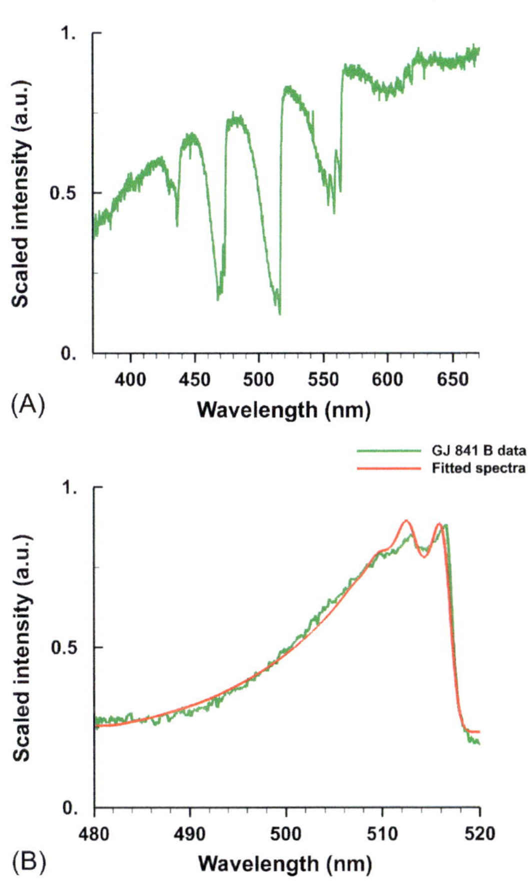

FIG. 7 White dwarf GJ 841 B. (A) C_2 Swan absorption spectrum. (B) $\Delta v = 0$, $T = 7.1$ kK, $\delta\lambda = 1.8$ nm [64].

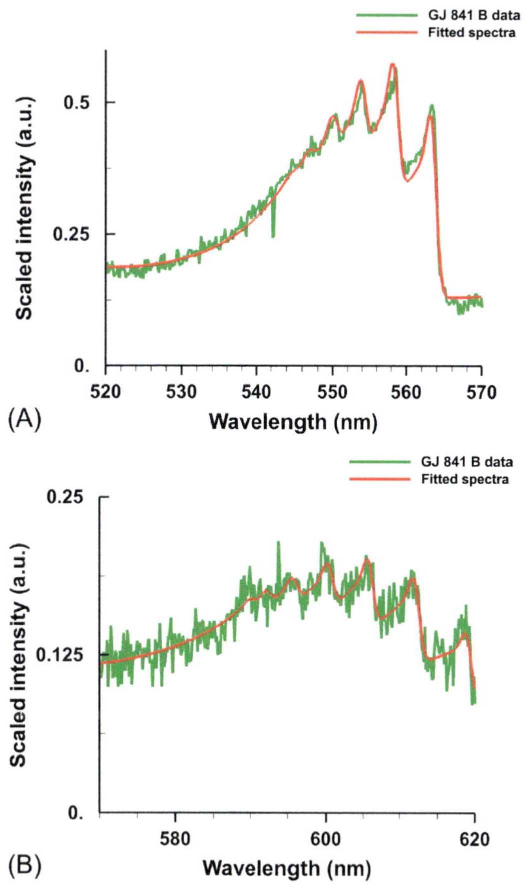

FIG. 8 (A) $\Delta v = -1$: $T = 4.5$ kK, $\delta\lambda = 1.5$ nm. (B) $\Delta v = -2$: $T = 4.2$ kK, $\delta\lambda = 1.5$ nm [64].

than that from fitting all bands $\Delta v = +2, +1, 0, -1, -2$. The temperature inferred from the $\Delta v = 0$ band is higher than that from the other bands.

Atomic white dwarf (WD) spectroscopy of the Balmer series hydrogen beta line, and for WDs hotter than typically 20 kK, frequently reveals narrow absorption lines at the central wavelength position of 486.14 nm. This particular narrow absorption may be explained by allowing for the presence of C, N, and O in the white dwarf atmospheres. Clearly, analysis of molecular spectra will help to infer details of the cooling phenomena. For several of these investigations, databases that include isotopologues will be instrumental, e.g., the ExoMol database [66, 67] or spectra for the visible region from the HITRAN database [68] that can be explored and compared with PGOPHER [69, 70] predictions.

5 Plasma temperature distributions from AlO measurements

This section presents the results of molecular spectroscopy of laser-produced plasma for the extraction of temperature distributions along the plasma plume. The ability to extract temperatures is useful in a number of applications, including monitoring nanoparticle synthesis [71], as well as understanding the shockwave diagnostics [72–74] and plasma chemical kinetics of the expanding laser-produced plasma [75, 76]. Differences in temperature along the plasma plume provide the opportunity to understand the shock interaction and propagation with the plasma and the ambient gaseous material in a laser ablation configuration. In particular, due to the lower temperatures at which molecular recombination occurs relative to atomic processes, molecular emissions allow one to investigate the late time dynamics of the plasma plume. Depending on the diatomic molecule, laser energy, gaseous breakdown versus focused laser ablation, and ambient gas pressures, the time one can investigate can range from 1 to 10 s of microseconds and up to 100 μs for large irradiances following the onset of laser-induced optical breakdown. We investigate the use of diatomic molecular aluminum monoxide (AlO) spectra to extract plasma temperature distributions along the plasma plume expansion along the direction of laser incidence.

Diatomic molecular spectra are calculated from detailed tables of line strengths [56]. Recent wavelength position and transition strength computations [77] include AlO for stellar astrophysics applications. Calculating diatomic line strengths is an involved process that depends on access to accurate diatomic constants and line positions. Here, we provide a brief overview of how diatomic spectra are calculated and where further details on these tabulations may be found. In general, the line strength, S_{ul}, is one of the more fundamental spectroscopic quantities as it can be calculated directly from the dipole moment of a transition from upper state, u, to lower state, l:

$$S_{ul} = \sum_{u}\sum_{l}\left|\left(\left\langle u,|T_k^{(q)},|l\right)\right\rangle\right|^2 \quad (3)$$

where $T_k^{(q)}$ is an irreducible tensor of rank q that represents the electric dipole operator. One of the properties of line strength is that it can be factored into components that account for the electronic, vibrational, and rotational structure of the diatomic molecule. Under this scheme, the diatomic line strength is given by

$$S_{ul} = S(J', J)S(n', n)q(v', v) \quad (4)$$

where the first term is the Hönl-London line strength that accounts for rotational aspects of the spectrum due to spin, the second term is an electronic line strength, and the third term is the so-called Frank-Condon factor. The primed coordinates indicate upper states. When calculated, the electronic line strength and

Frank-Condon factors are combined into a single electronic vibrational line strength:

$$S(n'v', nv) = \left|a_0 + a_1\bar{r}(v', v) + a_2\bar{r}^2(v', v) + \ldots\right|^2 q(v', v) \quad (5)$$

where the Born-Oppenheimer approximation has been made. The first term of Eq. (5) is the square of the electronic transition moment given as a Taylor expansion that is calculated at the r-centroids, $\bar{r}^n(v', v)$, which account for positional effects along the internuclear axis of the diatomic molecule. Details on how to calculate the diatomic line strength, how it is factored, and how each of the factored components is tabulated can be found in [56, 78–82]. Details on the electronic structure of molecules can be found in most books on atomic molecular physics, such as Chapters 10 and 11 of [83].

The line strength is used to calculate the intensity of spectral lines by assuming that the vibrational and rotational elements are in thermal equilibrium and the populations are described using Boltzmann statistics. Under this assumption, the intensity that reaches the pixels of a spectroscopic detector is given for a transition as:

$$I_{ul} = \frac{16\pi^3 c(a_0 e)^2 N_0 C}{3\epsilon_0 Q} v_{ul}^4 S_{ul} e^{-hF_u/k_B T} \quad (6)$$

where a_0 is the Bohr radius, c is the speed of light, Q is the partition function, e is the electronic charge, ϵ_0 is the vacuum permittivity, N_0 is the total species population, h is Planck's constant, C is a calibration constant, F_u is the term value of the upper energy state, T is the temperature, k_B is Boltzmann's constant, and v_{ul} is the transition frequency. Provided one has access to an accurate table of line strengths with upper energy term values and line positions, the diatomic spectra can be modeled as a function of temperature.

In the present effort, the results concerned focused laser ablation on the surface of a 96% pure alumina (Al_2O_3) target under the presence of a hydrogen/nitrogen gas mixture at a ratio of 9:1 under an ambient pressure of 1.12×10^5 Pa (840 Torr). The plasma was initiated by focusing the beam of an Nd:YAG laser onto the surface of the alumina target operating at its fundamental wavelength using an ultraviolet-fused silica (UV-FS) lens with a focal length of 125 mm. The laser energy per pulse was 120 mJ and the temporal pulse width was 14 ns. The plasma was imaged onto the slit of a Jobin-Yvon, Czerny-Turner spectrometer installed with an 1800-grooves/mm grating. Light from the plasma was focused onto the spectrometer slit using a UV-FS, 100-mm focal length lens. The plasma dynamics were imaged as a function of time from the onset of the plasma emission through the use of a two-dimensional chip on a gated ICCD (Andor iStar). The chip on the detector was binned such that 1024 horizontal pixels were used for spectral resolution and 128 pixels were used along the height of the plasma plume. The effective spatial resolution of the system was 0.1 mm and the spectral resolution was approximately 0.1 nm. The measurement system was calibrated using low-density hydrogen gas lamps for the wavelength response and a thermal continuum (Ocean Optics DH2000Bal) source for the spectrometer-detector intensity response.

The AlO transition under consideration is the $B^2\Sigma^+ \rightarrow X^2\Sigma^+$ $\Delta v = 0$ transition. For a primer on molecular spectroscopic notation, please refer to Chapter 11 of [83]. The AlO spectra were measured over the wavelength region between 480 and 500 nm for the purpose of measuring the temperature distribution within the expanding plasma along the plane of axial expansion parallel to the direction of laser incidence and perpendicular to the surface. Measurements were recorded for times ranging between 2.5 and 20 μs from the plasma onset in 2.5-μs steps. The gate width of the measurements was 1.25 μs. Fig. 9 displays these results in contour plots. One notices that there are early (2.5 μs) significant interferences from the H_β line indicating

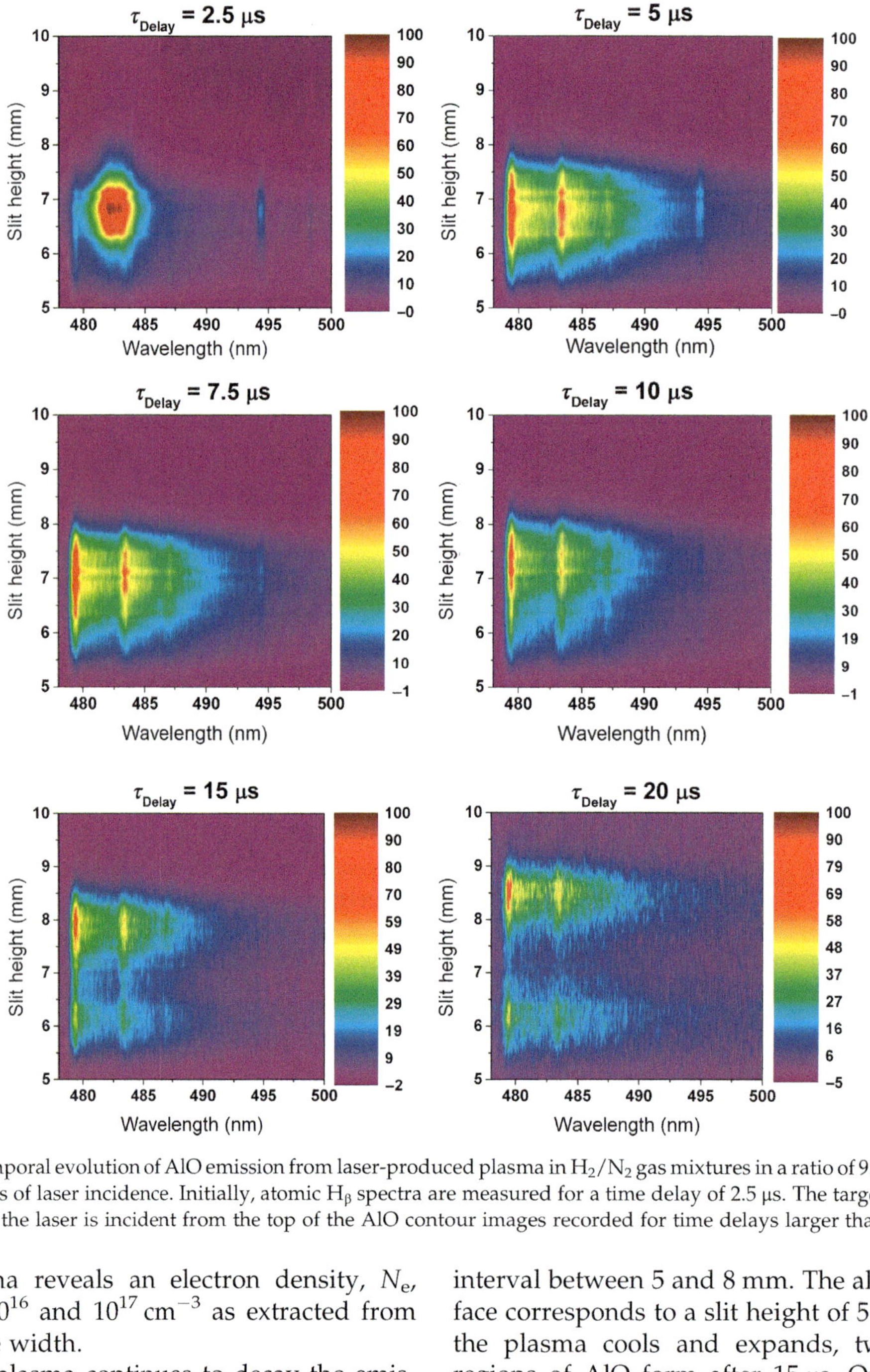

FIG. 9 Temporal evolution of AlO emission from laser-produced plasma in H_2/N_2 gas mixtures in a ratio of 9:1. The vertical axis is the axis of laser incidence. Initially, atomic H_β spectra are measured for a time delay of 2.5 μs. The target surface is at 5.44 mm and the laser is incident from the top of the AlO contour images recorded for time delays larger than 5 μs [74].

that plasma reveals an electron density, N_e, between 10^{16} and 10^{17} cm^{-3} as extracted from the H_β line width.

As the plasma continues to decay the emissions are dominated by AlO on the spatial interval between 5 and 8 mm. The alumina surface corresponds to a slit height of 5.44 mm. As the plasma cools and expands, two distinct regions of AlO form after 15 μs. One remains close to the surface and another, larger region appears to continue to expand with the shockwave between 8 and 9 mm in the 20-μs contour plots. It is important to note that the spatial units provided in the contour plots are in the image plane of the spectrometer and do not necessarily map one to one to the plasma. The magnification of the imaging system was approximately 1.1, but one needs to know the modulation transfer function of the imaging system to map the image plane heights back to the original plasma.

To make further inferences about AlO distributions at later times in the plasma expansion, the temperature profile along the axis of laser incidence was extracted. This was completed using a fitting routine according to Eq. (4) in which the intensity of each transition is modeled as having a Gaussian line profile. Fitting was performed using a trust-region fitting routine with fitting parameters of temperature, line amplitude, and linear offset [84, 85]. Additionally, the spectral resolution is a variable parameter that is used to establish the uncertainty of the inferred temperature from diatomic line profile fitting. Following initial fitting of each spectrum, the linear offset and spectral resolution parameters were randomly varied 1000 times and subsequently refitted to generate a distribution of possible temperatures for each AlO spectrum. An additional fit was performed in which the line amplitude, temperature, linear offset, and spectral resolution were fit parameters. A total of 1002 inferred temperatures were used to establish $1-\sigma$ uncertainties in which the spectral resolution, linear offset, and temperature are the fit parameters. Further details on fitting and uncertainty assessments can be found in [74, 82, 86, 87].

Fig. 10 shows the extracted temperatures at times >5 μs. One notices immediately that the uncertainties increase dramatically through the plasma evolution. This is a consequence of the weakening of the AlO signal relative to the detector noise. This effect is also amplified due to the loss of signal strength associated with the spatial resolution. The amplitude of the rotational structure of the diatomic emission becomes comparable with the detector noise, the signal quality degrades, and the uncertainty in the temperature that is extracted increases.

The decrease in signal quality is evident in the 20-μs image of Fig. 9 from the distinct fuzziness that is present in the image as compared to the early time delays. Despite the rather large uncertainties, some trends can still be observed. The initial temperature of the plasma near the surface is approximately 5500 K and subsequently drops to a temperature of ~3000 K by the 20-μs delay. The temperature of the intense region of AlO far from the surface remains hotter for longer, maintaining a temperature of ~4000–4500 K near the region between 8 and 8.5 mm in the 15-μs image of Fig. 10. By 20 μs the temperature of the outer region has also dropped to ~3000 K, though the rather large uncertainties are the main characteristic of this time delay. Previous results of aluminum laser ablation in air in which spatially averaged AlO spectra were measured indicate that temperatures can be as high as 5500 K at 10 μs following breakdown and decay to 3000 K by 80 μs using a similar incident laser energy per pulse [88, 89]. This temperature trend seems to indicate that the presence of a shockwave increases the AlO temperature as the plasma expands. After the shockwave has left the plasma region, the observed temperature cools as well.

6 Laser ablation molecular isotopic spectrometry

Emission spectra from molecules reveal distinct differences due to isotopic composition. LAMIS [44, 45] measures molecular emission as the plasma cools. In laser plasma, isotopic elements produce electronically, vibrationally, and rotationally excited isotopologues of dimers, oxides, nitrides, or halides in plasma reactions of the atomized sample with ambient

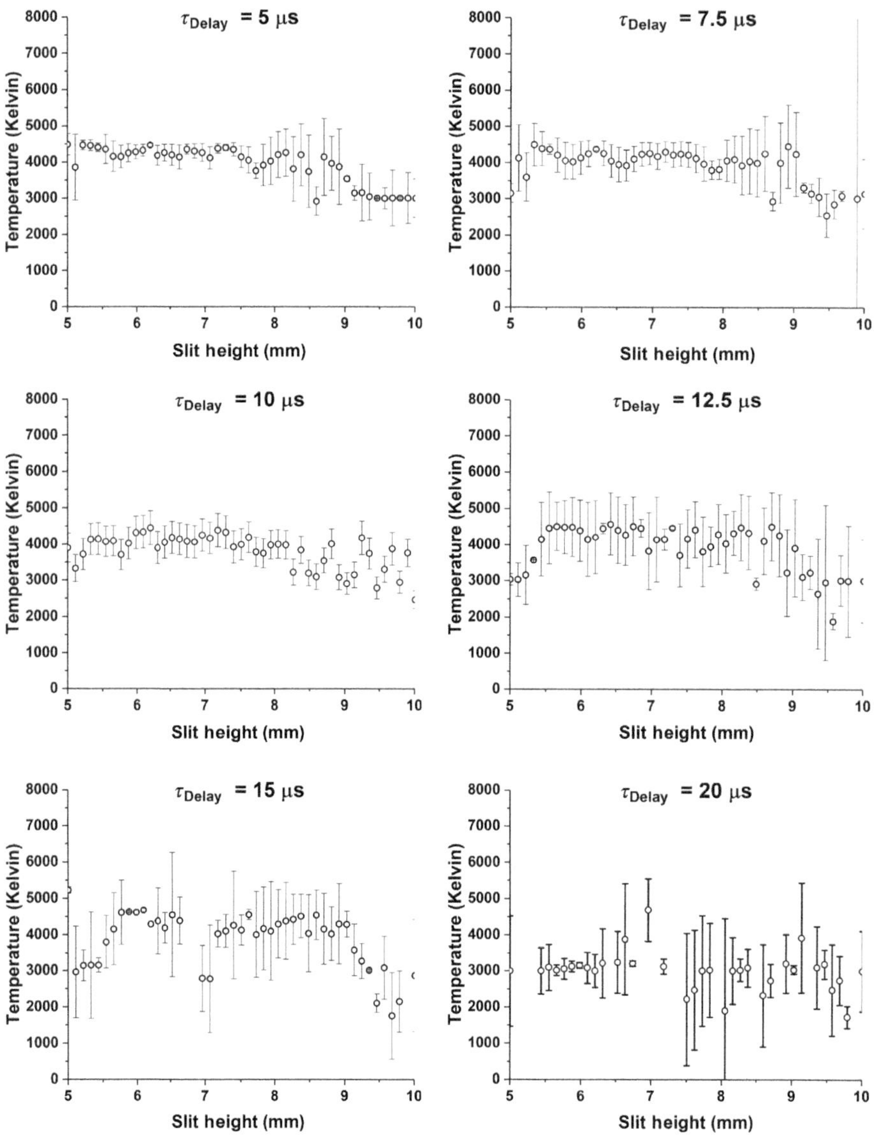

FIG. 10 Temperature decay of the AlO-containing plasma along the axis of laser incidence as a function of time following laser-induced breakdown [74].

atmosphere. Measurement of molecular spectra allows one to infer the presence of isotopes, including hydrogen, boron, carbon, nitrogen, oxygen, chlorine, strontium, zirconium, and uranium. This section communicates recent results of isotope detection with LAMIS [45].

6.1 Hydrogen LAMIS

Hydrogen and deuterium are among the most often studied isotopes in laser-induced plasma. The isotopic separation of 0.18 nm between H and D atomic lines at 656.29 and

656.11 nm makes it possible to measure these isotopes using their atomic spectra [90–94]. However, broadening of these spectral lines due to the Stark effect in laser plasma is larger than 1 nm for the Balmer series hydrogen alpha line at electron densities of the order of $10^{17}\ cm^{-3}$. Practice shows that it is appreciably easier to measure spectra of hydroxyl isotopologues using LAMIS because molecular spectra are significantly less prone to Stark broadening than atomic lines of H and D.

LAMIS of hydrogen isotopes [44, 45, 95–97] initially used a 1064-nm Nd:YAG laser device generating 4-ns pulses with an adjustable pulse energy up to 150 mJ. The laser was operated at a repetition rate of 10 Hz. Ablation plasma was projected through a lens onto an entrance slit of a spectrograph in the direction perpendicular to the incident laser beam. Czerny-Turner or echelle spectrographs were used interchangeably to disperse wavelengths onto an ICCD. The spectrum acquisition gate widths and delays were optimized to measure molecular emission from the OH and OD species.

Streams of vapors of ordinary water (H_2O) and heavy water (D_2O) were ablated to demonstrate hydrogen isotopic measurements by LAMIS [44]. The OH and OD spectra of the hydrogen and deuterium $A^2\Sigma^+ \rightarrow X^2\Pi$ (0-0) transitions from the plasma plume were recorded using a low-resolution spectrograph. This transition consists of six main branches because of its doublet multiplicity. The gate width of the ICCD was 60 μs with a time delay of 25 μs, compared to delays of ~1 μs, typically used for atomic detection in LIBS. The most prominent spectral features were observed at ~306 (R_{11}, R_{22} branch heads) and ~309 nm (Q_{22} branch head) with partially resolved individual rotational lines. The experimental shift between the Q_{22} branch heads of OH and OD was approximately 0.65 nm.

LAMIS was also used to analyze light and heavy water introduced with a moist argon flow that was produced by bubbling argon through several water samples with different D/H ratios [95]. A partial least squares regression (PLSR) model built on spectral responses from a set of eight reference samples was employed for isotopic calibration. Several PLSR models were constructed over spectra collected under different conditions in two spectral intervals with resolution ~15 pm. Statistically better results for D/H determination were achieved in a 311.5–315.5-nm interval that included several overlapping branches with relatively high J-rotational quantum numbers belonging to vibrational bands (0-0) and (1-1) of OH and OD. Another tested spectral interval was 306–310 nm that included rotational lines with relatively low J numbers. The isotopic shifts between OH and OD typically increase with increasing J numbers, and thus differentiation among the spectra was better for the high J interval. Branches R_{11} and R_{22} undergo a red shift from OH to OD, while the other four branches are shifted to the lower wavelength.

The possibility of using LAMIS for simultaneous determination of H and O isotopes in ice was recently demonstrated [96]. Emissions from hydroxyl radicals at the $A^2\Sigma^+ \rightarrow X^2\Pi$ transitions of the isotopologues were recorded during ablation of ordinary water ice (H_2O), heavy water ice (D_2O), and ice with 90% atomic fraction of ^{18}O. Argon flow was applied to displace ambient air. The isotopic shifts between ^{16}OH and ^{18}OH are smaller than those between OH and OD. The spectra of hydroxyl isotopologues ^{16}OH, ^{18}OH, and ^{16}OD were resolved using a compact spectrograph (IsoPlane-320, Acton/PI). The wavelength region 311.8–314.0 nm that included rotational lines with relatively high J numbers (branches Q_{11} and Q_{22} with quantum numbers from $J = 13.5$ to $J = 18.5$) was chosen for analysis. These results confirmed that a compact spectrograph is sufficient for LAMIS. Isotopic calibration for ice analysis can be accomplished using PLSR, previously realized for water vapor [95].

While spectra of the three hydroxyl isotopologues (^{16}OH, ^{18}OH, and ^{16}OD) were resolved in the region 311.5–314.0 nm using IsoPlane SCT-320 [96], the more intense rotational lines characterized by the lower J numbers were not resolved (below 311.5 nm), particularly for molecules ^{16}OH and ^{18}OH because of the small isotopic shifts formed in the transitions involving no change in the vibrational quantum number ($\Delta v = 0$). The isotopic shifts increase significantly when the vibrational quantum number changes. Thus it is logical to explore the vibrational band (1-0) of the hydroxyl molecules that stretches into the wavelength region 281–287 nm for ^{16}OH and ^{18}OH, and is shifted into the region 287–291 nm for ^{16}OD. Detailed lists of the rotational lines of the 1-0 bands $\{A^2\Sigma^+ \rightarrow X^2\Pi\}_i$ for the ith hydroxyl isotopologues are available in Refs. [98–102].

A plasma-reheating electric discharge after laser ablation (secondary cross-excitation) provided large enhancement in the ultraviolet emission of OH radicals for all bands with $\Delta v = 0, 1$ of the transitions $\{A^2\Sigma^+ \rightarrow X^2\Pi\}_i$ relative to that from the single-pulse laser ablation [97]. The electric discharge required optical breakdown by laser ablation. The applied voltage was 1 or 2 kV with a ballast resistor of 5 kΩ. The front edge of the electric pulse was delayed by 5 μs after the laser pulse, and the discharge duration was 500 μs. The ICCD gate was 500 μs delayed by 2 μs relative to the laser. The results demonstrated a significant increase in the signal-to-noise ratio in all spectral measurements of OH. The bands with $\Delta v = 0$ became identifiable in a larger interval of 305–324 nm. Relatively strong intensity of the OH bands $\{A^2\Sigma^+ \rightarrow X^2\Pi\}_i$ (1-0) within 281–287 nm was observed, but no discernible OH emission in this range could be obtained from laser ablation alone. The ablation laser pulse energy in these measurements was 116 mJ.

Fig. 11 displays spectra of the bands $\{A^2\Sigma^+ \rightarrow X^2\Pi\}_i$ (1-0) for ^{16}OH and ^{16}OD recorded in a flow of moist helium after bubbling through a bottle of deionized natural or heavy water. The experimental conditions were optimized for the best possible quality of the spectra (high intensity and resolution). The ablation laser pulse energy was reduced to 34 mJ. The ICCD gate width was 500 μs, but acquisition delay relative to the laser pulse was increased to 20 μs. The electric discharge voltage was 2 kV, and other parameters were kept the same as in the preceding experiment. The positions of individual lines of the six main rotational branches of the 1-0 band are indicated for both ^{16}OH

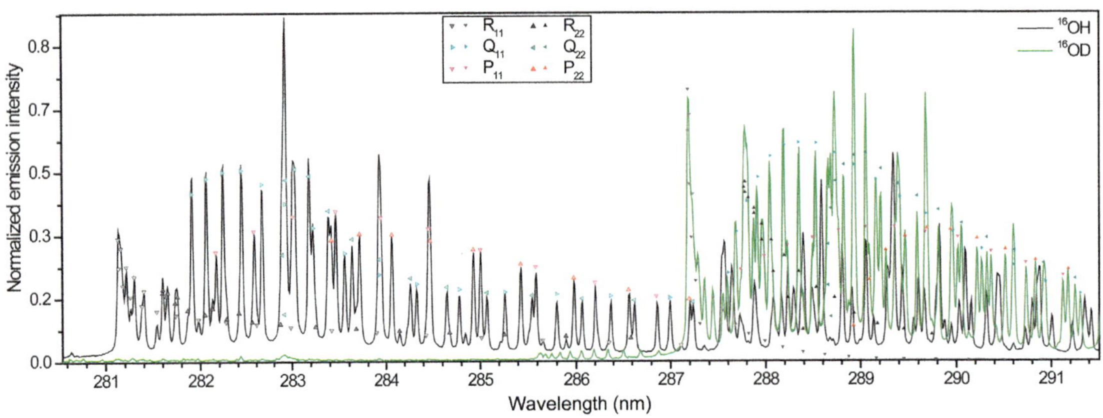

FIG. 11 Spectra of bands $\{A^2\Sigma^+ \rightarrow X^2\Pi\}_i$ ($\Delta v = 1$) of ^{16}OH and ^{16}OD acquired in a moist He flow after laser ablation reheated by an electric pulse. The line positions in the main branches of the band (1-0) are indicated for ^{16}OH (*open symbols*) and ^{16}OD (*solid symbols*) [97]. *Reproduced with permission of the Royal Society of Chemistry (RSC).*

(open symbols) and ^{16}OD (solid symbols) [98, 100]. The heads of branches R_{11} are shifted by 6.05 nm relative to each other (281.13 nm in ^{16}OH and 287.18 nm in ^{16}OD). This isotopic shift is an order of magnitude larger than the shifts found in the $\Delta v = 0$ bands.

Fig. 11 also reveals minor satellite branch $^{S}R_{21}$ for both hydroxyl isotopologues, i.e., below 281 nm for ^{16}OH and below 287 nm for ^{16}OD. At least 11 rotational lines of this branch with J values from 1.5 to 11.5 (not labeled) are clearly visible in the spectrum of ^{16}OD. The corresponding lines of this branch with $J = 1.5$, 2.5 are illustrated in Fig. 12 for ^{16}OH and ^{18}OH. Because of low-transition probabilities in this satellite branch, the individual rotational lines of the branch $^{S}R_{21}$ were not reported prior to obtaining these spectra in 1987 with tunable laser excitation [101]. The 2-1 band of ^{16}OH is also displayed in Fig. 11 with its branch R_{11} starting at 287.53 nm and spanning to longer wavelengths (not labeled). These results indicate the high excitation efficiency of laser ablation with electric discharge reheating. It is worth mentioning that the isotope-resolved bands $\{A^3\Pi \rightarrow X^3\Sigma^-\}_i$ (0-0), (1-1) of NH and ND radicals around 335–337 nm emerge when air leaks into the line of moist helium. The latter spectra are similar to those obtained by femtosecond laser ablation of deuterated solids [103, 103a].

The isotopic shifts between the rotational lines in the 1-0 band of ^{18}OH and ^{16}OH are apparent in the spectra reproduced in Fig. 12. Shifts are significantly smaller than those between OH and OD. The spectrum of ^{16}OH is identical to the one in Fig. 11, while the spectrum of ^{18}OH is recorded for the same conditions. Both spectra are similar, but the features of the heavier isotopologue are shifted by about 0.06 nm toward the red. The known positions of the individual lines are indicated for the branches Q_{11}, Q_{22}, P_{11}, and P_{22} of the ^{18}OH radical [99]. The R_{11} and R_{22} branches are congested with overlapping lines, and their labels are omitted in Fig. 12 (see Fig. 11 for the line positions in all main branches of ^{16}OH).

Molecular emission of OH and NH radicals was used for hydrogen isotopic analysis by LAMIS of liquid and frozen water samples

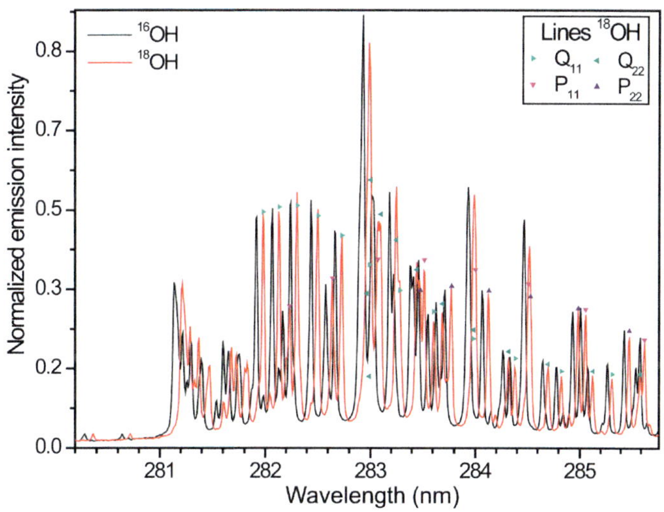

FIG. 12 Spectra of bands $\{A^2\Sigma^+ \rightarrow X^2\Pi\}_i$ (1-0) of ^{16}OH and ^{18}OH obtained under the same conditions as the spectra in Fig. 11. The line positions are indicated only for ^{18}OH [97]. *Reproduced with permission of the Royal Society of Chemistry.*

[103a]. The optimal laser pulse energy at 532 nm (5 ns) was 80 mJ for ice and 100 mJ for liquid water. The analytical ranges for PLSR quantification of the isotope-selective spectra of these radicals were set at 306–322 and 329–345 nm for the OH and NH isotopologues, respectively. Emission of both radicals reached their respective maxima and then decayed in liquid water faster than in ice. Furthermore, the OH emission persisted longer and had higher integrated intensity than that of NH for both water phases. Accordingly, the signal-to-noise ratio, accuracy and precision of the OH/OD measurements were better than those for NH/ND. The lowest relative standard deviation (0.5%) was attained for liquid water in LAMIS of hydroxyl isotopologues. Such precision and accuracy are sufficient to surveil the quality of coolant water in pressurized heavy-water reactors.

Formation processes of diatomic hydrogenated radicals using femtosecond laser-induced plasmas in air at atmospheric pressure were recently analyzed [103]. Each of the three tested substances, namely, urea, terephthalic acid, and anthracene, were either of naturally low isotopic ratio D/H or highly enriched in deuterium, thus making up six samples in total. The spectra of OH and OD were measured in the spectral range of 305–315 nm with resolution of ~60 pm using ablation of H- and D-urea with 60-fs laser pulses with 3.3 mJ per pulse, namely, a Ti:sapphire laser operating at 800 nm. The ICCD gate delay was 0.5 μs and the gate width was 25 μs. The experimental spectra were compared with simulations of the sequences $\{A^2\Sigma^+ \rightarrow X^2\Pi\}_i$ $(\Delta v = 0)$ computed for both OH and OD using LIFBASE software [104]. The fitting process yields average vibrational, T_v, and rotational, T_r, temperature, namely, integration over the gate-open time of 25 μs. The deduced values were $T_v = 5300$ (5500) K and $T_r = 3200$ (3500) K for OH (OD), respectively. Such similarity between the OH and OD internal energy distributions (T_v, T_r) implies common mechanisms of their formation.

The results from laser ablation of terephthalic acid $C_6H_4(COOH)_2$ and $C_6D_4(COOH)_2$ indicated that the extracted atoms H or D from the sample reacting with atmospheric oxygen were responsible for the formation of the OH or OD radicals in the laser plasma, without direct detachment of the intact hydroxyl groups from the sample [103]. Moisture in air also did not supply any noticeable amount of OH as evidenced by the absence of the OH emission during laser ablation of deuterated urea $CO(ND_2)_2$. These conclusions are important for the accurate implementation of LAMIS because complete atomization and equilibration of the ablated material preserves the stoichiometry of the sample. However, complete atomization does not always occur in laser ablation, and therefore careful considerations and optimization are necessary prior to analysis.

Femtosecond ablation of urea and terephthalic acid yielded the isotope-resolved spectra of NH and ND featuring the band 0-0 and 1-1 heads of the system $A^3\Pi_i \rightarrow X^3\Sigma^-$. These two band heads were observed at 335.99 and 337.10 nm for NH and at 335.66 and 336.39 nm for ND [103]. The sample of urea produced more intense NH emission than did the terephthalic acid sample. The H atoms extracted from terephthalic acid reacting with nitrogen from air were producing NH, as no nitrogen was in the sample. However, the native NH fragments were probably released from urea contributing to higher intensity of the NH spectrum relative to that from terephthalic acid. The natural and deuterium-enriched samples also exhibited CH and CD spectra. The observed features were the band 0-0 and 1-1 heads of the system $C^2\Sigma^+ \rightarrow X^2\Pi_r$ at 314.38 and 315.74 nm for CH and at 314.23 and 315.05 nm for CD. Also, the Q-branch heads of the 0-0 band $A^2\Delta \rightarrow X^2\Pi_r$ were observed at 430.99 and 430.31 nm for CH and CD, respectively [103].

6.2 Boron LAMIS

The stable isotopes of boron ^{10}B and ^{11}B with an average natural abundance ratio $^{10}B/^{11}B$ of about 0.247 were measured most often among all other isotopes in LAMIS. This is in part because ^{10}B is a crucial element for the nuclear industry due especially to the high neutron absorbing efficiency of ^{10}B. Furthermore, there are many applications in which the $^{10}B/^{11}B$ ratio is measured as a natural tracer in geochemistry, hydrology, oceanography, and environmental and other sciences. These important applications combined with easily obtainable emission of boron monoxide during laser ablation in air motivated a strong scientific interest in LAMIS.

Laser ablation chemical analysis has been studied since the invention of lasers, and isotope-specific molecular spectra have been known for more than a century. However, resolved isotopologue emission measurements in ablation plasma were first reported in 1998 [105]. A low-pressure chamber was filled with air at 400 Pa to ablate a boron disk of the natural isotopic ratio. At 400 Pa, the ablation plumes were hemispherical and 15 mm in diameter. Twelve bands of BO were observed using gate delays of 1–30 μs, within 330–560 nm at a resolution of 0.4 nm. A change of 5% in isotopic composition could be determined by fitting the calculated and experimental spectra of the 0-2 band $A^2\Pi_i \rightarrow X^2\Sigma^+$ of ^{10}BO and ^{11}BO. The size of the low-pressure plumes precluded collection of light from the entire plume. Subsequently, the authors suggested using a detector with spatial resolution to improve precision [105].

Recently, boron isotopes ^{10}B and ^{11}B were measured by laser ablation in SATP air using three chemically pure substances: isotope-enriched $^{10}B_2O_3$ and $^{11}B_2O_3$, and natural BN [106–108]. Spectral resolution better than 0.23 nm was found unnecessary for boron isotopic determination with LAMIS [106]. Spectra were accumulated over 1000 laser pulses with an analysis time of 100 s for a 10-Hz laser pulse repetition rate. Precision can be further improved to errors within only 1‰ using an increased repetition rate and a total number of up to 120,000 laser pulses, but with an analysis time of 2 min at 1 kHz. The results indicated that LAMIS can be calibrated using spectral normalization followed by PLSR for measurements of the isotopic ratio with a precision characterized by relative deviations within 9‰ from the mean value of the $^{10}B/^{11}B$ ratio with 95% confidence, namely, 2σ confidence [107].

Industrial applicability of LAMIS was demonstrated by measuring neutron absorber materials from 3M-Ceradyne (Canada) [96]. Two samples of aluminum-alloyed composites with different boron isotope ratios were ablated in air and analyzed using a compact IsoPlane-320 spectrograph. One of the samples was a Boral core, i.e., aluminum alloy consolidated with 56% boron carbide of natural boron isotope abundance. The other sample was BorAluminum, a regular unalloyed aluminum (grade 1100) with 4.5% addition of elemental boron, enriched in the lighter isotope ^{10}B up to 95%. Fig. 13 illustrates the ablation spectra of these samples recorded between 591 and 608 nm. The displayed spectra were acquired with an ICCD gate delay of 10 μs and gate width 100 μs. The spectra show rotational lines of a BO molecule in the 0-3 vibrational band of the $A^2\Pi_i \rightarrow X^2\Sigma^+$ system [96].

For research work in LAMIS, the BO $A^2\Pi_i \rightarrow X^2\Sigma^+$ emission was used in the range of the visible 0-2 band [107–109] and the 0-3 band [106, 107, 109, 110]. Ultraviolet emission from the $B^2\Sigma^+ \rightarrow X^2\Sigma^+$ 0-2 band [44, 106, 110a, 111–113] was also explored. Ultraviolet emission can be beneficial to reduce sunlight interference in outdoor standoff applications. However, usually a choice of the spectral region was driven by the quality of the spectra for multivariate quantification. Visible bands for the selected instrumentation appear several hundred times more intense than the ultraviolet bands of BO.

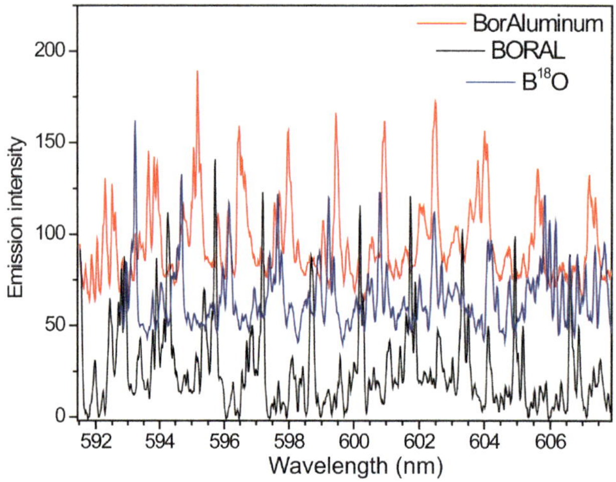

FIG. 13 Spectra of BO A → X (0-3) emission formed by ablating in air of two aluminum-based composite materials containing different $^{11}B/^{10}B$ ratios. Spectrum of $B^{18}O$ obtained by ablating BN in $H_2{}^{18}O$ vapor. Spectra are shifted up for clarity (continuum background subtracted) [45]. *Reproduced with permission of the Royal Society of Chemistry.*

The data in Fig. 13 illustrate a significant isotopic shift between spectra of Boral (80% ^{11}B) and Bor-Aluminum (95% ^{10}B), requiring only modest spectral resolution. Because of the optimized time delay of 10 μs, these spectra did not encounter interferences from atomic aluminum or chemical impurities in the samples.

A double-pulse approach, in which a second laser pulse arrives into a laser plasma at a small delay after the ablation pulse, was shown to increase BO emission [106, 110a]. The intensified emission can be attributed to higher electronic and collisional excitation of molecules in the double-pulse method as compared to the single-pulse LAMIS. An enhancement factor of ~3 was reported for a combination of 355 and 1064 nm laser pulses [106] similarly to a factor of ~2 for a combination of 532 nm and 10.6 μm laser pulses [110a]. However, the standard deviations in the double-pulse LAMIS measurements were higher than expected, and therefore the detection limits were only slightly improved relative to the single-pulse LAMIS [110a].

A femtosecond pulsed laser was used to ablate four boron carbide samples in a femtosecond regime of LAMIS. Reported relative accuracy values were within 2% in three-point calibration for determination of the $^{10}B/^{11}B$ isotopic ratio [110] by measuring $^{10}B/^{11}B$ ratios from BO emissions in the 532–540-nm region. The spectra were deconvolved using a multivariate regression function over 155 cubic splines to reconstruct the two constituents for the boron isotopes. Femtosecond ablation resulted in low continuum emission background, enabling nongated time-integrated measurements that were sufficient for accurate isotopic characterization. A nonradioactive B_4C sample from the Breazeale Nuclear Reactor was analyzed for ^{10}B content using 200 ablation pulses and confirmed <5% accuracy in ^{10}B determination with LAMIS. Precision and accuracy within a few percent are often satisfactory for the nuclear industry.

Recently, femto- and nanosecond ablation were compared in LAMIS while using nongated acquisition of spectra in both approaches [109]. It was shown that preprocessing of LAMIS with numerical spectral filtering, including low-pass or high-pass filters, and normalization prior to creating spline-approximated PLSR can effectively reduce the influence of significant continuum background in nanosecond ablation. Application of optimized numerical filters decreased the variances of calibration points up to two orders of magnitude, which resulted in improved accuracy of isotopic determination. The work [109] emphasized that neither femtosecond radiation nor ICCDs are necessary for boron isotopic measurements with LAMIS, therefore compact instruments with nongated CCD detectors are sufficient and can be developed for this LAMIS application.

Matrix effects were investigated for the determination of ^{10}B atomic fraction using LAMIS and for analysis of several boron-containing substances and mixtures [111]. The following materials were tested: pure metallic boron, boron nitride, solid boric acid, lanthanum borate, anhydrous borax, borax detergent additive, Pyrex glass, and two blends of borax with aluminum silicate and boric acid with cellulose. The experimental LAMIS parameters were optimized for BO measurements during BN ablation in air, and then applied unchanged for all other samples. The matrix effect was minimal for ^{10}B determination for most samples, except $LaBO_3$, Pyrex, and a silicate blend. The $LaBO_3$ spectrum was vastly affected by atomic and ionic lines of lanthanum, while Pyrex and a silicate blend did not produce discernible BO emission for the experimental arrangement. Native oxygen contained in most substances did not affect BO emission, implying domination of plasma reactions with air to form BO in the ablation plume. These results indicated that LAMIS can accurately measure ^{10}B in different substances with a relative uncertainty of ~3% as long as BO emissions were assertively recorded.

Similar to other analytical techniques, LAMIS is subject to chemical and physical matrix effects. The previous results on matrix effects [111] demonstrated that LAMIS can be remarkably robust regarding measurements of boron isotopes in very different environments even without optimizing the experimental conditions for a specific matrix. A proposed algorithm [96] automatically fits a Lorentzian profile to each interfering spectral line, even when the interference is 1000× stronger than the molecular feature of interest. Subsequently, the interferences are numerically subtracted from the molecular spectrum. The algorithm can be applied to "digitally purge" the spectrum of BO from the lanthanum lines that previously precluded ^{10}B determination. This algorithm was recently implemented to evaluate the fundamental analytical accuracy and precision attainable by LAMIS for minor isotopes in the presence of interferences [114]. Lorentzian or Gaussian profiles can also be used to convolve multiple molecular lines with instrument prompts, and thus digitally reduce experimental noise.

A possibility of LAMIS application for geological analysis [96] was examined using two tourmaline samples of different color and composition. Otherwise crowded LIBS spectra had the least atomic interferences in a region 572–585 nm. At a time delay of 3 μs, only two spectral lines of atomic iron were persistent in this spectral interval in one of the tourmaline samples. The positions of these interfering lines could be masked from the spectrum to determine boron isotopic composition using PLSR. Provided that the interfering lines are sparse, they do not necessarily preclude the isotopic analyses. However, appropriate isotopic standards of tourmaline would be required for accurate calibration.

Precision in determining $^{10}B/^{11}B$ ratios [107] with LAMIS at an uncertainty of 9‰ is comparable with that obtained in SIMS and better than that offered by inductively coupled plasma-optical emission spectrometry, inductively coupled plasma-quadrupole mass spectrometry, and laser ablation time-of-flight mass spectrometry for boron isotopic ratio measurements. This level of precision is reasonably adequate for geochemical analysis, because natural variation of the $^{10}B/^{11}B$ ratio in the terrestrial environment is >90‰. Precision of LAMIS can be further improved by increasing spectral accumulation. In this course, it will probably be necessary to consider two isotopes of oxygen ^{16}O and ^{18}O, because both participate in BO emission. The spectrum of $B^{18}O$ was obtained by ablating a BN sample in argon with $H_2{}^{18}O$ water vapor (Fig. 13).

Higher sensitivity of LAMIS measurements is certainly desirable. A combination of LAMIS with laser-induced fluorescence of molecular radicals [112a, 113] can enhance spectral radiance of particular emission features. A tunable OPO laser source can be used for additional selective excitation of molecular energy levels in individually chosen isotopologues. The enhancement factors achieved were 34 for ^{11}BO and 16 for ^{10}BO [113]. This approach resulted in improved detection limits (0.98 and 0.85 at.%) for determination of ^{11}B and ^{10}B atomic fractions using ^{11}BO and ^{10}BO spectra. An important advantage of this technique is that it facilitates isotopic measurements in standoff operation from a distance, and resonant laser-induced fluorescence is applicable to many molecular radicals [115]. A disadvantage is the high cost of an OPO source. A secondary electric pulsed discharge is a less expensive option [97], but a discharge does not allow for standoff operation.

LAMIS of triatomic radicals $^{10}BO_2$ and $^{11}BO_2$ was also investigated recently [116]. This is the first and so far the only work on LAMIS of triatomic isotopologues. A sample of B_4C was ablated in several ambient gases. Higher intensity of BO_2 emission was observed in an oxygen atmosphere relative to that for laser ablation in air. The intensities of BO emission were almost the same in air and in oxygen. The latter effect was attributed to direct association of atomic B with molecule O_2. Emission of BO_2 was considerably more intense relative to emission of BO in the experiments [116]. While progression bands of BO_2 emission for the $A^2\Pi_u \rightarrow X^2\Pi_g$ transition were broad (~10 nm), the observed isotope shifts were also significant, namely, 0.55, 1.35, and 1.93 nm for bands 000-000, 100-000, and 200-000, respectively. The experimental isotope shifts were in reasonable agreement with the predicted shifts.

Vibrational bands of boron nitride ^{11}BN and ^{10}BN were observed in plasma produced by pulsed laser ablation of a boron nitride target in low-pressure (130 Pa) nitrogen or argon atmospheres [117]. The ICCD gate delay was 1.4 μs. The observed emission bands were identified to be the $\Delta v = -1, 0, +1$ sequences of the triplet transition $A^3\Pi \rightarrow X^3\Pi$ in the spectral range 342–385 nm. The isotopic shifts in bands 0-1 and 1-2 of this transition were found to be 0.61 and 0.69 nm, respectively, in agreement with calculated values. In contrast to efficient formation of oxides, the nitride species (except for CN) are scarcely produced in laser ablation plasma because nitridation is generally a strongly endothermic process. Accordingly, the low emission intensity of the ^{10}BN band heads was masked by noise and overlapping emission bands from N_2 and N_2^+. The density of N_2 molecules in the plasma was always several orders higher than that of BN even in an argon environment because the boron nitride target supplied nitrogen.

6.3 Carbon LAMIS

Carbon isotopes are indicative of primary bioproductivity and energy cycling and are most important for studying biochemistry. The relative biological enhancement of ^{12}C over ^{13}C can be up to 5% (e.g., in methanogen bacteria). Carbon isotope measurements can facilitate tracing the origins of anthropogenic CO_2 and can reveal environmental challenges related to subsurface carbon sequestration and shale fracking. Variation of the natural isotopic ratio $^{13}C/^{12}C$ in different materials ranges from 0.96% to 1.15%. These variations can be measured using LAMIS. The ability to distinguish between spectra of the three C_2 isotopologues in laser ablation of graphite has been asserted using computed Swan spectra [118]. Laser ablation of carbon-containing samples in air typically produces C_2 and CN radicals in comparable quantities. Both species have been used for measuring carbon isotope ratio [119].

Utilization of spectra of carbon dimers C_2 for LAMIS measurements was thoroughly studied [44, 96, 108, 114, 119–122, 122a]. Kinetic isotope fractionation was observed in C_2 emission, suggesting that carbon isotopes ^{12}C and ^{13}C can behave differently in laser ablation plumes [120–122]. Fragmentation of aromatic rings from benzoic acid samples into $^{12}C_2$ dimers was found more effective than formation of molecules $^{13}C^{12}C$ [121]. Consequently, determined isotopic ratios $^{13}C/^{12}C$ deviated from the ratio in the sample. The latter effect was observed relatively close to the sample surface within 2 mm, therefore these results can be attributed to preferential evaporation of the native $^{12}C_2$ fragments from a sample surface subsequent to laser pulse excitation. Evaporation of carbon in the molecular versus atomic form would be thermodynamically favored.

Incomplete dissociation of the double bond $^{13}C=^{13}C$ in laser ablation of a ^{13}C-labeled fumaric acid pellet was recently studied [120]. It was found that in the beginning the ablation plasma was mostly atomized. Up to a gate delay of 4 μs, with a gate width of 2 μs, the majority of C_2 in the plasma was formed by association of free carbon atoms. At longer time delays the main supply of the gas-phase dimers $^{13}C_2$ was from partial dissociation of the original molecules of fumaric acid of the sample. A switch in the main formation mechanism was concurrent with the change in a plume shape from prolate to oblate at longer time delays. An oblate plasma plume adjacent to the sample surface within ~0.6 mm suggested prevalence of the molecular evaporation and partial fragmentation mechanisms over the free atom association. Contributions of partially dissociated molecules directly from the sample to the gas-phase C_2 were significantly less in an argon than in a nitrogen atmosphere. Only after a delay of 40 μs did the role of fragmentation in argon become comparable to that in nitrogen at delays of 5–10 μs.

The ambient gas obviously affects the mechanisms and their dynamics in the ablation plumes. Laser ablation plasma in argon remains excited for a longer time than that in air or nitrogen because argon is inert, has low thermal conductivity, and shows metastable states. Nitrogen consumes energy for dissociation or dissipates it through numerous intramolecular channels and becomes chemically reactive in its atomic form. Because of these effects, the plasma plume cools faster in nitrogen (or air), but also because the thermal conductivity of nitrogen and air is higher than that of argon. At long time delays from plasma initiation, the surface of the sample may remain at elevated temperatures to continue releasing molecular fragments into an already cold but still ionized plume. Since formation of C_2 does not require reactions with atmospheric constituents, argon is a better environment in which to sustain enough energy in the gaseous phase of the plume for complete dissociation of species and for sufficient time for equilibration.

Other noble gases can also be useful in LAMIS. Laser ablation plumes in helium have larger size and lower gas temperature than plumes in argon or air. A helium environment can be used to reduce Stark and collisional broadening of spectral lines. A pellet of mixed benzoic acid and amorphous carbon was ablated in a chamber filled with neon [122], and significant deviations were reported for the observed isotopic ratio $^{13}C/^{12}C$ from the nominal one at distances within 2 mm of the sample surface. At larger distances from the sample (2–5 mm) and gate delays between 4 and 10 μs, the observed isotopic ratio $^{13}C/^{12}C$ was close to the actual ratio in the sample.

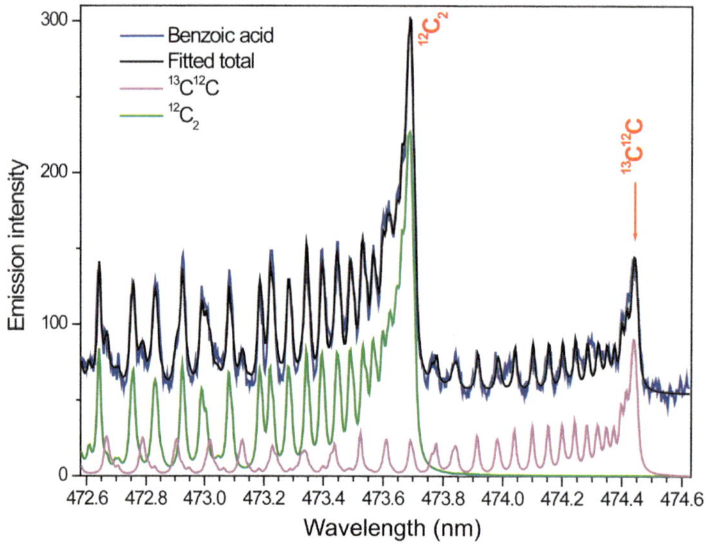

FIG. 14 Swan bands (1-0) of the two isotopologues $^{13}C^{12}C$ and $^{12}C_2$. Synthetic spectrum fitted to the experimental spectrum acquired from ablating a ^{13}C-labeled benzoic acid pellet ($^{13}C/^{12}C = 16.7\%$) in air at standard ambient temperature and pressure [45]. *Reproduced with permission of the Royal Society of Chemistry.*

Experimental conditions in LAMIS should be selected so that the ablated material becomes completely atomized, then equilibrates and associates into excited diatomic molecules. Use of the fourth harmonic for laser ablation at 266 nm that would directly break chemical bonds is probably preferable over ablation at 1064 nm. However, all studies on LAMIS for carbon isotopes to date have been performed using 1064-nm pulsed radiation. Operating at longer delays facilitates equilibration of the atomized matter within the plume but can yield erroneous results due to partial molecular fragmentation. For the LAMIS analysis of organic samples using C_2 emission, an argon atmosphere assists in full atomization of the ablation plume. Remaining fractionation effects may be minimized when volume-integrated emission is collected from the whole plume, contrary to the studies [120–122], in which fractionation was purposely refined and investigated.

The ratio of densities of the free dimers $^{13}C^{12}C$ and $^{12}C_2$ was determined to elucidate the isotopic effects in the ablation plumes in air using simulated Swan spectra [121]. Fitting the synthesized to an experimental spectrum from a benzoic acid pellet with the ratio ^{13}C:$^{12}C = 1:6$, or 16.7%, was performed for isotopic quantification. Because of the significant enrichment in the ^{13}C isotope, Fig. 14 displays results from simulating only the 1-0 bands $d^3\Pi_g \rightarrow a^3\Pi_u$ for both isotopologues of $^{13}C^{12}C$ and $^{12}C_2$, which was sufficient. A spectral resolution of 10 pm, delay time of 1 μs, and gate width of 0.75 μs were used. Assuming carbon dimers originated only from association of free atoms, the statistical relation between the number of the resulting isotopologue molecules should be $^{13}C^{12}C:^{12}C_2 = 1:3$, or 33.3%. A relatively slow release of the native $^{12}C_2$ fragments directly from the aromatic rings without their atomization was suggested as a probable cause.

Samples with the natural abundance of carbon isotopes required a more sophisticated spectrum simulation due to the significantly higher relative intensity in the tails of several nearby bands from the major isotopologue $^{12}C_2$. A synthetic spectrum of the Swan system of C_2 was computed in the region where the band head of $^{13}C^{12}C$ $d^3\Pi_g \rightarrow a^3\Pi_u$ (1-0) at 474.45 nm appears among other lines of the $^{12}C_2$ band system [119]. Fitting of the computed to an experimental spectrum from the ablation of graphite was performed. The best fit was achieved with a $^{13}C/^{12}C$ ratio of 1.08% and rotational temperature of 5155 K. The isotopic ratio obtained by LAMIS agreed well with the value of 1.0804%, measured from the same graphite sample by isotope ratio mass spectrometry [119]. The individual contribution of $^{13}C^{12}C$ and the final simulated total sum of five components from both isotopologues along with the experimental spectrum are shown in Fig. 15. The intensity scale was normalized on the maximum of the 1-0 band head of $^{12}C_2$ $d^3\Pi_g \rightarrow a^3\Pi_u$. The gate width was 30 μs at a time delay of 2 μs. Spectral resolution was 10 pm, and data were averaged over 1000 laser pulses.

Formation of CN radicals in LAMIS of organic and inorganic samples was found less prone to isotope fractionation compared to C_2 [119]. Moreover, carbon and nitrogen isotopes can be determined simultaneously using spectra of CN isotopologues [44, 96, 108, 119]. Two spectral intervals of CN emission were used for LAMIS measurements: 354–362 and 411–422 nm [96, 119]. These intervals include two band sequences with opposite changes in vibrational quantum numbers $\Delta v = \pm 1$ of the CN transition $B^2\Sigma^+ \rightarrow X^2\Sigma^+$. Spectra from a natural graphite sample and two benzamide pellets enriched in ^{13}C and ^{15}N (Fig. 16) were measured by a compact IsoPlane-320 spectrograph around the bands with $\Delta v = +1$. The rotational structure of these bands was barely resolved but isotopic shifts between isotopologues of $^{12}C^{14}N$, $^{13}C^{14}N$, and $^{12}C^{15}N$ were clearly observed.

Isotopic shifts are larger for the other sequence ($\Delta v = -1$) between 411 and 422 nm, but in this case the bands of the minor isotopologues sit on top of the relatively intense tails from the major isotopologue. Fig. 17 illustrates the isotopic shifts.

The 0-1 bands $B^2\Sigma^+ \rightarrow X^2\Sigma^+$ of $^{12}C^{14}N$ and $^{13}C^{14}N$ in the interval 419.8–422.0 nm were computed and measured at 10-pm spectral

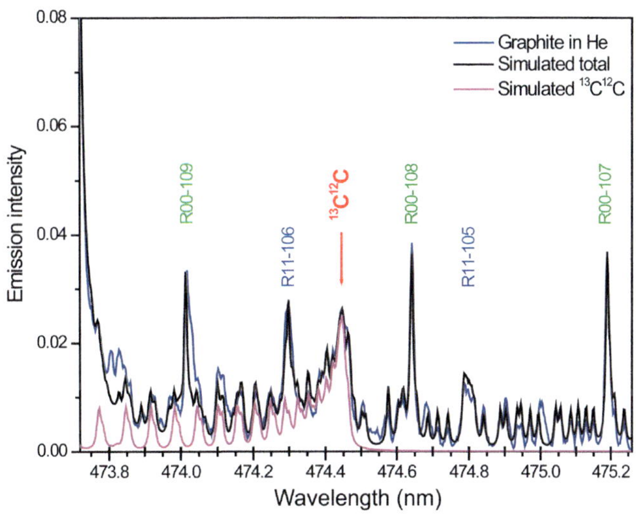

FIG. 15 Part of the Swan system with band (1-0) of $^{13}C^{12}C$ among rotational lines of $^{12}C_2$. Synthetic spectrum fitted to that recorded during ablation of graphite of natural abundance in helium. Designations of $^{12}C_2$ lines are simplified, e.g., "R00-109" stands for coincident triplet lines: $R_1(J = 109)$, $R_2(J = 108)$, and $R_3(J = 107)$ of $^{12}C_2$ (0-0) [45]. *Reproduced with permission of the Royal Society of Chemistry.*

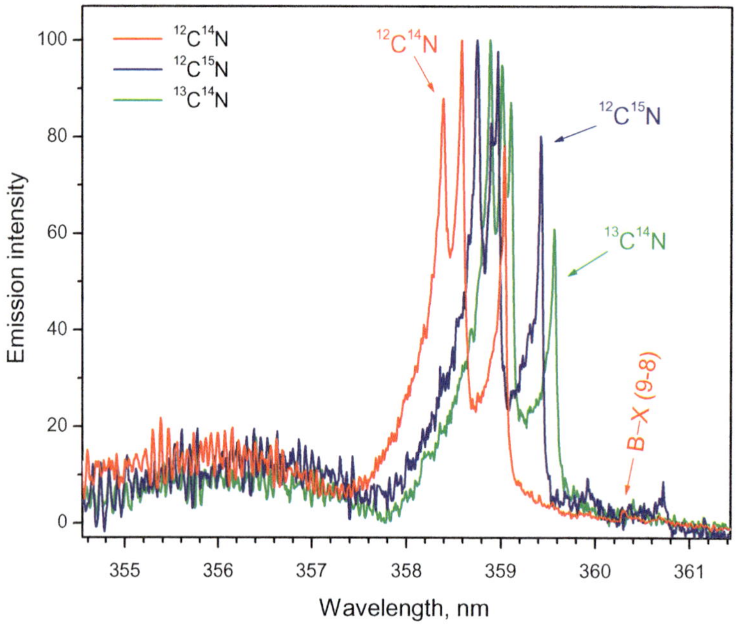

FIG. 16 Normalized spectra of cyanide vibrational band progression (B → X; $\Delta v = +1$) formed during ablation of natural graphite and benzamide pellets enriched in ^{13}C and ^{15}N isotopes. Spectra were averaged over 100 laser pulses. Bands of $^{12}C^{15}N$ were recorded in a helium flow [45]. *Reproduced with permission of the Royal Society of Chemistry.*

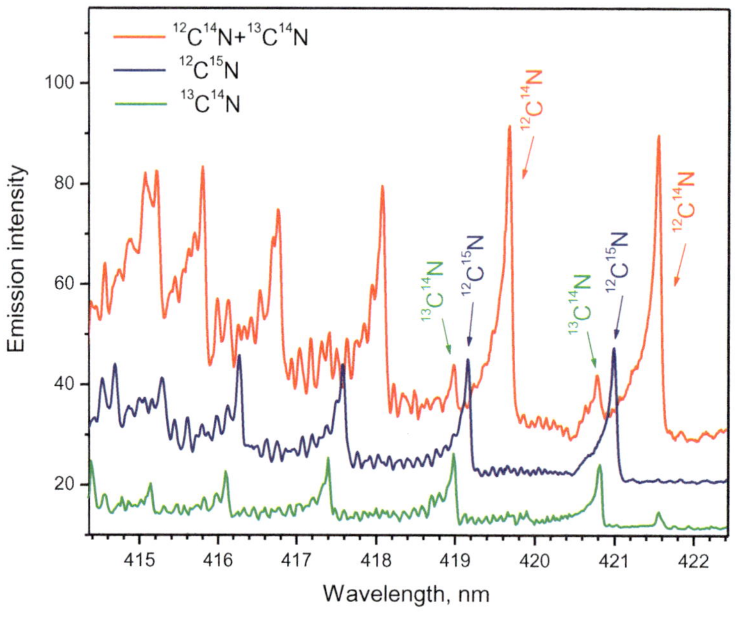

FIG. 17 Spectra of cyanide vibrational band progression (B → X; $\Delta v = -1$) formed during ablation of benzamide pellets enriched in ^{13}C and ^{15}N isotopes. Spectra were averaged over 100 laser pulses. Bands of $^{12}C^{15}N$ were recorded in a helium flow [45]. *Reproduced with permission of the Royal Society of Chemistry.*

resolution [121]. Pelletized benzoic acid powder was used as an ablation target in air. Fitting the experimental spectrum with the computed one provided relative densities of these isotopologues in the state $B^2\Sigma^+$ ($v = 0$). The derived data were used to calculate a dependence of the ratio $^{13}C/^{12}C$ on the time delay up to 8 μs. Within experimental uncertainty, the determined $^{13}C/^{12}C$ ratio was equal to the stoichiometric ratio at the delay of 0.6 μs and dropped to slightly lower values for longer delays. Fragmentation of the benzene rings into C_2 dimers that prevailed over full dissociation into free atoms was offered as an explanation of nonstoichiometric ratios. However, the ratio $^{13}C/^{12}C$ derived from CN spectra deviated significantly less from the sample stoichiometry than was the deviation of this ratio derived from C_2 spectra [121].

Quantitative LAMIS of carbon isotopes in samples of enriched urea was demonstrated [113] using CN bands ($B \rightarrow X$; $\Delta v = +1$) in the spectral region of 357.5–361.0 nm. Two PLSR calibration models were build for each of the ^{12}C and ^{13}C atomic fractions. The detection limit of ^{13}C determination was 5.3 at.% that is higher than natural abundance of this isotope. A combination of LAMIS with additional selective excitation was applied using an OPO laser tuned at the respective band heads $B \rightarrow X$ (1-0) of ^{12}CN (359.04 nm) and ^{13}CN (359.57 nm). The enhancement factors were 36 and 30 for ^{12}CN and ^{13}CN emission, respectively. This combination of LAMIS and laser induced fluorescence resulted in the detection limits lowered down to 1.6 and 1.2 at.% for ^{12}C and ^{13}C, respectively [113]. A review [115] listed 5 carbon-bearing radicals including CN that can be measured by laser-induced fluorescence in plasma or gaseous phase.

The temporal evolution of the band heads of ^{13}CN at 420.9 nm and ^{12}CN at 421.6 nm of the 0-1 band of the $B^2\Sigma^+ \rightarrow X^2\Sigma^+$ system was analyzed to study the isotope-specific formation of CN during laser ablation in nitrogen of ^{13}C-labeled fumaric acid [120]. The intensity ratio for the band heads $^{13}CN/^{12}CN$ increased from 0.76 to 1.3 for delays from 2.5 to 40 μs, while the value matching the stoichiometric ratio $^{13}C/^{12}C = 1$ occurred at about 8 μs. The temporal increase of the ratio $^{13}CN/^{12}CN$ was tentatively linked to a depleted supply of ^{12}C in the late afterglow because of preferential fragmentation of fumaric acid into ^{12}CO or $^{12}CO_2$ molecules, because the most stable bond in fumaric acid is the double bond $^{12}C{=}O$. The fragments $^{13}C_2$ might still be partially atomized to form ^{13}CN, or to react directly with N in the plasma to form CN.

Laser ablation of nitrogenous organic compounds in air or nitrogen results in formation of the free CN radicals that include nitrogen either from the sample or from the atmosphere. The contribution of the ambient nitrogen dominated over the native intramolecular nitrogen in the formation of CN molecules [120]. Mixing of the ambient gas and the target material in the plasma plume was incomplete at the early stage of plume development. For delays longer than ~20 μs, the contributions from sample and ambient nitrogen equilibrated [120].

Spectra of the CN molecule were calibrated using PLSR for quantitative determination of the ^{13}C isotope in solid samples [96, 119]. A set of reference samples was prepared by blending powders of natural and ^{13}C-enriched decanoic acid. A natural atomic fraction of ^{13}C was calculated from these measurements as $1.09\% \pm 0.14\%$ in good agreement with the known average atomic fraction of ^{13}C. Averaging over 10 measurements (each one made of 100 pulse accumulations) can reduce an estimated random error in the average ^{13}C fraction down to $\pm 0.044\%$. Precision and accuracy of carbon isotope measurements could be improved if better homogenized reference standards were available for calibration. This empirical calibration method reduced possible errors from unpredicted isotope fractionation effects.

The success of the empirical calibration of the CN spectra for quantitative LAMIS measurements was also based on the volume-integrated collection of light from the whole ablation plume, thus minimizing an impact of spatial fractionation of isotopes within the plume. It is possible that simultaneous measurements and analysis of the spectra of both CN and C_2 isotopologues will further improve the accuracy of LAMIS because these two radicals seemingly behave complementary to each other in regard to the isotopic fractionation.

Spectra of cyanide radicals (mostly ^{12}CN) forming in laser ablation of human exhale were compared with spectra of ^{13}CN forming in ablation of highly enriched gaseous $^{13}CO_2$ [122a]. The results of this study indicated that LAMIS can be a promising method for online detection of carbon isotopes in human breath or smoke from burning coal, wood and paper.

LAMIS was tested for isotopic analysis of coal, graphite, diamond [119], and soil samples [96]. However, practical applications were hampered by interfering metallic atomic lines, particularly iron and chromium that were persistent in the region of CN $B \rightarrow X$ ($\Delta v = +1$) bands even at long delays from the ablation pulse. It was concluded that an algorithm of fitting a Lorentzian or Voigt profile to every interfering spectral line and digitally subtracting them from the spectrum can be applied to suppress the metallic lines in the analysis. The efficiency of this algorithm was assessed in numerical simulations of the C_2 bands $d^3\Pi_g \rightarrow a^3\Pi_u$ ($\Delta v = +1$) in the presence of noise and spectral interference from four elements (Ti, Fe, Mn, Sr) [114]. The study claimed that the precision of LAMIS would degrade about 10 (20) times in the presence of an interfering spectral peak that is $100\times$ (or $1000\times$) stronger than the band head of the measured isotopologue.

In the absence of spectral interference, the theoretical detection limit of ^{13}C atomic fraction (equal to triple the standard deviation) was estimated at 0.003% (30 ppm) for LAMIS measurements when using an accumulation of C_2 spectra from 100 laser pulses [114]. Therefore in samples with ^{13}C depleted 300 times below natural abundance, the band head of $^{13}C^{12}C$ could still be supposedly detected. Theoretical accuracy in this case would be within 0.01%, but with a precision of only 30%. For samples with natural ^{13}C abundance, the precision in measuring ^{13}C atomic fraction was estimated at about 1‰ [114]. These results represented the ideal analytical situation. Accuracy depends on the quality of spectral simulation or the quality and homogeneity of the reference standards and improves with increasing number of calibration standards.

6.4 Nitrogen LAMIS

The stable isotope ^{15}N is frequently used as a tracer complementing ^{13}C because measurement of only one isotope is often insufficient in studying complex biotic and abiotic processes relevant to ecology, the biosphere, and geochemistry, both organic (fossil) and inorganic. Analytical techniques based on measurements of the ^{15}N tracer have been utilized and comprehensively studied particularly in agronomy, since nitrogen is a critical plant resource. Fertilizers enriched typically above 40% in ^{15}N relative to ^{14}N are broadly used to track the efficiency of plant uptakes, fertilizer losses, and nitrogen turnover in soil. The ^{15}N is used to trace nitrate leaching from agricultural fields as it causes both groundwater pollution and losses of fertilizer at an additional economic cost.

Studies of the interaction of laser-ablated organics, namely, enriched 98% ^{15}N urea, with ambient $^{14}N_2$ atmosphere [120] showed that $C^{15}N$ in the plume amounted to <10% of the total produced CN. The majority were $C^{14}N$ molecules resulting from reactions of the ablated matter with ambient nitrogen, evidently because ambient gas provided an excess of N in the plasma. The pronounced anisotropy of the $C^{15}N$-to-$C^{14}N$ ratio across the plume diameter was observed in the early plasma, indicating poor initial mixing of the ablated material with the ambient gas. Steep gradients of the distribution of ^{15}N originating from the target material and ^{14}N from the ambient gas were observed over the diameter of the plasma. For plasma at time delays longer than 20 μs, contributions from sample and ambient nitrogen isotopes equilibrated. When the ^{15}N-urea sample was ablated in argon, the resulting CN radicals were almost exclusively $C^{15}N$.

Calibration of LAMIS for measurements of ^{15}N was performed with reference samples prepared by pelletizing blended powders of natural and ^{15}N-enriched benzamide [96]. Atomic fractions of ^{15}N in the reference samples were 0.37, 1.0, 2.0, 5.0, 10, 20, and 99%. The samples were ablated in a helium flow to preclude nitrogen entrainment from air. The CN spectra (Fig. 17) of the vibrational band sequence $B^2\Sigma^+ \rightarrow X^2\Sigma^+$ ($\Delta v = -1$) were collected within the range 414–422 nm using a compact spectrograph IsoPlane-320. Ten replicates of spectra accumulating 100 laser pulses from each sample were acquired using a 10 μs gate width delayed by 1.5 μs from the ablation pulse. A multivariate PLSR calibration was built straightforwardly, but the resulting precision was insufficient for determination of natural variations of the isotopic ratio $^{15}N/^{14}N$.

Precision and accuracy of quantification were again limited by microscale isotopic inhomogeneity of the reference samples that were prepared. Inaccuracies in nominal isotopic content in the prepared reference mixtures also increased the calibration errors. Application of LAMIS for measuring the ratio $^{15}N/^{14}N$ at the current level of development is therefore confined to artificially enriched samples, such as found in agronomic studies on fertilizer efficiency or therapeutic research based on isotope-labeled drugs. Similarly to determining the ratio $^{13}C/^{12}C$ in coal and soil samples [96], there will be a need to address persistent interferences caused by atomic lines from metals that are present in all geo- or ecological sample spectra.

In a recent study [123], the content of ^{15}N in enriched urea, a nitrogenous compound containing the carbonyl group, was determined by LAMIS in an argon flow and applying PLSR calibration. Standard solutions of known isotopic content were prepared by mixing urea enriched in ^{15}N (98%) and urea of natural isotopic ratio. A volume of 30 μL of each solution was dispensed to filter paper discs that, after drying, produced uniformly distributed analytes. The filter paper also served as an additional source of carbon atoms that facilitated the formation of the CN radicals. Spectra were accumulated from 100 laser pulses and averaged over 12 replicates for each sample using a 3-μs time delay and a 50-μs gate width. Two spectral regions of the CN violet system, namely, $B \rightarrow X$ ($\Delta v = \pm 1$) from 357 to 360 nm and 417 to 422 nm, were included in a multivariate regression model for determination of the isotopic content of the samples. Each spectrum was normalized by the CN emission intensity of the pure electronic ($\Delta v = 0$) band head 1-1 at 387.1 nm. The isotopic shift of the latter band is only ~0.03 nm, while the bands with $\Delta v = \pm 1$ are shifted by 0.4 and 0.6 nm. The method can determine the ^{15}N content in the range 5%–50% with a root mean square error of prediction equal to 1.5% as compared to the true mass spectrometry data. Absolute repeatability and the detection limit were estimated at 0.6% and 2.0%, respectively. The analytical performance is adequate for evaluation of the quality of expensive enriched materials that are used for fraud detection, e.g., adulteration, and in scientific research.

6.5 Oxygen LAMIS

Oxygen isotopic analysis of ice is the main tool in paleoclimatology and glaciology studies. Measuring $^{18}O/^{16}O$ ratios in rocks and minerals is important for geochemistry. Oxygen isotope ratios between mineral pairs commonly serve as geothermometers. Enriched water-^{18}O is used as an important precursor in the radiopharmaceutical industry and for various medical and biological tracer studies. Measurements of ^{18}O with LAMIS using hydroxyl and boron monoxide spectra have already been discussed in this chapter.

Several intense lines of ^{18}OH appear at the spectral locations where emission of ^{16}OH is relatively low, because the former lines are situated in between or apart from the stronger lines of ^{16}OH. Such attributes of the spectra can facilitate LAMIS measurements of the $^{18}O/^{16}O$ ratio. Approximately twice smaller isotopic shifts ($\sim$0.03 nm) were reported previously between the lines of the 0-0 band of ^{16}OH, ^{18}OH, and ^{16}OD [45, 96]. The hydroxyl isotopologue measurements indicated a mediocre signal-to-noise ratio despite the higher transition probabilities in the 0-0 band than those in the 1-0 band. The spectra presented in Figs. 11 and 12 demonstrate a remarkable enhancement of the signal-to-noise ratio in the hydroxyl emission when reheating by electric discharge was applied. The ability to measure all isotopologues ^{16}OH, ^{18}OH, and ^{16}OD simultaneously for the 1-0 transitions $\{A^2\Sigma^+ \rightarrow X^2\Pi\}_i$ using a compact spectrograph is important for LAMIS applications.

Aluminum and calcium oxides are very common constituents in many minerals. Accordingly, molecular spectra of AlO and CaO are easily detectable in the laser ablation of rocks and sediments. Five band progressions of AlO $B^2\Sigma^+ \rightarrow X^2\Sigma^+$, namely, $\Delta v = 0, \pm1, \pm2$, were observed within a broad interval of 445–545 nm during ablation of several rocks, including various feldspars, tourmalines, and rock salts [96]. This spectral range is usually populated with many atomic spectral lines, and therefore LAMIS measurements required choosing a smaller segment of the spectrum and using relatively long delays from plasma initiation. The natural abundance ratio for $^{18}O/^{16}O$ is about 0.2%, and to date only ^{18}O-enriched samples have been measured by LAMIS.

Natural and isotope-enriched Al_2O_3 pellets were ablated in a helium flow to demonstrate that $Al^{18}O$ and $Al^{16}O$ can be spectrally resolved in LAMIS. The spectra of AlO band sequence $B^2\Sigma^+ \rightarrow X^2\Sigma^+$ ($\Delta v = +1$) were measured for both isotopologues indicating that AlO emission can be used to measure the oxygen isotopic ratio $^{18}O/^{16}O$ in rock samples [96]. Rotational and vibrational temperatures of AlO derived from the 0-0 band sequence $B \rightarrow X$ behaved differently relative to each other across the plasma plume, especially at small delays after the ablation pulse. With the increase in time delays for acquisition, rotational and vibrational temperature profiles tend to flatten and then equilibrate after about 20 μs.

Femtosecond LAMIS combined with fast photography were used to study the spatiotemporal evolution of isotopic species in the plasma generated by laser ablation of ^{18}O-enriched Al_2O_3 pellet [34]. A ratio of number densities $Al^{16}O/Al^{18}O$ was found higher in the area that is closer to the sample surface relative to that in the elongated tail of the plasma plume. These results verified that chemical reactions between plasma species and ambient air are more intense in the tail of plasma. The results also demonstrated that direct association of free Al and O atoms is the main mechanism for the formation of AlO molecules at the early stage of plasma evolution. Contrarily, reactions of plasma species with ambient oxygen and isotope exchange effect are the dominant mechanisms of the formation of AlO and the evolution of $Al^{16}O/Al^{18}O$ number density ratio at later stages of plasma (about 2–10 μs). The incomplete dissociation of specific bonds in the original molecular structure was shown to be negligible among AlO formation mechanisms [34].

A spectral interval containing emission from several bands of the system $B^2\Sigma^+ \rightarrow X^2\Sigma^+$ of $Al^{18}O$ and $Al^{16}O$ recorded during ablation of the ^{18}O-enriched pellet of Al_2O_3 is shown in Fig. 18. The 0-1 band head of the minor isotopologue appears at 507.04 nm between rotational lines of the transitions with $\Delta v = 0$ of the main

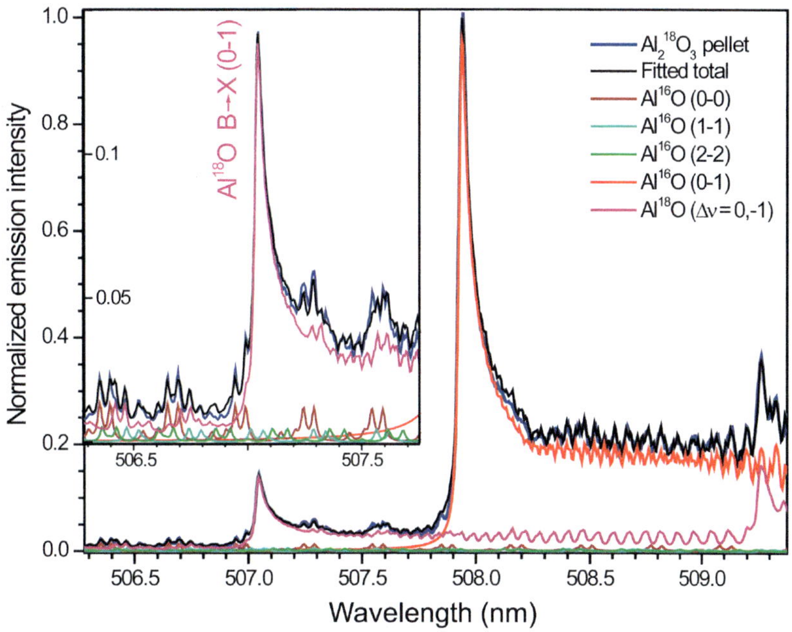

FIG. 18 Part of emission B → X ($\Delta v = 0, -1$) of $Al^{16}O$ and $Al^{18}O$ formed during ablation of the ^{18}O-enriched Al_2O_3 pellet. The simulation was based on ExoMol data [77]. The *inset* shows an expanded intensity scale around the band head (0-1) of $Al^{18}O$. Experimental spectra were acquired with a 20-μs delay and averaged over 100 laser pulses [45]. *Reproduced with permission of the Royal Society of Chemistry.*

isotopologue. The corresponding 0-1 band head of $Al^{16}O$ lies at 507.93 nm, while the next 1-2 band head of $Al^{18}O$ is observed at 509.26 nm. The simulated spectra of these bands and the tails of the $\Delta v = 0$, B → X sequence of $Al^{16}O$ are also included in Fig. 18. The total sum of these computed spectra was fitted to the experimental spectrum returning the ratio $^{18}O/^{16}O = 10\%$ in this experiment.

The inset in Fig. 18 displays a portion of the data around the band head (0-1) of $Al^{18}O$ at the expanded intensity scale. The background emission near this isotopic band head consists of series of doublets that belong to the tails of the $Al^{16}O$, $\Delta v = 0$, B → X sequence; most branches are P_{11} and P_{22} of the 0-0 band despite their very high J values up to $J = 147.5$, as seen in the inset of Fig. 18.

Several emission bands of CaO $A^1\Sigma^+ \rightarrow X^1\Sigma^+$ (2-0, 1-0, 0-0, 0-1) were observed within 765–960 nm during ablation of calcite ($CaCO_3$) and calcium chloride ($CaCl_2$) in SATP air [96]. The spectra of the 1-0 band of $A^1\Sigma^+ \rightarrow X^1\Sigma^+$ for both $Ca^{18}O$ and $Ca^{16}O$ were measured with IsoPlane-320 by ablating a calcite sample (Fig. 19). For the rare isotope, a Peltier cooler was used to freeze a thin layer of $H_2{}^{18}O$ water ice on top of the sample. Argon was flown to displace ambient air. In this case, $Ca^{18}O$ radicals were formed by association of Ca from calcite and ^{18}O from surface ice. The spectrum of $Ca^{18}O$ was blended with a lower contribution of $Ca^{16}O$ that originated from calcite itself. The spectrum of the main isotopologue $Ca^{16}O$ was obtained by ordinary ablation of calcite in air. Both spectra were collected using an acquisition gate of 100 μs delayed for 50 μs after ablation.

Studying the possibilities of the $^{18}O/^{16}O$ measurements by LAMIS demonstrated that the spectra of $Al^{18}O$, $Ca^{18}O$, and $B^{18}O$ can be

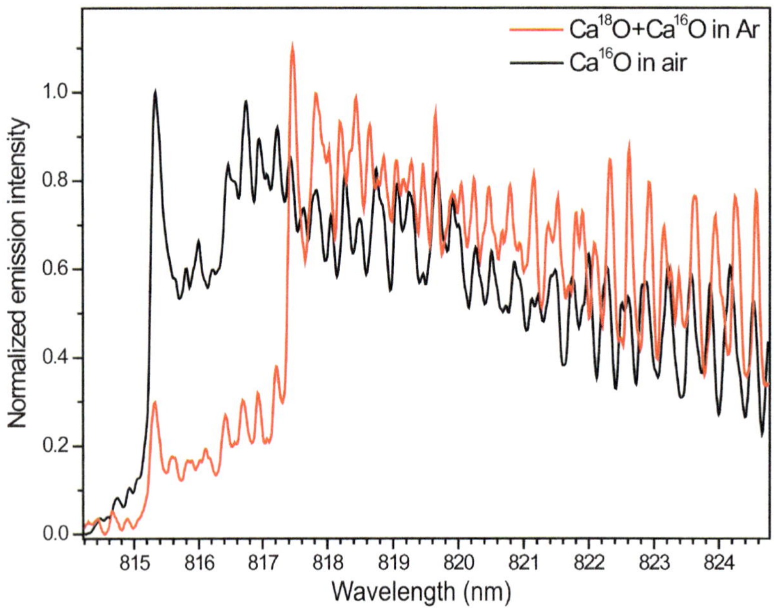

FIG. 19 Spectra of the A → X (1-0) band of $Ca^{16}O$ and $Ca^{18}O$ during laser ablation of a calcite sample. Spectra were acquired with a 50-μs delay and averaged over 100 laser pulses [45]. *Reproduced with permission of the Royal Society of Chemistry.*

used to complement each other for measuring the ratio of oxygen isotopes. The potential combination of LIBS and LAMIS techniques in one portable instrument for in-field geological analysis appears possible and beneficial. Such an instrument would alleviate expenses and logistics related to collection, delivery, and management of the samples from the field to an analytical laboratory. Molecular emission spectra of SrO, ZrO, and UO oxides substituted with ^{16}O and ^{18}O were used to study high-temperature chemistry in laser ablation plumes that can induce chemical fractionation during condensation [124, 125]. Spectra acquired after ablation of an $SrZrO_3$ target experimentally demonstrated the formation of multiple molecular species (ZrO and SrO) as a function of time as the laser ablation plume evolved. Although the stable oxides SrO and ZrO_2 are both refractory, emission from the ZrO intermediate at earlier times than SrO was observed. The timescale of oxygen entrainment into the laser ablation plume using an $^{18}O_2$ environment was deduced by observing the growth of $Zr^{18}O$ in the emission spectra relative to $Zr^{16}O$, which was formed by reaction of Zr with ^{16}O from the target itself. Time-resolved plume images demonstrated that ZrO is formed more readily at early times, volumetrically in the plume, while SrO is formed later in time, around the periphery. This technique is effective for identifying (and assigning) $^{238}U^{16}O$ and $^{238}U^{18}O$ emission lines for investigation of high-temperature uranium chemistry using LAMIS [125].

6.6 Chlorine LAMIS

Biomediated alteration of the $^{37}Cl/^{35}Cl$ ratio is distinctive in the microbial reduction of anthropogenic perchlorates, biphenyls, freons, and other chlorinated compounds. The values of ^{37}Cl showed an increasing trend with depth in oil-field brines. The diffusion-driven isotope fractionation in hydrogeological systems occurs because ^{35}Cl is more mobile than

^{37}Cl. However, the range of variations of chlorine isotopes in natural materials is smaller than the isotopic variation of boron, carbon, nitrogen, and oxygen.

Laser ablation of $CaCl_2$ produced emission of the CaCl $B^2\Sigma^+ \rightarrow X^2\Sigma^+$ band progressions with $\Delta v = +1$ and $\Delta v = 0$ within 579–585 and 585–600 nm, respectively. Spectra of the CaCl $\Delta v = 0$, $B^2\Sigma^+ \rightarrow X^2\Sigma^+$ system were recorded using two spectrographs [96]: one was a 1.25-m Czerny-Turner spectrograph (Horiba JY) and the other was a compact echelle spectrograph EMU-65 (Catalina Scientific). The natural relative isotopic abundances of chlorine isotopes ^{37}Cl and ^{35}Cl are approximately 24.2% and 75.8%, respectively. The isotopic shifts due to these two isotopes are apparent in the spectra that are illustrated in Fig. 20. While the 1.25-m, high-resolution spectrograph recorded highly resolved spectra, the compact echelle spectrograph yielded sufficient resolution for LAMIS measurements.

6.7 Strontium LAMIS

Strontium is a relatively abundant element in geological materials with total content up to ~3000 ppm in marine and magmatic carbonates. Although the major isotopes ^{88}Sr, ^{87}Sr, and ^{86}Sr are stable, a continuous increase in ^{87}Sr occurs naturally over time as a result of radioactive β-decay of ^{87}Rb. Therefore a ratio of Sr isotopes (usually ^{87}Sr/^{86}Sr) is used for age dating in geochronology, oceanography, archeology, and also as a provenance tracer for defining the origin of historic or forensic samples.

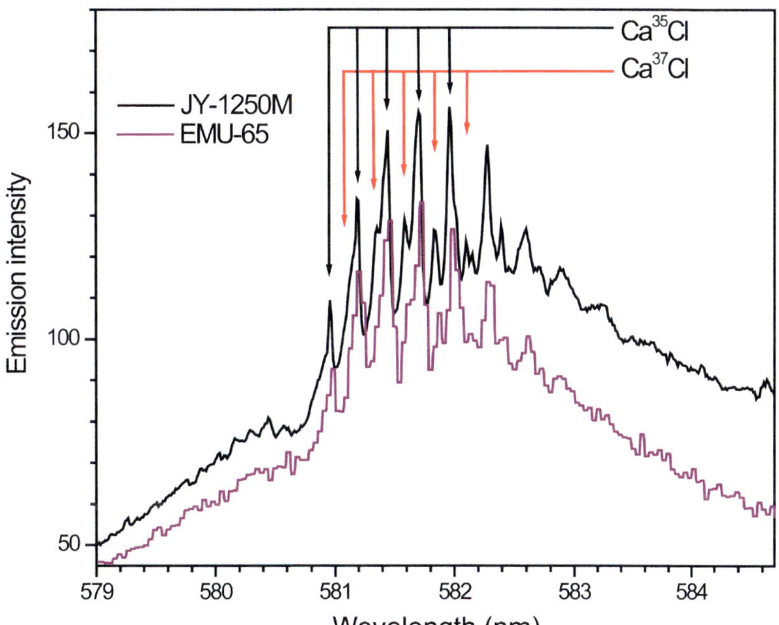

FIG. 20 Spectra of CaCl band sequence B → X ($\Delta v = +1$) formed during ablation of natural $CaCl_2$ pellets. Spectra were acquired with a 20-μs delay and averaged over 20 laser pulses [45]. *Reproduced with permission of the Royal Society of Chemistry.*

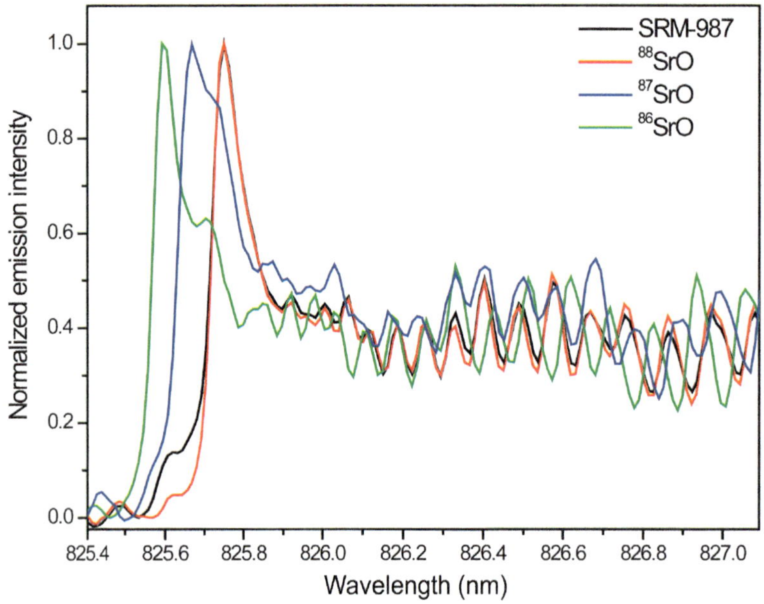

FIG. 21 Spectra of the band A → X (2-0) of SrO measured from enriched $^{88}SrCO_3$, $^{87}SrCO_3$, and $^{86}SrCO_3$ samples, and from a NIST (SRM-987) sample with natural isotopic abundance. The signal was normalized to the intensity in the band heads [45]. *Reproduced with permission of the Royal Society of Chemistry.*

Emission from the $A^1\Sigma^+ \rightarrow X^1\Sigma^+$ system of the isotopologues ^{88}SrO, ^{87}SrO, and ^{86}SrO within 780–960 nm was used for LAMIS of strontium isotopes (^{88}Sr, ^{87}Sr, and ^{86}Sr) [126]. Fig. 21 illustrates spectra of the SrO 2-0 band of the A → X system generated from natural material (NIST SRM-987) and isotope-enriched $SrCO_3$ pellets. A high-resolution Horiba JY-1250 M spectrograph resolved all three individual isotopes ^{86}Sr, ^{87}Sr, and ^{88}Sr that were quantified using a set of isotope-enriched standards. Emission from the SRM-987 was a sum of the Sr isotopes with their natural abundances, dominated by ^{88}Sr. Similar spectra of ^{88}SrO and ^{86}SrO of four bands of the A → X system, namely, 2-0, 1-0, 0-0, and 0-1, were measured using a compact spectrograph EMU-65 within 820–990 nm [108]. The results demonstrated that the isotopic shifts of three band heads were from 0.08 to 0.15 nm, with an exception for the 0-0 band. The spectra were easily resolvable, particularly in the regions of high rotational quantum numbers, *J*, in the band tails.

An empirical PLSR model was built to calibrate the spectra of the SrO $A^1\Sigma^+ \rightarrow X^1\Sigma^+$ transition within the tail region of the 1-0 band between 870 and 880 nm [126]. Calibration was based on a library of 400 spectra, including 100 spectra collected from four reference samples. Time delay was 10 μs and gate width was 30 μs. Isotopic composition of the SRM-987 was predicted to be 9.8% ± 2.8% (^{86}Sr), 9.1% ± 3.5% (^{87}Sr), and 81.6% ± 3.4% (^{88}Sr). These values were consistent with the certified isotopic abundances of 9.86% (^{86}Sr), 7.00% (^{87}Sr), and 82.58% (^{88}Sr) in the interrogated sample.

Diatomic strontium monohalide radicals formed during ablation of strontium salts (SrF_2, $SrCl_2$, $SrBr_2$, and SrI_2) in air were also investigated to provide spectral signatures for the naturally occurring strontium isotopes. Laser ablation spectra of the $B^2\Sigma^+ \rightarrow X^2\Sigma^+$ SrF transition were measured between 561 and 573 nm, using an SrF_2 pellet as a target [126]. A portion of this spectrum (561–564.5 nm) was simulated using the weighted sum of strontium isotopes $^{88}Sr \times 0.8258 + {}^{87}Sr \times 0.07 + {}^{86}Sr \times 0.0986 + {}^{84}Sr \times 0.0056$ in accordance with their natural abundances. Good visual agreement occurred between experimental and simulated SrF spectra.

Structured materials in industry and highly heterogeneous natural samples in geological studies usually require localized microanalysis. For example, strontium titanate films in memory capacitors are typically ~20 to ~250 nm thick, while the grain sizes in strontium ceramics as well as in natural rocks are on the submillimeter scale. Surface scanning and depth profiling are in the realm of commercial LIBS instruments that can be adjusted for a lateral resolution of ~30 μm and a depth resolution of ~20 nm. Similarly, LAMIS instruments can be implemented for rapid isotopic microanalysis.

The possibility of using LAMIS of strontium isotopes in rocks for radiometric isochron age determination has been discussed [108, 126]. Measurement accuracy within ±2% in the $^{87}Rb/^{86}Sr$ ratio and ± 0.2% in the $^{87}Sr/^{86}Sr$ ratio using LAMIS can result in determination of the geological age of rocks with uncertainty less than ±500 million years. Current age estimates for Mars geological components have uncertainties of billions of years; moreover, the validity of these estimates is unknown. There is no other means to make direct age-dating measurements on Mars or other planets, but a compact LAMIS instrument for robotic space exploration can be realized [108].

6.8 Zirconium LAMIS

Zirconium alloys are used as nuclear reactor core components and cladding materials for nuclear fuels because four stable isotopes, ^{90}Zr, ^{91}Zr, ^{92}Zr, ^{94}Zr, show low cross-sections for neutron capture, with ^{90}Zr being the lowest neutron absorbent. Rates of fuel burn-up and neutron fluxes can be determined by measuring Zr isotope ratios. Zirconium isotopes are among the products from the fission of ^{235}U and ^{239}Pu, and therefore their analysis is important in nuclear forensics.

Femtosecond LAMIS of zirconium isotopes was carried out by ablating natural Zr metal and an enriched $^{94}ZrO_2$ pellet in SATP air with a laser pulse energy of 160 μJ and a repetition rate of 1 kHz [127]. The singlet 0-1 band of ZrO $E^1\Sigma^+ \rightarrow X^1\Sigma^+$ at 381.8 nm and six triplet bands of the $\Delta v = 0, \pm 1$ bands of the ZrO $d^3\Delta_{1,2,3} \rightarrow a^3\Delta_{1,2,3}$ transition within 446–485 nm were observed with the isotopic band head shifts up to 74 pm. Molecular emission from femtosecond ablation peaked at delays of about 0.75 μs, while a ratio of molecular-to-atomic intensities increased up to a delay of ~3 μs. The 0-1 band of $d^3\Delta_3 \rightarrow a^3\Delta_3$ was used for Zr isotope determination by fitting simulated to experimental spectra at an optimized delay of 2 μs and a gate width of 4 μs. Fig. 22 displays spectra that were recorded with a JY-1250 M spectrograph accumulating 5000 ablation pulses as a single measurement with a total acquisition time of 5 s at 1 kHz.

The ratios of four Zr isotopes in the form of a proportion $^{90}Zr{:}^{91}Zr{:}^{92}Zr{:}^{94}Zr$ amounted to (2.93 ± 0.08):(0.47 ± 0.03):(1.16 ± 0.06):1. Experimental uncertainties were determined from 10 replicates. The measured ratios corresponded reasonably well to the proportion of natural isotopic abundances for zirconium, namely, 2.96:0.65:0.99:1. The discrepancies were attributed to unavailable data on perturbations and hyperfine splitting in ZrO energy levels that precluded more accurate simulation of the spectra. Nevertheless, feasibility was demonstrated of obtaining semi-quantitative isotopic information without isotopic standards for an element with a complex emission spectrum, such as zirconium.

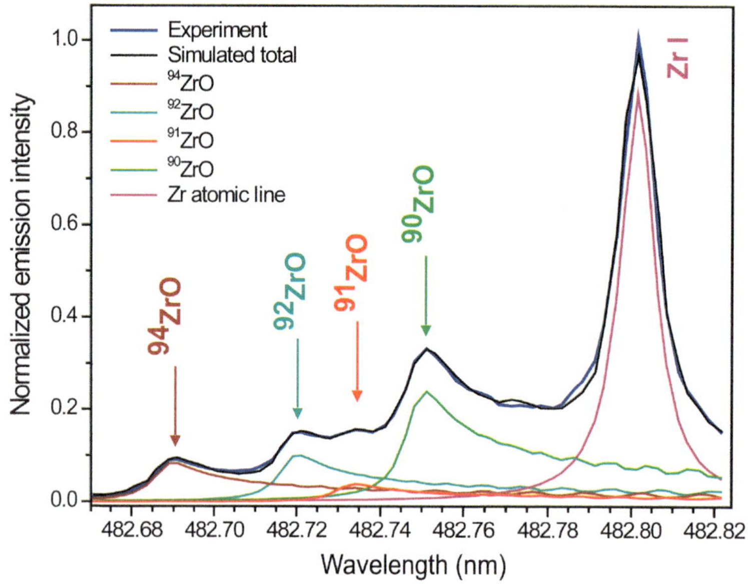

FIG. 22 Experimental and fitted synthetic spectra of ZrO $d^3\Delta_3 \rightarrow a^3\Delta_3$ (0-1) during femtosecond ablation of a Zr metal sample. The simulation includes an atomic line of Zr at 482.804 nm. Laser pulse duration was 500 fs operating at a wavelength of 343 nm and repetition rate of 1 kHz [45]. *Reproduced with permission of the Royal Society of Chemistry.*

At low laser pulse energy, the femtosecond ablation plasma was cooler than that from usual nanosecond ablation. This resulted in enhanced rates for association of free atoms into molecules (ideally after complete atomization of the ablated material) versus corresponding rates of molecular association in nanosecond laser plasmas. At the same laser irradiance delivered on a surface of the zirconium sample, signals in LAMIS with femtosecond ablation were sixfold stronger than that from nanosecond ablation.

A demonstration of the remotely excited LAMIS at distances of 5–7.8 m between a laser beam focusing lens and the analyzed sample surface was recently communicated [128]. This new technique combined a self-focusing, propagating laser beam through a contracted plasma channel (filament) in SATP air with LAMIS [128]. The technique was named femtosecond filament-induced laser ablation molecular isotopic spectrometry (F^2-LAMIS). A femtosecond laser pulse with adequate peak irradiance can initiate nonlinear refractive index effects in air causing the laser beam to establish a plasma filament with a dynamic balance of diffraction and self-focusing [30]. Thus F^2-LAMIS can remotely ablate the sample to enable all-optical isotopic measurements.

Recently, a Ti:sapphire laser was employed to generate filaments [128] using the center wavelength of 800 nm, pulse duration of 100 fs, and energy of 7 mJ per pulse, and at a repetition rate of 10 Hz. The laser filaments propagated over several meters before impinging on a Zr metal plate. Isotope-specific molecular emission of ZrO at the 0-1 band of the $d^3\Delta_3 \rightarrow a^3\Delta_3$ system within 482.6–483 nm was collected using a lens that was coupled to an optical fiber; both were directly mounted on the sample base to keep identical light collection conditions for different filament propagation distances. Spectra were measured using a JY-1250 M spectrograph and averaged over 500 pulses. The ICCD gate delay was set at 3 μs, and the gate width was 20 μs. The proportion of zirconium isotopes ^{90}Zr:^{91}Zr:^{92}Zr:^{94}Zr was deduced as 2.5:0.35: 1.15:1 by fitting simulated to the experimental spectra. These deduced ratios deviated from the actual isotopic composition of the sample but remained unaffected by a filament propagation distance within the tested range. These results indicated minimal influence of the filament propagation distance on isotopic quantification. More accurate analysis can be expected with the use of reference standards for calibration.

6.9 Uranium LAMIS

F^2-LAMIS was applied for isotope-selective standoff detection of uranium using samples of depleted (0.4% ^{235}U) and highly enriched (98% ^{235}U) uranium metal foil [129]. The uranium oxide isotope shift of 50 ± 7 pm was observed at 593.6 nm between emission features of ^{235}UO and ^{238}UO. The prominent $^{238}U^{16}O$ emission peak at 593.55 nm was earlier identified at a temperature of 130 K in supersonic expansion [130]. However, LAMIS spectra were broadened (for a gate delay of 1 μs), hence possible contributions from higher oxide U_xO_y emission could not be excluded at this wavelength [128]. Both spectroscopic and acoustic diagnostics were implemented to characterize the properties of uranium plasma at different filament-uranium interaction points. The resulting uranium oxide emission exhibited a nearly constant signal-to-background ratio over the length of the filament, unlike the uranium atomic and ionic emission, for which the signal-to-background ratio varies significantly along the filament propagation. The results provided a basis for the optimal use of filaments for standoff detection and analysis of uranium isotopes and indicated the potential of the technique for a wider range of remote sensing applications that require isotopic composition.

The reported isotope shift of 50 pm in emission band heads of uranium oxide was twice larger than the 25-pm isotopic shift of the commonly used ionic U II 424.43-nm spectral line. Therefore it is easier to resolve the uranium isotopes in LAMIS relative to resolving isotopes in atomic/ionic spectra using conventional LIBS, as has been consistently demonstrated for many elements from hydrogen to uranium. However, more important is that molecular, atomic, and ionic spectral information can be combined to improve accuracy and precision of optical isotopic analysis for difficult elements, such as uranium. The analytical performance of the LAMIS-LIBS combined approach for uranium isotopic analysis was recently demonstrated in a numerical model [131]. The use of multiple atomic emission lines combined with molecular spectral features improved 3.5 times the precision of determining the ^{235}U absolute atomic fraction as compared to precision attained with only one U line at 593.38 nm. The best precision was 0.42%, computed as a relative standard deviation of the ^{235}U atomic fraction measurement. The spectral window was 591.6–596.5 nm, which included 32 uranium atomic lines and the intense uranium oxide band at 593.55 nm. Further work continues to accomplish multiple spectral line fitting for uranium isotopic analysis [132].

7 Conclusions

Measured CN recombination spectra following LIBS show spatial temperature variations. Of note is the early occurrence of CN for time delays <1 μs. The presence of the C I 193.09-nm atomic line in second order indicates higher electron density near the edges of the plasma. Recorded absorption spectra from astrophysical objects such as WDs require accurate databases for making temperature inferences. In turn, well-maintained databases that include diatomic isotopologues are instrumental for LIBS and LAMIS applications.

Detailed studies of AlO molecular spectroscopy reveal the significance of shockwave and expansion dynamics in laser plasma. The theoretical basis for the computation of molecular spectra includes the use of *r*-centroids, over and above the usual Frank-Condon and Hönl-London line strengths, for accurate prediction of all lines that contribute to spectral regions separated from the vibrational band heads, i.e., primarily "rotational" lines that add up to exhibit an undulant spectrum. Details of the methodology for temperature inferences indicate the importance of the availability of accurate line positions and values of the amplitude for hundreds of lines that may be within the spectral resolution of the experimental system composed of collection optics, spectrometer, and gated detector.

Clearly, molecular laser plasma spectroscopy shows new pathways for the detection of isotopes or isotopic composition, especially of diatomic molecules. LAMIS reveals significant enhancement of molecular band edge shifts for isotopologues. New developments of femtosecond laser ablation for remote analysis, termed F^2-LAMIS, rely on the balance of ambient medium self-focusing and diffraction for delivery of plasma filaments.

Experimental investigations of molecular emission spectra require reasonable excitation of the molecules following formation by recombination. Consequently, nanosecond LIBS appears favorable for the sustenance of both reasonable excitation and a local thermodynamic equilibrium environment. Usually, diatomic molecular spectra are easily discernible following decay of the free-electron background radiation, and typically for time delays of the order of several tens of microseconds after optical breakdown.

For femtosecond excitation, molecular spectra reveal significantly less free electron background, thereby allowing analysis of the emission spectra for smaller time delays than that for nanosecond excitation. However, for laser ablation of materials, critical electron density is almost instantaneously reached, namely, electron densities of the order of one-tenth of solid densities, and reflection of femtosecond radiation occurs. Reflection would diminish the absorption and transfer of radiation to the plasma. For nanosecond radiation, critical density arguments are not important because further excitation occurs near the target material.

References

[1] L.J. Radziemski, D.A. Cremers (Eds.), Laser-Induced Plasmas and Applications, Marcel Dekker, New York, 1989.

[2] A.W. Miziolek, V. Palleschi, I. Schechter (Eds.), Laser Induced Breakdown Spectroscopy, Cambridge University Press, New York, 2006.

[3] D.E. Cremers, L.J. Radziemski, Handbook of Laser-Induced Breakdown Spectroscopy, John Wiley, New York, 2006.

[4] J.P. Singh, S.N. Thakur (Eds.), Laser Induced Breakdown Spectroscopy, Elsevier Science, New York, 2007.

[5] P. Mulser, D. Bauer, High Power Laser-Matter Interaction, Springer, Heidelberg, 2010.

[6] R. Noll, Laser-Induced Breakdown Spectroscopy Fundamentals and Applications, Springer, Heidelberg, 2012.

[7] S. Musazzi, U. Perini (Eds.), Laser-Induced Breakdown Spectroscopy, Springer, Heidelberg, 2014.

[8] A.A. Bol'shakov, n.d. LIBS at Work on Mars, https://appliedspectra.com/mars-libs.html. (Published on-line June 3, 2016; Accessed 28 March 2019).

[9] V.N. Ochkin, Spectroscopy of Low Temperature Plasma, Wiley-VCH, Weinheim, 2009.

[10] I.V. Hertel, C.-P. Schulz, Atoms, Molecules and Optical Physics 1, Atoms and Spectroscopy, Springer, Heidelberg, 2015.

[11] I.V. Hertel, C.-P. Schulz, Atoms, Molecules and Optical Physics 2, Molecules and Photons—Spectroscopy and Collisions, Springer, Heidelberg, 2015.

[12] J.O. Hornkohl, L. Nemes, C.G. Parigger, Spectroscopy of carbon containing diatomic molecules (Chapter 4), in: L. Nemes, S. Irle (Eds.), Spectroscopy, Dynamics and Molecular Theory of Carbon Plasmas and

Vapors—Advances in the Understanding of the Most Complex High-Temperature Elemental System, World Scientific, New Jersey, 2011, , pp. 113–165 Ch. 4.

[13] P.V. Minorsky, On the inside, Plant Physiol. 155 (2011) 169–170.

[14] T. Mudder, M. Botz, A. Smith, Chemistry and Treatment of Cyanidation Wastes, second ed., Mining Journal Books, London, 2001.

[15] N. Gupta, C. Balomajumder, V.K. Agarwal, Enzymatic mechanism and biochemistry for cyanide degradation: a review, J. Hazard. Mater. 176 (2010) 1–13.

[16] J. Ma, P.K. Dasgupta, Recent developments in cyanide detection: a review, Anal. Chim. Acta 673 (2010) 117–125.

[17] D. Hahn, N. Omenetto, Laser-induced breakdown spectroscopy (LIBS), part II: review of instrumental and methodological approaches to material analysis and applications to different fields, Appl. Spectrosc. 66 (2012) 347–419.

[18] J.O. Hornkohl, C. Parigger, J.W.L. Lewis, Temperature measurements from CN spectra in a laser-induced plasma, J. Quant. Spectrosc. Radiat. Transf. 46 (1991) 405–411.

[19] A. Kushwaha, R.K. Thareja, Dynamics of laser-ablated carbon plasma: formation of C_2 and CN, Appl. Optics 47 (2008) G35–G71.

[20] C.G. Parigger, Atomic and molecular emissions in laser-induced breakdown spectroscopy, Spectrochim. Acta B At. Spectrosc. 79–80 (2013) 4–16.

[21] V. Motto-Ros, Q.L. Ma, S. Grégoire, W.Q. Lei, X. C. Wang, F. Pelascini, F. Surma, V. Detalle, J. Yu, Dual-wavelength differential spectroscopic imaging for diagnostics of laser-induced plasma, Spectrochim. Acta B At. Spectrosc. 74–75 (2012) 11–17.

[22] J. Merten, M. Jones, S. Hoke, S. Allen, Differential spectral imaging of the CN violet band in laser-induced plasmas on TNT simulant molecules, J. Phys. Conf. Ser. 548 (2014) 012042.

[23] C.A. Zuhlke, J. Bruce, T.P. Anderson, D. Alexander, C. G. Parigger, A fundamental understanding of the dependence of the laser-induced breakdown spectroscopy (LIBS) signal strength on the complex focusing dynamics of femtosecond laser pulses on either side of the focus, Appl. Spectrosc. 68 (2014) 1021–1029.

[24] M. Braun, H. Metiu, V. Engel, Molecular femtosecond excitation described within the Gaussian wave packet approximation, J. Chem. Phys. 108 (1998) 8983–8988.

[25] E.N. Rao, S. Sreedhar, S.P. Tewari, G.M. Kumar, S. V. Rao, CN, C2, molecular emissions from pyrazole studied using femtosecond LIBS, in: Proceedings International Conference on Fibre Optics and Photonics, paper TPo.3, OSA International Conference on Fibre Optics and Photonics, 9–12 December 2012, Chennai, IN, 2012.

[26] E.N. Rao, P. Mathi, S.A. Kalam, S. Sreedhar, A. K. Singh, B.N. Jagatap, S.V. Rao, Femtosecond and nanosecond LIBS studies of nitroimidazoles: correlation between molecular structure and LIBS data, J. Anal. At. Spectrom 31 (2016) 737–750.

[27] F.C. De Lucia, J.L. Gottfried, A.W. Miziolek, Evaluation of femtosecond laser-induced breakdown spectroscopy for explosive residue detection, Opt. Express 17 (2009) 419–425.

[28] E.N. Rao, S. Sunku, S.V. Rao, Femtosecond laser-induced breakdown spectroscopy studies of nitropyrazoles: the effect of varying nitro groups, Appl. Spectrosc. 69 (2015) 1342–1354.

[29] S.A. Kalam, N.L. Murthy, P. Mahti, N. Kommu, A. K. Singh, S.V. Rao, Correlation of molecular, atomic emissions with detonation parameters in femtosecond and nanosecond LIBS plasma of high energy materials, J. Anal. At. Spectrom 32 (2017) 1535–1546.

[30] S.L. Chin, Femtosecond Laser Filamentation, Springer, Heidelberg, 2010.

[31] Y. Shuai, C.S. Leang, Z. He-Ping, Femtosecond filamentation induced fluorescence technique for atmospheric sensing, Chin. Phys. B 24 (2015) 014208.

[32] A.A. Ilyin, S.S. Golik, K.A. Shmirko, Absorption and emission characteristics of femtosecond laser plasma filaments in the air, Spectrochim. Acta B At. Spectrosc. 112 (2015) 16–22.

[33] H. Huang, L.-M. Yang, J. Liu, Femtosecond fiber-laser-based laser-induced breakdown spectroscopy, in: Proceedings of the Society of Photo-Optical Instrumentation Engineers (SPIE) Defense, Security, and Sensing (2012) 8358-43, April 23–27, 2012. Baltimore, Maryland, US.

[34] H. Hou, X. Mao, V. Zorba, R.E. Russo, Laser ablation molecular isotopic spectrometry for molecules formation chemistry in femtosecond-laser ablated plasmas, Anal. Chem. 89 (2017) 7750–7757.

[35] M. Kotzagianni, S. Couris, Femtosecond laser induced breakdown spectroscopy of air–methane mixtures, Chem. Phys. Lett. 561–562 (2013) 36–41.

[36] M. Baudelet, L. Guyon, J. Yu, J.-P. Wolf, T. Amodeo, E. Fréjafon, P. Laloi, Spectral signature of native CN bonds for bacterium detection and identification using femtosecond laser-induced breakdown spectroscopy, J. Appl. Phys. 99 (2006) 063901.

[37] M. Baudelet, L. Guyon, J. Yu, J.-P. Wolf, T. Amodeo, E. Fréjafon, P. Laloi, Femtosecond time-resolved laser-induced breakdown spectroscopy for detection and identification of bacteria: a comparison to the nanosecond regime, J. Appl. Phys. 99 (2006) 084701.

[38] C.G. Parigger, G. Gautam, A.C. Woods, D.M. Surmick, J. O. Hornkohl, Atomic and molecular emission spectroscopy measurements for characterization of laser-induced plasma, Trends Appl. Spectrosc. 11 (2014) 1–13.

[39] A. Thorne, U. Litzén, S. Johansson, Spectrophysics, Principles and Applications, Springer, New York, 1999.
[40] C.G. Parigger, J.O. Hornkohl, L. Nemes, Time-resolved spectroscopy diagnostic of laser-induced optical breakdown, Int. J. Spectrosc. 2010 (2010) 593820.
[41] C.G. Parigger, C.M. Helstern, G. Gautam, Molecular emission spectroscopy of cyanide in laser-induced plasma, Int. Rev. At. Mol. Phys. 8 (2017) 25–35.
[42] I.G. Dors, C.G. Parigger, J.W.L. Lewis, Spectroscopic temperature determination of aluminum monoxide in laser ablation with 266-nm radiation, Opt. Lett. 23 (1998) 1778–1780.
[43] C.G. Parigger, K.A. Drake, C.M. Helstern, G. Gautam, Laboratory hydrogen-beta emission spectroscopy for analysis of astrophysical white dwarf spectra, Atoms 6 (2018) 030036.
[44] R.E. Russo, A.A. Bol'shakov, X. Mao, C.P. McKay, D. L. Perry, O. Sorkhabi, Laser ablation molecular isotopic spectrometry, Spectrochim. Acta B At. Spectrosc. 66 (2011) 99–104.
[45] A.A. Bol'shakov, X. Mao, J.J. González, R.E. Russo, Laser ablation molecular isotopic spectrometry (LAMIS): current state of the art, J. Anal. At. Spectrom 31 (2016) 119–134.
[46] G. Gautam, C.M. Helstern, K.A. Drake, C.G. Parigger, Imaging of laser-induced plasma expansion dynamics in ambient air, Int. Rev. At. Mol. Phys. 7 (2016) 45–51.
[47] G. Gautam, C.G. Parigger, C.M. Helstern, K.A. Drake, Emission spectroscopy of expanding laser-induced gaseous hydrogen-nitrogen plasma, Appl. Optics 33 (2017) 9277–9284.
[48] C. Parigger, Y. Tang, D.H. Plemmons, J.W.L. Lewis, Spherical aberration effects in lens–axicon doublets: theoretical study, Appl. Optics 36 (1997) 8214–8221.
[49] C.G. Parigger, Laser-induced breakdown in gases: experiments and simulation, chapter 4, in: A. W. Miziolek, V. Palleschi, I. Schechter (Eds.), Laser Induced Breakdown Spectroscopy, Cambridge University Press, New York, 2006, , pp. 171–193 Ch. 4.
[50] Y.-L. Chen, J.W.L. Lewis, C. Parigger, Spatial and temporal profiles of pulsed laser-induced air plasma emissions, J. Quant. Spectrosc. Radiat. Transf. 67 (2000) 91–103.
[51] H.R. Griem, Spectral Line Broadening by Plasmas, Academic Press, New York, 1974.
[52] N. Konjević, A. Lesage, J.R. Fuhr, W.L. Wiese, Experimental stark widths and shifts for the spectral lines of neutral and ionized atoms (a critical review of selected data for the period 1989 through 2000), J. Phys. Chem. Ref. Data Monogr. 31 (2002) 819–927.
[53] G.I. Taylor, The formation of a blast wave by a very intense explosion—II. The atomic explosion of 1945, Proc. R. Soc. Lond. A Math. Phys. Sci. 201 (1950) 175–186.
[54] C.M. Helstern, C.G. Parigger, Time-resolved plasma spectroscopy of diatomic molecular cyanide, J. Phys.: Conf. Ser. 1289 (2019). 012016.
[55] C.G. Parigger, A.M. El Sherbini, R. Splinter, Selected laser-induced plasma spectroscopy: from medical to astrophysical applications, J. Phys.: Conf. Ser. 1253 (2019) 012001 (in press).
[56] C.G. Parigger, A.C. Woods, D.M. Surmick, G. Gautam, M.J. Witte, J.O. Hornkohl, Computation of diatomic molecular spectra for selected transitions of aluminum monoxide, cyanide, diatomic carbon, and titanium monoxide, Spectrochim. Acta B At. Spectrosc. 107 (2015) 132–138.
[57] B.J. McBride, S. Gordon, Computer Program for Calculating and Fitting Thermodynamic Functions, NASA RP-1271; on-line 2005 version at https://cearun.grc.nasa.gov/, 1992. (Accessed 26 November 2017).
[58] C.G. Parigger, G. Gautam, D.M. Surmick, Radial electron density measurements in laser-induced plasma from Abel inverted hydrogen Balmer beta line profiles, Int. Rev. At. Mol. Phys. 6 (2015) 43–55.
[59] C.G. Parigger, D.M. Surmick, G. Gautam, Self-absorption characteristics of measured laser-induced plasma line shapes, J. Phys. Conf. Ser. 810 (2017) 012012.
[60] G. Pretzler, A new method for numerical Abel-inversion, Z. Naturforsch. 46a (1991) 639–641.
[61] G. Pretzler, H. Jäger, T. Neger, H. Philipp, J. Woisetschläger, Comparison of different methods of Abel inversion using computer simulated and experimental side-on data, Z. Naturforsch. 47a (1992) 955–970.
[62] H.-J. Kunze, Introduction to Plasma Spectroscopy, Springer, Heidelberg, 2009.
[63] S. Eschlböck-Fuchs, A. Demidov, I. Gornushkin, T. Schmid, R. Rössler, N. Huber, U. Panne, J. Pedarnig, Tomography of homogenized laser-induced plasma by radon transform technique, Spectrochim. Acta B At. Spectrosc. 123 (2016) 59–67.
[64] C.G. Parigger, C.M. Helstern, G. Gautam, K.A. Drake, Atomic and molecular line shapes in laboratory and selected astrophysical plasma, J. Phys.: Conf. Ser. 1289 (2019) 012001.
[65] P. Dufour, S. Blouin, S. Coutu, M. Fortin-Archambault, C. Thibeault, P. Bergeron, G. Fontaine, The Montreal White Dwarf database: a tool for the community, in: P.-E. Tremblay, B. Gaensicke, T. Marsh (Eds.), Astronomical Society of the Pacific (ASP) Conference Series 509: 20th European White Dwarf Workshop, Univ. Warwick, 25–29 July 2016, UK, Utah Valley University, Orem, UT, US, 2016.
[66] J. Tennyson, The ExoMol project: molecular opacity calculations at University College London, workshop on astrophysical opacities, in: C. Mendoza, S. Turck-Chiéze, J. Colgan (Eds.), Astronomical Society of the

Pacific (ASP) Conference Series 515, Michigan University, Kalamazoo, 1–4 August 2017, Michigan, US, Astronomical Society of the Pacific, 2018, p. 137.

[67] J. Tennyson, S.N. Yurchenko, A.F. Al-Refaie, E. J. Barton, K.L. Chubb, P.A. Coles, S. Diamantopoulou, M.N. Gorman, C. Hill, A. Z. Lam, L. Lodi, L.K. McKemmish, Y. Na, A. Owens, O.L. Polyansky, C. Sousa-Silva, D.S. Underwood, A. Yachmenev, E. Zak, The ExoMol Database: Molecular Line Lists for Exoplanet and Other Hot Atmospheres, arXiv, 2018. 1603.05890 [astro-ph.GA].

[68] L.S. Rothman, I.E. Gordon, Y. Babikov, A. Barbe, D. C. Benner, P.F. Bernath, M. Birk, L. Bizzocchi, V. Boudon, L.R. Brown, A. Campargue, K. Chance, E.A. Cohen, L.H. Coudert, V.M. Devi, B.J. Drouin, A. Fayt, J.-M. Flaud, R.R. Gamache, J.J. Harrison, J.-M. Hartmann, C. Hill, J.T. Hodges, D. Jacquemart, A. Jolly, J. Lamouroux, R.J. Le Roy, G. Li, D.A. Long, O.M. Lyulin, C.J. Mackie, S.T. Massie, S. Mikhailenko, H.S.P. Müller, O.V. Naumenko, A.V. Nikitin, J. Orphal, V. Perevalov, A. Perrin, E.R. Polovtseva, C. Richard, M.A.H. Smith, E. Starikova, K. Sung, S. Tashkun, J. Tennyson, G.C. Toon, G. Tyuterev VI, G. Wagner, The HITRAN2012 molecular spectroscopic database, J. Quant. Spectrosc. Radiat. Transf. 130 (2013) 4–50.

[69] C.M. Western, PGOPHER: a program for simulating rotational, vibrational and electronic spectra, J. Quant. Spectrosc. Radiat. Transf. 186 (2017) 221–242.

[70] B.J. Wimberly, J.O. Hornkohl, C.G. Parigger, Measurement of strontium monoxide in methane–air flames, Appl. Spectrosc. 71 (2017) 267–278.

[71] J. Lam, D. Amans, F. Chaput, M. Diouf, G. Ledoux, N. Mary, K. Masenelli-Varlot, V. Motto-Ros, C. Dujardin, γ-Al_2O_3 nanoparticles synthesized by pulsed laser ablation in liquids: a plasma analysis, Phys. Chem. Chem. Phys. 16 (2016) 963–973.

[72] S. Harilal, G. Milshevsky, P. Diwakar, N. LaHaye, A. Hassanein, Experimental and computational study of complex shockwave dynamics in laser ablation plumes in argon atmospheres, Phys. Plasmas 19 (2012) 083504.

[73] C. Kimblin, R. Trainahm, G. Capelle, X. Mao, R. Russo, LIBS as a Compliment to Large-Scale Detonations and Means to Characterize Intermediates, Special Technologies Laboratory, LLC, 2015.

[74] D.M. Surmick, Spectroscopic Imaging of Aluminum Containing Plasma, unpublished dissertation, University of Tennessee, Knoxville, TN, US, 2016.

[75] S. Harilal, B. Brumfield, B. Cannon, M. Phillips, Shock wave mediated plume chemistry for molecular formation in laser ablation plasmas, Anal. Chem. 88 (2016) 2296–2302.

[76] I.G. Dors, C.G. Parigger, Computational fluid-dynamic model of laser-induced breakdown in air, Appl. Optics 42 (2003) 5978–5985.

[77] A.T. Patrascu, S.N. Yurchenko, J. Tennyson, ExoMol molecular line lists—IX. The spectrum of AlO, Mon. Not. R. Astron. Soc. 449 (2015) 3613–3619.

[78] J.O. Hornkohl, C.G. Parigger, Angular momentum states of the diatomic molecule, Am. J. Phys. 64 (1996) 623–632.

[79] J.O. Hornkohl, C.G. Parigger, L. Nemes, Diatomic Hönl-London factor computer program, Appl. Optics 44 (2005) 3687–3695.

[80] J.O. Hornkohl, A.C. Woods, C.G. Parigger, An algorithm for computation of diatomic spectra, J. Phys.: Conf. Ser. 548 (2014) 012033.

[81] C.G. Parigger, J.O. Hornkohl, Computation of AlO $B^2\Sigma^+ \rightarrow X^2\Sigma^+$ emission spectra, Spectrochim. Acta A Mol. Biomol. Spectrosc. 81 (2011) 404–411.

[82] A.C. Woods, C.G. Parigger, J.O. Hornkohl, Measurement and analysis of titanium monoxide spectra in laser-induced plasma, Opt. Lett. 37 (2012) 5139–5141.

[83] B.H. Bransden, C.J. Joachain, Physics of Atoms and Molecules, second ed., Prentice Hall, Harlow, 2003.

[84] R. Byrd, R. Schnabel, G. Schultz, A trust region algorithm for nonlinearly constrained optimization, SIAM J. Numer. Anal. 24 (1987) 1152–1170.

[85] T.F. Coleman, Y. Li, An interior trust region approach for nonlinear minimization subject to bounds, SIAM J. Optim. 6 (1996) 418–445.

[86] A.C. Woods, Atomic and Molecular Laser-Induced Breakdown Spectroscopy Above a Titanium Target, unpublished dissertation, University of Tennessee, Knoxville, TN, 2014.

[87] A.C. Woods, C.G. Parigger, J.O. Hornkohl, Time-resolved temperature inferences utilizing the TiO $A^3\Phi \rightarrow X^3\Delta$ band in laser-induced plasma, Int. Rev At. Mol. Phys. 3 (2012) 103–111.

[88] D.M. Surmick, C.G. Parigger, Aluminum monoxide emission measurements in a laser-induced plasma, Appl. Spectrosc. 68 (2014) 992–996.

[89] D.M. Surmick, C.G. Parigger, Time-resolved aluminium laser-induced plasma temperature measurements, J. Phys. Conf. Ser. 548 (2014) 012046.

[90] A. D'Ulivo, M. Onor, E. Pitzalis, R. Spiniello, L. Lampugnani, G. Cristoforetti, S. Legnaioli, V. Palleschi, A. Salvetti, E. Tognoni, Determination of the deuterium/hydrogen ratio in gas reaction products by laser-induced breakdown spectroscopy, Spectrochim. Acta B At. Spectrosc. 61 (2006) 797–802.

[91] F.R. Doucet, G. Lithgow, R. Kosierb, P. Bouchard, M. Sabsabi, Determination of isotope ratios using laser-induced breakdown spectroscopy in ambient air at atmospheric pressure for nuclear forensics, J. Anal. At. Spectrom 26 (2011) 536–541.

[92] Z.S. Lie, M. Pardede, E. Jobiliong, H. Suyanto, D. P. Kurniawan, R. Hedwig, M. Ramli, A. Khumaeni, T.J. Lie, K.H. Kurniawan, K. Kagawa, H-D analysis employing low-pressure microjoule picosecond laser-induced breakdown spectroscopy, Anal. Chem. 89 (2017) 4951–4957.

[93] R. Fantoni, S. Almaviva, L. Caneve, F. Colao, G. Maddaluno, P. Gasior, M. Kubkowska, Hydrogen isotope detection in metal matrix using double-pulse laser-induced breakdown spectroscopy, Spectrochim. Acta B At. Spectrosc. 129 (2017) 8–13.

[94] M. Suchoňová, P. Veis, J. Karhunen, P. Paris, M. Pribula, K. Piip, M. Laan, C. Porosnicu, C. Lungu, A. Hakola, Determination of deuterium depth profiles in fusion-relevant wall materials by nanosecond LIBS, Nucl. Mater. Energy 12 (2017) 611–616.

[95] A. Sarkar, X. Mao, G.C.-Y. Chan, V. Zorba, R.E. Russo, Laser ablation molecular isotopic spectrometry of water for ${}_1D^2/{}_1H^1$ ratio analysis, Spectrochim. Acta B At. Spectrosc. 88 (2013) 46–53.

[96] A.A. Bol'shakov, X. Mao, D.L. Perry, R.E. Russo, Laser ablation molecular isotopic spectrometry for rare isotopes of the light elements, Spectroscopy 29 (6) (2014) 30–39.

[97] A.A. Bol'shakov, X. Mao, R.E. Russo, Spectral emission enhancement by an electric pulse for LIBS and LAMIS, J. Anal. At. Spectrom 32 (2017) 657–670.

[98] G.H. Dieke, H.M. Crosswhite, The ultraviolet bands of OH. Fundamental data, J. Quant. Spectrosc. Radiat. Transf. 2 (1962) 97–199.

[99] A.S.-C. Cheung, C.M.-T. Chan, N.S.-K. Sze, The $A^2\Sigma^+$–$X^2\Pi$ bands of ^{18}OH, J. Mol. Spectrosc. 174 (1995) 205–214.

[100] H. Oura, On the band-system of neutral OD, Low Temp. Sci. 6 (1951) 41–70.

[101] S.-R. Lin, S.-T. Lee, Y.-P. Lee, The S_{21} lines of the $A^2\Sigma^+(v'=1) \leftarrow X^2\Pi(v''=0)$ transitions of OH and OD, J. Quant. Spectrosc. Radiat. Transf. 38 (1987) 163–166.

[102] C.G. Parigger, G. Guan, J.O. Hornkohl, Measurement and analysis of OH emission spectra following laser-induced optical breakdown in air, Appl. Optics 42 (2003) 5986–5991.

[103] J. Serrano, J. Moros, J.J. Laserna, Exploring the formation routes of diatomic hydrogenated radicals using femtosecond laser-induced breakdown spectroscopy of deuterated molecular solids, J. Anal. At. Spectrom 30 (2015) 2343–2352.

[103a] S.-U. Choi, S.-C. Han, J.-I. Yun, Hydrogen isotopic analysis using molecular emission from laser-induced plasma on liquid and frozen water, Spectrochim. Acta B At. Spectrosc. 162 (2019) 105716.

[104] J. Luque, D.R. Crosley, LIFBASE: Database and Spectral Simulation Program, SRI International Report MP, 99–009, 1999.

[105] H. Niki, T. Yasuda, I. Kitazima, Measurement technique of boron isotopic ratio by laser-induced breakdown spectroscopy, J. Nucl. Sci. Technol. 35 (1998) 34–39.

[106] X. Mao, A.A. Bol'shakov, D.L. Perry, O. Sorkhabi, R. E. Russo, Laser ablation molecular isotopic spectrometry: parameter influence on boron isotope measurements, Spectrochim. Acta B At. Spectrosc. 66 (2011) 604–609.

[107] A. Sarkar, X. Mao, R.E. Russo, Advancing the analytical capabilities of laser ablation ablation molecular isotopic spectrometry for boron isotopic analysis, Spectrochim. Acta B At. Spectrosc. 92 (2014) 42–50.

[108] A.A. Bol'shakov, X. Mao, C.P. McKay, R.E. Russo, Laser ablation—optical cavity isotopic spectrometer for Mars rovers, Proc. SPIE 8385 (2012) 83850C.

[109] P. Ko, I. Jovanovic, Boron isotopic measurements from spectrally filtered non-gated molecular spectra induced by laser ablation, Spectrochim. Acta B At. Spectrosc. 90 (2013) 68–71.

[110] B. Yee, K.C. Hartig, P. Ko, J. McNutt, I. Jovanovic, Measurement of boron isotopic ratio with non-gated molecular spectroscopy of femtosecond laser-produced plasma, Spectrochim. Acta B At. Spectrosc. 79–80 (2013) 72–76.

[110a] S. Brown, A. Ford, C.A. Akpovo, L. Johnson, Mid-IR enhanced laser ablation molecular isotopic spectrometry, Spectrochim. Acta B At. Spectrosc. 122 (2016) 178–187.

[111] S. Brown, A. Ford, C.C. Akpovo, J. Martinez, L. Johnson, Matrix effects in laser ablation molecular isotopic spectrometry, Spectrochim. Acta B At. Spectrosc. 101 (2014) 204–212.

[112] C.D. Harris, L.T.M. Profeta, C.A. Akpovo, L. Johnson, A.C. Stowe, Laser ablation molecular isotopic spectroscopy (LAMIS) towards the determination of multivariate LODs via PLS calibration model of ^{10}B and ^{11}B boric acid mixtures, Proc. SPIE 10192 (2017). paper 101920P.

[112a] C.A. Akpovo, L. Helms, L.T.M. Profeta, L. Johnson, Multivariate determination of ^{10}B isotopic ratio by laser-induced breakdown spectroscopy using multiple BO molecular emissions, Spectrochim. Acta B At. Spectrosc. 162 (2019) 105710.

[113] Z.H. Zhu, J.M. Li, Z.Q. Hao, S.S. Tang, Y. Tang, L. B. Guo, X.Y. Li, X.Y. Zeng, Y.F. Lu, Isotopic determination with molecular emission using laser-induced breakdown spectroscopy and laser-induced radical fluorescence, Opt. Express 27 (2019) 470–482.

[114] X. Mao, G.C.-Y. Chan, V. Zorba, R.E. Russo, Reduction of spectral interferences and noise effects in laser ablation molecular isotopic spectrometry with partial least square regression—a computer simulation study, Spectrochim. Acta B At. Spectrosc. 122 (2016) 75–84.

[115] A.A. Bol'shakov, N.V. Golovenkov, S.V. Oshemkov, A. A. Petrov, Laser—fluorescent analysis of inorganic gases and vapors (review), J. Appl. Spectrosc. 51 (1989) 741–753.

[116] S.H. Amiri, S.M.R. Darbani, H. Saghafifar, Detection of BO_2 isotopes using laser-induced breakdown spectroscopy, Spectrochim. Acta B At. Spectrosc. 150 (2018) 86–91.

[117] C. Dutouquet, S. Acquaviva, J. Hermann, Detection of boron nitride radicals by emission spectroscopy in a laser-induced plasma, Spectrochim. Acta B At. Spectrosc. 56 (2001) 629–635.

[118] C.G. Parigger, J.O. Hornkohl, A.M. Keszler, L. Nemes, Measurement and analysis of atomic and diatomic carbon spectra from laser ablation of graphite, Appl. Optics 42 (2003) 6192–6198.

[119] A.A. Bol'shakov, X. Mao, J. Jain, D.L. McIntyre, R. E. Russo, Laser ablation molecular isotopic spectrometry of carbon isotopes, Spectrochim. Acta B At. Spectrosc. 113 (2015) 106–112.

[120] R. Glaus, J. Riedel, I. Gornushkin, Insight into the formation of molecular species in laser-induced plasma of isotopically labeled organic samples, Anal. Chem. 87 (2015) 10131–10137.

[121] M. Dong, G.C.-Y. Chan, X. Mao, J.J. González, J. Lu, R. E. Russo, Elucidation of C_2 and CN formation mechanisms in laser-induced plasmas through correlation analysis of carbon isotopic ratio, Spectrochim. Acta B At. Spectrosc. 100 (2014) 62–69.

[122] M. Dong, X. Mao, J.J. González, J. Lu, R.E. Russo, Carbon isotope separation and molecular formation in laser-induced plasmas by laser ablation molecular isotopic spectrometry, Anal. Chem. 85 (2013) 2899–2906.

[122a] Q. Zhang, Y. Liu, W. Yin, Y. Yan, Q. Tang, G. Xing, The online detection of carbon isotopes by laser-induced breakdown spectroscopy, J. Anal. At. Spectrom. 35 (2020) 341–346.

[123] M.S.R. dos Santos, C. Pasquini, Determination of the isotopic composition of enriched materials using Laser Ablation Molecular Isotopic Spectrometry: partial least squares and multivariate curve resolution for the determination of ^{15}N content in enriched urea, Anal. Bioanal. Chem. (2020). https://doi.org/10.1007/s00216-020-02656-y.

[124] D.G. Weisz, J.C. Crowhurst, M.S. Finko, T.P. Rose, B. Koroglu, R. Trappitsch, H.B. Radousky, W. J. Siekhaus, M.R. Armstrong, B.H. Isselhardt, M. Azer, D. Curreli, Effects of plume hydrodynamics and oxidation on the composition of a condensing laser-induced plasma, J. Phys. Chem. A 122 (2018) 1584–1591.

[125] D.G. Weisz, J.C. Crowhurst, W.J. Siekhaus, T.P. Rose, B. Koroglu, H.B. Radousky, J.M. Zaug, M. R. Armstrong, B.H. Isselhardt, M.R. Savina, M. Azer, M.S. Finko, D. Curreli, Formation of $^{238}U^{16}O$ and $^{238}U^{18}O$ observed by time-resolved emission spectroscopy subsequent to laser ablation, Appl. Phys. Lett. 111 (2017) 034101.

[126] X. Mao, A.A. Bol'shakov, I. Choi, C.P. McKay, D. L. Perry, O. Sorkhabi, R.E. Russo, Laser ablation molecular isotopic spectrometry: strontium and its isotopes, Spectrochim. Acta B At. Spectrosc. 66 (2011) 767–775.

[127] H. Hou, G.C.-Y. Chan, X. Mao, V. Zorba, R. Zheng, R. E. Russo, Femtosecond laser ablation molecular isotopic spectrometry for zirconium isotope analysis, Anal. Chem. 87 (2015) 4788–4796.

[128] H. Hou, G.C.-Y. Chan, X. Mao, R. Zheng, V. Zorba, R. E. Russo, Femtosecond filament-laser ablation molecular isotopic spectrometry, Spectrochim. Acta B At. Spectrosc. 113 (2015) 113–118.

[129] K.C. Hartig, I. Ghebregziabher, I. Jovanovic, Standoff detection of uranium and its isotopes by femtosecond filament laser ablation molecular isotopic spectrometry, Sci. Rep. 7 (2017) 43852.

[130] L.A. Kaledin, M.C. Heaven, Electronic spectroscopy of UO, J. Mol. Spectrosc. 185 (1997) 1–7.

[131] X. Mao, G.C.-Y. Chan, I. Choi, V. Zorba, R.E. Russo, Combination of atomic lines and molecular bands for uranium optical isotopic analysis in laser induced plasma spectrometry, J. Radioanal. Nucl. Chem. 312 (2017) 121–131.

[132] J. Song, G.C.-Y. Chan, X. Mao, J.D. Woodward, R. W. Smithwick, T.G. Schaaff, A.C. Stowe, C.D. Harris, R. Zheng, V. Zorba, R.E. Russo, Multivariate nonlinear spectral fitting for uranium isotopic analysis with laser-induced breakdown spectroscopy, Spectrochim. Acta B At. Spectrosc. 150 (2018) 67–76.

PART III

LIBS applications

CHAPTER

8

Combustion applications of laser-induced breakdown spectroscopy

Fang Y. Yueh[a], *Markandey M. Tripathi*[a], *Jagdish P. Singh*[a], *Banu Sezer*[b], *Kemal Efe Eseller*[c]

[a]Institute for Clean Energy Technology, Mississippi State University, Starkville, MS, United States [b]Department of Food Engineering, Hacettepe University, Ankara, Turkey [c]Electrical and Electronics Engineering Department, Atilim University, Ankara, Turkey

1 Introduction

Combustion has been the fundamental way of energy generation for different industrial processes such as heating, production of electrical power, and automotive and aircraft/space shuttle propulsion for decades. Energy generation by combustion technologies provides 85% of worldwide energy with significant impacts on our daily lives. Even though combustion technologies provide a sufficient amount of energy in a wide range of industrial areas, today's world is exploring new ways to upgrade combustion technologies in terms of efficiency, stability, and environmental impact. Implementation of governmental regulations on emission control strategies and decreasing number of fuel sources are the two major problems encountered in industrial combustion technologies. As a natural consequence of this situation, a new field of research, combustion diagnostics, has emerged. Combustion diagnostics mainly focuses on understanding the nature of highly complex multistage combustion processes, leading to improved combustion efficiency and reduction of pollutants and greenhouse gases. In combustion diagnostics the main challenges are limitations of the analysis of the environment and the analyte itself. To surpass these limitations without compromising the accuracy and efficiency of analysis, advanced technologies with unique features are required. Recent advancements in laser-based technologies have brought considerable advantages in combustion diagnostics such as its nondestructive and in situ nature, remote detection option, capability for simultaneous temporal and spatial resolution, and independence from temperature. Laser-based techniques, including Laser Doppler Velocimetry (LDV), Laser Raman Spectroscopy, Coherent Anti-stokes Raman Scattering (CARS), Laser-Induced Florescence (LIF), Laser-Induced Incandescence (LII), etc., have been used for combustion diagnostics

https://doi.org/10.1016/B978-0-12-818829-3.00008-3

[1, 2]. Previous applications of combustion diagnostics were applied for the measurement of species concentrations, velocity, temperature, particle size, composition, and other thermodynamic/fluid dynamic parameters by using laser-based technologies [1, 2].

In addition to these technologies, laser-induced breakdown spectroscopy (LIBS) with unique features is an important technique that can be used to meet the requirements of combustion diagnostics. LIBS is a spectroscopy-based atomic emission technique in which a high-energy laser pulse is focused on a small area (typically a few microns in diameter) of a sample. This high-energy laser pulse delivers its energy to the surface of the sample and excites, ionizes, and vaporizes the sample into a plasma plume. Later, plasma decays and emits specific radiation to elements [3–5]. Since the laser was focused on a small area, LIBS provides both qualitative and quantitative results for a wide range of elements. Following the collection of emissions and monitoring the LIBS spectra, one can attain information on the fuel content and determine the chemical composition, flame parameters, and fuel type of combustible mixtures. The major advantage of the LIBS technique compared to other methods includes the capability to perform multielemental, remote, online, in situ analysis with no/minimum sample preparations. Considering these features, LIBS has created a great deal of interest in combustion diagnostics research. Early studies dealing with the potential use of LIBS in combustion processes were conducted in the 1980s. One of the first studies was the detection of toxic metal concentrations in off-gases. Later, a potential use for LIBS has been the testing of continuous monitoring in different test sites. The first feasibility study of the combustion process with LIBS was the characterization of an upstream region of a large magnetohydrodynamic (MHD) coal-fired flow facility by Singh et al. [3]. In addition to being the first, this study was of great importance because it was carried out in a harsh, turbulent, and highly luminous coal-fired MHD combustion environment and has presented successful results. After this study, especially in recent years, new studies have been put forward that develop different perspectives on combustion diagnostics such as measurements of flame temperature, fuel type, fuel concentration, equivalence ratio, etc. [4–8]. The main objective of this chapter is to present a comprehensive conspectus of the applications of LIBS in combustion diagnostics, including combustion products and flame characteristics. To cover the subject in broad terms, experimental parameters, data processing methods, and other combustion applications are discussed throughout the chapter.

2 Continuous monitoring of combustion emission

Combustion pollution is one of the major problems in terms of public health, environmental protection, and climate change. The combustion of coal, oil, gasoline, diesel, and natural gas results in major pollutants, including carbon monoxide, nitrogen dioxide, polycyclic aromatic hydrocarbons, and both fine (particulate matter (PM) 2.5) and ultrafine particles, etc. Air quality suffers due to these pollutants, which have simultaneous adverse effects on human health [9]. To reduce the adverse effects of these pollutants on public health and the environment, strict regulations and continuous monitoring of pollutant ratios in exhaust gases caused by combustion are essential requirements. Previously, conventional analytical techniques were applied for pollution measurement, which took time and effort for sample collection and transport to laboratories. Furthermore, these techniques faced problems in off-gas systems for sample collection, which is the core of the analysis. To overcome these disadvantages, laser-based methods, especially LIBS, were developed. LIBS with remote sensing capability and no sampling requirement can be used in off-gas measurement by focusing the laser beam on the gas stream and collecting the emission signals. Due to these advantages, LIBS provides a remote sensing alternative in in situ environments as a continuous emission monitor (CEM) for air pollution.

Evaluation of the LIBS system as a CEM for toxic metals in off-gases in thermal treatment facilities has been presented by different scientists [9–13]. Buckley et al. used LIBS in two incinerators and in a burns facility belonging to the US Department of Defense to evaluate the potential of LIBS as a CEM for toxic metals. In this LIBS system, a laser beam of 1064-nm Nd: YAG laser source was passed through a hole in the mirror and was focused on the gas stream. Emission of the plasma was reflected by the mirror and directed into a gated detector via optical fiber. To improve the emission signal, especially at low concentration levels, a new method called "single-shot detection" was applied. In this approach, "hits" of single-shot particles were separated from "misses" of single-shot particles. As a result of the applied method, spectral noise was reduced, resulting in increased signal-to-noise ratios. For the stack gas, a calibration study was carried out using LIBS parameters such as lens focal length, laser power, etc. used in the field measurement. For each analyte emission line, the integrated emission line peak was divided into the surrounding continuous background intensity level. Reported field test results of this study revealed detection limits between 2 and 100 mg/dscm for Be, Cd, Cr, Hg, and Pb. Yoshiie et al. used LIBS as a CEM to monitor the transient change in cadmium emission in the flue gas from waste incineration [12].

Later, Zhang et al. used a mobile LIBS system to monitor toxic metal concentrations in stack emissions [13]. In this study, a frequency-doubled 532-nm Nd:YAG laser source was used to focus the laser beam and collect the signal through an optical port. The port was designed to have the focusing lens close to the spark. To keep the lens clean and cool, the port was purged with nitrogen. An emission signal was collected with an ultraviolet optical fiber bundle, which was separated into two bundles from the output end and coupled with two intensified charge-coupled detector spectrographs. For data acquisition/analysis, two computers were interfaced to control the detectors. To obtain the toxic metal concentration, the system required an in situ calibration model. Since it is difficult to prepare the known concentration standards for gas streams, LIBS calibration standards were prepared under laboratory conditions and transferred to the field test with correction factors. A calibration study was conducted by injecting known concentrations of dry aerosols from an ultrasonic nebulizer through air as a carrier gas into two different systems: a sample cell acting as a closed system and air as acting as an open system. Prepared calibration standards obtained from an ultrasonic nebulizer were diluted from an analytical grade standard metal solution. In the open system, the laser beam was focused very close to the nozzle. To reduce the interference from the surrounding air, the sample injection tube was enclosed. In the sample cell system (closed system), continuous injection of metal aerosol was applied. Data collection was started after the gas composition reached equilibrium in the cell. In the data analysis, a calibration curve was obtained by using the peak area of the analyte line from the LIBS spectra. The concentration of metals was obtained after normalizing the peak area of the analyte line from the test spectrum with the slope of the linear calibration curve. Although the calibration performed in the laboratory was quite successful, there were some differences in the field tests. This showed that there were significant difficulties in transporting the calibration developed under laboratory conditions to the field environment. To correct the LIBS signal variation caused by shot-to-shot fluctuation of the laser and laser-matter interactions, a method that correlated the LIBS signal with a continuum background was applied.

LIBS measurements were performed with three sets of metal concentrations at the EPA Rotary Kiln Incinerator Simulator (RKIS) [9]. During the shakedown test at RKIS, an on-site calibration study with similar setup was performed for the detection of Cr, Pb, Cd, and Be. In the on-site calibration study, an injection probe was used to inject the metal aerosol into the RKIS gas stream from the opposite port. The gas stream parameters, such as temperature, flue gas flow rate, and particle loading in the test environment, were ~232°C, 3.4 scm/min, and 25–50 mg/dscm, respectively. Since the injected sample was quickly diluted by the gas flow, it was difficult to make an accurate estimation of the metal concentration in near focal volume. Because of this, the on-site calibration was used only to test the system response. In addition to the calibration study, the effect of gas stream parameters on LIBS calibration was also studied. The data collection conditions were with and without fly ash, as well as full flow and reduced flow. When compared to laboratory tests, the RKIS on-site study revealed much stronger CN band emission, which was related to the dust particles in the gas stream. In addition, the LIBS signal was found to be 1.1 times higher in the presence of fly ash. In the conditions of slightly reduced flow and temperature, LIBS signals showed density dependence as was expected. In this near-real-time test, concentrations of Cr, Pb, Cd, and Be were successfully measured with the LIBS system. There was good agreement between averaged LIBS concentrations and the results of the EPA method. However, large standard deviations from the sampling volume and widespread particulate conditions of metal in the gas stream were observed in the results. To correct the signal variations caused by optical damage, variation in laser power, and conditions of gas flow, a background normalization was also applied.

In addition to successfully demonstrating the potential of using LIBS as a CEM, the accuracy of the measurements depended largely on calibration and was greatly influenced by gas, particle size, and optical conditions. This situation clearly demonstrated the need for different calibration methods for the measurement of stack gas such as self-calibration, multivariate calibration, etc. In addition, the proposed LIBS method should be developed in terms of sensitivity and selectivity to be used as a routine analysis method for toxic metal measurement.

Another combustion product that must be monitored as much as toxic metals is PM. Particulate substances are, by definition, a mixture of solid particles and liquid droplets in the air. These PMs, named in accordance to their particle diameter such as particle diameters <2.5 μm and between 2.5 and 10 μm, are called PM2.5 and PM10, respectively. Both PM2.5 and PM10 contain toxic heavy metals and organic compounds, which can travel to the respiratory system and even the blood stream and cause significant health problems. As a result, accurate detection and quantification of particulate substances has an emerging importance. In this respect, LIBS with single-shot analysis has been used for the sizing and elemental analysis of detached, submicrometer to micrometer-sized aerosol particles in laboratory conditions [11, 14, 15]. These studies are an important indicator that the LIBS method can be used in the analysis of combustion products for size distribution, number densities, and composition of species both qualitatively and quantitatively. However, as an important note, it should be kept in mind that the quantification of particle size with high accuracy in the field condition involves considerable difficulties, which can be overcome by fully understanding the complex physical processes taking place. Reviews examining this issue in more detail are available in Refs. [16, 17].

3 Equivalence ratio

The universal target of combustion processes is continuous monitoring of the toxic pollutants and increasing efficiency. During the combustion process, continuous monitoring and control of the fuel-air equivalence ratio (φ) can contribute to the reduction of pollutant emissions such as oxides of nitrogen, carbon monoxide, unburned hydrocarbons, PM2.5, and PM10. Real-time analysis of fuel-air equivalence ratios can help to monitor combustion and closed-loop control.

The equivalence ratio (φ) of a burning fuel-oxidizer mixture is defined as [18, 19]:

$$\text{Equivalence ratio}\,(\varphi) = \text{AFR}_s/\text{AFR}_a$$

where AFR_s is the stoichiometric mass-based air-fuel ratio and AFR_a is the actual mass-based air-fuel ratio.

The main advantages of real-time monitoring in combustion can be obtained by control of the equivalence ratio, which prevents serious accidents, such as blow-off, flashback, or damage to the combustor due to combustion oscillations. As a result, various researchers have studied the potential of LIBS for measurements of fuel-air equivalence ratio in premixed combustion systems. One of the preliminary studies was reported by Phouc et al. In this study, the equivalence ratio of a methane-air combustion mixture in an ignition cell was measured by producing laser sparks [20]. The results of this study show that the equivalence ratio increases linearly with the intensity ratio of elemental H (656.3 nm) to O (777 nm) for equivalence ratios of 0.1–5.0. In a different study, real-time measurement of the equivalence ratio was measured by LIBS in spark-ignited engines [21–23]. Ferioli et al. applied the LIBS method to a General Motors LD-9 dual overhead cam four-cylinder engine exhaust gas, which was sampled in an optical cell. To quantify the equivalence ratio in the range from 0.8 to 1.2, spectral lines of atomic C (711.3 nm), atomic O (776.6 nm), atomic N (746.3 and 743.8 nm), and CN (broad emission 707–734 nm) were used. Ferioli et al. also studied direct measurement of atomic species over a wide range of mixture fractions of C_3H_8, CH_4, and CO_2 in air by LIBS [24]. In this study, spatial resolution was 1 mm, and the equivalence ratio was determined from a single shot of the laser. To quantify the mixture fraction of C_3H_8 in air, a C/(N + O) atomic ratio was used. The LIBS spectrum of a mixture of air and hydrocarbons can be seen in Fig. 1. The standard deviation was calculated as 3% of the mean for mixtures of 0%, 1%, and 2% propane in air for a single breakdown. The data belong to a single-shot measurement acquired at a 10-Hz, 3-μs delay and a 15-μs gate, and is presented in Fig. 2. As one can see from Fig. 2, quantitatively, LIBS has good potential for the determination of an unburned mixture of air and hydrocarbon.

Another application of LIBS was reported by Stavropoulos et al. [25, 26]. In this study, local equivalence ratio was measured in laminar premixed methane-air Bunsen flames by measuring the total emission intensities of atomic H (656.3 nm) and atomic O (777 nm). Later, the H/O intensity ratio was correlated with the flame equivalence ratio. Application of LIBS for a small premixed turbulent burner and a lean premixed combustor was studied by Zimmer et al., which revealed that H excitation of laser-induced plasma is the main factor in the correct equivalence ratio measurement [27]. Furthermore, Michalakou et al. showed that the intensity ratios H (656.3 nm)/O (777 nm) and C (833.5 nm)/O (844.6 nm) obtained from Bunsen burner laminar flames by LIBS in methane-, ethylene-, and propane-air mixtures have a very strong correlation with the equivalence ratio [28].

On the other hand, fuel-to-air ratios at the point of ignition by laser-induced gas breakdown have been studied by many researchers [6, 29, 30]. In this concept, measurement of CH_4-air and H_2-air mixtures in a high-pressure ignition cell was accomplished by Phuoc [29]. This study showed that there is a linear dependency of H (656.3 nm)/O (777 nm) LIBS spectral peak ratio on equivalence ratio. In another study, a combination of laser-induced ignition and plasma spectroscopy was investigated for the measurement of equivalence ratio in a

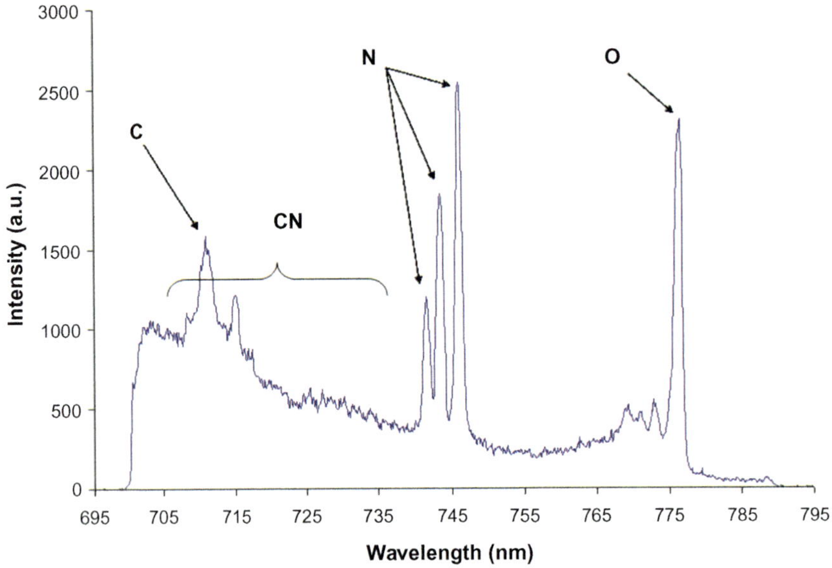

FIG. 1 Laser-induced breakdown spectroscopy spectrum of a mixture of air and hydrocarbons in the region between 690 and 790 nm.

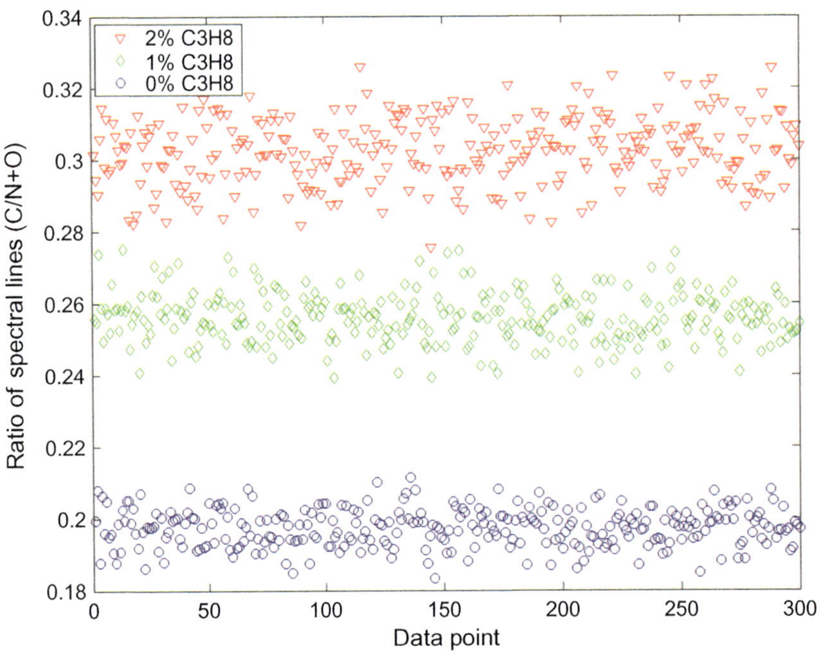

FIG. 2 Ratio of the integrated spectral peak area of 711-nm C to the combined signal from the 746-nm N line and 777-nm O line obtained in different mixtures of propane and air: 0% C_3H_8 (*circles*), 1% C_3H_8 (*diamonds*), 2% (*triangles*).

partially premixed hydrogen-air burner [31]. In this study, the ratios of H (486 nm)/O (777 nm) and H (656 nm)/N (744 nm) were simultaneously measured. In a different study, in-cylinder equivalence ratio measurement was accomplished by Joshi et al. [32]. This study was conducted in laser-ignited natural gas engines by using a 1064-nm Nd:YAG laser source to start the combustion. The optical emission was produced during the initiation processes of combustion. In this study, the equivalence ratios were measured by a wideband universal exhaust gas oxygen sensor. According to the results, H (656.3 nm)/O (777 nm), H (656.3 nm)/N (746 nm), and H (656.3 nm)/N (746 nm + 744 nm + 742 nm) ratios showed good correlation.

The general approach of the LIBS method used gated detection, which gave a delay between laser pulse and signal collection from the plasma. The main reason for this selection was to avoid initial continuum, improve the signal-to-background ratio, and block the elastically scattered high-intensity laser beam. However, in recent years, there have been certain efforts to use ungated detectors with an appropriate design to block scattered laser beams for LIBS measurements [4, 6, 33]. The main underlying reason for using an ungated system was its advantages of providing a simple, robust, and cost-effective environment when compared to time-gated systems. With this perspective, LIBS measurement of a methane-air premix flame to determine equivalence ratio and application of biodiesel diffusion flame using an ungated LIBS system was explored by Eseller et al. [4]. Furthermore, a polarization filter alongside online background signal correction was used with an ungated LIBS system by Kiefer et al. [6]. Later, they used this setup to explore the potential of LIBS in methane and dimethyl ether flames. Tripathi et al. approached this situation from a different point of view and studied the difference between a notch filter and a polarizer to suppress elastically scattered laser light [33]. This study showed that a polarizer has better blocking performance when compared to a notch filter. Moreover, Tripathi et al. used multivariate calibration for measurement of equivalence ratio, resulting in better performance when compared to intensity ratio-based calibration. There are different LIBS applications for the measurement of equivalence ratio in Refs. [31, 34–36].

4 Online coal analysis

Coal is one of the world's most abundant and leading electrical energy sources as a heterogeneous mineral, which is composed of carbon, hydrogen, oxygen, nitrogen, sulfur, and various minor/trace elements. Coal composition consists of complex mixtures of both organic and inorganic components. Many factors affect coal composition such as geographical location of mining area, depth, coalification, etc. The importance of the determination of the quality of coal may be attributed to many reasons, including environmental and industrial reasons such as the formation of toxic pollutants, boiler slagging and fouling, the requirement of expensive treatments in desulfurization, and disposal of ash and CO_2. However, the major impact of coal quality is net power output. Considering the amount and importance of coal use in both industry and society, determination of coal composition from direct mining operations has great importance. When coal is burned, most of the minerals and minor/trace elements generally form ash; however, some minerals break down into gaseous compounds that are expelled via the furnace's flue. Ash led to many problems, including inefficient combustion, machine downtime, and operational difficulties by depositing on the heated surface area.

The determination of coal quality has been studied for many years, including its coal rank, moisture, ash yield, sulfur content, and volatile matter. Electrical power plants, coke facilities, and various commercial laboratories employ calorific value [37], proximate (including moisture content, ash content, volatile matter, and fixed carbon) [38], and ultimate analyses (determination of C, H, O, N, and S concentrations) [39], as well as other related analyses. Currently, there are several analytical methods for the determination of coal analysis with certain disadvantages, including use of hazardous chemicals, long analysis time, sampling problems (a few grams of sample were used in the analysis despite the tons of coal transported on the conveyors), etc. So far, certain physical methods such as X-ray fluorescence (XRF) [40], prompt gamma neutron activation analysis (PGNAA) [41], near-infrared spectrometry (NIRS) [42], inductively-coupled plasma atomic emission spectrometry [43], atomic absorption spectrometry (AAS) [44], Fourier transform infrared spectrometry [45], and microwave analysis [46] have been used to accomplish coal analysis. Among these methods, XRF, PGNAA, NIRS, and microwave stand out with their ability to provide real-time measurement. PGNAA and XRF are two of the most applied methods particularly for online coal ore and ash analysis [47, 48]. However, these methods do not meet the demands of the coal industry and suffer from low accuracy, high cost, and radiation risk. As a result, there is a need for safe, real-time, relatively low-cost analytical methods that can help to sustain better quality control for production and easy implementation of regulations in many samples. At this point, the LIBS method is the best alternative for coal analysis. LIBS has the ability to be applied in both production plant and mining area. When compared to other coal analysis methods, LIBS has frontier advantages, including high-level safety with remote analysis, high sensitivity to low atomic number elements, detection of trace elements, and relatively low-cost portable system options. LIBS is rapid, less bulky, and safer (no need for radioactive source) when compared to PGNAA.

Initial studies were done by Ottesen et al. for in situ direct quantitative measurements of Si, Al, Mg, Ca, Fe, and Ti of coal particulates in combustion conditions by using LIBS [49, 50]. Later, Chadwick et al. demonstrated the potential of LIBS for the determination of elemental composition of lignite coals [51, 52]. In another study, Noda et al. used LIBS for the determination of caloric heat values using four different coals under high-pressure and -temperature conditions [53]. In this study, an automated LIBS unit was used to determine the monitoring capability of LIBS in power plants. The typical LIBS spectra of fly ash are presented in Fig. 3. According to the results of this study, the developed LIBS system had a 1-min detection time, which is significantly better when compared to over 30 min of sampling and analysis time in conventional methods. In Figs. 4 and 5, the correlation between LIBS measurements and conventional measurement results (Japanese Industrial Standards: JIS-8814 for calorific value and JIS-8815 for unburned carbon) are presented. Results of the unburned carbon measured in fly ash are presented in Fig. 4 under conditions of T_s 570 K, P_s 0.1 MPa, and a gas composition of N_2, 74%; H_2O, 8%; CO_2, 15%; and O_2, 3%, which is a typical exhaust gas composition for a pulverized coal power plant. As one can see from the results, LIBS results were in a good agreement with the conventional method. In Fig. 5, the correlation of unburned carbon and calorific value measurement under high-pressure conditions (98% N_2; 2% O_2; temperature 720 K) for a gasification power plant is presented. The results also show good agreement with the conventional methods. In conclusion, it can be said that LIBS has a high potential for high-pressure and -temperature plant conditions.

Furthermore, Kurihara et al. used LIBS for real-time detection of unburnt carbon in fly ash in the Haramachi thermal power plant to show that LIBS can be used in boiler control. In this study, different types of coal (>20) were used and the LIBS system was designed to measure the unburnt carbon in the range of 0%–10%. This study revealed that LIBS results were

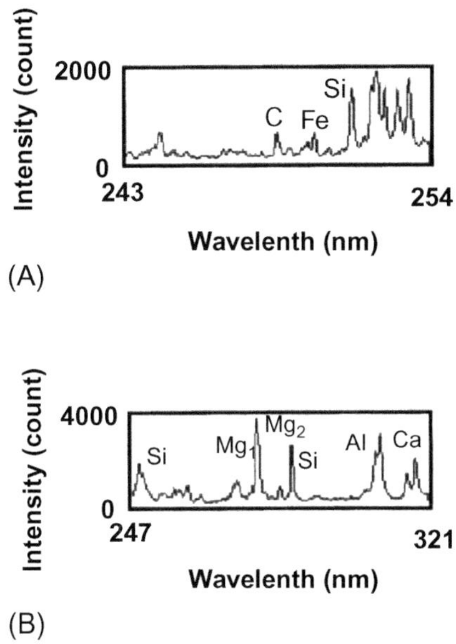

FIG. 3 Laser-induced breakdown spectroscopy spectra of fly ash: (A) high-resolution and (B) wide-range spectrum.

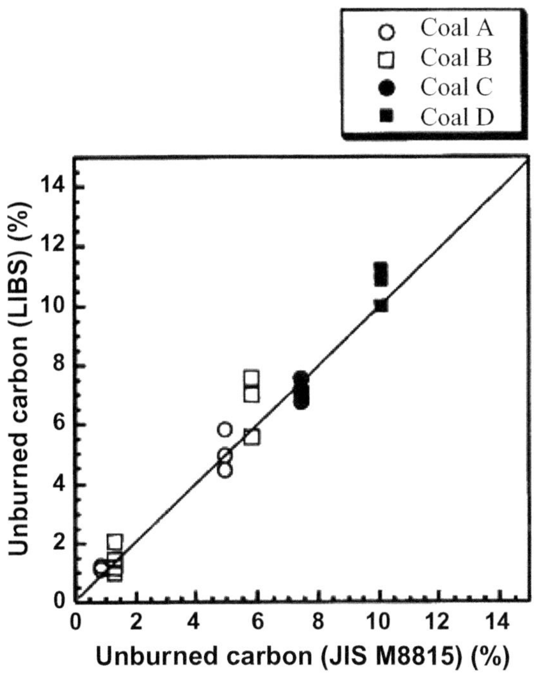

FIG. 4 Unburned carbon measurement results: pressure, 0.1 MPa; gas composition, N_2 74%, H_2O 8%, CO_2 15%, O_2 3%; temperature, 570 K.

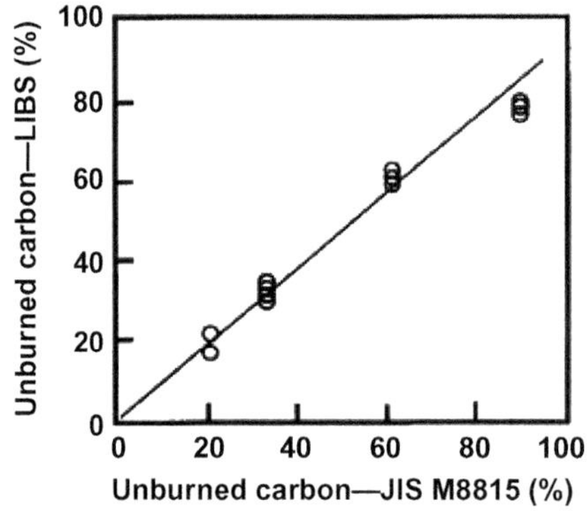

FIG. 5 Unburned carbon measurement results: pressure, 3.0 MPa; gas composition, N_2 98%, O_2 2%; temperature, 720 K.

compatible with the Japanese Industrial Standard 8815 test resulting in a standard deviation of 0.27% [54]. Blevins et al. used the LIBS method in coal-biomass boilers at 1170 K and 18 L/min of N_2 flow. In this study, a novel probe that was improved for severe conditions was used with an echelle spectrometer to determine Na, K, Ca, Mg, C, B, Si, Mn, Al, Fe, Rb, Cl, and Ti [55]. In inline coal analysis, Gaft et al. used LIBS for the analysis of coal ash content for 4 months on a conveyor belt. In this study, the potential of the LIBS system coupled with a moving belt conveyor was evaluated. This system was calibrated using five samples (ash content ranging from 12.5% to 53.9% in conventional laboratory analysis) with moisture content. The results of this study showed that the correlation between LIBS measurements and laboratory ash analysis has a standard deviation of 0.21 and $R^2 = 0.99$, and the average standard error was 0.32% between LIBS measurements and PGNAA [47, 48]. In a different study done by Yin et al., a fully software-controlled LIBS system was designed, including LIBS and sampling attachment for online coal analysis [56]. This system was designed to operate both multipoint sampling and sample preparation modes. In this study, elemental analysis (including C, Ca, Mg, Ti, Si, H, Al, Fe, etc.), net calorific value analysis, and proximate analysis of pulverized coal were achieved with 101% measurement error in elemental analysis. According to this result, it can be concluded that the designed system can be used in online analysis of pulverized coal with acceptable accuracy.

Another major objective of coal analysis was the determination of combustion efficiency. Coal combustion products include fly ash, bottom ash, boiler slag, and flue gas materials, with fly ash having the major proportion. In terms of coal combustion, the unburned fly ash content stands as a measure of the combustion efficiency in coal power plants. As a consequence, many researchers have focused on the determination of fly ash content by using LIBS. In this respect, Ctvrtnickova et al. used LIBS as a rapid and easy method to determine elemental content of fly ash (namely, C, Ba, Mg, Ti, Fe, and Si) in a pelletized form [57, 58]. In this study, two different binders (namely, KBr and polyvinyl alcohol) were used to prepare pellet samples. Moreover, they used carbon tapes to stick the fly ash samples, which resulted in speedy and easy detection of interference problems in the analysis. The result of this study showed that LIBS is a rapid and easy technique for combustion process control, especially when compared to inductively coupled plasma-optical emission spectroscopy. Zhang et al. designed an at-line

LIBS system for unburned fly ash measurement [59]. The results were obtained by using second-order polynomial multivariate inverse regression and showed 0.18% standard deviation, R^2 of 0.997, 0.26% measurement accuracy, and 3.81% average relative error. Yao et al. achieved better results in unburned carbon measurement by using multivariate calibration models compared to univariate calibration [60, 61]. In this study, the results of thermogravimetric analysis were correlated with LIBS spectra using the partial least square regression (PLSR) method. The obtained results were in good agreement in terms of the prediction ability of unknown samples. This study showed that LIBS coupled with PLSR has good potential for quantitative determination of coal ash content. It should be noted that multivariate analysis techniques, especially PLSR, have been increasingly applied in LIBS applications. It should not be forgotten that in addition to all the advantages it provides, it may not always produce an accurate estimate due to nonlinearity caused by spectral interferences. The PLSR technique surpasses the physical background of the LIBS analysis and, dependent on curve fitting, may result in noise overfitting. As a result, Feng et al. presented a new PLS method based on a dominant factor (including nonlinear correlation into PLSR) for better correlation for the determination of carbon concentration of bituminous coal [62].

In the process of coal combustion, coal ash fusion temperature (AFT) determines coal ash behavior. The coal AFT is an important parameter in terms of coal ash fouling and slagging since it may cause slagging problems and a decline in heat transfer efficiency. Hence, it is necessary to study coal AFT both theoretically and in practice. Determination of coal AFT can be useful for the prediction of slag potential. In a recent study, Romero et al. applied the LIBS method combined with an artificial neural network (ANN) for the determination of AFT of 16 coal samples under both laboratory and plant conditions [63]. According to the comparison of the LIBS-ANN result and ASTM methods, both methods were compatible with each other in terms of repeatability and reproducibility. Hence, it can be concluded that LIBS with ANN can be a useful decision control tool for slagging problems in boilers.

Oxygen in coal occurs in both organic and inorganic forms. Organic oxygen is one of the chief substances in coal. Real-time measurement of oxygen content in boilers of combustion plants has significant importance. Therefore Zhang et al. used LIBS for the measurement of organic oxygen content to build a calibration curve for eight different anthracite coal samples [64]. In this study, organic oxygen content was obtained by subtracting inorganic oxygen from the total oxygen content. To normalize the oxygen emission lines, Zhang et al. used a nitrogen atomic emission line as an internal standard. The results revealed that there was a linear correlation, while R was higher than 0.98, accuracy was in the range of 1.15%–1.37%, and absolute relative error was calculated as 19.39%. In this study, quantification of organic oxygen was demonstrated for coal for the first time.

As previously mentioned, ultimate coal analysis is a complex and wide-ranging examination of elements that exist in coal samples. It includes minor/trace element analysis alongside C, H, O, N, and S. Mateo et al. characterized the inorganic structure (Fe, Mg, Si, Al, and Ca) of coal samples under different operating conditions in thermal power utilization by using LIBS [65]. In this study, a 1064-nm and 355-nm laser source was used to generate laser plasma. The results obtained from the LIBS analysis were validated with the results of AAS and a good correlation was found between them. According to the results, the 355-nm laser source revealed better correlation when compared to the 1064-nm laser source in eight coal mine samples.

5 Flame temperature

Laser-based techniques such as laser-induced fluorescence, laser Raman, and coherent anti-Stokes spectroscopy have been used to determine flame temperatures for many years [66–69]. Temperature measurements with LIBS are based on the principle that gas densities are inversely proportional to temperature. Since the laser breakdown threshold is a function of the gas density, determination of the laser breakdown threshold can be used to *estimate* the local gas temperature. The temperature dependence of the laser breakdown threshold pulse energy was studied for flame thermometry [6, 7]. The effects of the sample composition on the disintegration threshold were determined by the threshold laser pulse energy dependence measured at different locations of the flame.

Molecular and atomic concentrations in flame are related to flame temperature. Lee and Hedge [8] used LIBS spectra for in situ flame temperature measurements by assuming that the nitrogen signal was inversely proportional to the flame temperature. Measurements of nitrogen LIBS signals taken at flame endpoints were related to absolute temperature values. The results of LIBS temperature measurements at these points were verified by thermocouple and Rayleigh scattering measurements.

6 Engine health

At very high temperatures, rocket engines have been shown to produce intense optical emissions from the wear of metals of engine parts. Evaporation of metal parts in the engine during combustion at temperatures such as 2000 K reduces the motor efficiency of the result. This can be observed by atomic emission spectrometry. Hydrocarbon-driven engines contain various species in the form of atomic carbon, C_2^+, and other carbon free radicals in their feathers that significantly increase background emission from the hair and the main OH band observed in oxygen and hydrogen in general. Even scattering from unburned carbon creates a strong background, which increases the opacity of the feather, which is another disadvantage of the atomic emission spectroscopic technique in detecting the presence of metal corrosion and engine wear in the hydrocarbon-fired rocket engine.

The performance of LIBS had been evaluated in detecting trace elements in the fuel plume of a hybrid rocket engine simulator at Stennis Space Center, Mississippi, USA. The hybrid rocket engine simulator used plexiglass as its fuel [36]. An adequate flow of oxygen was maintained for the proper burning of plexiglass. Initially, a spark was started in a small chamber between the electrode and the body of the ignition chamber using an electric current from a 12 V battery. This initial spark started the ignition of plexiglass as a main fuel, which generated a high-speed, luminous plume of ~5 cm from a 3–4-mm diameter exit nozzle. The laser was focused at various locations of the plume to record the LIBS spectra at different spatial locations with a lens of 10 cm focal length. The copper and stainless-steel wires were used as the seeding sample. Copper and stainless-steel wires were kept axially inside the ignition chamber extending up to the exit nozzle. The sample metals melted and vaporized due to the high temperature of the fuel and then exited with the burnt fuel as a plume.

LIBS spectra of the rocket engine simulator plume were recorded when a 316L stainless-steel wire of diameter 1.76 mm was inserted into the ignition chamber. The laser was focused 3 in. away from the exit of the nozzle. Stainless-steel 316L contains ~17% of Cr. LIBS spectra of the plume have shown a significant amount of Cr present in the plume. No iron lines could be identified in the spectra at this location, which is probably due to the low concentrations near the exit channel where the plume speed is very high. The presence of chromium lines seems to be due to their high transition probabilities. Fig. 6 shows the LIBS spectrum of the plume ranging from 305 to 350 nm. This spectrum shows the presence of strong OH and NH bands, as well as the two strong lines from atomic copper. OH and NH bands appear as a result of the reaction of hydrogen and oxygen as well as atmospheric nitrogen and hydrogen, respectively. Strong copper lines were recorded in the emission from the plume when copper wire was kept out of contact with the plume near the exit channel. Copper could be detected for a fraction of a second as it vaporized and escaped with the high-speed burnt fuel plume.

The LIBS spectra of the plume were recorded at different spatial locations from the exit nozzle. Fig. 7 shows the LIBS spectra at different spatial locations of the plume when copper wire was used as the seeded sample in the ignition chamber. Spectra were recorded at 13/8, 2, 3, and 5 in. from the exit nozzle. The presence of copper was detected strongly near the nozzle exit during an initial fraction of a second when the burnt fuel plume started building up. The Cu signal decreased with time when the plume attained its full length (>2 in.), high temperature, and high speed.

It was noted that as the measurement point moved away from the plume exit channel (luminous zone), the copper lines appeared throughout the plume. The background emission from the plume was also decreased in this measurement location. This was likely due to the better mixing of metal vapor in the plume away from the exit channel. In this location, the plume has a lower speed and a lower gas temperature, as compared with the plume near the exit channel. The data from the preliminary test showed that the measurements made away from the luminous part of the plume can provide more meaningful information about the health of the rocket engine. This test demonstrated that LIBS

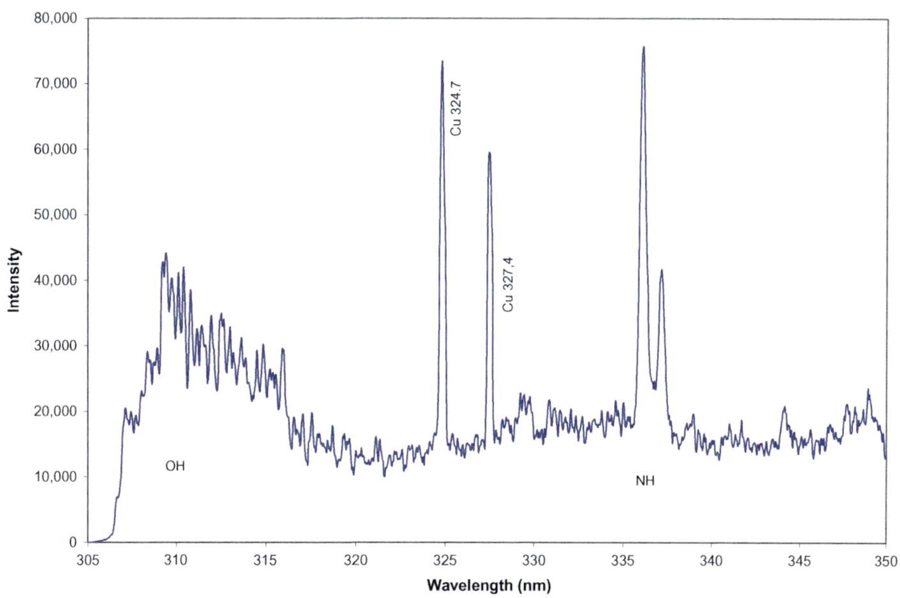

FIG. 6 Laser-induced breakdown spectroscopy spectra of the rocket engine simulator plume.

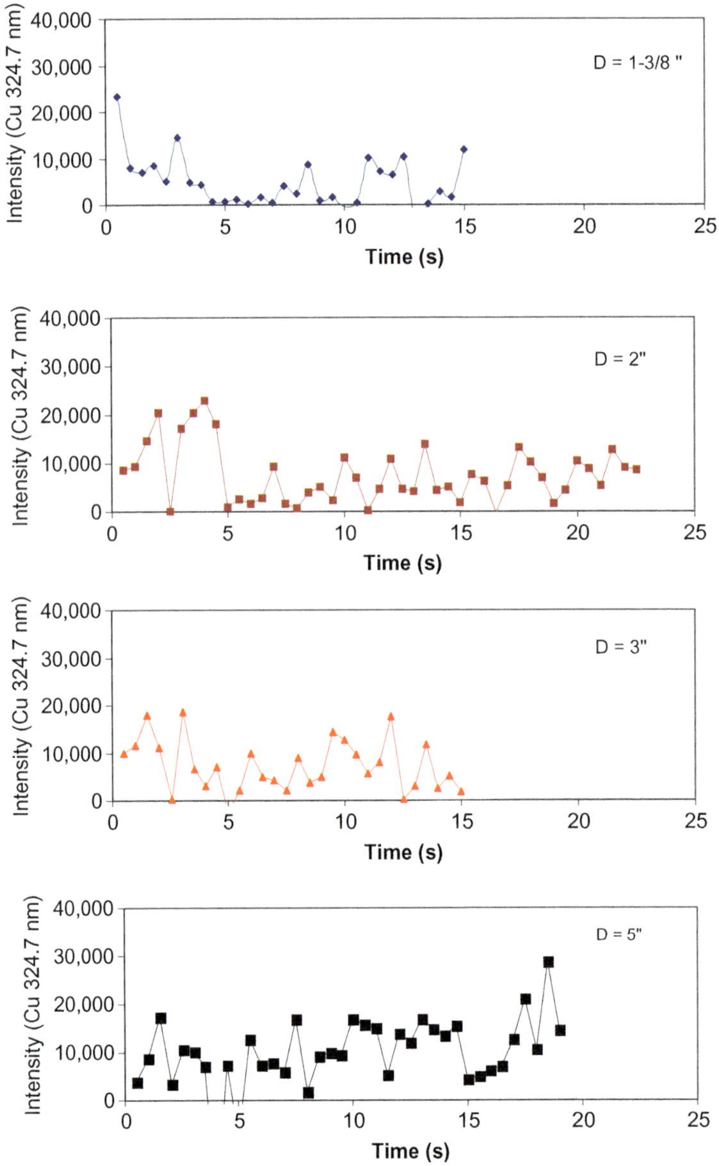

FIG. 7 Variation of laser-induced breakdown spectroscopy signal with time at different spatial locations.

was capable of being used as an engine health monitor to detect a trace amount of metal emerging from any part of a rocket's engine. The proper calibration and metal seeding techniques are now under investigation.

7 Conclusions

LIBS has been identified as a diagnostic tool in combustion applications. Online, in situ, and non-intrusive measurements can be provided easily by LIBS. The range of applications to which LIBS is applied varies between combustion monitoring, equivalence ratio, flame temperature, and determination and monitoring of the quality of flammable materials (such as coal) to measure air pollutants (toxic metals, PM, etc.). Current limitations that prevent broader adaptation of LIBS in combustion diagnostics include variability in the LIBS signal due to the pulse-to-pulse change in laser-induced plasmas, the effects of the matrix on the LIBS signal, and the need for proper calibration prior to insertion for practical applications. Research has been developed to improve the reproducibility of LIBS results, improve signal-to-noise and signal-to-background ratios, and develop robust calibration techniques that can lead to major improvements in LIBS-based diagnostic methods, including combustion applications.

References

[1] A. Eckbreth, Laser Diagnostics for Combustion Temperature and Species, CRC Press, Cambridge, MA, 1988.

[2] L.J. Radziemski, D.A. Cremers, Laser-Induced Plasmas and Applications, Marcel Dekker, Inc., New York, NY, 1989.

[3] J.P. Single, H. Zhang, F.-Y. Yueh, R.L. Cook, Laser-induced breakdown spectra in a coal-fired MHD facility, in: The 28th Intersociety Energy Conversion Engineering Conference, Atlanta, GA, USA, 08/08–13/93, 1993, pp. 995–1000.

[4] K.E. Eseller, F.Y. Yueh, J.P. Singh, Laser-induced breakdown spectroscopy measurement in methane and biodiesel flames using an ungated detector, Appl. Optics 47 (31) (2008) G144–G148.

[5] N. Kawahara, E. Tomita, T. Oka, Y. Ikeda, Fuelconcentration measurement of premixed mixture in a spark-ignition engine using spark-induced breakdown spectroscopy, in: 15th International Symposium on Applications of Laser Techniques to Fluid Mechanics, 2010.

[6] J. Kiefer, J. Tröger, Z. Li, M. Aldén, Laser-induced plasma in methane and dimethyl ether for flame ignition and combustion diagnostics, Appl. Phys. B 103 (1) (2011) 229–236.

[7] J. Kiefer, J.W. Tröger, Z. Li, T. Seeger, M. Alden, A. Leipertz, Laser-induced breakdown flame thermometry, Combust. Flame 159 (12) (2012) 3576–3582.

[8] T.-W. Lee, N. Hegde, Laser-induced breakdown spectroscopy for in situ diagnostics of combustion parameters including temperature, Combust. Flame 3 (142) (2005) 314–316.

[9] H. Zhang, F.-Y. Yueh, J.P. Singh, Laser-induced breakdown spectrometry as a multimetal continuous-emission monitor, Appl. Optics 38 (9) (1999) 1459–1466.

[10] S.G. Buckley, H.A. Johnsen, K.R. Hencken, D.W. Hahn, Implementation of laser-induced breakdown spectroscopy as a continuous emissions monitor for toxic metals, Waste Manag. 20 (5–6) (2000) 455–462.

[11] D. Hahn, W. Flower, K. Hencken, Discrete particle detection and metal emissions monitoring using laser-induced breakdown spectroscopy, Appl. Spectrosc. 51 (12) (1997) 1836–1844.

[12] R. Yoshiie, A. Kojima, S. Uemiya, M. Nishimura, S. Kambara, H. Moritomi, Monitoring of volatile cadmium in flue gas from the waste incineration process using libs, J. Chem. Eng. Jpn. 38 (7) (2005) 528–534.

[13] H. Zhang, F.-Y. Yueh, J.P. Singh, Performance evaluation of laser-induced breakdown spectrometry as a multimetal continuous emission monitor, J. Air Waste Manage. Assoc. 51 (5) (2001) 681–687.

[14] D. Hahn, Laser-induced breakdown spectroscopy for sizing and elemental analysis of discrete aerosol particles, Appl. Phys. Lett. 72 (23) (1998) 2960–2962.

[15] D. Hahn, M. Lunden, Detection and analysis of aerosol particles by laser-induced breakdown spectroscopy, Aerosol Sci. Tech. 33 (1–2) (2000) 30–48.

[16] P.K. Diwakar, K.H. Loper, A.-M. Matiaske, D.W. Hahn, Laser-induced breakdown spectroscopy for analysis of micro and nanoparticles, J. Anal. At. Spectrom 27 (7) (2012) 1110–1119.

[17] D.W. Hahn, Laser-induced breakdown spectroscopy for analysis of aerosol particles: the path toward quantitative analysis, Spectroscopy 24 (9) (2009).

[18] S.R. Turns, Thermodynamics: Concepts and Applications, Cambridge University Press, 2006.

[19] M.M. Tripathi, S.R. Krishnan, K.K. Srinivasan, F.-Y. Yueh, J.P. Singh, Chemiluminescence-based multivariate sensing of local equivalence ratios in premixed atmospheric methane–air flames, Fuel 93 (2012) 684–691.

[20] T.X. Phuoc, F.P. White, Laser-induced spark for measurements of the fuel-to-air ratio of a combustible mixture, Fuel 81 (13) (2002) 1761–1765.

[21] F. Ferioli, S. Buckley, P. Puzinauskas, Real-time measurement of equivalence ratio using laser-induced breakdown spectroscopy, Int. J. Eng. Res. 7 (6) (2006) 447–457.

[22] F. Ferioli, P.V. Puzinauskas, S.G. Buckley, LIBS for real-time equivalence ratio measurements in spark-ignited engines, in: Second International Conference on Laser-Induced Plasma Spectroscopy, 2002.

[23] F. Ferioli, P.V. Puzinauskas, S.G. Buckley, Laser-induced breakdown spectroscopy for on-line engine equivalence ratio measurements, Appl. Spectrosc. 57 (9) (2003) 1183–1189.
[24] F. Ferioli, S.G. Buckley, Measurements of hydrocarbons using laser-induced breakdown spectroscopy, Combust. Flame 144 (3) (2006) 435–447.
[25] P. Stavropoulos, A. Michalakou, G. Skevis, S. Couris, Laser-induced breakdown spectroscopy as an analytical tool for equivalence ratio measurement in methane–air premixed flames, Spectrochim. Acta B At. Spectrosc. 60 (7–8) (2005) 1092–1097.
[26] P. Stavropoulos, A. Michalakou, G. Skevis, S. Couris, Quantitative local equivalence ratio determination in laminar premixed methane–air flames by laser induced breakdown spectroscopy (LIBS), Chem. Phys. Lett. 404 (4–6) (2005) 309–314.
[27] L. Zimmer, S. Tachibana, Laser induced plasma spectroscopy for local equivalence ratio measurements in an oscillating combustion environment, Proc. Combust. Inst. 31 (1) (2007) 737–745.
[28] A. Michalakou, P. Stavropoulos, S. Couris, Laser-induced breakdown spectroscopy in reactive flows of hydrocarbon-air mixtures, Appl. Phys. Lett. 92 (8) (2008). p. 081501.
[29] T.X. Phuoc, Laser-induced spark for simultaneous ignition and fuel-to-air ratio measurements, Opt. Lasers Eng. 44 (6) (2006) 520–534.
[30] L. Zimmer, K. Okai, Y. Kurosawa, Combined laser induced ignition and plasma spectroscopy: fundamentals and application to a hydrogen–air combustor, Spectrochim. Acta B At. Spectrosc. 62 (12) (2007) 1484–1495.
[31] M. Dackman, J.W. Lewis, Y.-L. Chen, C.G. Parigger, Laser-induced breakdown spectroscopy for measurement of fuel/oxygen mixing in combustion, in: Digital Holography and Three-Dimensional Imaging, Optical Society of America, 2008, p. JMA22.
[32] S. Joshi, D.B. Olsen, C. Dumitrescu, P.V. Puzinauskas, A.P. Yalin, Laser-induced breakdown spectroscopy for in-cylinder equivalence ratio measurements in laser-ignited natural gas engines, Appl. Spectrosc. 63 (5) (2009) 549–554.
[33] M.M. Tripathi, K.K. Srinivasan, S.R. Krishnan, F.-Y. Yueh, J.P. Singh, A comparison of multivariate LIBS and chemiluminescence-based local equivalence ratio measurements in premixed atmospheric methane–air flames, Fuel 106 (2013) 318–326.
[34] M.S. Mansour, H. Imam, K.A. Elsayed, W. Abbass, Local equivalence ratio measurements in turbulent partially premixed flames using laser-induced breakdown spectroscopy, Spectrochim. Acta B At. Spectrosc. 64 (10) (2009) 1079–1084.
[35] A.E. Majd, A.S. Arabanian, R. Massudi, M. Nazeri, Spatially resolved laser-induced breakdown spectroscopy in methane—air diffusion flames, Appl. Spectrosc. 65 (1) (2011) 36–42.
[36] S. Zhang, X. Yu, F. Li, G. Kang, L. Chen, X. Zhang, Laser induced breakdown spectroscopy for local equivalence ratio measurement of kerosene/air mixture at elevated pressure, Opt. Lasers Eng. 50 (6) (2012) 877–882.
[37] D.M. Mason, K.N. Gandhi, Formulas for calculating the calorific value of coal and coal chars: development, tests, and uses, Fuel Process. Technol. 7 (1) (1983) 11–22.
[38] C.J. Donahue, E.A. Rais, Proximate analysis of coal, J. Chem. Educ. 86 (2) (2009) 222.
[39] A. White, J. Whittingham, Ultimate analysis of coals and cokes using instrumental techniques, Fuel 62 (9) (1983) 1058–1061.
[40] Z. Yan, et al., Online X-ray fluorescence (XRF) analysis of heavy metals in pulverized coal on a conveyor belt, Appl. Spectrosc. 70 (2) (2016) 272–278.
[41] J. Marshall, J. Zumberge, On-line measurements of bulk coal using prompt gamma neutron activation analysis, Nucl. Geophys. 3 (4) (1989) 445–459.
[42] M. Kaihara, T. Takahashi, T. Akazawa, T. Sato, S. Takahashi, Application of near infrared spectroscopy to rapid analysis of coals, Spectrosc. Lett. 35 (3) (2002) 369–376.
[43] S. Caroli, A.F. Mazzeo, A. Laurenzi, O. Senofonte, N. Violante, Determination of sulphur in coal products by inductively coupled plasma atomic emission spectrometry, J. Anal. At. Spectrom 3 (1) (1988) 245–248.
[44] A. Baysal, S. Akman, A practical method for the determination of sulphur in coal samples by high-resolution continuum source flame atomic absorption spectrometry, Talanta 85 (5) (2011) 2662–2665.
[45] Z. Li, P.M. Fredericks, C.R. Ward, L. Rintoul, Chemical functionalities of high and low sulfur Australian coals: a case study using micro attenuated total reflectance–Fourier transform infrared (ATR–FTIR) spectrometry, Org. Geochem. 41 (6) (2010) 554–558.
[46] E. Binner, et al., A review of microwave coal processing, J. Microw. Power Electromagn. Energy 48 (1) (2014) 35–60.
[47] M. Gaft, E. Dvir, H. Modiano, U. Schone, Laser induced breakdown spectroscopy machine for online ash analyses in coal, Spectrochim. Acta B At. Spectrosc. 63 (10) (2008) 1177–1182.
[48] M. Gaft, I. Sapir-Sofer, H. Modiano, R. Stana, Laser induced breakdown spectroscopy for bulk minerals online analyses, Spectrochim. Acta B At. Spectrosc. 62 (12) (2007) 1496–1503.
[49] D. Ottesen, J. Wang, L. Radziemski, Real-time laser spark spectroscopy of particulates in combustion environments, Appl. Spectrosc. 43 (6) (1989) 967–976.

[50] D.K. Ottesen, L. Baxter, L. Radziemski, J. Burrows, Laser spark emission spectroscopy for in-situ, real-time monitoring of pulverized coal particle composition, Energy Fuel 5 (2) (1991) 304–312.

[51] D. Body, B.L. Chadwick, Simultaneous elemental analysis system using laser induced breakdown spectroscopy, Rev. Sci. Instrum. 72 (3) (2001) 1625–1629.

[52] B.L. Chadwick, D. Body, Development and commercial evaluation of laser-induced breakdown spectroscopy chemical analysis technology in the coal power generation industry, Appl. Spectrosc. 56 (1) (2002) 70–74.

[53] M. Noda, Y. Deguchi, S. Iwasaki, N. Yoshikawa, Detection of carbon content in a high-temperature and high-pressure environment using laser-induced breakdown spectroscopy, Spectrochim. Acta B At. Spectrosc. 57 (4) (2002) 701–709.

[54] M. Kurihara, K. Ikeda, Y. Izawa, Y. Deguchi, H. Tarui, Optimal boiler control through real-time monitoring of unburned carbon in fly ash by laser-induced breakdown spectroscopy, Appl. Optics 42 (30) (2003) 6159–6165.

[55] L.G. Blevins, C.R. Shaddix, S.M. Sickafoose, P.M. Walsh, Laser-induced breakdown spectroscopy at high temperatures in industrial boilers and furnaces, Appl. Optics 42 (30) (2003) 6107–6118.

[56] W. Yin, L. Zhang, L. Dong, W. Ma, S. Jia, Design of a laser-induced breakdown spectroscopy system for on-line quality analysis of pulverized coal in power plants, Appl. Spectrosc. 63 (8) (2009) 865–872.

[57] T. Ctvrtnickova, M. Mateo, A. Yañez, G. Nicolas, Laser induced breakdown spectroscopy application for ash characterisation for a coal fired power plant, Spectrochim. Acta B At. Spectrosc. 65 (8) (2010) 734–737.

[58] T. Ctvrtnickova, M.-P. Mateo, A. Yanez, G. Nicolas, Characterization of coal fly ash components by laser-induced breakdown spectroscopy, Spectrochim. Acta B At. Spectrosc. 64 (10) (2009) 1093–1097.

[59] L. Zhang, et al., Development of an apparatus for on-line analysis of unburned carbon in fly ash using laser-induced breakdown spectroscopy (LIBS), Appl. Spectrosc. 65 (7) (2011) 790–796.

[60] S. Yao, J. Lu, M. Dong, K. Chen, J. Li, J. Li, Extracting coal ash content from laser-induced breakdown spectroscopy (LIBS) spectra by multivariate analysis, Appl. Spectrosc. 65 (10) (2011) 1197–1201.

[61] S. Yao, J. Lu, J. Zheng, M. Dong, Analyzing unburned carbon in fly ash using laser-induced breakdown spectroscopy with multivariate calibration method, J. Anal. At. Spectrom 27 (3) (2012) 473–478.

[62] J. Feng, Z. Wang, L. West, Z. Li, W. Ni, A PLS model based on dominant factor for coal analysis using laser-induced breakdown spectroscopy, Anal. Bioanal. Chem. 400 (10) (2011) 3261–3271.

[63] C.E. Romero, R.D. Saro, J. Craparo, A. Weisberg, R. Moreno, Z. Yao, Laser-induced breakdown spectroscopy for coal characterization and assessing slagging propensity, Energy Fuel 24 (1) (2009) 510–517.

[64] L. Zhang, L. Dong, H. Dou, W. Yin, S. Jia, Laser-induced breakdown spectroscopy for determination of the organic oxygen content in anthracite coal under atmospheric conditions, Appl. Spectrosc. 62 (4) (2008) 458–463.

[65] M. Mateo, G. Nicolas, A. Yanez, Characterization of inorganic species in coal by laser-induced breakdown spectroscopy using UV and IR radiations, Appl. Surf. Sci. 254 (4) (2007) 868–872.

[66] C.E. Baukal Jr., Industrial Combustion Testing, CRC Press, 2010.

[67] C. Chan, J.W. Daily, Measurement of temperature in flames using laser induced fluorescence spectroscopy of OH, Appl. Optics 19 (12) (1980) 1963–1968.

[68] T. Cheng, S. March, R. Pitz, J. Wehrmeyer, J.-Y. Chen, Laser raman measurements of temperature and species concentration in swirling lifted hydrogen jet diffusion flames, Int. J. Hydrogen Energy 37 (9) (2012) 7900–7911.

[69] T. Lee, W.G. Bessler, H. Kronemayer, C. Schulz, J. B. Jeffries, Quantitative temperature measurements in high-pressure flames with multiline NO-LIF thermometry, Appl. Optics 44 (31) (2005) 6718–6728.

CHAPTER

9

LIBS application to liquid samples

Chet R. Bhatt[a], Christian L. Goueguel[b], Jinesh C. Jain[a], Dustin L. McIntyre[a], Jagdish P. Singh[c]

[a]National Energy Technology Laboratory, U.S. Department of Energy, Morgantown, WV, United States [b]Logiag Inc., Chateauguay, QC, Canada [c]Institute for Clean Energy Technology, Mississippi State University, Starkville, MS, United States

1 Introduction

Liquid samples are usually classified as aqueous, nonaqueous, and mixtures. Examples of aqueous liquids are most often surface water, groundwater, drinking water, and wastewater, whereas nonaqueous liquids include organic solvents, oils, and many other hydrophobic liquids. Mixtures may be referred to as the combination of aqueous and nonaqueous liquids. Although most laser-induced breakdown spectroscopy (LIBS) research has been focused on the characterization of solid samples, the analysis of liquid samples by LIBS offers a huge potential in a number of applications such as surface and groundwater monitoring, oceanography, carbon dioxide (CO_2) leak detection in geologic carbon storage, quality control in pharmaceutical and food industries, medical diagnostics, and industrial process monitoring [1–4]. Traditionally, the analysis of liquid samples has been performed by well-established laboratory-based analytical techniques such as inductively coupled plasma atomic emission spectroscopy and inductively coupled plasma mass spectrometry. However, LIBS offers numerous advantages over the aforementioned techniques, including in situ and online measurements. Indeed, LIBS analysis is relatively fast, the instrumentation can be miniaturized to make the system portable, and the coupling of fiber optics allows LIBS measurements in harsh environments and places that are inaccessible by traditional lab-based techniques. In addition, only optical access to the sample is needed to perform a LIBS measurement to allow a standoff analysis.

LIBS analysis can be performed in a bulk liquid, in static or laminar flow of the liquid surface, in jets, and in a stream of aerosols or droplets [5–10]. Although bulk liquid analysis is relatively simple to set up, it is only usable for transparent liquids. The sensitivity of analysis in bulk liquid remains the main issue since most of the plasma energy is lost in vaporization of the surrounding liquid and in generation of a shockwave and cavitation bubble accompanying breakdown. Only a small fraction of

Laser-Induced Breakdown Spectroscopy
https://doi.org/10.1016/B978-0-12-818829-3.00009-5

the plasma energy is converted into radiative energy. Furthermore, plasma formation in bulk water generally suffers from short lifetimes as the emission intensity falls steeply after approximately 1 μs. On the other hand, liquid surface analysis is typically more sensitive than bulk analysis because of plasma expansion in air. However, strong mechanical effects arising from the laser-liquid surface interaction cause several experimental difficulties (i.e., splashing, generation of bubbles and aerosol, and ripples on the liquid surface), which greatly affect the accuracy, repeatability, and reproducibility of the measurement. Interestingly, most of these difficulties can be minimized by using various experimental configurations to convert liquid into jets, aerosols, or droplets. However, LIBS setup complexity increases due to the use of additional instrumentation (e.g., peristaltic pump or various types of nebulizers) to transform the liquid sample and to ensure that there is a continuous exchange of the sample in the flow channel. The limits of detection (LODs) obtained using the aforementioned configurations are mostly in the parts per million (ppm) or high parts per billion (ppb) level. Even though the sensitivity of these configurations can be satisfactory for a number of applications, research is under way to improve the detection limits and to expand LIBS applicability to a higher number of applications that require detection at trace level. To this end, many researchers have investigated various alternative approaches involving sample preparation or pretreatment [11–15] or the use of two successive laser pulses (double-pulse LIBS [DP-LIBS]) [16–19], including the use of hyphenated approaches (e.g., LIBS-laser-induced fluorescence [LIF]) [20–23]. Another approach proposed to enhance LIBS sensitivity is the use of an ArF laser at 193 nm as an excitation source instead of the commonly used 1064-nm YAG laser [24–26]. This chapter presents some general aspects and research trends on liquid sample analysis using LIBS. General procedures for preparing and analyzing liquid samples are described.

2 Experimental details

2.1 Samples

The main goal of LIBS study is to characterize a sample for its elemental composition qualitatively and quantitatively. LIBS has extensively been used for the analysis of liquids, particularly aqueous samples. In principle, sample preparation is not required for test samples; however, for quantifying the elemental concentrations a reliable calibration of the instrument is required. The calibration is generally achieved by using analyte solutions of known concentrations. These solutions are known as calibration standards and are prepared by mixing pure analyte liquid standards or dissolving analyte salts in desired concentrations. For construction of the calibration curves in their study of metal ions, Goueguel et al. [27] prepared stock solutions by dissolving pure powders of $CaCl_2{\cdot}2H_2O$, $MnCl_2{\cdot}4H_2O$, $SrCl_2{\cdot}6H_2O$, $MgCl_2{\cdot}6H_2O$, and $BaCl_2{\cdot}2H_2O$ in ultrapure water. The stock solutions were then used to prepare sets of calibration standards for quantification of Ca, Mn, Sr, and Mg. Ba solution served as an internal standard in this study. Bhatt et al. [28, 29] used solutions of Eu (10,017 ± 29 μg/mL Eu, 7% HNO_3) and Yb (9978 ± 27 μg/mL Yb, 7% HNO_3), which were diluted with deionized water containing 2% HNO_3 to obtain the desired concentrations of solutions for generating the calibration curves for Eu and Yb. In a separate study, Bhatt et al. [30] utilized calibration standards of Ca, Mg, Mn, and Sr for investigating the leaching of these metals from their carbonate pellets under rising CO_2 pressure. The released amount of Ca, Mg, Mn, and Sr from their corresponding metal carbonates was measured by LIBS and quantified by developing calibration curves using the calibration standards. A similar study by Goueguel et al. with $CaCO_3$ is reported in [31].

2.2 Experimental apparatus

In general, the LIBS apparatus consists of three units: excitation source (laser), optics (lens, optical fiber), and detection system coupled with a computer (camera, spectrometer) for analysis. For the LIBS study of liquids, the laser pulse is focused either on the surface or into the liquid. The generation of a laser spark on the liquid surface sometimes results in splashing and surface agitation. The excitation capacity, on the other hand, may be compromised due to lower temperature spark in the liquid. The experimental setup used by Goueguel et al. [31] for liquid LIBS analysis is shown in Fig. 1. Laser radiation at 1064 nm was produced in pulses of 6-ns duration by a solid-state Nd:YAG laser (Q-smart 850, Quantel). The laser beam was focused 2 cm inside the liquid using two ultraviolet-grade fused-silica planoconvex lenses of 180 and 70 mm focal length, and 25.4 cm diameter. The lenses were separated by 1 mm, resulting in a focal length of approximately 50 mm. In these experiments, the laser repetition rate was held at 5 Hz. Pulse energy levels were attenuated by means of two pairs of half-wave plates and polarizing beam splitter cubes. The monitoring of the pulse-to-pulse energy was carried out using a high-energy pyroelectric head detector (70,273, Oriel Instruments) connected to a computer through an Ophir compact Juno USB interface. The liquid samples were kept in a 0.5-L stainless-steel high-pressure vessel to perform measurements under pressurized conditions. Similar LIBS setups have been reported by other research groups with some changes in laser sources, optics, and detection systems depending on the nature of the study and availability of the equipment [32, 33]. Joseph and David created laser sparks on the surface of the solution to determine uranium [34]. If DP-LIBS is to be used, a second source of excitation can be added to the system in various geometrical configurations, such as collinear, orthogonal, and oblique [16, 18, 35, 36]. Rai et al. [18, 37] focused the laser pulse on the jet orthogonally.

3 Optimization of experimental parameters

Major experimental parameters that come into play in a single-pulse LIBS (SP-LIBS) study of liquid samples are laser energy, gate time delay,

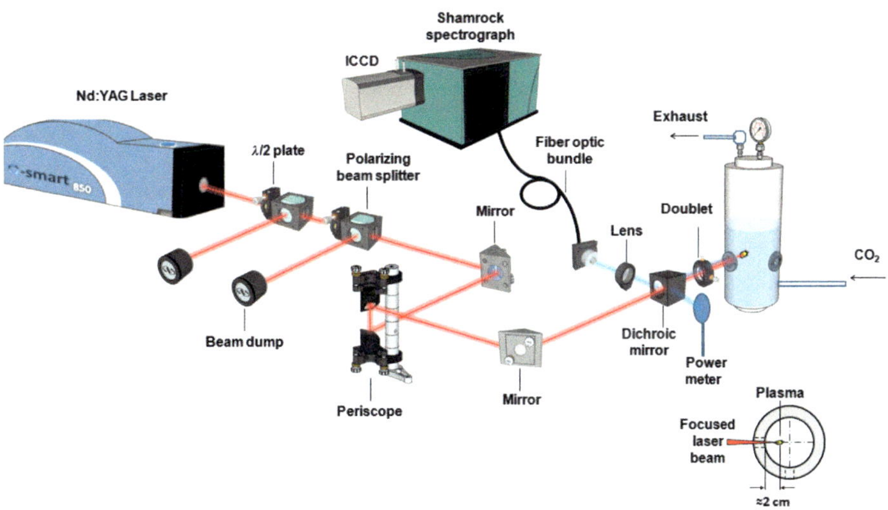

FIG. 1 Experimental setup for laser-induced breakdown spectroscopy.

gate width, and sample-to-the-focusing lens distance. Interpulse delay adds up when two lasers are used for sample excitation in DP-LIBS. For generating accurate and precise results, all these parameters need to be optimized. LIBS study primarily depends on emission lines of analyte elements. The intensity profile of spectral lines provides hints as to whether there are any detrimental effects, like self-absorption, phase reversals, and unequal temperature in the center and edges of the plasma. For quantitative analysis, emission lines should be free from these effects. Moreover, for better sensitivity, emission intensity of analyte lines should be maximum with minimum background effects. Considering favorable signal-to-noise ratio (SNR) values is one of the more widely followed practices for optimization of experimental parameters.

Generally, signal intensity is higher in the beginning and decreases as time delay increases. Since the background emission is higher at early times, the SNR consideration would be better for the determination of optimum time delay. Goueguel et al. [38] optimized gate time delay and laser energy by using SNR as shown in Fig. 2 to study the matrix effect of sodium compounds on metal ions in aqueous solutions. Optimum values of experimental parameters are not always those producing the largest SNR; some adverse effects may be associated with them. For instance, higher laser energy may produce larger SNR, but it sometimes causes plasma shielding and moving the breakdown along the laser beam pathway. Therefore moderate energy is preferred rather than the energy values producing the largest SNR [39].

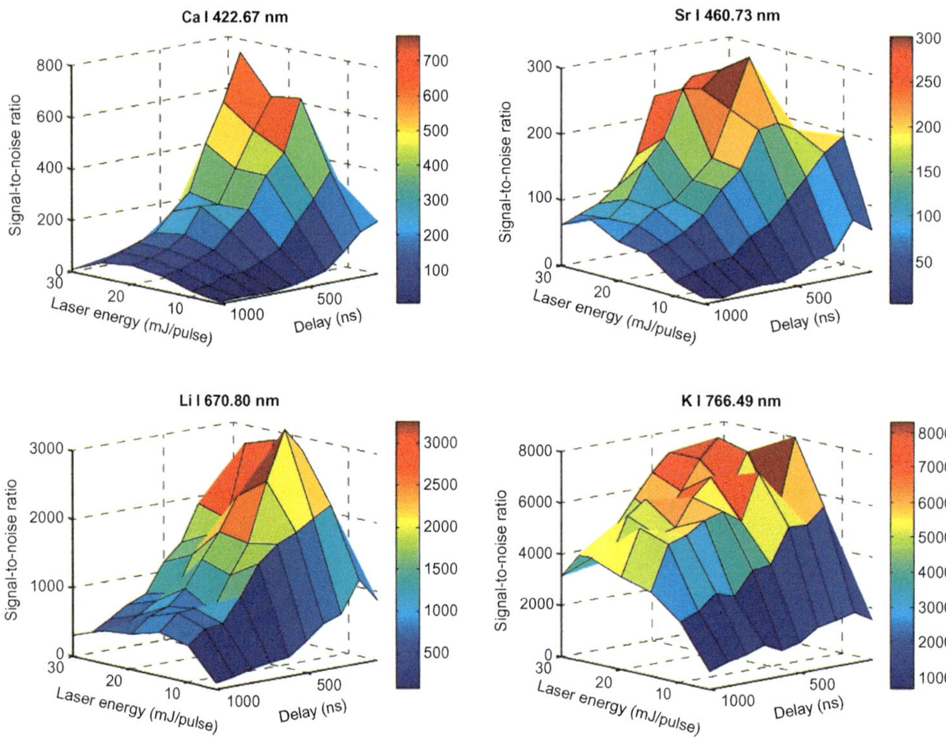

FIG. 2 3D display of the signal-to-noise ratio for the Ca I 422.67-nm, Sr I 460.73-nm, Li I 670.80-nm, and K I 766.49-nm emission lines versus laser energy versus gate delay [38].

In DP-LIBS, an appropriate interpulse delay is required to maximize sample ablation and plasma production. With larger plasmas, both plasma temperature and electron density will be increased producing stronger emission signals in the LIBS spectra. Optimum interpulse delay for underwater LIBS measurements is generally analyte and pressure dependent. Michel et al. [40] reported the importance of shorter interpulse delay at high pressures.

4 LIBS signal enhancement techniques

Laser-induced plasma formation and its emission mechanism are different in an aqueous media than in gas. This is due primarily to the difference in compressibility of liquids compared to gases. The formation and evolution of plasma and emission spectra from a plasma created in liquids generally suffer from matrix, pressure, plasma shielding, line broadening due to increased collisions, and Stark-broadening effects [38, 39, 41–46]. Moreover, the plasma loses its energy more rapidly to the surrounding liquid, resulting in reduction of persistence time and lower temperature. All these effects in the liquid environment contribute to weakening of the emission signals and compromising the sensitivity, accuracy, reproducibility, and stability of analysis. Although these effects are inevitable while using LIBS in real field applications, it is important to explore ways of enhancing the emission signals. To enhance the emission signals, several enhancement techniques have been reported by various research groups. Some of the common enhancement techniques are briefly described in the following sections.

4.1 Substrate technique

In this pretreatment technique, liquid samples are transformed into solid samples by using different solid substrates. Chen et al. [47] used a wood slice as the solid substrate in their experiment for fast and sensitive trace heavy metal analysis and claimed two- to threefold improvement in LODs for Cr, Mn, Cu, Cd, and Pb. Likewise, carbon planchet, paper, $Ca(OH)_2$, and ion exchange polymer membrane have also been used as solid substrates [48–53]. When calcium oxide (CaO) is added to aqueous solution, calcium hydroxide ($Ca(OH)_2$) is formed and the resulting precipitate in the form of solid can be used for LIBS analysis. The ion exchange polymer containing an iminodiacetic acid functional group has also been used for preconcentrating liquid samples for LIBS analysis. As carboxylate groups are negatively charged, the membrane exchanges cations at a $pH > 5$ [49].

In their recent study, He et al. [54] reported LOD in low ppb (18.4 ppb) for Li detection in a liquid sample by using filter paper as the adsorption substrate. To analyze Mn in aqueous solution and seawater, the use of a paper substrate as an absorber is demonstrated in [55]. Alamelu et al. and Sarkar et al. were able to detect low levels of some lanthanides and actinides in aqueous solutions by LIBS using the substrate techniques [56, 57].

4.2 Dual-pulse technique

Another useful and widely used technique for enhancing emission signals is DP-LIBS. In DP-LIBS, two laser pulses either from two different sources or from the same source in various geometrical configurations are employed to induce sample ablation and subsequent plasma expansion and emission. Various research groups have used different geometrical configurations, like collinear, orthogonal, prespark, orthogonal preheating, and dual-pulse crossed-beam modes for signal enhancement. In double-pulse mode, the first pulse induces the plasma, which starts cooling down and decaying. Depending on the interpulse delay, the second pulse will reheat the plasma produced by the first laser pulse or a certain portion will be absorbed by the plasma and the rest will be used for further ablation of the sample. The second pulse could also ablate the sample again to create a plasma. Therefore interpulse delay has a significant role in signal enhancement in the double-pulse regime.

Cremers et al. [58] were the first to demonstrate bulk liquid analysis by using multiple laser sparks. They studied the emission lines of Be, Mg, Ca, B, Al, and five alkali metals (Li, Na, K, Rb, Cs) without taking pressure effects into account. Nakamura et al. [59] reported detection of a 20-ppb concentration of Fe in FeO(OH) colloidal suspensions by using two laser pulses at optimum interpulse delay and gate delay time. Laser pulses were focused perpendicularly onto the laminar jet of aqueous Cr solution to study the temporal enhancement of Cr emission by Rai et al. [37]. These authors observed that the material ablation in DP-LIBS was $\geq$3–5 times that of SP-LIBS. Their experimental results were found to qualitatively match with theoretical calculation of the enhancement. By combining DP- and crossed-beam LIBS, the detection limit of Na was brought down to 0.1 ppb by Kuwako et al. [60]. DP-LIBS has also been used for the study of submerged materials in water and the analytical performances were evaluated for sensitivity enhancement [35, 61, 62].

4.3 Hyphenated techniques

In hyphenated techniques, two different analytical techniques are combined or coupled to form a more powerful integrated system for material analysis. Hyphenated techniques may include separation-separation, separation-identification, and identification-identification techniques. Nakane et al. [20] reported the detection of trace metal atoms in water by combining LIBS and laser-excited atomic fluorescence spectroscopy and claimed LODs as low as10 ppb for Fe in water. Loudyi et al. [21] detected traces of metallic impurities in acid solutions using LIBS combined with LIF; the LODs obtained were 39 and 65 ppb for Pb and Fe, respectively. The authors claimed that the use of a flow cell improved the reproducibility of analysis. Similarly, other groups have also combined LIF with LIBS for improvement in sensitivity and analytical performance [22, 23].

5 Underwater LIBS measurements

Underwater LIBS has been demonstrated as a powerful tool for the analysis of both dissolved materials and submerged materials. This is becoming a very useful technique for a number of applications, including environmental and industrial waste monitoring, and geological and marine research activities [63–65]. Although problems related to splashing of a sample by a liquid surface are eliminated in underwater LIBS, the liquid pressure comes into play when the plasma is created at depths below the surface. This becomes more prominent when LIBS analysis is performed in oceanic and hydrothermal vent fluids [33, 39]. In the complex natural aqueous samples, matrix effects also affect the LIBS analysis. It is therefore important to analyze and quantify matrix and pressure effects on the LIBS signal to assist in designing appropriate instrumentation for underwater LIBS measurements.

5.1 Matrix effects

Matrix effects are simply an influence of constituents other than analyte of interest on the analyte signal. In LIBS, major matrix effects can be explained as spectral, chemical, and physical effects [66]. When weak emission lines of an element of interest and strong emission lines of matrix elements overlap, spectral interference takes place. If emission characteristics of analyte are altered due to the presence of matrix elements, it is known as the chemical matrix effect. Lastly, because of differences in physical properties of elements present in the sample the plasma behavior can be affected resulting in a physical matrix effect. The matrix effect could bring increased false-positive or false-negative LIBS response to compromise analytical precision. For quantitative measurements in complex matrix samples, multivariate analysis is generally preferred to achieve better accuracy. In oceanic environments, deep saline waters contain Na as the most abundant element constituting approximately 70%–90% of the total cation mass [67, 68]. Other major constituents of saline brines include Ca, Mg, and K; however, brine composition varies by location. Therefore understanding the matrix effect of NaCl on LIBS measurements becomes crucial. In general, matrix effects cause suppression of analyte emission signals; however, in the case of an NaCl matrix, the outcome is different. While atomic emission signals are enhanced the suppression of ionic spectral lines was observed in the presence of a sodium chloride (NaCl) matrix. This effect on spectral lines may be because of the higher plasma temperature and electron number density caused by NaCl. Significant increase in the intensity of atomic emission (Ca I 422.67 nm) and no notable change in ionic emission (Ca II 393.366 nm and Ca II 396.847 nm) were observed by Michel et al. [32] with NaCl addition. Likewise, Thornton et al. [69] also observed more intense Ca signals in an Na matrix when they compared the spectra recorded from solutions with and without NaCl. While decrease in intensity ratio of Ca(II)/Ca (I) was reported by Cremers et al. [58], Michel et al. [32] did not observe any change in Mn I 403 and K I (766.491-nm and 769.897-nm) intensities by NaCl. Goueguel et al. [38, 41] carried out a detailed study of the effect of NaCl concentration and other Na compounds on underwater LIBS measurements of common metal ions. For determining the effect of NaCl on Ca and K the authors used three concentrations (0.5 M, 2 M, and 3 M) of NaCl. Although NaCl enhanced both Ca and K signals the magnitude of enhancement varied. Intensity of a Ca I 422.67-nm line increased two-, four-, and eightfold in the presence of 0.5 M, 2 M, and 3 M NaCl, respectively (Fig. 3A). In the case of a K I 766.49-nm line the magnitude of enhancement was fivefold with 0.5 M NaCl and 10-fold with 3 M NaCl added to the solution (Fig. 3B). Signal enhancement by NaCl in this study was attributed to the increase in plasma temperature and electron density.

In an another study, Gouguel et al. [38] investigated the effect of Na compounds (NaCl, Na_2CO_3, and Na_2SO_4) on the LIBS measurements of Li. The analyte emission intensity was found to increase in the presence of NaCl and Na_2SO_4 and decreased with addition of Na_2CO_3 as shown in Fig. 4. This behavior appears to be related to a change in Bremsstrahlung process efficiency with increasing amount of impurities and difference in the thermal properties (boiling point and thermal conductivity) of the impurities.

Gouguel et al. [38] also generated calibration curves of Li and K in the presence of three different concentrations (0.1, 1, and 10 wt%) of NaCl, Na_2SO_4, and Na_2CO_3 (Fig. 5). The authors pointed out that it is feasible to perform quantitative analysis by LIBS in complex matrix samples such as groundwater.

5.2 Pressure effects

The laser-induced plasma formation process is basically the same in gaseous and liquid media; however, there are some differences in plasma expansion and emission processes. Plasmas formed in liquid are more confined than those in air and are relatively smaller in size. As surrounding pressure increases, both plasma evolution and emission process are affected significantly [70]. It should be noted that the extent of pressure effect on emission line intensity and width depends on several factors, including emission lines, elemental concentration, sample matrix, and experimental conditions (temperature, laser energy, gate delay). Effect of pressure

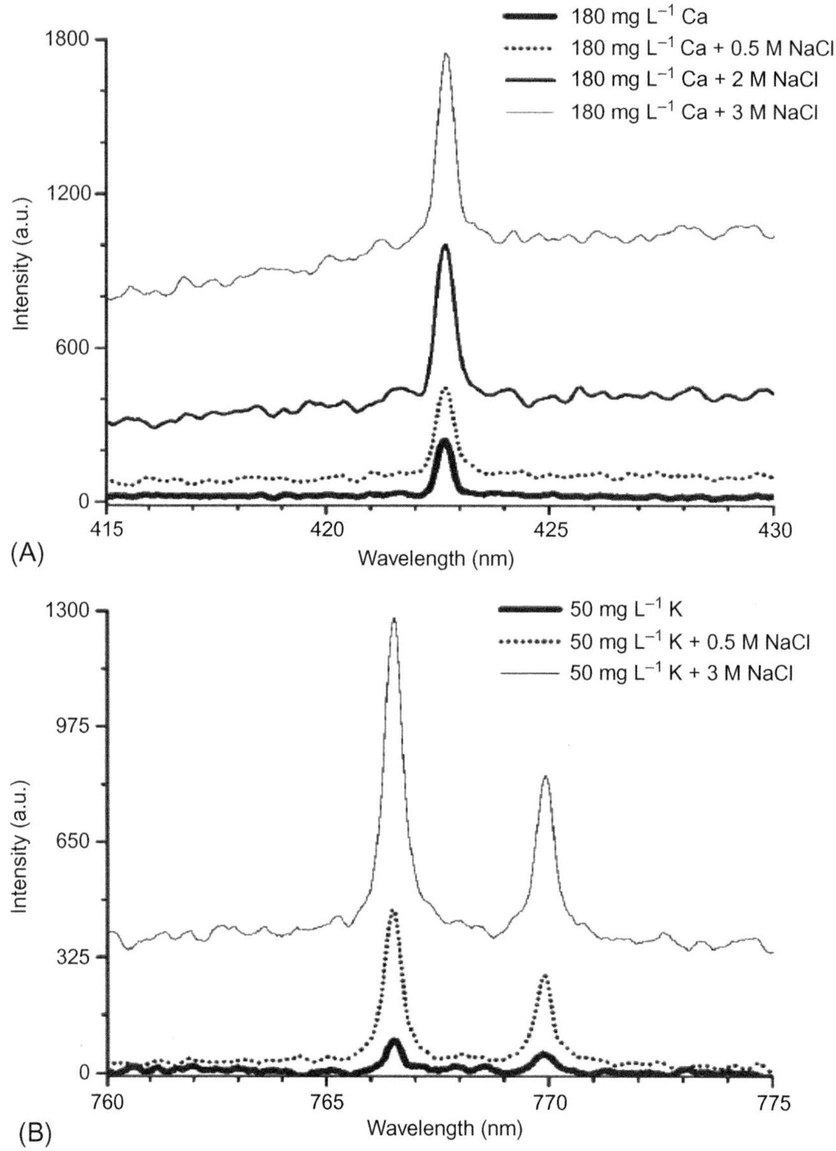

FIG. 3 Effect of NaCl concentration on emission spectra recorded using samples containing 180 mg L^{-1} Ca and 50 mg L^{-1} - K. (A) Ca I 422.67 nm, (B) K I doublet (766.49 and 769.90 nm) [41].

in underwater LIBS measurements was first taken into account by Lawrence-Snyder et al. [46]. The authors reported larger pressure effects on the emission line width and minor effects on the intensity of the Li I 670-nm line, mostly during late-stage plasma evolution. In 2014, Hou et al. [71] observed the pressure effects on the peak width, intensity, and integrated intensity of Ca, K, and Mg in seawater. The intensity of three atomic lines, Ca I 422 nm, K I 766 nm, and Mg I 518 nm, increased with increasing pressure until 15 MPa and then

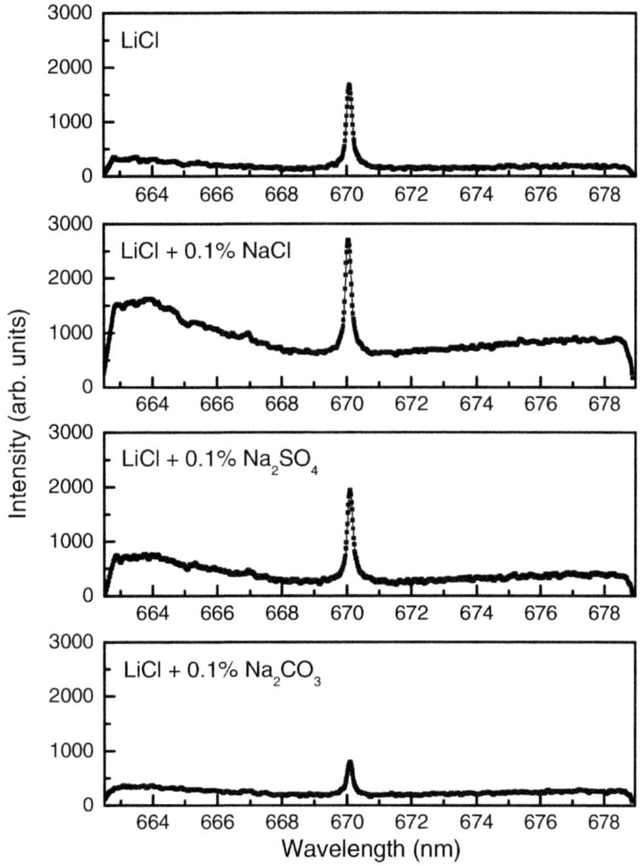

FIG. 4 Li emission spectra acquired using (*from top to bottom*) LiCl, LiCl-NaCl, LiCl-Na_2SO_4, and LiCl-Na_2CO_3 aqueous samples [38].

decreased. It was, however, noticed that the atomic and ionic lines of Ca behaved differently and followed a different trend in variation of their intensity as the pressure was increased. While the intensity of the atomic line (Ca I 422 nm) increased until 15 MPa and then decreased, the intensity of the ionic line (Ca II 393 nm) increased steadily with increasing pressure up to 40 MPa except a marginal decrease observed in the pressure range of 15–20 MPa. In contrast, the integrated intensity and the line width of both the Ca lines increased gradually with the increasing pressure. The integrated intensity of the Mg I 518-nm line also gradually increased with increasing pressure; however, in the case of the K I 766-nm line, the intensity increased sharply up to 15 MPa and then decreased steadily.

In separate studies, Thornton et al. [9, 69, 72] reported no significant effect of pressure on Zn and Cu spectra recorded from metal plates immersed in water. However, decrease in emission intensity of Zn I lines and broadening of Cu lines in DP-LIBS with rising pressure are reported in [73, 74], respectively. LIBS has also been demonstrated for the study of sediments in aqueous medium. Lazic et al. [75] used DP excitation from a single laser source to study natural and certified sediments by underwater LIBS.

FIG. 5 Comparison of the effect of NaCl, Na_2CO_3, and Na_2SO_4 on the calibration curves of Li and K in aqueous solutions [38].

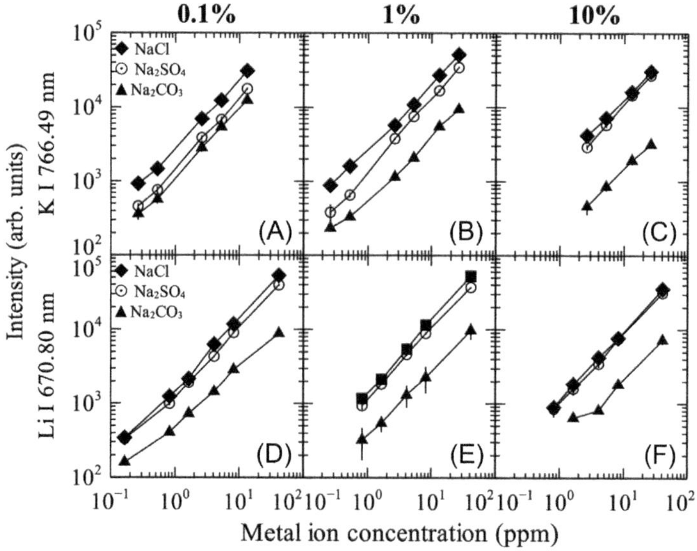

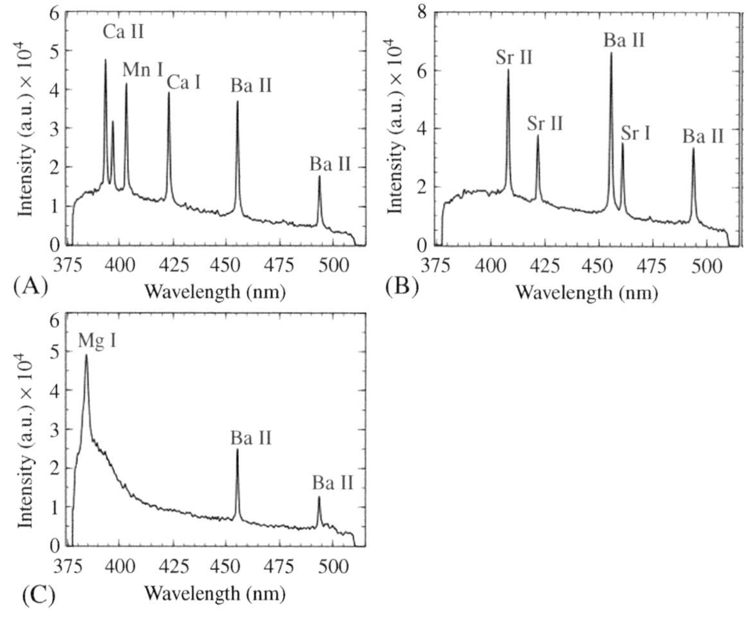

FIG. 6 Underwater laser-induced breakdown spectroscopy spectra of Ca, Mn, Ba, Sr, and Mg obtained in CO_2–water solutions. The spectra were acquired at a CO_2 pressure of 400 bar using (A) $CaCl_2$–$MnCl_2$–$BaCl_2$, (B) $SrCl_2$–$BaCl_2$, and (C) $MgCl_2$–$BaCl_2$ aqueous samples [27].

5.3 LIBS study in a compressed gas-water mixture

Pressure conditions in oceanic and downhole environments generally change due to increasing depth and hydrothermal activities. It is therefore important to evaluate the effect of increasing pressure on the LIBS measurement in compressed gas-water mixtures. Goueguel et al. [76] performed underwater LIBS measurements to detect Ca under rising CO_2 and N_2 pressures. Their study demonstrated that Ca emission line (Ca I 422.67 nm) intensity decreased with increasing CO_2 and N_2 pressures. However, the gas-dependent differences were evident as the Ca line intensity in CO_2 was found to be higher at 120 bar but lower at 10 bar than that in N_2. LIBS measurements of common metal ions (Ba, Ca, Mg, Mn, Sr) in a compressed CO_2-water mixture are also reported by Goueguel et al. [27]. In these experiments, the Ba emission line was used as an internal standard; LIBS spectra in the wavelength range of 375–510 nm are shown in Fig. 6.

Underwater LIBS study in a compressed CO_2–water mixture could be useful in geological carbon storage (GCS). In GCS, CO_2 is captured and injected deep underground for permanent storage. In case of leakage, upward migration of CO_2 could react with carbonate minerals of storage formations to release specific metals to shallow groundwater aquifers. These metals could be useful to provide early indication of CO_2 leakage. Goueguel et al. [31] demonstrated underwater LIBS for studying the dissolution of calcium carbonate under elevated CO_2 pressures up to 350 bar. In this study a $CaCO_3$ pellet was inserted into an aqueous $BaCl_2$ solution in a pressure vessel, and Ba was added to use it as an internal standard. CO_2 was injected into the liquid to raise the pressure to 50, 150, 250, and 350 bar and LIBS spectra were recorded at each pressure setting (Fig. 7). Ca emission lines

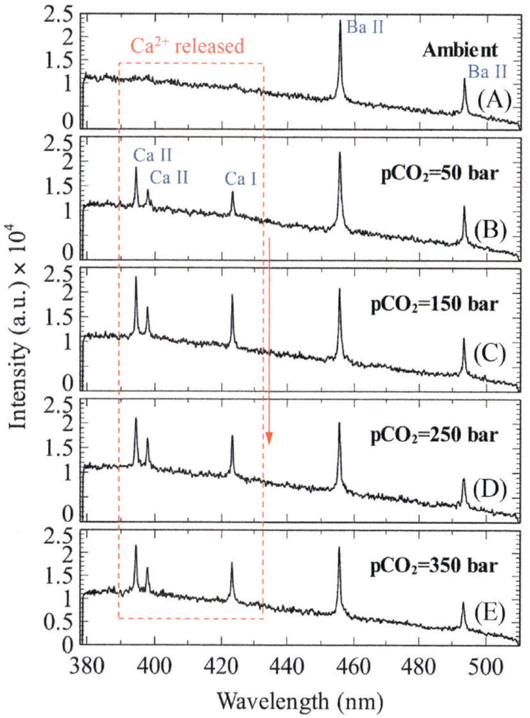

FIG. 7 Laser-induced breakdown spectroscopy spectra showing leaching of Ca^{2+} cation from $CaCO_3$ by CO_2-enriched water [31].

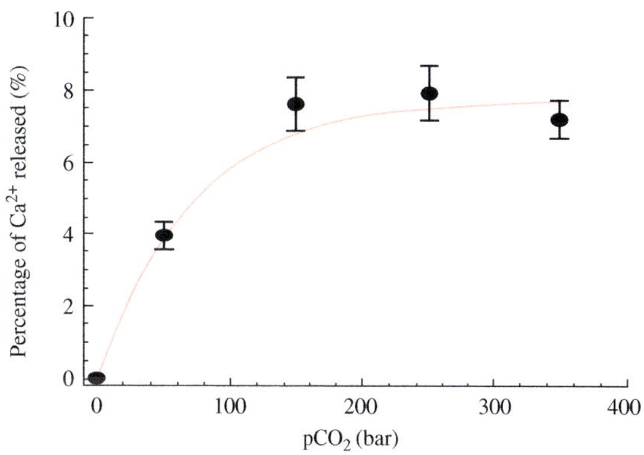

FIG. 8 Ca ion released to the water as a function of CO_2 pressure (pCO_2) [31].

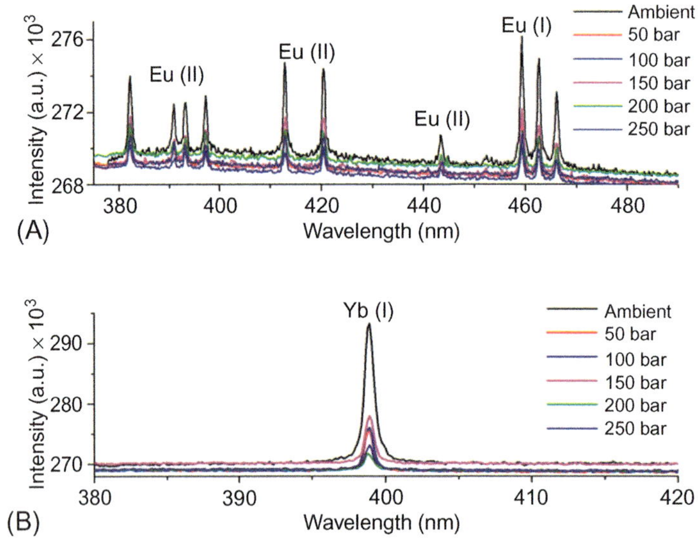

FIG. 9 Emission lines of Eu and Yb identified in the spectra obtained from (A) 4000-ppm Eu solution, (B) 4000-ppm Yb solution.

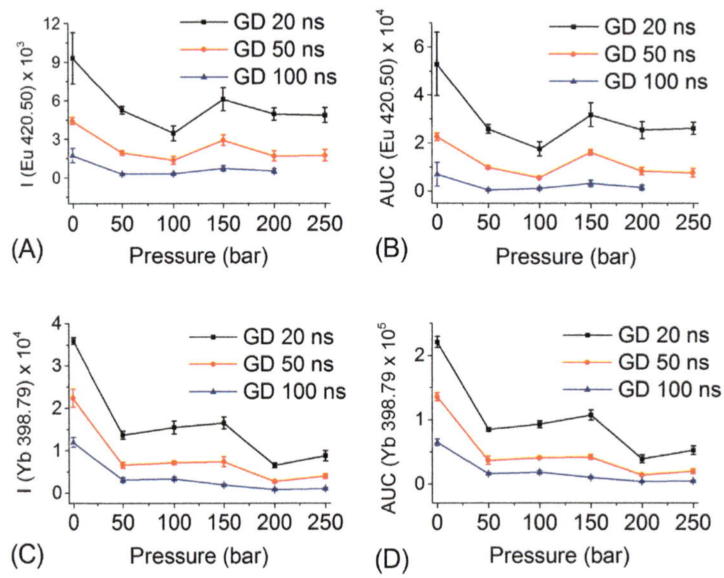

FIG. 10 Pressure effect on Eu and Yb emission. (A, C) Effect on intensity of emission lines and (B, D) effect on integrated intensity (AUC).

(which were not present at ambient conditions) were detected at higher pressures confirming the pressure-induced leaching of Ca ion from the carbonate by CO_2-enriched water. Increasing intensity of Ca emission lines with increase in pressure revealed the proportionate dissolution of $CaCO_3$ with applied CO_2 pressure. The amount of dissolved Ca at each pressure setting was quantified by generating calibration curves at the corresponding CO_2 pressure (Fig. 8). It is clear from Fig. 8 that the released amount of Ca increased up to 150 bar and then remained almost constant up to 350 bar.

Bhatt et al. [28] analyzed two rare earth elements (REEs), Eu and Yb, by underwater LIBS in a compressed gas–water mixture. LIBS spectra were recorded at six different pressure settings (ambient, 50, 100, 150, 200, 250 bar) from solutions of Eu and Yb and the detected emission lines are shown in Fig. 9. This study demonstrated the capability of LIBS to identify REE liquid sources present in high-pressure environments.

Effect of pressure and gate delays on the intensity and integrated intensity (area under curve) of Eu II 420.50-nm and Yb I 398.79-nm lines shown in Fig. 10. Intensity and integrated intensity of both elements declined with increase in gate delay. On the other hand, a steep decline in intensity and integrated intensity was observed in the beginning of CO_2 injection until the pressure reached 50 bar. No significant changes in both parameters were seen after 50 bar and the influence of pressure on the emission of Eu and Yb remained negligible up to 250 bar. The suppression of emission intensity in this study could be attributed to the confinement of laser-induced plasma by pressure in the beginning of CO_2 injection. Michel et al. [32] and Thornton et al. [9] also reported the weak effect of pressure on emission of Ca, Na, and Zn.

6 Conclusions

This chapter partially reviewed the applications of LIBS for the analysis of liquid samples. The technique could be used for in situ analysis and quantification of elements in both ambient and high-pressure conditions. Major challenges in underwater LIBS measurements and possible solutions to generate reliable data were discussed. Improvement in analytical performance could be achieved to some extent by using an appropriate experimental setup, proper gating parameters, excitation energy, and averaging the emission line intensity for several laser shots. LIBS signals in liquid media are generally weaker than in solids, therefore the signal enhancement techniques were briefly described. Lastly, matrix and pressure effects play a major role in underwater LIBS analysis that could bring challenges in quantitative analysis. Based on the recently published research, these effects were explained to improve the accuracy of analysis in liquids.

References

1. M. Yao, et al., Detection of chromium in wastewater from refuse incineration power plant near Poyang Lake by laser induced breakdown spectroscopy, Appl. Opt. 51 (10) (2012) 1552–1557.
2. L. St-Onge, et al., Rapid analysis of liquid formulations containing sodium chloride using laser-induced breakdown spectroscopy, J. Pharm. Biomed. Anal. 36 (2) (2004) 277–284.
3. N. Melikechi, et al., Laser-induced breakdown spectroscopy of whole blood and other liquid organic compounds, in: Optical Diagnostics and Sensing VIII, International Society for Optics and Photonics, 2008.
4. L. Zheng, et al., On the performance of laser-induced breakdown spectroscopy for direct determination of trace metals in lubricating oils, Spectrochim. Acta B At. Spectrosc. 99 (2014) 1–8.
5. C.L. Goueguel, D.L. McIntyre, J.C. Jain, Influence of CO 2 pressure on the emission spectra and plasma parameters in underwater laser-induced breakdown spectroscopy, Opt. Lett. 41 (23) (2016) 5458–5461.
6. T. Sakka, et al., Use of a long-duration ns pulse for efficient emission of spectral lines from the laser ablation plume in water, Appl. Phys. Lett. 88 (6) (2006) 061120.
7. T. Sakka, et al., Emission spectroscopy of laser ablation plume: composition analysis of a target in water, Appl. Surf. Sci. 255 (24) (2009) 9576–9580.
8. T. Sakka, et al., Effects of pulse width on nascent laser-induced bubbles for underwater laser-induced breakdown spectroscopy, Spectrochim. Acta B At. Spectrosc. 97 (2014) 94–98.

9. B. Thornton, et al., Spectroscopic measurements of solids immersed in water at high pressure using a long-duration nanosecond laser pulse, Appl. Phys. Express 6 (8) (2013) 082401.
10. B. Thornton, et al., Long-duration nano-second single pulse lasers for observation of spectra from bulk liquids at high hydrostatic pressures, Spectrochim. Acta B At. Spectrosc. 97 (2014) 7–12.
11. Y. Wang, et al., Sensitive analysis of copper in water by LIBS–LIF assisted by simple sample pretreatment, J. Appl. Spectrosc. 86 (2) (2019) 353–359.
12. F. Al-Adel, et al., Optimization of a laser induced breakdown spectroscopy method for the analysis of liquid samples, J. Appl. Spectrosc. 80 (5) (2013) 767–770.
13. D. Zhu, et al., Laser-induced breakdown spectroscopy for determination of trace metals in aqueous solution using bamboo charcoal as a solid-phase extraction adsorbent, Anal. Methods 4 (3) (2012) 819–823.
14. M. Bukhari, et al., Development of a method for the determination of chromium and cadmium in tannery wastewater using laser-induced breakdown spectroscopy, J. Anal. Methods Chem. 2012 (2012).
15. X. Yang, et al., Sensitive determinations of Cu, Pb, Cd, and Cr elements in aqueous solutions using chemical replacement combined with surface-enhanced laser-induced breakdown spectroscopy, Opt. Express 24 (12) (2016) 13410–13417.
16. K. Rifai, et al., Quantitative analysis of metallic traces in water-based liquids by UV-IR double-pulse laser-induced breakdown spectroscopy, J. Anal. At. Spectrom. 27 (2) (2012) 276–283.
17. W. Pearman, J. Scaffidi, S.M. Angel, Dual-pulse laser-induced breakdown spectroscopy in bulk aqueous solution with an orthogonal beam geometry, Appl. Opt. 42 (30) (2003) 6085–6093.
18. V.N. Rai, F.-Y. Yueh, J.P. Singh, Study of laser-induced breakdown emission from liquid under double-pulse excitation, Appl. Opt. 42 (12) (2003) 2094–2101.
19. A. Kumar, F.Y. Yueh, J.P. Singh, Double-pulse laser-induced breakdown spectroscopy with liquid jets of different thicknesses, Appl. Opt. 42 (30) (2003) 6047–6051.
20. M. Nakane, et al., Analysis of trace metal elements in water using laser-induced fluorescence of laser-breakdown plasma, in: Laser Plasma Generation and Diagnostics, International Society for Optics and Photonics, 2000.
21. H. Loudyi, et al., Improving laser-induced breakdown spectroscopy (LIBS) performance for iron and lead determination in aqueous solutions with laser-induced fluorescence (LIF), J. Anal. At. Spectrom. 24 (10) (2009) 1421–1428.
22. Y. Godwal, et al., Determination of lead in water using laser ablation–laser induced fluorescence, Spectrochim. Acta B At. Spectrosc. 62 (12) (2007) 1443–1447.
23. S.L. Lui, et al., Detection of lead in water using laser-induced breakdown spectroscopy and laser-induced fluorescence, Anal. Chem. 80 (6) (2008) 1995–2000.
24. C. Ng, W. Ho, N. Cheung, Spectrochemical analysis of liquids using laser-induced plasma emissions: effects of laser wavelength on plasma properties, Appl. Spectrosc. 51 (7) (1997) 976–983.
25. S. Ho, N. Cheung, Sensitive elemental analysis by ArF laser-induced fluorescence of laser ablation plumes: elucidation of the fluorescence mechanism, Appl. Phys. Lett. 87 (26) (2005) 264104.
26. X. Pu, N. Cheung, ArF laser induced plasma spectroscopy of lead ions in aqueous solutions: plume reheating with a second Nd: YAG laser pulse, Appl. Spectrosc. 57 (5) (2003) 588–590.
27. C.L. Goueguel, et al., Quantification of dissolved metals in high-pressure CO2-water solutions by underwater laser-induced breakdown spectroscopy, Opt. Laser Technol. 108 (2018) 53–58.
28. C.R. Bhatt, J.C. Jain, D.L. McIntyre, Investigating the CO2 pressure effect on underwater laser-induced plasma emission of Eu and Yb, Spectrochim. Acta B At. Spectrosc. 149 (2018) 42–47.
29. C.R. Bhatt, et al., Measurement of Eu and Yb in aqueous solutions by underwater laser induced breakdown spectroscopy, Spectrochim. Acta B At. Spectrosc. 137 (2017) 8–12.
30. C.R. Bhatt, et al., Mineral carbonate dissolution with increasing CO2 pressure measured by underwater laser induced breakdown spectroscopy and its application in carbon sequestration, Talanta (2019) 120170.
31. C.L. Goueguel, et al., In situ measurements of calcium carbonate dissolution under rising CO2 pressure using underwater laser-induced breakdown spectroscopy, J. Anal. At. Spectrom. 31 (7) (2016) 1374–1380.
32. A.P. Michel, et al., Laser-induced breakdown spectroscopy of bulk aqueous solutions at oceanic pressures: evaluation of key measurement parameters, Appl. Opt. 46 (13) (2007) 2507–2515.
33. V. Lazic, S. Jovićević, Laser induced breakdown spectroscopy inside liquids: processes and analytical aspects, Spectrochim. Acta B At. Spectrosc. 101 (2014) 288–311.
34. J.R. Wachter, D.A. Cremers, Determination of uranium in solution using laser-induced breakdown spectroscopy, Appl. Spectrosc. 41 (6) (1987) 1042–1048.
35. A. De Giacomo, et al., Double-pulse LIBS in bulk water and on submerged bronze samples, Appl. Surf. Sci. 247 (1–4) (2005) 157–162.
36. V. Babushok, et al., Double pulse laser ablation and plasma: laser induced breakdown spectroscopy signal enhancement, Spectrochim. Acta B At. Spectrosc. 61 (9) (2006) 999–1014.
37. V.N. Rai, F.Y. Yueh, J.P. Singh, Time-dependent single and double pulse laser-induced breakdown

spectroscopy of chromium in liquid, Appl. Opt. 47 (31) (2008) G21–G29.
38. C. Goueguel, et al., Matrix effect of sodium compounds on the determination of metal ions in aqueous solutions by underwater laser-induced breakdown spectroscopy, Appl. Opt. 54 (19) (2015) 6071–6079.
39. P.K. Kennedy, D.X. Hammer, B.A. Rockwell, Laser-induced breakdown in aqueous media, Prog. Quantum Electron. 21 (3) (1997) 155–248.
40. A.P. Michel, A.D. Chave, Double pulse laser-induced breakdown spectroscopy of bulk aqueous solutions at oceanic pressures: interrelationship of gate delay, pulse energies, interpulse delay, and pressure, Appl. Opt. 47 (31) (2008) G131–G143.
41. C. Goueguel, et al., Effect of sodium chloride concentration on elemental analysis of brines by laser-induced breakdown spectroscopy (LIBS), Appl. Spectrosc. 68 (2) (2014) 213–221.
42. J.-S. Huang, C.-B. Ke, K.-C. Lin, Matrix effect on emission/current correlated analysis in laser-induced breakdown spectroscopy of liquid droplets, Spectrochim. Acta B At. Spectrosc. 59 (3) (2004) 321–326.
43. T. Sakka, et al., Spectral profile of atomic emission lines and effects of pulse duration on laser ablation in liquid, Spectrochim. Acta B At. Spectrosc. 64 (10) (2009) 981–985.
44. J. Aguilera, C. Aragon, F. Penalba, Plasma shielding effect in laser ablation of metallic samples and its influence on LIBS analysis, Appl. Surf. Sci. 127 (1998) 309–314.
45. O. Samek, et al., Application of laser-induced breakdown spectroscopy to in situ analysis of liquid samples, Opt. Eng. 39 (8) (2000) 2248–2262.
46. M. Lawrence-Snyder, et al., Laser-induced breakdown spectroscopy of high-pressure bulk aqueous solutions, Appl. Spectrosc. 60 (7) (2006) 786–790.
47. Z. Chen, et al., Fast and sensitive trace metal analysis in aqueous solutions by laser-induced breakdown spectroscopy using wood slice substrates, Spectrochim. Acta B At. Spectrosc. 63 (1) (2008) 64–68.
48. R.L. Vander Wal, et al., Trace metal detection by laser-induced breakdown spectroscopy, Appl. Spectrosc. 53 (10) (1999) 1226–1236.
49. N.E. Schmidt, S.R. Goode, Analysis of aqueous solutions by laser-induced breakdown spectroscopy of ion exchange membranes, Appl. Spectrosc. 56 (3) (2002) 370–374.
50. C.R. Dockery, J.E. Pender, S.R. Goode, Speciation of chromium via laser-induced breakdown spectroscopy of ion exchange polymer membranes, Appl. Spectrosc. 59 (2) (2005) 252–257.
51. J. Cáceres, et al., Quantitative analysis of trace metal ions in ice using laser-induced breakdown spectroscopy, Spectrochim. Acta B At. Spectrosc. 56 (6) (2001) 831–838.
52. P. Yaroshchyk, et al., Quantitative determination of wear metals in engine oils using laser-induced breakdown spectroscopy: a comparison between liquid jets and static liquids, Spectrochim. Acta B At. Spectrosc. 60 (7–8) (2005) 986–992.
53. D.D. Pace, et al., Analysis of heavy metals in liquids using laser induced breakdown spectroscopy by liquid-to-solid matrix conversion, Spectrochim. Acta B At. Spectrosc. 61 (8) (2006) 929–933.
54. Y. He, et al., Lithium ion detection in liquid with low detection limit by laser-induced breakdown spectroscopy, Appl. Opt. 58 (2) (2019) 422–427.
55. J. Xiu, et al., Quantitative determination of manganese in aqueous solutions and seawater by laser-induced breakdown spectroscopy (LIBS) using paper substrates, Appl. Spectrosc. 68 (9) (2014) 1039–1045.
56. D. Alamelu, A. Sarkar, S. Aggarwal, Laser-induced breakdown spectroscopy for simultaneous determination of Sm, Eu and Gd in aqueous solution, Talanta 77 (1) (2008) 256–261.
57. A. Sarkar, D. Alamelu, S.K. Aggarwal, Determination of thorium and uranium in solution by laser-induced breakdown spectrometry, Appl. Opt. 47 (31) (2008) G58–G64.
58. D.A. Cremers, L.J. Radziemski, T.R. Loree, Spectrochemical analysis of liquids using the laser spark, Appl. Spectrosc. 38 (5) (1984) 721–729.
59. S. Nakamura, et al., Determination of an iron suspension in water by laser-induced breakdown spectroscopy with two sequential laser pulses, Anal. Chem. 68 (17) (1996) 2981–2986.
60. A. Kuwako, Y. Uchida, K. Maeda, Supersensitive detection of sodium in water with use of dual-pulse laser-induced breakdown spectroscopy, Appl. Opt. 42 (30) (2003) 6052–6056.
61. S. Koch, et al., Detection of manganese in solution in cavitation bubbles using laser induced breakdown spectroscopy, Spectrochim. Acta B At. Spectrosc. 60 (7–8) (2005) 1230–1235.
62. A. De Giacomo, et al., Elemental chemical analysis of submerged targets by double-pulse laser-induced breakdown spectroscopy, Anal. Bioanal. Chem. 385 (2) (2006) 303–311.
63. M. Simileanu, R. Radvan, N. Puscas, Underwater LIBS investigations setup for metals' identification, Univ. Pol. Bucharest Sci. Bull.-Ser. A-Appl. Math. Phys. 72 (4) (2010) 209–216.
64. C. Fabre, et al., Palaeofluid chemistry of a single fluid event: a bulk and in-situ multi-technique analysis (LIBS, Raman Spectroscopy) of an Alpine fluid (Mont-Blanc), Chem. Geol. 182 (2–4) (2002) 249–264.
65. M. Saeki, et al., Development of a fiber-coupled laser-induced breakdown spectroscopy instrument for analysis of underwater debris in a nuclear reactor core, J. Nucl. Sci. Technol. 51 (7–8) (2014) 930–938.
66. T. Takahashi, B. Thornton, Quantitative methods for compensation of matrix effects and self-absorption in laser induced breakdown spectroscopy signals of solids, Spectrochim. Acta B At. Spectrosc. 138 (2017) 31–42.

67. Y.K. Kharaka, J. Hanor, Deep fluids in the continents: I. Sedimentary basins, in: Treatise on Geochemistry, vol. 5, Elsevier, 2003, p. 605.
68. K.J. Breen, et al., Chemical and isotopic characteristics of brines from three oil- and gasproducing sandstones in eastern Ohio, with applications to the geochemical tracing of brine sources, in: Water-Resources Investigation Report, 1985, p. 84–4314.
69. B. Thornton, et al., Development and field testing of laser-induced breakdown spectroscopy for in situ multi-element analysis at sea, in: 2012 Oceans, IEEE, Piscataway, NJ, 2012.
70. A. De Giacomo, et al., Effects of the background environment on formation, evolution and emission spectra of laser-induced plasmas, Spectrochim. Acta B At. Spectrosc. 78 (2012) 1–19.
71. H. Hou, et al., Study of pressure effects on laser induced plasma in bulk seawater, J. Anal. At. Spectrom. 29 (1) (2014) 169–175.
72. B. Thornton, T. Ura, Effects of pressure on the optical emissions observed from solids immersed in water using a single pulse laser, Appl. Phys. Express 4 (2) (2011) 022702.
73. M. Lawrence-Snyder, et al., Sequential-pulse laser-induced breakdown spectroscopy of high-pressure bulk aqueous solutions, Appl. Spectrosc. 61 (2) (2007) 171–176.
74. T. Takahashi, B. Thornton, T. Ura, Investigation of influence of hydrostatic pressure on double-pulse laser-induced breakdown spectroscopy for detection of Cu and Zn in submerged solids, Appl. Phys. Express 6 (4) (2013) 042403.
75. V. Lazic, et al., Underwater sediment analyses by laser induced breakdown spectroscopy and calibration procedure for fluctuating plasma parameters, Spectrochim. Acta B At. Spectrosc. 62 (1) (2007) 30–39.
76. C. Goueguel, et al., Laser-induced breakdown spectroscopy (LIBS) of a high-pressure CO2–water mixture: application to carbon sequestration, Appl. Spectrosc. 68 (9) (2014) 997–1003.

CHAPTER

10

LIBS application to powder samples

Chet R. Bhatt[a], *Herve K. Sanghapi*[b], *Fang Y. Yueh*[c], *Jagdish P. Singh*[c]

[a]National Energy Technology Laboratory, U.S. Department of Energy, Morgantown, WV, United States
[b]Alcorn State University, Lorman, MS, United States
[c]Institute for Clean Energy Technology, Mississippi State University, Starkville, MS, United States

1 Introduction

Laser-induced breakdown spectroscopy (LIBS) is a robust analytical technique with great prospects and with the capability of analyzing very light to heavy elements of the periodic table. Several books have been published on the application of LIBS for qualitative and quantitative analysis [1–4]. LIBS consists of generating a plasma by focusing a laser pulse on the sample. The optical emission from the plasma is collected with a spectrometer; the resulting signal is then used for qualitative analysis (identification of elements) and for quantitative analysis (sample composition through calibration curves). The use of temporally and spatially resolved plasma provides great advantages over other standard analytical techniques. Without being exhaustive, among other advantages are its ability for in situ and standoff analysis due to the use of fiber optics; multielemental analysis at low and high atomic number; real-time and fast analysis; and little sample preparation. These features have made LIBS an interdisciplinary workhorse. It can be used in material science, forensic science, and biological science just to name a few. It is being used extensively on solids, liquids, and aerosols [4, 5]. With established quantitative and qualitative results, coupled to more powerful computers nowadays, industrial applications of LIBS have gained attention more than ever. Some of these applications are (but not limited to) metals and alloys processing, scrap sorting, and characterization of highly radioactive nuclear waste using glass batch [6]. In industrial sectors like chemical, pharmaceutical, and metallurgy, the raw material used is often a composite powder that necessitates prior elemental analysis to optimize the purity of end products and minimal waste. LIBS can provide online analysis of the raw material and multielemental and in situ analysis for industrial applications where the sample is often in a hard-to-reach or harsh environment and sample manipulation is not needed. In this chapter, attention will be given to powder samples with case studies of glass batch, rare earth elements (REEs), and pharmaceutical powders.

https://doi.org/10.1016/B978-0-12-818829-3.00010-1

2 Experimental

2.1 Powder samples

Despite the versatility of LIBS, its direct application to powder sample analysis has been a challenge due to samples blowing off and scattering during laser-matter interaction. Many approaches have been developed to overcome such undesirable effects. Among these is the operation of LIBS under low-ambient pressure. Spectrochemical analyses have been performed directly on the powder samples [7–9]. Suyanto et al. [7] developed a technique to bend and confine plasma in a sample holder and detected plasma emission on the bent shockwave plasma outside the sample holder for powder analysis. The use of a subtarget-supported micromesh sample holder in low-pressure ambient gas has been demonstrated for powder analysis [10]. These results showed a practical approach to a small quantity of sample and low background emission, but the use of low pressure is a limiting factor and cannot be used for in situ analysis. Researchers have used a thin film with an ultraviolet Nd:YAG laser at 355 nm but films suffered from inhomogeneity [11]. Double-pulse systems have also been reported for direct analysis of powder samples. Two lasers with different wavelengths are used to control the production of dust in a low-pressure environment [12].

For the LIBS study of powder samples, generally, pellets are prepared and used to obtain LIBS spectra; however, depending upon the powder sample nature and available amount of analyte, the double-sided taping method has also been found useful. In the double-sided taping method, a finely ground powder sample is uniformly distributed on a glass slide with double-sided tape [13, 14]. This method is especially useful when the sample is available in limited amounts and needs to be away from other contaminants. To prepare pellets, depending upon the sample, different binders are used; polyvinyl alcohol, starch, potassium bromide, silver, sodium metasilicate, boric acid, and aluminum are some common binders used by various research groups to prepare pellets for LIBS study [15–18]. To achieve analytical results comparable to those of solids, it is imperative to have pellets that have certain characteristics. Pellets should be homogeneous, which requires grinding of powder grains to fine powder. Also, the mechanical strength of the pellets is another factor to be considered to reduce fractionation caused by the interaction of laser beam and pellets during laser ablation. To optimize the homogeneity and mechanical strength of the pellets, the binder should be carefully chosen by taking into consideration its bond strength; this is the strength between particles that bind a sample and resist ablation [15].

In principle, the binders are expected not only to bind the powder sample particles together but also to have no interference in elemental composition of the major analyte. Therefore selection of a binder is also critical to prepare pellets for the LIBS analysis of powder samples. The role of various binders in LIBS for trace elemental analysis of powder samples is presented in [15] where the authors have reported potassium bromide as a better binder for Mg analysis.

Yual et al. [18] evaluated use of sodium metasilicate as a binder and reported reduction of the relative standard deviation (RSD) value by an internal calibration method to 12.1% from the original intensity average value of 30.7% without binders. The importance of optimizing the binder percentage is highlighted in [17] to reduce matrix effects with the case study of carbon in coal. For nine samples, the plasma temperature and its fluctuation (represented by RSD) of the four groups of pellet samples (KBr 0 wt%, 30 wt%, 60 wt%, and 90 wt%) are reported where the minimum RSD is obtained by mixing coal with 60 wt% KBr binder. In a study performed by Shi et al. [19], epoxy resin glue has been reported as a significantly better binder compared to traditional polyethylene because of higher signals, better R^2 values of calibration curves, and lower limits of detection for Fe, Ca, and Cr in soil samples. A sample-to-binder ratio of 1:9 was used in the analysis of brick clay powder with cellulose as the binder [20]. Impurities in the binder may increase the matrix effect and subsequently can compromise the sensitivity of LIBS. Gondal et al. [15] studied the matrix effect with different binders on a LIBS signal by correlating the effect of laser energy and crater depth with the ablated mass, which determines the intensity of emission lines.

In their study, the deepest craters, maximum ablated material, and highest LIBS signal intensity and signal-to-noise ratio (SNR) were recorded for KBr (Figs. 1 and 2). Hence, they concluded that the greatest LIBS sensitivity may be achieved by using KBr as a binder.

During laser pulse-material interaction, a shockwave and mechanical vibration are produced leading to the material crumbling away and powder splashing, which interferes with the stability of the plasma's appearance. A comparative study of the epoxy resin and traditional polyethylene pellets showed that epoxy resin glue as a binder of powder samples improved performance in LIBS technique as a quantitative analytical tool [19, 20].

Although pellets are common in the LIBS analysis of powder samples, the density of the powder compact is often not reported and its effect on the analytical figure of merit is unknown. Pandey et al. [21] studied the influence of the aluminum oxide powder compact density of these pellets on the calibration curve of nickel contaminant. They obtained an increase in the sensitivity of LIBS when the grain size increases for a given density and its RSD decreases with the density of the powder compacts without any grain size effect. They concluded that both the powder compact density and the ultimate particle size of the powders must be specified in the protocol for building the calibration curve. Regarding the particle size, Carvalho et al. [22] noted a 50% enhancement in emission signal intensity and attained better measurement precision when the particle size was smaller. Also, they were able to reduce matrix effects by analyzing pellets prepared from <75-μm sieved fractions.

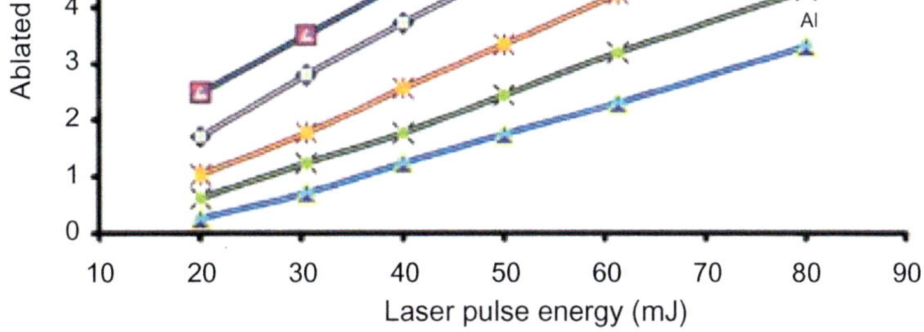

FIG. 1 Influence of incident laser pulse energy of ablated mass for different binders [15].

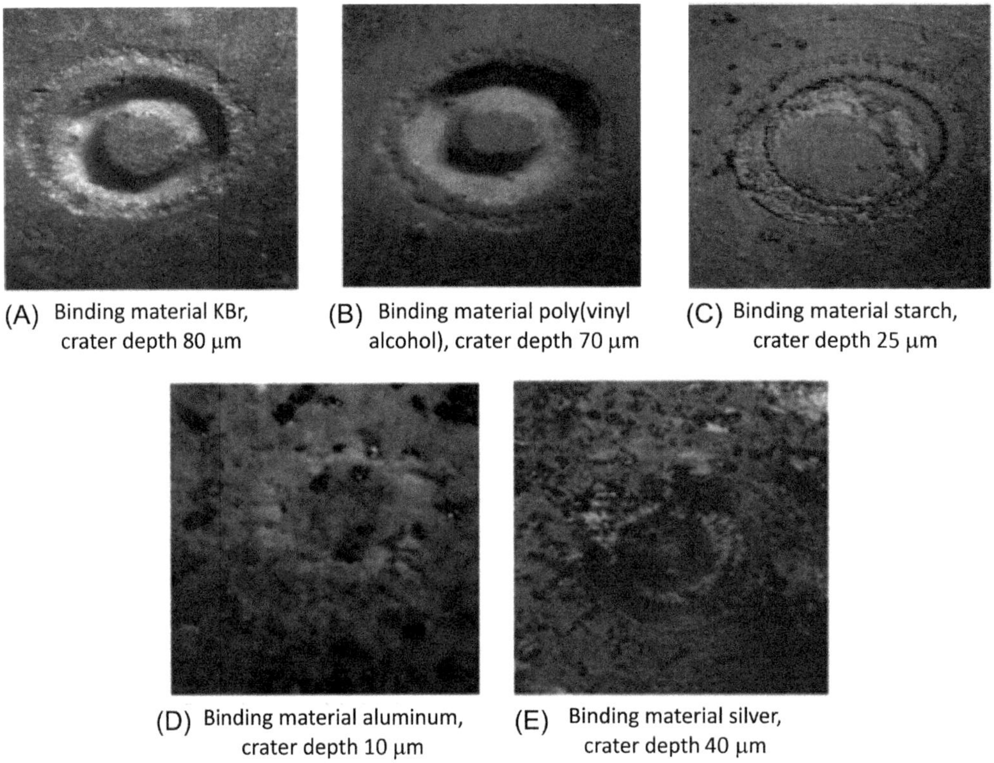

FIG. 2 Optical scanning microscopy images of the laser-ablated crater using different binding materials: (A) KBr, (B) poly(vinyl alcohol), (C) starch, (D) aluminum, and (E) silver [15].

2.2 Experimental setup

As in other samples, laser pulses of different wavelengths are used as excitation source to produce LIBS spectra of powder samples. A typical LIBS setup for powder samples consists of laser(s), optics, spectrometer, and computer as shown in Fig. 3. Laser pulses of known pulse width, pulse duration, and wavelength at the desired energy and frequency are focused onto the analyte sample surface orthogonally. Generally, the sample is placed on a platform rotating with uniform speed so that the laser pulse will hit at different spots and the effect of differences in craters produced will be minimized. To remove the aerosols produced by a laser-matter interaction, a sweeping gas flow can also be applied [23]. Plasma emission is collected by using certain collecting optics in different geometry modes supplied to the detection system through an optical fiber. Geometry modes like orthogonal, parallel, and oblique (at an angle) to the sample surface can be used to collect LIBS spectra. Both echelle and Czerny-Turner spectrometers coupled with gated intensified detectors can be used for detection. To enhance the spectral signals and for better limit of detection, a second laser pulse with a certain pulse time gap (double-pulse LIBS) could also be used. The second laser pulse can be focused onto the analyte sample at various geometrical configurations such as collinear, orthogonal, and oblique (at an angle).

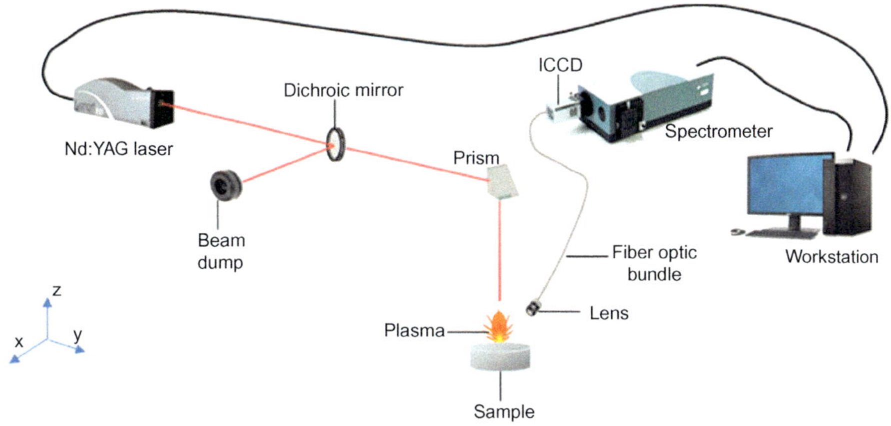

FIG. 3 Laser-induced breakdown spectroscopy experimental setup used for powder samples.

2.3 Experimental parameters

Major experimental parameters in LIBS study are gate delay, laser energy, gate width, and distance of the sample to the focusing lens. Gate delay is the time gap between onset of laser pulse and data acquisition, while gate width is the time duration for which data are acquired. For accurate and precise analysis, these parameters are optimized. Optimization may include (but is not limited to) maximizing signal intensity and minimizing background and noise effects. Generally, researchers look for the most favorable signal-to-background ratio (SBR) or SNR for optimization of these experimental parameters (Fig. 4). Selection of emission lines of analyte elements is also crucial in LIBS. As different lines have different excitation and ionization levels, optimum experimental parameters vary according to the selection of analyte emission lines. Analytical results of LIBS rely on plasma emission line intensity; therefore emission lines should be free from self-absorption and spectral interference.

3 Analytical methods

LIBS analytical figures of merit rely on the quality of the emission lines in the spectra collected from the test sample. Although LIBS is capable of multielemental analysis, spectral interference is common and affects the peak identification and calibration curves; however, plasmas can be temporally and spatially resolved to mitigate this. Data processing techniques have also been developed to minimize spectral interference and background effect to improve the quantitative analysis of LIBS [25–27]. Modified simple linear regression (SLR) with line intensity normalization and internal standardization and multivariate analysis are used to improve the quantitative results of LIBS. Partial least squares regression (PLSR), principal component analysis (PCA), and multiple linear regression are common types multivariate analysis [28–33].

Internal standardization is used to improve the precision of the measurements and minimize the effect of laser intensity shot-to-shot variation. An internal standard is generally a known substance with equal known concentration used in all test samples. Signals of both analyte and internal standard are assumed to be affected equally by any possible affecting factor. Therefore the ratio of analyte signal and internal standard signal is generally used for quantitative analysis, which also improves the linearity of the calibration curves. For best results, it is required that the energy of the upper level of

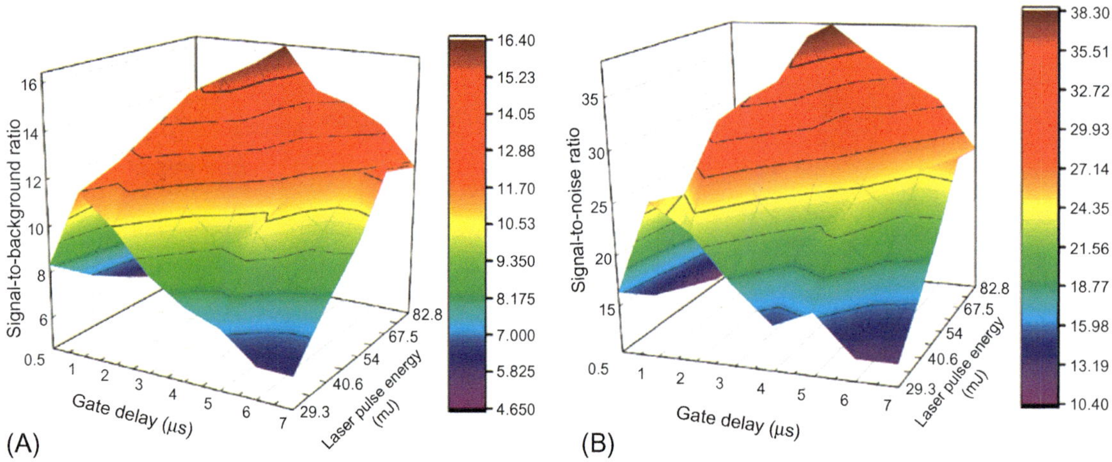

FIG. 4 3D plots showing variation of (A) signal-to-background ratio and (B) signal-to-noise ratio of an Sr II 430.54-nm line with gate delay and laser pulse energy [24].

the internal standard element be close to that of the analyte of interest [1].

PLSR provides a calibration model for the relationship between a set of predictor variables X (spectra) and a set of response variables Y (concentrations). If the spectral data contain information about the properties (concentrations) of interest, a reliable calibration model can be constructed. The samples with known elemental concentrations are used to create a model relating Y to X that is used to predict the concentrations of unknown specimens. More information on PLSR can be found in *Basic Chemometric Techniques in Spectroscopy* [34]. PCA, which is a more qualitative approach, is used principally for classification purposes [28, 35, 36]. PCA seeks the hidden structure within a large set of complex data, such as spectroscopic data. It provides a visual representation of the relationships between samples and variables and provides insights into how measured variables cause some samples to be similar to or different from others. The most significant information carried by the original variables are projected onto a smaller number of latent variables called principal components [37]. The first two principal components are generally those containing the most sources of information.

4 Applications

4.1 Application to glass batch analysis

Although modern glass production and its utilization have reached different levels, glass production has been an integral part of human beings' socioeconomic evolution since ancient times [38, 39]. During medieval days, glass was used in pottery and doors to allow light in; now it has unlimited applications. Modern high-tech equipment and domestic kitchen utensils are incomplete without glass. Major constituents of glass include silica, alkalis, lead, and calcium; however, oxides like alumina and phosphorus oxide are also used in modern glasses for different purposes such as for increasing durability, colorization, and refractive index.

The raw materials mixed in an appropriate proportion to prepare glass are collectively termed glass batch. These materials come from a variety of sources (some are mined from the earth and used only after processing, while others are already refined, like metal oxide powders) and have various functions. Common glass batch ingredients are listed in Table 1.

The integrated approach (batching, melting, and forming) of glass production is driven by economics. Forming a glass melt requires considerable energy, so direct formation of the product from the melt is preferred. The alternative to the integrated route is to create a melt followed by forming a glass powder instead of a product. The glass powder is then shipped as a starting material to customers who carry out the forming processes for desired glass items. Glass batches are thus prepared such that they meet the requirements of the end-products. To achieve this, a prior analysis of the elemental composition of the raw material is needed before entering the furnace.

Industrial production of glass is determined by its demand and use. Various analytical techniques such as X-ray fluorescence and inductively coupled plasma mass spectrometry (ICP-MS) are used for the compositional study of glass. For these techniques, sample preparation and data processing can be time consuming. With its in situ, multielemental, and little to no sample preparation requirement capabilities, LIBS can be a better option as compared to traditional analytical methods for glass batch study. Reports have shown that LIBS can be used for glass study at different stages (before, during, and after) of its production.

TABLE 1 Common raw materials for glass batches [40].

Material name	Purpose
Barite ($BaSO_4$)	Flux and fining agent (help remove bubbles during melting), source of barium oxide
Anhydrous borax (Na_2O_2, B_2O_3)	Source of B_2O_3
Chromite (FeO, Cr_2O_3)	Colorant for green bottles
Cullet	Crushed or powdered glass may be internal from the plant or from other sources
Anorthite (CaO, Al_2O_3, $2SiO_2$)	Sources of alumina, which improves durability of silicate glasses
Limestone ($CaCO_3$)	Source of CaO, an important ingredient for the durability of soda-lime silica glasses
Silica sources/sand/quartz	Source of silica (glass former)

Buckley et al. [41] used LIBS to measure sodium concentration in an oxygen-natural gas-fired soda-lime silica glass melting furnace. More precisely, they measured time-resolved sodium in the flue gas for which they used a LIBS monitor in the exhaust duct downstream from the tank.

Szelagowska et al. [42] utilized LIBS to study historic stained glasses from historical buildings and churches of the 18th20th centuries. The authors claimed the possibility of determining the manufacturing time of glasses. They identified characteristic lines of major elements, such as Si, Ca, K, Mg, and Na, in their LIBS spectra and used this information to identify modern and historic glasses.

Lal et al. [43] employed LIBS to study the batch used for soda-lime glass production and that used as surrogate for the batch used for the trapping of radioactive waste. They employed both Czerny-Turner and echelle spectrograph systems to obtain LIBS spectra from the glass batches. Spectra showing emission lines of the Fe, Si, Ca, and Al around the 390-nm wavelength range are shown in Fig. 5. The authors developed calibration curves for selected elements using spectral data obtained from both detection systems and compared analytical performances. Comparable figures of merit from both systems are reported; however, the echelle system was suggested for its wider spectral range, while the Czerny-Turner system showed faster data acquisition. In a separate study, Lal et al. [16] reported LIBS analytical performance

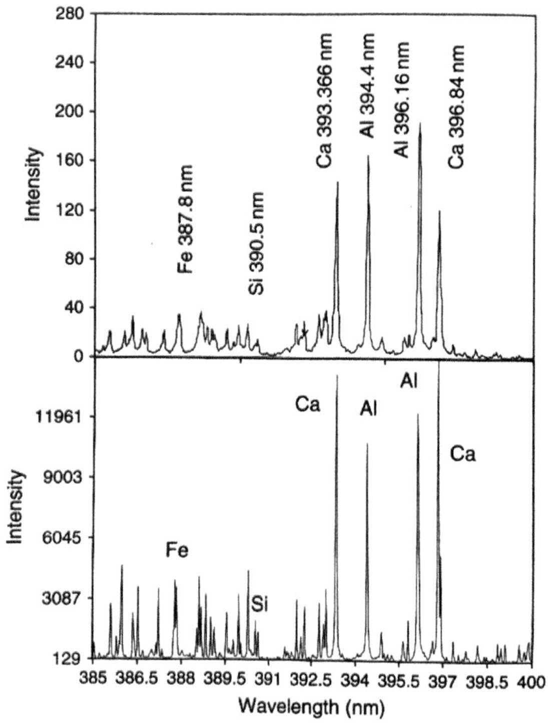

FIG. 5 Laser-induced breakdown spectroscopy spectra of a surrogate glass batch around 390 nm. *Top*: Recorded by a 0.5 Czerny-Turner spectrometer. *Bottom*: Recorded by an intensified charge-coupled device detector fitted to an ESA 3000 echelle spectrometer [43].

for a surrogate glass batch and suggested using pellets for batch analysis.

Zheng et al. [44] presented an evaluation of LIBS for the monitoring of plutonium feed composition before combining it for glass formation. CeO_2 was used as Pu surrogate and the elemental composition of a CeO_2 composite powder was determined during lanthanide borosilicate glass fabrication to verify process control. Analytical figures of merit of calibration curves developed using LIBS data obtained from pellet and powder samples were compared for various elements (Ce, Cr, Si, Fe, Ta, Ni, Zn, Al Mg, Gd, and W). Pellet samples yielded slightly better precision results than the direct analysis of powder samples from double-sided tape.

The aforementioned works of Lal et al. and Zheng et al. used either normalization or intensity ratio to improve the analytical figures of merit in their LIBS studies with univariate calibrations. Tripathi et al. [31], on the other hand, used both univariate as well as multivariate calibrations to analyze plutonium oxide surrogate. Principal component regression and partial least squares were used for multivariate calibration. The results were then compared with univariate calibrations (Fig. 6).

Multivariate calibration results showed significant improvement in the coefficient of determination from 0.87 to 0.97 for Ce compared to univariate calibration. The root mean square error was also reduced from 7.46% to 2.93%. An analytical method for remote monitoring of the elemental composition of glass during a vitrification process for fly and bottom ashes from waste incineration by laser-induced plasma spectroscopy is presented in [45, 46]. Yun et al. [47] presented an application of LIBS for glass analysis in a high-level liquid waste vitrification process. Optical glasses doped with REEs (Nd, Eu, Er, Ho) were studied using the LIBS technique by Dwivedi et al. [48]. Many researchers have presented LIBS applications for the forensic characterization of glass samples [49–53].

4.2 Application to REE analysis

REEs are 17 chemical elements of the periodic table, including 15 lanthanides (lanthanum-57 to lutetium-71) along with two other elements, scandium-21 and yttrium-39. These elements are critical as they are used in modern high-tech products like plasma screens, lasers, computing components, batteries, cell phones, and many more. The main sources of REEs are bastnaesite, monazite, and xenotime [54]. As REEs are available in low concentrations in their original sources, in situ techniques capable of being used in the field appear more useful for source identification and initial estimation of their abundance. LIBS is a promising in situ analytical technique that has been demonstrated for REE

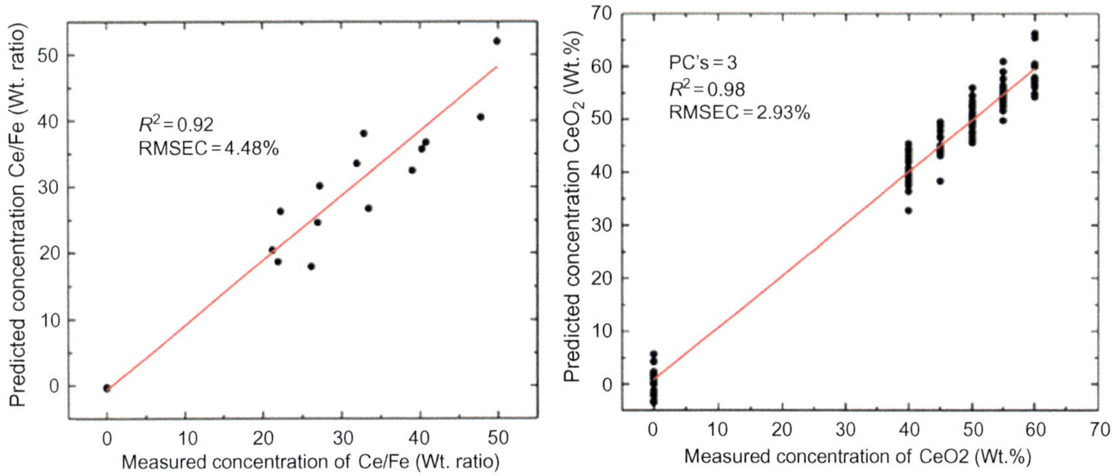

FIG. 6 *Left*: Univariate calibration plot developed for Ce/Fe (413.38 nm/373.50 nm). *Right*: Multivariate calibration plot developed for CeO_2 concentration with partial least squares regression [31].

study by many researchers [55–58]. Some REEs in aqueous solutions are also studied with LIBS [59–62]. Bhatt et al. [56, 58] reported qualitative and quantitative measurements of REEs in powder samples. Powder samples of six REE oxides (CeO_2, Eu_2O_3, Gd_2O_3, Nd_2O_3, Sm_2O_3, and Y_2O_3) in an Al_2O_3 batch were utilized to produce LIBS spectra in [56]. SLR calibration models were developed using selected emission lines for each element as shown in Fig. 7.

For multivariate analysis, LIBS spectra were recorded from the mixture of all six REE oxides. Reduced spectral ranges containing most of the strong emission lines of elements were taken into account for PLSR. Reduced wavelength ranges from 300 to 700 nm for Ce, 275 to 700 nm for Eu, 280 to 600 nm for Gd, 345 to 575 nm for Nd, 325 to 530 nm for Sm, and 290 to 680 nm for Y were used to obtain calibration models. Unscrambler X (version 10.3) CAMO software was used for PLSR. Calibration and validation curves are shown in Fig. 8. The slopes, offsets, root mean square error, and regression coefficients for both calibration and validation sets were excellent and slopes were close to one, which showed a strong correlation between predicted and reference values. This demonstrated that PLSR models for REEs can be developed to predict the concentration of unknown samples.

After analyzing pure powders of REE oxides, geological samples were studied by LIBS [58].

Eight natural samples containing REEs (Y, La, Ce, Pr, Nd, Sm, Eu, Gd, Dy, Ho, Er, Tm, Yb, and Lu) were taken for experiment. ICP-MS measurements of these samples showed that the REEs were in the 1–6000 ppm concentration range. To enhance the signals, an Ar atmosphere was created near the sample-laser pulse interaction by providing a continuous supply of Ar gas. Significant increments in signal were obtained with Ar use as shown in Fig. 9.

Spectral peaks with intensity levels higher than three times the noise (i.e., 3σ using International Union of Pure and Applied Chemistry criterion) were identified in the LIBS spectra using a NIST database and spectra from the standard samples as references to confirm the presence of corresponding elements in the samples. Multivariate analysis was performed for quantification of three elements (Ce, La, and Nd). PLSR models were developed by using spectra obtained from seven samples with added concentrations of Ce, La, and Nd. By

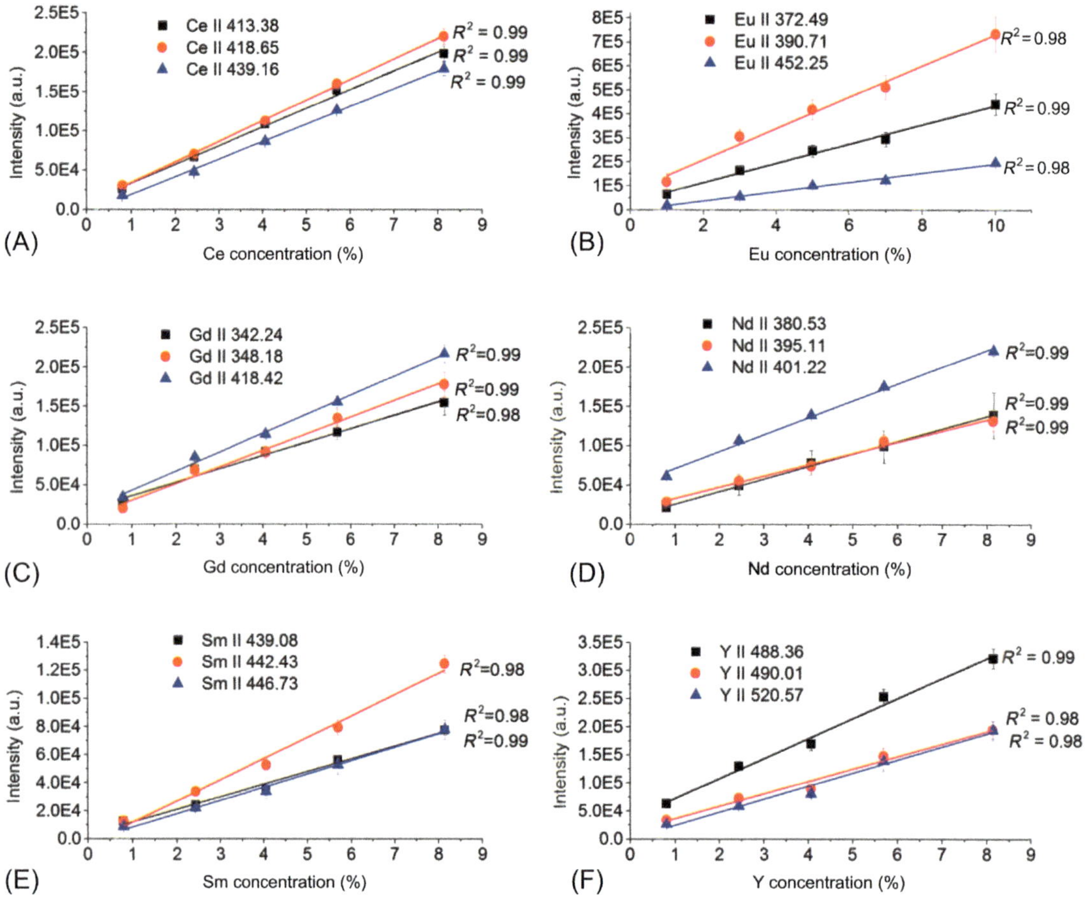

FIG. 7 Simple linear regression calibration curves for (A) Ce, (B) Eu, (C) Gd, (D) Nd, (E) Sm, and (F) Y [56].

using these models, concentrations of Ce, La, and Nd in the remaining eighth sample were predicted and obtained values were found comparable to ICP-MS values.

4.3 Application to pharmaceutical powder analysis

LIBS has been applied for the quality control of pharmaceutical powders, namely talc and starch-based cosmetic powders. Most cosmetic powders are essentially made from talc because of its softness. However, the potential presence of asbestos in talc is of concern in human health and pharmaceutical products [63]. The production and use of asbestos-containing products are regulated and in some countries, like Japan, a content of >1% is prohibited [64]. Nowadays, alternative sources, like cornstarch, are found on the market as a replacement for talc. There is a need for careful chemical analysis by manufacturers to make sure that talc used in cosmetics is asbestos free.

Analytical techniques such as transmission electron microscopy (TEM), scanning electron microscopy equipped with energy dispersive

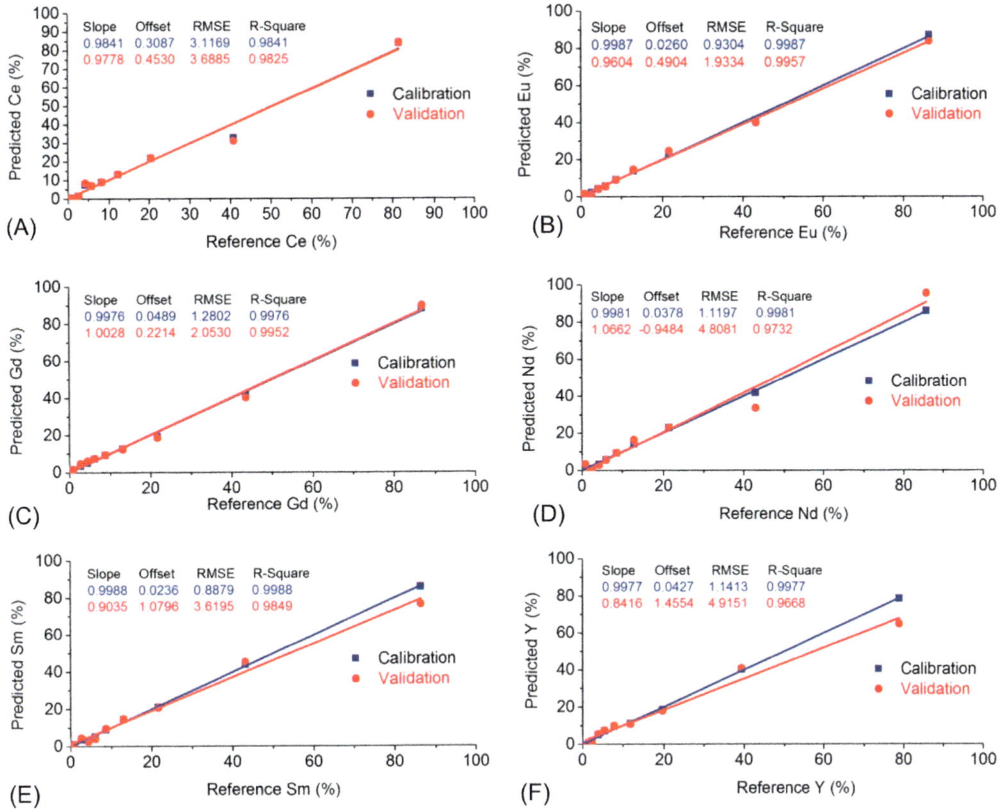

FIG. 8 Partial least squares regression calibration models for (A) Ce, (B) Eu, (C) Gd, (D) Nd, (E) Sm, and (F) Y [56].

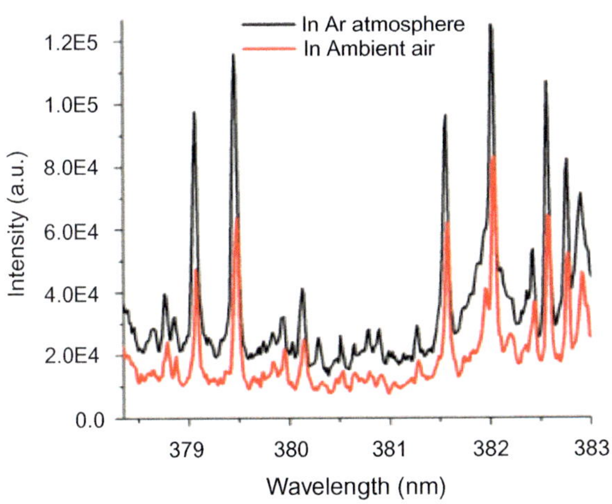

FIG. 9 Laser-induced breakdown spectroscopy spectra in an Ar atmosphere and ambient air [58].

X-ray spectroscopy, and polarized light microscopy have been used to investigate the presence asbestos. Light scattering and Raman spectroscopy are also considered for the same purpose [65–67]. Gordon et al. [68] reported the presence of asbestos in cosmetic talcum powder as a cause of mesothelioma in women, using TEM in conjunction with energy dispersive X-ray spectroscopy and selected-area electron diffraction to determine asbestos fiber number and type.

However, these techniques are cumbersome, time consuming, and require hands-on experience; moreover, they are sometimes suitable only for elements with atomic number $Z > 28$. Low concentrations and small dimension of

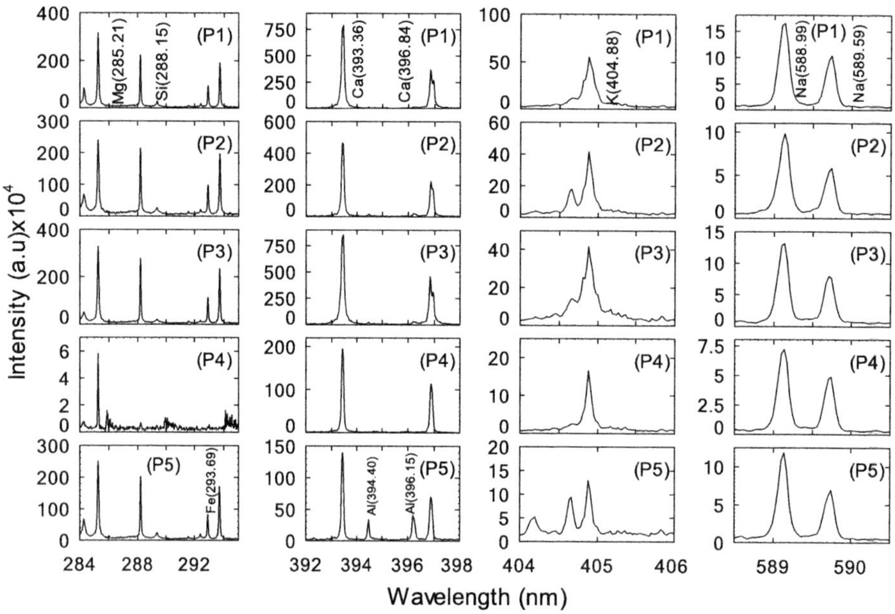

FIG. 10 Selected emission lines from talcum (P1, P2, P3, P5) and starch (P4) baby powders [72].

asbestos fibers also make their characterization difficult [69]. Therefore there is a need for a simplified, fast, and sensitive technique capable of analyzing all elements, and LIBS is such a technique with a detection limit on the order of ppm for most of the elements. LIBS has been proposed to characterize various types of asbestos (chrysotile, crocidolite, amosite, anthophyllite, actinolite, and tremolite). Intensity ratios of Mg/Si and Fe/Si were used as intrinsic characteristics of asbestos [70, 71].

Double-sided tape was used to prepare samples and a microplasma was created using a pulsed laser to record emission spectra [72]. By using these spectra, elemental analysis, screening, and classification based on the raw material used were carried out using multivariate data analysis. Major elements (Si, Fe, Mg) as well as minor elements (Ca, Na, K) were qualitatively detected as shown in Fig. 10. Similarly, a CN/CH molecular band was identified in the starch sample (Fig. 11).

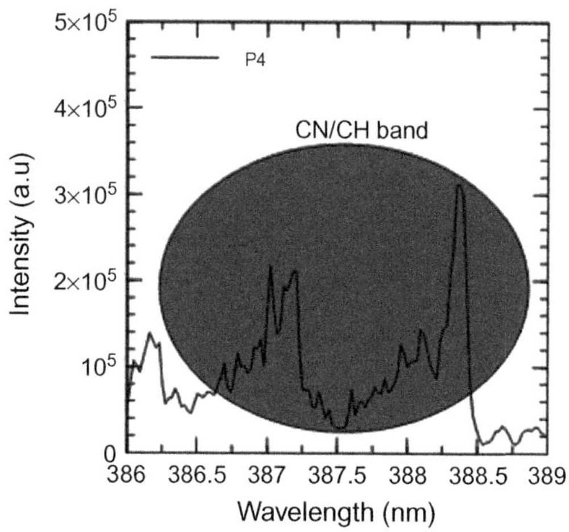

FIG. 11 Molecular band of CN/CH in starch-based powder sample [72].

With the aid of PCA, talc-based cosmetic powder was differentiated from the starch-based powder as shown in Fig. 12.

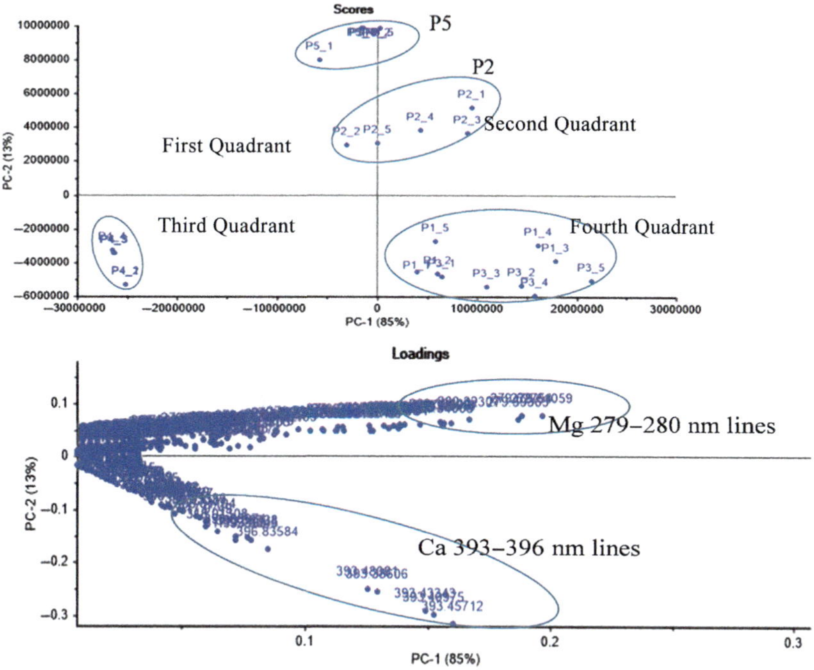

FIG. 12 Principal component analysis results. *Top*: Scores plot. *Bottom*: 2D scatter loading plot of PC1 versus PC2 [72].

A scatter plot of the scores of the main principal components gives information about the possible trends in the samples. From the scores plot, two principal components are sufficient to explain the data set. The points plotted here represent the objects which are samples in this case. These two components together explain 98% of the collected emission spectra with PC1 representing 85% and PC2 13%. In the scores plot, the talcum powders seem to have a vertical distribution along the axis of PC2. On the other hand, as one moves from left to right of PC1 one successively finds P4 (starch-based cosmetic), P5, P2, and P1, P3 (talc-based cosmetics). Since P1, P2, P3, and P5 are all talc based, this suggests there is a possible subgrouping in talc-based powders. The main difference in composition might be understood from the loadings plots. Loadings plots give insight into the variables behind the main differences between starch and talc-based samples, which could be used as screening factors. From the scores plot, one observes that the samples are grouped horizontally from cornstarch (left) to talc based (right). Since Mg has the highest contribution, the major grouping can be attributed to Mg content. Thus the principal component that carries the most information in this study is Mg and is distributed along the PC1 axis. On the other hand, looking at the PC2 line loadings plot, the only lines contributing are the calcium lines. From the scores plot observation, it is realized that there is a subgrouping within the talc-based cosmetics along the PC2 axis. We can infer from the observation of PC2 loadings plot that the calcium content is responsible of the subgrouping of talcum.

5 Conclusions

Powder material is widely used in the industrial sector for material processing. Prior knowledge of the elemental composition of the powder samples is needed to meet the specific purposes and requirements of the end-product. In this chapter, we presented LIBS application to powder sample analysis. Its application to glass batch, REEs, and pharmaceutical powder analyses was discussed. LIBS appeared to be useful for continuous online monitoring during every stage of glass manufacturing. The possibility of using LIBS for source characterization and abundance assessment for critical elements, like REEs, was demonstrated. Similarly, how pharmaceutical products in powder form can be examined with LIBS was also discussed.

In addition, possible ways of improving data acquisition and analysis were presented. LIBS analytical performance for powder samples was found to be better with pellets compared to direct analysis. However, pellets require homogeneity of particle size and appropriate binder. For data analysis, normalization of signals to be used in calibration and applying the multivariate analysis approach are recommended for complex sample analysis to increase the sensitivity and precision of LIBS for powder sample study.

References

1. D.A. Cremers, L.J. Radziemski, Handbook of Laser Induced Breakdown Spectroscopy, John Wiley & Sons Ltd, Chiehester, England, 2006.
2. R. Noll, Laser-induced breakdown spectroscopy, in: Laser-Induced Breakdown Spectroscopy, Springer, Berlin, Heidelberg, 2012, pp. 7–15.
3. J.P. Singh, S.N. Thakur (Eds.), Laser-Induced Breakdown Spectroscopy, Elsevier, Amsterdam, The Netherlands, 2007.
4. A.W. Miziolek, V. Palleschi, I. Schechter, Laser-Induced Breakdown Spectroscopy, Cambridge University Press, Cambridge, UK, 2006.
5. F. Anabitarte, A. Cobo, J.M. Lopez-Higuera, Laser-induced breakdown spectroscopy: Fundamentals, applications, and challenges, ISRN Spectrosc. 2012 (2012).
6. R. Noll, et al., Laser-induced breakdown spectroscopy expands into industrial applications, Spectrochim. Acta B At. Spectrosc. 93 (2014) 41–51.
7. H. Suyanto, et al., Application of laser plasma confinement and bending effect for direct analysis of powder sample, Spectrochim. Acta B At. Spectrosc. 57 (8) (2002) 1325–1332.
8. N. Idris, et al., Analysis of heavy metal pollution in soil using transversely excited atmospheric CO2 laser-induced plasma by trapping the soil in microstructured holes on metal subtargets, Appl. Spectrosc. 61 (12) (2007) 1344–1351.
9. Z.S. Lie, et al., Spectrochemical analysis of powder using 355 nm Nd-YAG laser-induced low-pressure plasma, Anal. Bioanal. Chem. 390 (7) (2008) 1781–1787.
10. J. Iqbal, et al., Shock wave plasma generation in low pressure ambient gas from powder sample using subtarget supported micro mesh as a sample holder and its potential applications for sensitive analysis of powder samples, Microchem. J. 142 (2018) 108–116.
11. R. Hedwig, et al., Film analysis employing subtarget effect using 355 nm Nd-YAG laser-induced plasma at low pressure, Spectrochim. Acta B At. Spectrosc. 61 (12) (2006) 1285–1293.
12. M. Suliyanti, et al., Direct powder analysis by laser-induced breakdown spectroscopy utilizing laser-controlled dust production in a small chamber, J. Korean Phys. Soc. 58 (5) (2011) 1129.
13. C.R. Bhatt, et al., Study of atomic and molecular emission spectra of Sr by laser induced breakdown spectroscopy (LIBS), Appl. Optics 54 (34) (2015) 10264–10271.
14. H.K. Sanghapi, et al., Analysis of slags using laser-induced breakdown spectroscopy, Spectrochim. Acta B At. Spectrosc. 115 (2016) 40–45.
15. M. Gondal, et al., The role of various binding materials for trace elemental analysis of powder samples using laser-induced breakdown spectroscopy, Talanta 72 (2) (2007) 642–649.
16. B. Lal, et al., Parametric study of pellets for elemental analysis with laser-induced breakdown spectroscopy, Appl. Optics 43 (13) (2004) 2792–2797.
17. S. Yao, et al., Optimizing the binder percentage to reduce matrix effects for the LIBS analysis of carbon in coal, J. Anal. At. Spectrom 32 (4) (2017) 766–772.
18. T. Yuan, et al., Quantitative carbon measurement in anthracite using laser-induced breakdown spectroscopy with binder, Appl. Optics 51 (7) (2012) B22–B29.
19. L. Shi, Q. Lin, Y. Duan, A novel specimen-preparing method using epoxy resin as binding material for LIBS analysis of powder samples, Talanta 144 (2015) 1370–1376.
20. T. Čtvrtníčková, et al., Comparison of double-pulse and single-pulse laser-induced breakdown spectroscopy techniques in the analysis of powdered samples of silicate raw materials for the brick-and-tile industry, Spectrochim. Acta B At. Spectrosc. 63 (1) (2008) 42–50.
21. S. Pandey, et al., Effect of powder compact density on the LIBS analysis of Ni impurities in alumina powders, Spectrochim. Acta B At. Spectrosc. 148 (2018) 99–104.

22. G.G.A. de Carvalho, et al., Influence of particle size distribution on the analysis of pellets of plant materials by laser-induced breakdown spectroscopy, Spectrochim. Acta B At. Spectrosc. 105 (2015) 130–135.
23. C.R. Bhatt, et al., Analysis of charcoal blast furnace slags by laser-induced breakdown spectroscopy, Appl. Optics 56 (28) (2017) 7789–7795.
24. H.K. Sanghapi, et al., Laser-Induced Breakdown Spectroscopy: Application to Powder Samples, National Energy Technology Lab.(NETL), Pittsburgh, PA, and Morgantown, WV, 2018.
25. D. Body, B. Chadwick, Optimization of the spectral data processing in a LIBS simultaneous elemental analysis system, Spectrochim. Acta B At. Spectrosc. 56 (6) (2001) 725–736.
26. I. Gornushkin, et al., Automatic correction of continuum background in laser-induced breakdown and Raman spectrometry, Appl. Spectrosc. 57 (2) (2003) 197–207.
27. I. Gornushkin, et al., Curve of growth methodology applied to laser-induced plasma emission spectroscopy, Spectrochim. Acta B At. Spectrosc. 54 (3–4) (1999) 491–503.
28. S.M. Clegg, et al., Multivariate analysis of remote laser-induced breakdown spectroscopy spectra using partial least squares, principal component analysis, and related techniques, Spectrochim. Acta B At. Spectrosc. 64 (1) (2009) 79–88.
29. K.K. Ayyalasomayajula, et al., Quantitative analysis of slurry sample by laser-induced breakdown spectroscopy, Anal. Bioanal. Chem. 400 (10) (2011) 3315–3322.
30. C.B. Stipe, et al., Laser-induced breakdown spectroscopy of steel: a comparison of univariate and multivariate calibration methods, Appl. Spectrosc. 64 (2) (2010) 154–160.
31. M.M. Tripathi, et al., Multivariate calibration of spectra obtained by laser induced breakdown spectroscopy of plutonium oxide surrogate residues, Spectrochim. Acta B At. Spectrosc. 64 (11 – 12) (2009) 1212–1218.
32. J.W.B. Braga, et al., Comparison of univariate and multivariate calibration for the determination of micronutrients in pellets of plant materials by laser induced breakdown spectrometry, Spectrochim. Acta B At. Spectrosc. 65 (1) (2010) 66–74.
33. H.K. Sanghapi, et al., Determination of elemental composition of shale rocks by laser induced breakdown spectroscopy, Spectrochim. Acta B At. Spectrosc. 122 (2016) 9–14.
34. J.M. Andrade-Garda, Royal Society of Chemistry, Basic Chemometric Techniques in Atomic Spectroscopy, RSC Publishing, Cambridge, 2009.
35. S. Yao, et al., Study of laser-induced breakdown spectroscopy to discriminate pearlitic/ferritic from martensitic phases, Appl. Surf. Sci. 257 (7) (2011) 3103–3110.
36. Q. Godoi, et al., Laser-induced breakdown spectroscopy and chemometrics for classification of toys relying on toxic elements, Spectrochim. Acta B At. Spectrosc. 66 (2) (2011) 138–143.
37. K.H. Esbensen, et al., Multivariate data analysis: in practice: an introduction to multivariate data analysis and experimental design, in: Multivariate Data Analysis, CAMO, Oslo, Norway, 2002.
38. J. Henderson, The raw materials of early glass production, Oxf. J. Archaeol. 4 (3) (1985) 267–291.
39. P. Marson, Glass and Glass Manufacture, Sir Isaac Pitman & Sons, Ltd., Bath, Melbourne, 1949.
40. De Saro, R., A. Weisberg, and J. Craparo. In situ, real time measurement of aluminum, steel, and glass melt chemistries using laser induced breakdown spectroscopy. in Energy Research Company. Conference Paper at the ACEEE Summer Study on Energy Efficiency in Industry. Available at: *http://www.eceee.org/conference_proceedings/ACEEE_industry/2005/Panel_1/p1_3/paper*. 2005.
41. S.G. Buckley, et al., Measurements of sodium in an oxygen-natural gas fired soda-lime-silica glass melting furnace, in: A Collection of Papers Presented at the 60th Conference on Glass Problems: Ceramic Engineering and Science Proceedings, 2000. Wiley Online Library.
42. K. Szelagowska, et al., Comparative study of historic stained glass by LIBS and SEM/EDX, in: 7th International Conference on Lasers in the Conservation of Artworks, LACONA 7; Madrid; Spain; 17 September 2007 through 21 September 2007, 2008.
43. B. Lal, F.-Y. Yueh, J.P. Singh, Glass-batch composition monitoring by laser-induced breakdown spectroscopy, Appl. Optics 44 (18) (2005) 3668–3674.
44. H. Zheng, et al., Analysis of plutonium oxide surrogate residue using laser-induced breakdown spectroscopy, Spectrochim. Acta B At. Spectrosc. 63 (9) (2008) 968–974.
45. U. Panne, et al., Analysis of glass and glass melts during the vitrification of fly and bottom ashes by laser-induced plasma spectroscopy. Part II. Process analysis, Spectrochim. Acta B At. Spectrosc. 53 (14) (1998) 1969–1981.
46. U. Panne, et al., Analysis of glass and glass melts during the vitrification process of fly and bottom ashes by laser-induced plasma spectroscopy. Part I: Normalization and plasma diagnostics, Spectrochim. Acta B At. Spectrosc. 53 (14) (1998) 1957–1968.
47. J.-I. Yun, R. Klenze, J.-I. Kim, Laser-induced breakdown spectroscopy for the on-line multielement analysis of highly radioactive glass melt simulants. Part II: Analyses of molten glass samples, Appl. Spectrosc. 56 (7) (2002) 852–858.
48. Y. Dwivedi, S. Thakur, S. Rai, Laser induced breakdown spectroscopy diagnosis of rare earth doped optical glasses, Appl. Optics 49 (13) (2010) C42–C48.
49. C. Barnett, E. Cahoon, J.R. Almirall, Wavelength dependence on the elemental analysis of glass by laser induced

breakdown spectroscopy, Spectrochim. Acta B At. Spectrosc. 63 (10) (2008) 1016–1023.
50. M.Z. Martin, et al., High resolution applications of laser-induced breakdown spectroscopy for environmental and forensic applications, Spectrochim. Acta B At. Spectrosc. 62 (12) (2007) 1426–1432.
51. E. Rodriguez-Celis, et al., Laser induced breakdown spectroscopy as a tool for discrimination of glass for forensic applications, Anal. Bioanal. Chem. 391 (5) (2008) 1961.
52. C.M. Bridge, et al., Forensic comparative glass analysis by laser-induced breakdown spectroscopy, Spectrochim. Acta B At. Spectrosc. 62 (12) (2007) 1419–1425.
53. M.M. El-Deftar, et al., Assessment and forensic application of laser-induced breakdown spectroscopy (LIBS) for the discrimination of Australian window glass, Forensic Sci. Int. 241 (2014) 46–54.
54. J. Voncken, The rare earth elements—a special group of metals, in: The Rare Earth Elements, Springer, Switzerland, 2016, pp. 1–13.
55. S.J. Rehse, C.A. Ryder, Laser-induced breakdown spectroscopy for branching ratio and atomic lifetime measurements in singly-ionized neodymium and gallium, Spectrochim. Acta B At. Spectrosc. 64 (10) (2009) 974–980.
56. C.R. Bhatt, F.Y. Yueh, J.P. Singh, Univariate and multivariate analyses of rare earth elements by laser-induced breakdown spectroscopy, Appl. Optics 56 (8) (2017) 2280–2287.
57. M. Martin, et al., Quantification of rare earth elements using laser-induced breakdown spectroscopy, Spectrochim. Acta B At. Spectrosc. 114 (2015) 65–73.
58. C.R. Bhatt, et al., Determination of rare earth elements in geological samples using laser-induced breakdown spectroscopy (LIBS), Appl. Spectrosc. 72 (1) (2018) 114–121.
59. D. Alamelu, A. Sarkar, S. Aggarwal, Laser-induced breakdown spectroscopy for simultaneous determination of Sm, Eu and Gd in aqueous solution, Talanta 77 (1) (2008) 256–261.
60. C.R. Bhatt, et al., Measurement of Eu and Yb in aqueous solutions by underwater laser induced breakdown spectroscopy, Spectrochim. Acta B At. Spectrosc. 137 (2017) 8–12.
61. C.R. Bhatt, J.C. Jain, D.L. McIntyre, Investigating the CO2 pressure effect on underwater laser-induced plasma emission of Eu and Yb, Spectrochim. Acta B At. Spectrosc. 149 (2018) 42–47.
62. J.-I. Yun, et al., Selective determination of europium (III) oxide and hydroxide colloids in aqueous solution by laser-induced breakdown spectroscopy, Appl. Spectrosc. 55 (3) (2001) 273–278.
63. D.W. Hahn, N. Omenetto, Laser-induced breakdown spectroscopy (LIBS), part II: Review of instrumental and methodological approaches to material analysis and applications to different fields, Appl. Spectrosc. 66 (4) (2012) 347–419.
64. T. Yamamoto, et al., Development of a testing method for asbestos fibers in treated materials of asbestos containing wastes by transmission electron microscopy, Waste Manag. 34 (2) (2014) 536–541.
65. Z. Ulanowski, et al., Respirable asbestos detection using light scattering and magnetic alignment, J. Aerosol Sci. 29 (1998) S13–S14.
66. E. Hirst, P.H. Kaye, J. Hoskins, Potential for recognition of airborne asbestos fibres from spatial laser scattering profiles, Ann. Occup. Hyg. 39 (5) (1995) 623–632.
67. C. Rinaudo, E. Belluso, D. Gastaldi, Assessment of the use of Raman spectroscopy for the determination of amphibole asbestos, Mineral. Mag. 68 (3) (2004) 455–465.
68. R. Gordon, S. Fitzgerald, J. Millette, Asbestos in commercial cosmetic talcum powder as a cause of mesothelioma in women, Int. J. Occup. Environ. Health 21 (4) (2015) 347.
69. Y. Suzuki, S.R. Yuen, Asbestos fibers contributing to the induction of human malignant mesothelioma, Ann. N. Y. Acad. Sci. 982 (1) (2002) 160–176.
70. L. Caneve, et al., Laser-induced breakdown spectroscopy analysis of asbestos, Spectrochim. Acta B At. Spectrosc. 60 (7–8) (2005) 1115–1120.
71. D.M. Benton, Identification of Asbestos using laser-induced breakdown spectroscopy: A viable alternative to the conventional approach? ISRN Spectrosc. 2013 (2013).
72. H.K.-N. Sanghapi, Laser-induced breakdown spectroscopy. [electronic book]: simultaneous multi-elemental analysis and geological applications, Mississippi State University, 2017.

CHAPTER

11

LIBS of slurry samples

K.K. Ayyalasomayajula[a], Fang Y. Yueh[b,c], Jagdish P. Singh[b,c]

[a]Center for Informatics and Analytics, University of Mississippi Medical Center, Jackson, MS, United States [b]Institute for Clean Energy Technology & Department of Physics, Mississippi State University, Starkville, MS, United States [c]JPS Advanced Technology R & D LLC, Starkville, MS, United States

1 Introduction

The laser-induced breakdown spectroscopy (LIBS) technique is widely used for qualitative and quantitative analysis of the elemental composition of solids, liquids, and gases [1–3]. One of the main advantages of using LIBS is the possibility of rapid multielemental analysis with little or no sample preparation. However, the use of LIBS for the direct analysis of liquid samples is still very challenging due to problems associated with splashing, aerosols above the surface, and bubbling inside the liquid. To overcome these challenges, researchers have implemented two different approaches: one is finding an appropriate experimental configuration and the second is transforming the liquid specimen into a solid form [4–10]. The selection of experimental and sampling configuration is found to greatly affect the stability and reproducibility of the LIBS measurement for liquid samples. Generally, liquid samples are used in the form of a static liquid pool, droplets or aerosols, bulk flowing liquids, liquid streams, or jets, with a double-pulsed configuration in LIBS analysis [11–17]. Cremers et al. employed a double-pulse LIBS configuration for bulk liquid analysis and observed signal enhancement [4]. Ito et al. used laminar flows of a liquid jet configuration for the detection of colloidal iron in a turbid solution of FeO in water and obtained an estimated limit of detection (LOD) of 0.6 ppm [18]. Nakumara et al. used a liquid jet configuration along with double-pulse excitation in the presence of purging gas and reported enhancement in the LOD of iron [19]. St. Onge et al. performed LIBS analysis by considering various liquid configurations, including bulk liquid in a closed glass bottle, flowing horizontal surface, and nonflowing surface in a container [20], and concluded that the conversion of liquid to a flowing horizontal surface was the most promising approach.

Applications of LIBS to slurry samples not only involve the challenge associated with liquid sample analysis but also the additional challenge of obtaining a valid representative sample of the slurry. Previous studies have shown that the presence of solid materials suspended in the liquid and dissolved gases can influence the laser focus and decrease the intensity of the LIBS signal, giving rise to limited success in the quantitative analysis of such samples. To reduce

Laser-Induced Breakdown Spectroscopy
https://doi.org/10.1016/B978-0-12-818829-3.00011-3

splashing and provide better LODs, Schmidt et al. performed elemental analysis of solutions via LIBS by concentrating and immobilizing metal ions into an ion exchange polymer membrane [21]. Michaud et al. applied LIBS to the analysis of iron ore slurries [22–24]. They used sampler geometry with a slurry recirculator to produce a column of free-falling slurry to overcome the difficulties of LIBS measurement [23]. They also studied the influence of particle size and mineral phase on the analysis of iron ore slurries and discovered the LIBS intensity to be dependent upon the particle size and the magnetite content [22]. An empirical correction equation for the calibration model was needed to correct the intensity of either iron or silicon lines due to variation of the average particle size within the ore sample. Application of LIBS as a remote, real-time analytical technique to analyze liquid radioactive wastes during the vitrification process was explored using direct sampling and slurry circulation systems [25,26]. The problems encountered in the direct analysis of wet slurry samples were sedimentation, splashing, and surface turbulence. Also, water could quench the laser plasma and suppress the LIBS signal, resulting in poor sensitivity. Investigations on the effect of water from slurries on LIBS spectra indicated that as the water content decreased, the LIBS signal was enhanced with the decrease in standard deviation.

To reduce the water content in liquid samples, researchers have implemented the second approach mentioned earlier, namely, transforming the liquid specimen into a solid form. Pace et al. analyzed a liquid sample by adding calcium oxide to form a precipitate that was pressed into pellets after stirring and drying [27]. Yaroshchyk et al. analyzed used engine oil on an absorbent paper substrate and the results concluded that the paper substrate method provided better LODs than those obtained from flowing liquid jets [28]. Gondal et al. analyzed liquid oil samples in solid tablet form to perform multielemental analysis and the results were in agreement with conventional techniques [29]. Xiu et al. analyzed an oil sample by coating it as a thin layer on the polished surface of a pure aluminum target and achieved LODs of sub-ppm level [30].

A preliminary study of the slurry samples coated on substrates for elemental analysis has shown good correlation between spectral intensity and elemental concentration [31]. For this study, a thin-coating procedure called "spin-on-glass" is applied for LIBS analysis of slurry samples. In this approach, the moisture in the slurry sample is almost completely removed before analysis, transforming it into an almost solid sample [32]. Several experimental configurations and detection conditions have been studied to obtain optimum experimental conditions for slurry measurements. To evaluate the quantitative measurement and reproducibility capabilities of the LIBS technique, the data of an unknown slurry sample were analyzed and compared with the results of inductively coupled plasma optical emission spectroscopy (ICP-OES) analysis.

2 Experimental

2.1 Apparatus

The two experimental setups used to record the LIBS signal from slurry samples are shown in Fig. 1. A frequency doubled, Q-switched Nd:YAG laser (Continuum Surelite III, 10 Hz, 5-ns pulse width, 9 mm diameter, 425 mJ maximum) was used as an excitation source. The laser light at 532 nm was focused on the sample surface by using a fused silica convex lens of 30-cm focal length. The size of the focused spot is about 50 μm and the energies of the laser pulse used in this work were between 8 and 44 mJ, corresponding to peak intensities of approximately 8.2×1010 and 4.5×1011 W/cm^2, respectively. The emitted light from the laser-induced plasma was collected either in a backward direction or

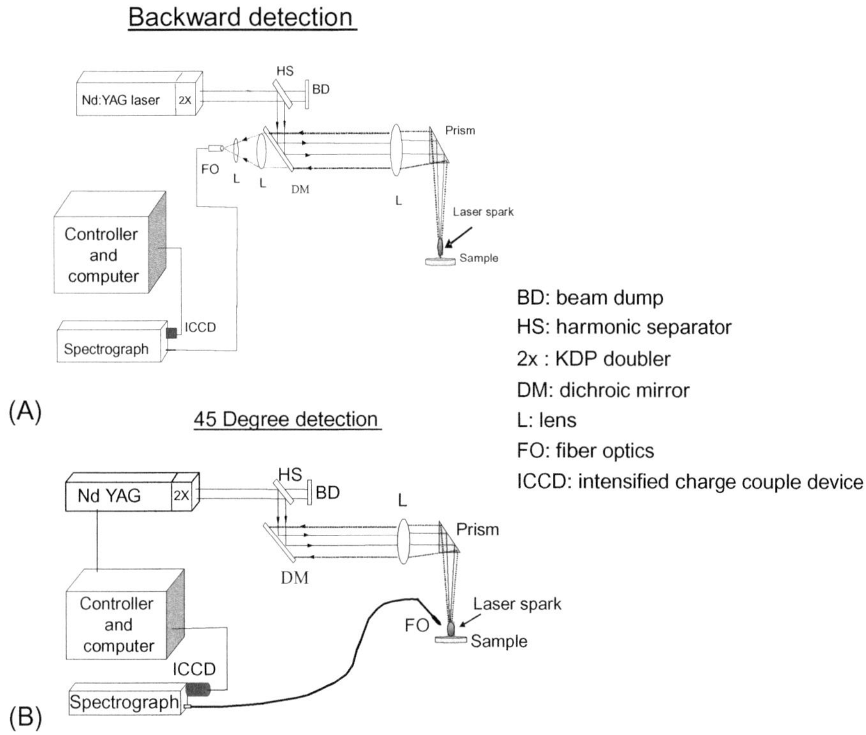

FIG. 1 Schematic diagram of two experimental setups of laser-induced breakdown spectroscopy.

with a pickup lens (Ocean Optics Inc. Part No. 74-UV) aligned at a 45-degree angle with respect to the laser beam into a 100-μm optical fiber. The distance of the pickup lens was 6 cm from the focal point of the laser beam. The collected signal was sent to a broadband spectrograph (Andor Mechelle 5000), which was integrated with an intensified charge-coupled detector. The detector head has a built-in digital delay generator (DDG) (DG535, Stanford Research Systems) that controls the gate delay and gate width. To synchronize the data acquisition with the laser pulse, the DDG is activated with a trigger from the laser Q-switch output. Data acquisition and analysis were performed by using a personal computer and Andor iStar software. To obtain better signal-to-noise data, each spectrum recorded was an average of five shots. Ten spectra were collected from each sample. Measurement was performed on a rotating platform to ensure data collection from various parts of the prepared sample. LIBS data were collected for various rotation speeds from 1 to 3 rotations/min.

2.2 Sample preparation

Standard procedures available in the literature were followed to evaluate LIBS calibration for slurry analysis, and five sludge simulants were prepared labeled as 8F, 10F, 12F, 14F, and unknown [33]. Five analytes, aluminum, calcium, nickel, iron, and silicon, were selected for weight percent adjustments based upon original percentages found in 8F simulant composition. A sample from each simulant was submitted for ICP-OES analysis. Samples from 8F, 10F, 12F, and 14F simulants were utilized

TABLE 1 Weight percent solids for the simulants.

	Sample				
Analyte	8F	10F	12F	14F	Unknown
Al	9.52	10.3	13.4	4.61	8.09
Fe	24.6	20.6	22.4	28.4	27.1
Ca	0.65	3.78	0.78	3.11	2.22
Si	0.93	0.74	1.89	3.86	1.54
Ni	2.73	4.46	1.74	4.12	3.22

for instrument calibrations for each analyte. Table 1 contains the weight percent (dry) of the five constituent elements obtained from the sludge simulants. The unknown simulant sample contained each of the selected five elements differing only in the weight percentages, but within the calibration ranges used in the LIBS experiments. LIBS data of the four slurry samples (i.e., 8F, 10F, 12F, and 14F) were collected and used to build calibration models and the data of the unknown sample were analyzed by these calibration models.

As described earlier, the LIBS experiments with liquid samples create splashing and shockwaves, which cause poor signal reproducibility. The slurry samples used in the present study contained mostly water (~80%), ferric oxide, alumina, sodium oxide, and small quantities of oxides of carbon, silicon, chromium, manganese, magnesium, etc. [33]. To solve problems with liquid samples, we have evaluated the method of "spin-on-glass" with double-sided tape to reduce water content. The slurry sample was first deposited on a substrate (glass slide with double-sided tape), then placed in a spin-coater machine (Chemat KW-4A) for spinning. Samples were coated using 380 rpm for 18 s and 2060 rpm for 60 s. By adjusting the rotation speeds, coating thickness, and sample distribution on the substrate moisture removal was controlled. When the substrate was at the lowest rotation speed, the coating thickness was adjusted, whereas at a higher rotation speed, distribution of the sample on the substrate and moisture removal were controlled. For the average of samples, upon completion of the spin coating, excess slurry was removed, leaving the total slurry present on the substrate about 7.8 mg. Finally, the slurry-coated substrate sample was dried on a hotplate (55 °C) for 90 s to remove the moisture from the sample completely. The remaining weight of the dried slurry sample on the substrate after moisture removal was about 2.5 mg. Two advantages of sample preparation using this method are to ensure that: (1) uniform thickness (2.5–3.5 μm) of the slurry sample on the glass substrate is maintained and (2) moisture content or wetness of the sample is so reduced that low laser energy is sufficient to obtain a good signal.

2.3 Data analysis

Forty spectra from each calibration sample were used to develop the calibration model, and 10 additional spectra of each calibration sample were used for validation of the calibration model. Three calibration models evaluated in this work were univariate calibration (UC), multiple linear regression (MLR), and partial least square regression (PLS1) [34,35]. UC is based on a plot of the signal of an analyte spectral line as a function of analyte concentration. The calibration curves were obtained from the LIBS data of a series of known concentration samples and then used to determine the concentration of an unknown sample. To account for various factors that could affect the LIBS signal, such as interference effects caused by other constituents of the sample matrix or random experimental errors, the MLR were also evaluated for LIBS calibration. The peak area of each selected analyte line was extracted from the spectral data of each slurry sample and normalized to the total emission intensity. The calibration curve was obtained by fitting the linear equation to a set of experimental data consisting of the measured peak area and known concentrations of analyte. The MLR procedures estimated a linear equation of the form:

$$C_x = b_0 + \sum_i^n (b_i A_i) \tag{1}$$

where C_x is the concentration of the element x and A_i is the peak areas of the selected analyte lines. A multiple line scan be used in the regression to improve the measurement accuracy by correcting for the matrix effect. PLS1, a multivariate analysis using whole spectrum or selected spectral regions to develop a calibration model, was also tested in this work. Some preprocessing methods, including multiple scattering correction, normalization, and baseline correction, could be applied to the calibration data sets. This analysis method relies on simultaneously fitting the intensity of the selected spectral region and species concentration to minimize the variance of both the spectral intensity and species concentration. GRAMS/AI PLSplus/IQ (Thermo Scientific) software was used to perform multivariate analysis for slurry data.

The predictive quality of the calibration models was evaluated by calculating the correlation coefficient R^2, root mean square error prediction (RMSEP), and relative error of prediction (REP) in the validation step with independent samples. The RMSEP values provide an estimate of the absolute error of prediction for each component:

$$\text{RMSEP} = \left[\frac{1}{n}\sum_{i=1}^n (\hat{x}_i - x_i)^2\right]^{0.5} \tag{2}$$

The REP is calculated as:

$$\text{REP}(\%) = \frac{100}{\overline{x}}\left[\frac{1}{n}\sum_{i=1}^n (\hat{x}_i - x_i)^2\right]^{0.5} \tag{3}$$

where x_i is the true concentration of the analyte in the sample i, $\hat{x}_i$ represents the estimated concentration of the analyte in the sample i, $\overline{x}$ is the mean of true concentrations in the prediction set, and n is the total number of samples used in the prediction sets.

3 Results and discussion

3.1 Effects of laser energy and gate delay

The backward detection (Fig. 1A) and 45-degree detection (Fig. 1B) configurations were used in the study of experimental parameters and LIBS calibration. Backward detection allows remote measurement but demands a complex optical geometry. In comparison with backward detection, 45-degree detection configuration needs simple alignment and gives high sensitivity, but it cannot be employed for remote detection applications. In a 45-degree detection configuration, a pickup lens aligned at 45 degrees with the laser beam was used to directly couple the LIBS signal into the optical fiber. The LIBS spectra of spin-coated samples with both detection configurations were recorded, and it was found that the sensitivity was improved (~5–10 times) with the 45-degree detection configuration by using low laser energy. The influence of laser energies and gate delays was studied to determine the optimum experimental parameters for quantitative measurement. The relation between the intensity of analyte lines and laser energy at a constant gate delay of 0.5 μs is shown in Fig. 2A. Initially, the LIBS signal gradually increases with the increase in laser energy and becomes constant at higher laser energy. The relation between the intensity of analyte lines and gate delay at a constant laser pulse energy of 22 mJ is shown in Fig. 2B.

The LIBS signal decreases dramatically with the increase in gate delay time for gate delays of <5 μs and decreases slowly with the increase of gate delay to higher delay values. To achieve good reproducibility, the laser-induced plasma during the measurement time frame needs to be as stable as possible. Plasma temperatures were evaluated for different experimental conditions and are shown in Fig. 3. Plasma temperatures were estimated using a Boltzmann plot of intensities of 14 spectral lines of iron from 400 to 500 nm. As the pulse energy increases above

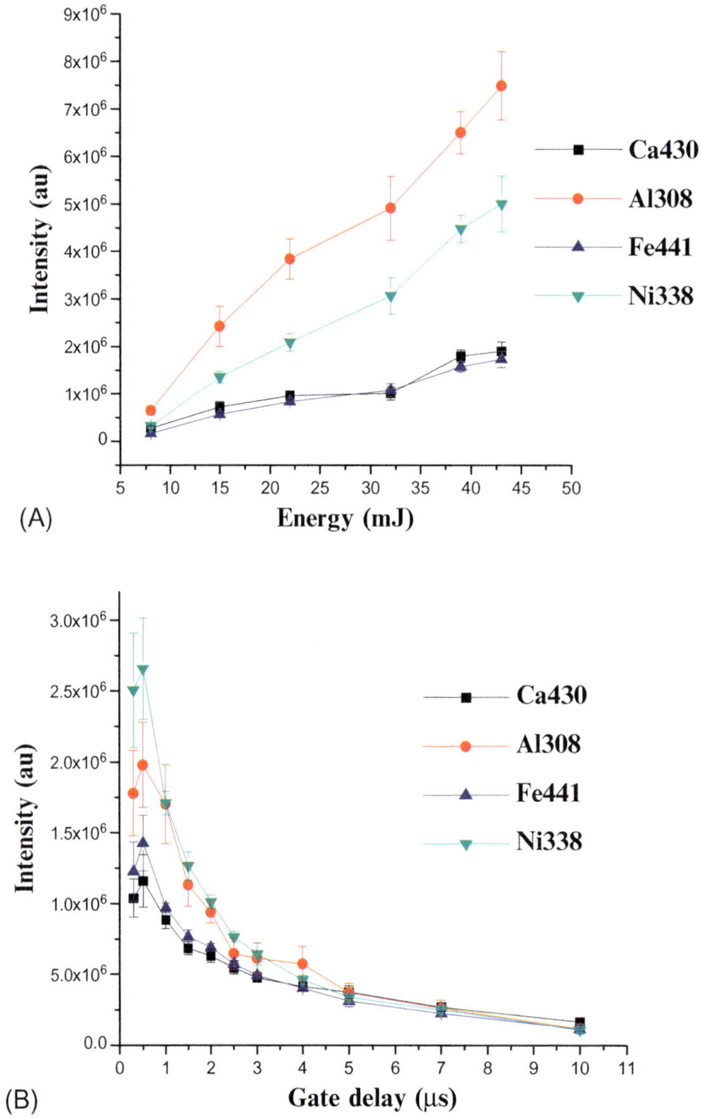

FIG. 2 (A) The relation between intensity of a laser-induced breakdown spectroscopy (LIBS) signal and laser energy at a constant gate delay (0.5 μs). (B) The relation between intensity of a LIBS signal and gate delay at constant laser energy (22 mJ).

22 mJ, the plasma temperature was found to be less influenced by variations of laser pulse energy. This study indicates that laser pulse energy above 22 mJ is required to obtain reproducible measurement in LIBS analysis.

To further improve the reproducibility of LIBS measurement and overcome the effect of pulse-to-pulse laser fluctuation, the recorded spectra were normalized by integral plasma emission [25]. This method significantly

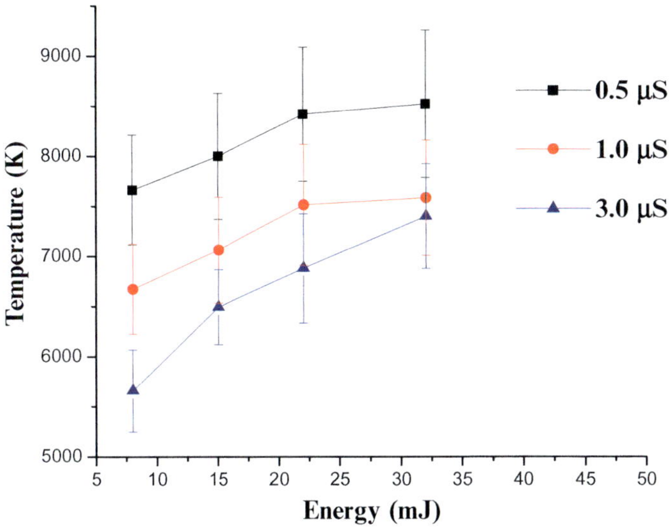

FIG. 3 Plasma temperature at different gate delays and laser pulse energies.

reduced the matrix influence on quantitative analyses. Measurement accuracy was also improved due to the reduction of data scattering on the calibration graphs. To test the selected analyte lines for measurement reproducibility, the five slurry samples were prepared every day prior to running the experiment. The LIBS data of these five slurry samples were collected and normalized with the integral plasma emission. Fig. 4 shows the normalized data of the Ni line at 313.4 nm for the five slurry samples taken over a period of 4 days using experimental parameter #1 (Table 2). As a result of this approach, day-by-day reproducibility improved and the relative standard deviation from the measurements of 4 days was $<4\%$ ($n = 4$, each measurement was made on a different day).

3.2 Calibration

3.2.1 Model validation

Three calibration models were developed and tested of which the linear regression model assumes a linear relationship between the peak intensity (area) of the analyte line and the elemental concentration. This calibration is easy to develop; however, the relationship between signal and concentration is not always linear due to spectral interference, matrix effect, and variation of laser-induced plasma, and a normalization technique was applied to the calibration data to reduce these effects. In this work, more than one spectral line for each element was used to check the linear behavior of the calibration curves. The observed Si signals from the calibration samples have very poor signal-to-noise ratio; therefore they were not used for this calibration. The calibration results for test data are shown in Table 3, where the high R^2 values indicate that the predicted concentrations of Ni and Ca would be more reliable compared to those of Fe and Al. The model has given a relatively high error margin due to the effect of self-absorption for predicted concentrations of Fe and Al.

To account for various factors that could affect the LIBS signal, such as interference effects caused by other components of the sample matrix or random experimental errors, the MLR was also evaluated for LIBS calibration [36]. The peak area of each selected analyte line was extracted from the spectral data of each

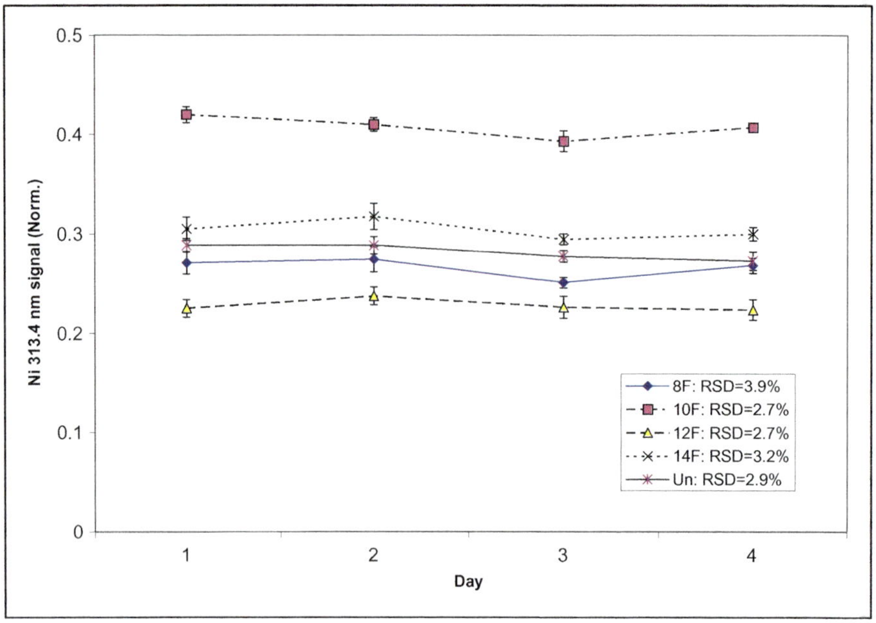

FIG. 4 Reproducibility of the normalized intensity of the Ni 313.4-nm line for an extend period of time.

TABLE 2 Experimental parameters in slurry measurements (gate width: 5.0 μs).

Experiment	Laser energy (mJ/pulse)	Gate delay (μs)
EXP_1	22	0.5
EXP_2	22	2.0
EXP_3	22	3.0
EXP_4	8	0.5

slurry sample and normalized to the total emission intensity using the function provided by AndoriStar software. The selected lines for the linear regression are given in Table 3. The MLR would be a good approach to predict the unknown concentration if the elemental concentration could be well correlated with the intensity of a few spectral lines. However, this method would be inefficient or inappropriate if many spectral lines are needed or if there is a significant collinear relation between the spectral lines. The spectral interference, such as self-absorption, can affect the relation between spectral intensity and elemental concentration. Therefore the selection of analyte lines in the MLR is also important. The RMSEP and REP values for validation data using the MLR model are given in Table 4 for comparison.

In a multivariate calibration one can use either the whole spectrum or a selected spectral region to develop a calibration model. There are some preprocessing methods, such as multiple scattering correction, normalization, and baseline correction, which could be applied to the calibration data sets. PLS1 was used in this work and relies on simultaneously fitting the intensity of the selected spectral region and the species concentration to minimize the variance of both the spectral intensity and species concentration. Because of the high Fe concentration in the calibration sample (>20 wt%), self-absorption was found in many Fe lines. To minimize the effect of self-absorption, only selected spectral regions were used in the PLS regressions. Multiplicative scatter correction was applied to the training set

TABLE 3 Comparison of univariate calibration results with different analyte lines.

Analyte	RMSEP	REP (%)	R^2
Al 308.2 nm	0.65	7.06	0.957
Al 394.4 nm	1.03	11.23	0.893
Al 396.2 nm	1.02	11.10	0.895
Ca 431.9 nm	0.17	9.15	0.984
Ca 612.2 nm	0.09	5.98	0.996
Ca 616.1 nm	0.07	5.21	0.997
Fe 561.6 nm	0.82	2.72	0.920
Fe 492.1 nm	0.80	2.54	0.925
Ni 349.3 nm	0.09	3.97	0.994
Ni 338.1 nm	0.06	3.01	1.000
Ni 345.8 nm	0.03	1.85	0.999

REMSEP, root mean square error prediction; *REP*, relative error of prediction.

to compensate for additive and/or multiplicative effects in the spectra. The optimum number of factors in PLS to model the system without overfitting was determined to be six from a series of tests. Table 5 shows the RMSEP and REP values for validation data obtained using a PLS1 model with a desired number of factors. This model gave a reasonably good prediction for the validation set. The REP values for Al, Ca, Fe, and Ni from the model were all below 7%. The REP values for Si were 9.45%. Therefore the prediction error for Si was expected to be higher as compared to other elements. A comparison of the test results from different calibration methods shows that the prediction error for Al, Fe, and Ni using UC, PLS1, and MLR are comparable [37]. However, the PLS1 model gives much less prediction error for Ca.

3.2.2 Comparison of LIBS and ICP-OES

To evaluate the quantitative ability of LIBS, the calibration models mentioned in the previous section were tested with the data of the unknown sample taken under the same experimental conditions as the calibration data. First, the effects of experimental parameters on the LIBS analysis were evaluated with the multivariate calibration. The test experimental parameters are given in Table 2. The LIBS data of the four calibration samples were taken with these four experimental conditions and used to build calibration models. The analysis results of unknown samples from the three different calibration models are shown in Table 6.

The results of Al, Fe, Ca, and Ni between different experiments were tested with F statistics and indicate that no statistically significant difference exists. The results of Si between the different experiments show larger variation as compared to other elements. To investigate this variation, the Si spectral data at 288.15 nm taken from sample 14F at different gate delay times and laser energies were compared. The Si

TABLE 4 RMS error of MLR model for the elemental concentration of the validation set.

		MLR		
		RMSEP	REP (%)	R^2
Al	308.21, 309.28, 394.39, 343.36	0.35	3.66	0.991
Ca	612.20, 616.85, 854.21, 396.16, 492.05	0.28	13.43	0.979
Fe	492.05, 561.56, 545.50, 396.16, 612.20	0.31	1.27	0.993
Ni	343.36, 396.16	0.08	2.42	0.997
Si	288.15, 492.05	0.46	22.59	0.921

MLR, multiple linear regression; *REMSEP*, root mean square error prediction; *REP*, relative error of prediction.

TABLE 5 Analysis results of validation data using the PLS1 model.

	RMSEP	REP (%)	R^2
Al	0.31	3.22	0.991
Ca	0.14	6.72	0.997
Fe	0.30	1.27	0.983
Ni	0.10	2.94	0.996
Si	0.18	9.45	0.980

PLS1, partial least square regression; *REMSEP*, root mean square error prediction; *REP*, relative error of prediction.

TABLE 6 Comparison of the LIBS analysis using different calibration models.

Analyte	UC (wt%)	MLR (wt%)	PLS1 (wt%)
Al	8.84 ± 0.15	8.88 ± 0.10	8.74 ± 0.16
Ca	2.71 ± 0.19	2.77 ± 0.12	2.39 ± 0.18
Fe	24.89 ± 0.41	24.39 ± 0.17	24.91 ± 0.62
Ni	3.25 ± 0.10	3.20 ± 0.06	3.28 ± 0.10
Si	1.18 ± 0.13	1.93 ± 0.27	2.21 ± 0.36

LIBS, laser-induced breakdown spectroscopy; *MLR*, multiple linear regression; *PLS1*, partial least square regression; *UC*, univariate calibration.

TABLE 7 Comparison between LIBS analysis and ICP-OES analysis.

Analyte	ICP-OES value	LIBS value[a] ± estimated uncertainty (wt%)	Relative error
Al	8.09	8.74 ± 0.16	0.08
Ca	2.22	2.39 ± 0.18	0.08
Fe	27.1	24.91 ± 0.62	0.08
Ni	3.22	3.28 ± 0.1	0.02
Si	1.54	2.21 ± 0.36	0.43

ICP-OES, inductively coupled plasma optical emission spectroscopy; *LIBS*, laser-induced breakdown spectroscopy; *PLS1*, partial least square regression.
[a] *Values from PLS1 model.*

spectral line 288.15 nm at a different gate delay time clearly shows a spectral interference from an unidentified line near 288.09 nm. The intensity ratio of the Si 288.15-nm line and the interference line 288.09 nm varies with the gate delay time. It shows that the Si detection sensitivity can be improved with longer gate delay time. Because the spectral interference depends on the gate delay time and laser energy, it can explain the larger difference found in Si measurements between the different experiments.

LIBS analyses of the unknown sample were performed with UC, MLR, and PLS1 regression and compared with ICP-OES analysis. The results from the PLS1 model from different days and experimental conditions were found to be more consistent as compared to the results obtained from the UC and MLR models. The comparison between PLS1 analysis of the unknown sample and ICP-OES analysis is shown in Table 7. The results for Ni agreed well with ICP-OES analysis (percent difference <1%). The percent differences between ICP-OES and LIBS analyses for Al, Fe, and Ca are all <8%. The LIBS-predicted Fe concentration was found to be consistently lower than ICP-OES analysis because all the calibration samples contain high Fe concentrations (>20 wt%), and most Fe emission lines suffered from self-absorption. The small sensitivity was obtained from a less intense line for the Fe calibration curve. For the determination of high concentration elements, the measurement conducted under reduced pressure of atmospheres can significantly reduce the effect of self-absorption and improve measurement quality.

The inconsistent result between LIBS and ICP-OES for Si was further investigated by examining the data of all samples. One possible reason is that the samples prepared for LIBS measurements are not representative samples due to the mixing process (i.e., sample might not be thoroughly mixed), but the repeated measurements with newly prepared samples all show the same trend. Another possible reason may be the sample viscosity that affected the sample deposition on the substrate slide during the spin-coating method. The low Si intensity from sample 12F still needs further investigation.

4 Conclusions

The application of LIBS to slurry samples has achieved limited success and has some inherent disadvantages, and the quantitative analysis of slurry samples is crucial and challenging. A new sampling method, "spin-on-glass," was evaluated in this work. It converts slurry measurement to almost solid-sample measurement to avoid the problems associated with liquids. The results of this work show improvements in the reproducibility and sensitivity of LIBS for slurry sample analysis. The inconsistent results for Si between the LIBS and ICP-OES analyses indicate the need for further study to investigate the effect of particulate size on this sampling method.

Acknowledgments

This work is supported by US Department of Energy Cooperative Agreement no. DE-FC26-98FT 40395. The authors would like to thank Mr. Markandey M. Tripathi, Mr. Tracy Miller, and Dr. Kemal E. Eseller for help in the laboratory, and Dr. Jeff Lindner for useful discussion.

References

[1] D.A. Cremers, L.J. Radziemski, Handbook of Laser-Induced Breakdown Spectroscopy, John Wiley & Sons, 2006.

[2] A.W. Miziolek, V. Palleschi, I. Schechter, Laser-Induced Breakdown Spectroscopy (LIBS): Fundamentals and Applications, Cambridge University Press, 2006.

[3] J.P. Singh, S.N. Thakur (Eds.), Laser Induced Breakdown Spectroscopy, Elsevier, 2007.

[4] D.A. Cremers, L.J. Radziemski, T.R. Loree, Spectrochemical analysis of liquids using the laser spark, Appl. Spectrosc. 38 (1984) 721–729.

[5] M.S. Cheri, S.H. Tavassoli, Quantitative analysis of toxic metals lead and cadmium in water jet by laser-induced breakdown spectroscopy, Appl. Opt. 50 (2011) 1227–1233.

[6] S.B. Mirov, R.E. Pitt, A.Y. Dergachev, W. Lee, D.V. Martyshkin, O.D. Mirov, J.J. Randolph, L.J. DeLucas, C.G. Brouillette, T.T. Basiev, Y.V. Orlovskii, O.K. Alimov, I.N. Vorob'ev, Novel laser breakdown spectrometer for environmental monitoring, Proc. SPIE Int. Soc. Opt. Eng. 3855 (1999) 34–41.

[7] E.M. Cahoon, J.R. Almirall, Quantitative analysis of liquids from aerosols and microdrops using laser induced breakdown spectroscopy, Anal. Chem. 84 (2012) 2239–2244.

[8] M.A. Aguirre, S. Legnaioli, F. Almodóvar, M. Hidalgo, V. Palleschi, A. Canals, Elemental analysis by surface-enhanced laser-induced breakdown spectroscopy combined with liquid–liquid microextraction, Spectrochim. Acta B 79–80 (2001) 88–93.

[9] A. De Giacomo, M. Dell'Aglio, A. Casavola, G. Colonna, O. De Pascale, M. Capitelli, Elemental chemical analysis of submerged targets by double-pulse laser-induced breakdown spectroscopy, Anal. Bioanal. Chem. 385 (2006) 303–311.

[10] S.K. Ho, N.H. Cheung, Sub-part-per-billion analysis of aqueous lead colloids by ArF laser induced atomic fluorescence, Anal. Chem. 77 (2005) 193–199.

[11] V.N. Rai, A.K. Rai, F.Y. Yueh, J.P. Singh, Optical emission from laser induced breakdown plasma of solid and liquid samples in the presence of magnetic field, Appl. Opt. 42 (2003) 2085–2093.

[12] O. Samek, D.C.S. Beddows, J. Kaiser, S.V. Kukhlevsky, M. Liska, H.H. Telle, J. Young, Application of laser induced breakdown spectroscopy to in situ analysis of liquid samples, Opt. Eng. 39 (2000) 2248–2262.

[13] S.L. Lui, Y. Godwal, M.T. Taschuk, Y.Y. Tsui, R. Fedosejevs, Detection of lead in water using laser-induced breakdown spectroscopy and laser-induced fluorescence, Anal. Chem. 80 (6) (2008) 1995–2000.

[14] Z. Chen, H. Li, M. Liu, R. Li, Fast and sensitive trace metal analysis in aqueous solutions by laser-induced breakdown spectroscopy using wood slice substrates, Spectrochim. Acta B 63 (2008) 64–68.

[15] N.K. Rai, A.K. Rai, LIBS—an efficient approach for the determination of Cr in industrial wastewater, J. Hazard. Mater. 150 (2008) 835–838.

[16] V. Lazic, S. Jovicevic, R. Fantoni, F. Colao, Efficient plasma and bubble generation underwater by an optimized laser excitation and its application for liquid analyses by laser-induced breakdown spectroscopy, Spectrochim. Acta B 62 (2007) 1433–1442.

[17] S.C. Jantzi, V. Motto-Ros, F. Trichard, Y. Markushin, N. Melikechi, A.D. Giacomo, Sample treatment and preparation for laser-induced breakdown spectroscopy, Spectrochim. Acta B 115 (2016) 52–63.

[18] Y. Ito, O. Ueki, S. Nakamura, Determination of colloidal iron in water by laser-induced breakdown spectroscopy, Anal. Chim. Acta 299 (1995) 401.

[19] S. Nakamura, K.S.Y. Ito, H. Hiraga, K.I. Kanedo, Determination of an iron suspension in water by laser-induced breakdown spectroscopy with two sequential laser pulses, Anal. Chem. 68 (1996) 2981–2986.

[20] L. St-Onge, E. Kwong, M. Sabsabi, E.B. Vadas, Rapid analysis of liquid formulations containing sodium chloride using laser-induced breakdown spectroscopy, J. Pharm. Biomed. Anal. 36 (2004) 277–284.

[21] N.E. Schmidt, S.R. Goode, Analysis of aqueous solutions by laser-induced breakdown spectroscopy of ion exchange membranes, Appl. Spectrosc. 56 (2002) 370–374.

[22] D. Michaud, R. Leclerc, É. Proulx, Influence of particle size and mineral phase in the analysis of iron ore slurries by laser-induced breakdown spectroscopy, Spectrochim. Acta B 62 (2007) 1575–1581.

[23] D. Michaud, E. Proulx, J.G. Chartrand, L. Barrette, Shooting slurries with laser-induced breakdown spectroscopy: sampling is the name of the game, Appl. Opt. 42 (2003) 6179–6183.

[24] L. Barrette, S. Turmel, On-line iron-ore slurry monitoring for real-time process control of pellet making processes using laser-induced breakdown spectroscopy: graphitic vs. total carbon detection, Spectrochim. Acta B 56 (2001) 715–723.

[25] S.Y. Oh, F.Y. Yueh, J.P. Singh, C.C. Herman, K. Zeigler, Preliminary evaluation of laser induced breakdown spectroscopy for slurry samples, Spectrochim. Acta B 64 (2009) 113–118.

[26] S.Y. Oh, T. Miller, F.Y. Yueh, J.P. Singh, Comparative study of laser-induced breakdown spectroscopy measurement using two slurry circulation systems, Appl. Opt. 46 (2007) 4020–4025.

[27] M.D.D. Pace, C.A. D'Angelo, D. Bertuccelli, G. Bertuccilli, Analysis of heavy metals in liquids using laser induced breakdown spectroscopy by liquid-to-solid matrix conversion, Spectrochim. Acta B 61 (2006) 929–933.

[28] P. Yaroshchyk, R.J.S. Morrison, D. Body, B.L. Chadwick, Quantitative determination of wear metals in engine oils using LIBS: the use of paper substrates and a comparison between single- and double-pulse LIBS, Spectrochim. Acta B 60 (2005) 1482–1485.

[29] M.A. Gondal, T. Hussain, Z.H. Yamani, M.A. Baig, Detection of heavy metals in Arabian crude oil residue using laser induced breakdown spectroscopy, Talanta 69 (2006) 1072–1078.

[30] J. Xiu, V. Motto-Ros, G. Panczer, R. Zheng, J. Yu, Feasibility of wear metal analysis in oils with parts per million and sub-parts per million sensitivities using laser-induced breakdown spectroscopy of thin oil layer on metallic target, Spectrochim. Acta B 91 (2014) 24–30.

[31] K.E. Eseller, M.M. Tripathi, F.Y. Yueh, J.P. Singh, Elemental analysis of slurry samples with laser induced breakdown spectroscopy, Appl. Opt. 49 (2010) C21–C26.

[32] K.K. Ayyalasomayajula, V. Dikshit, F.Y. Yueh, J.P. Singh, L.T. Smith, Quantitative analysis of slurry sample by laser-inducedbreakdown spectroscopy, Anal. Bioanal. Chem. 400 (2011) 3315–3322.

[33] M.R. Poirier, Recipe for Simulated Tank 8F Sludge Containing No RCRA Metals or Halides, WSRC-TR-2005-00045 Rev.0, Westinghouse Savannah River Company, Aiken, SC, 2005.

[34] S. Weisberg (Ed.), Applied Linear Regression, third ed., John Wiley & Sons, 2005.

[35] K.H. Esbensen, Multivariate Data Analysis in Practice, fifth ed., Camo Inc., 2004.

[36] E.C. Ferreira, J.M. Anzano, D.M.B.P. Milori, E. J. Ferreira, R.J. Lasheras, B. Bonilla, B. Montull-Ibor, J. Casas, L.M. Neto, Multiple response optimization of laser-induced breakdown spectroscopy parameters for multi-element analysis of soil samples, Appl. Spectrosc. 63 (2009) 1081–1088.

[37] J.W.B. Braga, L.C. Trevizan, L.C. Nunes, I.A. Rufini, D. Santos Jr., F.J. Krug, Comparison of univariate and multivariate calibration for the determination of micronutrients in pellets of plant materials by laser induced breakdown spectrometry, Spectrochim. Acta B 65 (2010) 66–74.

CHAPTER

12

Defense applications

Jennifer L. Gottfried

CCDC Army Research Laboratory, Weapons and Materials Research Directorate, Aberdeen Proving Ground, MD, United States

1 Introduction

Laser-induced breakdown spectroscopy (LIBS) is an ideal candidate for numerous defense-related applications due to its inherent advantages such as rapid data collection, versatility, and configurability (Fig. 1). These applications include the detection of hazardous materials in both civilian and military populations, the characterization of materials of interest to the defense community, and the identification of foreign munitions. For hazardous material detection, the ability to obtain single-shot, broadband LIBS spectra from minute quantities of suspicious material is essential. Many of these applications require the development of customized LIBS systems and software; some example man-portable and standoff systems developed and tested with the Combat Capabilities Development Command Army Research Laboratory (ARL) are shown in Fig. 1. Other configurations such as remote/fiber-optic (Chapter 17), hand-held (Chapter 23), suitcase, and robotic- (Chapter 20), vehicle-, ship-, or submersible-mounted LIBS systems have been developed for other applications as well. Since an overview of LIBS instrumentation was provided in Chapter 5, here we will only discuss in detail LIBS systems specifically developed for defense applications, e.g., portable systems for biological/chemical agent or landmine detection and standoff LIBS systems for explosive residue detection.

In addition to ruggedized LIBS hardware, chemometric algorithms designed for each application are essential for successful discrimination of chemical, biological, radiological, nuclear, and explosives (CBRNE)-related materials. As a subset of machine learning, chemometrics is a term encompassing a variety of techniques for the automated extraction of chemical information from raw data. Because the data obtained from LIBS consists of high-dimensional spectral data with a significant amount of chemical information (e.g., potentially thousands of intensity channels with multiple emission features from each observed species), advanced data analysis methods are required to determine the most important spectral features (atomic, ionic, and/or molecular) for the desired application. The development of chemometric algorithms for the discrimination of materials using LIBS has been reviewed in detail elsewhere [1–3].

This chapter presents example LIBS systems developed to address defense-related problems, with a focus on the custom hardware and

Laser-Induced Breakdown Spectroscopy
https://doi.org/10.1016/B978-0-12-818829-3.00012-5

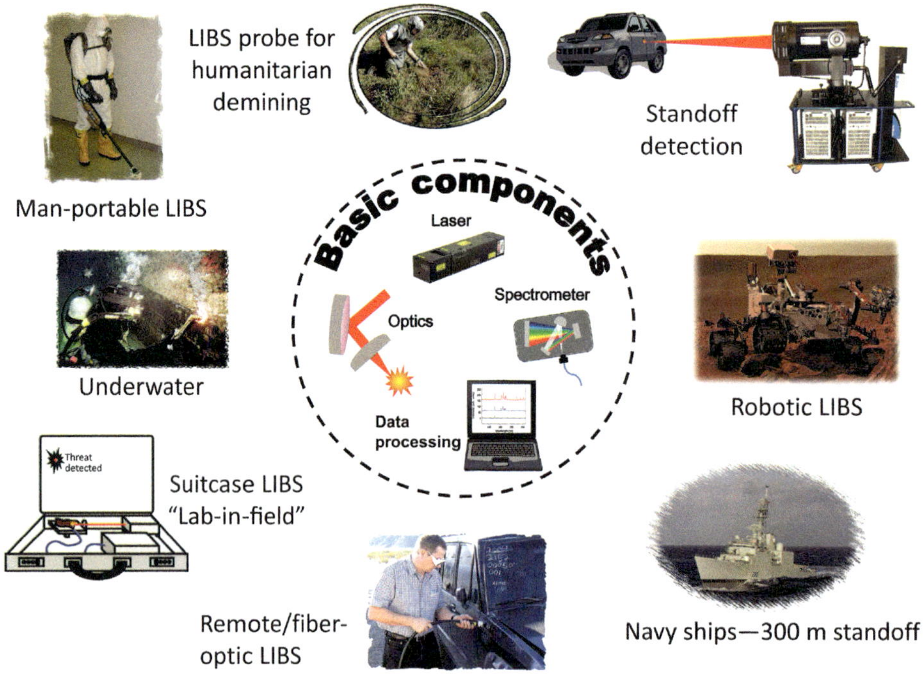

FIG. 1 Some example configurations of laser-induced breakdown spectroscopy systems that can be used for defense applications.

chemometric models developed for each application. The results from several types of chemometric techniques will be discussed in this chapter. Principal components analysis (PCA) is a technique for reducing the dimension of the data so that it can be easily visualized in two or three dimensions [4]. In PCA, a training set consisting of LIBS data is used to generate principal components (PCs); the scores and loadings for the PCs describe how the samples or variables, respectively, relate to each other. Because no class information is used to develop the model, unsupervised chemometric methods such as PCA are only suitable when the variability between classes (i.e., sample types) is significantly larger than the variability within groups. Soft independent method of class analogy (SIMCA) models consist of a collection of PCA models, one for each class in the data set [5]. The PCs for each PCA model therefore describe the variance within a particular sample class. Partial least squares discriminant analysis (PLS-DA) is a multivariate inverse least squares discrimination method used to classify samples [6]. PLS-DA is a supervised chemometric technique that maximizes the variance between classes (e.g., by emphasizing minor impurities that enable discrimination of two otherwise similar materials). The latent variables generated by the model consist of the linear combination of independent variables (e.g., emission intensities) that best describe the variance in the data. Variable importance in projection scores can be used to investigate which variables contribute the most to each class in the model.

This chapter describes several LIBS applications of interest to the military and homeland security, including the identification of CBRNE threats, weapons-related materials (explosive fills, bullets, landmines), residual environmental hazards associated with the use of explosives, characterization of novel explosives, and troubleshooting military systems. The following sections provide examples of each of these application areas.

2 Identification of materials

The threat of terrorists using weapons of mass destruction has led to the need for reliable sensors to detect a potential attack. While strictly regulated, the use of real explosives to develop sensors in research laboratories is relatively common since in very small quantities they can be handled safely by trained personnel. Because handling chemical and biological agents requires even more stringent safety precautions and specialized laboratories, proof-of-principle LIBS systems typically rely on common simulant materials. A number of configurations of LIBS systems have been used for these applications, including laboratory, portable, and standoff instruments. Most of these systems focus on identifying surface contamination (e.g., suspicious powders or liquids), although vapor detection is also important for chemical agents with high vapor pressures. Aerosol detection is another important approach for biological agent detection, and is described in Chapter 22. For most chemical and biological sensors, the surface (or air) must be first sampled by a direct contact method, then the sample is analyzed ex situ in a laboratory—a process that could take many hours (or longer). LIBS offers the ability to directly sample the suspicious material (or continuously monitor the air) and provide an immediate identification of the material in situ by testing the acquired spectra against the appropriate chemometric model.

2.1 Chemical agent detection

Chemical warfare involves the use of the toxic properties of nonliving, chemical substances such as botulinum toxin, ricin, and sarin. Chemical warfare agents (CWAs) may be in liquid, gas, or solid form and are primarily composed of carbon (C), hydrogen (H), nitrogen (N), and/or oxygen (O). They often contain one or more of the following elements, which are detectable by LIBS in ppm (or lower) quantities: arsenic (As), phosphorus (P), fluorine (F), chlorine (Cl), and sulfur (S). In 2005, De Lucia et al. [7] analyzed the LIBS spectra of organophosphate liquid nerve agent simulants. They showed a direct correlation between the C/P emission intensity ratio and the stoichiometry of the simulants, even when embedded in soil. Munson et al. [8] used linear correlation and PCA to identify five different nerve agent simulants; however, not all the simulants could be discriminated from each other with these methods due to significant shot-to-shot variation in the LIBS spectra.

A double-pulse standoff LIBS (ST-LIBS) system (shown in the top right of Fig. 1) designed for hazardous materials detection was used to obtain spectra of the nerve agent simulants dimethyl methylphosphonate (DMMP) and diisopropyl methylphosphonate (DIMP) (Fig. 2), successfully demonstrating detection of the phosphorous emission lines at standoff distances for the first time, which is important for the safety of first responders [10]. A subsequent study by the same group obtained ST-LIBS spectra of thin residues of five nerve agent simulants (DMMP, DIMP, diethyl methylphosphonate, diethyl ethylphosphonate, and triethyl phosphate) and two interferent liquids (laundry detergent and fertilizer) on aluminum (Al) foil at 20 m [9]. Using a PLS-DA model trained with four of the simulants and the fertilizer (with both summed intensities and complex intensity ratios as input variables), a test set consisting of the fifth type of simulant, additional fertilizer samples, and the detergent resulted in 100% true positives for the nerve agent simulant with no false positives (Fig. 3). A combined chemical, biological, and explosive (CBE) model was also used to correctly classify the nerve agent simulant with 100% accuracy.

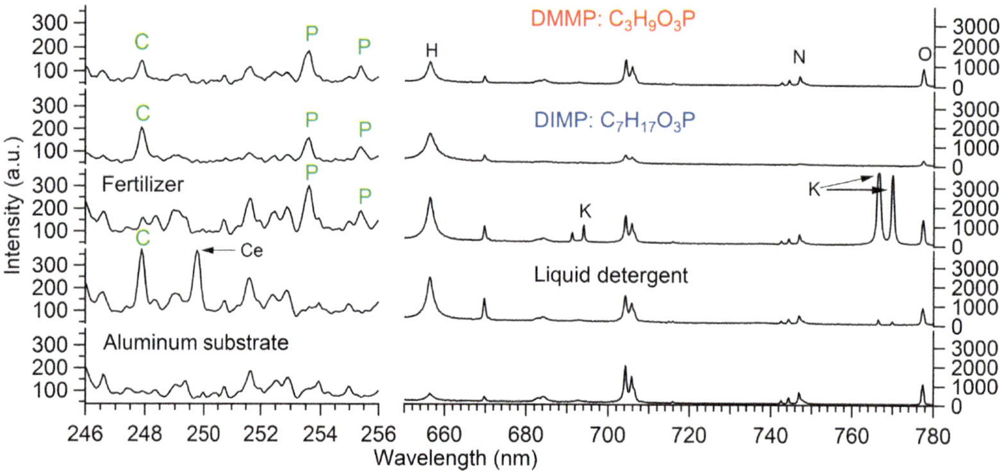

FIG. 2 Spectra of chemical agent simulants and interferents acquired at 20 m with a double-pulse standoff laser-induced breakdown spectroscopy system developed at the Army Research Laboratory [9].

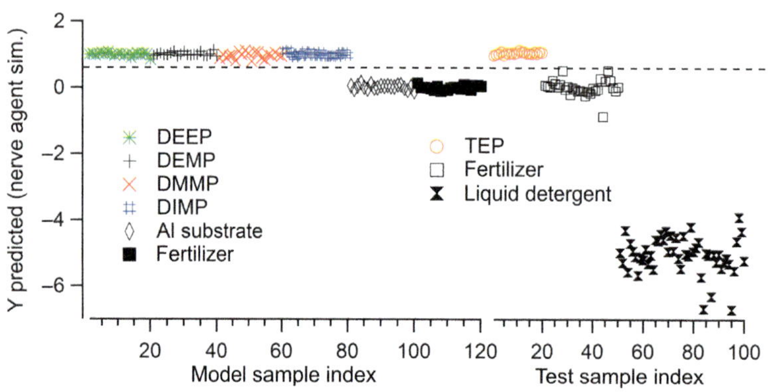

FIG. 3 Partial least squares discriminant analysis model prediction for nerve agent simulant class membership; both the samples used to train the model and the test samples are shown. Each symbol represents a single-shot laser-induced breakdown spectroscopy spectrum acquired at 20 m; data points that fall above the Bayesian threshold determined by the model (*dashed horizontal line*) were identified by the model as a nerve agent simulant [9].

One concern with residue detection using LIBS is that the ablation of the substrate can contribute to the emission spectra; not only can substrates with similar elemental compositions add to intensity signals from the residue, but matrix effects can influence the relative emission intensities as well. Landstrom et al. [11] analyzed the LIBS and Fourier transform infrared (FTIR) spectra of contaminants such as sarin, soman, and organophosphates on various substrates such as silicon (Si) wafers and chemical agent resistant coating (CARC) paint systems. While they were able to use the phosphorous signals from LIBS for depth profile analysis of the substrates and monitoring diffusion processes, FTIR was used to monitor residual contaminants on the different substrates. L'Hermite et al. [12] used a portable LIBS system to detect sarin, lewisite, mustard gas, and ethyl ({2-[bis(propan-2-yl)amino]ethyl}sulfanyl)(methyl)phosphinate

(VX) surface contamination on wood, concrete, CARC paint, gloves, and ceramic. Identification of the CWAs was based on the presence of emission lines from As, Cl, P, F, and/or S; sensitivities for those elements ranged from 15 μg/cm^2 (for Cl) up to 140 μg/cm^2 (for P). No attempt was made to discriminate the CWAs from other, nonhazardous surface contaminates with similar elemental compositions.

Gottfried [13] combined the results from two PLS-DA models, one trained with the full spectra and one trained with selected emission intensities and ratios, to discriminate a mustard agent simulant (2-chloroethyl ethyl sulfide) and an organophosphate nerve agent simulant from biological agent simulants and inert interferents on multiple substrates (aluminum, steel, and polycarbonate). The full spectra models depended on certain substrate emission lines for classification, resulting in higher correct classification rates (for that specific substrate), while the intensity/ratio models had lower overall correct classification rates (but depended much less on the substrate). By fusing the probabilities for class membership determined by each model, relatively high correct classification rates (>85%) could be obtained with lower misclassification (<1%) and unclassified (<15%) rates than either individual model. Individual components in mixtures of the simulants (with ovalbumin and limestone interferents) were more accurately identified with the intensity/ratio model. Training the PLS-DA model with multiple substrates reduced the dependence of the model on substrate emission features, but the percentage of single-shot spectra that could not be classified by the model was relatively high (35%).

The use of swipe materials to sample the unknown residue prior to LIBS analysis is one possible approach to mitigate this challenge. In this way, the chemometric model can be trained on spectra of residues applied to a fixed substrate, and any test materials would be applied to the same substrate, improving the discrimination. Gottfried [14] demonstrated 100% true-positive identification of a nerve agent simulant on two types of swipe materials, although overall classification rates for hazardous materials were higher for the rayon/polyester pad than the cotton twill wipe. Another approach is to detect vapor-phase chemical agents. A study by Stearns et al. [15] demonstrated 99% successful discrimination of two vapor-phase nerve agent simulants and three rocket fuels using PLS-DA with 16 intensities and ratios as input variables. More recently, Brown et al. [16] obtained long-wave infrared (LWIR) LIBS emission spectra of a nerve agent simulant on asphalt. They suggested that the combination of conventional LIBS spectra (covering wavelengths spanning the ultraviolet/visible/near-infrared) and LWIR signatures could be used to improve the discrimination of different chemicals on complex substrates.

2.2 Biological agent detection

Biological weapons consist of disease-causing, living organisms such as bacteria, viruses, and fungi deployed for the purpose of causing harm to humans, livestock, or crops and inciting fear in the population. Incidents such as those involving postal letters containing anthrax spores demonstrate the willingness and ability of terrorists to use these materials. Since the detection of bioaerosols is covered in Chapter 22, here we focus on the use of LIBS for interrogation of suspicious powders (or liquids) on various surfaces. As with chemical agent detection, the biggest advantage of LIBS for this application is real-time, in situ analysis that provides first responders and investigators with timely information critical for public safety. The LIBS signatures of biological agent surrogates (e.g., bacterial spores such as *Bacillus globigii*, *Bacillus thurengensis*, and *Escherichia coli*) were obtained by multiple groups in 2003 [17–19]. PCA was used for the initial discrimination studies [17, 19]. De Lucia et al. [7]

discriminated between three bacillus spores, two molds, and a protein using a simple linear correlation technique. Leone et al. [20] investigated the effect of the nutrient media on the bacterial growth and found that the use of a mineral medium influenced the bacterial metabolic activity and subsequent LIBS emission features. Discrimination of three strains of *E. coli*, one strain of mold, and one strain of *Candida albicans* yeast was demonstrated by Diedrich and Rehse [21] using the chemometric technique discriminant function analysis, which is similar to PCA. The full LIBS spectra were used for the analysis, and good separation of the strains was achieved. A subsequent study investigated bacteria colonies grown in three different media: a trypticase soy agar plate, a blood agar plate, or an agar plate containing bile salts [22]. Merdes et al. [23] used PCA, linear discriminant analysis, and hierarchical cluster analysis to identify bacterial spores on painted surfaces with low false-positive (1%) and false-negative (3%) rates. Snyder et al. [24] used a neural network model for discrimination between biological agent surrogates and interferent compounds, with false-negative rates of 0% for *Bacillus atrophaeus*; the limit of detection was 12 colony-forming units. LIBS was also used to differentiate strains of live pathogens *Bacillus anthracis* and *Francisella tularensis*, as well as ultraviolet-killed *hantavirus* [25].

Munson et al. [8] showed good discrimination of bacterial spores and interferents such as pollen, mold, and dust using SIMCA; classification of bacteria and dust mixtures was also demonstrated. The identification of biological agent surrogates (including the anthrax simulant *B. atrophaeus*, a Gram-negative bacterial biothreat simulant *E. coli*, a smallpox simulant MS-2 bacteriophage, and a biological toxin simulant α-hemolysin from *Staphylococcus aureus*) was demonstrated on polycarbonate, stainless steel (SS), and aluminum foil substrate coupons [13]. In addition to the pure residues, LIBS spectra from mixtures of the surrogates and interferents on each substrate were obtained. A chemometric algorithm was developed to discriminate the threat materials with reasonable success (misclassification rates as low as 1.8%); however, the number of single-shot spectra that could not be classified by the model was likely too high for real-world application, suggesting an approach based on a single substrate (such as a swipe material [14]) may be more practical.

The use of femtosecond LIBS enabled Baudelet et al. [26] to distinguish between the contribution of native CN bonds released from the sample and CN formed through recombination of carbon with atmospheric nitrogen; they suggested this approach could be used to improve the discrimination of organic and biological samples. Remote femtosecond filament-induced breakdown spectroscopy was proposed for the identification of biological agents based on the spectral differences observed between yeast and egg white [27].

For the detection of biological powders on indoor surfaces, a fieldable backpack LIBS system was developed for ARL (shown in the top left of Fig. 1) [28]. This configuration enables a fully suited first responder to immediately enter the area with possible contamination (e.g., suspicious powder on a desk) and evaluate the threat in situ. The man-portable LIBS (MP-LIBS) prototype system consists of a backpack containing the spectrometer, computer, and power supply and a hand-held wand containing the laser head (1064 nm, 30 mJ/pulse). The system also has the capability to flow argon across the sample surface, and the tip of the wand is designed to block the laser radiation when placed on a surface. Spectra of three bacillus spores (the biological agent surrogates), molds, and common powders (sugar, salt, coffee creamer, flour, baking powder, etc.) were first acquired on powders affixed to double-sided tape. Ten office surfaces (e.g., desk, carpet, filing cabinet, paper, etc.) were contaminated with one of the surrogate powders and the subsequent LIBS spectra were compared to those obtained by using 10 different swipe materials (e.g., tackcloth, paper towel, etc.) to sample the powder off the surface (Fig. 4). While linear correlation was not sufficient to identify surface contaminants, PLS-DA models enabled the identification of the surrogates from the confusant materials on both the office surfaces and the swipe materials.

Standoff detection and discrimination of biological agent surrogates was also achieved by the ARL group using the ST-LIBS instrument in Fig. 1 [9]. Spectra from a large sample set of 25 surrogates, biological confusants, and environmental or household interferents were obtained at a standoff distance of 20 m. Surrogates from different commercial sources were prepared on various surfaces (aluminum, glass) and mixed with dust in varying concentrations. All of the surrogates in the test set were 100% correctly identified by the PLS-DA model trained using preselected, summed emission intensities and ratios from powdered samples on double-sided tape affixed to white cardboard. The true-positive and false-positive rates for the mixtures varied from 95%–89% to 20%–7.3%, respectively, depending on the user-defined threshold for class membership. In other words, the true positives could be increased at the expense of false positives. However, the determination of whether a single-shot spectrum from the mixtures was classified correctly depends on whether the laser actually sampled the surrogate with each shot—this might not be the case, especially for mixtures with only 25% surrogate and 75% dust. In a combined PLS-DA model for CBE detection, the surrogate (on aluminum) was correctly identified 100% of the time.

To facilitate the use of LIBS for biological agent detection (or other applications), ARL worked closely in conjunction with researchers

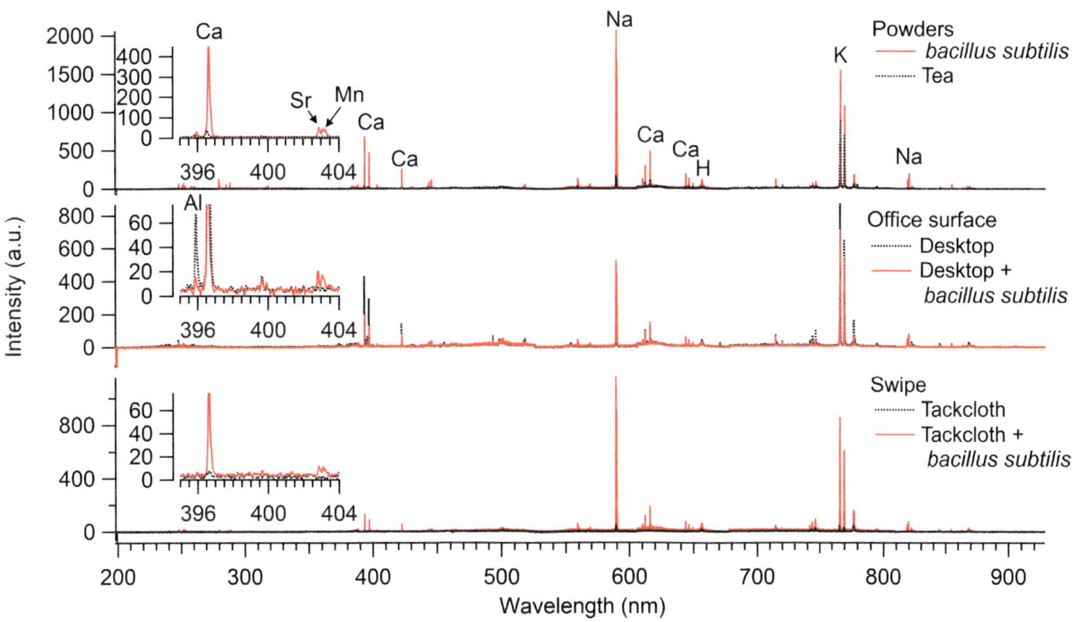

FIG. 4 Spectra of the biological agent simulant *Bacillus subtilis* in powder form (compared to tea), in trace amounts on a desktop surface, and on a tackcloth swipe material; the spectra were acquired directly on the surfaces with the man-portable laser-induced breakthrough spectroscopy system developed for the Army Research Laboratory [28].

at Duke University [29] to automate the development of the PLS-DA model (including training data collection, feature extraction, classifier development, and model evaluation) based on the chemometric processes for LIBS data analysis developed by ARL. The laser-induced breakdown spectroscopy graphical development tool is a graphical user interface that enables real-time automatic processing of new LIBS test data as it is acquired once the PLS-DA model has been developed based on LIBS training data. The positive identification of a threat material is indicated by the red light on a stop sign; green indicates no threat was detected, and yellow indicates additional data must be collected to confirm identification. The user-defined thresholds for class membership can be adjusted "on-the-fly" to increase or decrease the sensitivity for threat detection. Fig. 5 shows a screen capture from a demonstration video of the real-time detection of a biological agent surrogate (*B. subtilis*) with a commercial LIBS system; the interferents used for the demonstration included dust, fertilizer, soot, and mold. The surrogate was identified correctly 100% of the time within a few seconds, with no false positives. For raster scans, the test results appear as a color-coded matrix of red/yellow/green squares indicating the classification of the sample at each single-shot location.

2.3 Radiological/nuclear detection

The threat of nuclear terrorism (e.g., dirty bombs) is of grave concern for homeland security and necessitates new detection tools, since not all radiation detection devices can differentiate between radioactive isotopes, some are not capable of measuring very high radiation rates, and others require sample preparation in a laboratory. LIBS has been demonstrated for the detection of radiological and nuclear isotopes, although one primary disadvantage is the need for line-of-sight measurements (thus limiting the potential applications). A standoff LIBS system was used to demonstrate the detection of uranium under various conditions, including in soil and on surfaces such as aluminum, plastic, and ceramic tiles [30]. The observed detection limits fell within the

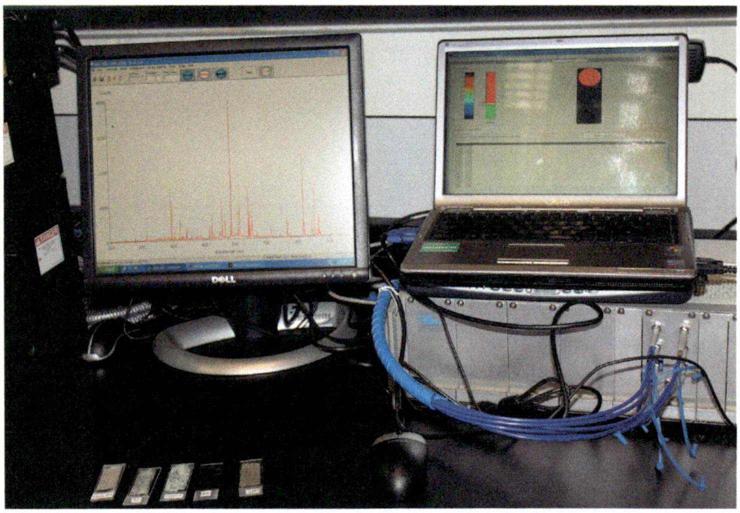

FIG. 5 Screen capture from a video demonstrating biological agent simulant identification with a commercial laser-induced breakthrough spectroscopy system and custom-designed analysis software with real-time threat analysis.

guidelines established for safe decontamination of facilities and equipment. Doucet et al. [31] demonstrated the determination of isotopic ratios for uranium-235/uranium-238 and deuterium/hydrogen with a portable LIBS system. In addition, the technique of laser ablation molecular isotopic spectrometry has enabled the use of LIBS to determine isotope ratios for lighter elements such as boron, carbon, strontium, and zinc [32]. Gaona et al. [33] demonstrated the detection of radioactive surrogates (several hundred nanograms per square centimeter) on the surface of common urban materials at 30 m using LIBS.

LIBS has also been demonstrated for various applications related to the nuclear industry because of its ability to remotely confirm material composition in areas not safely accessible by personnel (e.g., with a fiber optic system or robotic platform) and because of the ability to perform rapid, in situ analysis with no sample preparation. These applications include the characterization of nuclear-related materials such as steel [34, 35], depleted powders containing uranium and thorium [36], origin determination of yellow cake samples (powdered uranium concentrate used as the first step in the fabrication of nuclear fuel) [37], and establishment of detection limits for common nuclear fission products such as strontium, cesium, and cerium (a common surrogate for radioactive plutonium) [38]. LIBS has also been suggested for online monitoring of fusion devices [39]. For more details on the topic of LIBS for radiological/nuclear detection, the reader is referred to Chapter 17.

2.4 Explosives detection

Following the terrorist attacks on September 11, 2001 and the subsequent military campaigns in the Middle East, an urgent need for fast, accurate explosives detection arose. In particular, the threat of improvised explosive devices (IEDs) to civilian and military personnel was of significant concern. For IED detection, standoff methods for explosives detection are required for the protection of personnel and equipment. The advantages of LIBS such as the capability for rapid analysis, strong sensitivity to trace elemental concentrations (including light elements such as C and H), and the ability to focus the laser beam at long distances while collecting the plasma emission with a telescope made LIBS an ideal candidate for this application. For a more comprehensive discussion of LIBS spectra of energetic materials, the reader is referred to Chapter 18. Here, we concentrate on the development of standoff LIBS by ARL for IED detection.

Explosives detection with LIBS began at ARL in 2002. Spectra of 10 different bulk energetic materials (including explosives and propellants) were acquired with a laboratory LIBS system [40]. Observed emission features included C, H, N, O, and impurities such as potassium (K), calcium (Ca), sodium (Na), and magnesium (Mg). Most military explosives contain C, H, N, and O, with significantly more O and N than nonexplosive organic materials. Unfortunately, the entrainment of air into the laser-induced plasma with single-pulse LIBS complicates the discrimination of explosives due to the presence of O and N emission from the excited air. A single-pulse, standoff LIBS system designed by the Laserna group at the University of Málaga was field tested by a team of researchers from Málaga, ARL, Ocean Optics, Inc., and Applied Photonics, Ltd. [41]. Using a decision-making flowchart accounting for the presence of specific emission lines and ratio values, 13/15 known residue samples were correctly identified as explosive or nonexplosive with this system (86.7%), and all six test samples were correctly identified as part of a blind test. Based on the success of this preliminary demonstration for standoff explosives detection with LIBS, ARL began a program to develop an improved standoff LIBS system for IED detection.

The first breakthrough for the application of LIBS to explosives detection was the development of double-pulse LIBS, in which two laser pulses separated in time (Δt is usually a few microseconds or less) are used to sample the material. Double-pulse LIBS not only increases the intensity of the spectral signatures (compared to single-pulse LIBS with the same pulse energy), it also reduces the amount of air entrained in the plasma since the shockwave from the first laser pulse produces a rarified environment [42]. Although many configurations of double-pulse LIBS are possible, the most applicable for standoff detection is when the two laser pulses are collinear. The improved discrimination of explosives and nonexplosives with double-pulse LIBS was first demonstrated in the laboratory [43] and then at a standoff distance of 20 m [10]. Fig. 6 compares the standoff double- and single-pulse spectra of an explosive residue (cyclotrimethylenetrinitramine, RDX) on an Al substrate. The overall intensities for the double-pulse spectra are higher and the O/N ratios are more representative of the sample rather than just the air. For example, the O/N ratios are indistinguishable for Al and RDX with single-pulse LIBS, but O/N is significantly higher for RDX than Al using double-pulse LIBS. These features of double-pulse LIBS spectra enable better discrimination of explosives.

Because of the advantages of double-pulse LIBS for explosives detection, most of the ST-LIBS systems developed for ARL incorporated a collinear double-pulse laser configuration (Fig. 7). The Gen 1 system suffered from poor beam quality at far-field and inefficient light collection, limiting the signal-to-noise ratio of the LIBS spectra. Gen 2 had better, more powerful lasers, a larger telescope, and a three-channel charge-coupled device (CCD) grating spectrometer with broadband coverage from the ultraviolet to the near infrared. The Gen 3 system was designed to operate at 1540 nm (a

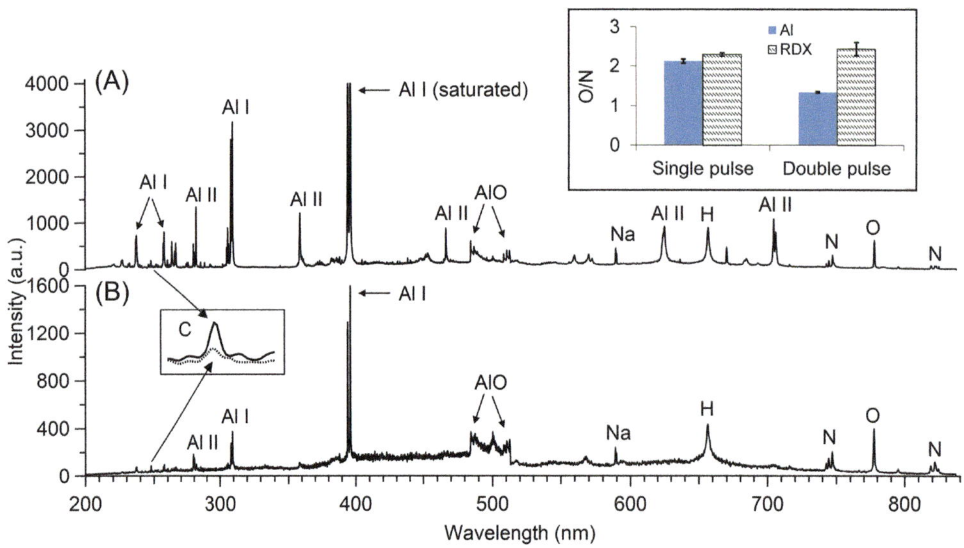

FIG. 6 Laser-induced breakthrough spectroscopy (LIBS) spectra of cyclotrimethylenetrinitramine (RDX) residue on Al acquired at 20 m using the Gen 2 system with (A) a double laser pulse with 275-mJ total energy ($\Delta t = 3\ \mu s$) and (B) a single 275-mJ laser pulse. The *left inset* directly compares the C intensity for double (black) and single (gray) pulse and the *top right inset* compares the O/N ratios for pure Al and RDX on Al for single- and double-pulse LIBS [10].

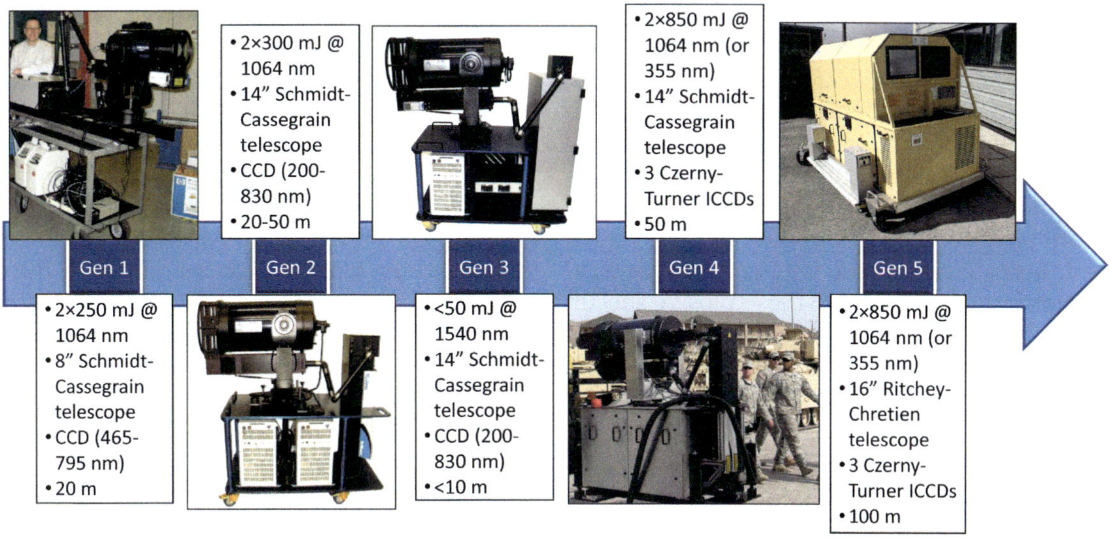

FIG. 7 Five generations of standoff laser-induced breakthrough spectroscopy systems developed for the Army Research Laboratory for explosives detection. The laser pulse energies and wavelengths, telescope size/type, spectrometer type and wavelength range, and maximum practical standoff distance for each system are specified. The Gen 2 system provided most of the data (acquired in indoor ranges) for our published results. The Gen 4 system was repeatedly field tested and the Gen 5 system was developed to be extremely ruggedized.

more eye-safe wavelength [44]), but suffered from poor conversion efficiency and low laser pulse energy. Gen 4 was designed for field testing and performed reasonably well at up to 50 m in desert conditions. The lessons learned from field testing of the Gen 4 system led to the design of a more ruggedized Gen 5 ST-LIBS with a military-grade telescope, completely automated focusing, firing and data collection, and built-in air conditioning. However, even the ruggedized Gen 5 system still required routine maintenance and alignment by qualified personnel, especially under extreme field conditions. An expensive (and fragile) spectrometer consisting of three Czerny–Turner grating spectrometers with intensified-CCD detectors was used for field testing to provide nearly broadband coverage, i.e., the C, CN, C_2, H, N, and O emission lines from explosives could all be observed simultaneously. This system was tested outdoors in the winter on an active test range and obtaining useful spectra at 100 m proved challenging. Because of their higher energy lasers, both the Gen 4 and Gen 5 systems could be converted to 355-nm operation for increased eye safety (at the expense of alignment stability and laser pulse energy). More details regarding the ARL ST-LIBS system capabilities can be found elsewhere [45].

Another important breakthrough for explosives detection with LIBS was the application of chemometrics to LIBS [1]. Until the use of chemometrics to analyze LIBS data became widespread, LIBS was considered nothing more than an elemental detector and most people did not believe it could be used for the discrimination of molecular species with similar stoichiometries. Chemometrics enables not only the discrimination of explosives from nonexplosives with LIBS [46, 47], but also the discrimination of different types of explosives [46, 48], organic residues on metallic [49] or organic [50, 51] substrates, and complex mixtures of explosives [46, 47]. PLS-DA models provided the fewest false positives or false negatives for explosives discrimination in practical applications, i.e., when unknown residues and/or substrates are likely to be present. The choice of input variable selection significantly improved the ability to discriminate explosive residues in the presence of interferents on complex substrates. In most cases, using the entire broadband spectrum did not perform the best. Instead, combinations of emission intensities for features relevant to the explosive spectra and ratios of those emission features provided the best discrimination (e.g., [46, 48–51]). In some cases, the fusion of model results from PLS-DA models built using full spectra or intensities/ratios improved the explosive classification rate [49–51].

Since one of the primary goals at ARL for ST-LIBS was vehicle-borne IED detection (i.e., car bombs), our focus was to demonstrate the discrimination of explosive residues on vehicle surfaces. Trace explosives are easily transferred via the fingerprints of a person who has been handling explosive devices [52]. Using ST-LIBS to sample surfaces that have a high degree of probability to contain fingerprint residues (e.g., on or near door handles, trunk latches, etc.) thus provides a viable route for standoff detection of hidden explosive devices at a vehicle checkpoint, for example.

Car panels cut from vehicles of different makes, models, years, and colors were obtained from local junkyards [51]. The optimal timing for the Gen 2 ST-LIBS system based on minimizing the error in a PLS-DA model to discriminate explosive residues on the car panels at 30 m was determined: 1-μs gate delay with a 500-ns interpulse delay. Although combinations of different laser pulse energies were investigated, full laser power for both lasers produced the best signal-to-noise ratio and the most effective explosive discrimination. A photo of a teal car panel from a 1993 Volkswagen Jetta with laser ablation marks, spectra of various explosive and nonexplosive residues acquired on the teal panel at 30 m, and two emission ratios averaged over 50 shots for each of the different residue samples is shown in Fig. 8.

In total, over 4000 single-shot spectra of various explosive (RDX, Composition-B) and nonexplosive (fingerprint oil, diesel fuel, lubricant oil, road dust, and sand) residues were acquired on seven different car panels for this study [51]. Although the spectra of the residues are all very similar to the blank car panels (Fig. 8B), differences in the relative intensities of the emission features were used to discriminate the samples with PLS-DA. Using just two emission ratios—$(O + N)/(C + H)$ and C_2/CN—the explosive residues were separated from the nonexplosive residues (Fig. 8C). However, the relative standard deviations of the ratios for each sample ranged from 10% to 50%, so many additional features from the residue spectra are required to correctly classify single-shot test spectra. The combination of 132 emission intensities and ratios from the explosive residue spectra were used as inputs for the PLS-DA model and compared to model results using the whole spectra. While the true-positive rate was slightly lower for the ratio model and the false-positive rate was slightly higher for test sets containing sample types included in the model, by fusing the results of the two models the overall false-positive rate was significantly lowered compared to the whole spectra model (0.16% vs. 1.9%), at the expense of the true-positive rate (97% vs. 99.5%). Overall, the correct classification rate of numerous residues on multiple car panels (including residues and car panels not incorporated in the model) from ST-LIBS spectra with PLS-DA was very promising.

Although concerns about the fieldability of ST-LIBS systems (to be operated by military personnel) and the minor unavoidable damage to vehicle surfaces following laser ablation (Fig. 8A) were issues raised during the development of the ST-LIBS systems, the biggest limitation to deploying the ST-LIBS systems concerned laser eye safety. Even "eye-safe" wavelengths such as 1540 nm [44] can cause

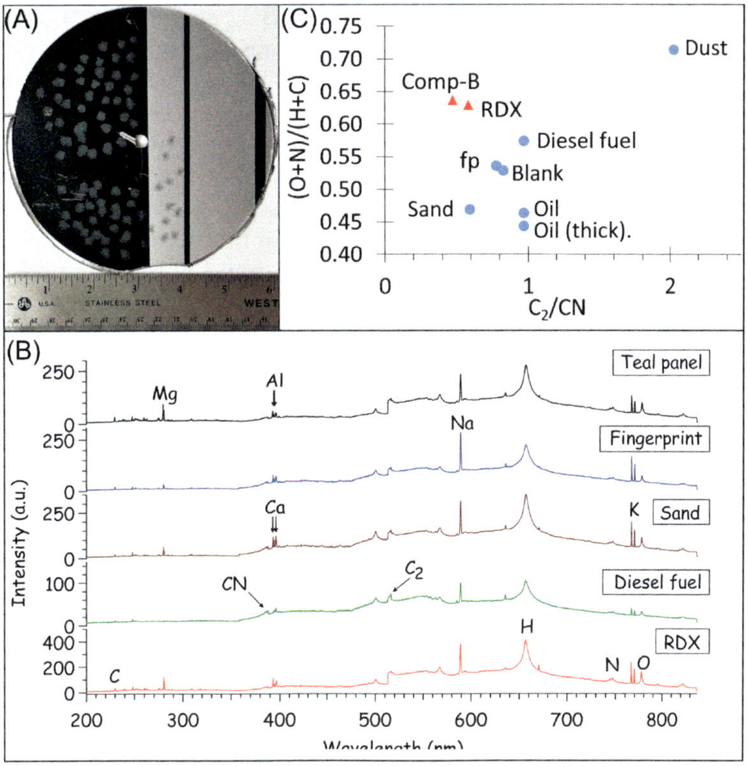

FIG. 8 (A) Photo of a teal car panel with laser ablation marks, (B) spectra of the car panel with and without various residues acquired at 30 m, and (C) separation of the samples based on two emission ratios: (O + N)/(C + H) and C_2/CN [51].

permanent eye damage at the laser focus, especially at the laser pulse energies required to obtain good LIBS signal at standoff distances. While none of these issues were insurmountable, the wind-down of military operations in the Middle East eventually decreased urgent military interest in standoff IED detection.

2.5 Landmine detection

The identification of buried landmines is an important capability for US forces in many foreign theaters and for humanitarian demining to protect civilian populations in areas of current and former conflict. In conjunction with a manual probe (i.e., a mine prodder), the hand-held metal detector has been the US military's primary landmine detection tool since World War II. Since most soils contain trace metals [53, 54], metal detectors can lead to a high number of false alarms. In addition, the widespread use of plastic mines with little to no metal content also poses a significant problem for metal detectors. ARL first demonstrated the identification of landmines with a commercial LIBS system and suggested the use of a portable LIBS sensor with a hand-held probe through which the laser could be focused to directly interrogate the surface of a buried object [55]. Linear correlation was subsequently used to discriminate landmines and clutter items, with 92%–93% correct mine/no-mine classification and 78%–80% correct mine identification [7]. The MP-LIBS system (top left of Fig. 1) was used to acquire single-shot spectra of 15 types of antipersonnel and antitank landmines, as well as eight plastic and wood mine confusants [56]. The mine/no-mine classification success was 90% using linear correlation, with 95% of the samples identified as landmine casings correctly identified as a specific mine type. Bohling et al. [57] demonstrated a compact LIBS system with a low-pulse energy laser (600 μJ) delivered to the target via a fiber optic cable. They compared the time evolution of the CN emission for two plastic landmine casings, an aluminum can, a brass bullet casing, a wood mine casing, and a rock. The system was field tested on pure metals and coated metals (including an antipersonnel mine) in a follow-up study [58], with a correct mine identification rate of 80% and false-positive rates <5% based on the use of PCA and emission decay times to train a neural network.

To improve the identification of specific landmine types, PCA and PLS-DA models were developed and tested using 3000 single-shot LIBS spectra of 39 inert mine casings (14 PMA-1A, 9 M14, 6 PMN-1, 2 TS-50, 2 VS-MK2, 1 TNG-M80, 1 VS-HCT2, 1 VS 1.6, 1 TMA-3, 1 TM-62P3, and 1 POM) and seven confusant materials (1 wood, 2 black plastic, 4 nylon) [59]. Two of the landmines (PMN-1 and VS-MK2) had casings with distinctly different caps (rubber) and bottom (plastic) pieces, so both areas were sampled. A custom laboratory LIBS setup was required to accommodate the large antitank mines (Fig. 9). This system used a 1064-nm, 150-mJ laser with a broadband spectrometer (200–940 nm). For each sample, one cleaning shot followed by 25 successive laser shots was acquired for at least two different locations. Example single-shot spectra are shown in Fig. 10.

The loadings plot for the PCA model built using 36 different landmines and five confusant samples showed that the strongest contributions to the loadings were due to the organic species (C, CN, C_2, and H), contribution from air (O, N), and Ca. Since all of the samples have these emission features, PCA could not be used to discriminate the mines or confusants. Because the PLS-DA model generates latent variables that maximize the interclass variance (i.e., the differences in the spectra based on the assigned

FIG. 9 Photo of a custom laser-induced breakthrough spectroscopy setup with sufficient clearance for large antitank landmines (two small antipersonnel landmines are shown). The laser pulse is directed through a pierced mirror down onto the mine surface. The emission from the laser-induced plasma is focused by the off-axis parabolic mirror (*left*) onto the fiber optic bundle (*right*) for the seven-channel spectrometer (*background*) [59].

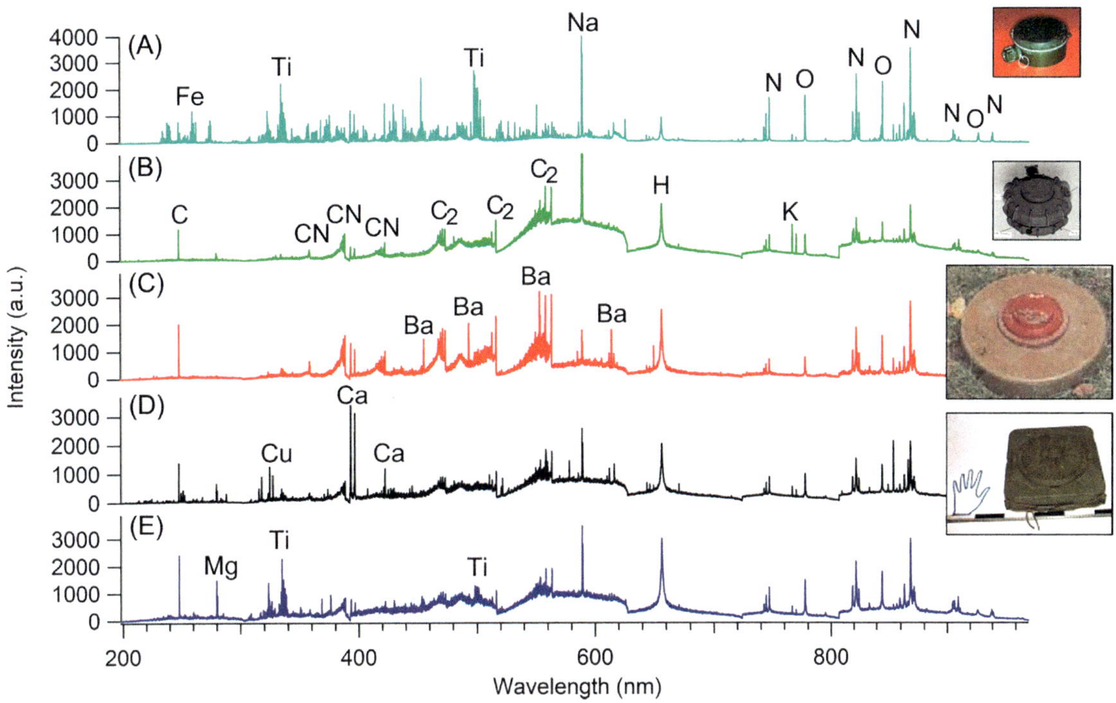

FIG. 10 Single-shot laser-induced breakthrough spectroscopy spectra of antipersonnel mines (A) PMN-1 (Yugoslavia) and (B) TS-50 (Italy), antitank mines (C) TM-62P3 (Russia) and (D) TNG-M80 (USA), and (E) the black plastic confusant [59].

sample class), a model trained on 17 sample classes with the full spectra from 42 individual mines (including caps and bases where applicable) and six confusants was able to identify the trace impurities important for each mine/confusant type. For example, the titanium (Ti) lines contribute to the PMN-1 class (Fig. 10A), the barium (Ba) lines (Fig. 10C) contribute to the TM-62P3 class, and the copper (Cu) lines (Fig. 10D) contribute to the TNG-M80 class.

A test set consisting of 1500 additional spectra from each mine and confusant in the model resulted in 99.0% correct classifications, with 1.8% misclassifications. Of the misidentifications, 2.6% resulted in a confusant being identified as a mine (false positive) and 1.6% resulted in a mine being identified as a confusant (false negative). A second validation set consisting of 450 spectra from six mines and two confusants not included in the model and spectra of the POM mine acquired at 90 mJ instead of 150 mJ was also tested against the model. A correct classification rate of 98.6% was achieved, with 4.6% misclassification. If more restrictive criteria were applied to determine class membership (i.e., a sample was only assigned class membership if the probability was >50% for only one class in the model), the misclassification rate was reduced to 0.4%. While 99.3% of the mines in the validation set were classified with other mines in the model, only 15/100 confusant spectra were correctly identified as not mines (84 were unclassified). Increasing the number of confusant classes in the model would significantly improve these results.

Overall, we have demonstrated that landmine detection is an excellent application for LIBS. The combination of field portability with the ability to deliver the laser beam via a mine prodder, distinctive emission features from plastic landmines of various origins, and well-trained chemometric algorithms can result in extremely high classification rates with relatively low false positives and negatives. Additional studies are needed to test the ability of LIBS to discriminate mines buried in soils of different types; however, the judicious use of cleaning shots should negate any possible environmental influences on the emission spectra.

2.6 Toxic gas detection

LIBS has also been developed to detect toxic gases during fire suppression tests of halon alternatives for use in military applications. Because of environmental concerns due to the destruction of the ozone layer by halons (halogenated hydrocarbons) containing bromine and chlorine (e.g., CF_3Br and CF_2ClBr), new fire suppressants were required for military vehicles that quickly suppress flames, are compatible with storage materials, meet environmental safety standards, and produce a minimum of toxic gases. Alternative candidates investigated with LIBS included C_2F_5H, CF_4, CF_3H, and CF_2H_2 [60]. Hydrofluorocarbons, as potential halon replacements, tend to produce HF, which is toxic and corrosive. Single-shot LIBS measurements were essential for determining the relative distribution of species in the flame environment with nanosecond time resolution (using a monochromator-photomultiplier tube detection system). The LIBS system enabled real-time measurement of HF concentration down to the ppm level. A subsequent study expanded the list of potential candidates to include C_3F_7H and $C_2F_4H_2$ and optimized an intensified photodiode array spectrograph that could be fielded [61]. After assembly and laboratory testing, the LIBS sensor was transitioned to field use in a hostile measurement environment with intense flames.

2.7 Lead content in solder

Due to increasing environmental and health concerns of lead-containing solders, new "green" replacements have been developed. These "lead-free" solders predominantly contain tin (Sn) alloyed with trace amounts of Cu or silver (Ag). The mixing of lead (Pb) and "lead-free" solders may result in premature failure from brittle or cracked joints. Thus the continued reliability of fielded military devices depends on ensuring that repairs are performed using compatible solder materials. LIBS provides a fast, simple, and, most importantly, in situ method for determining whether a solder material (e.g., from a printed circuit board) contains Pb and/or Sn [62, 63].

LIBS spectra of multiple lead-containing (2.2-mm and 1.1-mm diameter leaded solder from unknown manufacturers, 1/32-in. silver-bearing solder from an unknown manufacturer, and a 60/40 Sn/Pb solder from Bow Solder Products) and "lead-free" solders (rosin core, acid core, and no flux Dutch Boy lead-free solders) were acquired with a commercial LIBS system using a 72-mJ Nd:YAG laser pulse (1064 nm) and a broadband CCD spectrometer (200–960 nm, 0.1-nm resolution, 250-μs delay, 2.1-ms gate width) [62, 63]. The solders were prepared in two ways: by mechanically flattening the solder and by soldering the material onto a copper gasket. The spectra of the solders are shown in Fig. 11. Besides Pb, other elements detected in the lead-containing solders included Al, Ag, Ba, bismuth (Bi), cerium (Ce), lanthanum (La), and Sn. The "lead-free" solders contained Al, indium (In), and Sn. Note that all three of the "lead-free" solders contained trace amounts of Pb; subsequent investigation of two commercial lead-free solders with LIBS by Chen et al. [64] measured Pb concentrations of 160 and 230 ppm.

Soldering the material to a gasket resulted in significant spectral differences (Fig. 12). Overall, the signal intensity was lower due to the organic

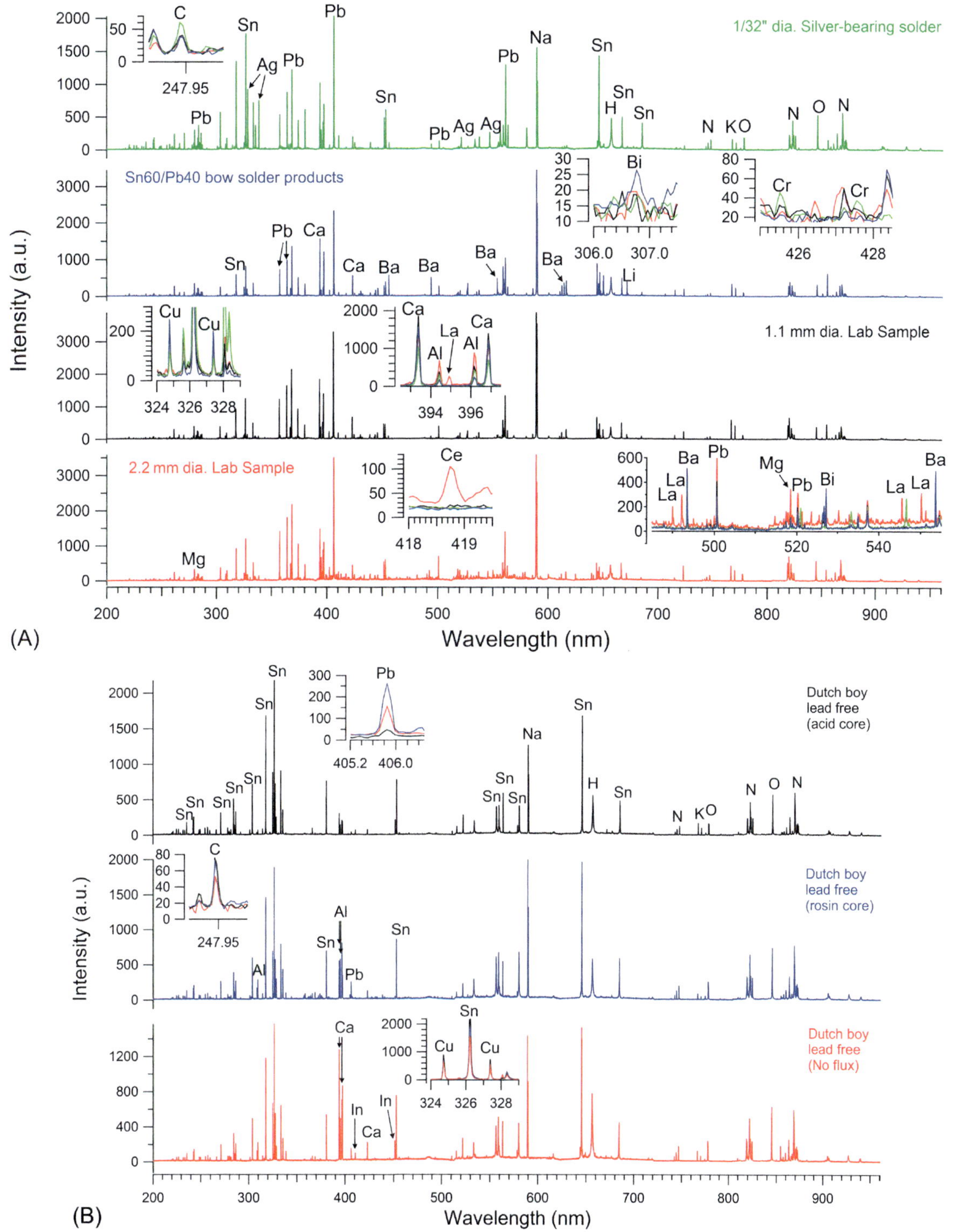

FIG. 11 Laser-induced breakthrough spectroscopy spectra of (A) lead-containing and (B) lead-free solders prepared by flattening the solder. The insets compare the emission intensities from specific elements [62].

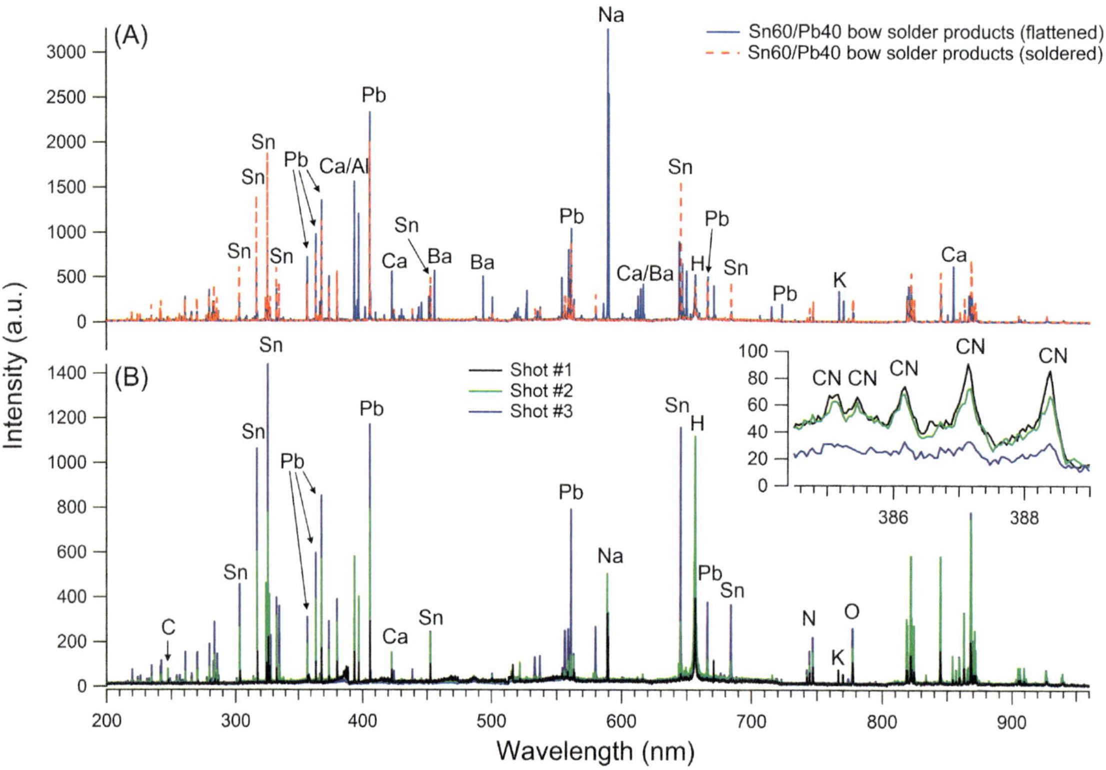

FIG. 12 Laser-induced breakthrough spectroscopy spectra of an Sn/Pb solder (A) flattened versus soldered on a gasket and (B) after successive laser shots at the same location [62].

flux material suppressing the laser–material interaction. Some emission features observed in the presoldered spectra disappeared (e.g., Al, Ba, Ca, Na, K) and some new features appeared or increased in intensity (Sn I and Sn II). In particular, the Pb/Sn ratio decreased after soldering; although the difference in their melting temperatures is <100°C, the boiling temperature of Pb is almost 900°C lower. Strong CN bands were present in the soldered material due to the presence of an organic coating (flux) on the solidified solder surface. The CN bands disappeared after several sequential laser shots at the same position. These results emphasize the importance of considering the type of data (presoldered vs. soldered spectra) input in the training set for the chemometric model, since easily vaporized species may become depleted and other diffusion or stratification effects [65] may occur. In addition, accurate analysis of solder may require multiple successive laser shots to penetrate through surface flux layers. Future studies should investigate the effect of soldering conditions (e.g., temperature and duration) and the target surface on the LIBS signal.

To evaluate the ability to discriminate solder materials with LIBS, 95 emission features from Ag, Al, Ba, Bi, C, Ce, chromium (Cr), Cu, iron (Fe), In, La, lithium (Li), Mg, Pb, Si, and Sn were identified. Individual emission lines for each element were background subtracted and summed to yield 16 emission intensities. The summed intensities were then used to create 231 ratios to be used as input variables for the model. A PCA model of the samples demonstrated that they could be discriminated based on lead content (i.e., lead vs. no lead), but individual samples could not be discriminated. A PLS-DA model was trained on 332 single-shot spectra (both flattened and soldered material) and tested with the remaining 82 spectra. The training and test set data for the 60/40 Sn/Pb Bow Solder Products material are shown in Fig. 13. Flattened and soldered spectra are shown as circles and triangles, respectively; the model data is represented by closed symbols, while the test data is shown as open symbols. The overall correct identification rate for each specific solder type in the test set was 96.3%, with 2.8% incorrect identifications across all classes (samples can be classified with more than one model class based on the Bayesian threshold, or none). With a two-class PLS-DA model, 100% of the test samples were correctly classified as lead containing or lead free. Future models should include types of solder materials not used to train the model in the test set, and solder affixed to different components under varying conditions.

Subsequently, Bol'shakov et al. [66] used LIBS to perform quantitative determination of Pb in 12 different solder materials, with single-shot detection limits at or below 10 ppm. Ismail and Yoo [67] compared LIBS with other analytical techniques for monitoring Pb in solder products. They found that the accuracy and sensitivity of LIBS were comparable to the industry standard inductively coupled plasma optical emission spectroscopy, without the need for time-consuming sample preparation. Galbacs et al. [68] were able to quantitatively determine the Sn and Pb in flattened solder materials using multipulse LIBS with accuracies of 1%–5%. Carvalho et al. [69] used LIBS to investigate the metal composition of printed circuit boards from a cell phone and a computer mouse; the

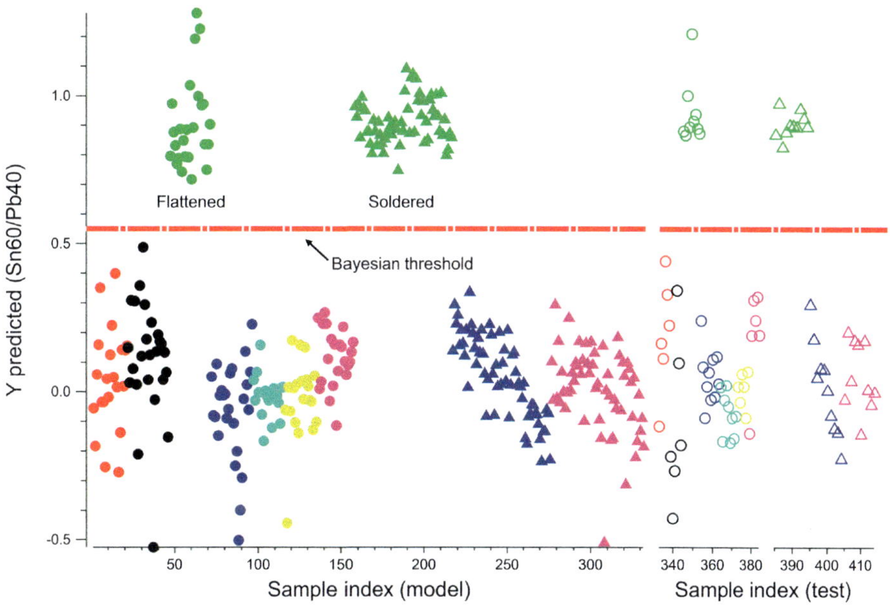

FIG. 13 Predicted class membership for the Sn/Pb solder based on a partial least squares discriminant analysis model trained using laser-induced breakthrough spectroscopy spectra of seven different solders (either flattened [*circles*] or soldered [*triangles*] onto a gasket). All of the Sn/Pb solder test spectra (*open symbols*) were correctly classified by the model, with no false identifications from the other test samples [62].

mouse circuit board was found to contain Pb from a soldered connection. Zdunek et al. [70] reviewed machine-learning methods for analyzing LIBS spectra, with example data from solder spectra. Most recently, Speranca et al. [71] used LIBS to perform semiquantitative analysis of the chemical species present in solder masks, which are polymers (with many additives) used to provide a thin, protective coating on printed circuit boards.

2.8 Steel alloy identification

Benét Laboratories, part of the US Army Armament Research, Development and Engineering Center (now the Combat Capabilities Development Command Armament Center), designs, develops, and provides engineering and field support for the US Army's larger caliber armament systems (including cannons, mortars, and tank gun mounts). Based on their need to differentiate steel samples with small compositional variations in C and S content using a commercial LIBS system, ARL performed a proof-of-principle study [72]. The main challenge was to avoid the experimental complexity necessary to use the less convoluted atomic emission lines in the vacuum ultraviolet region (e.g., [73]) and quickly determine whether the provided steel sample was 4130 steel. Spectra from a commercial LIBS system (Nd:YAG laser, 300 mJ, with a seven-channel CCD spectrometer integrated over 1 ms after a 1.25-μs delay) were acquired under argon (Ar). Fourteen 4130 steel samples (plus two base plate samples made up of an unidentified steel alloy) were provided by Benét Laboratories for analysis, and compared to three standard reference materials (4130 steel, 1761a low-alloy steel, and C1296 SS) from the National Institute of Standards and Technology. Averaged spectra are shown in Fig. 14. The first laser pulse at each

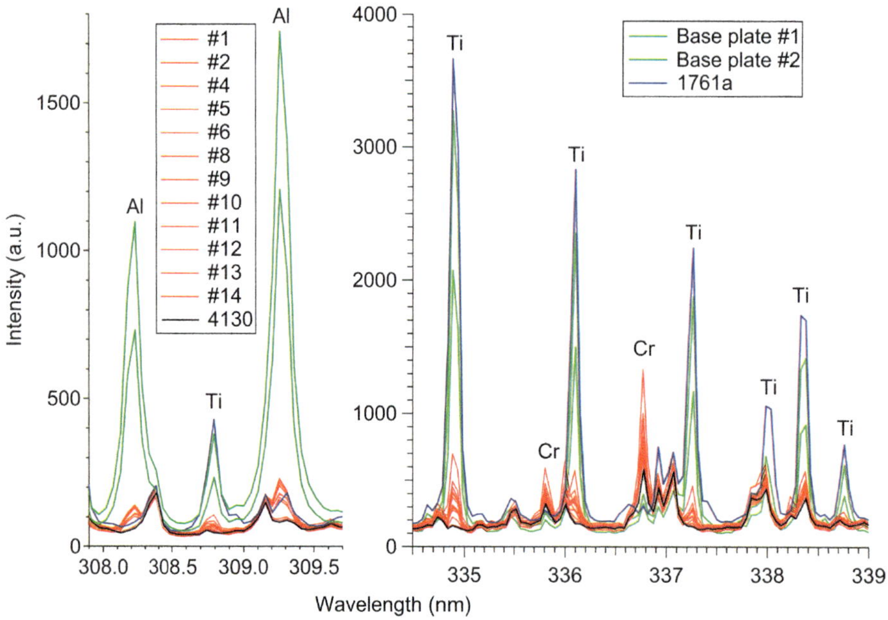

FIG. 14 Selected regions from the laser-induced breakthrough spectroscopy spectra of Benét Laboratories steel samples and National Institute of Standards and Technology standard steel alloys 4130 and 1761a acquired under Ar [72].

spot was used as a cleaning shot to remove surface contamination.

Linear correlation was used to confirm if the samples provided by Benét Laboratories were 4130 steel; however, most of the numbered samples correlated well with both the 4130 (average $r^2 = 0.95$) and 1761a (average $r^2 = 0.94$) standards. None correlated with the C1296 steel (average $r^2 = 0.72$), which has 27.9% Cr compared to 0.91% in 4130 and 0.222% in 1761a. The base plates correlated slightly better with 1761a steel (average $r^2 = 0.83$) than 4130 steel (average $r^2 = 0.80$). The trace Ti and Cr content was ultimately used to differentiate the 14 Benét 4130 samples and two base plate steel samples (which also contain Al and Mg impurities not present in the 1761a steel). The only C emission line detected in the 200–1000-nm region (247.8 nm) was convoluted with an Fe line, as expected, precluding calibration of the C concentration with the commercial LIBS system. In addition, none of the prominent S emission lines (869.47, 921.29, 922.81, and 923.75 nm) were observed. Despite the lack of spectral information for the C and S emission, the steel alloys could be differentiated with a commercial LIBS system using other trace impurities, a process easily amenable to chemometric analysis with a supervised technique such as PLS-DA.

3 Material characterization

In addition to CBRNE detection and other material identification applications discussed in Section 2, LIBS has been applied to characterize materials of interest to the military and the processes used to synthesize or modify them. The following sections describe examples of the characterization of energetic material composition and performance with LIBS or the monitoring of various processes involved in energetic materials synthesis or ordinance disposal.

3.1 Energetic material characterization

As an extension of the LIBS-related research performed at ARL for explosives detection, we have begun using the processes involved in the laser–material interaction initiated with a nanosecond-pulsed laser to characterize novel energetic materials [74]. The conditions (temperature, pressure, electron density, etc.) in the laser-induced plasma are very similar to those found in the chemical reaction zone behind the detonation wave in an explosive event. In addition to the formation of the laser-induced plasma, relevant processes also include the high-temperature chemistry of the ablated material in the plasma as the plasma is cooling, the formation of a laser-induced shockwave, which expands into the air above the sample, and the laser-induced deflagration (i.e., fast combustion) of particles ejected off the sample surface (Fig. 15). The time-resolved LIBS emission provides information about the composition of the energetic material and the high-temperature chemical reactions [75–80], the velocity of the laser-induced shockwave provides information about the energy content of the material (since exothermic reactions in the plasma increase the plasma temperature and accelerate the shockwave) [81], and the laser-induced deflagration reactions provide information about the combustion behavior of the energetic material [82].

When the laser system and diagnostics have been properly calibrated for the experimental conditions, the laser-induced shock velocities can be directly correlated to the measured detonation velocities of explosives from large-scale testing, which is significantly more expensive and hazardous than laser ablation experiments [81]. This technique is called laser-induced air shock from energetic materials (LASEM). The subsequent laser-induced deflagrations also provide information about the extended energy release of energetic materials on the millisecond timescale, analogous to the detonation fireball

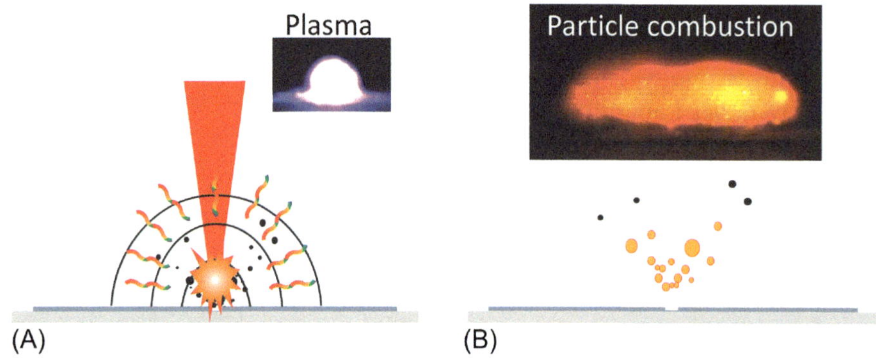

FIG. 15 The interaction of a focused, nanosecond-pulsed laser with a thin residue of energetic material produces (A) the laser-induced plasma consisting of the highly ionized and excited ablated material that emits radiation and laser-induced shockwaves, which expand into the air above the sample on the microsecond timescale; (B) on the millisecond timescale, large particles ejected from the sample surface by the laser-induced shockwave can begin combusting as they pass through the air heated by the plasma and the passage of the shockwave. The photos are snapshots from the high-speed video of the laser excitation event.

[82]. Generally, the most powerful explosives have very high laser-induced shock velocities (i.e., detonation velocities) and weaker, shorter deflagration reactions because most of their energy has been expended on the microsecond timescale. A key advantage of LASEM is that it requires very small quantities of material (tens of milligrams or less), and thus serves as a useful prescreening tool to evaluate the potential performance of new energetic materials under development prior to scale-up for detonation testing.

In addition to calibrating and validating LASEM with conventional military explosives [81], LASEM has been used to evaluate the performance of novel energetic materials currently under development [83]. One of the energetic materials investigated with LASEM in that study, dihydroxylammonium 5,5′-bistetrazole-1,1′-diolate (TKX-50), had an estimated detonation velocity of 9.56 ± 0.28 km/s based on the measured laser-induced shock velocity. LASEM testing of a TKX-50 sample produced via an alternative synthesis process designed to facilitate larger-scale production determined that the estimated detonation velocity was only 9.29 ± 0.25 km/s, nearly 3% lower than expected [84]. Along with the lower laser-induced shock velocity, the LIBS spectra obtained during the LASEM experiment showed the presence of an unusually large quantity of Na; the 589 nm Na emission was even visible in the high-speed video of the laser excitation event (Fig. 16). Analytical characterization of the scaled-up TKX-50 sample revealed that the alternate synthesis method resulted in the presence of a 5% disodium 5,5′-bis(tetrazole-1-oxide) tetrahydrate salt impurity, which reduced the explosive performance despite being slightly energetic itself.

In addition to conventional organic explosives, LASEM has been applied to explosive formulations with metal additives, including either Al or boron (B) [85]. Al and B both have very high heats of combustion and are therefore of interest for enhancing explosive performance. The addition of Al to military explosives results in enhanced blast effects, but with a slight reduction in detonation velocity. LASEM results confirmed this behavior for micron-sized Al particles (m-Al), which resulted in lower laser-induced shock velocities and stronger

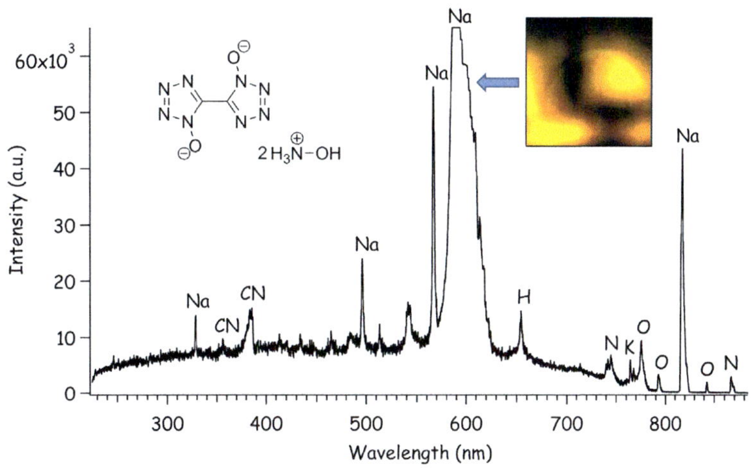

FIG. 16 Emission spectrum of laser-excited TKX-50 (molecular structure shown) contaminated by a disodium bistetrazole salt impurity. The first frame from the high-speed video shows strong Na emission partially obscuring the laser-induced shockwave [84].

laser-induced deflagration reactions. Commercial m-Al particles do not react fast enough to increase the detonation performance of explosives; while nanoaluminum (nano-Al) is expected to react faster due to its large surface-to-volume ratio, its reaction is inhibited by the presence of the native oxide layer that forms on Al particles (alumina, Al_2O_3). Recently, we have been using LASEM to investigate methods for catalyzing the fast reaction of Al in explosive formulations via novel synthesis or surface treatment methods to improve the detonation performance of explosives (Fig. 17) [86–89]. While coating the Al in a material like graphene oxide (GO) that undergoes exothermic reactions upon rapid heating can increase the microsecond timescale energy release from Al [86], the most successful methods so far involve replacing the Al_2O_3 layer with another passivating material like C [87] or adding an oxidizing material such as aluminum iodate hexahydrate (AIH) to the nano-Al surface [88, 89].

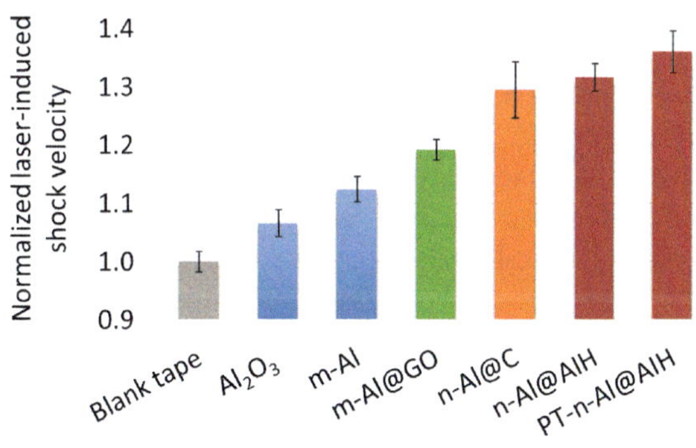

FIG. 17 Laser-induced shock velocities (normalized to the blank tape substrate) for Al_2O_3, micron-sized Al (m-Al), m-Al wrapped in graphene oxide [86], nanoaluminum (nano-Al) with the oxide layer replaced by carbon (n-Al@C) [87], commercial nano-Al coated in aluminum iodate hexahydrate (AIH) [88], and plasma-treated nano-Al coated in AIH [89].

Other groups are also using LIBS to evaluate explosive performance. Farhardian et al. [90] used LIBS to study the chemical structure of composite solid propellants with performance-enhancing additives. Rezaei et al. [76–78] demonstrated the use of LIBS emission features such as the Al/O or AlO/Al ratios to determine the detonation velocity and pressure of aluminized explosive formulations. Kalam et al. [80] studied the influence of energetic material functional groups on the formation pathways for CN and C_2, and found that the $(CN + C_2)/(C + H + N + O)$ ratio correlated well with detonation parameters such as oxygen balance, detonation velocity, detonation pressure, and detonation energy. Finally, Kimblin et al. [79] demonstrated that the chemical reactions in laser-induced plasma involving Al and AlO can be used to understand the chemical reactions of detonating explosives under controlled laboratory conditions.

Laser ablation has also been used to investigate the aging of explosives. Gottfried et al. [91] used LASEM to evaluate the performance of composite energetic materials containing hexanitrohexaazaisowurtzitane (CL-20) that had been artificially aged for 5 years at either 70 or 100°C. Not only did aging result in decreased (estimated) detonation performance, it also resulted in increased thermal sensitivity for the formulations. Analytical characterization revealed that the evaporation of included solvents in the material produced significant damage to the energetic crystals. Farhadian et al. [92, 93] used LIBS to compare C_2, CN, and AlO intensities in aged composite propellants. Similarly, Ahmadi et al. [94] used LIBS to study the effect of accelerated aging on the mechanical properties of plastic-bonded explosives.

3.2 Monitoring catalyst deactivation

LIBS has been used to investigate the deactivation or concentration of catalysts for many applications, including automotive catalytic converters [95, 96], online monitoring of industrial processes [97], oil or petroleum refining [98], selective hydrogenation [99], process control/design [100], and other industrial applications (Chapter 19). Recently, a new application for LIBS related to the chemical synthesis of the most powerful military explosive, CL-20, was introduced. Belyani et al. [101] used a commercial LIBS device to monitor the deactivation of a $Pd/(OH)_2/C$ catalyst involved in the synthesis of CL-20. The spent catalysts had lower palladium (Pd) and H intensities and stronger C_2 bands. LIBS thus provides a fast, cost-effective method for monitoring catalyst efficiency after each use to determine when it needs to be replaced, without the need for extensive sample preparation. Since catalyst deactivation is one of the most important problems associated with the notoriously costly synthesis of CL-20, LIBS could potentially alleviate some of the expenses associated with CL-20 production.

3.3 Monitoring decontamination processes

As discussed in Section 2.1, LIBS has been used to monitor decontamination processes such as the removal of CWAs from CARC panels. The Non-Stockpile Chemical Materiel Project funded ARL to assess the capability of LIBS for the detection of As surface contamination on the walls of a blast chamber at Idaho National Laboratory used to demilitarize chemical weapons, before and after laser cleaning [102]. For this application, the detection of As in surface concentrations of approximately 1.0 μg/cm^2 on steel was required. A commercial LIBS system (100 mJ at 1064 nm, 7 ns) with the sample chamber bathed in Ar was used to obtain LIBS spectra from As-spiked 316 SS and mild steel samples with concentrations ranging from 0 μg/cm^2 (blank) to 27.6 μg/cm^2. Only a few As lines within the spectrometer coverage (238–1000 nm) were relatively free from interference by nearby

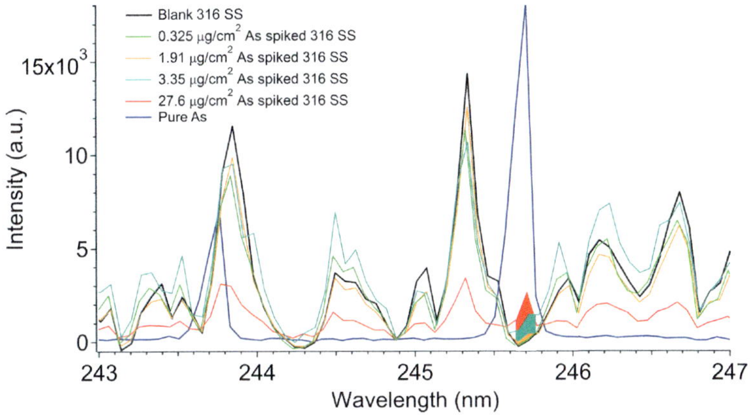

FIG. 18 Laser-induced breakthrough spectroscopy spectra of 316 stainless steel (SS), pure As, and 316 SS spiked with four different concentrations of As. The atomic emission line for As at 245.7 nm was used to determine the presence of As on the 316 SS sample; the *shaded areas* highlight the As emission from the spiked substrates [102].

Fe, Cr, or nickel (Ni) lines. Fig. 18 shows a region of the spectrum near the As line at 245.7 nm; this line is still partially convoluted with an Fe line at 245.9 nm.

A mild steel sample containing fewer alloyed metals had significantly fewer emission features; the As line at 286 nm was free from interference. Arsenic concentrations as low as 0.13 μg/cm^2 were observed on this substrate. Spectra from a contaminated educator tube removed from the Idaho National Laboratory blast chamber (As concentration unknown) indicated the presence of As. The educator tube was made of mild steel, thus demonstrating the feasibility of LIBS for monitoring As surface contamination on "real-world" substrates. Using a higher resolution spectrometer with a broader wavelength range (below 238 nm, where numerous As lines free of interference are observable) would enable extension of this application to more complex substrates such as 316 SS.

4 Forensic analysis

LIBS has increasingly been applied to different forensic analysis problems for detecting and investigating security incidents or violations of organizational rules or procedures. The following sections describe specific defense-related applications, including the provenance determination of weapons-related hardware and residues, evaluation of soil contamination from test ranges, and monitoring or troubleshooting of military equipment.

4.1 Provenance determination of spent cartridge casings

The identification of the source of origin (or provenance) of many different materials has been demonstrated with LIBS, including artwork [103], archeological samples [104, 105], gems [106, 107], geological samples [108, 109], nuclear-related materials [37, 110], and consumer goods [111, 112]. This application is of particular interest to the military for weapons-related materials that may be supplied via illicit markets. For example, LIBS has also been demonstrated for the provenance determination of unmarked cartridge casings [113]. LIBS spectra for six different types of spent cartridges (with 10 replicate samples) were acquired with a commercial LIBS system (100 mJ, 1064 nm, 7 ns). Unspent cartridges (i.e., before being fired out of a weapon) consist of the case, primer,

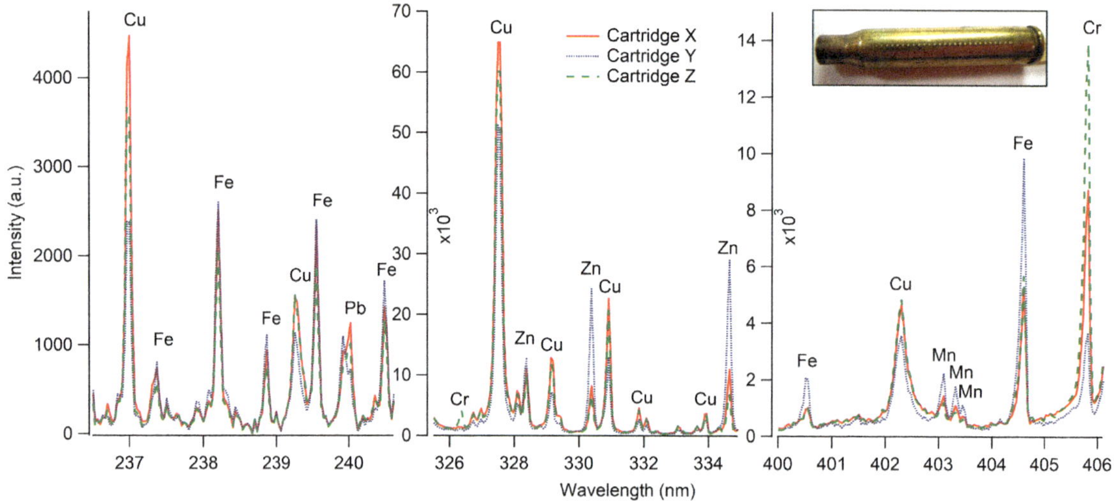

FIG. 19 Selected regions of the laser-induced breakthrough spectroscopy (LIBS) spectra of three different cartridge casings (after being fired from a gun). Subtle spectral differences can be used to discriminate the type/origin of the cartridge using chemometric analysis. The *inset photo* shows a sample cartridge analyzed with LIBS [113].

propellant, and projectile. Cartridge brass contains the main elements of Cu and zinc (Zn) in the 70:30 mass ratio range, with many other elements such as Pb, Sn, Ni, antimony (Sb), manganese (Mn), Bi, Al, beryllium (Be), P, cadmium (Cd), and Ag present in trace quantities. Fig. 19 shows example LIBS spectra of spent bullet cartridges.

A PLS-DA model was generated using the entire broadband spectra (240 total) as training inputs for the model; a validation set of additional laser shots (60) was tested against the model. For the six bullet types, 93.3% were positively identified by their origin, with a false-positive rate of 5.8% [113]. In an unpublished follow-up experiment, 10 cleaning shots were acquired on each cartridge case prior to saving the LIBS spectra; a total of 600 spectra were obtained, with 360 used to train a PLS-DA model and the rest used for the validation set. If at least 7 of the 10 spectra were classified with a particular cartridge type in the model, the cartridge was categorized as a match. Using this methodology, a 100% true-positive rate was achieved, with a 5.8% false-positive rate, despite the very subtle differences in the LIBS spectra (Fig. 19). Additional preprocessing of the data to remove saturated emission features and down-select to key intensity ratios may result in reduction of the false-positive rate.

4.2 Gunshot residues

Trace residues produced by the discharge of a firearm have been used to provide valuable information for forensic investigations for many years [114]. Gunshot residue contains organic and inorganic material originating from the primer, propellant, bullet, bullet jacket, cartridge case, and gun barrel. Goode et al. [115] demonstrated the detection of gunshot residue on a shooter's hand by directly analyzing double-sided tape pressed to the skin of the shooter with LIBS. The principal elements detected were Ba from the barium nitrate oxidant in the primer and Pb from the lead styphnate explosive in the primer. Antimony, the other main component of gunshot residues due to the presence of antimony sulfide fuel in the primer, was not detected. Even after washing the shooter's hand with soap and water, the gunshot residue was detectable. A subsequent study concluded that identifying gunshot residue on a person who fired a single shot from a clean gun was more challenging, but possible with the use of statistics to minimize false positives and false negatives [116]. Successful detection of gunshot residue with a commercial LIBS system >5 days after a firearm discharge was also demonstrated [117].

Silva et al. [118] used SIMCA to classify residue samples acquired from the hands of shooters (firing different types of guns with various ammunition) and nonshooters (including an automechanic, a profession known to cause false positives for gunshot residue detection due to regular exposure to Ba, Pb, and Sb from brake pads). While one of the three samples collected from the automechanic resulted in a false positive, no false negatives were obtained using LIBS. Rinke et al. [119] demonstrated the classification of residues surrounding bullet holes based on the type of bullet jacket (copper, metal alloy, or nonjacketed lead projectiles) using the combination of LIBS and target factor analysis. Hondrogiannis et al. [120] were able to detect and discriminate gunshot residue based on the manufacture of the smokeless powder using a field-portable, battery-operated commercial LIBS system. Dona-Fernandez et al. [121] developed a portable LIBS system for gunshot residue detection and confirmed its performance (100% sensitivity for samples with at least three particles) by comparison to scanning electron microscopy/energy dispersive X-ray spectroscopy. The feasibility of LIBS for rapid screening of lead-free gunshot residues was also recently demonstrated [122, 123]. Finally, the use of LIBS to spectrochemically map gunshot residue patterns on clothing has been demonstrated [124, 125].

A related LIBS defense application is the detection of firing residues from larger weapons. The firing of mortars and rockets can result in the accumulation of propellant residue in deposition rates of up to 70% [126]. In a "proof-of-concept" study at ARL [63], firing residue samples were collected on double-sided tape following shots from a rocket-propelled grenade and an 82 mm mortar. Strong emission features from Al, Ba, Mg, Fe, Cu, Li, and Ca were observed in the LIBS spectra; minor impurities included Ti and Cr (Fig. 20). The mortar residue showed more evidence of organic material (molecular C_2 and CN emission features and stronger H emission). Future work could explore the feasibility of identifying the weapon origin/manufacturer.

4.3 Soil contamination

One of the earliest LIBS applications investigated at ARL was the issue of toxic metal contamination in soil, an environmental concern for both industrial and military sites [127]. LIBS was first demonstrated for analysis of heavy metals in soils in the early 1990s, with detection limits in the 10–100-μg/g range [128, 129]; subsequent field-portable LIBS systems were developed for that application (e.g., [130–133]). As discussed in Section 4.2, residues from small-arms and other weapons testing and live-fire exercises can be deposited throughout the test range. Since most toxic metal-monitoring techniques recognized by governmental oversight organizations (e.g., the Environmental Protection Agency) are laboratory-based methodologies, there exists a need for direct, in-field measurement for continuous monitoring. Current field-portable instruments such as X-ray fluorescence have their own limitations, such as the use of a radioactive source and poor sensitivity to lighter elements such as Be. A field-portable LIBS instrument specifically designed to detect Pb in paint and soil was developed as part of an Army-funded Small Business Innovation Research program and tested at both Fort Carson Military Reservation in Colorado and

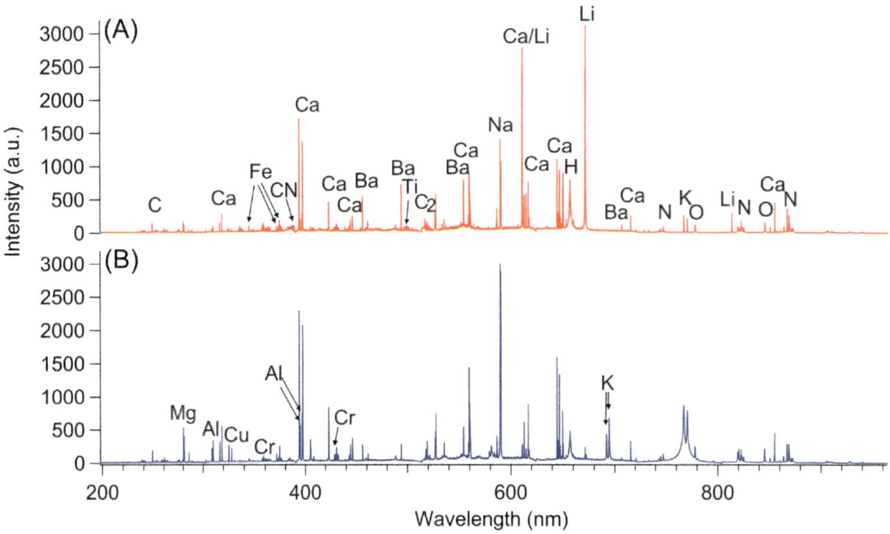

FIG. 20 Laser-induced breakthrough spectroscopy spectra of firing residue collected on double-sided tape from (A) an 82-mm mortar and (B) a rocket-propelled grenade.

the Sierra Army Depot in California. Wainner et al. [134] compared the results from the field tests using the portable LIBS system to the results from a benchtop LIBS system, and found that the results were comparable despite the limitations of the portable system.

The detection of trace contaminants in a complex matrix such as soil can be challenging due to the effects of the matrix on the LIBS signals for the various chemical species [135, 136]. Environmental factors such as sample moisture and multiple types of contaminants can also complicate the interpretation of the LIBS results. Chemometric analysis of LIBS spectra from soil samples has enabled a significant improvement in the classification (e.g., [137]) and quantification (e.g., [138]) of soils with and without heavy metal contaminants. For example, Gottfried et al. [53] used PLS-DA to distinguish 19 soil samples without any sample preparation, including 10 international soil standards and 9 soils collected from Yuma (AZ), Aberdeen Proving Ground (MD), the National Training Center (CA), and the Sierra Army Depot (CA). Example spectra are shown in Fig. 21. Significant improvement in the classification results was achieved with double-pulse LIBS compared to single-pulse LIBS. Using a threshold of >50% predicted probability for class membership with averaged spectra, 100% correct classification with 1.1% misclassification was achieved from the test spectra. With a more restrictive >95% predicted probability threshold, 89.5% of the test spectra were correctly classified, with no misclassifications (10.5% of the spectra were unclassified). A more comprehensive review of trace metal detection in soils can be found in the literature [139].

4.4 Engine wear analysis

Engine wear analysis is an important component in the maintenance and repair of military vehicles and aircraft. The concentration of metals in the engine lubricants can be used to evaluate whether a system is near failure (e.g., damaged cylinders, worn or broken bearings,

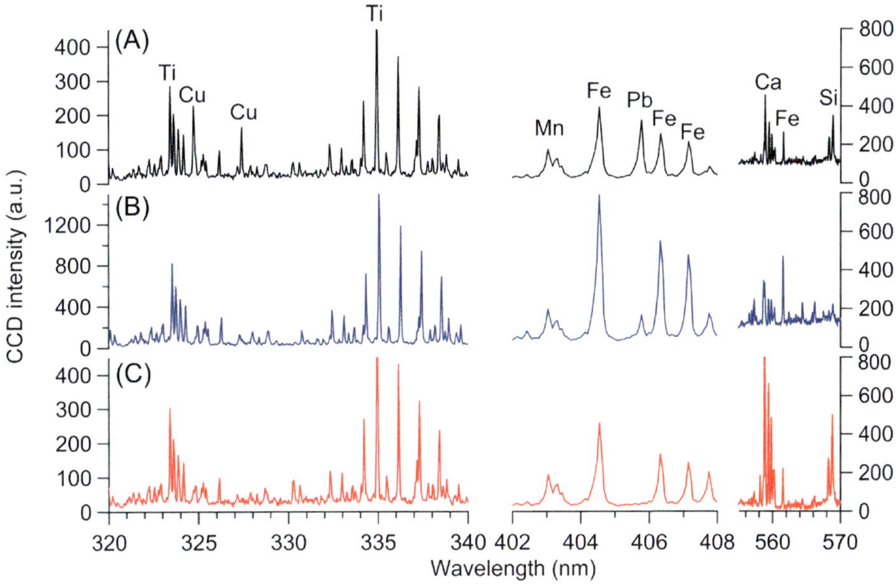

FIG. 21 Double-pulse laser-induced breakthrough spectroscopy spectra of loose soils from (A) National Institute of Standards and Technology standard 2587, (B) Aberdeen Proving Ground, and (C) the National Training Center.

misaligned gears, etc.) and requires immediate maintenance or repair to avoid further damage. Typically, this requires sending samples off-site for laboratory analysis, a time-consuming process. LIBS has been demonstrated as a faster alternative for this application. Yaroshchyk et al. [140] determined the limits of detection (LODs) for Na, Mg, Al, Ca, Ti, V, Cr, Mn, Ni, Fe, Cu, Zn, molybdenum (Mo), Ag, Cd, and Ba in engine oils and found that they were mostly comparable to those measured using the widely accepted standard inductively coupled plasma atomic emission spectroscopy. Improved LODs (2–4 ×) were obtained by using paper substrates for the analysis rather than direct sampling of the liquids [141]. Compact LIBS instruments for this application are commercially available.

Similarly, LIBS has been used to identify metallic contaminants in rocket engine plume flows for early identification of component failures [142] or wall erosion [143]. Majewski et al. [144] used LIBS to monitor the laser cleaning of thermal barrier coatings on turbine blades contaminated during engine operation. In a related application, LIBS has been demonstrated for online engine equivalence ratio measurements of the air-to-fuel ratio, which affects engine power efficiency and exhaust gas emission [145–147].

5 Future outlook

Numerous defense-related LIBS applications have been demonstrated by various groups. The ability to obtain real-time results in situ is a key advantage of LIBS for many of these applications. Commercial LIBS systems designed for specific applications (including analysis software) have only recently started to become available. As their availability spreads, LIBS is likely to become much more routine in many different application areas. Currently fielded LIBS systems for defense-related applications are generally designed for very specific applications, where the chemometric analysis models do not have to be as rigorous. While standoff detection of CBRNE threats with LIBS is also quite promising, commercial systems for this application are not yet available, and the development of sophisticated chemometric algorithms for acceptable classification rates (i.e., no false negatives with very few false positives) under real-world conditions continues. Other concerns such as eye safety, fieldability under extreme conditions, and damage to the target surface must also be resolved before LIBS becomes routine for standoff threat detection.

Acknowledgments

The author would like to thank Dr. Frank C. De Lucia, Jr. and Dr. Chase A. Munson for their assistance locating the original data used to prepare some of the figures in this chapter.

References

[1] J.L. Gottfried, Chemometric analysis in LIBS, in: D. A. Cremers, L.J. Radziemski (Eds.), Handbook of Laser-Induced Breakdown Spectroscopy, John Wiley & Sons Ltd, Singapore, 2013, pp. 223–255.

[2] P. Torrione, L.M. Collins, K.D. Morton, Multivariate analysis, chemometrics, and machine learning in laser spectroscopy, in: M. Baudelet (Ed.), Laser Spectroscopy for Sensing, Woodhead Publishing, Cambridge, UK, 2014, pp. 125–164.

[3] T. Zhang, H. Tang, H. Li, Chemometrics in laser-induced breakdown spectroscopy, J. Chemometr. 32 (2018) e2983.

[4] S. Wold, K. Esbensen, P. Geladi, Principal component analysis, Chemom. Intel. Lab. Syst. 2 (1987) 37–52.

[5] S. Wold, Pattern-recognition by means of disjoint principal components models, Pattern Recogn. 8 (1976) 127–139.

[6] M. Barker, W. Rayens, Partial least squares for discrimination, J. Chemometr. 17 (2003) 166–173.

[7] F.C. De Lucia Jr., A.C. Samuels, R.S. Harmon, R. A. Walters, K.L. McNesby, A. LaPointe, R. J. Winkel Jr., A.W. Miziolek, Laser-induced breakdown spectroscopy (LIBS): a promising versatile chemical sensor technology for hazardous material detection, IEEE Sensors J. 5 (2005) 681–689.

[8] C.A. Munson, F.C. De Lucia Jr., T. Piehler, K. L. McNesby, A.W. Miziolek, Investigation of statistics strategies for improving the discriminating power of laser-induced breakdown spectroscopy for chemical and biological warfare agent simulants, Spectrochim. Acta B 60 (2005) 1217–1224.

[9] J.L. Gottfried, F.C. De Lucia Jr., C.A. Munson, A. W. Miziolek, Standoff detection of chemical and biological threats using laser-induced breakdown spectroscopy, Appl. Spectrosc. 62 (2008) 353–363.

[10] J.L. Gottfried, F.C. De Lucia Jr., C.A. Munson, A. W. Miziolek, Double-pulse standoff laser-induced breakdown spectroscopy for versatile hazardous materials detection, Spectrochim. Acta B 62 (2007) 1405–1411.

[11] L. Landstrom, L. Orebrand, K. Svensson, P. O. Andersson, Spectroscopic investigation of substrates contaminated by chemical warfare agents, J. Anal. At. Spectrom 30 (2015) 2394–2402.

[12] D. L'Hermite, E. Vors, T. Vercouter, G. Moutiers, Evaluation of the efficacy of a portable LIBS system for detection of CWA on surfaces, Environ. Sci. Pollut. Res. 23 (2016) 8219–8226.

[13] J.L. Gottfried, Discrimination of biological and chemical threats in residue mixtures on multiple surfaces, Anal. Bioanal. Chem. 400 (2011) 3289–3301.

[14] J.L. Gottfried, Comparison of swipe materials for chemical, biological, and explosive residue discrimination with laser-induced breakdown spectroscopy, report no. ARL-TR-5231, US Army Research Laboratory, Aberdeen Proving Ground, MD, 2010.

[15] J.A. Stearns, S.E. McElman, J.A. Dodd, Identification of vapor-phase chemical warfare agent simulants and rocket fuels using laser-induced breakdown spectroscopy, Appl. Optics 49 (2010) C8–C15.

[16] E.E. Brown, C.S.C. Yang, U. Hommerich, F. Jin, S. B. Trivedi, A.C. Samuels, Comparative studies of long-wave laser-induced breakdown spectroscopy emissions excited at 1.064 nm and eye-safe 1.574 nm, Opt. Express 25 (2017) 7238–7250.

[17] J.D. Hybl, G.A. Lithgow, S.G. Buckley, Laser-induced breakdown spectroscopy detection and classification of biological aerosols, Appl. Spectrosc. 57 (2003) 1207–1215.

[18] S. Morel, N. Leone, P. Adam, J. Amouroux, Detection of bacteria by time-resolved laser-induced breakdown spectroscopy, Appl. Optics 42 (2003) 6184–6191.

[19] A.C. Samuels, F.C. De Lucia Jr., K.L. McNesby, A. W. Miziolek, Laser-induced breakdown spectroscopy of bacterial spores, molds, pollens, and protein: initial studies of discrimination potential, Appl. Optics 42 (2003) 6205–6209.

[20] N. Leone, G. D'Arthur, P. Adam, J. Amouroux, Detection of bacterial deposits and bioaerosols by time-resolved laser-induced breakdown spectroscopy (TRELIBS), High Temp. Mater. Process. 8 (2004) 1–22.

[21] J. Diedrich, S.J. Rehse, S. Palchaudhuri, *Escherichia coli* identification and strain discrimination using

nanosecond laser-induced breakdown spectroscopy, Appl. Phys. Lett. 90 (2007) 163901.

[22] S.J. Rehse, J. Diedrich, S. Palchaudhuri, Identification and discrimination of Pseudomonas aeruginosa bacteria grown in blood and bile by laser-induced breakdown spectroscopy, Spectrochim. Acta B 62 (2007) 1169–1176.

[23] D.W. Merdes, J.M. Suhan, J.M. Keay, D.M. Hadka, W. R. Bradley, The investigation of laser-induced breakdown spectroscopy for detection of biological contaminants on surfaces, Spectroscopy 22 (2007) 28–35.

[24] E.G. Snyder, C.A. Munson, J.L. Gottfried, F.C. De Lucia Jr. B. Gullett, A. Miziolek, Laser-induced breakdown spectroscopy for the classification of unknown powders, Appl. Optics 47 (2008) G80–G87.

[25] R.A. Multari, D.A. Cremers, M.L. Bostian, The use of laser-induced breakdown spectroscopy for the differentiation of pathogens and viruses on substrates, Appl. Optics 51 (2012) B57–B64.

[26] M. Baudelet, L. Guyon, J. Yu, J.P. Wolf, T. Amodeo, E. Frejafon, P. Laloi, Femtosecond time-resolved laser-induced breakdown spectroscopy for detection and identification of bacteria: a comparison to the nanosecond regime, J. Appl. Phys. 99 (2006) 84701.

[27] H.L. Xu, W. Liu, S.L. Chin, Remote time-resolved filament-induced breakdown spectroscopy of biological materials, Opt. Lett. 31 (2006) 1540–1542.

[28] C.A. Munson, J.L. Gottfried, E.G. Snyder, F.C. De Lucia Jr. B. Gullett, A.W. Miziolek, Detection of indoor biological hazards using the man-portable laser induced breakdown spectrometer, Appl. Optics 47 (2008) G48–G57.

[29] P. Torrione, K. Morton, *Laser Induced Breakdown Spectroscopy (LIBS) Graphical Development Tool*, Signal Innovations Group, Research Triangle Park, NC, 2008.

[30] R.C. Chinni, D.A. Cremers, L.J. Radziemski, M. Bostian, C. Navarro-Northrup, Detection of uranium using laser-induced breakdown spectroscopy, Appl. Spectrosc. 63 (2009) 1238–1250.

[31] F.R. Doucet, G. Lithgow, R. Kosierb, P. Bouchard, M. Sabsabi, Determination of isotope ratios using laser-induced breakdown spectroscopy in ambient air at atmospheric pressure for nuclear forensics, J. Anal. At. Spectrom 26 (2011) 536–541.

[32] A.A. Bol'shakov, X.L. Mao, J.J. Gonzalez, R.E. Russo, Laser ablation molecular isotopic spectrometry (LAMIS): current state of the art, J. Anal. At. Spectrom 31 (2016) 119–134.

[33] I. Gaona, J. Serrano, J. Moros, J.J. Laserna, Evaluation of laser-induced breakdown spectroscopy analysis potential for addressing radiological threats from a distance, Spectrochim. Acta B 96 (2014) 12–20.

[34] C.M. Davies, H.H. Telle, A.W. Williams, Remote in situ analytical spectroscopy and its applications in the nuclear industry, Fresenius J. Anal. Chem. 355 (1996) 895–899.

[35] A.I. Whitehouse, J. Young, I.M. Botheroyd, S. Lawson, C.P. Evans, J. Wright, Remote material analysis of nuclear power station steam generator tubes by laser-induced breakdown spectroscopy, Spectrochim. Acta B 56 (2001) 821–830.

[36] E.J. Judge, J.E. Barefield, J.M. Berg, S.M. Clegg, G. J. Havrilla, V.M. Montoya, L.A. Le, L.N. Lopez, Laser-induced breakdown spectroscopy measurements of uranium and thorium powders and uranium ore, Spectrochim. Acta B 83–84 (2013) 28–36.

[37] J.-B. Sirven, A. Pailloux, Y. M'Baye, N. Coulon, T. Alpettaz, S. Gosse, Towards the determination of the geographical origin of yellow cake samples by laser-induced breakdown spectroscopy and chemometrics, J. Anal. At. Spectrom 24 (2009) 451–459.

[38] M.Z. Martin, S. Allman, D.J. Brice, R.C. Martin, N. O. Andre, Exploring laser-induced breakdown spectroscopy for nuclear materials analysis and in-situ applications, Spectrochim. Acta B At. Spectrosc. 74–75 (2012) 177–183.

[39] C. Li, C.L. Feng, H.Y. Oderji, G.N. Luo, H.B. Ding, Review of LIBS application in nuclear fusion technology, Front. Physiol. 11 (2016) 16.

[40] F.C. De Lucia Jr., R.S. Harmon, K.L. McNesby, R. J. Winkel Jr., A.W. Miziolek, Laser-induced breakdown spectroscopy analysis of energetic materials, Appl. Optics 42 (2003) 6148–6152.

[41] C. Lopez-Moreno, S. Palanco, J. Javier Laserna, F. De Lucia Jr., A.W. Miziolek, J. Rose, R.A. Walters, A. I. Whitehouse, Test of a stand-off laser-induced breakdown spectroscopy sensor for the detection of explosive residues on solid surfaces, J. Anal. At. Spectrom 21 (2006) 55–60.

[42] V.I. Babushok, F.C. De Lucia Jr., J.L. Gottfried, C. A. Munson, A.W. Miziolek, Double pulse laser ablation and plasma: laser induced breakdown spectroscopy signal enhancement, Spectrochim. Acta B 61 (2006) 999–1014.

[43] F.C. De Lucia Jr., J.L. Gottfried, C.A. Munson, A. W. Miziolek, Double pulse laser-induced breakdown spectroscopy of explosives: initial study towards improved discrimination, Spectrochim. Acta B 62 (2007) 1399–1404.

[44] J.L. Gottfried, F.C. De Lucia Jr., C.A. Munson, A. W. Miziolek, Laser-induced breakdown spectroscopy for detection of explosives residues: a review of recent advances, challenges, and future prospects, Anal. Bioanal. Chem. 395 (2009) 283–300.

[45] F.C. De Lucia Jr., J.L. Gottfried, C.A. Munson, A. W. Miziolek, Current status of standoff LIBS security applications at the United States Army Research Laboratory, Spectroscopy 24 (2009) 1–7.

[46] J.L. Gottfried, F.C. De Lucia Jr., C.A. Munson, A. W. Miziolek, Strategies for residue explosives detection using laser-induced breakdown spectroscopy, J. Anal. At. Spectrom 23 (2008) 205–216.
[47] F.C. De Lucia Jr. J.L. Gottfried, C.A. Munson, A.W. Miziolek, Multivariate analysis of standoff laser-induced breakdown spectroscopy spectra for classification of explosive-containing residues, Appl. Optics 47 (2008) G112–G121.
[48] F.C. De Lucia Jr., J.L. Gottfried, Influence of variable selection on partial least squares discriminant analysis models for explosive residue discrimination, Spectrochim. Acta B 66 (2011) 122–128.
[49] J.L. Gottfried, Influence of metal substrates on the detection of explosive residues with laser-induced breakdown spectroscopy, Appl. Optics 52 (2012) B10–B19.
[50] J.L. Gottfried, F.C. De Lucia Jr. A.W. Miziolek, Discrimination of explosive residues on organic and inorganic substrates using laser-induced breakdown spectroscopy, J. Anal. At. Spectrom 24 (2009) 288–296.
[51] F.C. De Lucia Jr. J.L. Gottfried, Classification of explosive residues on organic substrates using laser induced breakdown spectroscopy, Appl. Optics 51 (2012) B83–B92.
[52] D.J. Phares, J.K. Holt, G.T. Smedley, R.C. Flagan, Method for characterization of adhesion properties of trace explosives in fingerprints and fingerprint simulations, J. Forensic Sci. 45 (2000) 774–784.
[53] J.L. Gottfried, R.S. Harmon, F.C. De Lucia Jr. A.W. Miziolek, Multivariate analysis of laser-induced breakdown spectroscopy chemical signatures for geomaterial classification, Spectrochim. Acta B 64 (2009) 1009–1019.
[54] M.E. Essington, G.V. Melnichenko, M.A. Stewart, R.A. Hull, Soil metals analysis using laser-induced breakdown spectroscopy (LIBS), Soil Sci. Soc. Am. J. 73 (2009) 1469–1478.
[55] R.S. Harmon, F.C. De Lucia, R.J. Winkel Jr., A. LaPointe, S.L. Grossman, K.L. McNesby, A.W. Miziolek, LIBS: a new versatile field-deployable real-time detector system with potential for landmine detection, Proc. SPIE Int. Soc. Opt. Eng. 5089 (2003) 1065.
[56] R.S. Harmon, F.C. De Lucia Jr., A. LaPointe, R.J. Winkel Jr., A.W. Miziolek, LIBS for landmine detection and discrimination, Anal. Bioanal. Chem. 385 (2006) 1140–1148.
[57] C. Bohling, D. Scheel, K. Hohmann, W. Schade, M. Reuter, G. Holl, Fiber-optic laser sensor for mine detection and verification, Appl. Optics 45 (2006) 3817–3825.
[58] C. Bohling, K. Hohmann, D. Scheel, C. Bauer, W. Schippers, J. Burgmeier, U. Willer, G. Holl, W. Schade, All-fiber-coupled laser-induced breakdown spectroscopy sensor for hazardous materials analysis, Spectrochim. Acta B 62 (2007) 1519–1527.
[59] J.L. Gottfried, R.S. Harmon, A. LaPointe, Progress in LIBS for landmine detection, report no. ARL-TR-5127, U.S. Army Research Laboratory, Aberdeen Proving Ground, MD, 2010.
[60] C.K. Williamson, R.G. Daniel, K.L. McNesby, A.W. Miziolek, Laser-induced breakdown spectroscopy for real-time detection of Halon alternative agents, Anal. Chem. 70 (1998) 1186–1191.
[61] E.D. Lancaster, K.L. McNesby, R.G. Daniel, A.W. Miziolek, Spectroscopic analysis of fire suppressants and refrigerants by laser-induced breakdown spectroscopy, Appl. Optics 38 (1999) 1476–1480.
[62] C.A. Munson, J.L. Gottfried, F.C. Delucia Jr., A.W. Miziolek, Characterization of soldering materials using laser-induced breakdown spectroscopy (LIBS), NASLIBS2007, New Orleans, LA, 2007.
[63] J.L. Gottfried, F.C. De Lucia Jr. Laser-induced breakdown spectroscopy: capabilities and applications, report no. ARL-TR-5238, US Army Research Laboratory, Aberdeen Proving Ground, MD, 2010.
[64] B. Chen, H. Kano, M. Kuzuya, Quantitative analysis of trace lead in tin-base lead-free solder by laser-induced plasma spectroscopy in air at atmospheric pressure, Anal. Sci. 24 (2008) 289–291.
[65] K. Loebe, A. Uhl, H. Lucht, Microanalysis of tool steel and glass with laser-induced breakdown spectroscopy, Appl. Optics 42 (2003) 6166–6173.
[66] A.A. Bol'shakov, J.H. Yoo, C. Liu, J.R. Plumer, R.E. Russo, Laser-induced breakdown spectroscopy in industrial and security applications, Appl. Optics 49 (2010) C132–C142.
[67] N. Ismail, J. Yoo, IEEE, Determination of lead (Pb) concentration level in solder finished product using laser induced breakdown spectroscopy (LIBS), in: 2010 12th Electronics Packaging Technology Conference, IEEE, Singapore, 2010, pp. 456–461.
[68] G. Galbacs, N. Jedlinszki, A. Metzinger, Analysis and discrimination of soldering tin samples by collinear multi-pulse laser induced breakdown spectrometry, supported by inductively coupled plasma optical emission and mass spectrometry, Microchem. J. 107 (2013) 17–24.
[69] R.R.V. Carvalho, J.A.O. Coelho, J.M. Santos, F.W. B. Aquino, R.L. Carneiro, E.R. Pereira, Laser-induced breakdown spectroscopy (LIBS) combined with hyperspectral imaging for the evaluation of printed circuit board composition, Talanta 134 (2015) 278–283.
[70] R. Zdunek, M. Nowak, E. Plinski, Statistical classification of soft solder alloys by laser-induced breakdown spectroscopy: review of methods, J. Eur. Opt. Soc. Rapid Publ. 11 (2016) 20.

[71] M.A. Speranca, A. Virgilio, E.R. Pereira, F.W.B. de Aquino, Determination of elemental content in solder mask samples used in printed circuit boards using different spectroanalytical techniques, Appl. Spectrosc. 72 (2018) 1205–1214.

[72] F.C. De Lucia Jr., J.L. Gottfried, A.W. Miziolek, Analysis of carbon and sulfur in steel samples using benchtop laser-induced breakdown spectroscopy (LIBS), report no. ARL-TR-5000, U.S. Army Research Laboratory, Aberdeen Proving Ground, MD, 2009.

[73] I. Radivojevic, C. Haisch, R. Niessner, S. Florek, H. Becker-Ross, U. Panne, Microanalysis by laser-induced plasma spectroscopy in the vacuum ultraviolet, Anal. Chem. 76 (2004) 1648–1656.

[74] J.L. Gottfried, Influence of exothermic chemical reactions on laser-induced shock waves, Phys. Chem. Chem. Phys. 16 (2014) 21452–21466.

[75] J.L. Gottfried, Laser-induced plasma chemistry of the explosive RDX with various metallic nanoparticles, Appl. Optics 51 (2012) B13–B21.

[76] A.H. Rezaei, M.H. Keshavarz, M. Kavosh Tehrani, S.M.R. Darbani, A.H. Farhadian, S.J. Mousavi, A. Mousaviazar, Approach for determination of detonation performance and aluminum percentage of aluminized based explosives by laser-induced breakdown spectroscopy, Appl. Optics 55 (2016) 3233–3240.

[77] A.H. Rezaei, M.H. Keshavarz, M.K. Tehrani, S.M. R. Darbani, Quantitative analysis for the determination of aluminum percentage and detonation performance of aluminized plastic bonded explosives by laser-induced breakdown spectroscopy, Laser Phys. 28 (2018) 065605.

[78] A.H. Rezaei, M.H. Keshavarz, M.K. Tehrani, M.R. Darbani, Assessment of detonation performance and characteristics of 2,4,6-trinitrotoluene based melt cast explosives containing aluminum by laser induced breakdown spectroscopy, Cent. Eur. J. Energ. Mater. 16 (2019) 3–20.

[79] C. Kimblin, R. Trainham, G.A. Capelle, X. Mao, R.E. Russo, Characterization of laser-induced plasmas as a complement to high-explosive large-scale detonations, AIP Adv. 7 (2017) 095208.

[80] S.A. Kalam, N.L. Murthy, P. Mathi, N. Kommu, A.K. Singh, S.V. Rao, Correlation of molecular, atomic emissions with detonation parameters in femtosecond and nanosecond LIBS plasma of high energy materials, J. Anal. At. Spectrom 32 (2017) 1535–1546.

[81] J.L. Gottfried, Laboratory-scale method for estimating explosive performance from laser-induced shock waves, Propellants Explos. Pyrotech. 40 (2015) 674–681.

[82] E.S. Collins, J.L. Gottfried, Laser-induced deflagration for the characterization of energetic materials, Propellants Explos. Pyrotech. 42 (2017) 592–602.

[83] J.L. Gottfried, T.M. Klapötke, T.G. Witkowski, Estimated detonation velocities for TKX-50, MAD-X1, BDNAPM, BTNPM, TKX-55 and DAAF using the laser-induced air shock from energetic materials technique, Propellants Explos. Pyrotech. 42 (2017) 353–359.

[84] J.L. Gottfried, Laser-induced air shock from energetic materials (LASEM) method for estimating detonation performance: challenges, successes and limitations, AIP Conf. Proc. 1979 (2018) 100014.

[85] J.L. Gottfried, E.J. Bukowski, Laser-shocked energetic materials with metal additives: evaluation of chemistry and detonation performance, Appl. Optics 56 (2017) B47–B57.

[86] Y. Jiang, S. Deng, S. Hong, J. Zhao, S. Huang, C.-C. Wu, J.L. Gottfried, K.-i. Nomura, Y. Li, S.C. Tiwari, R.K. Kalia, P. Vashishta, A. Nakano, X. Zheng, Energetic performance of optically activated aluminum/graphene oxide composites, ACS Nano 12 (2018) 11366–11375.

[87] S.A. Davari, J.L. Gottfried, C. Liu, E.L. Ribeiro, G. Duscher, D. Mukherjee, Graphitic-coated Al nanoparticles manufactured as superior energetic materials via laser ablation synthesis in organic solvents, Appl. Surf. Sci. 473 (2019) 156–163.

[88] J.L. Gottfried, D.K. Smith, C.-C. Wu, M.L. Pantoya, Improving the explosive performance of aluminum nanoparticles with aluminum iodate hexahydrate (AIH), Sci. Rep. 8 (2018) 8036.

[89] K.K. Miller, J.L. Gottfried, S.D. Walck, M.L. Pantoya, C.-C. Wu, Plasma surface treatment of aluminum nanoparticles for energetic material applications, Combust. Flame 206 (2019) 211–213.

[90] A.H. Farhadian, M.K. Tehrani, M.H. Keshavarz, S.M.R. Darbany, M. Karimi, A.H. Rezayi, Application of laser-induced breakdown spectroscopy as nondestructive and safe analysis method for composite solid propellants, Chem. Bulg. J. Sci. Educ. 24 (2015) 734–744.

[91] J.L. Gottfried, R.A. Pesce-Rodriguez, D. Farrow, J. Dellinger, Laboratory-scale investigation of the influence of ageing on the performance and sensitivity of an explosive containing ε-CL-20, Propellants Explos. Pyrotech. 43 (2018) 616–625.

[92] A.H. Farhadian, M.K. Tehrani, M.H. Keshavarz, M. Karimi, S.M. Reza Darbani, Relationship between the results of laser-induced breakdown spectroscopy and dynamical mechanical analysis in composite solid propellants during their aging, Appl. Optics 55 (2016) 4362–4369.

[93] A.H. Farhadian, M.K. Tehrani, M.H. Keshavarz, M. Karimi, S.M.R. Darbani, A.H. Rezayi, A novel approach for investigation of chemical aging in composite propellants through laser-induced breakdown spectroscopy (LIBS), J. Therm. Anal. Calorim. 124 (2016) 279–286.

[94] S.H. Ahmadi, M.H. Keshavarz, H.R. Hafizi Atabak, Introducing laser induced breakdown spectroscopy (LIBS) as a novel, cheap and non-destructive method to study the changes of mechanical properties of plastic bonded explosives (PBX), Z. Anorg. Allg. Chem. 644 (2018) 1667–1673.

[95] P. Lucena, J.M. Vadillo, J.J. Laserna, Mapping of platinum group metals in automotive exhaust three-way catalysts using laser-induced breakdown spectrometry, Anal. Chem. 71 (1999) 4385–4391.

[96] M.A. Palacios, M.M. Gomez, M. Moldovan, G. Morrison, S. Rauch, C. McLeod, R. Ma, J. Laserna, P. Lucena, S. Caroli, A. Alimonti, F. Petrucci, B. Bocca, P. Schramel, S. Lustig, M. Zischka, U. Wass, B. Stenbom, M. Luna, J.C. Saenz, J. Santamara, J.M. Torrens, Platinum-group elements: quantification in collected exhaust fumes and studies of catalyst surfaces, Sci. Total Environ. 257 (2000) 1.

[97] F.A. Barreda, F. Trichard, S. Barbier, N. Gilon, L. Saint-Jalmes, Fast quantitative determination of platinum in liquid samples by laser-induced breakdown spectroscopy, Anal. Bioanal. Chem. 403 (2012) 2601–2610.

[98] M. Hornackova, Z. Grolmusova, M. Hornacek, J. Rakovsky, P. Hudec, P. Veis, Calibration analysis of zeolites by laser induced breakdown spectroscopy, Spectrochim. Acta B 74–75 (2012) 119–123.

[99] F. Trichard, L. Sorbier, S. Moncayo, Y. Blouet, C.P. Lienemann, V. Motto-Ros, Quantitative elemental imaging of heterogeneous catalysts using laser-induced breakdown spectroscopy, Spectrochim. Acta B 133 (2017) 45–51.

[100] J.E. Jaine, M.R. Mucalo, Rapid determination of rhodium, palladium, and platinum in supported metal catalysts using multivariate analysis of laser induced breakdown spectroscopy data, Spectrochim. Acta B 145 (2018) 58–63.

[101] S. Belyani, M.H. Keshavarz, S.M.R. Darbani, M.K. Tehrani, Application of laser-induced breakdown spectroscopy to assess palladium catalyst deactivation, Appl. Optics 58 (2019) 794–799.

[102] F.C. De Lucia Jr., A.W. Miziolek, Determination of surface arsenic contamination on steel substrates using laser induced breakdown spectroscopy, Report no. ARL-TR-6161, Aberdeen Proving Ground, MD, (2012).

[103] D. Anglos, K. Melesanaki, V. Zafiropulos, M.J. Gresalfi, J.C. Miller, Laser-induced breakdown spectroscopy for the analysis of 150-year-old daguerreotypes, Appl. Spectrosc. 56 (2002) 423–432.

[104] A.J. Lopez, A. Ramil, G. Nicolas, M.P. Mateo, V. Pinon, A. Yanez, Provenance characterization of archaeological ceramics by laser induced plasma spectroscopy, Mater. Sci. Forum 514–516 (2006) 1638–1642.

[105] A. Erdem, A. Cilingiroglu, A. Giakoumaki, M. Castanys, E. Kartsonaki, C. Fotakis, D. Anglos, Characterization of iron age pottery from eastern Turkey by laser-induced breakdown spectroscopy (LIBS), J. Archaeol. Sci. 35 (2008) 2486–2494.

[106] N. McMillan, C. McManus, R. Harmon, F. De Lucia, A. Miziolek, Laser-induced breakdown spectroscopy analysis of complex silicate minerals—beryl, Anal. Bioanal. Chem. 385 (2006) 263–271.

[107] D.C. Alvey, K. Morton, R.S. Harmon, J.L. Gottfried, J.J. Remus, L.M. Collins, M.A. Wise, Laser-induced breakdown spectroscopy-based geochemical fingerprinting for the rapid analysis and discrimination of minerals: the example of garnet, Appl. Optics 49 (2010) C168–C180.

[108] J.J. Remus, J.L. Gottfried, R.S. Harmon, A. Draucker, D. Baron, R. Yohe, Archaeological applications of laser-induced breakdown spectroscopy: an example from the Coso volcanic field, California, using advanced statistical signal processing analysis, Appl. Optics 49 (2010) C120–C131.

[109] R.S. Harmon, K.M. Shughrue, J.J. Remus, M.A. Wise, L. J. East, R.R. Hark, Can the provenance of the conflict minerals columbite and tantalite be ascertained by laser-induced breakdown spectroscopy? Anal. Bioanal. Chem. 400 (2011) 3377–3382.

[110] B. Bhatt, K.H. Angeyo, A. Dehayem-Kamadjeu, LIBS development methodology for forensic nuclear materials analysis, Anal. Methods 10 (2018) 791–798.

[111] E.R. Schenk, J.R. Almirall, Elemental analysis of cotton by laser-induced breakdown spectroscopy, Appl. Optics 49 (2010) C153–C160.

[112] M.M. Tan, S. Cui, J. Yoo, S.H. Han, K.S. Ham, S.H. Nam, Y. Lee, Feasibility of laser-induced breakdown spectroscopy (LIBS) for classification of sea salts, Appl. Spectrosc. 66 (2012) 262–271.

[113] A.W. Miziolek, F.C. De Lucia Jr., A spectroscopic tool for identifying sources of origin for materials of military interest, Proc. SPIE 9101 (2014) J1–J10.

[114] M. Maitre, K.P. Kirkbride, M. Horder, C. Roux, A. Beavis, Current perspectives in the interpretation of gunshot residues in forensic science: a review, Forensic Sci. Int. 270 (2017) 1–11.

[115] S.R. Goode, C.R. Dockery, M.F. Bachmeyer, A.A. Nieuwland, S.L. Morgan, Detecting gunshot residue by laser induced breakdown spectroscopy, Trends Opt. Photonics 81 (2002) 175–177.

[116] C.R. Dockery, S.R. Goode, Laser-induced breakdown spectroscopy for the detection of gunshot residues on the hands of a shooter, Appl. Optics 42 (2003) 6153–6158.

[117] M.B. Rosenberg, C.R. Dockery, Determining the lifetime of detectable amounts of gunshot residue on the hands of a shooter using laser-induced breakdown spectroscopy, Appl. Spectrosc. 62 (2008) 1238–1241.

[118] M.J. Silva, J. Cortez, C. Pasquini, R.S. Honorato, A.P. S. Paim, M.F. Pimentel, Gunshot residues: screening analysis by laser-induced breakdown spectroscopy, J. Braz. Chem. Soc. 20 (2009) 1887–1894.

[119] C.N. Rinke, M.R. Williams, C. Brown, M. Baudelet, M. Richardson, M.E. Sigman, Discriminant analysis in the presence of interferences: combined application of target factor analysis and a Bayesian soft-classifier, Anal. Chim. Acta 753 (2012) 19–26.

[120] E. Hondrogiannis, D. Andersen, A.W. Miziolek, The evaluation of a new technology for gunshot residue (GSR) analysis in the field, Proc. SPIE 8726 (2013) 87260P.

[121] A. Dona-Fernandez, I. de Andres-Gimeno, P. Santiago-Toribio, E. Valtuille-Fernandez, F. Aller-Sanchez, A. Heras-Gonzalez, Real-time detection of GSR particles from crime scene: a comparative study of SEM/EDX and portable LIBS system, Forensic Sci. Int. 292 (2018) 167–175.

[122] L.A. Fambro, E.T. Miller, D.D. Vandenbos, C.R. Dockery, Characterization of lead-free gunshot residue analogs, Anal. Methods 8 (2016) 3132–3139.

[123] L.A. Fambro, D.D. Vandenbos, M.B. Rosenberg, C.R. Dockery, Laser-induced breakdown spectroscopy for the rapid characterization of lead-free gunshot residues, Appl. Spectrosc. 71 (2017) 699–708.

[124] M. Lopez-Lopez, C. Alvarez-Llamas, J. Pisonero, C. Garcia-Ruiz, N. Bordel, An exploratory study of the potential of LIBS for visualizing gunshot residue patterns, Forensic Sci. Int. 273 (2017) 124–131.

[125] C.V. Pyl, O. Ovide, M. Ho, B. Yuksel, T. Trejos, Spectrochemical mapping using laser induced breakdown spectroscopy as a more objective approach to shooting distance determination, Spectrochim. Acta B 152 (2019) 93–101.

[126] M.R. Walsh, M.E. Walsh, G. Ampleman, S. Thiboutot, S. Brochu, T.F. Jenkins, Munitions propellants residue deposition rates on military training ranges, Propellants Explos. Pyrotech. 37 (2012) 393–406.

[127] J.P. Singh, F.Y. Yueh, V.N. Rai, R. Harmon, S. Beaton, P. French, F.C. DeLucia, B. Peterson, K.L. McNesby, A.W. Miziolek, Civilian and military environmental contamination studies using LIBS, in: A.W. Miziolek, I. Schechter, V. Palleschi (Eds.), Laser Induced Breakdown Spectroscopy, Cambridge University Press, Cambridge, 2006, pp. 368–399.

[128] A. Koskelo, D.A. Cremers, RCRA metals analysis by laser-induced breakdown spectroscopy: detection limits in soils, report no. LA-UR-94-1544, Los Alamos National Laboratory, 1994.

[129] R. Wisbrun, I. Schechter, R. Niessner, H. Schroder, K.L. Kompa, Detector for trace elemental analysis of solid environmental-samples by laser-plasma spectroscopy, Anal. Chem. 66 (1994) 2964–2975.

[130] K.Y. Yamamoto, D.A. Cremers, M.J. Ferris, L.E. Foster, Detection of metals in the environment using a portable laser-induced breakdown spectroscopy instrument, Appl. Spectrosc. 50 (1996) 222–233.

[131] A. Ciucci, V. Palleschi, S. Rastelli, R. Barbini, F. Colao, R. Fantoni, A. Palucci, S. Ribezzo, H.J.L. van der Steen, Trace pollutants analysis in soil by a time-resolved laser-induced breakdown spectroscopy technique, Appl. Phys. B Lasers Opt. 63 (1996) 185–190.

[132] B.H. Miles, J. Cortes, E.R. Cespedes, Laser induced breakdown spectroscopy (libs) detection of heavy metals using a cone penetrometer: system design and field investigation results, in: Field Analytical Methods for Hazardous Wastes and Toxic Chemicals, Air & Waste Management Assoc, Pittsburgh, 1997, pp. 671–680.

[133] G.A. Theriault, S. Bodensteiner, S.H. Lieberman, A real-time fiber-optic LIBS probe for the in situ delineation of metals in soils, Field Anal. Chem. Technol. 2 (1998) 117–125.

[134] R.T. Wainner, R.S. Harmon, A.W. Miziolek, K.L. McNesby, P.D. French, Analysis of environmental lead contamination: comparison of LIBS field and laboratory instruments, Spectrochim. Acta B 56 (2001) 777–793.

[135] A.S. Eppler, D.A. Cremers, D.D. Hickmott, M.J. Ferris, A.C. Koskelo, Matrix effects in the detection of Pb and Ba in soils using laser-induced breakdown spectroscopy, Appl. Spectrosc. 50 (1996) 1175–1181.

[136] V. Lazic, R. Barbini, F. Colao, R. Fantoni, A. Palucci, Self-absorption model in quantitative laser induced breakdown spectroscopy measurements on soils and sediments, Spectrochim. Acta B 56 (2001) 807–820.

[137] B. Bousquet, G. Travaillé, A. Ismaël, L. Canioni, K. Michel-Le Pierrès, E. Brasseur, S. Roy, I. le Hecho, M. Larregieu, S. Tellier, M. Potin-Gautier, T. Boriachon, P. Wazen, A. Diard, S. Belbèze, Development of a mobile system based on laser-induced breakdown spectroscopy and dedicated to in situ analysis of polluted soils, Spectrochim. Acta B 63 (2008) 1085–1090.

[138] J.-B. Sirven, B. Bousquet, L. Canioni, L. Sarger, Laser-induced breakdown spectroscopy of composite samples: comparison of advanced chemometrics methods, Anal. Chem. 78 (2006) 1462–1469.

[139] R.S. Harmon, R.E. Russo, R.R. Hark, Applications of laser-induced breakdown spectroscopy for geochemical and environmental analysis: a comprehensive review, Spectrochim. Acta B 87 (2013) 11–26.

[140] P. Yaroshchyk, R.J.S. Morrison, D. Body, B.L. Chadwick, Quantitative determination of wear metals in engine oils using laser-induced breakdown spectroscopy: a comparison between liquid jets and static liquids, Spectrochim. Acta B 60 (2005) 986.

[141] P. Yaroshchyk, R.J.S. Morrison, D. Body, B.L. Chadwick, Quantitative determination of wear metals in engine oils using LIBS: the use of paper substrates and a comparison between single- and double-pulse LIBS, Spectrochim. Acta B 60 (2005) 1482–1485.

[142] K.W. Whitaker, Analyzing rocket plume spectral data with neural networks for condition monitoring, in: RTO AVT Course on Intelligent Systems for Aeronautics, iRhode-Saint-Genese, Belgium, 2002.

[143] M. Thomas, S. Deaconu, J. Lewis, E. Coy, Laser induced breakdown spectroscopy (LIBS) applied to reacting gases for mixture ratio measurement and detection of metallic species, report no. AFRL-PR-ED-TP-2007-170, Air Force Research Laboratory, Edwards Air Force Base, CA, 2007.

[144] M.S. Majewski, C. Kelley, W. Hassan, W. Brindley, E. H. Jordan, M.W. Renfro, Laser induced breakdown spectroscopy for contamination removal on engine-run thermal barrier coatings, Surf. Coat. Technol. 205 (2011) 4614–4619.

[145] F. Ferioli, P.V. Puzinauskas, S.G. Buckley, Laser-induced breakdown spectroscopy for on-line engine equivalence ratio measurements, Appl. Spectrosc. 57 (2003) 1183.

[146] F. Ferioli, S.G. Buckley, P.V. Puzinauskas, Real-time measurement of equivalence ratio by using laser-induced breakdown spectroscopy for combustion performance in gasoline engine, Int. J. Engine Res. 7 (2006) 447–457.

[147] S. Joshi, D.B. Olsen, C. Dumitrescu, P.V. Puzinauskas, A.P. Yalin, Laser-induced breakdown spectroscopy for in-cylinder equivalence ratio measurements in laser-ignited natural gas engines, Appl. Spectrosc. 63 (2009) 549–554.

CHAPTER

13

Applications of LIBS in drug analysis

Pravin Kumar Tiwari[a,b], *Pradeep Kumar Rai*[c], *Awadhesh Kumar Rai*[a]

[a]Department of Physics, University of Allahabad, Pragyagraj, India [b]Institute for Plasma Research, Gandhinagar, Gujarat, India [c]Department of Nephrology, Opal Hospital, Varanasi, India

1 Introduction

Pharmaceutical drugs or medicines are kinds of chemical compounds used for the prevention or cure of any disease/disorder in humans and animals. The drugs come in many forms and types (i.e., capsules, tablets, lozenges, syrups, pastilles, powders, gels, ointments, drops, lotions, injections/vaccines etc.) and may have different compositions, shapes, sizes, colors, and phases [1, 2]. In local markets, both branded drugs (i.e., drugs manufactured by companies under patent) and generic drugs (manufactured without patent) are available for sale with or without a prescription. A generic drug claims to have the same clinical benefits as an equivalent branded drug but is cheaper [3, 4]. A pharmaceutical drug would be effective in curing a disease only when it is free from impurities and adulteration. Counterfeit drugs are detrimental to human health and may cause severe complications during treatment and sometimes may result in death of the patient [2, 5]. Thus drug safety and authenticity have become major concerns in the worldwide pharmaceutical industry due to the availability of counterfeit medicines in the market.

A number of chemical and physical methods to identify fake drugs have been suggested by different pharmacopoeia and drug industries [1–3, 5]. Generally, the drugs are manufactured by compacting a combination of organic composites, i.e., hydrogen (H), carbon (C), nitrogen (N), and oxygen (O), as an active pharmaceutical ingredient (API) [2, 3]. Therefore a suitable technique is required that offers fast identification of complete chemical composition, including APIs, excipients, and impurities present in the medicinal product. The most preferred analytical techniques for drug analysis in the manufacturing industry are high-performance liquid chromatography, atomic absorption spectroscopy (AAS), and near-infrared and Raman spectroscopy [5–7]. The other analytical techniques, e.g., X-ray fluorescence (XRF), laser ablation inductively coupled plasma mass spectrometry (LA-ICP-MS), inductively coupled plasma atomic emission spectroscopy (ICP-AES), and neutron activation analysis (NAA), have also been used for the elemental analysis of samples [6–17].

Laser-Induced Breakdown Spectroscopy
https://doi.org/10.1016/B978-0-12-818829-3.00013-7

At the manufacturing unit, a technique is required, which is capable of real-time, in situ, in vivo, standoff/remote, and point detection analysis of pharmaceutical drugs. LIBS satisfies the foregoing constraints and can analyze samples under any atmospheric condition irrespective of their phase, i.e., solid, liquid, gas, powder, aerosol, biological material, polymer, etc. [7–9]. LIBS is a safe and environment-friendly technique and although other analytical techniques (e.g., XRF, NAA, AAS, XPS, ICP-AES, and LA-ICP-MS) are more precise, they are less popular because of the complexity of instrumentation or constraints of sample preparation before measurements [7–12].

Conventional techniques are costly, time consuming, and may generate chemical waste; therefore LIBS is being adopted as a novel analytical technique for elemental analysis. A comparative study of different analytical techniques with their merits and demerits [11] is summarized in Table 1. It is evident from Table 1 that LIBS offers several unique features over other elemental analysis techniques. All elements present in the pharmaceutical sample can be simultaneously detected in a LIBS experiment in a single step with minimal sample preparation [7–9, 16, 17].

LIBS has become an established laboratory technique with a wide range of applications involving quantitative as well as qualitative analysis of materials [7–9]. Several research publications illustrate the strength of LIBS for the fast and online monitoring of pharmaceutical manufacturing processes [6, 7, 17–19]. Current developments of LIBS aim at enhancing the limit of detection up to the parts per billion level that has great significance in the identification of trace elements in a sample [6–8].

LIBS has been used for the detection of atomic as well as molecular constituents of a variety of material samples by several research groups [18, 19]. The spectral lines emanating from the laser-produced plasma at the target sample constitute the signatures of atomic and molecular species present in the plasma plume. The spectral emission from the plasma plume is dependent on its temporal evolution in addition to the chemical composition of the target [8]. A typical duration of plasma plume ignited by a nanosecond laser pulse is a few microseconds [7, 8]. Laser-induced plasma is very hot in the initial stages of its ignition and emits continuum radiation. However, with the passage of time the plasma plume expands and cools down leading to the emission of spectral lines by neutral atoms and ions after about 1 μs of its ignition. At a further delay from the time of plasma ignition, along with the ionic and atomic lines, some spectral features of molecular bands have also been observed in the LIBS spectra of organic materials [7, 8]. The molecular fingerprints of CN, C_2, CaO, and NH could be observed in the LIBS spectra of organic materials, e.g., polycyclic aromatic hydrocarbons, polymers, aromatic carboxylic acid, amides, etc. [17–21]. In the present work a correlation has been established between the presence of different organic compounds in the samples and the observation of C_2 Swan bands and CN violet bands in their LIBS spectra.

In this study, selected pharmaceutical drugs categorized into two groups, i.e., Group-1 (antidiabetes drugs) and Group-2 (pain-relieving drugs), are studied. The uncoated drug samples of Group-1 contain a voglibose ($C_{10}H_{21}NO_7$) molecule as the API, and are classified as samples of A, B, C, D, and E brands [18]. World Health Organization (WHO)-listed pain-relieving drugs of different brands of Group-2 drugs are taken for elemental and molecular analysis and are termed S1, S2, S3, S4, S5, and S6. Samples S1–S5 contain paracetamol and S6 ibuprofen [21].

2 Material and methods

The LIBS technique is based on laser-matter interaction, and it is necessary that all the experimental parameters in the application of LIBS be

TABLE 1 Comparisons between different elemental analytical techniques.

Analytical techniques	XRF	AAS	LIBS	XPS [13–15]	LA-ICP-MS [9]	ICP-AES [9]	NAA
Type of elements	Most elements with $Z \geq 12$	Most elements with $Z \geq 6$	Most elements	Metals with $Z \geq 3$	Most elements	Most elements	Utmost elements
Sample preparation	Minimal	Required	No/Minimal	Minimal	Required	Required	No/Minimal
Analysis speed	Seconds	10–15 min	Seconds	10–12 min	Minutes	Minutes	Minutes
Detection limits	10–100 ppm	Subppb	10–100 ppm	1000 ppm	Sub-ppb	1–10 ppb	1–10 ppm
Operation cost	Low	Medium	Low	High	High	High	High
Instrumentation	Complex	Complex	Simple	Complex	Complex	Complex	Complex
Quantitative analysis	Yes	Yes	Yes	Semi-Q	Semi-Q	Yes	Yes
Standard required	Yes	Yes	No	Yes	Yes	Yes	Yes
Sample phase	Solid/Liquid	Solid,[a] liquid/gas	Any phase	Solid	Liquid	Liquid/Gas	Any phase
Destructibility	Minimal	Yes	No/Minimal	Minimal	Minimal	Yes	No/Minimal
Sample can be reused	Yes	No	Yes	No	No	No	No
Remote analysis	Possible [15]	No	Possible	No	No	No	Yes
Molecular detections	No	No	Possible	Possible [14]	No	No	No

[a] *Graphite furnace atomic absorption spectroscopy [9].*
AAS, Atomic absorption spectroscopy; ICP-AES, inductively coupled plasma atomic emission spectroscopy; LA-ICP-MS, laser ablation inductively coupled plasma mass spectrometry; LIBS, laser-induced breakdown spectroscopy; NAA, neutron activation analysis; ppb, parts per billion; ppm, parts per million; XPS, X-ray photoelectron spectroscopy; XRF, X-ray fluorescence; Z, atomic number.

optimized to obtain noise-free clean spectral lines for the sample under investigation. In addition to the concentration of an element in the sample, the intensity of its spectral lines depends on the various components of the LIBS detection system. The input energy of the laser pulse that ignites the plasma plume is of greatest importance, as is the efficient collection of light emission to form the spectrum. The distance between the sample and the lens for collecting light and the focal length of the latter have to be selected with great care; the angle between the laser beam and the collecting optics and the time delay between the ignition of laser-plasma and its collection play major roles in the quality of detection.

2.1 Optimization of laser energy

To optimize some of the experimental parameters, LIBS spectra of a solid target made up of Al, Mg, and Ca were recorded, and spectral intensities of lines of Mg at 285.2 nm and of Al at 358.6 nm were monitored. The signal strength of the spectral lines increased linearly with the increase in laser pulse energy from 15 mJ initially and then saturated at higher

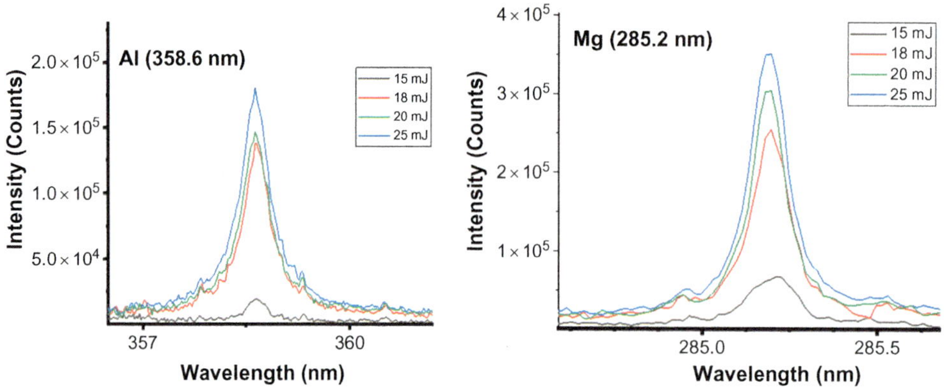

FIG. 1 Spectral line intensity of Al (358.6 nm) and Mg (285.2 nm) at different laser energies.

energies. The power of the pulsed laser was measured by a Genetec-E (model UP19K-30H-VM-DO) energy meter. The saturation effect was observed in the spectral lines of Mg at 20 and 25 mJ, whereas in the same energy range the lines of Al were not saturated (Fig. 1). However, the effect of saturation in the spectral lines for both elements was not observed at 18 mJ of laser energy and hence it was used to record the LIBS spectra of pharmaceutical drugs in the present work.

2.2 Optimization of the angle between collection optics and laser beam

Experiments were carried out to optimize the angle between the laser beam and the optical axis of the assembly to get the best results. The signal-to-background (S/B) and signal-to-noise (S/N) ratios were greatly improved for this angle in the range of ~35–45 degrees. Hence, an ≈45-degree angle was chosen to record the LIBS spectra of all drug samples (Fig. 2).

2.3 Optimizations of gate delay and gate width

The gate width of the spectrometer represents the actual exposure time for which the light from the laser-plasma plume is used to record the spectrum. Similarly, the gate delay refers to the interval of time between the ignition of the plasma plume and the beginning of its light being sensed by the charge-coupled device detectors in the spectrometer. The intensity of the spectral lines in the LIBS spectra depends on the plasma temperature and it is necessary to optimize the gate delay and gate width of the spectrometer to record noise-free LIBS spectra. In the ambient atmosphere, the laser-produced plasma lifetime is about 10 μs, and the continuum emissions are dominated at the early stage of the plasma. When the plasma cools down, after a few microseconds of its ignition, ionic and atomic spectral lines are observed with minimal continuum background [7, 8]. To determine the optimum condition for the best S/N ratio, lines in the LIBS spectra are monitored with different combinations of gate width and gate delay (Fig. 3). It was found that for 1-μs gate width, the best S/B and S/N ratio is observed at about a 0.7-μs gate delay in the spectrometer.

2.4 Accumulation of LIBS spectra

Once the LIBS experimental parameters are optimized, spectra of any sample can be recorded by using a single laser shot or by

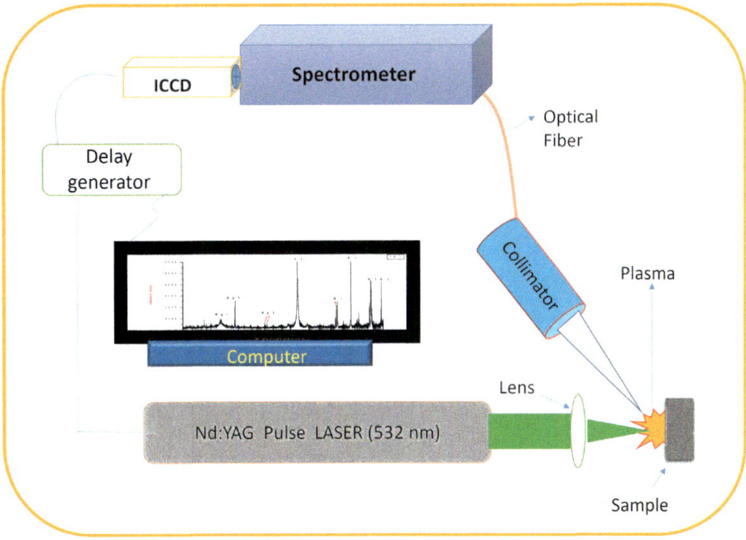

FIG. 2 Orientation of collection optics with respect to the incident laser beam.

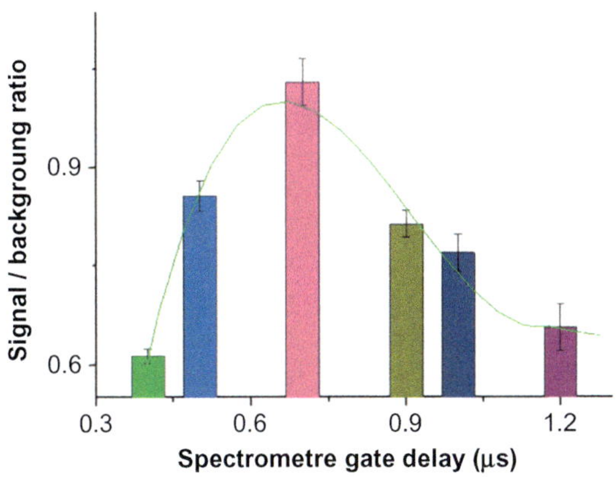

FIG. 3 Optimization of gate delay for Ca (422.6 nm) emission line.

accumulation of spectra from multiple laser shots. The accumulation of the LIBS spectra involves the addition of several single laser shot spectra. The LIBS spectral intensity and S/B ratio increase with an increase in the number of single laser shot spectra in the accumulation process. Thus the sensitivity and limit of detecting constituent elements in the sample can be enhanced by accumulation of LIBS spectra.

A schematic diagram of the LIBS experimental setup is shown in Fig. 4, which consists of an Nd:YAG laser (Continuum Surelite III-10) and wavelength separator, which is used as a filter for residual wavelengths other than 532 nm from the source. The delay generator is used to synchronize the laser and spectrometer. The spectrometer (spectral range 200–850 nm, $\lambda/\Delta\lambda \approx 6000$; Mechelle ME5000, Andor Technology), equipped with an intensified charge-coupled device (iStar 734, Andor Technology) and Andor Solis software, is used to record LIBS spectra.

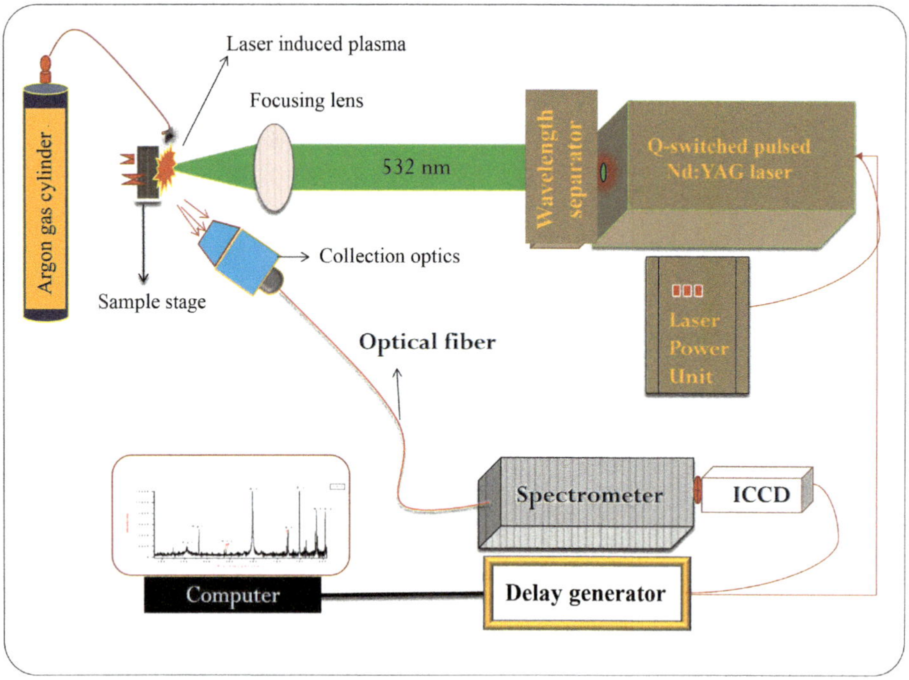

FIG. 4 A schematic block diagram of the experimental setup to record LIBS spectra.

The atmospheric interference (contribution of nitrogen and oxygen) in the LIBS spectra cannot be avoided completely, therefore the LIBS spectra are also recorded in an argon atmosphere by purging the argon gas across the sample surface to reduce the interference of ambient atmosphere [9–12] as shown in Fig. 4. To produce the laser-induced plasma on the target material, under the ambient environments, high-peak laser irradiance of the order of 100 GW/cm^2 is employed.

2.5 Chemometric analysis

In the pharmaceutical industry, LIBS is used to inspect and compare a huge amount of drug samples, thus creating vast spectral datasets that involve high complexity in the analysis. To overcome such complexity and to achieve visual patterns in data, chemometric analysis can be employed on variables of LIBS, i.e., spectral intensities corresponding to different wavelengths [18, 21]. Two different types of chemometric methods, i.e., principal component analysis (PCA) and partial least squares (PLS) regression analysis, have been used in the present work and details of these methods are described in the following paragraphs.

PCA is an approach in chemometric methods used for discrimination/classification of a large number of datasets. It enables visual representation of the dataset through the projection of variables [18, 21]. For a large number of spectra of various samples having similar compositions, it would become a complex and difficult task to discriminate them. In such cases, PCA is preferred for the classification of multiple samples. For the quantitative analysis of a very large number of samples, multivariate regression analysis (i.e., PLS regression analysis) is preferred. In PLS regression analysis, the information is extracted from the matrix having any standard/known values that correspond to

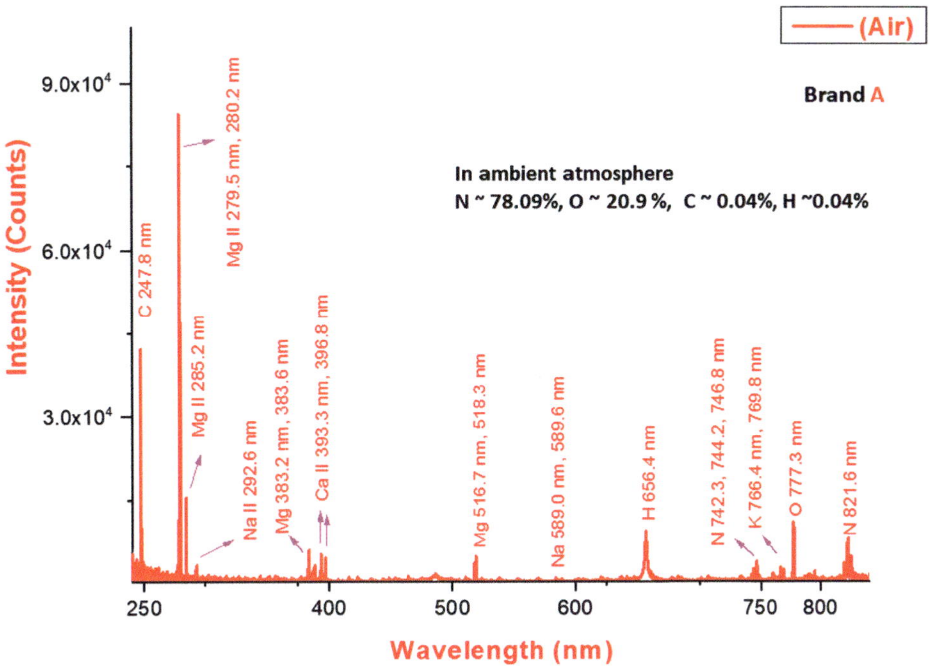

FIG. 5 Laser-induced breakdown spectroscopy spectrum of Group-1 drug (Brand A) recorded in ambient atmosphere.

establishing a robust calibration model [18]. PLS regression analysis is used on the LIBS spectral dataset along with reference data to build up the calibration models that can predict the relative concentrations of different elements in unknown samples [17, 18]. The linear regression approach is no longer practical if the number of predictors/variables is large compared to the number of observations. In the present work, PLS regression analysis is used to illustrate the capability of LIBS techniques for the rapid quantitative analysis of bulk materials.

3 Results and discussion

3.1 Analysis of Group-1 (antidiabetes) drugs of different brands

LIBS spectra of Group-1 drugs of different brands have been recorded under the same experimental conditions in the ambient atmosphere. Six LIBS spectra of each tablet of Group-1 drugs are recorded, and each spectrum is the accumulation of 50 laser shots. Five replicates of each drug of Group-1 that contains voglibose are chosen for the present study [18]. Voglibose, discovered in Japan in 1981, is a kind of organic compound that is typically used in type-2 antidiabetic drugs [22]. A typical LIBS spectrum of Brand A drug of Group-I (antidiabetes drug) is shown in Fig. 5, which exhibits the spectral signatures of organic elements (e.g., carbon [C; 247.8 nm], hydrogen [H; 656.4 nm], nitrogen [N; 742.3, 744.2, 746.8, and 821.6 nm], oxygen [O; 777.1, 77.4, 777.5, and 844.8 nm]) as well as inorganic elements (e.g., Mg; 279.5(II), 280.2(II), 285.2, 382.9(II), 383.2(II), 516.7, 517.3, 518.3 nm), silicon (Si; 250.7, 251.4, 251.6, 252.8, and 288.1 nm), calcium (Ca; 393.3 (II), 396.8(II), 397.2(II), and 422.6 nm), and potassium (K; 766.4 and 769.8 nm). Similar spectral signatures are observed for all brands of Group-1 drugs (Figs. 5–7). The presence of a

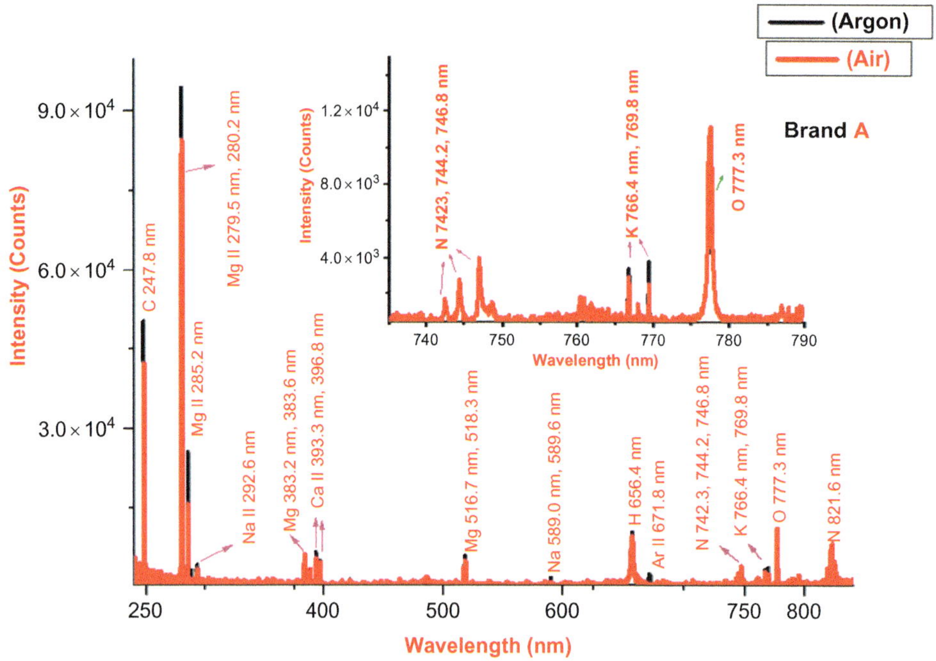

FIG. 6 Laser-induced breakdown spectroscopy spectra of Group-1 drug (Brand A) recorded in argon and ambient atmospheres.

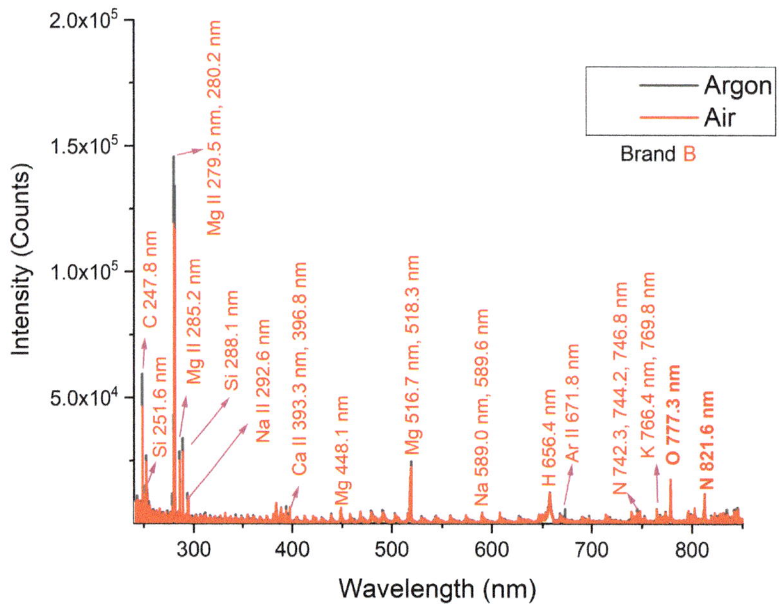

FIG. 7 Laser-induced breakdown spectroscopy spectrum of Group-1 drug (Brand B) recorded in argon and ambient atmospheres.

TABLE 2 Laser-induced breakdown spectroscopy spectral intensity ratios of organic elements[a] of Group-1 drugs.

Group-1 sample (Brand)	H/C (argon)	H/C (air)	O/N (argon)	O/N (air)	H/N (argon)	H/N (air)
A	3.35 ± 0.16	3.73 ± 0.29	6.61 ± 0.16	6.32 ± 0.67	14.32 ± 0.20	12.66 ± 0.39
B	2.57 ± 0.22	2.79 ± 0.35	8.12 ± 0.44	7.59 ± 0.48	16.38 ± 0.43	13.90 ± 0.75
C	3.52 ± 0.19	3.89 ± 0.32	9.26 ± 0.24	8.95 ± 0.33	16.83 ± 0.29	15.27 ± 0.50
D	4.21 ± 0.14	4.72 ± 0.43	9.18 ± 0.66	8.17 ± 0.39	17.64 ± 0.53	14.55 ± 0.40
E	4.01 ± 0.10	4.04 ± 0.33	9.10 ± 0.46	9.36 ± 0.57	15.95 ± 0.37	15.39 ± 0.77

[a] *The generating lines are taken for H at 656.4 nm, C at 247.8 nm, N at 746.8 nm, and O at 777.2 nm.*

spectral signature of organic elements represents the detection of organic compounds like voglibose in the drug sample. The presence of spectral lines of inorganic elements may be due to inactive ingredients of drug such as binders, excipients, etc. or it may also be because of impurities during the manufacturing process.

For the LIBS spectra recorded in an ambient atmosphere, the appearance of the spectral lines of H, C, O, and N from the atmosphere in the drug sample cannot be avoided. To overcome this interference, we have also recorded LIBS spectra of Group-1 drugs in an argon atmosphere, and the spectral signatures of organic elements (H, C, O, and N) are clearly seen (Fig. 6). Thus the presence of spectral signatures of organic elements in the LIBS spectra recorded in the argon atmosphere confirms the presence of organic compounds like voglibose in the drug (Fig. 6).

Similarly, we have recorded the LIBS spectra of all other brands of Group-1 drug samples (i.e., Brand B, Brand C, Brand D, and Brand E). A typical LIBS spectrum of Brand B, shown in Fig. 7, clearly exhibits the fact that the composition of Brand B is similar to that of Brand A. An additional spectral signature of silicon (Si) is observed in the LIBS spectra of Brand B and Brand D. Intensities of the spectral lines of all the elements except N and O recorded in an argon atmosphere are enhanced compared to the ambient atmosphere (Figs. 6 and 7). In general, a laser-produced plasma plume generated in an argon atmosphere is more confined compared to the plasma generated in the ambient atmosphere as the mass density of argon is higher than that of air, and it leads to enhancement of spectral intensity [6, 7, 18–20].

With the help of LIBS spectral data, we now demonstrate that the present drug sample is composed of an organic compound (voglibose). We have calculated the intensity ratio of spectral lines of a pair of organic elements in the LIBS spectra recorded in ambient air as well as in an argon atmosphere as shown in Table 2, using the measured intensity of the spectral lines of C at 247.8 nm, N at 746.8 nm, H at 656.4 nm, and O at 777.2 nm [18]. These estimated ratios are correlated from stoichiometric ratios, e.g., ratios of two organic elements present in the molecular composition of voglibose, i.e., H/C is 2.1, H/N is 21, O/N is 7, etc. Based on experimental intensity ratios of spectral lines and stoichiometric ratios, univariate regression analyses have been drawn and are shown in Fig. 8.

It is clear from Fig. 8 that the experimental intensity ratios are close to the stoichiometric ratio, therefore this analysis confirms the presence of voglibose in Group-1 drugs. The observed regression coefficient (R^2) value is 0.93 for argon and 0.86 for an ambient atmosphere in above-univariate linear regression plots. The R^2 value is close to one, which shows reasonably good correlations for both ambient and argon atmospheres. The present linear regression analyses clearly demonstrate

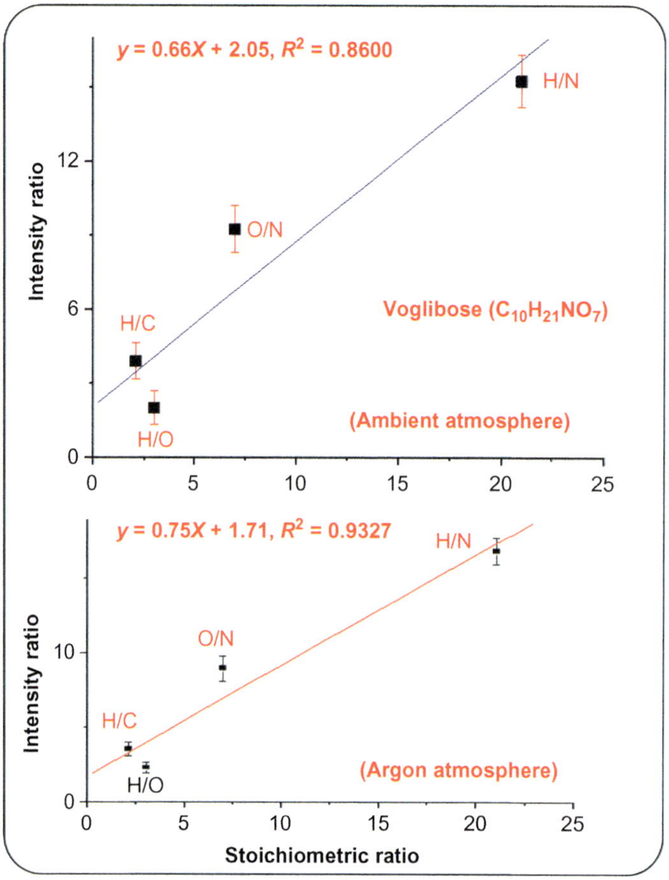

FIG. 8 Ratio-based linear regression analysis of Brand A of Group-1 drug.

the potential use of the LIBS technique for identification of molecular compositions in various drugs (Fig. 8).

3.2 Statistical analysis of LIBS spectral data of Group-1 drugs

3.2.1 Principal component analysis of Group-1 drugs

LIBS spectra of Group-1 drugs (Figs. 6 and 7) reveals that the compositions of the drugs may be similar. If we have a huge number of samples with nearly similar compositions, it will be very difficult to discriminate them from each other. PCA is a multivariate method that helps in discriminating such samples into different clusters on the basis of slight differences in their compositions. PCA has been performed on Group-1 drugs by generating matrices of order $30 \times 30{,}456$ with the help of LIBS spectral data. The score plot of Fig. 9 represents classification of Group-1 samples about 98% (96% by PC-1 and 2% by PC-2) using PCA.

Fig. 9 illustrates the compositional variations in samples A, B, C, D, and E by clustering each brand in color-coded circles: A (blue data), B (red data), C (dark green data), D (black data), and E (light brown data). Samples A, B, and E are clustered close to each other, which indicates that these samples may have a similar

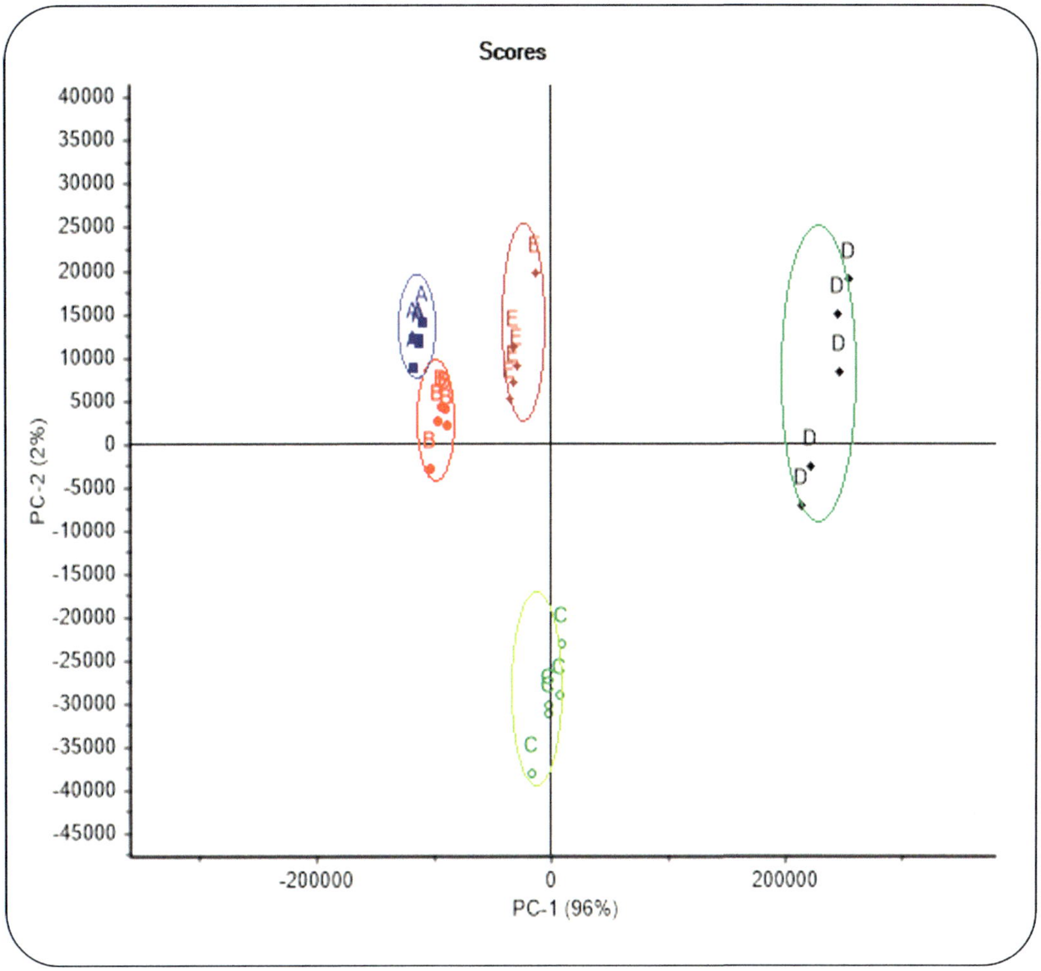

FIG. 9 Two-dimensional principal component analysis score plot of Group-1 drugs.

composition, which is also evident from the pattern of spectral lines in Figs. 6 and 7.

3.2.2 PLS regression analysis of Group-1 drugs

Using a reference dataset, PLS regression analysis can construct a calibration model to predict the concentrations of the constituents in materials having the same matrix as the reference material. In general, it is not easy to have a reference material having the same matrix as the material that has to be analyzed. In the present case, we do not have any reference dataset of Group-1 drugs. However, in Table 1, we have demonstrated that LIBS spectral intensity ratios of organic elements represent the presence of voglibose molecules in the sample. Therefore we have taken these calculated ratios as the reference dataset for PLS regression analysis to construct an empirical ratio-based calibration model for Group-1 drugs.

A typical wavelength region containing spectral emission lines of organic elements is used to create the elemental ratio-based PLS calibration model. Among 30 LIBS spectra of Group-1 drugs, 20 spectra are chosen as the training set (known sample), and 10 spectra are chosen as the test set (unknown sample) to prepare the PLS regression model [18]. The performance of the PLS calibration model (test sets) can be relied upon if the predicted values are very close to the reference values. The reliability can also be judged by the root mean square error (RMSE) and R^2 values [17–19]. The R^2 value represents a direct linear relationship between the LIBS spectra and the calculated elemental ratios; however, for a good correlation, RMSE values should be close to zero, while R^2 values should be almost equal to one [18].

For the prediction of unknown samples (test samples), a ratio-based library dataset is constructed and the result is represented in a calibration curve which has two different regression lines, i.e., calibration line and validation line (Fig. 10). The blue line in the figure stands for the calibration line and the red line represents the validation line. The validation line deviates very little from the calibration line, which infers that calibration is close to validation. However, the developed model cannot be trusted if the calibration and the validation lines differ substantially. The calibration set can be evaluated for several other matrixes of similar composition, once the PLS regression model is precisely developed for the given matrix [18]. The H/C ratio-based PLS regressions have been performed for air and argon atmospheres and are shown in Fig. 10A and B, respectively.

It is seen from Fig. 10 that the R^2 value is close to unity (0.85 in argon and 0.89 in air) and the RMSE value is close to zero (0.2 in argon and 0.3 in air). This provides a reasonably good agreement between calibration and validation and thus the developed model is found to be authentic. Similarly, we have also performed PLS regression analysis for other elemental ratios like O/N, H/N, etc., and the results show that the H/C ratio-based regression model has better accuracy and reliability because its lower value of RMSE and R^2 value are close to unity. Using the developed regression models, the predicted spectral intensity ratios are compared with the observed experimental values in Table 3.

It is clear from Table 3 that the experimentally calculated (two times repeated for each) spectral

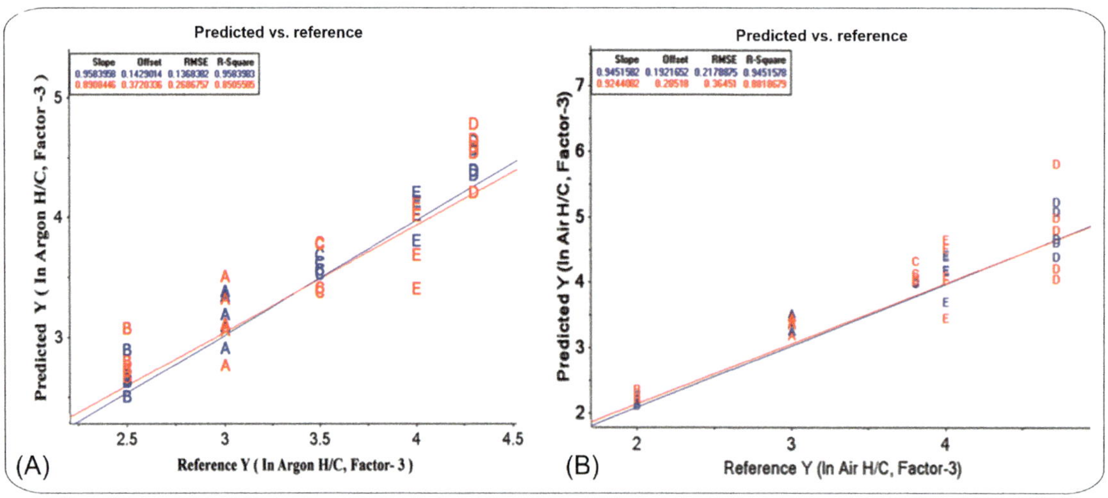

FIG. 10 (A, B) PLS regression analysis of Group-1 drugs using H/C ratio in argon and ambient atmospheres.

TABLE 3 Calculated intensity ratios of H/C and O/N[a] of Group-1 drugs using partial least squares regression.

	Ambient atmosphere		Argon atmosphere		Ambient atmosphere		Argon atmosphere	
Group-1 (Brand)	Experimental H/C	Predicted H/C	Experimental H/C	Predicted H/C	Experimental O/N	Predicted O/N	Experimental O/N	Predicted O/N
A	3.73 ± 0.29	3.16 ± 0.26	3.35 ± 0.16	3.08 ± 0.18	6.61 ± 0.67	6.15 ± 0.65	6.32 ± 0.16	6.65 ± 0.69
A	3.73 ± 0.29	3.29 ± 0.28	3.35 ± 0.16	3.06 ± 0.19	6.61 ± 0.67	6.45 ± 0.68	6.32 ± 0.16	6.86 ± 0.68
B	2.79 ± 0.35	2.08 ± 0.30	2.57 ± 0.22	2.55 ± 0.20	8.12 ± 0.48	8.40 ± 0.58	7.59 ± 0.44	7.76 ± 0.63
B	2.79 ± 0.35	2.13 ± 0.24	2.57 ± 0.22	2.54 ± 0.24	8.12 ± 0.48	8.50 ± 0.55	7.59 ± 0.44	7.46 ± 0.59
C	3.89 ± 0.32	3.88 ± 0.30	3.52 ± 0.19	3.71 ± 0.23	9.26 ± 0.33	9.30 ± 0.64	8.95 ± 0.24	8.27 ± 0.73
C	3.89 ± 0.32	3.85 ± 0.25	3.52 ± 0.19	3.40 ± 0.25	9.26 ± 0.33	9.40 ± 0.57	8.95 ± 0.24	9.08 ± 0.61
D	4.72 ± 0.43	4.53 ± 0.26	4.21 ± 0.14	4.56 ± 0.30	9.18 ± 0.39	9.10 ± 0.62	8.17 ± 0.66	8.29 ± 0.65
D	4.72 ± 0.43	4.34 ± 0.33	4.21 ± 0.14	4.08 ± 0.26	9.18 ± 0.39	9.01 ± 0.65	8.17 ± 0.66	8.81 ± 0.75
E	4.04 ± 0.33	4.00 ± 0.35	4.01 ± 0.10	3.78 ± 0.25	9.10 ± 0.57	9.10 ± 0.60	9.36 ± 0.46	9.25 ± 0.64
E	4.04 ± 0.33	3.26 ± 0.25	4.01 ± 0.10	3.98 ± 0.24	9.10 ± 0.57	9.10 ± 0.60	9.36 ± 0.46	9.11 ± 0.62

[a] *Wavelength of spectral lines used in ratios C at 247.8 nm, H at 656.4 nm, N at 746.8 nm, and O at 777.2 nm.*

intensity ratios are very close to the predicted PLS regression values [18]. In addition to this, the regression analyses for H/C ratios are unswerving in both air and argon atmospheres as compared to other ratios. Therefore it is cost effective to perform H/C ratio-based regression analysis in the ambient atmosphere. Our study shows that the statistical data processing methods (PCA and PLS regression) of LIBS spectral datasets for large amounts of samples lead to more productive and rapid results than those from the univariate approach.

The present experimental results show that the presence of voglibose composition in unknown drug samples can be predicted rapidly with the developed PLS regression model. Hence, this approach can be applied to identify organic compounds in any drug samples by developing PLS regression analysis of elemental ratios of organic elements. The PLS regression analysis of the LIBS spectral dataset can be effectively used for the in situ and online analysis of bulk amounts of drugs.

3.3 LIBS analysis of Group-2 (pain-relieving) drugs

A study of atomic, ionic, and molecular signatures from laser-produced plasma can provide valuable information about sample compositions. If the organic samples contain ample amounts of carbon and nitrogen, then for a longer gate delay in the spectrometer (concerning the plasma ignition), the spectral bands of CN and C_2 can be observed in LIBS spectra of this sample [18–21]. The molecular signatures in the laser-induced plasma are expected to form due to recombination of native carbon-carbon and carbon-nitrogen species present in the sample [19–21, 23]. Thus the presence of molecular bands of CN and C_2 in the LIBS spectra indicates that the target material contains organic compounds [19–21, 23]. Therefore molecular composition in the sample can be predicted from the observed spectral signature of the CN violet band system ($B_2\Sigma^+ - X_2\Sigma^+$) and C_2 Swan band system ($d_3\Pi g$–$a_3\Pi u$) in the LIBS spectra.

However, the interaction of atmospheric carbon and nitrogen with the laser-produced plasma may also result in the appearance of CN bands in the LIBS spectra. To avoid atmospheric interference, LIBS spectra of six different brands of Group-2 drugs are also recorded in the argon atmosphere (Fig. 11A, B, and C). In the present study this LIBS technique has been tried to identify the molecular compositions of Group-2 drugs, which (reported by the manufacture) are: S1 (ibuprofen and paracetamol), S2 (aceclofenac and paracetamol), S3 (diclofenac sodium and paracetamol), S4 (paracetamol), S5 (paracetamol), and S6 (ibuprofen) [21].

In the LIBS spectra recorded in an argon atmosphere, the spectral signatures of the atomic species (C [247.8 nm], Hα [656.3 nm], O [777.1, 777.4, 777.5 nm], N [742.46, 744.28, 746.85 nm], Na [589.0, 589.5 nm], Mg [279.5(II), 280.2 (II), 285.2, 382.9(II), 383.2(II), 516.7, 517.3, 518.3 nm], Ca [393.3(II), 396.8(II), 397.2(II), 422.6 nm], and Si [250.7, 251.4, 251.6, 252.8, 288.1 nm]) are observed in the LIBS spectra as shown in Fig. 11A. In addition to this, the electronic bands of CN corresponding to (0,0), (1,1), (2,2), (3,3), and (4,4) transitions at 388.2, 387.0, 386.1, 385.4, and 385.0 nm, respectively, are also observed in the drug samples (S1–S5) recorded in an argon atmosphere (Fig. 11B). However, the CN bands are absent in the LIBS spectra of sample S6 (ibuprofen), which reflects the absence of nitrogen in the composition of sample S6 [21].

Mousavi et al. demonstrated that in the absence of C—C or C=C bonds in the molecular structure of the organic compounds, C_2 spectral bands will not be observed in their LIBS spectra of this compound/sample [24]. Therefore the strongest molecular bands of C_2 (0,0) at 516.4 nm and CN (0,0) at 388.32 nm are calculated to identify the presence of different molecules in the drug samples (Figs. 12 and 13). The molecular bands of C_2 Swan band systems, at 471.5, 469.7, 468.4, 467.8, 516.4, and 512.8 nm corresponding to (2,1), (3,2), (4,3), (5,4), (0,0), and (1,1) vibrational transitions, respectively, are also observed in the LIBS spectra of Group-2 drugs (Fig. 11C).

The foregoing results clearly demonstrate that the presence of the molecular bands of C_2 and CN molecules in the LIBS spectra provides a great deal of information about the molecular compositions of the target materials (Figs. 12 and 13). The CN and C_2 band spectral intensity in LIBS spectra clearly divulge the presence of organic compounds/molecules in each sample. The variation in the intensity of the bands of CN and C_2 molecules in the LIBS spectra represents the presence of different kinds of molecular compositions in the matrix of pharmaceutical drugs (Figs. 12 and 13). As reported by the manufacturer, drug samples S4 and S5 have similar compositions (paracetamol) and this is also confirmed by the present work as their LIBS results are very similar. The spectral signature of organic as well as inorganic elements can be detected/monitored using the LIBS technique, which is useful for the detection of the matrix of pharmaceutical drugs. With the help of the LIBS technique, one can simultaneously monitor the atomic lines of the elements as well as molecular bands, thus the variation in their intensities in the LIBS spectra can be utilized at the time of drug manufacturing.

3.4 Principal component analysis of Group-2 drugs

In Fig. 14, the three-dimensional PCA of the LIBS spectral datasets of Group-2 drugs is shown with the principal component scatter plot. To make a PCA, a library is created in the form of a matrix of certain order, i.e., (60 × 24,806), with the help of the LIBS spectral dataset of Group-2 drugs in terms of intensity and wavelength.

The representations in PCA are displayed in three different PCs (PC-1, PC-2, and PC-3) to elucidate the variations of the spectra along with their scores for classification of the samples. In

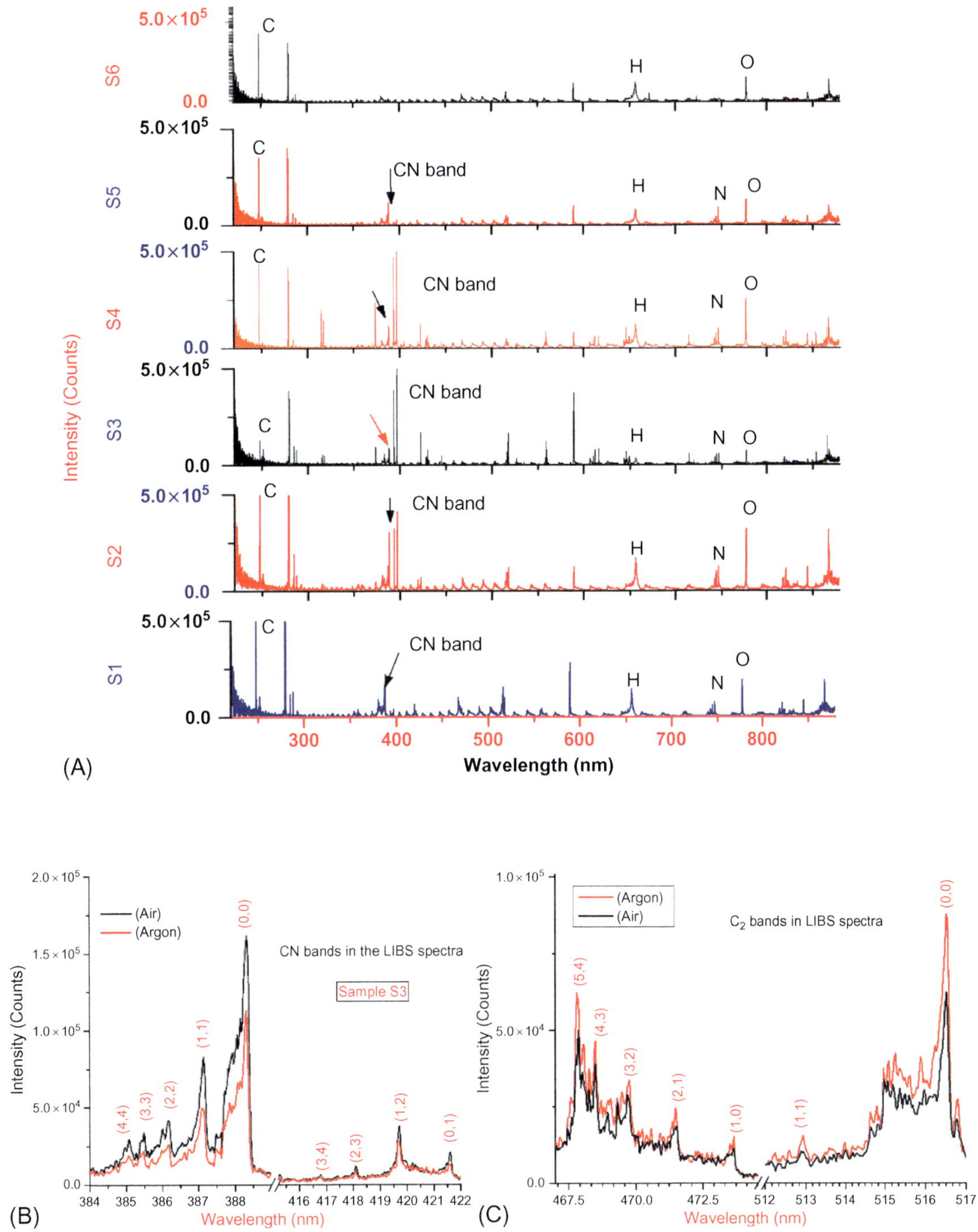

FIG. 11 (A) Laser-induced breakdown spectroscopy (LIBS) spectra of Group-2 drugs recorded in an argon atmosphere (240–850 nm); (B, C) CN and C_2 bands in LIBS spectra of an S3 sample recorded in argon and ambient atmospheres.

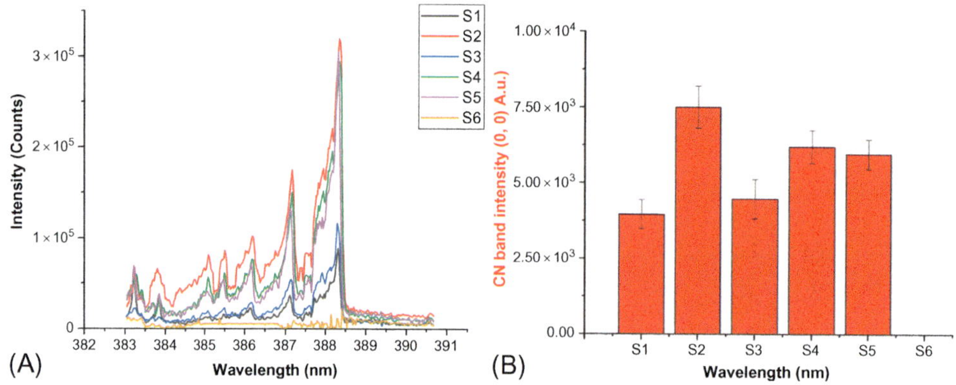

FIG. 12 (A) Vibrational bands of a CN molecule in the laser-induced breakdown spectroscopy (LIBS) spectra of Group-2 drugs; (B) CN band intensities of (0,0) band present in the LIBS spectra of Group-2 drugs.

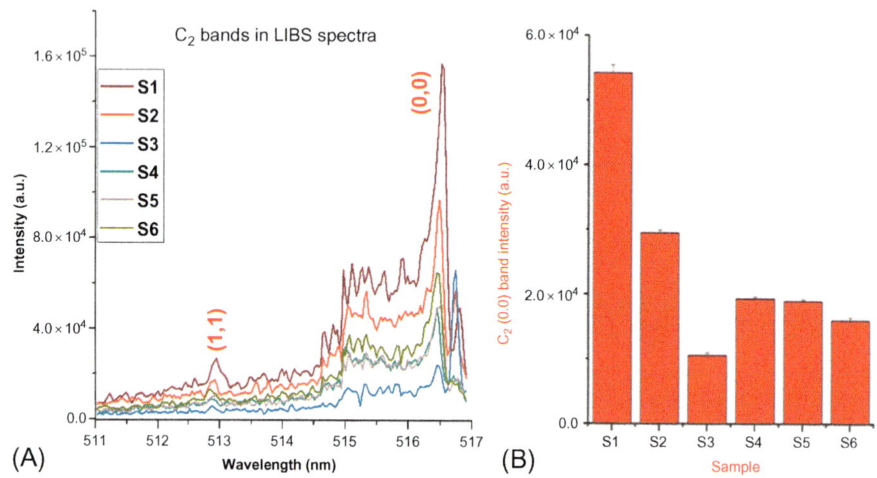

FIG. 13 (A) Laser-induced breakdown spectroscopy (LIBS) spectra of Group-2 drugs showing the presence of the vibrational band of a C_2 molecule; (B) C_2 band intensities of the (0,0) band present in the LIBS spectra of Group-2 drugs.

the present study, 10 LIBS spectra for each brand of Group-2 drugs are recorded and these data are included for PCA analysis. Fig. 14 reveals the apparent clustering of each brand of Group-2 drugs, e.g., S1 (blue data), S2 (red data), S3 (dark green data), S4 (black data), S5 (light green data), and S6 (brown data) in a 3D score plot. The analysis infers that information extracted from the LIBS spectral dataset despite having similar compositions reflects significant discrimination of Group-2 drugs with the help of PCA.

If there is a bulk amount of sample of either similar or slightly varying compositions, then it is a most tedious job to classify them. The present finding claims that PCA can help in the discrimination/classification of an immense number of drugs having nearly similar or different compositions.

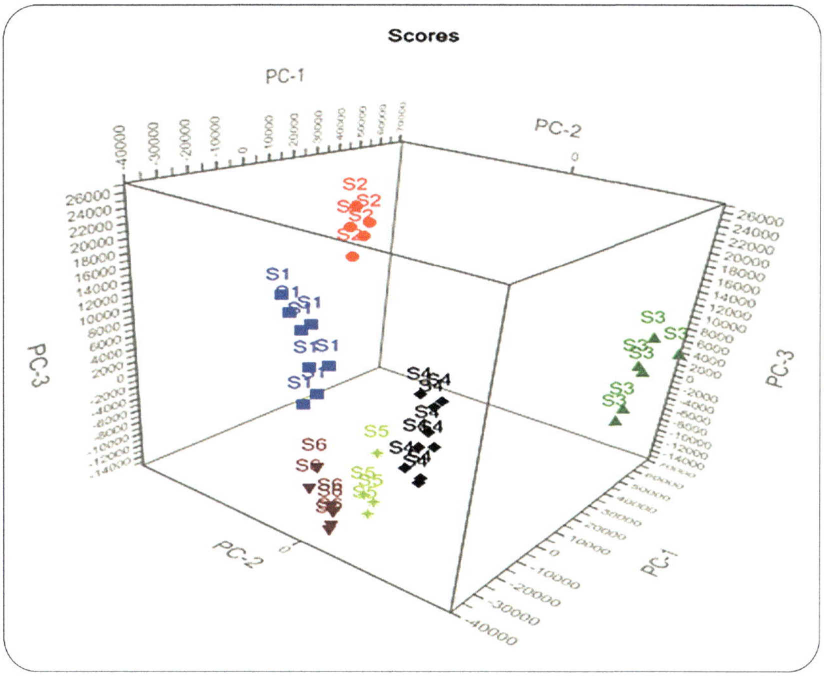

FIG. 14 Three-dimensional principal component scatter plot of Group-2 drugs where the PC-3 axis is perpendicular to the plane of this page.

4 Conclusion and future outlook

Online LIBS analysis in the pharmaceutical industry remains a challenge, but with the help of chemometrics on LIBS spectral data, it can be a useful tool in drug compositional analysis. For the classification/discrimination of bulk amounts of drugs, the PCA of a LIBS dataset emerges as a valuable approach/method. The developed calibration model on the LIBS spectral dataset using PLS regression with the help of known reference data or estimated values using different approaches could be one of the significant methods for rapid investigation of impurities detection or counterfeit drugs. Once a calibration model is devolved, it could be a user-friendly interface for both nontechnical and technical users to make their job easier. The CN and C_2 band intensity shows the evidence of simultaneous identification of molecular signature together with elemental fingerprints present in the LIBS spectra. The sample can be reused after LIBS analysis and it can be the most attractive way to perform real-time, online, and in situ analysis of drugs for industrial as well as other medical applications. Handheld LIBS integrated with chemometric models could be a rapid and nondestructive counterfeit detection tool for online sampling/inspection/identification of unknown substances in versatile phases in just a single unit.

Acknowledgment

Dr. Pravin Kumar Tiwari thanks the Board of Research in Nuclear Sciences, India, for providing Senior Research Fellowship support from the project Sanction No. 39/14/30/2016-BRNS/34434, 24.01.17. The authors are grateful to Prof. S. N. Thakur, Ex-Dean Faculty of Science and Ex-HOD, Physics,

Banaras Hindu University, Varanasi, India, for his valuable suggestions. We are also thankful to Dr. A. K. Pathak, Ewing Christian College, Prayagraj, India, for his feedback.

References

[1] Definition and Classification of Drug or Pharmaceutical Regulatory Aspects of Drug Approval, http://www.pharmacistspharmajournal.org, 2013. (Accessed 30.12.13).

[2] Food and Drug Administration (FDA), USA, https://www.fda.gov/drugs/information-healthcare-professionals-drugs/drug-safety-information, 2019.

[3] Generic Drug Facts, Center for Drug Evaluation and Research, U.S. Food and Drug Administration, 2017. 2017-10-06 retrieved (2017).

[4] What's the Difference Between Brand-Name and Generic Prescription Drugs? Scientific American. Retrieved (2017).

[5] R. Martino, M. Malet-Martino, V. Gilard, S. Balayssac, Counterfeit drugs: analytical techniques for their identification, Anal. Bioanal. Chem. 398 (1) (2010) 77–92.

[6] L. Nancy, The use of atomic spectroscopy in the pharmaceutical industry for the determination of trace elements in pharmaceuticals, J. Pharm. Biomed. Anal. 55 (2011) 653–661.

[7] A.W. Miziolek, V. Palleschi, I. Schechter, Laser Induced Breakdown Spectroscopy: Fundamental and Application, Cambridge University Press, New York, 2006. ISBN: 978-0-511-24529-9.

[8] J.P. Singh, S.N. Thakur, Laser-Induced Breakdown Spectroscopy, Elsevier, Amsterdam, UK, 2007. ISBN: 978-0-444-51734-0.

[9] J.D. Winefordner, N. Omenetto, et al., Comparing several atomic spectrometric methods to the super stars: special emphasis on laser induced breakdown spectrometry, LIBS, a future super star, J. Anal. At. Spectrom. 19 (2004) 1061–1083.

[10] J. Willis, K. Turner, G. Pritchard, XRF in the Workplace: A Guide to Practical XRF Spectrometry, PANalytical, Chipping Norton, NSW, Australia, 2011. ISBN (9780620506656).

[11] Q. Lin, G. Niu, Q. Wang, Q. Yu, Y. Duan, Combined laser-induced breakdown with Raman spectroscopy: historical technology development and recent application, Appl. Spectrosc. Rev. 48 (6) (2013) 487–508.

[12] P.S. Remya, A. Devi, A. Dalvi, K.K. Swain, R. Verma, Comparison and statistical evaluation of neutron activation methodologies for the determination of gold in copper concentrate, Anal. Methods 7 (2015) 3833–3840.

[13] D.R. Baer, M.H. Engelhard, XPS analysis of nanostructured materials and biological surfaces, J. Electron Spectrosc. Relat. Phenom. 178–179 (2010) 415–432.

[14] D.L. Perry, Z. Ma, A. Olson, E. Topp, Molecular characterization of gadolinium-doped zinc telluride films by X-ray photoelectron, Spectroscopy 30 (6) (2015) 38–43.

[15] D.F. Blake, Remote X-Ray Diffraction and X-Ray Fluorescence Analysis on Planetary Surfaces, https://ntrs.nasa.gov/archive/nasa/casi.ntrs.nasa.gov/20000102383.pdf, 1999.

[16] D.W. Hahn, N. Omenetto, Laser-induced breakdown spectroscopy (LIBS), part II: review of instrumental and methodological approaches to material analysis and applications to different fields, Appl. Spectrosc. 66 (4) (2012) 347–419.

[17] R. Noll, et al., LIBS analyses for industrial applications—an overview of developments from 2014 to 2018, J. Anal. At. Spectrom. 33 (2018) 945–956.

[18] P.K. Tiwari, S. Awasthi, R. Kumar, R.K. Anand, P. K. Rai, A.K. Rai, Rapid analysis of pharmaceutical drugs using LIBS coupled with multivariate analysis, Lasers Med. Sci. 33 (2) (2018) 263–270.

[19] A. Portnov, S. Rosenwaks, I. Bar, Emission following laser induced breakdown spectroscopy of organic compounds in ambient air, Appl. Opt. 42 (2003) 2835–2842.

[20] L. St-Onge, R. Sing, S. Béchard, M. Sabsabi, Carbon emissions following 1064 μm laser ablation of graphite and organic samples in ambient air, Appl. Phys. A Mater. Sci. Process. 69 (1999) S913–S916.

[21] P.K. Tiwari, N.K. Rai, R. Kumar, C.G. Parigger, A.K. Rai, Atomic and molecular laser-induced breakdown spectroscopy of selected pharmaceuticals, Atoms 7 (3) (2019) 2019060095.

[22] A.S. Dabhi, N.R. Bhatt, M.J. Shah, Voglibose: an alpha glucosidase inhibitor, J. Clin. Diagn. Res. (2013), https://doi.org/10.7860/JCDR/2013/6373.3838.

[23] M. Dong, X. Mao, J.J. Gonzalez, J. Lu, R.E. Russo, Time resolved LIBS of atomic and molecular carbon from coal in air, argon, and helium, J. Anal. At. Spectrom. 27 (2012) 2066–2075.

[24] S.J. Mousavi, M.H. Farsani, S.M.R. Darbani, N. Asadorian, M. Soltanolkotabi, A.E. Majd, Identification of atomic lines and molecular bands of benzene and carbon disulfide liquids by using LIBS, Appl. Opt. 54 (2015) 1713–1720.

CHAPTER

14

LIBS imaging applications

V. Motto-Ros[a], S. Moncayo[b], C. Fabre[c], B. Busser[d,e]

[a]Light and Matter Institute (UMR 5306), Lyon 1 University—CNRS, Villeurbanne, France [b]Ramón y Cajal Health Research Institute (IRYCIS), Madrid, Spain [c]GeoRessources Laboratory (UMR 7359), University of Lorraine—CNRS, Vandoeuvre-les-Nancy, France [d]Grenoble Alpes University Hospital, CHUGA, Grenoble, France [e]Institute for Advanced Biosciences (Inserm U 1209 / CNRS 5309), Grenoble Alpes University, Grenoble, France

1 Introduction

In the last few years, the application of laser-induced breakdown spectroscopy (LIBS) to elemental imaging has attracted increasing attention. The currently established technique is a promising axis for the future development of LIBS. In LIBS-based imaging (or mapping), laser-induced plasma is generated at different positions on the sample surface with a predefined sequence covering the region of interest. Elemental maps are then obtained after extracting the line intensity associated with the elements of interest from each recorded spectrum [1, 2]. This approach has several advantages, such as multielement capability, ease of use, and ambient atmosphere operation. This LIBS approach is an all optical elemental analysis method that can reach ppm-scale sensitivity with ~10 μm (or lower) resolution [3–5]. In addition, fast acquisition speeds of up to kilohertz per pixel have already been demonstrated [6]. The high acquisition speed combined with complete compatibility with optical microscopy and ease of use provide LIBS with unique features and allow it to serve as a good complement to gold standard elemental imaging methods, such as laser ablation inductively coupled plasma mass spectrometry (LA-ICP-MS), high-resolution secondary ion mass spectrometry (SIMS), synchrotron radiation microanalysis, and electron probe microanalysis (EPMA) [7–10].

To the best of our knowledge, the first works reporting the capability of LIBS to create elemental maps were published in the 1990s [11–17]. In the years that followed, space-resolved LIBS has been further developed, and several reference works have established the basis of LIBS mapping while demonstrating its potential in terms of applications. The possibility of reaching micrometer-scale resolution was demonstrated by Menut et al. [18], a kilohertz acquisition rate was proposed by Bette et al. [6], and the possibility of scanning a large surface was proposed by the group of Javier Laserna in several publications on various application fields [17, 19–22]. In the early 2010s, the appearance of new

Laser-Induced Breakdown Spectroscopy
https://doi.org/10.1016/B978-0-12-818829-3.00014-9

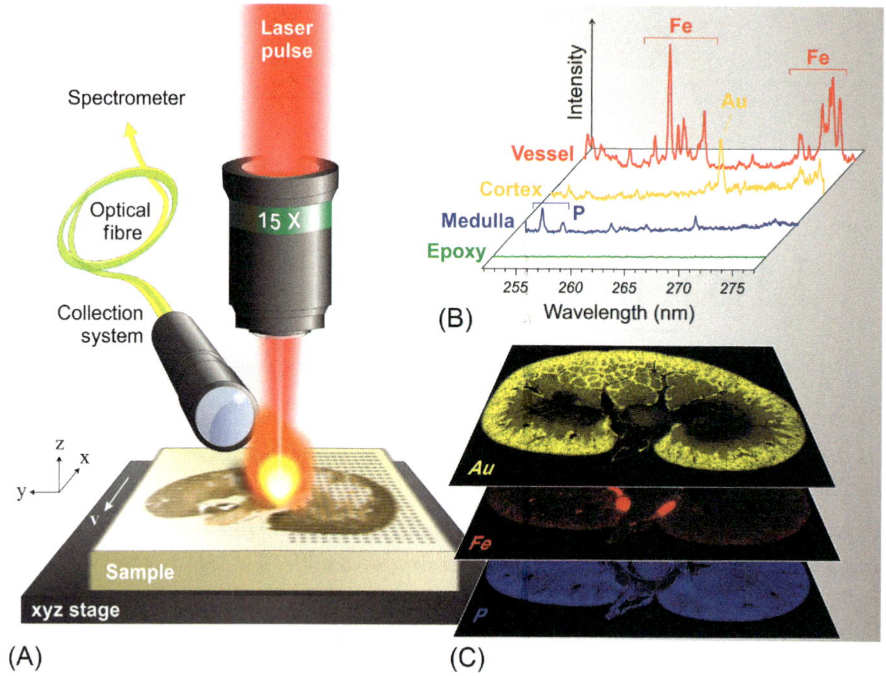

FIG. 1 General protocol for laser-induced breakdown spectroscopy (LIBS) imaging. (A) Schematic view of the LIBS imaging instrument showing the major components: the microscope objective used to focus the laser pulse, the motorized platform supporting the sample, and the detection system connected to a spectrometer via an optical fiber. (B) Examples of single-shot emission spectra in the spectral range between 250 and 280 nm. (C) Sample relative-abundance images of Au (yellow), Fe (red), and P (blue) represented using false color scales for a 2000 × 1000-pixel sequence (resolution of 10 μm).

detectors, both fast and sensitive, on the market boosted the development of LIBS imaging, allowing the emergence of new applications, such as the possibility of obtaining elemental images on biological tissues [23, 24].

Here we aim to review recent progress on LIBS-based imaging with a particular focus on applications, including the characterization of geological, industrial, or biomedical materials, among others.

2 Experimental

2.1 Preamble

The principle of any LIBS imaging instrumentation is to generate plasma at different positions on the sample of interest. Each spectrum is then processed to extract the relevant signals and build elemental maps (cf. Fig. 1). In a LIBS imaging setup, the sample is generally positioned on a motorized XY stage, allowing automated sequences to be performed. The step size (i.e., distance between consecutive laser shots) will then define the lateral resolution of the imaging measurement. In general, the accessible resolution is ultimately related to the crater size.

There are mainly two possible configurations for scanning acquisition: continuous or step by step. In the continuous scan mode, the movement of the sample is realized line by line while being synchronized with laser firing. In such cases, the lateral resolution is set by adjusting the speed of the X translation stage with regard to the laser frequency rate. At the end of each line, the sample is moved along Y by the step size, and the operation is repeated [24, 25]. In the second acquisition mode, the measurement is performed step by step (i.e., position by position) [26–29]. At the initial position, the laser performs a certain number of laser shots while the position of the sample remains fixed. The sample is then moved along X (or Y) by the step size, and the operation is repeated. Compared to the step-by-step mode, continuous scanning benefits from a higher operating speed and better accessible resolution since less damage is done to the sample by one shot than by a series of shots; however, the analysis must be performed in a single-shot configuration (no possibility of accumulating the signal at a fixed position).

2.2 Instrumental configurations

A large number of experimental configurations can be found in the literature. These configurations use various types of laser sources, spectrometers, detectors, and focusing properties, among others. These configurations have been recently described and reviewed [1], and we will mainly focus here on the main trends.

There are basically two types of focusing configuration: macroscopic and microscopic. In the macroscopic mode, focusing is generally performed by an optical lens with a focal distance of a few centimeters [30–32]. The crater size obtained with such a focal length is generally from one to several hundreds of micrometers [25, 30]. In the microscopic configuration, the microscope objectives used typically have a magnification higher than ×10 (i.e., a focal length of several millimeters) [33–35], allowing access to lateral resolutions in the range of 10 μm (or even lower) [18, 36]. However, with strong focusing, the amount of ablated mass is strongly reduced (nanograms or less per pulse), and argon gas is generally used to enhance the emission signal [28, 37–39]. In addition, in the microscopic configuration, the sample surface needs to be perfectly flat, and this usually requires a refined sample preparation.

When considering laser properties such as laser wavelength, pulse energy, and duration, various configurations have been proposed. In the microscopic mode, the laser energy generally does not exceed a few hundred microjoules [33, 40–42], while in the macroscopic mode, it is common to work with a laser energy of more than a few tens to hundreds of millijoules [31, 43, 44]. In terms of laser wavelength, the majority of cases use nanosecond Nd:YAG lasers, either in the fundamental configuration [27, 45–47], second harmonic [40, 48–50], or higher harmonics. In addition, several groups have developed femtosecond (fs) experiments using Ti:sapphire lasers operating at 800 nm [51] or the second harmonic [38], as well as frequency-tripled (343 nm) diode-pumped ytterbium laser [41, 52]. Several groups have also proposed using a double-pulse (DP) configuration with the aim of improving the signal-to-noise ratio with respect to the single-pulse approach [34, 53–56]. Finally, Abdelhamid et al. demonstrated the feasibility of imaging explosive residues in human fingerprints with an original approach called optical catapulting-LIBS, in which the first pulse causes the ejection of particles from the substrate and the catapulted particles are then analyzed by LIBS [29].

LIBS imaging performance (resolution, sensitivity, operating speed, and multielement capabilities) is also importantly related to the type of spectral detection apparatus (spectrometer and detectors). Depending on the application needs, various configurations of spectrometers have been used, such as échelle [52, 57–59] Paschen-Runge spectrometers [6, 60–62]. However, Czerny-Turner (CT) spectrometers coupled to solid-state multichannel detectors (charge-coupled device [CCD], intensified charge-coupled device [ICCD], electron multiplying charge-coupled device, etc.) are currently the most widespread spectral detection systems used for LIBS mapping experiments [25, 33, 40, 41, 46, 63–70]. Such spectrometers are bright (large entrance surface), fast (up to kHz), and versatile since they are generally equipped with several gratings. However, the drawback of CT spectrometers is undeniably their limited spectral range of detection, typically from ~20 nm with 2400 L/mm to ~80 nm with 600 L/mm grating [71, 72]. The use of compact CT spectrometers (Ocean Optics, Avantes, etc.) has also been reported in many studies [25, 58, 73–75].

2.3 Data processing

One of the challenging tasks in LIBS-based imaging is undoubtedly the data processing step. A series of LIBS spectra that constitute the mapping sequences can contain a large amount of spectral information, and several difficulties arise in extracting and representing the relevant information. In addition, unsupervised methods need to be implemented because the large number of spectra contained in the dataset does not allow conventional spectrum-by-spectrum processing. Despite the large number of papers published in this field, there is little consideration of processing such large spectral datasets. Our group has reported various methodologies to handle the processing of simple or complex matrices [76–79]. In addition, the use of chemometrics has been proposed by several groups and appears to be highly promising for the future [26, 27, 65, 80].

3 Applications

3.1 Applications for earth science

As LIBS-based imaging studies have increased in the last decade, the highly valuable potential of LIBS has been widely applied in the Earth science field. LIBS mapping provides important information about elemental distribution in geological samples through thousands to millions of LIBS spectra, permitting the characterization of heterogeneous mineral phases or the detection of major to trace elements down to the ppm level. Variations in elemental composition, pertinent ratios such as climate proxies, or the substitution of elements in specific phases (sulfur or oxide species) for economic aims have also been investigated. Numerous studies have examined the chemical heterogeneities of geological samples to assess their elemental quantification, classify minerals, and establish their origin and provenance.

3.1.1 *Paleoclimate studies*

Paleoclimate proxies are usually observed through Mg/Ca or Ca/Sr variations at the millimetric scale on carbonate phases, shells, or corals [81]. The first LIBS mapping experiment on this topic was reported by Fabre and Lathuiliere [82]. A sponge fossil was studied with a spatial resolution of 10 μm per laser shot for centimetric-scale surface analysis. However, automation of the X–Y LIBS mapping provided faster LIBS-based imaging analysis, such as that reported in Xu et al. [31] (carbonate rocks related to wet or dry climate periods), Hausmann et al. [83] (marine mollusk shells), and Sperança et al. [27] (correlation of the concentration of Ca and P on coprolite and fossils with the extinction environment). Recent studies were undertaken on large-scale speleothems and coral samples [66, 84] and megapixel elemental images were obtained with a resolution of 15 μm for up to three million spectra. The Ca, Mg, Sr, Fe, Mn, Al, and Si distributions revealed laminar structures related to vegetation or climatic variations.

3.1.2 *Mining industry*

Fast high-resolution multielemental LIBS-based mapping of geological samples has been developed in recent years. The repetition rate of up to 1 kHz with a spatial resolution from 15 to 50 μm provides elemental mapping of several square centimeters of surface for numerous elements, revealing rather complex elemental distributions. This elemental information can be related to the mineral composition of the sample, being faster than conventional and time-consuming techniques such as EPMA.

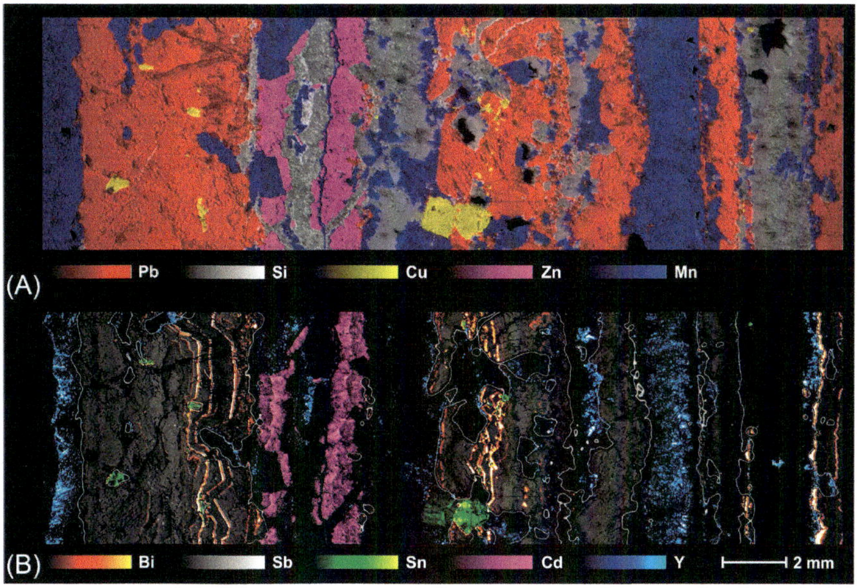

FIG. 2 Example of geological imaging of a rock section of sulfur minerals highlighting the dynamic range of measurements from major (A) to trace (B) elements. *Modified from C. Fabre et al., Elemental imaging by laser-induced breakdown spectroscopy for the geological characterization of minerals. J. Anal. Atom. Spectrom. 33 (2018) 1345–1353.*

A considerable amount of literature has also been published in the mining industry field over the last decade to characterize the spatial distribution of elements in ores, providing key information about their sources and deposition processes. These data assist in the decisions for further prospects in the field and in the evaluation of economic viability for extraction and refining. The first work was performed on granite by Novotný et al. [85]. The compositions of the drill cores were investigated by Rifai et al. [25]; their system worked at kilohertz speed and allowed real-time geochemical applications with special emphasis on gold content. Connors et al. [39] also performed on-site analyses of sulfide minerals in different gold-rich rocks. More recently, the contents of S, B, and As in a Canadian mine core composed of several mineral phases were explored by Trichard et al. [86]. In addition, the capability of LIBS-based imaging to obtain megapixel images was explored by Fabre et al. [77] when analyzing a complex multiphase hydrothermal ore sample. Five different mineral phases were identified on a sample surface of 5 cm^2. Minor and trace substituent elements were mapped with a spatial resolution of 15 μm using >2 million spectra (cf. Fig. 2).

3.1.3 Light elements and rare earth element analyses

Light elements ($Z < 6$) are hardly detectable by classical techniques, but they are of interest for economic purposes (battery or electronic devices). Sweetapple et al. [44] measured Li, Be, and B on altered spodumenes and other Li-rich minerals with an estimated limit of detection (LOD) of approximately 240 ppm for Li. LIBS-based imaging has also been used for the analysis of hydrocarbon-bearing shale samples [87]. The data collected from these experiments were used to construct 2D elemental maps, including hydrocarbon-forming elements, such as C and H. Fluorine has also been detected in rock with an LOD of 135 ppm on a millimetric map [65] and more recently in pellets [49]; the last study undertaken by Gaft et al. [88] presents

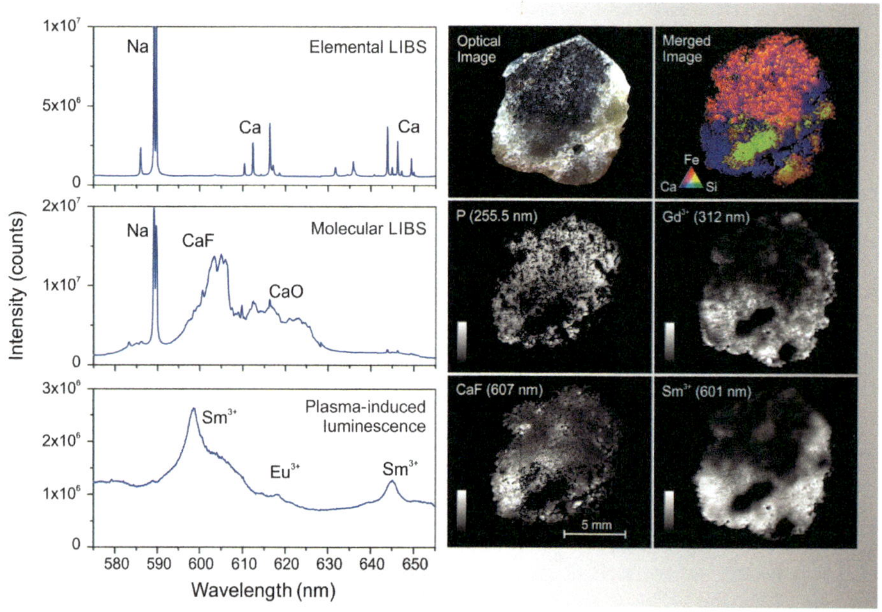

FIG. 3 Combined elemental, molecular, and plasma-induced luminescence imaging of Kovdor fluorapatite. *Modified from M. Gaft, Y. Raichlin, F. Pelascini, G. Panzer, V. Motto Ros, Imaging rare-earth elements in minerals by laser-induced plasma spectroscopy: molecular emission and plasma-induced luminescence. Spectrochim. Acta B 151 (2019) 12–19.*

imaging of rare earth elements (REEs) obtained by combining molecular emission (LaO and YO), atomic LIBS analyses, and plasma-induced luminescence using the same experimental setup and plasma source (cf. Fig. 3). The detection of REE contents in ores is a real challenge in economic studies. On millimetric surfaces, Chirinos et al. [68] used a tandem LIBS/LA-ICP-MS instrument to detect Ce, La, and Nd and their isotopes to create a 3D compositional mapping. More recently, Romppanen et al. [89] and Fabre et al. [77] investigated, on large surfaces of rock samples (larger than several square centimeters), Y and several other REE contents in various minerals at the ppm level.

3.1.4 Use of multivariate analyses

The complexity of rock structures generally imposes the use of multivariate analysis for processing the spectral dataset. One of the first geological works combining multivariate analysis and LIBS mapping was undertaken by Moncayo et al. [78]. In this work, a turquoise sample was analyzed to determine the elemental distribution of Al, Fe, P, Cu, and Si. El Haddad et al. [90] used multivariate curve resolution to image minerals on rock tiles, even in the presence of mixed mineral phases within the laser spot area. This study described a new method for mineral identification and quantification that could be scalable to automated mineralogy measurements in coarse rock streams. The analyzed mineral phases included sulfurs, molybdenite, quartz, chlorite, feldspars, fluorite, and calcite. Using a single laser shot, a root-mean-square-error below 10% was achieved for the main minerals with absolute errors below 3.5% for the major minerals and below 1% for the minor minerals. Finally, Pagnotta et al. [53] reported a fast μ-LIBS application for quantitative elemental mapping of highly inhomogeneous samples. The methodology combines the self-organizing map clustering method with calibration-free LIBS to analyze historical lime mortars. LIBS has also been used in the analysis of REE ores from the geological formation of the Norra Kärr Alkaline Complex in southern Sweden [89]. Yttrium was detected in eudialyte and catapleiite. Singular value decomposition (SVD) was used for the classification of the mineral rock samples based on their mineralogy. Based on the SVD classification, the percentage of yttrium-bearing ore minerals was calculated for the fine-grained rock samples.

3.1.5 3D geological mapping

When it is possible to combine LIBS maps obtained on rock samples, we may determine volumetric chemical variations at the microscale. The first account of 3D geological imaging was undertaken by Chirinos et al. [68] by simultaneous LIBS/LA-ICP-MS measurements. The authors presented 2D layer-by-layer mapping, 2D cross-sectional imaging, and 3D volume rendering of elements and isotopes in a Bastnäsite matrix. The approach proposed by Grassi et al. [34] and more recently by Senesi et al. [91] is based on microscale 3D compositional imaging using DP-μ-LIBS in conjunction with optical microscopy. The Fe, Si, and Na variations were analyzed to estimate the impact of environmental contamination on historical limestone rocks. In the context of planetary exploration, in Mars-simulated laboratory conditions, Lefebvre et al. [92] studied Ca/Fe, Mg/Fe, and Si/Fe elemental ratios for identifying and characterizing coatings on Martian rocks. All of these studies demonstrate the potential of LIBS for not only providing chemical profiles but also unambiguously detecting layer-to-layer elemental content variation in complex natural geological materials.

3.2 Industrial applications

This section will describe the past and current industrial applications of LIBS-based imaging. The scope includes both laboratory applications and real-world use of LIBS-based imaging in industrial settings. We have four sections according to the area of industry: (1) process control in the metallurgy industry, (2) catalyst industry, (3) nuclear field applications, and (4) miscellaneous industrial applications.

3.2.1 Process control in the metallurgy industry

The LIBS technique has become a reliable tool for metallurgical analysis because it is routinely used in various steel alloy plants. Among other state-of-the-art technologies, LIBS-based imaging has gained increasing attention for spatially resolved analysis in the iron and steel industries. In 2001, the group of R. Noll [60] published a paper giving an overview of LIBS R&D activities. This paper included a scanning microanalysis to study the homogeneity and cleanness characterization of steel. Kuss et al. [93, 94] reported a quick and powerful microanalytical methodology that provides qualitative and quantitative information on inclusions within different metals. These works describe a procedure to distinguish between the signals of the elements of interest in the inclusions and low signals from elements in the metallic matrix. Similarly, Bette et al. [95] demonstrated for the first time the detection of light elements such as C, N, O, and P in steel inclusions and compared LA-ICP-MS and LIBS using iron-base certified reference materials [96].

Another important application is the spatially resolved analysis of segregation and decarburization during the steelmaking process. The group of Boué-Bigne [61, 62] proposed a LIBS methodology to remove human subjectivity, improving reliability for the characterization of both defect types. Latkoczy et al. [97] developed a simultaneous LIBS-LA-ICP-MS setup to analyze major and trace elements in magnesium-based alloy samples. The influence of surface roughness on the LIBS signal was studied in stainless-steel samples with different surface finishes by López-Quintas et al. [57]. The same group proposed other industrial applications, such as the characterization of laser-produced clads [48] and the study of the dealloying process in brass [30]. Taparli et al. [98] performed in situ measurements on the heat-affected zone during tungsten inert gas welding processes on austenitic stainless steels. Elemental mapping and layer-by-layer contour elemental profiling (3D) of carbon in tungsten carbide was demonstrated by Lednev et al. [67] to control the coating quality of the composite material. Finally, Ahamer et al. [38] reported the spatially resolved analysis of $YBa_2Cu_3O_7$ thin films on MgO and yttria-stabilized zirconia substrates using fs-LIBS imaging.

3.2.2 *Catalyst industry*

Industry has an increasing demand for the development of new catalysts with applications in chemical, polymer, petrochemical, and automobile industries. Characterization of the local elemental composition of catalysts provides insight essential for their rational design. The group of Javier Laserna was the first to apply spatially resolved LIBS in the catalytic industry. In 1999, Lucena et al. [17] analyzed fresh gasoline catalytic converters for Pt, Pd, and Rh, proving the feasibility of LIBS for this type of elemental analysis. Several articles from the same group also demonstrated the possibility of performing 3D elemental reconstruction of these converters [20] to study the deterioration processes caused by poisoning elements [19] and the spatial distribution of catalytically active elements and deactivants [99]. More recently, Trichard et al. [79] demonstrated the capability of LIBS imaging to provide quantitative images of Pd in catalyst beads from % to the ppm scale (Fig. 4). These results were then used to estimate the catalyst crust thickness (related to the apparent activity) in egg shell-deposited catalysts [100]. Finally, the same authors analyzed the distribution of S, Ni, V, C, and Al in an alumina support with asphaltenes.

3.2.3 *Nuclear field applications*

Until recently, there has been little research on LIBS-based imaging applications in the nuclear industry. Li et al. [28] reported on the 2D distribution of Li deposition in the plasma-facing wall of a Tokamak fusion reactor (EAST). The maps contain only 78 laser spots with reduced spatial distribution (i.e., 0.4 mm); however, the measurements were performed in situ. A second study by these authors [37] on the same type of samples (plasma-facing components of EAST) showed results consistent with those from X-ray photoelectron spectroscopy analysis. Wang et al. reported μ-LIBS mapping of REEs in Mo-rich glass-ceramic. The results were validated by LA-ICP-MS, and atomic force microscopy topography was used to examine the morphology of the ablation craters [33]. LA-ICP-MS in tandem with LIBS was also investigated by Manard et al. to characterize uranium particles, showing the complementary ability of LIBS with other techniques [41]. A total of 51×503 points were scanned for different uranium particle concentrations in an Ni:Fe mixture held by double-sided tape. The combination of isotopic and elemental information is of particular interest for nuclear forensics to detect enrichment activities.

3.2.4 *Miscellaneous industrial applications*

The principal applications of LIBS-based imaging have already been described; however, other industrial sectors can benefit from elemental distribution analysis. Crystals are invaluable for a wide range of industries as they are present in our everyday electronic devices. Wiggins et al. [43] applied LIBS imaging to characterize $LiInSe_2$ crystals, which are used for thermal neutron detection. The authors generated an elemental map of Li, Se, In, Ca, Na, and other elements of interest. Yb-doped fibers have become a research hotspot in fiber lasers [40], and Li et al. proposed a LIBS-based imaging system to analyze ion

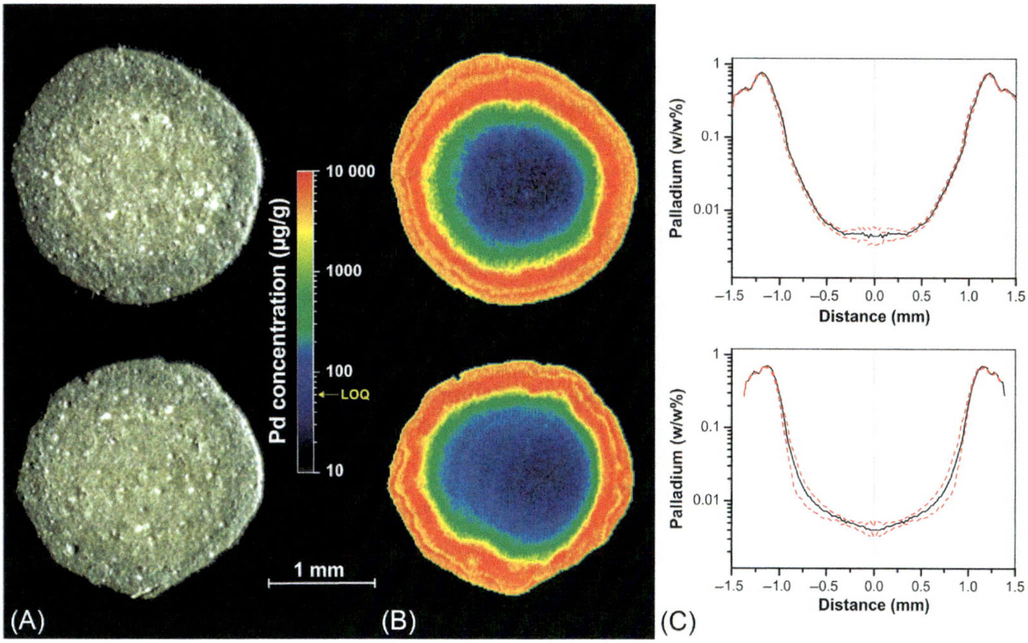

FIG. 4 Example of heterogeneous catalysis imaging. (A) Optical images, (B) quantitative elemental images, and (C) averaged profiles of palladium with their respective confidence interval (in dot red) for the two catalyst sections. *Extracted from F. Trichard et al., Quantitative elemental imaging of heterogeneous catalysts using laser-induced breakdown spectroscopy. Spectrochim. Acta B 133 (2017) 45–51.*

doping profiles in Yb-doped fiber preforms [40]; using the provided elemental information, the performance of the fiber can be predicted before its drawing. Alombert-Goget et al. [101] also applied LIBS-based imaging to crystal analysis; they investigated a Ti-doped sapphire for petawatt laser construction. LIBS and luminescence measurements were used to observe the titanium radial distribution on sapphire crystals grown by two methods, again highlighting the complementarity of LIBS with other optical techniques. The same group also analyzed the partition coefficients of Zr and Ca in a $BaTiO_3$-based piezoelectric crystal fiber grown at different velocities. The LIBS semiquantitative images were calibrated with respect to the EPMA measurements and performed on subcentimetric polycrystals with millimeter-sized grains [102].

Another application of LIBS is in the battery industry; lithium-ion battery technology is in constant development, and LIBS-based imaging can offer significant advantages for the spatially resolved analysis of battery elements. Zorba et al. [52] adopted fs-LIBS to chemically image the solid-state lithium-ion $Li_7La_3Zr_2O_{12}$ electrolyte to control the desired atomic ratio distribution between major (Li, La, Zr) and minor (Al) elements. In another work, Li distribution in an $LiCoO_2$ battery cathode was analyzed in a semiquantitative manner [103]. The authors were capable of detecting the precipitating decomposition products of the electrolyte with a methodology much simpler than conventional measurements by X-ray absorption spectroscopy. Finally, Smyrek et al. [45] explored LIBS mapping combined with depth-profiling techniques to improve lithium nickel manganese cobalt oxide cathodes by analyzing the Li-ion diffusion kinetics within the cathode.

Within the semiconductor industry, Darwiche et al. [64] analyzed elemental impurities in different solid silicon substrates to improve the production processes of photovoltaic-grade silicon. Lee et al. [104] analyzed a $Cu(In,Ga)Se_2$ thin-film solar cell using a commercial LIBS system. Finally, material sorting and recycling applications have also been developed using LIBS-based imaging (Fig. 5). Radivojevic et al. [105] illustrated the analysis of mineral bottom ash samples from a waste incineration process in the vacuum ultraviolet range, and Kim et al. [106] compared LIBS and LA-ICP-MS analyses for the determination of Zn in polycarbonate used for electrical appliances.

3.3 Applications for biology and medicine

There is major interest in studying the distribution of elements within vegetal, animal, and human tissues. Among the applications of multielemental imaging with LIBS, biomedical applications appear to be very promising. Two recent published reviews cover the increasing number of valuable research contributions in this recent expanding field of LIBS analysis [107, 108]. Biologists working on plant biology/animal pathophysiology and medical doctors attempting to understand metal-related diseases have taken advantage of the versatility of LIBS imaging to investigate the spatial distribution of chemical elements in tissues. Most of this work demonstrated how LIBS imaging contributed to uncovering some underlying mechanisms of the phenotypes they were studying. Since LIBS allows for the detection, imaging, and quantification of most of the elements in the periodic table, it appears to be an excellent candidate technology for studying both endogenous biological elements, i.e., elements naturally contained in tissues (P, Na, Mg, Ca, K, etc.), and exogenous elements, i.e., elements coming from necessary exposure to external agents (Au, Ag, Pt, Ti, and Hg). This latter category of foreign elements is very broad, and LIBS offers the ability to visualize the elemental distribution of chemical components from pesticides, dust, metal nanoparticles (NPs), or even microparticles, as well as metal ions in situ.

Among the growing number of valuable works already published in this field, we selected some of the most representative studies to describe in more detail the technical strategies and meaningful results that help illustrate the recent strong potential of LIBS imaging for vegetal and biomedical applications.

3.3.1 *Imaging vegetal tissues*

Some reviews underline the considerable importance of mapping essential or trace elements, as well as pollutants, toxic contaminants, nutrients, or organic substrates within plants [109, 110]. Accordingly, LIBS can be used to image elemental distribution in different areas of a plant, such as the roots, stem, fruit, or leaves. One of the main concerns related to human health is mapping the content of toxic metals within vegetal tissues. As an example, chronic exposure to metalloids such as As, Cd, Pb, or Hg in vegetables or cereals may induce severe adverse health effects in people cultivating or eating these plants [107, 111]. An in-field remote LIBS system was developed to directly monitor the mineral and pesticide content in maize leaves, both in 2D and 3D [111]. The spatial distribution of Ag ions in the tissues of treated sunflowers was determined with LIBS [112]. The content of toxic Pb was also imaged in the leaves of *Helianthus annuus* samples [113]. Several groups studied the absorption, transport, and exchange of elements among the different compartments and organs of plants. A low-energy LIBS system was successfully used to monitor micro- and macronutrients (Ca, Fe, Na, K) in the fresh leaves of *Citrus unshiu* and *Rhododendron obtusum* [114].

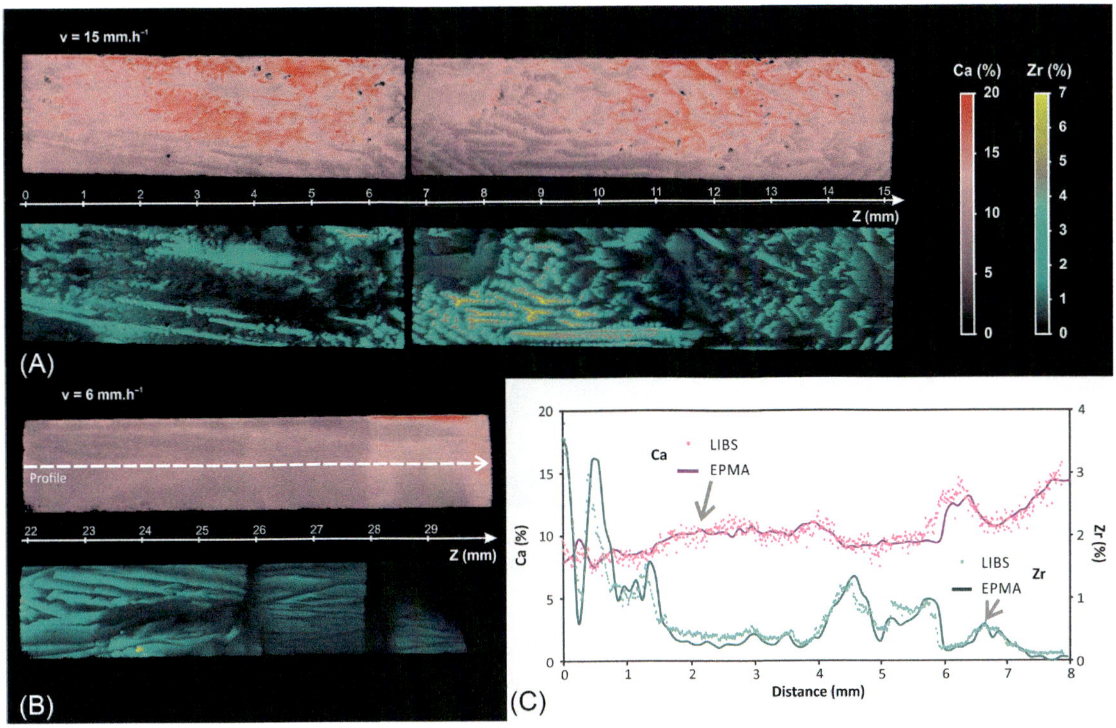

FIG. 5 Laser-induced breakdown spectroscopy longitudinal chemical mapping for Ca and Zr on $BaTiO_3$-based piezoelectric crystal fiber samples at different crystal growth speeds: (A) 15 and (B) 6 mm h^{-1}. (C) Molar contents normalized with respect to the electron probe microanalysis profile. *Extracted from P. Veber et al., Lead-free piezoelectric crystals grown by the micro-pulling down technique in the BaTiO3–CaTiO3–BaZrO3 system. Cryst. Eng. Comm. 21 (2019) 3844–3853.*

3.3.2 *Imaging animal tissues*

During the last 5 years, our group has been quite active in imaging metal-based NPs in the organs of laboratory animals. These preclinical studies allow visualization of the distribution of metal NPs inside living organisms. There is currently a broad interest in the design, synthesis, and functionalization of different metal NPs. When these NPs are used as contrast agents, a therapeutic drug, or carrier, preclinical studies are important to assess the kinetic parameters of these molecules. In 2012, we performed pioneering work in this field, and used LIBS imaging to map the distribution of Gd-NPs in fresh frozen murine kidneys with an achieved lateral resolution of 100 μm [23, 115]. We further refined the technical setup and, importantly, most of the preanalytical and experimental parameters. In 2014, we described how LIBS multielemental imaging was suitable for monitoring the kinetics of injected Gd-NPs [24, 116]. Kidneys containing Gd-NPs were fixed and further embedded in epoxy resin. This hardening step both helped preserve the samples and allowed for better laser shot-to-shot reproducibility. We managed to obtain a ppm-level detection limit and a typical resolution of ~30 μm. Higher resolutions in the range of 10 μm could be achieved but were associated with an altered sensitivity. We obtained elemental images of several elements (Na, P, Ca, Mg, Gd, and Si), and described the progressive accumulation of Gd in the cortex (external area) of the kidney, followed by progressive diffusion to the

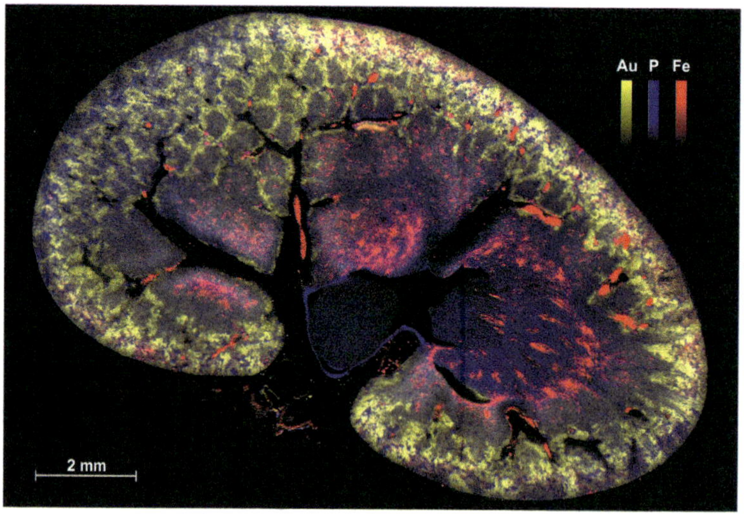

FIG. 6 Multielemental image of a murine kidney after intravenous injection of Au nanoparticles. Merged image composed of the combination of three false color elemental images for Au (yellow), P (blue) and Fe (red).

medulla (central area) of the kidney. One week after the injection, almost no remaining Gd signal was recorded, suggesting that the NPs were efficiently eliminated from the blood circulation of the animals. These studies were performed in 2D at the entire organ scale. The results highlight the versatility of LIBS imaging for collecting information about general localization, accumulation, and residence time of NPs within elimination organs, namely, the liver and kidneys. Importantly, we believe that elemental images obtained with an all-optical LIBS instrument are complementary to conventional tissue-imaging techniques such as electron microscopy or fluorescence microscopy [36]. In preclinical studies, most of the metal NPs are made with high Z elements, such as Gd [23, 117], Au (cf. Fig. 6), Pt, Ag, etc. LIBS can image the presence of these metals without requiring any further complementary labeling. Therefore LIBS technology appears to be a promising strategy for label-free imaging of metal NPs in tumors [118–120] and various organs [36, 116, 117]. A proof-of-concept study for 3D elemental imaging of Gd-NPs in the kidneys of animals was also performed at the entire organ level after a volume reconstruction was calculated from adjacent sections that were mapped with LIBS [121]. A second and more resolved strategy was based on continuous in-depth analysis. This real 3D elemental imaging did not require any interpolation; however, in this case the depth of the organ analyzed was technically limited to ~500 μm [121].

3.3.3 *Imaging human tissues*

Several pathophysiological conditions may result from the balance between the normal and abnormal distribution of chemical elements within tissues. Moncayo et al. demonstrated the feasibility of imaging bulk biological elements in different paraffin skin biopsies containing either normal skin or skin cancer tissues [71]. A LIBS multielemental analysis study of a melanoma, the deadliest form of skin cancer, and a squamous cell carcinoma or a Merkel-cell carcinoma helped the authors validate LIBS imaging as a relevant analytical tool, which may be helpful for pathologists and clinicians. The vast majority of human biopsies are generally fixed with buffered formaldehyde and subsequently embedded in paraffin. The elemental images recorded from skin tissues enabled the identification and visualization of the different physiological layers of the skin, perfectly matching the expected tissue architecture. The ability of combining classical histological microscopy images with the elemental images obtained by LIBS opens avenues for assisting pathologists in their quest for a better understanding of the heterogeneous distribution of elements in areas histologically homogeneous. For instance, areas with important levels of a given element (hotspots) or even gradients of concentration may help to identify actively proliferating tissues or necrotic areas. The other main application of LIBS imaging as a valuable tool for assisting pathologists is for investigating the origin or effects of disparate exogenous materials in human specimens. The absence of convenient or available techniques make this mission almost impossible as a routine analysis in pathology laboratories. Busser et al. used LIBS imaging to visualize exogenous chemical elements such as Al, Cu, Ti, and W within paraffin-embedded specimens [122]. The elemental images were stackable with histology, which helped to assess immune reactions to Al salts (cf. Fig. 7), but Ti particles were also stuck in a pigmented lymph node. The authors also showed the diversity of the chemical nature of metal particulates left behind tattoos, implants, or surgical procedures.

Importantly, LIBS imaging may be complementary to other elemental imaging technologies. As an example, fresh frozen tumor tissues can be scanned by a tandem LIBS/LA-ICP-MS instrument. This dual-elemental analysis allows imaging of more chemical elements than a single stand-alone modality [123].

The unique capabilities of LIBS imaging enable the analysis of major/minor/trace metal elements in biological specimens. Therefore we believe that LIBS is a novel technique for the

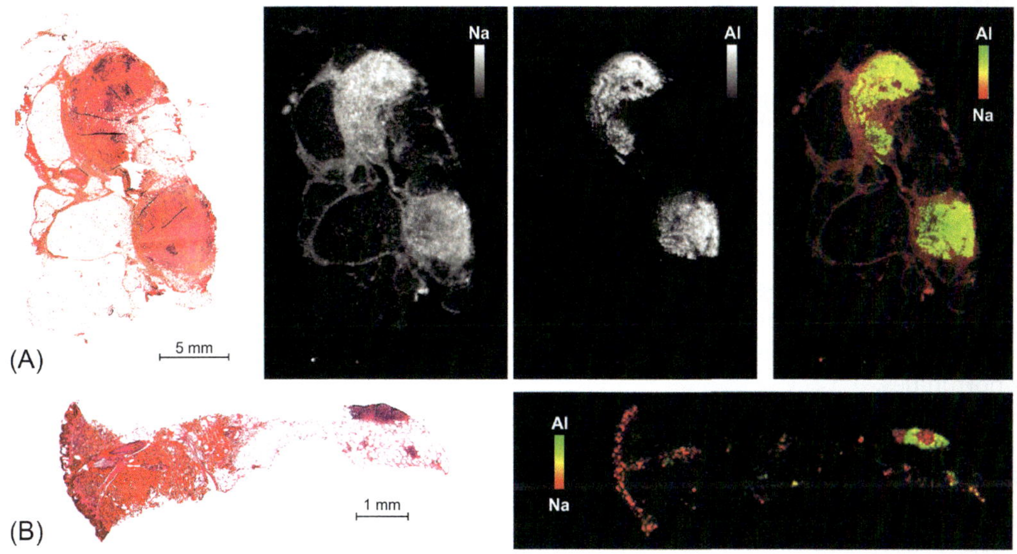

FIG. 7 Histological images of immune responses induced by Al salts contained in immunotherapies such as vaccines. Al salts are frequently used as adjuvants in these formulations, and some patients may develop sensitivity against them. The false color elemental images of Al in a granuloma (A) or a pseudolymphoma (B) indicate that Al was colocalized with the inflammatory and immune reactions in these skin biopsies. *Extracted from B. Busser et al., Characterization of foreign materials in paraffin-embedded pathological specimens using in situ multi-elemental imaging with laser spectroscopy. Mod. Pathol. 31 (2018) 378–384.*

observation of enriched or depleted metal distributions according to the location in an organism, or its exposure.

4 Summary and conclusion

We have reviewed the various instrumental configurations and the main applications of LIBS-based imaging. There is a large number of reports expanding through numerous and diverse applicative fields. Currently, LIBS imaging is experiencing important development with a strong increase in the number of published papers in recent years, and this approach is becoming increasingly more mature relative to the feasibility studies conducted in laboratories. This approach has the potential to become, in the near future, a reference approach for space-resolved elemental characterization. It has unique specificities, such as the possibility of attaining micrometric-scale resolution with limits of detection up to ppm for most elements (even for $Z < 6$), while working at fast acquisition rates in ambient conditions with table-top, all-optical instrumentation.

Despite the numerous advantages of LIBS-based imaging, there are still some technological barriers to overcome. In particular, the analysis of complex materials (i.e., composed of several matrices) still appears to be challenging both for quantification and data processing. Further studies dedicated to these issues need to be conducted, and the use of calibration-free LIBS and advanced chemometric tools represent interesting opportunities.

Acknowledgments

This work was partially supported by Pulsalys (#L0978-L1294), the French region Rhônes Alpes Auvergne (Optolyse, CPER2016), and the French ANR (projects Medi-LIBS and Imazinc). SM is the recipient of a Sara Borrell grant (CD17/00203) from Instituto de Salud Carlos III. In addition, we gratefully acknowledge Frédéric Pelascini from the Cetim Grand Est and Florian Trichard from Ablatom SAS for fruitful discussions.

References

[1] L. Jolivet, et al., Review of the recent advances and applications of LIBS-based imaging, Spectrochim. Acta B 151 (2019) 41–53.

[2] M.P. Mateo, G. Nicolas, V. Piñon, Laser-induced breakdown spectroscopy for chemical mapping of materials, Appl. Spect. Rev. 48 (2013) 357–383.

[3] M.T. Taschuk, I.V. Cravetchi, Y.Y. Tsui, R. Fedosejevs, Laser-Induced Breakdown Spectroscopy, Elsevier, Amsterdam, 2007 (Chapter 9).

[4] R. Noll, Laser-Induced Breakdown Spectroscopy, Springer-Verlag, Berlin, 2012.

[5] A. Miziolek, V. Palleschi, I. Schechter, Laser Induced Breakdown Spectroscopy, Cambridge University Press, Cambridge, 2006.

[6] H. Bette, R. Noll, High speed laser-induced breakdown spectrometry for scanning microanalysis, J. Phys. D Appl. Phys. 37 (2004) 1281.

[7] M. Reich, R. Large, A.P. Deditius, New advances in trace element geochemistry of ore minerals and accessory phases, Ore Geol. Rev. 81 (2017) 1215–1217.

[8] D. Pozebon, G.L. Scheffler, V.L. Dressler, M.A. G. Nunes, Review of the applications of laser ablation inductively coupled plasma mass spectrometry (LA-ICP-MS) to the analysis of biological samples, J. Anal. Atom. Spectrom. 29 (2014) 2204–2228.

[9] M.J. Pushie, I.J. Pickering, M. Korbas, M.J. Hackett, G.N. George, Elemental and chemically specific X-ray fluorescence imaging of biological systems, Chem. Rev. 114 (2014) 8499–8541.

[10] K.L. Moore, E. Lombi, F.J. Zhao, C.R. Grovenor, Elemental imaging at the nanoscale: NanoSIMS and complementary techniques for element localisation in plants, Anal. Bioanal. Chem. 402 (2012) 3263–3273.

[11] L. Moenke-Blankenburg, Laser Micro Analysis, John Wiley & Sons, New York, 1989.

[12] H.J. Häkkänen, J.E.I. Korppi-Tommola, UV-laser plasma study of elemental distributions of paper coatings, Appl. Spectrosc. 49 (1995) 1721–1728.

[13] C. Geertsen, J.L. Lacour, P. Mauchien, L. Pierrard, Evaluation of laser ablation optical emission spectrometry for microanalysis in aluminium samples, Spectrochim. Acta B 51 (1996) 1403–1416.

[14] J.M. Vadillo, S. Palanco, M.D. Romero, J.J. Laserna, Applications of laser-induced breakdown spectrometry (LIBS) in surface analysis, Anal. Bioanal. Chem. 355 (1996) 909–912.

[15] T. Kim, C.T. Lin, Y. Yoon, Compositional mapping by laser-induced breakdown spectroscopy, J. Phys. Chem. B 102 (1998) 4284–4287.

[16] D. Romero, J.J. Laserna, Surface and tomographic distribution of carbon impurities in photonic-grade silicon using laser-induced breakdown spectrometry, J. Anal. Atom. Spectrom. 13 (1998) 557–560.

[17] P. Lucena, J.M. Vadillo, J.J. Laserna, Mapping of platinum group metals in automotive exhaust three-way catalysts using laser-induced breakdown spectrometry, Anal. Chem. 71 (1999) 4385–4391.

[18] D. Menut, P. Fichet, J.L. Lacour, A. Rivoallan, P. Mauchien, Micro-laser-induced breakdown spectroscopy technique: a powerful method for performing quantitative surface mapping on conductive and nonconductive samples, Appl. Optics 42 (2003) 6063–6071.

[19] P. Lucena, J.M. Vadillo, J.J. Laserna, Compositional mapping of poisoning elements in automobile three-way catalytic converters by using laser-induced breakdown spectrometry, Appl. Spectrosc. 55 (2001) 267–272.

[20] P. Lucena, J.J. Laserna, Three-dimensional distribution analysis of platinum, palladium and rhodium in auto catalytic converters using imaging-mode laser-induced breakdown spectrometry, Spectrochim. Acta B 56 (2001) 177–185.

[21] D. Romero, J.J. Laserna, Multielemental chemical imaging using laser-induced breakdown spectrometry, Anal. Chem. 69 (1997) 2871–2876.

[22] D. Romero, J.J. Laserna, A microanalytical study of aluminium diffusion in photovoltaic cells using imaging-mode laser-induced breakdown spectrometry, Spectrochim. Acta B 55 (2000) 1241–1248.

[23] V. Motto-Ros, et al., Mapping nanoparticles injected into a biological tissue using laser-induced breakdown spectroscopy, Spectrochim. Acta B 87 (2013) 168–174.

[24] L. Sancey, et al., Laser spectrometry for multi-elemental imaging of biological tissues, Sci. Rep. 4 (2014) 6065.

[25] K. Rifai, F. Doucet, L. Özcan, F. Vidal, LIBS core imaging at kHz speed: paving the way for real-time geochemical applications, Spectrochim. Acta B 150 (2018) 43–48.

[26] R.R.V. Carvalho, et al., Laser-induced breakdown spectroscopy (LIBS) combined with hyperspectral imaging for the evaluation of printed circuit board composition, Talanta 134 (2015) 278–283.

[27] M.A. Sperança, et al., Application of laser-induced breakdown spectroscopy and hyperspectral images for direct evaluation of chemical elemental profiles of coprolites, Geostand. Geoanal. Res. 41 (2017) 273–282.

[28] C. Li, et al., In situ chemical imaging of lithiated tungsten using laser-induced breakdown spectroscopy, J. Nucl. Mater. 452 (2014) 10–15.

[29] M. Abdelhamid, F.J. Fortes, M.A. Harith, J.J. Laserna, Analysis of explosive residues in human fingerprints using optical catapulting–laser-induced breakdown spectroscopy, J. Anal. Atom. Spectrom. 26 (2011) 1445–1450.

[30] R. Cerrato, A. Casal, M.P. Mateo, G. Nicolas, Dealloying evidence on corroded brass by laser-induced breakdown spectroscopy mapping and depth profiling measurements, Spectrochim. Acta B 130 (2017) 1–6.

[31] T. Xu, et al., Multi-elemental surface mapping and analysis of carbonaceous shale by laser-induced breakdown spectroscopy, Spectrochim. Acta B 115 (2016) 31–39.

[32] M.P. Casado-Gavalda, et al., Quantification of copper content with laser induced breakdown spectroscopy as a potential indicator of offal adulteration in beef, Talanta 169 (2017) 123–129.

[33] X. Wang, et al., Mapping of rare earth elements in nuclear waste glass-ceramic using micro laser-induced breakdown spectroscopy, Spectrochim. Acta B 87 (2013) 139–146.

[34] R. Grassi, et al., Three-dimensional compositional mapping using double-pulse micro-laser-induced breakdown spectroscopy technique, Spectrochim. Acta B 127 (2017) 1–6.

[35] C. Schiavo, et al., High-resolution three-dimensional compositional imaging by double-pulse laser-induced breakdown spectroscopy, J. Instrum. 11 (2016) C08002.

[36] L. Sancey, et al., Long-term in vivo clearance of gadolinium-based AGuIX nanoparticles and their biocompatibility after systemic injection, ACS Nano 9 (2015) 2477–2488.

[37] R. Hai, et al., Characterization of Li deposition on the first wall of EAST using laser-induced breakdown spectroscopy, J. Nucl. Mater. 438 (2013) S1168–S1171.

[38] C.M. Ahamer, K.M. Riepl, N. Huber, J.D. Pedarnig, Femtosecond laser-induced breakdown spectroscopy: elemental imaging of thin films with high spatial resolution, Spectrochim. Acta B 136 (2017) 56–65.

[39] B. Connors, A. Somers, D. Day, Application of handheld laser-induced breakdown spectroscopy (LIBS) to geochemical analysis, Appl. Spectrosc. 70 (2016) 810–815.

[40] J. Li, et al., Analysis of ion doping profiles in Yb-doped fiber preforms using laser-induced breakdown spectroscopy, J. Anal. Atom. Spectrom. 31 (2016) 492–496.

[41] B.T. Manard, C. Derrick Quarles, E.M. Wylie, N. Xu, Laser ablation—inductively couple plasma—mass spectrometry/laser induced break down spectroscopy: a tandem technique for uranium particle characterization, J. Anal. Atom. Spectrom. 32 (2017) 1680–1687.

[42] V.N. Lednev, et al., Laser-induced breakdown spectroscopy for three-dimensional elemental mapping of composite materials synthesized by additive technologies, Appl. Optics 56 (2017) 9698–9705.

[43] B. Wiggins, et al., Investigation of non-uniformity and inclusions in 6LiInSe2 utilizing laser induced breakdown spectroscopy (LIBS), in: SPIE Optical Engineering + Applications (Ed.), Proceeding of Hard X-Ray, Gamma-Ray, and Neutron Detector Physics XV, San Diego, USA, 25–29 August (2013), 2013.

[44] M.T. Sweetapple, S. Tassios, Laser-induced breakdown spectroscopy (LIBS) as a tool for in situ mapping and textural interpretation of lithium in pegmatite minerals, Am. Mineral. 100 (2015) 2141–2151.

[45] P. Smyrek, Y. Zheng, H.J. Seifert, W. Pfleging, Post-mortem characterization of fs laser-generated micro-pillars in Li(Ni1/3Mn1/3Co1/3)O2 electrodes by laser-induced breakdown spectroscopy, in: SPIE Optical Engineering + Applications (Ed.), Proceedings of Laser-based Micro- and Nanoprocessing X, San Francisco, USA, 16–18 February (2016), 2016.

[46] P. Lucena, I. Gaona, J. Moros, J.J. Laserna, Location and detection of explosive-contaminated human fingerprints on distant targets using standoff laser-induced breakdown spectroscopy, Spectrochim. Acta B 85 (2013) 71–77.

[47] J.-H. Yang, S.-J. Choi, J.J. Yoh, Towards reconstruction of overlapping fingerprints using plasma spectroscopy, Spectrochim. Acta B 134 (2017) 25–32.

[48] J.A. Varela, et al., Characterization of hard coatings produced by laser cladding using laser-induced breakdown spectroscopy technique, Appl. Surf. Sci. 336 (2015) 396–400.

[49] P. Pořízka, et al., Detection of fluorine using laser-induced breakdown spectroscopy and Raman spectroscopy, J. Anal. Atom. Spectrom. 32 (2017) 1966–1974.

[50] A. Iqbal, Z. Sun, M. Wall, Z.T. Alwahabi, Sensitive elemental detection using microwave-assisted laser-induced breakdown imaging, Spectrochim. Acta B 136 (2017) 16–22.

[51] B.D. Strycker, K. Wang, M. Springer, A.V. Sokolov, Chemical-specific imaging of shallowly buried objects using femtosecond laser pulses, Appl. Optics 52 (2013) 4792–4796.

[52] H. Hou, et al., Three-dimensional elemental imaging of Li-ion solid-state electrolytes using fs-laser induced breakdown spectroscopy (LIBS), J. Anal. Atom. Spectrom. 30 (2015) 2295–2302.

[53] S. Pagnotta, et al., Micro-laser-induced breakdown spectroscopy (Micro-LIBS) study on ancient Roman mortars, Appl. Spectrosc. 71 (2017) 721–727.

[54] L. Krajcarová, et al., Mapping of the spatial distribution of silver nanoparticles in root tissues of Vicia faba by laser-induced breakdown spectroscopy (LIBS), Talanta 173 (2017) 28–35.

[55] P. Modlitbova, et al., Comparative investigation of toxicity and bioaccumulation of Cd-based quantum dots and Cd salt in freshwater plant Lemna minor L, Ecotoxicol. Environ. Saf. 147 (2018) 334–341.

[56] J. Novotný, et al., A versatile interaction chamber for laser-based spectroscopic applications, with the emphasis on laser-induced breakdown spectroscopy, Spectrochim. Acta B 101 (2014) 149–154.

[57] I. Lopez-Quintas, V. Piñon, M.P. Mateo, G. Nicolas, Effect of surface topography in the generation of chemical maps by laser-induced plasma spectroscopy, Appl. Surf. Sci. 258 (2012) 9432–9436.

[58] C. Beresko, P. Kohns, G. Ankerhold, Surface element-mapping of three dimensional structures by laser-induced breakdown spectroscopy, Spectrochim. Acta B 99 (2014) 20–27.

[59] M.B. Bueno Guerra, et al., Direct analysis of plant leaves by EDXRF and LIBS: microsampling strategies and cross-validation, J. Anal. Atom. Spectrom. 30 (2015) 1646–1654.

[60] R. Noll, et al., Laser-induced breakdown spectrometry—applications for production control and quality assurance in the steel industry, Spectrochim. Acta B 56 (2011) 637–649.

[61] Boué-Bigne, Laser-induced breakdown spectroscopy applications in the steel industry: rapid analysis of segregation and decarburization, Spectrochim. Acta B 63 (2008) 1122–1129.

[62] F. Boué-Bigne, Simultaneous characterization of elemental segregation and cementite networks in high carbon steel products by spatially-resolved laser-induced breakdown spectroscopy, Spectrochim. Acta B 96 (2014) 21–32.

[63] M. Hoehse, A. Paul, I. Gornushkin, U. Panne, Multivariate classification of pigments and inks using combined Raman spectroscopy and LIBS, Anal. Bioanal. Chem. 402 (2012) 1443–1450.

[64] S. Darwiche, M. Benmansour, N. Eliezer, D. Morvan, Laser-induced breakdown spectroscopy for photovoltaic silicon wafer analysis, Prog. Photovolt. 20 (2012) 463–471.

[65] C.D. Quarles, et al., Fluorine analysis using laser induced breakdown spectroscopy (LIBS), J. Anal. Atom. Spectrom. 29 (2014) 1238–1242.

[66] L. Bassel, et al., Laser-induced breakdown spectroscopy for elemental characterization of calcitic alterations on cave walls, Environ. Sci. Pollut. R. 24 (2017) 2197–2204.

[67] V.N. Lednev, et al., Combining Raman and laser induced breakdown spectroscopy by double pulse lasing, Anal. Bioanal. Chem. 410 (2018) 277–286.

[68] J.R. Chirinos, et al., Simultaneous 3-dimensional elemental imaging with LIBS and LA-ICP-MS, J. Anal. Atom. Spectrom. 29 (2014) 1292–1298.

[69] H. Wang, et al., Applying stand-off LIBS to paleoclimatic research: a case study on geochemical content of carbonate rocks, in: IEEE (Ed.), Proceedings of the Optoelectronics Global Conference Shenzhen, China, August 29–31 (2015), 2015, , pp. 1–3.

[70] J. Li, et al., Spatially selective excitation in laser-induced breakdown spectroscopy combined with laser-induced fluorescence, Opt. Express 25 (2017) 4945–4951.

[71] S. Moncayo, et al., Multi-elemental imaging of paraffin-embedded human samples by laser-induced breakdown spectroscopy, Spectrochim. Acta B 133 (2017) 40–44.

[72] J. Kaiser, et al., Utilization of selected laser ablation-based diagnostic methods for study of elemental distribution in various solid samples, in: SPIE Optical Engineering + Applications (Ed.), Proceeding of the

17th Slovak-Czech-Polish Optical Conference on Wave and Quantum Aspects of Contemporary Optics, December 16th (2010), Liptovsky Jan, Slovakia, 7746 1–9, 2010.

[73] C. Gottlieb, S. Millar, S. Grothe, G. Wilsch, 2D evaluation of spectral LIBS data derived from heterogeneous materials using cluster algorithm, Spectrochim. Acta B 134 (2017) 58–68.

[74] C. Zhao, D. Dong, X. Du, W. Zheng, In-field, in situ, and in vivo 3-dimensional elemental mapping for plant tissue and soil analysis using laser-induced breakdown spectroscopy, Sensors 16 (2016) 1764.

[75] J. Rakovský, et al., Testing a portable laser-induced breakdown spectroscopy system on geological samples, Spectrochim. Acta B 74-75 (2012) 57–65.

[76] V. Motto-Ros, S. Moncayo, F. Trichard, F. Pelascini, Investigation of signal extraction in the frame of laser induced breakdown spectroscopy imaging, Spectrochim. Acta B 155 (2019) 127–133.

[77] C. Fabre, et al., Elemental imaging by laser-induced breakdown spectroscopy for the geological characterization of minerals, J. Anal. Atom. Spectrom. 33 (2018) 1345–1353.

[78] S. Moncayo, et al., Exploration of megapixel hyperspectral LIBS images using principal component analysis, J. Anal. Atom. Spectrom. 33 (2018) 210–220.

[79] F. Trichard, et al., Quantitative elemental imaging of heterogeneous catalysts using laser-induced breakdown spectroscopy, Spectrochim. Acta B 133 (2017) 45–51.

[80] J. Klus, et al., Multivariate approach to the chemical mapping of uranium in sandstone-hosted uranium ores analyzed using double pulse laser-induced breakdown spectroscopy, Spectrochim. Acta B 123 (2016) 143–149.

[81] T.P. Guilderson, R.G. Fairbanks, J.L. Rubenstone, Tropical temperature variations since 20,000 years ago: modulating interhemispheric climate change, Science 263 (1994) 663.

[82] C. Fabre, B. Lathuiliere, Relationships between growth-bands and paleoenvironmental proxies Sr/Ca and Mg/Ca in hypercalcified sponge: a micro-laser induced breakdown spectroscopy approach, Spectrochim. Acta B 62 (2007) 1537–1545.

[83] N. Hausmann, et al., Elemental mapping of Mg/Ca intensity ratios in marine mollusc shells using laser-induced breakdown spectroscopy, J. Anal. Atom. Spectrom. 32 (2017) 1467–1472.

[84] J.O. Caceres, et al., Megapixel multi-elemental imaging by laser-induced breakdown spectroscopy, a technology with considerable potential for paleoclimate studies, Sci. Rep. 7 (2017) 5080.

[85] K. Novotný, et al., Mapping of different structures on large area of granite sample using laser-ablation based analytical techniques, an exploratory study, Spectrochim. Acta B 63 (2008) 1139–1144.

[86] F. Trichard, et al., Evaluation of a compact VUV spectrometer for elemental imaging by laser-induced breakdown spectroscopy: application to mine core characterization, J. Anal. Atom. Spectrom. 32 (2017) 1527–1534.

[87] J. Jain, et al., Elemental mapping and geochemical characterization of gas producing shales by laser induced breakdown spectroscopy, Spectrochim. Acta B 150 (2018) 1–8.

[88] M. Gaft, Y. Raichlin, F. Pelascini, G. Panzer, V. Motto Ros, Imaging rare-earth elements in minerals by laser-induced plasma spectroscopy: molecular emission and plasma-induced luminescence, Spectrochim. Acta B 151 (2019) 12–19.

[89] S. Romppanen, H. Häkkänen, S. Kaski, Singular value decomposition approach to the yttrium occurrence in mineral maps of rare earth element ores using laser-induced breakdown spectroscopy, Spectrochim. Acta B 134 (2017) 69–74.

[90] J. El Haddad, et al., Multiphase mineral identification and quantification by laser-induced breakdown spectroscopy, Miner. Eng. 134 (2019) 281–290.

[91] G.S. Senesi, et al., Elemental and mineralogical imaging of a weathered limestone rock by double-pulse micro-laser-induced breakdown spectroscopy, Spectrochim. Acta B 143 (2018) 91–97.

[92] C. Lefebvre, A. Catalá-Espí, P. Sobron, A. Koujelev, R. Léveillé, Depth-resolved chemical mapping of rock coatings using laser-induced breakdown spectroscopy: implications for geochemical investigations on Mars, Planet. Space Sci. 126 (2016) 24–33.

[93] H.-M. Kuss, H. Mittelstädt, G. Müller, C. Nazikkol, Fast scanning laser-OES. I. Characterization of non-metallic inclusions in steel, Anal. Lett. 36 (2003) 659–665.

[94] H.-M. Kuss, H. Mittelstaedt, G. Mueller, Inclusion mapping and estimation of inclusion contents in ferrous materials by fast scanning laser-induced optical emission spectrometry, J. Anal. Atom. Spectrom. 20 (2005) 730–735.

[95] H. Bette, et al., High-speed scanning laser-induced breakdown spectroscopy at 1000 Hz with single pulse evaluation for the detection of inclusions in steel, J. Laser Appl. 17 (2005) 183–190.

[96] G. Müller, F. Stahnke, D. Bleiner, Fast steel-cleanness characterization by means of laser-assisted plasma spectrometric methods, Talanta 70 (2006) 991–995.

[97] C. Latkoczy, T. Ghislain, Simultaneous LIBS and LA-ICP-MS analysis of industrial samples, J. Anal. Atom. Spectrom. 21 (2006) 1152–1160.

[98] U.A. Taparli, et al., In situ laser-induced breakdown spectroscopy measurements of chemical compositions in stainless steels during tungsten inert gas welding, Spectrochim. Acta B 139 (2018) 50–56.

[99] P. Lucena, J.M. Vadillo, J.J. Laserna, Spatial distribution of catalytically active elements and deactivants in diesel-engine automobile converters by laser-induced plasma spectrometry, J. Anal. Atom. Spectrom. 17 (2002) 548–551.

[100] L. Sorbier, F. Trichard, S. Moncayo, C.P. Lienemann, V. Motto-Ros, Calculation of catalyst crust thickness from full elemental laser-induced breakdown spectroscopy images, IOP Conf. Ser.: Mater. Sci. Eng. 304 (2018) 012016.

[101] G. Alombert-Goget, et al., Titanium distribution profiles obtained by luminescence and LIBS measurements on Ti: Al2O3 grown by Czochralski and Kyropoulos techniques, Opt. Mater. 65 (2017) 28–32.

[102] P. Veber, et al., Lead-free piezoelectric crystals grown by the micro-pulling down technique in the BaTiO3–CaTiO3–BaZrO3 system, Cryst. Eng. Comm. 21 (2019) 3844–3853.

[103] S. Imashuku, et al., Quantitative lithium mapping of lithium-ion battery cathode using laser-induced breakdown spectroscopy, J. Power Sources 399 (2018) 186–191.

[104] S.-H. Lee, J.-H. Choi, J.-H. In, S. Jeong, Fast compositional mapping of solar cell by laser spectroscopy technique for process monitoring, Int. J. Pr. Eng. Man-GT 6 (2019) 189–196.

[105] I. Radivojevic, et al., Microanalysis by laser-induced plasma spectroscopy in the vacuum ultraviolet, Anal. Chem. 76 (2004) 1648–1656.

[106] E. Kim, H.B. Lim, Quantification of Zinc in thermally conducting polycarbonate containing Fiber-type SiO2 fillers, Bull. Kor. Chem. Soc. 39 (2018) 341–346.

[107] B. Busser, S. Moncayo, J.-L. Coll, L. Sancey, V. Motto-Ros, Elemental imaging using laser-induced breakdown spectroscopy: a new and promising approach for biological and medical applications, Coord. Chem. Rev. 358 (2018) 70–79.

[108] R. Gaudiuso, et al., Laser-induced breakdown spectroscopy for human and animal health: a review, Spectrochim. Acta B 152 (2019) 123–148.

[109] D. Santos, et al., Laser-induced breakdown spectroscopy for analysis of plant materials: a review, Spectrochim. Acta B 71-72 (2012) 3–13.

[110] J. Kaiser, et al., Trace elemental analysis by laser-induced breakdown spectroscopy—biological applications, Surf. Sci. Rep. 67 (2012) 233–243.

[111] J. Peng, et al., Challenging applications for multi-element analysis by laser-induced breakdown spectroscopy in agriculture: a review, Trends Anal. Chem. 85 (2016) 260–272.

[112] V. Juvé, R. Portelli, M. Boueri, M. Baudelet, J. Yu, Space-resolved analysis of trace elements in fresh vegetables using ultraviolet nanosecond laser-induced breakdown spectroscopy, Spectrochim. Acta B 63 (2008) 1047–1053.

[113] J. Kaiser, et al., Mapping of lead, magnesium and copper accumulation in plant tissues by laser-induced breakdown spectroscopy and laser-ablation inductively coupled plasma mass spectrometry, Spectrochim. Acta B 64 (2009) 67–73.

[114] T. Ohta, M. Ito, T. Kotani, T. Hattori, Emission enhancement of laser-induced breakdown spectroscopy by localized surface plasmon resonance for analyzing plant nutrients, Appl. Spectrosc. 63 (2009) 555–558.

[115] V. Motto-Ros, et al., Mapping of native inorganic elements and injected nanoparticles in a biological organ with laser-induced plasma, Appl. Phys. Lett. 101 (2012) 223702.

[116] L. Sancey, et al., Laser-induced breakdown spectroscopy: a new approach for nanoparticle's mapping and quantification in organ tissue, J. Vis. Exp. 88 (2014).

[117] A. Moussaron, et al., Ultrasmall nanoplatforms as calcium-responsive contrast agents for magnetic resonance imaging, Small 11 (2015) 4900–4909.

[118] X. Le Guével, et al., Elemental and optical imaging evaluation of zwitterionic gold nanoclusters in glioblastoma mouse models, Nanoscale 10 (2018) 18657–18664.

[119] S. Kunjachan, et al., Nanoparticle mediated tumor vascular disruption: a novel strategy in radiation therapy, Nano Lett. 15 (2015) 7488–7496.

[120] A. Detappe, et al., Advanced multimodal nanoparticles delay tumor progression with clinical radiation therapy, J. Control. Release 238 (2016) 103–113.

[121] Y. Gimenez, et al., 3D imaging of nanoparticle distribution in biological tissue by laser-induced breakdown spectroscopy, Sci. Rep. 6 (2016) 29936.

[122] B. Busser, et al., Characterization of foreign materials in paraffin-embedded pathological specimens using in situ multi-elemental imaging with laser spectroscopy, Mod. Pathol. 31 (2018) 378–384.

[123] M. Bonta, et al., Elemental mapping of biological samples by the combined use of LIBS and LA-ICP-MS, J. Anal. Atom. Spectrom. 31 (2016) 252–258.

CHAPTER

15

Food science application

Kemal Efe Eseller[a], *Ismail Hakkı Boyaci*[b], *Banu Sezer*[b], *Gonca Bilge*[c], *Halil Berberoglu*[d]

[a]Electrical and Electronics Engineering Department, Atilim University, Ankara, Turkey [b]Department of Food Engineering, Hacettepe University, Ankara, Turkey [c]Department of Food Engineering, Konya Food and Agriculture University, Konya, Turkey [d]Physics Department, Ankara Haci Bayram Veli University, Ankara, Turkey

1 Introduction

Elements are essential components of human nutrition and biological function. Many of them such as Ca, Mg, and K are required for nutrition, while some elements such as Cd, Pb, and Hg may cause toxicity. In this respect, the elemental content of a food item refers to its nutritional value and toxicity directly and to its quality, authenticity, and possible adulteration indirectly. Determining the elemental content of a food product is vital, especially for certain consumer groups such as children, pregnant women, and people with allergies to specific elements. The elemental content of a food product can be determined by several analytical techniques. X-ray fluorescence spectroscopy, atomic absorption spectroscopy, inductively coupled optical emission spectroscopy/mass spectroscopy are the mainly used techniques. Due to the complex and nonhomogeneous nature of food products, determination of quality, authenticity, and adulteration requires certain techniques, which are time consuming, expensive, labor intensive, etc. Hence, there has been an increasing demand for basic, fast, accurate, reliable, and cost-effective techniques for food analysis. Furthermore, to enforce regulations and obtain diverse knowledge of food quality control and nutritional value, rapid and efficient analytical methods are required.

In recent years, laser-induced breakdown spectroscopy (LIBS) has been applied to numerous fields, including forensics, medicine, space exploration, metallurgy, mining, and environmental analysis. Especially since 2000, LIBS has become a rising star in the study of structural and compositional aspects of food analysis. Conventional techniques used in food analysis require long processes, certain hazardous chemicals, specialists, etc. However, LIBS provides numerous advantageous such as little or no sample preparation, in situ or stand-off detection, and the ability to analyze liquids, gases, and solids with multielemental analysis capability. Thanks to these features, LIBS has achieved great progress in food analysis alongside other fields. LIBS has been used for the determination

https://doi.org/10.1016/B978-0-12-818829-3.00015-0

of elemental composition, detection of food contamination and adulteration, geographical origin, and quality control analysis in many different food matrixes such as flour and bread [1, 2], milk [3], cheese [4], meat [5, 6], vegetables [7–9], etc.

Since food is a complex matrix that includes carbohydrates, lipids, proteins, and enzymes, and may also contain additives that are used to obtain a prolonged shelf life or better color, odor, taste, etc., problems may arise in quantitative analysis. A problem caused by a complex matrix is called the matrix effect, in which variations of physical and chemical properties of a food product change the signal. Also, inhomogeneity and freshness of food products are other problems. To solve these problems, further evaluations are needed to fully understand the interaction of lasers with food matrices, plasma formation, radiation emission, and signal analysis. Therefore experimental parameters such as laser energy, laser repetition rate, number of shots, sampling technique based on the physical state of the sample, etc. need to be optimized to improve detection sensitivity. Also, for reliable results, choice of sample preparation technique and data analysis technique is quite important.

In light of all this information and the increasing trend in LIBS-based food analysis, it can be concluded that LIBS is a critical monitoring tool in food science. In this context, the emphasis of this chapter will be on experimental design, sample preparations, and common applications of LIBS in food analysis.

2 Experimental

In the context of elemental or molecular analyzing techniques, LIBS as an analytical method comes to fore with a number of favorable features, some of which are little or no sample preparation, the possibility of real-time analysis, the detection capability of multiple elements, and minimal damage to samples.

Some of the components that the standard LIBS setup basically consists of are nanosecond pulsed laser, focusing optics, plasma collection optics, and spectrometry (Fig. 1). Additional equipment can be adopted such as digital delay generators to avoid continuum emission due to Bremsstrahlung and free-bound electronic recombination. Characteristic background signals can be eliminated by using a gated detection technique, which triggers LIBS spectra acquisition with a proper delay time after the laser pulse that produces plasma. Computer-controlled translational stages are used in the LIBS system to refresh or map the elemental composition of samples of interest. Sample

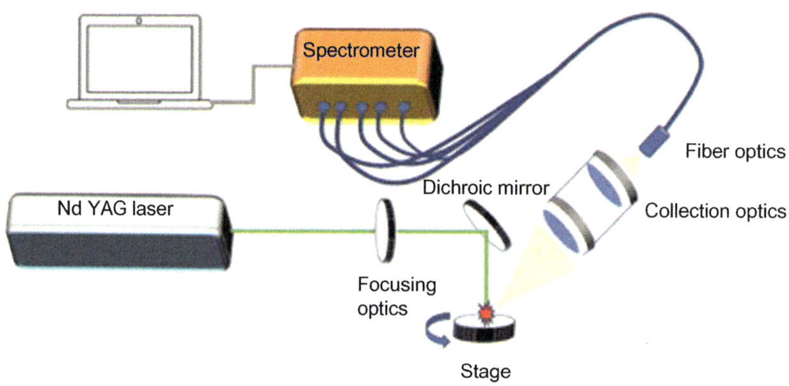

FIG. 1 Laser-induced breakdown spectroscopy setup.

chambers are also sometimes used in the LIBS system to reduce environmental or ambient air effects on laser-produced plasma and these chambers may be also purged with Ar gas to minimize oxidation on the surface.

To acquire a stable, noise-free, and clean spectrum (or emission intensity), some of the experimental parameters must be prepared for optimum signal-to-noise ratios, the first of which are laser parameters for ablation such as laser energy or fluence on samples, choice of wavelength, pulse duration, and repetition rates. A common laser to produce plasma on food samples is Q-switched Nd:YAG (yttrium-aluminum-garnet) or Nd:YVO_4 (yttrium-orthovanadate) solid-state lasers with a wavelength of 1064 nm. Most of the LIBS setup uses a lamp-pumped Q-switch or diode-pumped solid-state lasers with a pulse energy of a few to several hundred millijoules and with a pulse duration of about 10 ns. The fundamental wavelength can be converted to shorter wavelengths of 532, 355, and 266 nm with harmonic generation crystals. Pulses of around 10 ns are easily obtained by using the Q-switching mode with these lasers. The number of pulses generated by the laser is known as the repetition rate, which can usually be between 1 and 100 Hz for Nd:YAG laser types. A higher repetition rate shortens the measurement time. As is known, conventional LIBS systems use high pulse energy with a low repetition rate laser, which has very low duty cycles. On the other hand, microchip lasers have lower pulse energy with higher repetition rates of several kHz, which have two to four orders of magnitude higher duty cycles at comparable pulse durations [10].

Laser-produced plasma takes place when the laser irradiance, which depends on the spot size on the sample surface (typically 1 GW/cm^2), exceeds a specified threshold value to reach a breakdown that can be associated with the characteristic of materials such as its type, ionization energy, and concentration of the elements of interest. As a result of focusing a high irradiance laser light onto the surface of a sample within a very short time duration, immediately after exceeding that threshold a small amount of material of the target is heated and evaporated into the surrounding atmosphere. This material plume can continue to absorb laser energy and turn into a plasma state on the near surface of the material target, which then expands outward at supersonic speed and changes its parameters as it evolves. Then, a shockwave is formed by compressing the surrounding ambient atmosphere. Expansion of the plasma stops when the pressure of the plasma is similar to that of the internal wall of the shockwave [11]. At this stage, plasma is confined inside the contact wall and can be characterized by high electron number density (1016–1019 cm^{-3}). At the same time, all atoms and ions are excited in the plasma state, therefore LIBS spectra are mainly characterized by typical discrete and intense peaks that reflect the population of the excited levels of emitting species [12]. When continuum emission is strongly decreased, LIBS spectra can be easily recorded. It is generally assumed that this stage can be considered as local thermodynamic equilibrium (LTE), which can be described by Maxwell-Boltzmann distributions. At this stage, it is possible to determine the composition of plasma in a quantitative manner with a calibration-free approach. LTE is a necessary requirement for the calibration-free approach, which was proposed in 1999 [13]. Calibration-based methods may not always be applied for quantitative analysis because composition of unknown samples may not match that of the reference material. To use the calibration-free approach, plasma must be optically thin (less self-absorption), spatially homogeneous, and in LTE conditions [14]. Then, the famous Boltzmann equation can be applied to determine the elemental composition of the sample plasma. Improved sensitivity of calibration-free LIBS has been applied by recording two emission spectra with different delays between the laser pulse and the detector gate. A second measurement with improved limits of detection has been applied with a larger gate delay. To evaluate the concentrations of trace elements, LIBS is performed with a larger gate delay, which is a partial LTE condition to quantify minor and trace elements [15].

The next necessity is ambient conditions such as pressure and ambient gas flow. The laser pulse excites not only the sample but also the ambient gas. In the presence of an ambient gas, the plasma expansion dynamics are different to those in a vacuum because it results in denser plasma and an increase in frequency of collisions between ablated and ambient gas species, so the denser plasma partly shields the laser radiation and thus reduces the ablated target mass [16]. Therefore ambient conditions interact with the plasma and affect the emission spectra.

In some situations, the choice of wavelength for ablation might also be important. Organic samples, compared to the metallic samples when analyzed under atmospheric pressure, are much more sensitive to interaction between the plasma and the ambient air. It is difficult to distinguish the emission of CN radicals and other organic elements such as C, H, O, and N between those that come from the ablated sample and those that form due to the recombination of sample plasma and the ambient air, especially at high energy pulses [17]. To avoid this, in the same work [17], it is concluded that nanosecond infrared laser is not suitable for organic material analysis using molecular emission, because there is a low molecule production rate at low fluence; and at high fluence the absorption of laser energy by the plasma leads to shielding of the pulse and heating of the plasma. Therefore it can be concluded that ablation with a low fluence nanosecond ultraviolet laser, which is not shielded by the plasma, is quite suitable for efficient analysis of organic materials with LIBS under atmospheric pressure by providing efficient molecular ablation and allowing low interference due to recombination with ambient air.

LIBS collection optics is also of great importance since signal quality is directly determined by the design of collection optics. A collimator is usually needed between the plasma and the fiber that guides the signal into the spectrometer. For optimum collection of signals, the numerical aperture of the fiber and the collection lens parameters must be carefully taken into consideration.

The laser ablation mechanism is more complex for organic samples compared to metallic ones. Due to the large difference in ionization potentials between organic and metallic elements, it is much easier to obtain a partial stoichiometric ablation among metallic elements. For instance, ionization potentials for nitrogen and carbon atoms are 14.01 and 12.01 eV, but for metals like iron and aluminum they are almost half, 7.90 and 5.99 eV, respectively. Generally, single ionization takes places in the plasma for a typical LIBS signal.

To remove any potential contamination on the samples, it is suggested that a few cleaning pulses are sent to the focal spots where the measurements or spectra are being taken. Noticeable emission lines can be characterized by the NIST Atomic Spectra Database in LIBS spectra [18].

2.1 Sample preparation

One of the main advantages of LIBS is that it requires a minimum or no sample preparation step. Therefore this property makes LIBS fast and popular among other elemental analysis techniques and makes the spread of the LIBS application area wider, from explosives to biological materials. Performance of the LIBS technique strongly depends on laser-matter interaction and plasma parameters. Laser energy interaction on the sample surface is determinative for obtaining suitable spectra. In the LIBS technique, when the laser energy is transferred to the sample surface, the sample is ablated and atomic emission lines occur in the spectrum. As a consequence of plasma expansion, emission line intensities decrease until the plasma is extinguished. A longer emission time provides a better limit of detection (LOD) and signal-to-noise ratio values because a longer emission time means a higher amount of ablated material. Therefore adopting a sample is critical for effective energy transfer and better analytical performance.

In the LIBS technique the sample is ablated from only a several hundred micrometer laser spot size. Therefore only small amounts of sample can be analyzed and these amounts do not represent the total sample. To overcome this disadvantage, recording large numbers of spectra from different regions of sample is critical. Another important point is heterogeneity. This is a common problem, especially in food matrixes. Because food samples are complex matrixes consisting of different components such as carbohydrates, fats, proteins, vitamins, additives, and some aroma compounds, LIBS requires uniform and homogeneous surfaces. In the literature there are a number of solutions according to the phase of the food sample (solid, powder, liquid, and gel phase). To improve sensitivity, durability, repeatability, and reproducibility and to reduce the matrix effect, many specimen techniques are applied to food matrixes. Separation, dilution, internal standards, homogenization, milling, fusion, sol-gel, leaching and digestion, fixing, pellets, and use of adhesives are some of the solid specimen techniques [19].

Although solid food samples such as fruits and vegetables, meat, and cheese can be analyzed directly by slicing [20], for increased analytical performance of LIBS a smooth and uniform sample surface should be provided. At this point, particle size is also important because a smaller particle size provides easier vaporization and atomization in the plasma, which leads to uniform plasma. For this reason, some preprocessing methods are needed such as milling and forming of pellets. Gomes et al. used cryogenic grinding and planetary ball milling for sample preparation before pellet formation in plant analysis. Results showed that small particles reduced pellet porosity and may influence the ablation process with up to 50% emission of signal enhancement on LIBS measurements [21]. For the same reason, Lei et al. formed milk powder as pellets for measurements of Ca, Mg, and K [22] and Sezer et al. formed wheat flours as a sheet by using a Glutork machine (Fig. 2). These processes provided hard and uniform surfaces by mixing with water and cooking for 4 min [23]. Thanks to these methods, not only does it provide homogeneity, but it is also prevented from spreading around by giving a hard form to samples such as wheat flour. Cellulose and Ultrabind [24] are some of the binders used. On the other hand, water in solid foods is another important parameter that decreases the LIBS signal intensity. Therefore drying [7] and freezing are frequently used preprocess methods for foods. Like water, fat content is also problematic for LIBS analysis. Therefore Bilge et al. extracted fat from pork, beef, and chicken samples with the Soxhlet extraction method for LIBS analysis [6]. Another interesting approach is extracting nutrient elements from the food sample to obtain closely matrix-matched standards. Da Silva Gomes et al. performed acid extraction of sugar cane for standard samples and then added a proper amount of nutrients for a calibration dataset [25].

Foods with high water content such as sauces and beverages (fruit juices and milk) can also be analyzed by LIBS. However, in these types of samples, laser shots can cause some undesirable effects. Splashing, formation of aerosols, and formation of shockwaves and bubbles inside the liquid can be listed as undesirable effects. On the other hand, LIBS plasma in liquid causes lower plasma temperature emission intensity and shorter plasma lifetime, which affects the analytical performance of LIBS negatively. To overcome these undesirable effects, a number of configuration novelties and sample preparation techniques are used. Double pulse LIBS

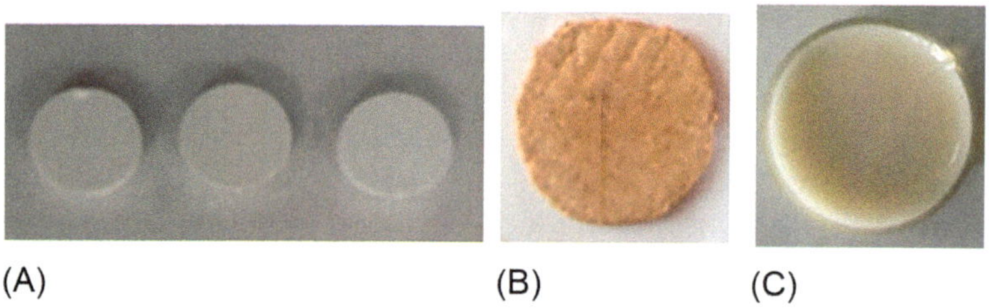

FIG. 2 Different sample forms, namely pellet (A), Gloturk (B), and gel form (C).

technique [26], aerosol formation [27], laminar flow, and jets or droplets are applied to food analysis by LIBS. Another common approach is transformation of liquid to solid by freezing [28] or adsorption to the solid surface. Cáceres et al. analyzed Na and Al metals in water by using liquid nitrogen for the transformation of water to ice [28]. In another study, absorbent paper was used to analyze Ca and Mg elements in aqueous solution [29]. For the transformation of a liquid to a solid sample, gel formation has been developed for liquid analysis by LIBS. Moncayo et al. analyzed red wine samples by using collagen to detect fake wines and to avoid unfair competition in the market. By this means, working difficulties in liquid analysis were overcome via transformation of liquid to gel form (Fig. 2) [30]. In another study, Sezer et al. transformed bovine, caprine, and ovine milk samples to form a gel by using gelatin and they classified pure and adulterated milk samples via chemometric techniques [31]. As an example, for the transformation of liquid to solid, chemometric techniques can be applied to dry microdroplets on metal surfaces, which is called surface-enhanced LIBS.

An internal standard can also be a good alternative to reduce the matrix effect. After subtraction of the background, an analytic signal is normalized to the internal standard signal. The emission line of the internal standard should be present in the interested spectral range. In some situations, known and fixed elements in analyte are used as internal standard, while in some situations, it is added individually.

Nanoparticles have been used in sample preparation methods for LIBS analysis. It was reported that nanoparticles improved ablation and plasma excitation by allowing exploitation of plasmonic resonance. By this means, lower LOD values could be obtained. This technique is called nanoparticle-enhanced LIBS by the authors. Ohta et al. were able to analyze *Citrus unshiu* and *Rhododendron obtusum* with improved LOD values [32]. Another new application in the LIBS technique is elemental mapping, which provides information on the elemental line intensity versus one or more spatial coordinates [33]. Although it is useful, especially for industrial applications in material characterization and medical applications, it can be applied to the elemental mapping of plants to show elemental distribution.

3 Food applications of LIBS

3.1 Determination of elemental composition

Measurement of nutritional and essential and toxic elements in foods has become a critical topic in recent years. While some elements such as Mg, Ca, and K are required by the human body, some of them (Pb, Hg, St, Mn, etc.) cause systematic diseases due to their toxicity. On the other hand, Na-like elements are not toxic but their levels are controlled routinely for preventing heart attacks, stroke, and coronary diseases.

Therefore amounts of elements in foods, especially toxic ones, should not exceed daily recommended amounts as recommended by the Food and Drug Administration and should be analyzed routinely. For other major elements, elemental composition indication on the label of the food product provides an educational tool for consumers. In conventional methods, elemental analyses are performed by inductively coupled plasma mass spectroscopy, X-ray fluorescence, or atomic absorption spectrometry techniques. Those methods need sample preparation and specialist personnel. The complex structure of food compounds can have an effect on some preprocesses for elemental analysis such as digestion and transformation of solid samples to liquid samples. As an alternative, LIBS provides rapid and multi-elemental analysis opportunities in several seconds for foods.

To determine major and minor elemental contents of foods, LIBS provides practical solutions by analyzing all elements at the same time. The only limitation of LIBS is for minor elements due to reduced sensitivity. Agrawal et al. analyzed the elemental content of food supplements for different brands by using the calibration-free LIBS approach and analyzed Ca, Mg, C, P, Zn, Fe, Cu, and Cr elements. In the study, nondetected elements, Na, K, and Ti, were found in some brands [34]. In another study, LIBS was used as an industrial at-line monitoring system for Ca analysis in comminuted poultry meats. The Ca element indicates the presence of bone particles in mechanically separated poultry meats, which is an important quality parameter for the meat industry [5]. In another meat application study, Dixit et al. quantified Rb as a trace element in beef, which is an important Rb source [35]. In the context of trace elements, Na is the most abundant element in many foods and is related to salt content in foods. In conventional methods, Na or Cl ions are determined for salt analysis and transformed into NaCl content. Because of the easy detection of Na with LIBS, Bilge et al. analyzed Na content in bakery products as a function of NaCl [36].

Because of the suitability of powder structures for forming pellets, powder-type foods are commonly used in LIBS analysis. Cho et al. determined trace elements such as Sr, Mg, Al, Cu, Cr, K, Mn, Rb, Cd, and Pb in starch-based flours with 18 μg/g of LOD for all elements [37]. Markiewicz-Keszycka et al. achieved analysis of K and Mg in gluten-free flours with 0.85 and 0.29 mg/g of LOD values, respectively. Furthermore, in the study of buckwheat, gluten-free flour blend, potato, rice, tapioca, and maize flours were able to be discriminated regarding elemental compositions [38]. Ca enrichment is a common approach in the flour industry to replace the reduced Ca amount during milling. Another reason for Ca addition is dietary fiber adulteration. Some manufacturers add Ca to show the increased ash amount, which increases with dietary fiber content. Bilge et al. used LIBS to identify Ca added flour based on the Ca/K ratio [1]. Milk powder is another example of powdered food samples. By using LIBS, K, Mg, and Ca analyses were performed in milk powder samples by Dos Santos Augusto [39]. Other performed elemental analyses by LIBS in powder-type foods are coffee [40] and tea [41]. In the study of coffee, Mg, Ca, Al, Cu, Na, Ba, Br, Co, Cr, Ce Mn, and Mb elements were quantified, while Br, Cr, and minerals like Fe, Ca, K, and Si were analyzed by LIBS in the tea samples.

Vegetables are another target food for LIBS analysis due to their high nutritional value. Beldjilali et el. analyzed minor elements in the flesh and skin of potatoes. According to the results, potato skin contains a large amount of Si and Al, whereas the flesh does not. Despite the limitation of LIBS in organic matrixes due to the matrix effect, in the study the measured spectrum was compared to the spectral radiance computed for a plasma in LTE. Thanks to this, difficulty in calibration for organic substances like vegetables could be overcome [42].

Rai et al. used the LIBS technique for identification of glycemic index-responsible elements in ripe and unripe fruit peel aqueous extracts of *Psidium guajava*. Results showed that Mg concentration is higher in unripe fruit peel aqueous extract than in ripe fruit peel aqueous extract, whereas K concentration is lower in the extract of unripe fruit peel than in the extract of ripe fruit peel [43]. Kim et al. determined Mg, Ca, Na, and K elements in spinach and unpolished rice samples [44].

Although liquids are problematic matrixes for LIBS, there are many applications for liquid foods in the literature. In this context, D'Angelo et al. analyzed Ca, Na, and Mg elements in water and Boussaidi et al. analyzed the water extracted from orange juice for the same elements [45]. Pb, Si, Ca, Na, Zn, Sn, Al, Cu, Ni, Fe, Mg, and Cr elements were studied in oil and water as liquid matrixes by Fichet et al. [46]. In another study, Bockova et al. analyzed majors (Na, Mg, K, and Ca), minors, and traces (Li, B, Si, P, Ti, Mn, Fe, Cu, Zn, Rb, Sr, Ba, and Pb) in wine samples [47].

3.2 Determination of food constituents

From the receiving of raw material to final food product, contamination can come from each step in production. Environmental contamination of raw material, storage and transportation of raw material, food processes such as thermal processes and cleaning, food packaging, and transport of final product can all be steps to be wary of. For this reason, contamination, which is the highest concern for consumers, should be prevented [48].

Contamination can be sourced from microbiological or chemical origins and commonly at trace levels. Therefore detection is still a challenge in many food matrixes. Although there are many analytical tools for the identification of contaminants, none of them provides a unique solution. Recently developed food applications of LIBS provide rapid solutions for the determination of contamination in foods.

A common source of food spoilage and food poisoning is microbiological contamination, especially for animal-sourced foods. LIBS cannot analyze the genetic difference between microorganisms but it can analyze the chemical composition of outer cell membranes. Mohaidat showed that LIBS has the potential to identify microorganisms based on the relative concentration of trace inorganic elements such as Mg, P, Ca, and Na [49]. A study performed by Kaiser et al. showed the responsible elements for identification of microorganisms in summary form [33]. Regarding this principle, food-borne pathogens can be analyzed with LIBS. Multari et al. differentiated *Escherichia coli* O157:H7 and *Salmonella enterica* on raw foods (eggshell, milk, bologna, ground beef, chicken, and lettuce) and food-contacted surfaces (metal drain strainer and cutting board) for food safety and streamlining the food-monitoring process [50]. Barnett et al. detected *S. enterica* serovar Typhimurium in liquid foods such as brain heart infusion, milk, or chicken broth according to Ca, Mg, and Na emission lines. Of course, LIBS is less sensitive than polymerase chain reaction (PCR) and quantitative polymerase chain reaction (qPCR), but LIBS was able to detect between 105 and 106 cfu/mL of *Salmonella* from milk, whereas the PCR and qPCR assays were unable to detect *Salmonella* from milk using DNA extracted by a simple boiling method [51]. Not only analyzed elemental distribution but also molecular bands can be useful for bacteria detection. Baudelet et al. analyzed *E. coli* using the strong CN emission bands from bacteria [52].

Heavy metals such as arsenic, cadmium, lead, and mercury are naturally occurring in chemical compounds, which are present at trace levels in water, the environment, and soil. Therefore foods can be contaminated with these metals during handling, processing, and storage. Because they are toxic to humans, their levels are controlled in foods. In this context, Huang et al. determined the heavy metal chromium in pork meat. Regarding statistical values, the characteristic line of Cr

I 425.43 gave better results for the calibration model than other characteristic lines of Cr I 425.43, Cr I 427.48, and Cr I 428.97 [53]. Fish are another heavy metal intake source. Rapid industrialization causes aquatic pollution and in this situation fish may be a source of metal contamination. Wan et al. analyzed fish for plumbum and barium elements, which are quite toxic. Results showed that heavy metal accumulation is high in fish liver, mouth, and gills, etc.; however, the heavy metal content in fish meat is very low [54]. In another fish study, Ponce et al. analyzed lead and copper in edible fish with 25 and 100 ppm of LOD values, respectively [55]. Yao et al. detected Cd in polluted fresh, leafy vegetables by using LIBS [8]. Another important heavy metal source is drinking water. Lee et al. analyzed water by soaking a filter paper for Pb and Cr content with 2.7 and 0.36 ppm of LOD, respectively. However, preconcentration by evaporation improved the LOD value to 75 and 18 ppb for Pb and Cr, respectively [56]. Hu et al. analyzed Cu contamination on orange peels [57].

Pesticides are another important contaminant in vegetables and fruits. Determination of pesticide residue in foods is important, especially for infants and children, because the amount of food they ingest per body weight is relatively high. Some pesticides contain chemicals disrupting the endocrine system, so the function of naturally occurring hormones can be impacted even at low concentrations. In conventional methods, gas chromatography mass spectrometry and liquid chromatography tandem mass spectrometry methods are used for pesticide detection. However, these methods are time consuming, expensive, and require specialist personnel. It has been shown that LIBS has potential in pesticide analysis as a practical solution. Some elements such as phosphorus, sulfur, and chlorine can be used as indicators for pesticides. In this context, Ma et al. employed LIBS for chlorpyrifos (a kind of pesticide) on apple surfaces [20] and Kim et al. determined pesticide residues (10 ppm of fosetyl-aluminum) in rice and spinach [44].

3.3 Detection of food adulteration

Detection of possible adulterants in food samples is an important task for governmental regulatory agencies, manufacturers, and consumers in terms of food safety, fair trade, and consumer health. Hence, scientists all over the world are trying to develop new methods to accomplish this task with certain requirements such as robust and easy-to-use instruments, portable options, no or minimum sample preparation, rapid response, qualitative and quantitative measurement, high specificity and precision, high detection limits, etc. In this respect, LIBS is a well-documented, rapidly evolving, in situ, nondestructive optical, elemental analysis technique, which meets several of these requirements. To this end, there are several applications in different food matrices such as meat, cereal, oil/fat, and dairy products.

To enhance profit, meat is one of the most adulterated food products in many ways. Meat adulteration has become common practice around the world. This proves to be a serious issue in terms of economics, ethics, health, commerce, trade, and religion. Allergic individuals and people holding religious beliefs restrict the allowable intake of certain types of meat. Furthermore, accurate labeling is important to support informed choice and fair trade. Due to this, there is a growing demand for reliable, accurate, and rapid methods to identify meat species. Hence, Bilge et al. achieved successful quantitative determination of beef meat addition to pork and chicken mixtures. LIBS spectra (Fig. 3) were evaluated by chemometric methods, and meat species were discriminated with principal component analysis (PCA) with an 83.37% ratio (Fig. 4). Pork/beef and chicken/beef meat mixtures were also analyzed with the partial least squares (PLS) method quantitatively (Fig. 5). Determination coefficient (R^2) and LOD values were found to be 0.994 and 4.4% for pork-adulterated beef, and 0.999 and 2.0% for chicken-adulterated beef, respectively [6].

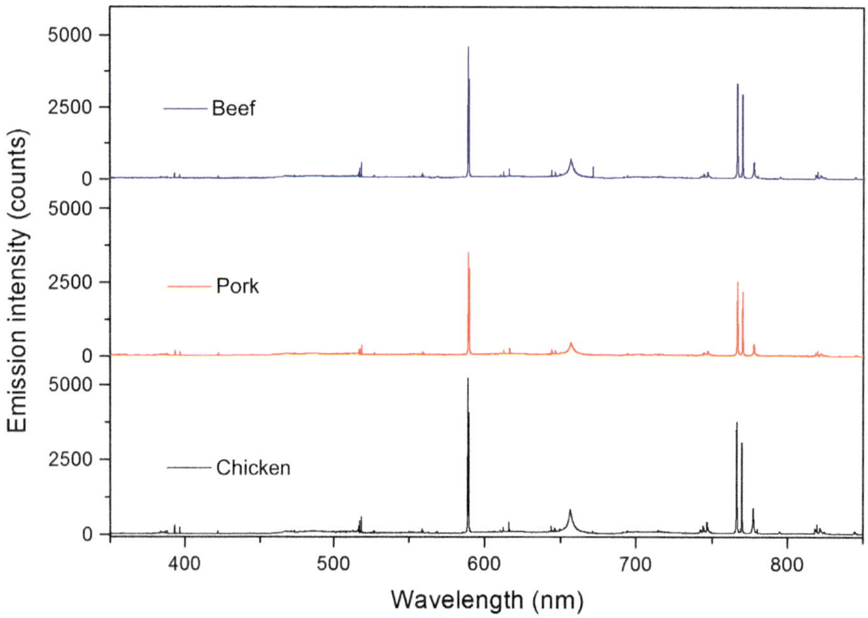

FIG. 3 Laser-induced breakdown spectroscopy spectra of meat samples.

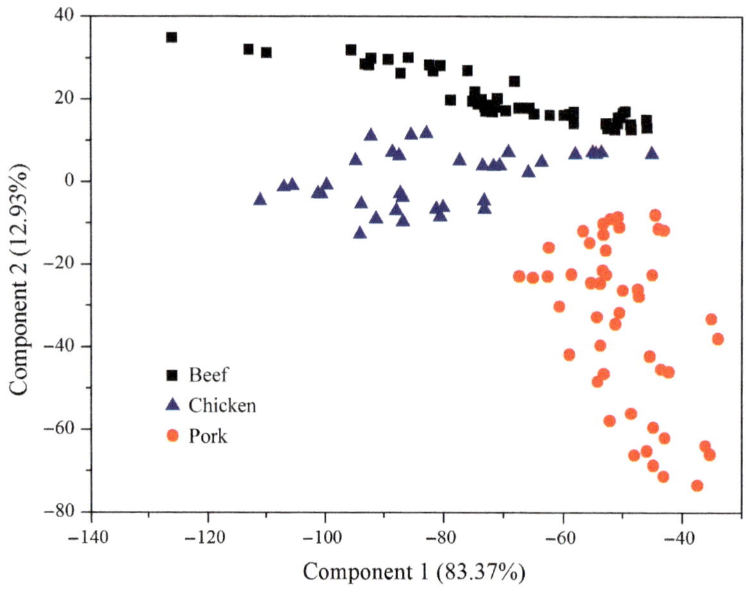

FIG. 4 Laser-induced breakdown spectroscopy spectra of meat samples.

Furthermore, Casado-Gavalda et al. approached this problem from a different angle [58]. They used LIBS for the analysis of copper content in minced beef samples as a potential indicator of offal adulteration. In this study, a calibration model with a high determination

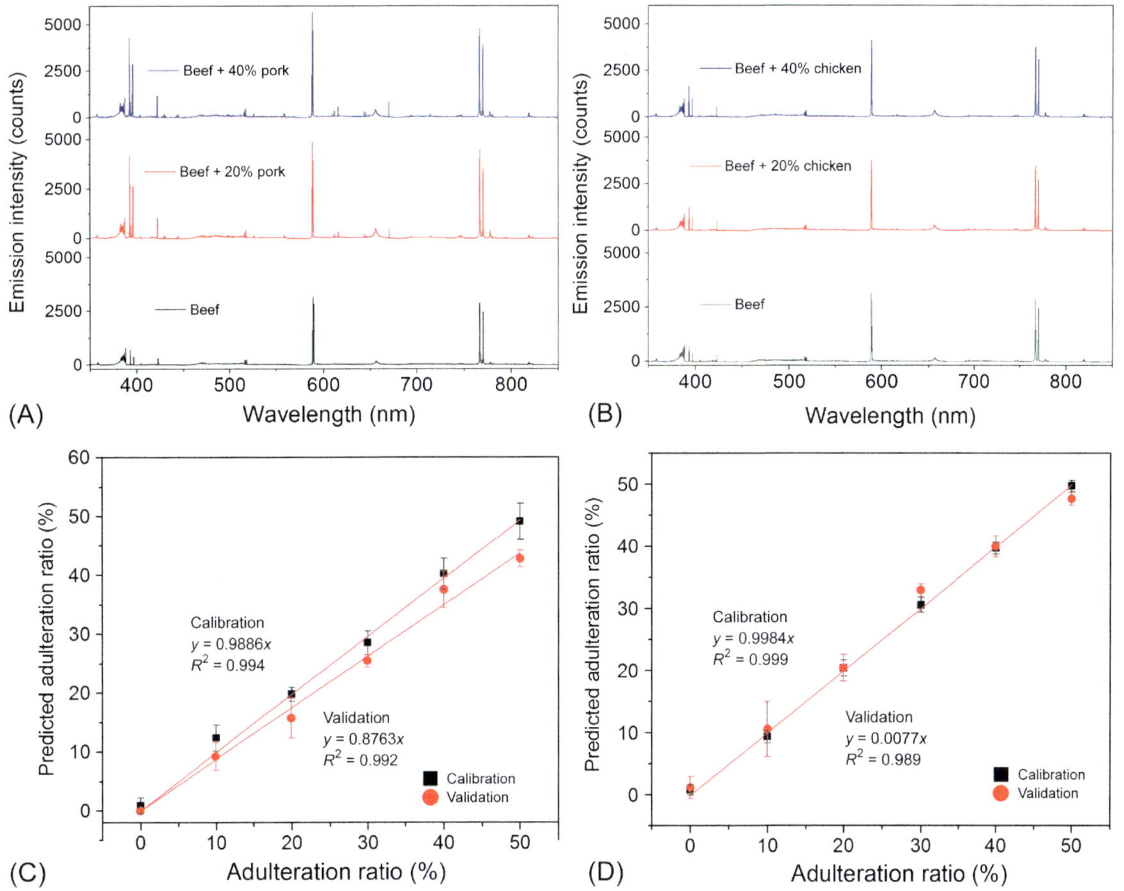

FIG. 5 Laser-induced breakdown spectroscopy spectra of adulterated beef with pork (A) and chicken (B), and partial least squares calibration models of adulterated beef with pork (C), and chicken (D).

coefficient of cross-validation of 0.85 and a root mean square error cross-validation of 43.5 ppm was obtained. The validation model showed a good prediction accuracy with a high determination coefficient for prediction of 0.85 and root mean square error of prediction of 36.8 ppm. In another study, Dixit et al. used LIBS to quantify beef kidney adulteration in beef meat by determining the Na and K concentrations. The models for Na and K showed calibration coefficients of determination (R_c^2) of 0.97 and 0.91, respectively. Similarly, a good calibration model was obtained for adulteration showing an R_c^2 of 0.97 [59]. On the other hand, Velioglu et al. applied the LIBS technique to discriminate beef meat and offal samples and to determine offal content in beef meats. In this study, 0.947 R^2 and 3.8% LOD values for offal mixture-adulterated beef samples were obtained [60].

Powdered samples have significant importance in terms of analysis success in LIBS applications. One good example of this is presented by Sezer et al. in ground-roasted coffee samples [61]. Coffee is one of the most widely consumed beverages all over the world. Due to the high economic importance for countries that produce, export, and import it, its purity and detection of adulterants have been of constant

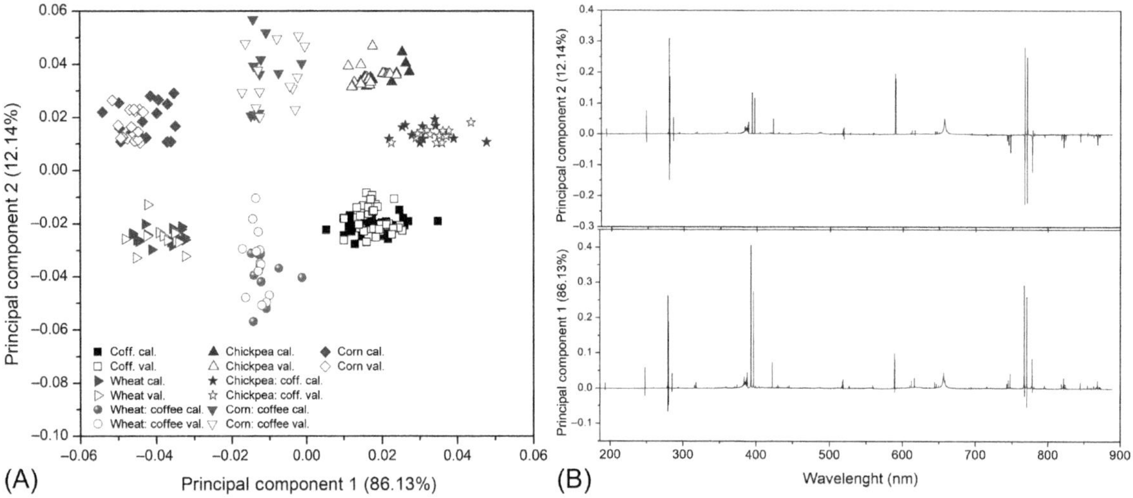

FIG. 6 Score plot (A) and loadings plot (B) of principal component analysis of *Coffee arabica*, chickpea, corn, and wheat samples.

concern. Sezer et al. demonstrated that LIBS can be an important tool for the determination of coffee adulteration with chickpea, corn, and wheat samples. In this study, PCA was used to discriminate *Coffee arabica* and wheat, corn, and chickpea samples (Fig. 6) and quantitative analysis was achieved using the PLS method (Fig. 7). According to the results, R^2 and LOD values for chickpea, corn, and wheat adulteration with *C. arabica* were found to be 0.996, 0.995, 0.995, and 0.56%, 0.52%, and 0.45%, respectively [61].

Another example put forward by Tiwari et al. [62], where four different turmeric samples were subjected to LIBS analysis and discrimination of pure turmeric, has been obtained in the PCA study. In another study, Bilge et al. applied the LIBS method to determine Ca concentration in wheat flour fortified or adulterated with the intention of obtaining higher ash values. In this study, $CaCO_3$-added wheat flour samples were analyzed for the discrimination of natural and Ca-added flour by determining Ca content and Ca/K ratio. Ca and Ca/K calibration graphs of the PLS method showed good linearity with R^2 0.999. LOD values for Ca and Ca/K analysis were calculated as 25.9 and 0.013 ppm, respectively [1]. In addition, Er et al. published a different study on saffron samples. Er et al. developed a LIBS-based method for the determination and quantification of saffron adulteration with certain plant adulterants (safflower, marigold, and turmeric). In this study, LOD and limit of quantification (LOQ) values were found to be 1.86% and 9.32%, respectively [63].

Dairy products are another important food commodity that have been subjected to significant amounts of adulteration due to their high nutritional value, great economic importance, and high consumption rates. There are different practices in terms of dairy product adulteration such as addition of water, external starch, whey, melanine, colorant, different types of milk, etc. Some of the practices result only in economic losses; however, some of the results show significant health concerns. Melamine is an important external adulterant for milk powders with negative health problems, including protein deficiency, kidney stones, and renal failure [64]. In this respect, Moncoya et al. employed LIBS to detect melamine adulteration in milk. In this study, univariate analysis using the CN emission band and multivariate calibration neural network (NN) model obtained R^2 values of

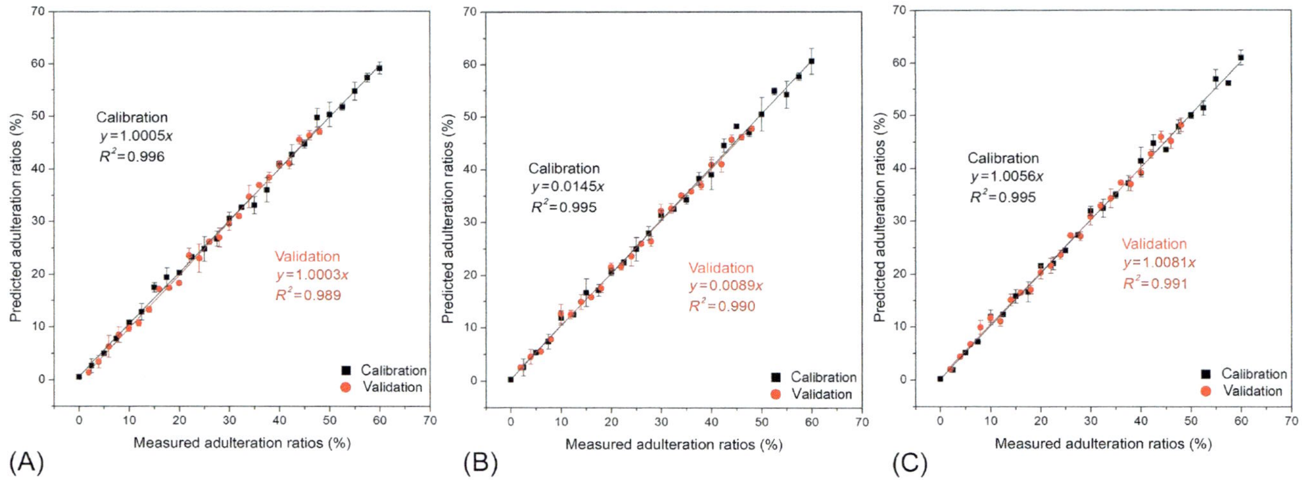

FIG. 7 Partial least squares calibration and validation models of chickpea- (A), corn- (B), and wheat- (C) adulterated *Coffee arabica*.

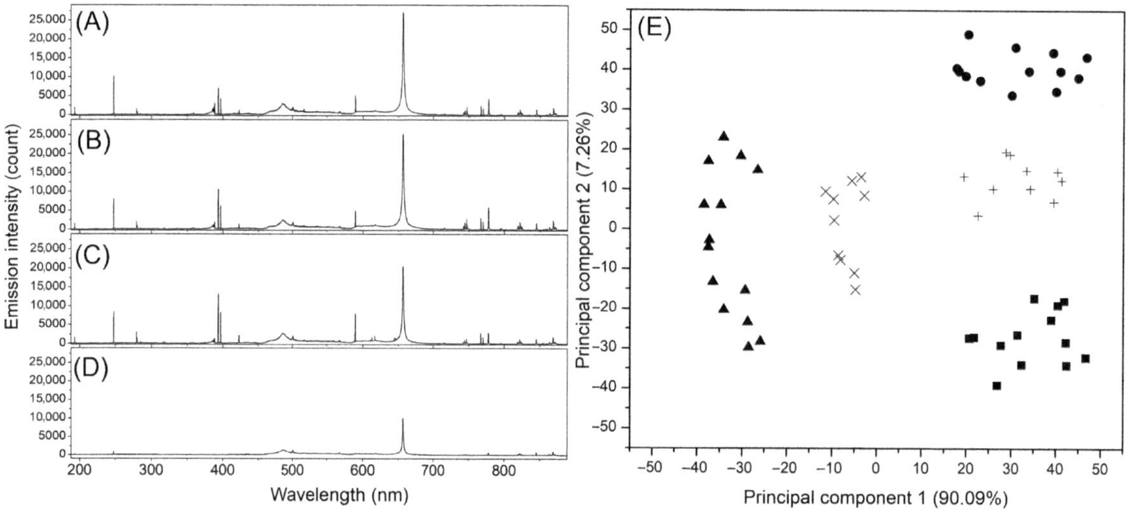

FIG. 8 Laser-induced breakdown spectroscopy spectra of pure (A) bovine, (B) caprine, and (C) ovine milk and gelatin (D) samples, and (E) principal component analysis score plot of pure bovine, caprine, and ovine milk and adulterated milk samples.

0.982 and 0.999, respectively [3]. Addition of whey to milk is one of the most frequently applied adulteration methods [65]. By using LIBS, Bilge et al. achieved quantitative and qualitative determination of whey in milk powder. R^2 and LOD values were 0.981 and 1.55% for adulteration with sweet whey powder, and 0.985 and 0.55% for adulteration with acid whey powder, respectively [66]. Sezer et al. employed LIBS to determine milk fraud by identifying species origin (bovine, ovine, and caprine) and determining bovine milk adulteration of ovine and caprine milk samples. In this study, LIBS spectra were evaluated with PCA for the classification of pure and adulterated milk samples (Fig. 8) and partial least squares regression was performed to determine the adulteration ratio (Fig. 9). The R^2 and LOD values for caprine milk adulteration with bovine milk were 0.995 and 1.39%, respectively; the corresponding values for ovine milk adulteration with bovine milk were 0.996 and 1.29%, respectively [31].

Adulteration and traceability are serious problems in the vegetable oil industry. For this, Caceres et al. used LIBS combined with NNs for identification, quality control, traceability, and adulteration detection in extra virgin olive oils. The results obtained allow the identification of the oils tested with a certainty of more than 95% [67]. Temiz et al. showed that LIBS can successfully determine margarine adulteration to butter samples, which is an important application due to its economic gain. Temiz et al. used PCA to classify butter and margarine samples (Fig. 10) and PLS to determine adulteration ratio (Fig. 11). In this study, LOD and LOQ values were calculated as 3.9% and 11.8%, respectively [68].

Products of plant origin have encountered mislabeling and adulteration often in the food industry. To prevent consumers from this kind of deception, LIBS provides an alternative solution with its multielemental nature. Sezer et al. used LIBS to determine *Pistachia vera* adulteration with green pea and spinach. The results showed that the calibration curve for green pea-adulterated pistachio had an R^2 of 0.995 and an LOD of 2.04%, whereas the calibration

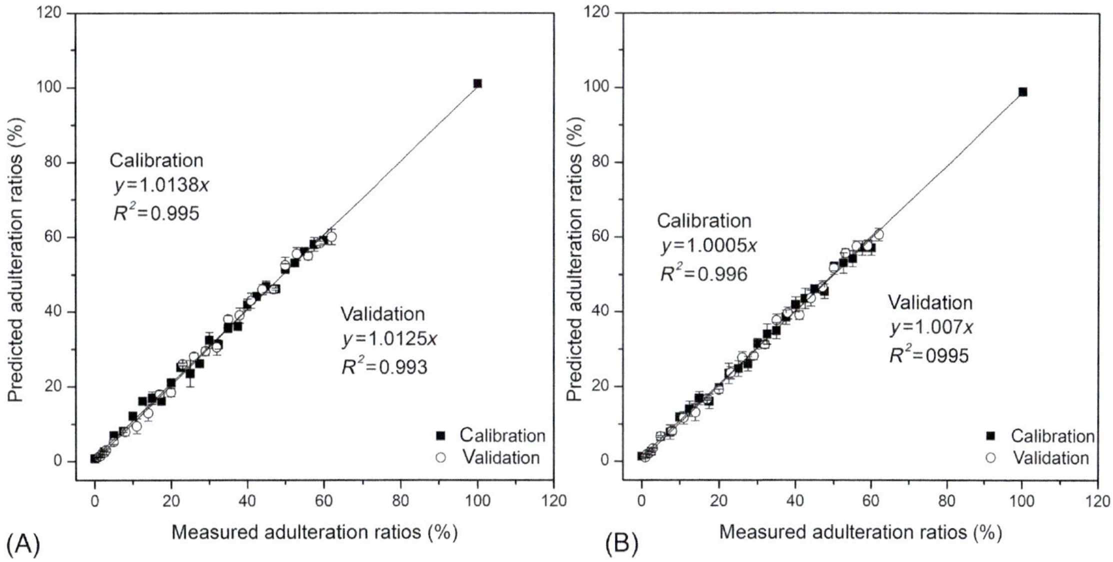

FIG. 9 Partial least squares regression calibration and validation models for quantitative determination of bovine milk-adulterated (A) caprine milk and (B) ovine milk.

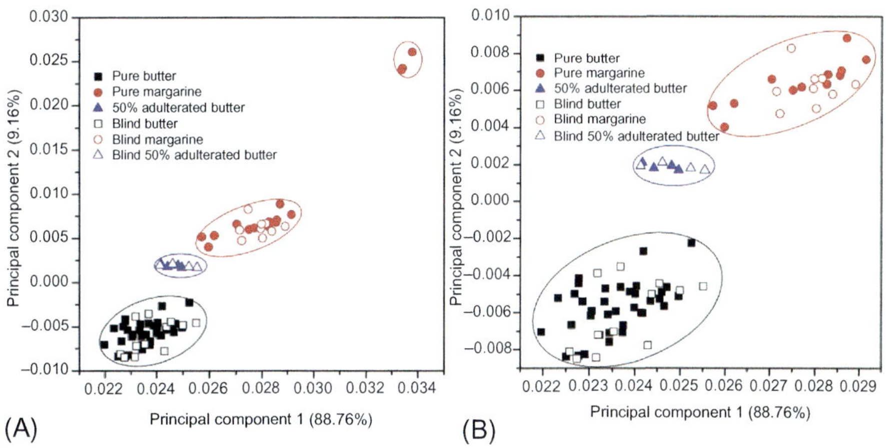

FIG. 10 (A) Principal component analysis (PCA) score plot based on laser-induced breakdown spectroscopy (LIBS) spectra. (B) Expanded graph of PCA score plot based on LIBS spectra.

curve for spinach-adulterated pistachio had an R^2 of 0.993 and an LOD of 1.64% [9]. The pistachio nut is in great demand globally. It is a popular ingredient in ice cream, desserts, confectionary, and bakery products, which makes it hard to detect in complex foods due to the suppressed taste of sugar.

3.4 Identification of geographical origin

Globalization of the food market has led to the transport of food commodities through and between countries and continents. This massive transportation around the world has created increased concern about geographical

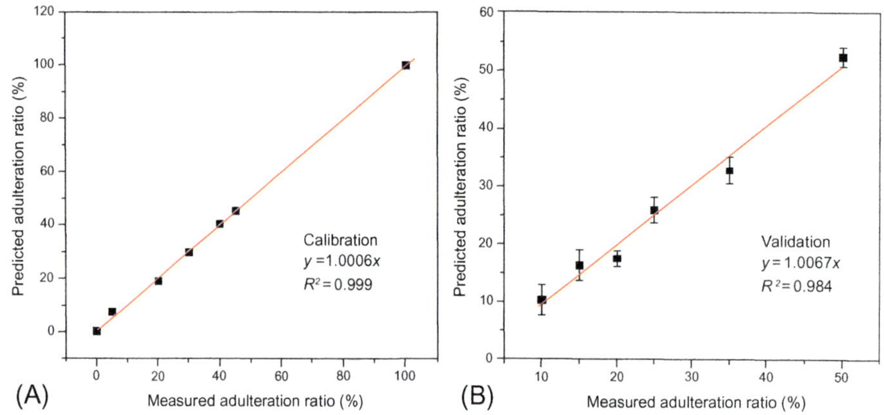

FIG. 11 Partial least squares (A) calibration and (B) validation models for quantitative determination of butter adulteration with margarine.

origins of consumed food items. On the other hand, accurate determination of protected designation of origin (PDO) of a food product gains importance to protect production and fare trade among manufacturers. Hence, there has been an exponential increase in research into the determination of geographical origin.

To this end, Moncayo et al. published early research into the classification and PDO of red wine samples [30]. In this study, wine samples belonging to different Spanish regions were analyzed by using LIBS combined with NNs. According to the study, all wines were correctly classified based on their origin. Furthermore, Tian et al. reported classification of 29 bottles of French wines according to their production sites (Alsace, Bourgogne, Beaujolais, Bordeaux, and Languedoc) by using LIBS [69]. Also, Tian et al. enabled a sub-ppm limit of detection and identification of at least 22 metal and nonmetal elements in this study.

Another study was reported by Yang et al. [70]. In this study, 20 different kinds of rice from different geographic origins were analyzed by LIBS and successfully classified with PCA and support vector machine.

In a different study, Akpovo et al. reported that LIBS is an attractive tool for geographical discrimination of oysters [71]. Oysters are high prize food items and one of the most harvested shellfish in the world. Therefore, geographical discrimination is significant in terms of price and species identification.

Furthermore, Eum et al. evaluated the performance of LIBS for geographical discrimination of milk vetch root, which is a medicinal herb commonly used in formulations due to its cardioprotective and antiinflammatory activities [72]. According to this study, near-infrared spectroscopy and LIBS provide useful information for successful discrimination of the geographical origin of milk vetch root samples.

3.5 Quality and control analysis

In food science, all applied analyses, whether chemical, biological, or sensory, have two major goals: financial profit for manufacturers and to ensure safety and quality for consumers. In this respect, certain attributes of the applied technique have gained significant importance such as speed, complexity, sample preparation, repeatability, detection limits, etc. Development of rapid analytical methods for food analysis have shown an exponential increase due to their ease of use, rapid and simple design, and use of fewer chemicals. LIBS is one of the most promising techniques to determine food quality and control due to its unique features such as rapid, no, or minimum sample preparation, applicability to solid, liquid, and gases, etc.

In this respect, Bilge et al., Sezer et al., and Markiewicz-Keszycka et al. applied the LIBS technique to determine ash content in wheat flour, wheat milling fractions, and gluten-free flours, respectively [2, 23, 38]. The ash content of a food product is a measure of total mineral content. Ash value represents an important quality parameter in terms of nutritional quality and process properties, and can discriminate between bran and germ. As a result, ash analysis is an important and commonly applied procedure. The conventional dry ash analysis procedure is lengthy (5–6 h) and causes energy waste because of the analysis of a limited number of samples in a single application. However, LIBS provides ash analysis in flour and wheat milling fractions with accuracy.

Protein content can be listed as one of the most important quality parameters in terms of nutritional value, price, accurate labeling, etc. Accurate determination of protein content in a high number of samples in a relatively short amount of time has been a very important task for food matrixes. In this respect, Sezer et al. successfully used LIBS to determine protein content in wheat flour and whole meal samples [73].

In a different study, the potential use of LIBS for the determination of chemical quality parameters in cheese samples was investigated. Ayvaz et al. were able to determine moisture, dry matter, salt, total ash, total protein, pH, fat, acidity, water-soluble nitrogen, trichloroacetic acid-soluble nitrogen, and phosphotungstic acid-soluble nitrogen in full fat Ezine cheese samples [4]. Another quality attribute is fatty acid content of food products. Determination of the fatty acid content and the classification of products based on this value have significant importance in terms of health, quality, and characterization of the food sample. Kongbonga et al. used LIBS to classify vegetable oils according to their fatty acid content [74].

Coffee is a widely consumed beverage around the world with high commercial value to the manufacturer countries. According to the International Coffee Organization, estimated global coffee consumption in 2014 was 149.2 million bags (each bag contains 60 kg of coffee) [75]. Because of this, its quality control is important for both consumers and manufacturers. Zhang et al. used LIBS to determine coffee varieties due to their elemental content with a classification accuracy of over 80% [76]. In another study, Siva et al. succeeded in determining defective beans in coffee blends by using LIBS to prevent the negative impact on product quality [77]. Ferreira et al. used the LIBS method for the classification of roasted coffees according to Global Quality (GQ) classes such as gourmet, superior, traditional, and inferior [78]. Furthermore, Anggraeni et al. used LIBS to evaluate coffee for quality and safety purposes [79].

Food products have external ingredients called additives for numerous reasons, such as taste, texture, storage, etc. In general, additives are food-grade chemicals and harmless for consumption. In some cases, the method of producing food additives can even change the class of an additive classified as Generally Recognized as Safe (GRAS), such as titanium dioxide (TiO_2). In some cases, the production method of food-grade additives even changes its use as safe such as titanium dioxide (TiO_2). TiO_2 is a well-known food additive used in foods and personal care products [80–82]. The International Agency for Research on Cancer conducted a risk assessment of TiO_2 cancer potential in 2010 and classified TiO_2 as human carcinogen group 2B. Because of this, the detection and determination of nano-TiO_2 in food is a critical issue. In this respect, Sezer et al. achieved a quantitative determination of TiO_2 in a food matrix [83].

4 Summary and conclusions

Presented reports show a significant increment in food applications of LIBS in the literature. LIBS has been used in different food matrixes and for the purpose of adulteration and geographical and biological origin detection via developments in sample preparation techniques and experimental setup. Because of the properties of LIBS such as little or no sample preparation and fast analysis, it has potential to become a process control tool in the food industry.

Although LIBS has huge potential in the industry for metal alloys, there is currently no known commercial LIBS system for food samples. This situation is related to complex matrixes of food compounds and difficulties in obtaining calibration models, because in real food samples, all components may change due to the concentration of analyte molecule and this causes the matrix effect. On the other hand, low concentrations of minor elements and heavy metals in a complex matrix are also problematic for sensitivity in analysis. However, because of improvements in the stability of systems and data analysis techniques like chemometrics, it seems that certain drawbacks will be overcome.

For online analysis in the industry, the main restrictions are the requirement of vacuum/inert atmosphere and sample preparation procedures for some food samples. Controlling of atmospheric conditions with huge systems is expensive at the industrial scale and without a controlled atmosphere organic molecules and molecular bands may not be analyzed. Therefore the analysis of inorganic molecules is more suited to food samples. On the other hand, depending on the inhomogeneity and physical state of the food sample, some preprocessing techniques may be required. Therefore solid and homogeneous samples such as flour and milk powder are better suited to industrial applications. The present state shows that further studies are needed for the industrialization of LIBS for food samples.

An exponentially increasing number of research articles have been published in the last 10 years dealing with the use of natural abundance isotope variations and multielemental concentrations as geographic markers to determine the origin of foodstuffs. Multielement and isotopic analyses have previously been applied to different foods to develop methods that will identify their geographical origins with varying degrees of certainty. Certain analytical techniques and parameters have been studied to verify the geographical location of foods, such as aroma, sugar, phenolic, and flavor compound profiling by gas and liquid chromatography; chemical profiling by ^{1}H NMR; and near infrared and fluorescence spectroscopy [84–86]. Measuring elemental concentrations and isotopic variation in regional products is arguably the best analytical strategy for accurately identifying geographical origin. In this respect, LIBS can be a powerful alternative for multielemental analysis techniques with its broad range of advantages. In food authentication studies, there is a significant requirement for a database. Similarly, in geographical origin determination, collecting and analyzing samples from a wide range of countries is time consuming and costly, especially with the aforementioned analytical techniques [87]. Furthermore, there is a significant requirement for a large number of independent variables to be measured and statistically analyzed to identify key markers that differentiate the regions or countries. In this respect, LIBS can perform multielemental analysis of a large number of samples with minimum sample preparation within seconds. This feature brings an important step to the research field of natural abundance isotope and geographical origin analysis for the future.

Developments in elemental mapping by LIBS are promising for food manufacturers. Elemental mapping has been used to observe the metal accumulations in different biological tissues [33, 88]. Applying this technology to the food industry will be ideal for the analysis of food products and food quality on a conveyor belt.

References

[1] G. Bilge, B. Sezer, K.E. Eseller, H. Berberoğlu, H. Köksel, İ.H. Boyacı, Determination of Ca addition to the wheat flour by using laser-induced breakdown spectroscopy (LIBS), Eur. Food Res. Technol. 242 (10) (2016) 1685–1692.

[2] G. Bilge, B. Sezer, K.E. Eseller, H. Berberoglu, H. Koksel, I.H. Boyaci, Ash analysis of flour sample by using laser-induced breakdown spectroscopy, Spectrochim. Acta B At. Spectrosc. 124 (2016) 74–78.

[3] S. Moncayo, S. Manzoor, J. Rosales, J. Anzano, J. Caceres, Qualitative and quantitative analysis of milk for the detection of adulteration by laser induced breakdown spectroscopy (LIBS), Food Chem. 232 (2017) 322–328.

[4] H. Ayvaz, B. Sezer, M.A. Dogan, G. Bilge, M. Atan, I. H. Boyaci, Multiparametric analysis of cheese using single spectrum of laser-induced breakdown spectroscopy, Int. Dairy J. 90 (2019) 72–78.

[5] M.B.S. Andersen, J. Frydenvang, P. Henckel, Å. Rinnan, The potential of laser-induced breakdown spectroscopy for industrial at-line monitoring of calcium content in comminuted poultry meat, Food Control 64 (2016) 226–233.

[6] G. Bilge, H.M. Velioglu, B. Sezer, K.E. Eseller, I. H. Boyaci, Identification of meat species by using laser-induced breakdown spectroscopy, Meat Sci. 119 (2016) 118–122.

[7] D. Rai, R. Agrawal, R. Kumar, A.K. Rai, G.K. Rai, Effect of processing on magnesium content of green leafy vegetables, J. Food Process. Technol. 80 (6) (2014) 878–883.

[8] M. Yao, H. Yang, L. Huang, T. Chen, G. Rao, M. Liu, Detection of heavy metal Cd in polluted fresh leafy vegetables by laser-induced breakdown spectroscopy, Appl. Opt. 56 (14) (2017) 4070–4075.

[9] B. Sezer, H. Apaydin, G. Bilge, I.H. Boyaci, Detection of Pistacia vera adulteration by using laser induced breakdown spectroscopy, J. Sci. Food Agric 99 (5) (2019) 2236–2242.

[10] P. Pořízka, B. Klessen, J. Kaiser, I. Gornushkin, U. Panne, J. Riedel, High repetition rate laser-induced breakdown spectroscopy using acousto-optically gated detection, Rev. Sci. Instrum. 85 (7) (2014) 073104.

[11] A. De Giacomo, J. Hermann, Laser-induced plasma emission: from atomic to molecular spectra, J. Phys. D: Appl. Phys. 50 (18) (2017) 183002.

[12] A. De Giacomo, M. Dell'Aglio, R. Gaudiuso, S. Amoruso, O. De Pascale, Effects of the background environment on formation, evolution and emission spectra of laser-induced plasmas, Spectrochim. Acta B At. Spectrosc. 78 (2012) 1–19.

[13] M. Gondal, A. Dastageer, M. Maslehuddin, A. Alnehmi, O. Al-Amoudi, Detection of sulfur in the reinforced concrete structures using a dual pulsed LIBS system, Opt. Laser Technol. 44 (3) (2012) 566–571.

[14] M. Xu, Q. Lin, G. Yang, T. Xu, T. Zhang, X. Wang, S. Wang, F. Bian, Y. Duan, A single-beam-splitting technique combined with a calibration-free method for field-deployable applications using laser-induced breakdown spectroscopy, RSC Adv. 5 (6) (2015) 4537–4546.

[15] C.-T. Chen, D. Banaru, T. Sarnet, J. Hermann, Two-step procedure for trace element analysis in food via calibration-free laser-induced breakdown spectroscopy, Spectrochim. Acta B At. Spectrosc. 150 (2018) 77–85.

[16] H. Yuan, A.B. Gojani, I.B. Gornushkin, X. Wang, Investigation of laser-induced plasma at varying pressure and laser focusing, Spectrochim. Acta B At. Spectrosc. 150 (2018) 33–37.

[17] M. Boueri, M. Baudelet, J. Yu, X. Mao, S.S. Mao, R. Russo, Early stage expansion and time-resolved spectral emission of laser-induced plasma from polymer, Appl. Surf. Sci. 255 (24) (2009) 9566–9571.

[18] Y. Ralchenko, A. Kramida, J. Reader, NIST Atomic Spectra Database (version 5.6.1), https://physics.nist.gov/asd (accessed).

[19] S.C. Jantzi, V. Motto-Ros, F. Trichard, Y. Markushin, N. Melikechi, A. De Giacomo, Sample treatment and preparation for laser-induced breakdown spectroscopy, Spectrochim. Acta Part B 115 (2016) 52–63.

[20] F. Ma, D. Dong, A measurement method on pesticide residues of apple surface based on laser-induced breakdown spectroscopy, Food Anal. Methods 7 (9) (2014) 1858–1865.

[21] M.d.S. Gomes, D. Santos, L.C. Nunes, G.G.A. de Carvalho, F. de Oliveira Leme, F.J. Krug, Evaluation of grinding methods for pellets preparation aiming at the analysis of plant materials by laser induced breakdown spectrometry, Talanta 85 (4) (2011) 1744–1750, https://doi.org/10.1016/j.talanta.2011.06.069.

[22] W. Lei, J. El Haddad, V. Motto-Ros, N. Gilon-Delepine, A. Stankova, Q. Ma, X. Bai, L. Zheng, H. Zeng, J. Yu, Comparative measurements of mineral elements in milk powders with laser-induced breakdown spectroscopy and inductively coupled plasma atomic emission spectroscopy, Anal. Bioanal. Chem. 400 (10) (2011) 3303–3313.

[23] B. Sezer, G. Bilge, T. Sanal, H. Koksel, I.H. Boyaci, A novel method for ash analysis in wheat milling fractions by using laser-induced breakdown spectroscopy, J. Cereal Sci. 78 (2017) 33–38.

[24] L.C. Peruchi, L.C. Nunes, G.G.A. de Carvalho, M.B. B. Guerra, E. de Almeida, I.A. Rufini, D. Santos, F. J. Krug, Determination of inorganic nutrients in wheat flour by laser-induced breakdown spectroscopy and energy dispersive X-ray fluorescence spectrometry, Spectrochim. Acta Part B: At. Spectrosc. 100 (2014) 129–136, https://doi.org/10.1016/j.sab.2014.08.025.

[25] M. da Silva Gomes, G.G.A. de Carvalho, D.S. Junior, F. J. Krug, A novel strategy for preparing calibration

standards for the analysis of plant materials by laser-induced breakdown spectroscopy: a case study with pellets of sugar cane leaves, Spectrochim. Acta Part B 86 (2013) 137–141.
[26] S. Beldjilali, W. Yip, J. Hermann, T. Baba-Hamed, A. Belasri, Investigation of plasmas produced by laser ablation using single and double pulses for food analysis demonstrated by probing potato skins, Anal. Bioanal. Chem. 400 (7) (2011) 2173.
[27] P. Devangad, V. Unnikrishnan, R. Nayak, M. Tamboli, K.M. Shameem, C. Santhosh, G. Kumar, D.K. Sardar, Performance evaluation of laser induced breakdown spectroscopy (LIBS) for quantitative analysis of rare earth elements in phosphate glasses, Opt. Mater. 52 (2016) 32–37.
[28] J. Cáceres, J.T. López, H. Telle, A.G. Ureña, Quantitative analysis of trace metal ions in ice using laser-induced breakdown spectroscopy, Spectrochim. Acta B 56 (6) (2001) 831–838.
[29] D. Zhu, L. Wu, B. Wang, J. Chen, J. Lu, X. Ni, Determination of Ca and Mg in aqueous solution by laser-induced breakdown spectroscopy using absorbent paper substrates, Appl. Opt. 50 (29) (2011) 5695–5699.
[30] S. Moncayo, J. Rosales, R. Izquierdo-Hornillos, J. Anzano, J. Caceres, Classification of red wine based on its protected designation of origin (PDO) using laser-induced breakdown spectroscopy (LIBS), Talanta 158 (2016) 185–191.
[31] B. Sezer, S. Durna, G. Bilge, A. Berkkan, A. Yetisemiyen, I.H. Boyaci, Identification of milk fraud using laser-induced breakdown spectroscopy (LIBS), Int. Dairy J. 81 (2018) 1–7.
[32] T. Ohta, M. Ito, T. Kotani, T. Hattori, Emission enhancement of laser-induced breakdown spectroscopy by localized surface plasmon resonance for analyzing plant nutrients, Appl. Spectrosc. 63 (5) (2009) 555–558.
[33] J. Kaiser, K. Novotný, M.Z. Martin, A. Hrdlička, R. Malina, M. Hartl, V. Adam, R. Kizek, Trace elemental analysis by laser-induced breakdown spectroscopy—biological applications, Surf. Sci. Rep. 67 (11) (2012) 233–243, https://doi.org/10.1016/j.surfrep.2012.09.001.
[34] R. Agrawal, R. Kumar, S. Rai, A.K. Pathak, A.K. Rai, G. K. Rai, LIBS: a quality control tool for food supplements, Food Biophys. 6 (4) (2011) 527.
[35] Y. Dixit, M.P. Casado-Gavalda, R. Cama-Moncunill, M. Markiewicz-Keszycka, X. Cama-Moncunill, P. Cullen, C. Sullivan, Quantification of rubidium as a trace element in beef using laser induced breakdown spectroscopy, Meat Sci. 130 (2017) 47–49.
[36] G. Bilge, İ.H. Boyacı, K.E. Eseller, U. Tamer, S. Çakır, Analysis of bakery products by laser-induced breakdown spectroscopy, Food Chem. 181 (2015) 186–190.
[37] H.-H. Cho, Y.-J. Kim, Y.-S. Jo, K. Kitagawa, N. Arai, Y.-I. Lee, Application of laser-induced breakdown spectrometry for direct determination of trace elements in starch-based flours, J. Anal. At. Spectrom. 16 (6) (2001) 622–627.
[38] M. Markiewicz-Keszycka, M.P. Casado-Gavalda, X. Cama-Moncunill, R. Cama-Moncunill, Y. Dixit, P. J. Cullen, C. Sullivan, Laser-induced breakdown spectroscopy (LIBS) for rapid analysis of ash, potassium and magnesium in gluten free flours, Food Chem. 244 (2018) 324–330.
[39] A. dos Santos Augusto, P.L. Barsanelli, F.M.V. Pereira, E.R. Pereira-Filho, Calibration strategies for the direct determination of Ca, K, and Mg in commercial samples of powdered milk and solid dietary supplements using laser-induced breakdown spectroscopy (LIBS), Food Res. Int. 94 (2017) 72–78.
[40] M. Gondal, U. Baig, M. Dastageer, M. Sarwar, Determination of elemental composition of coffee using UV-pulsed laser induced breakdown spectroscopy, in: AIP Conference Proceedings, vol. 1742 (1), AIP Publishing, 2016, p. 030007.
[41] M. Gondal, Y. Habibullah, U. Baig, L. Oloore, Direct spectral analysis of tea samples using 266 nm UV pulsed laser-induced breakdown spectroscopy and cross validation of LIBS results with ICP-MS, Talanta 152 (2016) 341–352.
[42] S. Beldjilali, D. Borivent, L. Mercadier, E. Mothe, G. Clair, J. Hermann, Evaluation of minor element concentrations in potatoes using laser-induced breakdown spectroscopy, Spectrochim. Acta B At. Spectrosc. 65 (8) (2010) 727–733.
[43] P.K. Rai, N.K. Rai, A. Rai, G. Watal, Role of LIBS in elemental analysis of *Psidium guajava* responsible for glycemic potential, Instrum. Sci. Technol. 35 (5) (2007) 507–522.
[44] G. Kim, J. Kwak, J. Choi, K. Park, Detection of nutrient elements and contamination by pesticides in spinach and rice samples using laser-induced breakdown spectroscopy (LIBS), J. Agric. Food Chem. 60 (3) (2012) 718–724.
[45] S. Boussaidi, Characterization of laser-induced plasma of water extracted from orange juice, Anal. Lett. 45 (13) (2012) 1928–1935.
[46] P. Fichet, P. Mauchien, J.-F. Wagner, C. Moulin, Quantitative elemental determination in water and oil by laser induced breakdown spectroscopy, Anal. Chim. Acta 429 (2) (2001) 269–278.
[47] J. Bocková, Y. Tian, H. Yin, N. Delepine-Gilon, Y. Chen, P. Veis, J. Yu, Determination of metal elements in wine using laser-induced breakdown spectroscopy (LIBS), Appl. Spectrosc. 71 (8) (2017) 1750–1759.
[48] C. Nerín, M. Aznar, D. Carrizo, Food contamination during food process, Trends Food Sci. Technol. 48 (2016) 63–68.
[49] Q.I. Mohaidat, Laser-Induced Breakdown Spectroscopy (LIBS): An Innovative Tool for Studying Bacteria, 2011.

[50] R.A. Multari, D.A. Cremers, J.A.M. Dupre, J. E. Gustafson, Detection of biological contaminants on foods and food surfaces using laser-induced breakdown spectroscopy (LIBS), J. Agric. Food Technol. 61 (36) (2013) 8687–8694.

[51] C. Barnett, C. Bell, K. Vig, A. Akpovo, L. Johnson, S. Pillai, S. Singh, Development of a LIBS assay for the detection of *Salmonella enterica* serovar Typhimurium from food, Anal. Bioanal. Chem. 400 (10) (2011) 3323–3330.

[52] M. Baudelet, L. Guyon, J. Yu, J.-P. Wolf, T. Amodeo, E. Fréjafon, P. Laloi, Spectral signature of native CN bonds for bacterium detection and identification using femtosecond laser-induced breakdown spectroscopy, Appl. Phys. Lett. 88 (6) (2006) 063901.

[53] L. Huang, T. Chen, X. He, H. Yang, C. Wang, M. Liu, M. Yao, Determination of heavy metal chromium in pork by laser-induced breakdown spectroscopy, Appl. Opt. 56 (1) (2017) 24–28.

[54] X. Wan, J.Y. Wang, J.H. Ye, P. Wang, Z.M. Zhang, Analysis of distribution and contents of heavy metal pollution in fish body with laser-induced breaddown spectroscopy, Spectrosc. Spectr. Anal. 33 (1) (2013) 206–209.

[55] L. Ponce, T. Flores, M. Sosa-Saldaña, F. Alvira, G. M. Bilmes, Laser-induced breakdown spectroscopy determination of toxic metals in fresh fish, Appl. Opt. 55 (2) (2016) 254–258.

[56] Y. Lee, S.-W. Oh, S.-H. Han, Laser-induced breakdown spectroscopy (LIBS) of heavy metal ions at the sub-parts per million level in water, Appl. Spectrosc. 66 (12) (2012) 1385–1396.

[57] H. Huiqin, H. Lin, L. Muhua, C. Tianbing, Y. Ping, Y. Mingyin, Nondestructive determination of Cu residue in orange peel by laser induced breakdown spectroscopy, Plasma Sci. Technol. 17 (8) (2015) 711.

[58] M.P. Casado-Gavalda, Y. Dixit, D. Geulen, R. Cama-Moncunill, X. Cama-Moncunill, M. Markiewicz-Keszycka, P.J. Cullen, C. Sullivan, Quantification of copper content with laser induced breakdown spectroscopy as a potential indicator of offal adulteration in beef, Talanta 169 (2017) 123–129.

[59] Y. Dixit, M.P. Casado-Gavalda, R. Cama-Moncunill, X. Cama-Moncunill, M. Markiewicz-Keszycka, P. Cullen, C. Sullivan, Laser induced breakdown spectroscopy for quantification of sodium and potassium in minced beef: a potential technique for detecting beef kidney adulteration, Anal. Methods 9 (22) (2017) 3314–3322.

[60] H.M. Velioglu, B. Sezer, G. Bilge, S.E. Baytur, I. H. Boyaci, Identification of offal adulteration in beef by laser induced breakdown spectroscopy (LIBS), Meat Sci. 138 (2018) 28–33.

[61] B. Sezer, H. Apaydin, G. Bilge, I.H. Boyaci, *Coffee arabica* adulteration: detection of wheat, corn and chickpea, Food Chem. 264 (2018) 142–148.

[62] M. Tiwari, R. Agrawal, A.K. Pathak, A.K. Rai, G.K. Rai, Laser-induced breakdown spectroscopy: an approach to detect adulteration in turmeric, Spectrosc. Lett. 46 (3) (2013) 155–159.

[63] S.V. Er, H.E. Kocak, H. Yetim, I.H. Boyaci, Novel spectroscopic method for determination and quantification of saffron adulteration, Food Anal. Methods 10 (5) (2017) 1547–1555.

[64] Y. Cheng, Y. Dong, J. Wub, X. Yang, H. Bai, H. Zheng, D. Ren, Y. Zou, M. Lia, Screening melamine adulterant in milk powder with laser Raman spectrometry, J. Food Compos. Anal. 23 (2) (2010) 199–202.

[65] M.A. De La Fuente, M. Juárez, Authenticity assessment of dairy products, Crit. Rev. Food Sci. Nutr. 45 (7–8) (2005) 563–585.

[66] G. Bilge, B. Sezer, K.E. Eseller, H. Berberoglu, A. Topcu, I.H. Boyaci, Determination of whey adulteration in milk powder by using laser induced breakdown spectroscopy, Food Chem. 212 (2016) 183–188.

[67] J.O. Caceres, S. Moncayo, J.D. Rosales, F.J.M. de Villena, F.C. Alvira, G.M.B. Caceres, Application of laser-induced breakdown spectroscopy (LIBS) and neural networks to olive oils analysis, Appl. Spectrosc. 67 (9) (2013) 1064–1072.

[68] H.T. Temiz, B. Sezer, A. Berkkan, U. Tamer, I.H. Boyaci, Assessment of laser induced breakdown spectroscopy as a tool for analysis of butter adulteration, J. Food Compos. Anal. 67 (2018) 48–54.

[69] Y. Tian, C. Yan, T. Zhang, H. Tang, H. Li, J. Yu, J. Bernard, L. Chen, S. Martin, N. Delepine-Gilon, Classification of wines according to their production regions with the contained trace elements using laser-induced breakdown spectroscopy, Spectrochim. Acta B At. Spectrosc. 135 (2017) 91–101.

[70] P. Yang, R. Zhou, W. Zhang, S. Tang, Z. Hao, X. Li, Y. Lu, X. Zeng, Laser-induced breakdown spectroscopy assisted chemometric methods for rice geographic origin classification, Appl. Opt. 57 (28) (2018) 8297–8302.

[71] C.A. Akpovo, J.A. Martinez Jr., D.E. Lewis, J. Branch, A. Schroeder, M.D. Edington, L. Johnson, Regional discrimination of oysters using laser-induced breakdown spectroscopy, Anal. Methods 5 (16) (2013) 3956–3964.

[72] C. Eum, D. Jang, J. Kim, S. Choi, K. Cha, H. Chung, Improving the accuracy of spectroscopic identification of geographical origins of agricultural samples through cooperative combination of near-infrared and laser-induced breakdown spectroscopy, Spectrochim. Acta B At. Spectrosc. 149 (2018) 281–287.

[73] B. Sezer, G. Bilge, I.H. Boyaci, Laser-induced breakdown spectroscopy based protein assay for cereal samples, J. Agric. Food Chem. 64 (49) (2016) 9459–9463.

[74] Y.G.M. Kongbonga, H. Ghalila, M.B. Onana, Z. B. Lakhdar, Classification of vegetable oils based on their concentration of saturated fatty acids using laser

induced breakdown spectroscopy (LIBS), Food Chem. 147 (2014) 327–331.
[75] C.M. Tucker, Coffee Culture: Local Experiences, Global Connections, Routledge, New York, 2017.
[76] C. Zhang, T. Shen, F. Liu, Y. He, Identification of coffee varieties using laser-induced breakdown spectroscopy and chemometrics, Sensors 18 (1) (2018) 95.
[77] T.V. Silva, S.Z. Hubinger, J.A.G. Neto, D.M.B.P. Milori, E.J. Ferreira, E.C. Ferreira, Potential of laser induced breakdown spectroscopy for analyzing the quality of unroasted and ground coffee, Spectrochim. Acta B At. Spectrosc. 135 (2017) 29–33.
[78] E. Ferreira, E. Ferreira, A. Delbem, D. Milori, Ensemble of predictors and laser induced breakdown spectroscopy for certifying coffee, Electron. Lett. 47 (17) (2011) 967–969.
[79] K. Anggraeni, A. Nasution, H. Suyanto, Recognition of spectral identifier from green coffee beans of arabica and robusta varieties using laser-induced breakdown spectroscopy, in: Second International Seminar on Photonics, Optics, and Its Applications (ISPhOA 2016), vol. 10150, International Society for Optics and Photonics, 2016, p. 1015019.
[80] M.C. Lomer, R.P. Thompson, J. Commisso, C. L. Keen, J.J. Powell, Determination of titanium dioxide in foods using inductively coupled plasma optical emission spectrometry, Analyst 125 (12) (2000) 2339–2343.
[81] D.B. Warheit, E.M. Donner, Risk assessment strategies for nanoscale and fine-sized titanium dioxide particles: recognizing hazard and exposure issues, Food Chem. Toxicol. 85 (2015) 138–147.
[82] A. Weir, P. Westerhoff, L. Fabricius, K. Hristovski, N. Von Goetz, Titanium dioxide nanoparticles in food and personal care products, Environ. Sci. Technol. 46 (4) (2012) 2242–2250.
[83] B. Sezer, G. Bilge, A. Berkkan, U. Tamer, I.H. Boyaci, A rapid tool for determination of titanium dioxide content in white chickpea samples, Food Chem. 240 (2018) 84–89.
[84] M.A. Brescia, G. Di Martino, C. Fares, N. Di Fonzo, C. Platani, S. Ghelli, F. Reniero, A. Sacco, Characterization of Italian durum wheat semolina by means of chemical analytical and spectroscopic determinations, Cereal Chem. 79 (2) (2002) 238–242.
[85] M. Brescia, G. Di Martino, C. Guillou, F. Reniero, A. Sacco, F. Serra, Differentiation of the geographical origin of durum wheat semolina samples on the basis of isotopic composition, Rapid Commun. Mass Spectrom. 16 (24) (2002) 2286–2290.
[86] J.-P. Renou, G. Bielicki, C. Deponge, P. Gachon, D. Micol, P. Ritz, Characterization of animal products according to geographic origin and feeding diet using nuclear magnetic resonance and isotope ratio mass spectrometry. Part II: Beef meat, Food Chem. 86 (2) (2004) 251–256.
[87] S. Kelly, K. Heaton, J. Hoogewerff, Tracing the geographical origin of food: the application of multi-element and multi-isotope analysis, Trends Food Sci. Technol. 16 (12) (2005) 555–567.
[88] X. Cama-Moncunill, M. Markiewicz-Keszycka, Y. Dixit, R. Cama-Moncunill, M.P. Casado-Gavalda, P.J. Cullen, C. Sullivan, Feasibility of laser-induced breakdown spectroscopy (LIBS) as an at-line validation tool for calcium determination in infant formula, Food Control 78 (2017) 304–310, https://doi.org/10.1016/j.foodcont.2017.03.005.

CHAPTER

16

LIBS study of geological samples coupled with chemometric methods

Abhishek Kumar Rai[a], *Jayanta Kumar Pati*[a,b], *Awadhesh Kumar Rai*[c]

[a]Department of Earth and Planetary Sciences, Nehru Science Centre, University of Allahabad, Allahabad, India [b]National Center of Experimental Mineralogy and Petrology, University of Allahabad, Allahabad, India [c]Department of Physics, University of Allahabad, Allahabad, India

1 Introduction

Earth is composed of rocks, one of the important Earth materials, and the rocks are aggregates of mineral(s). Hence, their characterization is a fundamental prerequisite to understand the structure and composition of Earth [1–4]. Identification of minerals involves various physical and chemical methods. Minerals have diagnostic physical properties (hardness, streak, diaphaneity, luster, color, cleavage, habit, magnetism, etc.), possess a distinct crystal structure, and have a definite range of chemical compositions. Mineral characterization, in general, requires extensive knowledge and experience [5–7]. The mineralogy of Earth's crust and mantle is variable, laterally as well as with depth [8, 9]. There are several chemically similar and yet texturally distinct rocks in nature formed under different geological settings and generated by various processes. The elemental and isotopic geochemistry of minerals and rocks helps in their identification, nomenclature, classification, and precise information regarding their genesis in space and time [10]. The commonly employed analytical techniques for the elemental analysis of rocks and minerals include atomic absorption spectroscopy, atomic emission spectrometry, X-ray fluorescence analysis, instrumental neutron activation analysis, laser ablation inductively coupled plasma atomic emission spectroscopy, laser ablation inductively coupled plasma mass spectrometry, and electron probe microanalysis (EPMA); however, each of these techniques has inherent limitations [9, 11, 12]. These analytical techniques are routinely used to quantitatively analyze chemical constituents of Earth material following specific sample preparation methods that are, in general, lengthy, expensive, mostly destructive, and laboratory based. In earth and planetary sciences, there is a need for real-time, in situ, field-based characterization techniques [13] having the capability to detect heavy as well as light elements.

The laser-induced breakdown spectroscopy (LIBS) technique has emerged as a rugged and

Laser-Induced Breakdown Spectroscopy
https://doi.org/10.1016/B978-0-12-818829-3.00016-2

fast analytical tool for the analysis of any type of material, including geological samples [14–16]. It is a versatile technique to detect and analyze the chemical constituents of any kind of sample: solid, liquid, or gas [17–19]. Some of the distinctive features of LIBS are: no sample preparation, less sample consumption, and rapid analysis of samples with minimal sample destruction [20]. It has the capability to measure all elements in the periodic table, including the light elements present in rocks, water, soils, etc., which is not possible with any other technique [21, 22]. LIBS provides a point detection/analysis capability by ablating a tiny region (~1 nm) on a rock sample. Portable LIBS systems are being successfully used in geosciences to identify minerals in the field and have many potential mineralogical applications [23, 24]. LIBS can be used in standoff detection. It is the first spectroscopic technique to be used on the surface of a planet other than Earth in the ChemCam equipment on board the Curiosity rover to explore the Gale Crater on Mars [25, 26]. It is vastly superior to X-ray fluorescence [27, 28] and alpha proton X-ray spectroscopy [29] for the analysis of planetary rocks and can perform in-depth profile analysis by repeated laser ablation at a single point to understand the composition of the underlying bulk as well as the weathered layers of rock [30].

2 LIBS experimental setup

A laboratory LIBS setup consists of a high-power pulse laser to create plasma on the surface of the sample, and an optical system to focus the laser beam on the sample, to collect the emission from laser-induced plasma, and to feed it into the spectrometer equipped with a suitable detector (CCD or ICCD) to record the spectra. The target sample is placed on the translational/rotational stage so that every time the laser is focused at fresh spot to avoid the crater formation on the sample surface.

In the present study, the second harmonic of an Nd:YAG laser (Continuum Surelite III-10) at 532 nm is used, having a maximum pulse energy of 425 mJ, a pulse width of 4 ns (full width at half maximum), and a repetition rate of 1–10 Hz as an excitation source. To achieve sufficient irradiance to exceed the breakdown threshold, the laser beam is focused using a convex lens of focal length 15 cm, which produces laser-induced plasma at the surface of the sample (Fig. 1). The sample-to-lens distance has been optimized to minimize the interference of atmospheric hydrogen, carbon, nitrogen, and oxygen as the LIBS spectra are recorded in an open atmosphere. The collection optics and optical fiber (CC52 collimator, Andor Technology) feed emission from the plasma into the Mechelle spectrograph (ME5000, Andor Technology) equipped with an ICCD camera (iStar 334, Andor Technology) having a spectral resolving power of 6000. In the early stages, plasma emission is dominated by continuum radiation, which needs to be eliminated to record the atomic spectral lines that appear at a later stage of plasma formation. The gated ICCD can be opened after a desired interval of time (a few nanoseconds) after plasma formation, and the gate width can also be adjusted. The optimized gate delay and gate width for the best signal-to-noise ratio, in the present case, are 0.7 and 4 μs, respectively.

3 Analysis of geological samples

The LIBS technique has been widely used for analyzing geological samples; Harmon et al. renamed it GEOLIBS [31]. The analysis of four mineral samples (vanadinite, pyrite, garnet, and quartz) was carried out for the first time in 1996 by Vadillo and Laserna [32] by recording the atomic emission spectra with the help of a single laser shot at 532 nm under ambient conditions using a CCD detector. They could detect

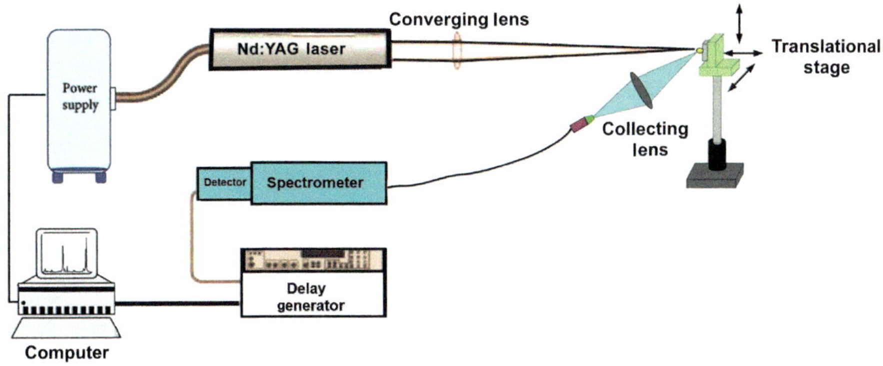

FIG. 1 Experimental setup of laser-induced breakdown spectroscopy.

several elements such as Fe, Mn, Mg, and Si, and concluded that LIBS could characterize and identify minerals. Subsequently, Haider et al. [33] and Abedin et al. [34] measured the elemental composition of beach sand samples from Bangladesh using the LIBS technique. Abedin et al. [34] showed that LIBS could detect rare earth elements in the sand samples. In 2007, McMillan et al. [35] applied the LIBS technique for the fast and precise identification of 52 minerals comprising 18 carbonates, 9 pyroxenes and pyroxenoids, 6 amphiboles, 8 phyllosilicates, and 11 feldspars. They prepared two databases, each containing the composite LIBS spectra having averages of 10 laser shots for each sample of the 52 minerals. Their results suggest that LIBS has the potential for in situ mineral identification to understand the process of formation. Gottfried et al. [36] used single- and double-pulse LIBS in close-contact benchtop and standoff (25 m) systems to examine a large suite of geological materials (natural carbonates, fluorite, and silicates). The configuration used in this experiment does not give any substantial result over the single-pulse LIBS system in the classification except in the case of the soil samples. Chemometric methods like principal components analysis (PCA) and partial least squares discriminant analysis (PLS-DA) were used to classify the geological samples [37].

4 Discrimination between rock types

The three main rock types (sedimentary, metamorphic, and igneous rocks) are routinely distinguished by geoscientists based on the characteristic mineral assemblage, diagnostic texture, and elemental chemistry using conventional physical and chemical analytical methods. LIBS coupled with PCA has been employed for in situ characterization of different rock types [38]. The LIBS spectra were recorded under ambient conditions in the wavelength range 200–900 nm and it is clear from Fig. 2 that the characteristic spectral lines of elements of the Earth's crust, such as Si, Ca, Mg, Fe, Na, Ti, and K along with lighter elements (C, H, N, and O), are present in the samples of different rocks. To analyze the spectra of these samples using PCA, a data matrix of 12×5849 was prepared using the LIBS spectral features of the sample. Fig. 3 shows that PC1 classifies these three samples into two groups: sedimentary and metamorphic rocks at the negative X axis and igneous rock samples at the positive X axis. Thus PC1 brings forth the similarity between sedimentary and metamorphic rocks and chemically isolates the igneous rock sample. The distinction between sedimentary and metamorphic rocks is established with the help of a two-dimensional PCA plot (Fig. 3A) where PC2

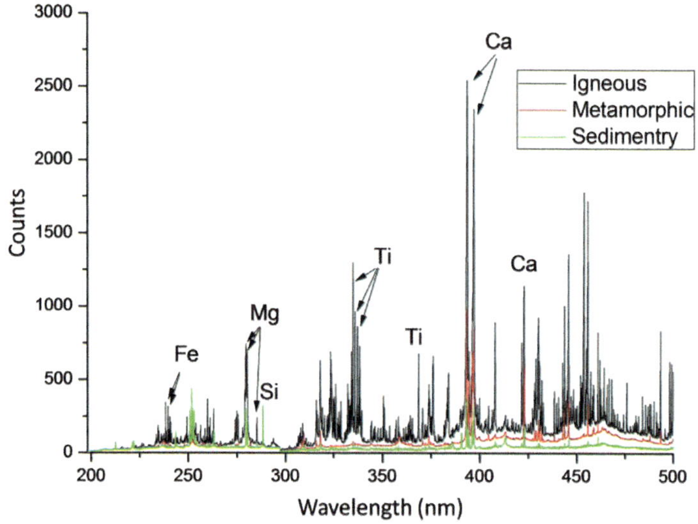

FIG. 2 Comparative laser-induced breakdown spectroscopy spectra of the igneous, metamorphic, and sedimentary rocks. *Reproduced from A. K. Rai, G.S. Maurya, R. Kumar, A.K. Pathak, J.K. Pati, A.K. Rai. Analysis and discrimination of sedimentary, metamorphic and igneous rocks using Laser Induced Breakdown Spectroscopy. J. Appl. Spectrosc. 83 (2017) 394–415, with the permission of Springer Nature.*

clearly differentiates between two distinct clusters: one at the positive *Y* axis and the other at the negative *Y* axis. The three-dimensional plot of the PCA in Fig. 3B is another way of representing PC1, PC2, and PC3 into three different groupings (present at three different planes) corresponding to igneous, metamorphic, and sedimentary rock samples. Fig. 4A and B shows the loading plots of PCA, which demonstrates the influence of spectral lines in the calculation of PC. It is clear from Fig. 4A that the elements Ca, Mg, Ti, Fe, and Sr have contributed to the calculation of PC1 as dominant components. Other spectral lines shown in Fig. 4B have contributed to the calculation of PC2.

The hardness of the rocks is compared by measuring the intensity ratio of the ionic-to-neutral atomic line, which is proportional to the compressive strength of the target sample [39]. The intensity ratio of the ionic line of Fe (238.2 nm) to the corresponding neutral atomic line of Fe (344.0 nm) is calculated for different rock samples. The calculated ratio is found to be 1.68 ± 0.07, 0.75 ± 0.12, and 1.23 ± 0.02 for igneous, metamorphic, and sedimentary rocks, respectively. These results reveal that the igneous rock is harder than the metamorphic and sedimentary rocks, and sedimentary rock is harder than the metamorphic rock. The spectral signatures for various elements (heavy and light) helped in distinguishing these three different rock samples and the change in diameter of laser-ablated spots suggested rheological variations, clearly supporting chemical discrimination [38].

Thus the present study demonstrates that LIBS coupled with PCA may be an important tool for rapid classification and in situ discrimination of rock samples in terrestrial as well as extraterrestrial environments.

5 LIBS study of moldavite, a possible Martian analog

Moldavite is a shocked silica-rich glass ejecta observed in parts of Central Europe associated with the Ries impact crater in Germany. Moldavites are translucent, exhibiting pale green to brown colors due to the presence of transition metal elements (Fe^{2+}, Fe^{3+}, and Mn^{2+} [40]), and display sculptured surface features. These impact ejecta were formed due to the collision of a kilometer-sized chondritic meteorite about

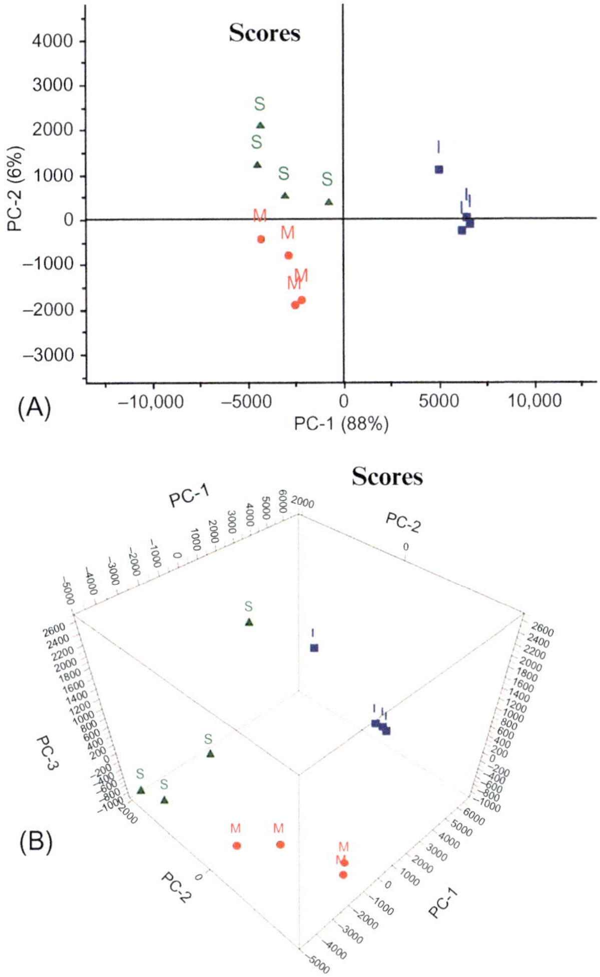

FIG. 3 (A) Two-dimensional score plot of three rock types. (B) Three-dimensional score plot of the first three principal components. *Reproduced from A. K. Rai, G.S. Maurya, R. Kumar, A.K. Pathak, J.K. Pati, A.K. Rai. Analysis and discrimination of sedimentary, metamorphic and igneous rocks using Laser Induced Breakdown Spectroscopy. J. Appl. Spectrosc. 83 (2017) 394–415, with the permission of Springer Nature.*

14.81 ± 0.2 Ma ago in parts of southern Germany forming twin impact structures (Ries and Steinheim). The physical and chemical properties of these glasses have been extensively studied to ascertain the chronology of impact event, nature of source material, and type of impactor (meteorite). It is interesting to note that the Bunte Breccia of the Ries crater constitutes both shocked and unshocked allochthonous clasts derived from the target rocks, and similar ejecta blankets are also observed on Mars [41]. The presence of well-preserved mafic impact glasses [42] and melts (melt-bearing ejecta deposit), also similar to the Ries crater, are spread over hundreds of square kilometers in an ~150-km diameter Bakhuysen crater on Mars [43–45]. Hence, it is possible that the silica-rich impact glass ejecta similar to moldavite associated with the twin

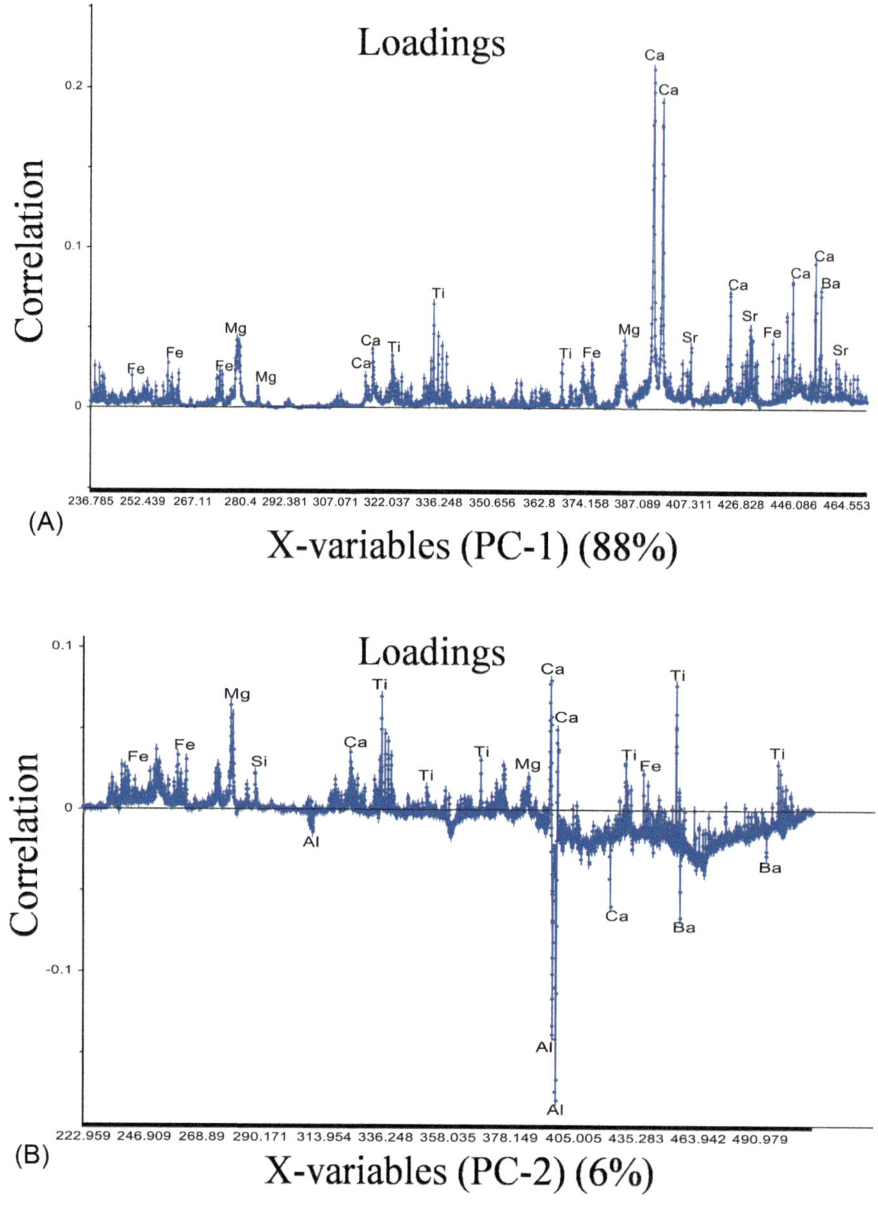

FIG. 4 Loadings plots of different rock samples. *Reproduced from A. K. Rai, G.S. Maurya, R. Kumar, A.K. Pathak, J.K. Pati, A.K. Rai. Analysis and discrimination of sedimentary, metamorphic and igneous rocks using Laser Induced Breakdown Spectroscopy. J. Appl. Spectrosc. 83 (2017) 394–415, with the permission of Springer Nature.*

craters of Ries and Steinheim could also be present on Mars. Here, the moldavite samples (Fig. 5) are analyzed for the first time using the LIBS technique to prepare a spectral dataset for comparison with elemental concentrations and possible identification of similar ejecta in the Martian environment using the Mars Science Laboratory database.

FIG. 5 (A) Moldavite sample (4204). (B) Moldavite sample (4205). (C) Microscope image of moldavite sample (4204). (D) Microscope image of moldavite sample (4205).

The LIBS spectra of two moldavite samples (4204 and 4205; Fig. 5) are generated (Fig. 6) using the experimental setup shown in Fig. 1. The main objective of this study is to compare the relative concentrations of the elements present in the two samples 4205 and 4205 by measuring the intensities of the spectral lines of the elements present (Fig. 6). Intensities of different spectral lines of the respective elements present in the LIBS spectra are calculated (Fig. 7) and it is found that the intensities of characteristic spectral lines of Na, Si, K, and Fe in moldavite (4205) are higher compared to those in moldavite (4204). The results obtained from the LIBS technique are in good agreement with the EPMA data.

One of the characteristic surficial features of moldavites is the presence of ridges and grooves [46] resulting from chemical heterogeneity. In such cases, LIBS provides an excellent in situ point detection capability to examine the chemical variability. This was established during the present study based on the change in the intensities of the spectral lines of the elements present in the LIBS spectra corresponding to ridge and groove portions of moldavite (4204). Similarly, the hardness of the ridges and grooves is compared using data derived from the intensity ratio of ionic-to-atomic spectral lines present in the spectra of the materials [39, 47, 48]. The ionic-to-atomic intensity ratios of the spectral lines of Mg II 280.2 nm/Mg I 285.2 nm, Si II 505.59/Si I 288.1 nm, and Fe II 238.2 nm/Fe I 344.0 nm in the LIBS spectra of the ridge and groove portions of moldavite (4204) are calculated and the corresponding values are 1.88, 1.79, and

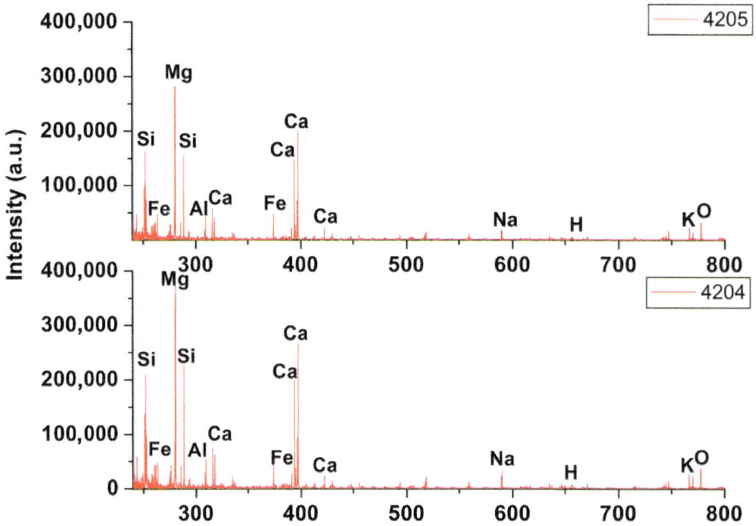

FIG. 6 Laser-induced breakdown spectroscopy spectra of moldavite (4204) and moldavite (4205). *Reproduced from A. K. Rai, J.K. Pati, R. Kumar. Spectro-chemical study of moldavites from Ries impact structure (Germany) using LIBS. Optics Laser Technol 117 (2019) 146–157, with the permission of Elsevier.*

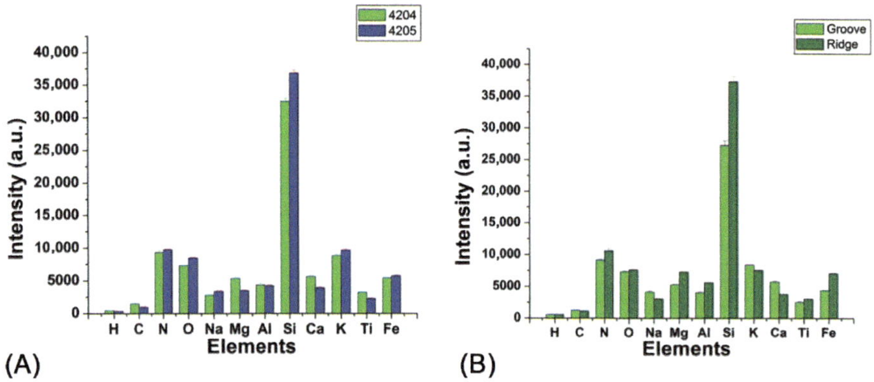

FIG. 7 Intensity of spectral lines of different elements present in the laser-induced breakdown spectroscopy spectra of (A) 4204 and 4205 and (B) the ridge and groove portion of 4204. *Reproduced from A. K. Rai, J.K. Pati, R. Kumar. Spectro-chemical study of moldavites from Ries impact structure (Germany) using LIBS. Optics and laser Technology, 117, 146–157, 2019, with the permission of Elsevier.*

1.98 (ridges), and 1.24, 1.11, and 1.23 (grooves), respectively. Since these ratios are higher for ridges as compared to grooves, the ridge portion is harder as well as more resistant compared to the groove part of moldavite. Also, the hardness of the material depends upon the bonding between the elements [49]. The intensities of the spectral lines of Al, Mg, Si, Fe, and Ti are also calculated and the results are shown in Fig. 7B where the spectral lines of Al, Mg, Si, Fe, and Ti are more intense in the ridge than in the groove portion and thus indicate that their concentrations are higher in the ridges in comparison to the grooves. However, the concentration of K is slightly elevated in the grooves as compared to the ridges since the intensity of the spectral line of K is about ~10% higher in the groove part as compared to the ridge part. Hence, it can be concluded that the concentrations of Al, Mg, Si, Fe, and Ti have an inverse relationship with K content in the sample. In addition, it is well known that the hardness of K is very low (0.4 on Mohr's scale of hardness [49]). The present results clearly exhibit the point detection capability of LIBS without any sample preparation based on the successful in situ analysis of a given sample. Based on relative elemental concentrations of these samples and by applying the multivariate analysis to the LIBS spectra, it is possible to discriminate between different moldavite samples having subtle chemical variability. It is clearly seen from Fig. 8A that two different clusters are formed corresponding to the LIBS spectra of moldavites (4205 and 4204) because of the subtle variation in the concentration of their constituents. Due to chemical homogeneity of the moldavite (4205) sample, the data points show a compact cluster (Fig. 8A). However, due to chemical heterogeneity corresponding to ridge and groove portions of the moldavite (4204) sample, the data in the plot show a scatter. From the loading plot of

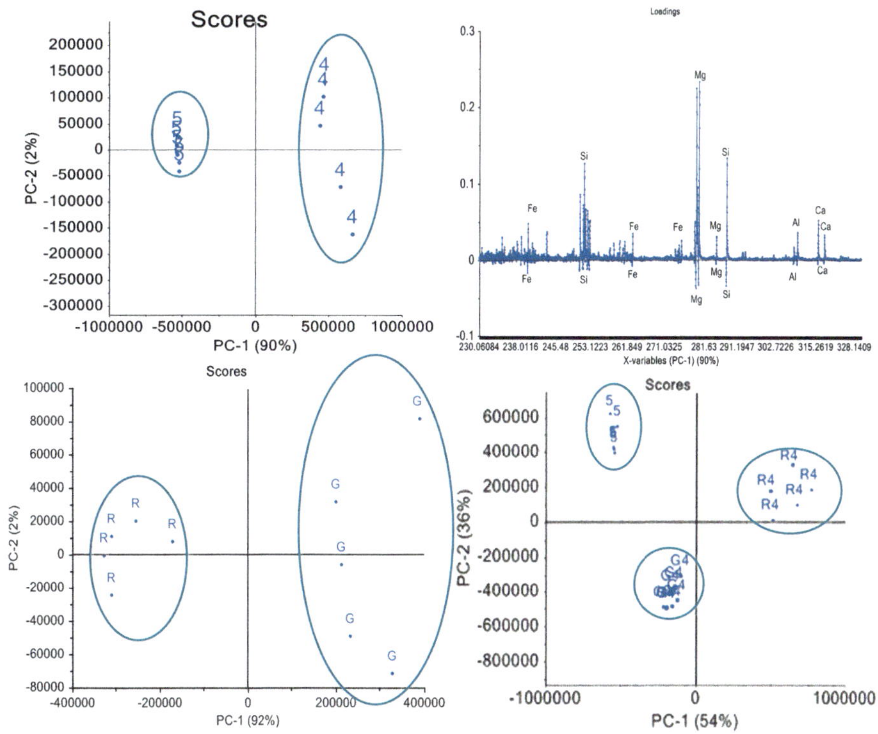

FIG. 8 (A) Two-dimensional score plot of 4204(4) and 4205(5). (B) Loading plot of principal components analysis carried out on the laser-induced breakdown spectroscopy spectral data obtained from 4204 and 4205 samples of moldavite. (C) Two-dimensional score plot of the ridge (R) and groove (G) portion of 4204. (D) Two-dimensional score plot of 4205(5), and ridge (R4) and groove (G4) portion of 4204. *Reproduced from A. K. Rai, J.K. Pati, R. Kumar. Spectro-chemical study of moldavites from Ries impact structure (Germany) using LIBS. Optics Laser Technol 117 (2019) 146–157, with the permission of Elsevier.*

PC1 (Fig. 8B) of the spectral data of moldavites (4205 and 4204), it is obvious that there is a positive correlation between the spectral lines corresponding to Si, Mg, Al, Ca, and Fe. However, the same elements exhibit negative correlation due to variable concentrations of these elements in both moldavites (4204 and 4205). Furthermore, to discriminate ridge and groove portions of moldavite (4204), the PCA to the LIBS data has been applied (Fig. 8C). The plot reveals a distinct compositional contrast within the sample having two different clusters corresponding to ridge and groove portions and concurring with earlier findings [46, 50]. The present study is also in agreement with previous reports [51, 52] pertaining to the development of a sculpted surface as a result of differential chemical erosion in response to internal chemical heterogeneity. The PCA shown in Fig. 8D clearly exhibits three distinct clusters corresponding to the sample 4205 and ridge and groove portions of 4204 suggesting overall compositional heterogeneity.

6 Topsoil analysis using LIBS as a regolith proxy

The soil found on Earth up to a depth of ~20 cm from the exposed surface is referred to as topsoil. This loose and unconsolidated Earth material is an important component to examine surficial contaminations in space and time and the role of operative *endo-* and exogenetic processes. On Earth, the soil horizon undergoes changes due to dynamic geological processes, whereas on other planetary surfaces (with an absence of atmosphere, unlike Earth), these sediments (regolith) bear the signatures of long-lost dynamic changes. The group at the University of Allahabad is using the LIBS technique to study the Ganga-Yamuna interfluve topsoil horizon as a proxy to planetary regolith horizon. The LIBS data has been evaluated to understand the distribution of heavy metals spatially and with varying depths. Representative soil samples from 49 discrete locations from different parts of the city of Allahabad, located on the confluence of the Ganga and Yamuna rivers, are collected. Three representative samples (321, 174, and 129) were chosen on the basis of mass magnetic susceptibility variations (64.84, 54.26, and $45.78 \times 10^{-8}\ m^3\ kg^{-1}$) at varying depths (2, 8, and 18 cm, respectively) as shown in Fig. 9. The LIBS spectra of the three representative soil samples (321, 174, and 129) are recorded in the spectral range from 300 to 800 nm using the experimental setup shown in Fig. 1. The identification of the spectral lines of elements present in the soil samples is done using NIST data [53] and is shown in Fig. 10A. The spectral lines suggest the presence of Li, Na, Mg, Al, Si,

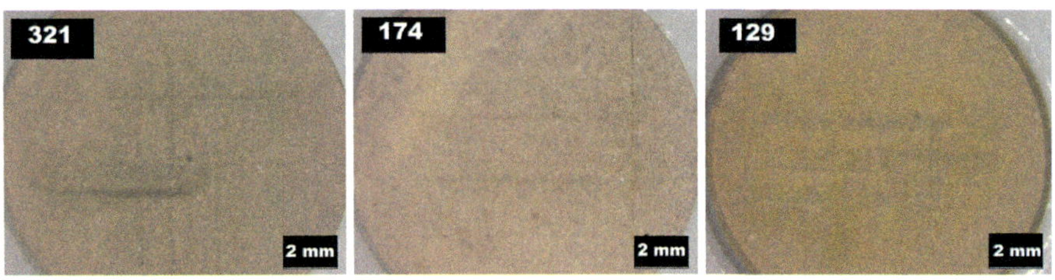

FIG. 9 Soil pellets of 321, 174, and 129, respectively.

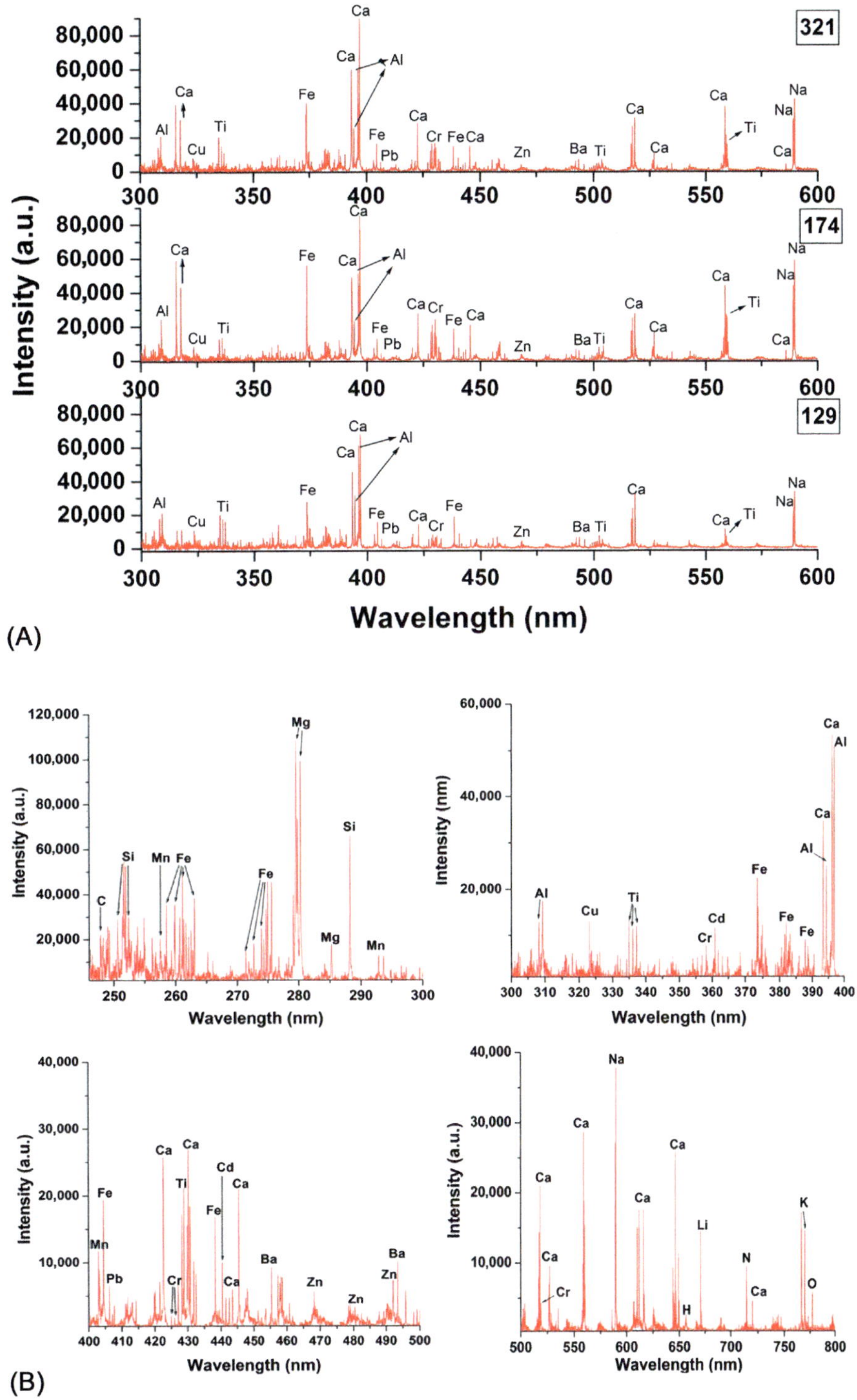

FIG. 10 (A) Comparative laser-induced breakdown spectroscopy (LIBS) spectra of soil samples 321, 174, and 129, respectively, for wavelength 300–600 nm. (B) Typical LIBS spectra of soil sample number 321 in wavelength range (1) 250–300 nm, (2) 300–400 nm, (3) 400–500 nm, and (4) 500–800 nm. *Reproduced from A. K. Rai, A. K. Singh, J. K. Pati, S. Gupta, M. Chakarvorty, A. Niyogi, A. Pandey, M. M. Dwivedi, K. Pandey, K. Prakash, Assessment of topsoil contamination in an urbanized interfluve region of Indo-Gangetic Plains (IGP) using magnetic measurements and spectroscopic techniques. Environ. Monitor. Assess. 2019 with the permission of Springer Nature.*

K, Ca, Ti, Cr, Mn, Fe, Cu, Zn, Cd, Ba, and Pb along with lighter elements (H, C, N, and O) in the analyzed soil samples, where N and O are also contributed by the atmospheric air. A chemometric method coupled with LIBS is more propitious for the qualitative, quantitative, and discriminative analysis of samples of very similar compositions [54]. The PCA of the LIBS data of the three soil samples is performed and it is clear from the three-dimensional score plots of Fig. 11 that three different clusters are formed corresponding to three soil samples, although their chemical compositions (Fig. 11) are almost the same but the concentrations of their constituents are different. PLSR has been used to determine the elemental concentrations present in these soil samples. To draw the calibration models for the different elements Al, Ca, and Mg, and for the toxic heavy metals Mn, Cr, Pb, and Zn, different wavelength ranges are chosen. Predicted versus reference plot of the PLSR model for Ca is shown in Fig. 12 where R^2 values are almost one, and RMSE values are close to zero, and the best fits of calibration (blue line) and validation (red line) deviate very little from each other. Thus the model shows a strong correlation between predicted and reference values. Similarly, the PLSR model of other elements when plotted shows a strong correlation between R^2 and RMSE. The predicted values of concentration of the elements Mg, Al, Ca, Cr, Mn, Zn, and Pb are shown in Table 1.

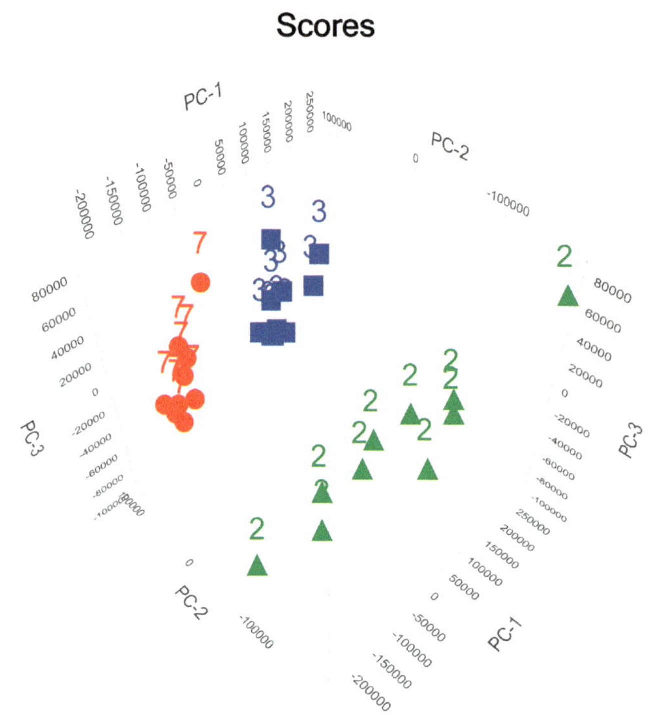

FIG. 11 Three-dimensional score plots of soil samples 3:321; 7:174; and 2:129. *Reproduced from A. K. Rai, A. K. Singh, J. K. Pati, S. Gupta, M. Chakarvorty, A. Niyogi, A. Pandey, M. M. Dwivedi, K. Pandey, K. Prakash, Assessment of topsoil contamination in an urbanized interfluve region of Indo-Gangetic Plains (IGP) using magnetic measurements and spectroscopic techniques. Environ. Monitor. Assess. 2019 with the permission of Springer Nature.*

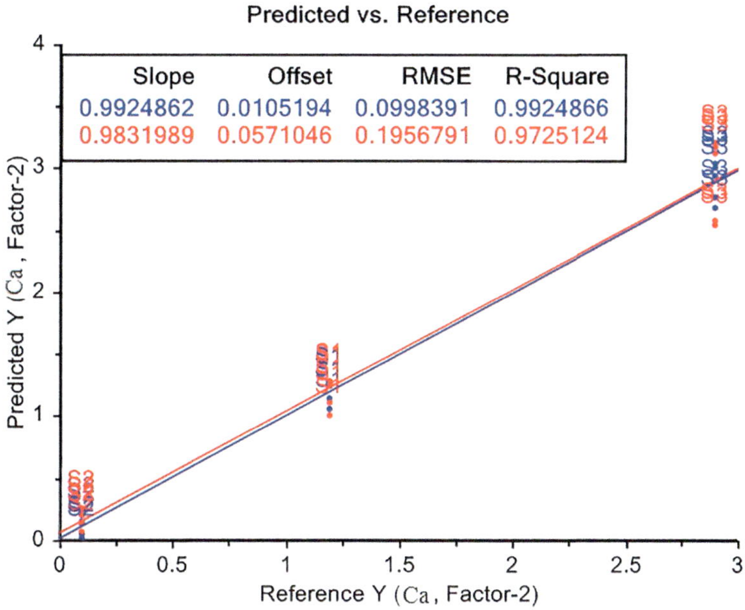

FIG. 12 Predicted versus reference plots for the concentration of Ca for soil certified reference materials. *Reproduced from A. K. Rai, A. K. Singh, J. K. Pati, S. Gupta, M. Chakarvorty, A. Niyogi, A. Pandey, M. M. Dwivedi, K. Pandey, K. Prakash, Assessment of topsoil contamination in an urbanized interfluve region of Indo-Gangetic Plains (IGP) using magnetic measurements and spectroscopic techniques. Environ. Monitor. Assess. 2019 with the permission of Springer Nature.*

TABLE 1 Concentration of samples in wt% from Ca to Mg and from Mn to Pb in ppm.

Element name	Sample 321	Sample 129	Sample 174
Ca	1.67 ± 0.07	1.21 ± 0.09	1.25 ± 0.02
Al	12.51 ± 0.64	14.15 ± 0.71	12.14 ± 0.35
Mg	0.86 ± 0.02	0.15 ± 0.09	0.56 ± 0.02
Mn	78.36 ± 0.18	38.00 ± 0.08	68.54 ± 0.21
Zn	122.58 ± 8.07	109.18 ± 6.09	114.09 ± 9.82
Cr	67.51 ± 4.78	34.15 ± 2.52	52.14 ± 3.15
Pb	45.86 ± 2.05	38.15 ± 1.09	47.56 ± 2.62

Reproduced from A. K. Rai, A. K. Singh, J. K. Pati, S. Gupta, M. Chakarvorty, A. Niyogi, A. Pandey, M. M. Dwivedi, K. Pandey, K. Prakash, Assessment of topsoil contamination in an urbanized interfluve region of Indo-Gangetic Plains (IGP) using magnetic measurements and spectroscopic techniques. Environ. Monitor. Assess. 2019 with the permission of Springer Nature.

7 Conclusions

Several studies have shown the unique capability of the LIBS technique to tackle difficulties in the compositional analysis of geological samples. In this chapter, the results of three investigations regarding earth and planetary sciences have been summarized. Discrimination of various rock types has been accomplished solely using LIBS data and respective ablated spot diameters. Both heavy and light elemental constituents were analyzed and rheological properties of respective samples were revealed. A large database of various rock types can be prepared under ambient and controlled conditions for rock identification in planetary environments. The detailed analyses of moldavite shows versatility of LIBS in establishing micro- and macrochemical heterogeneity in Earth materials, and its ability to decipher extraterrestrial components in impact ejecta samples. The recorded moldavite spectra are expected to aid in studies of planetary analogs, including the study of impactites littered on the Martian environment having felsic target rocks. Regolith analysis in planetary environments is a serious issue due to its unconsolidated nature and presence of fluid phases within and below the regolith horizon. The LIBS technique used in the characterization of select topsoil samples suggests that this spectroscopic method can establish the compositional heterogeneity of soil samples, detect heavy/toxic elements (Ti, Cr, Mn, Fe, Cu, Zn, Cd, and Pb) in them, and also detect the presence of various other elements (Li, Na, Mg, Al, Si, K, Ca, and Ba). The concentrations of toxic elements are very well predicted by the PLSR approach. These case studies strongly establish the versatility of the LIBS method and its unique capability to deal with fundamental earth and planetary science problems, including geoexploration and resource geology.

References

[1] C. Klein, B. Dutrow, Mineral Science, 23rd ed., John Wiley & Sons, New York, 2008.

[2] P.J. Potts, A Handbook of Silicate Rock Analysis, Blackie & Son Limited, Glasgow, 1987.

[3] W.D. Nesse, Introduction to Mineralogy, Oxford University Press, New York, 2000.

[4] W.A. Deer, R.A. Howie, J. Zussman, An Introduction to the Rock-Forming Minerals, second ed., Wiley, New York, 1992.

[5] H.R. Wenk, A. Bulakh, Minerals: Their Constitution and Origin, vol. 10, Cambridge University Press, 2004.

[6] F.C. Hawthorne, Reviews in Mineralogy 18, Mineralogical Society of America, Washington, DC, 1988.

[7] C. Klein, Minerals and Rocks: Exercise in Crystallography, Mineralogy, and Hand Specimen Petrology, Rev. ed., Wiley, New York, 1989.

[8] B.S. Kamber, Geochemical fingerprinting: 40 years of analytical development and real world applications, Appl. Geochem. 24 (6) (2009) 1074–1086, https://doi.org/10.1016/j.apgeochem.2009.02.012.

[9] B.J. Skinner, P.B. Barton, Genesis of mineral deposits, Annu. Rev. Earth Planet. Sci. 1 (1) (1973) 183–211, https://doi.org/10.1146/annurev.ea.01.050173.001151.

[10] J. Hoefs, Geochemical fingerprints: a critical appraisal, Eur. J. Mineral. 22 (1) (2010) 3–15.

[11] R. Tessadri, Introduction to the Mineralogical Sciences. Part 3: Analytical Techniques for Elemental Analysis of Minerals, UNESCO Encyclopedia of Life Support Systems EOLSS, 2003. on-line, publication, www.eolss.net.

[12] B.E. Naes, S. Umpierrez, S. Ryland, C. Barnett, J.R. Almirall, A comparison of laser ablation inductively coupled plasma mass spectrometry, micro X-ray fluorescence spectroscopy, and laser induced breakdown spectroscopy for the discrimination of automotive glass, Spectrochim. Acta B At. Spectrosc. 63 (10) (2008) 1145–1150, https://doi.org/10.1016/j.sab.2008.07.005.

[13] F. Smith, E. Arsenault, Microwave-assisted sample preparation in analytical chemistry, Talanta 43 (8) (1996) 1207–1268, https://doi.org/10.1016/0039-9140(96)01882-6.

[14] R.S. Harmon, F.C. DeLucia, C.E. McManus, N.J. McMillan, T.F. Jenkins, M.E. Walsh, et al., Laser-induced breakdown spectroscopy—an emerging chemical sensor technology for real-time field-portable, geochemical, mineralogical, and environmental applications, Appl. Geochem. 21 (5) (2006) 730–747, https://doi.org/10.1016/j.apgeochem.2006.02.003.

[15] R.S. Harmon, J. Remus, N.J. McMillan, C. McManus, L. Collins, J.L. Gottfried, et al., LIBS analysis of geomaterials: geochemical fingerprinting for the rapid

analysis and discrimination of minerals, Appl. Geochem. 24 (6) (2009) 1125–1141, https://doi.org/10.1016/j.apgeochem.2009.02.009.

[16] J. Rakovský, O. Musset, J. Buoncristiani, V. Bichet, F. Monna, P. Neige, et al., Testing a portable laser-induced breakdown spectroscopy system on geological samples, Spectrochim. Acta B At. Spectrosc. 74–75 (2012) 57–65, https://doi.org/10.1016/j.sab.2012.07.018.

[17] S. Qiao, Y. Ding, D. Tian, L. Yao, G.A. Yang, Review of laser-induced breakdown spectroscopy for analysis of geological materials, Appl. Spectrosc. Rev. 50 (1) (2014) 1–26, https://doi.org/10.1080/05704928.2014.911746.

[18] G.S. Senesi, Laser-induced breakdown spectroscopy (LIBS) applied to terrestrial and extraterrestrial analogue geomaterials with emphasis to minerals and rocks, Earth Sci. Rev. 139 (2014) 231–267, https://doi.org/10.1016/j.earscirev.2014.09.008.

[19] R. Gaudiuso, M. Dell'Aglio, O.D. Pascale, G.S. Senesi, A. D. Giacomo, Laser induced breakdown spectroscopy for elemental analysis in environmental, cultural heritage and space applications: a review of methods and results, Sensors 10 (8) (2010) 7434–7468, https://doi.org/10.3390/s100807434.

[20] F. Anabitarte, A. Cobo, J.M. Lopez-Higuera, Laser-induced breakdown spectroscopy: fundamentals, applications, and challenges, Int. Sch. Res. Not. Spectrosc. 2012 (2012) 1–12, https://doi.org/10.5402/2012/285240.

[21] M.A. Khater, Trace detection of light elements by laser-induced breakdown spectroscopy (LIBS): applications to non-conducting materials, Opt. Spectrosc. 115 (4) (2013) 574–590, https://doi.org/10.1134/S0030400X13100123.

[22] T. Hussain, M.A. Gondal, Laser induced breakdown spectroscopy (LIBS) as a rapid tool for material analysis, J. Phys. Conf. Ser. 439 (2013), https://doi.org/10.1088/1742-6596/439/1/012050.

[23] B. Lemiere, Y. Uvarova, Introduction: New developments in field portable geochemical techniques and site technologies and their place in mineral exploration, in: Exploration 17, Field Analysis Workshop, Oct 2017, Toronto, Canada, 2017.

[24] G.S. Senesi, Portable hand held laser-induced breakdown spectroscopy (LIBS) instrumentation for in-field elemental analysis of geological samples, Int. J. Earth Environ. Sci. 2 (2) (2017), https://doi.org/10.15344/2456-351x/2017/146.

[25] R.C. Wiens, S. Maurice, B. Barraclough, M. Saccoccio, W.C. Barkley, J.F. Bell, et al., The ChemCam instrument suite on the mars science laboratory (MSL) rover: body unit and combined system tests, Space Sci. Rev. 170 (1–4) (2012) 167–227, https://doi.org/10.1007/s11214-012-9902-4.

[26] J.P. Grotzinger, R.E. Arvidson, J.F. Bell, W. Calvin, B.C. Clark, D.A. Fike, et al., Stratigraphy and sedimentology of a dry to wet eolian depositional system, Burns formation, Meridiani Planum, Mars, Earth Planet. Sci. Lett. 240 (1) (2005) 11–72, https://doi.org/10.1016/j.epsl.2005.09.039.

[27] B.C. Clark, A.K. Baird, H.J. Rose, P. Toulmin, R.P. Christian, W.C. Kelliher, et al., The Viking X ray fluorescence experiment: analytical methods and early results, J. Geophys. Res. 82 (28) (1977) 4577–4594, https://doi.org/10.1029/JS082i028p04577.

[28] R. Rieder, T. Economou, H. Wänke, A. Turkevich, J. Crisp, J. Brückner, et al., The chemical composition of Martian soil and rocks returned by the mobile alpha proton X-ray spectrometer: preliminary results from the X-ray mode, Science 278 (5344) (1997) 1771–1774, https://doi.org/10.1126/science.278.5344.1771.

[29] B. Sallé, J.-L. Lacour, E. Vors, P. Fichet, S. Maurice, D.A. Cremers, et al., Laser-induced breakdown spectroscopy for Mars surface analysis: capabilities at stand-off distances and detection of chlorine and sulfur elements, Spectrochim. Acta B At. Spectrosc. 59 (9) (2004) 1413–1422, https://doi.org/10.1016/j.sab.2004.06.006.

[30] G.S. Senesi, G. Nicolodelli, D.M.B.P. Milori, O. De Pascale, Depth profile investigations of surface modifications of limestone artifacts by laser-induced breakdown spectroscopy, Environ. Earth Sci. 76 (16) (2017), https://doi.org/10.1007/s12665-017-6910-4.

[31] R.S. Harmon, R.E. Russo, R.R. Hark, Applications of laser-induced breakdown spectroscopy for geochemical and environmental analysis: a comprehensive review, Spectrochim. Acta B At. Spectrosc. 87 (2013) 11–26, https://doi.org/10.1016/j.sab.2013.05.017.

[32] J. Vadillo, J. Laserna, Laser-induced breakdown spectroscopy of silicate, vanadate and sulfide rocks, Talanta 43 (7) (1996) 1149–1154, https://doi.org/10.1016/0039-9140(96)01851-6.

[33] A.F.M.Y. Haider, M. Wahadoszamen, M.E. Sadat, K.M. Abedin, A.I. Talukder, Elemental profiling and determination of Ti content of the beach sand samples of Bangladesh using LIBS technique, Opt. Laser Technol. 42 (6) (2010) 969–974, https://doi.org/10.1016/j.optlastec.2010.01.017.

[34] K.M. Abedin, A.F.M.Y. Haider, M.A. Rony, Z.H. Khan, Identification of multiple rare earths and associated elements in raw monazite sands by laser-induced breakdown spectroscopy, Opt. Laser Technol. 43 (1) (2011) 45–49, https://doi.org/10.1016/j.optlastec. 2010.05.003.

[35] N.J. McMillan, R.S. Harmon, F.C. De Lucia, A.M. Miziolek, Laser-induced breakdown spectroscopy

analysis of minerals: carbonates and silicates, Spectrochim. Acta B At. Spectrosc. 62 (12) (2007) 1528–1536, https://doi.org/10.1016/j.sab.2007.10.037.

[36] J.L. Gottfried, R.S. Harmon, F.C. De Lucia, A.W. Miziolek, Multivariate analysis of laser-induced breakdown spectroscopy chemical signatures for geomaterial classification, Spectrochim. Acta B At. Spectrosc. 64 (10) (2009) 1009–1019, https://doi.org/10.1016/j.sab.2009.07.005.

[37] A.K. Rai, J.K. Pati, R. Kumar, Spectro-chemical study of moldavites from Ries impact structure (Germany) using LIBS, Opt. Laser Technol. 114 (2019) 146–157, https://doi.org/10.1016/j.optlastec.2019.01.028.

[38] A.K. Rai, G.S. Maurya, R. Kumar, A.K. Pathak, J.K. Pati, R. A.K. Analysis, Discrimination of sedimentary, metamorphic, and igneous rocks using laser-induced breakdown spectroscopy, J. Appl. Spectrosc. 83 (6) (2017) 1089–1095, https://doi.org/10.1007/s10812-017-0412-8.

[39] K. Tsuyuki, S. Miura, N. Idris, K.H. Kurniawan, T.J. Lie, K. Kagawa, Measurement of concrete strength using the emission intensity ratio between Ca(II) 396.8 nm and Ca(I) 422.6 nm in a Nd:YAG laser-induced plasma, Appl. Spectrosc. 60 (1) (2006) 61–64, https://doi.org/10.1366/000370206775382668.

[40] K. Žák, R. Skála, Z. Řanda, J. Mizera, A review of volatile compounds in tektites, and carbon content and isotopic composition of moldavite glass, Meteorit. Planet. Sci. 47 (6) (2012) 1010–1028, https://doi.org/10.1111/j.1945-5100.2012.01369.x.

[41] C.M. Caudill, G.R. Osinski, L.L. Tornabene, Ejecta deposits of Bakhuysen Crater, Mars, Icarus 314 (2018) 175–194, https://doi.org/10.1016/j.icarus.2018. 06.001.

[42] K.M. Cannon, J.F. Mustard, Preserved glass-rich impactites on Mars, Geology 43 (7) (2015) 635–638, https://doi.org/10.1130/G36953.1.

[43] J.-P. Bibring, Y. Langevin, J.F. Mustard, F. Poulet, R. Arvidson, A. Gendrin, et al., Global mineralogical and aqueous Mars history derived from OMEGA/Mars express data, Science 312 (5772) (2006) 400–404, https://doi.org/10.1126/science.1122659.

[44] C.M. Caudill, H.M. Sapers, G.R. Osinski, L. L. Tornabene, S. Chauvin, Mineralogical and geochemical study of ries ejecta deposits as a martian analogue for impact melt modification, in: 79th Annual Meeting of the Meteoritical Society, 1921, 2016.

[45] C.M. Caudill, R.N. Greenberger, H.M. Sapers, L.L. Tornabene, G.R. Osinski, B.L. Ehlmann, Mineralogical study and hyperspectral mapping of ries ejecta deposits as a martian analogue for impact melt alteration, in: Lunar and Planetary Science Confrence XLVIII, 2017.

[46] R. Rost, Sculpturing of moldavites and the problem of micromoldavites, J. Geophys. Res. 74 (27) (1969) 6816–6824, https://doi.org/10.1029/JB074i027 p06816.

[47] J.P. Singh, S.N. Thakur, Laser-Induced Breakdown Spectroscopy, Elsevier, 2007. https://doi.org/10.1016/B978-044451734-0.50015-6.

[48] M.M. ElFaham, A.M. Alnozahy, A. Ashmawy, Comparative study of LIBS and mechanically evaluated hardness of graphite/rubber composites, Mater. Chem. Phys. 207 (2018) 30–35, https://doi.org/10.1016/j.matchemphys.2017.12.036.

[49] G.V. Samsonov, M.E. Straumanis, Handbook of the physicochemical properties of the elements, Phys. Today 21 (9) (1968) 97, https://doi.org/10.1063/1.3035172.

[50] J. Konta, Tvarova Analyza Fragmentu Vltavínů, Acta Univ. Carol. Geol. 1–2 (1980) 45–67.

[51] K. Žák, R. Skála, Z. Řanda, J. Mizera, K. Heissig, L. Ackerman, et al., Chemistry of Tertiary sediments in the surroundings of the Ries impact structure and moldavite formation revisited, Geochim. Cosmochim. Acta 179 (2016) 287–311, https://doi.org/10.1016/j.gca.2016.01.025.

[52] I. Vetvicka, Inhomogeneous moldavites: the dependence of sculpture on chemical composition and glass flow structure, J. Non-Cryst. Solids 323 (1–3) (2003) 34–41, https://doi.org/10.1016/S0022-3093(03)00287-4.

[53] National Institute of Standards and Technology, National Institute of Standards and Technology Electronic database. http://physics.nist.gov/PhysRefData/ASD/linesform.html.

[54] K.-Q. Yu, Y.-R. Zhao, F. Liu, Y. He, Laser-induced breakdown spectroscopy coupled with multivariate chemometrics for variety discrimination of soil, Sci. Rep. 6 (1) (2016), https://doi.org/10.1038/srep27574.

Further reading

W.I. Rose, Introduction to optical mineralogy. William D. Nesse, J. Geol. 95 (2) (1987) 283, https://doi.org/10.1086/629128.

CHAPTER

17

Study of the different parts of a tokamak using laser-induced breakdown spectroscopy

Gulab Singh Maurya[a], *Pravin Kumar Tiwari*[b,c], *Rohit Kumar*[d], *Rajesh Kumar Singh*[c], *Awadhesh Kumar Rai*[b]

[a]Ultrafast Spectroscopy Lab, Department of Chemistry, IIT Kanpur, Kanpur, Uttar Pradesh, India [b]Laser Spectroscopy Research Lab, Department of Physics, University of Allahabad, Prayagraj, Uttar Pradesh, India [c]Institute for Plasma Research, Gandhinagar, Gujarat, India [d]Department of Physics, Chaudhary Mahadeo Prasad Degree College, Prayagraj, Uttar Pradesh, India

1 Introduction

The present energy requirement per person is much higher in comparison to the energy available from conventional sources and is increasing day by day [1]. In controlled nuclear reactions, a small amount of nuclear fuel is used and a huge amount of energy is released, which may fulfill our everyday requirements for energy. Nuclear energy may be obtained by two basic processes: nuclear fusion and nuclear fission, and both are cost effective. But nuclear fission has an inherent problem of producing long-lived radioactive waste and hence causes hazardous environments. Controlled nuclear fusion that occurs in a toroidal plasma machine called a tokamak is safe, and can be used as a major source of power supply without emission of greenhouse gases and without any long-lived radioactive waste [1].

Inside the tokamak, high-temperature plasma is generated and it revolves in the chamber by a strong magnetic field that causes erosion of the wall material. These eroded materials are redeposited as thin layers on different plasma-facing components (PFCs) of the tokamak such as optical window, mirror, flange, limiter, etc. [2]. Safety of PFCs of the tokamak is an important issue to maintain optimum plasma performance [3–8]. The deposited thin film decreases the performance of the optical components of the tokamak and hence it is essential to identify the deposited impurities so that we may protect and maintain the minimum quality of the optical components of the tokamak [3–6]. The *online* and in

Laser-Induced Breakdown Spectroscopy
https://doi.org/10.1016/B978-0-12-818829-3.00017-4

situ characterization of PFCs is of great importance for the operation of fusion devices with respect to safety [9]. Laser-induced breakdown spectroscopy (LIBS) is a suitable technique for *online* monitoring of the inner walls of the tokamak in current and next-generation fusion devices [6]. It can be used for the in situ analysis of deposited layers on the surface of fusion devices with micro-destructive characteristics [3–8]. To study the impurities deposited on PFCs of the tokamak, an experimental setup is required that can record the LIBS spectra without breaking the high vacuum of the tokamak chamber [10].

The LIBS technique is basically an atomic emission spectroscopic technique. A pulsed laser beam is focused on the sample of interest to produce a plasma plume on its surface. The excited ions and atoms of the plasma relax to their lower energy state and emit characteristic photons of the constituent elements of the sample [11–14]. The emitted radiation is collected and fed to a high-resolution spectrograph for identification of the sample constituents [15]. The merits of LIBS are its rapidity, ability to perform simultaneous multielemental analysis, including lighter elements, and requirement of very small amounts of sample (~1 μg) and for samples in any phase: solid, liquid, gas, aerosol, or molten state.

In this study, LIBS analyses of PFCs of the Aditya tokamak are presented. For analysis of different PFCs, a conventional LIBS setup in the laboratory [3–5] has been used, and for analysis of impurities deposited on the surface of the optical window a back-collection method [10] has been applied. After analysis of PFCs at the Department of Physics, Allahabad University, a prototype LIBS system has also been developed at the Institute of Plasma Research (IPR), Gandhinagar, India, to perform experiments at different atmospheric environments which can be utilized for *online* analysis of these PFCs. These investigations demonstrate the capability of our experimental setup to record the LIBS spectra of the contaminated surface of a window without dismantling the window or recycling the vacuum of the system.

2 Experimental

The experimental setup [3–5, 8] shown in Fig. 1 consists of a frequency-doubled Nd:YAG laser source (532 nm, Continuum Surelite III-10) capable of delivering a maximum energy of 425 mJ with a pulse duration of 4 ns (full width at half maximum [FWHM]) with a maximum pulse repetition rate of 10 Hz. A lens of

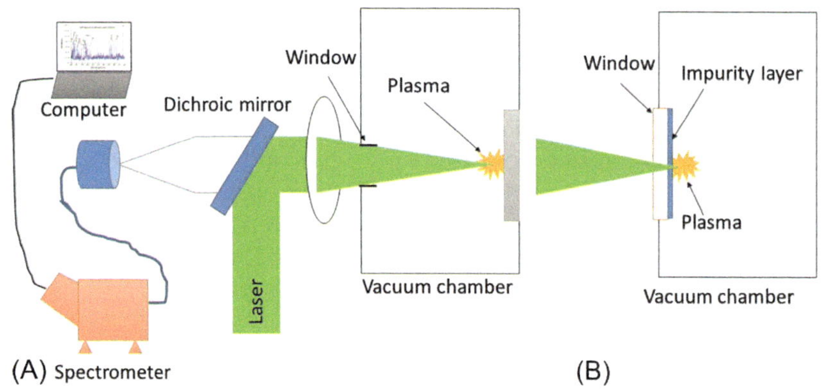

FIG. 1 Experimental setup of laser-induced breakdown spectroscopy for recording spectra of (A) samples present inside the vacuum chamber such as limiter and mirror using the conventional method and (B) an optical window installed in the tokamak using the back-collection system.

focal length 30 cm is used to focus the laser beam onto the surface of the optical window. The characteristic emission from the plasma plume is collected by a lens (fused silica) coupled to the optical fiber and fed into the spectrometer (ME5000, Andor Technology) equipped with an intensified charge-coupled device camera (iStar, Andor Technology). To get the best signal-to-background ratio and signal-to-noise ratio, the laser pulse energy is optimized at 15 mJ. The spectral resolution ($\lambda/\Delta\lambda$) of the spectrometer used in the present study is 6000.

A new experimental setup, termed as back-collection system, is shown in Fig. 1B with a detailed description in [8]. It has been designed for *online* recording of the LIBS spectra of the impurities deposited on the rear surface of the optical window of the Aditya tokamak at Institute for Plasma Research (IPR), Gandhinagar. In the back-collection method shown in Fig. 1B, the plasma is created at the back surface of the window. The radiation from the plasma plume travels the same path as the laser beam (but in the opposite direction), is transmitted through the optical window material, passes through a dichroic mirror (which has the property to transmit light of all wavelengths except 532 nm; for 532 nm at angle of 45 degrees, its reflectance is 99%), and finally is incident on the collection optics [10]. Within the collection optics, the radiation passes through a convex lens and the collimator assembly (CC52, Andor Technology), and is then guided to the spectrometer (ME5000, Andor Tecnology) through the optical fiber (as in Fig. 1A). The spectrometer is connected to an acquisition system (Andor Solis software) and a computer system, which give the LIBS spectra [10].

Finally, an experimental setup as shown in Fig. 2 has been developed at IPR, Gandhinagar, for the analysis of impurities deposited on PFCs of Aditya tokamak [16]. At IPR we do not have the facility of second-harmonic generation and a dichroic mirror for 1064 nm. Therefore we have to perform the experiment using Fig. 2. Our aim here is to record the LIBS spectra in a vacuum chamber without breaking its vacuum. The LIBS spectra of impurity deposited on the first wall of the Aditya tokamak (limiter and window) have been recorded at optimized gate delay and gate width in a vacuum chamber at different ambient pressures.

3 Results and analysis

To validate the satisfactory working of the experimental setup (Fig. 1A), i.e., to record the LIBS spectra of the PFCs of the Aditya tokamak through the optical window, an initial experiment was carried out by recording the LIBS spectra of a test sample (stainless-steel slab) through an optical window in the laboratory. Using the experimental setup shown in Fig. 1A, the LIBS spectra of stainless steel (SS304, as this material is available in the Department of Physics, University of Allahabad, India) was recorded by focusing the laser beam through the contaminated optical window to produce the plasma on its surface. This laboratory experiment is similar to the proposed LIBS experiment at the Aditya tokamak at atmospheric pressure. The characteristic emission from the plasma produced at the surface of the steel sample passes through the contaminated optical window in the opposite direction to that of the laser beam, and is fed to the spectrometer through the collection optics and optical fiber. The recorded LIBS spectra of stainless steel (SS304) is shown in Fig. 3, exhibiting the presence of the spectral signatures of Fe, Cr, Ni, Mn, and Cu, which are the constituents of the SS304 material, and this clearly demonstrates the working of the experimental setup shown in Fig. 1A.

After performing the experiment with the setup shown in Fig. 1A, we used another experimental setup with back-ablation geometry as shown in Fig. 1B. The initial experiment was performed using a thick glass plate (Fig. 4). The formation of the plasma plume at the back surface of the thick glass slab can be clearly seen in Fig. 4. For validation of experimental set up

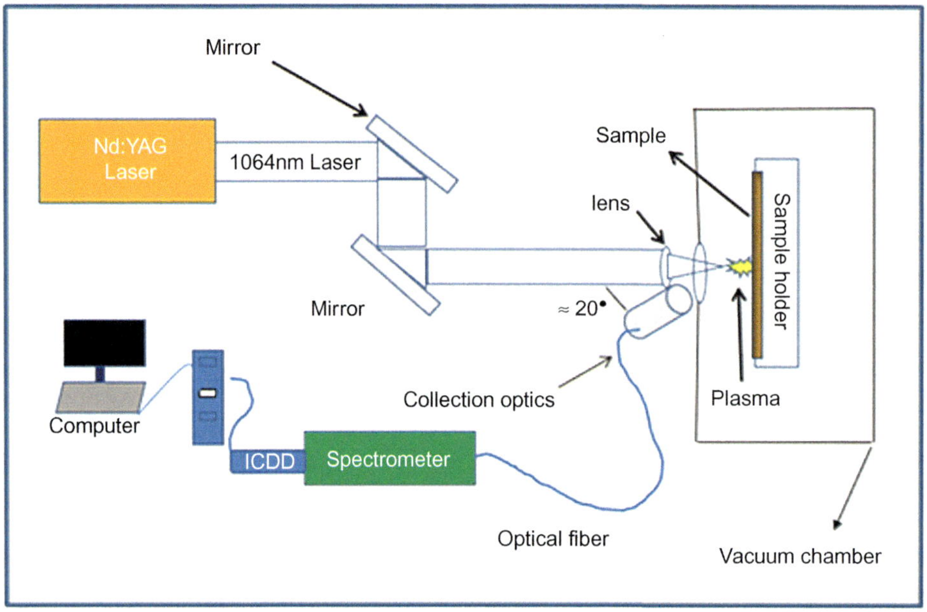

FIG. 2 Experimental setup for *online* analysis of the impurities deposited in the different parts of the Aditya tokamak.

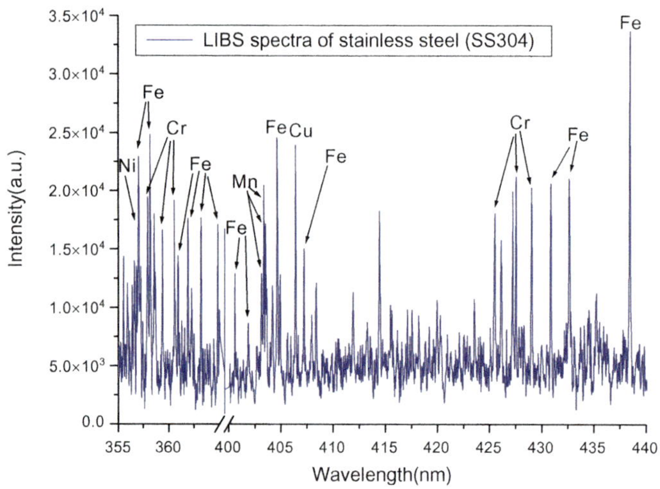

FIG. 3 Typical laser-induced breakdown spectroscopy spectra of a stainless-steel (SS304) sample recorded through an optical window. *Reproduced from G.S. Maurya, R. Kumar, A. Kumar, A.K. Rai, Rev. Sci. Instrum. 86 (2015) 123112, https://doi.org/10.1063/1.4938176, with the permission of AIP Publishing.*

shown in Fig. 1B, the laser beam is incident on the front surface of the quartz optical window of Aditya tokamak and after transmission through it, the laser beam is focused on the impurity layer deposited on its back surface. The emission from the plasma plume on the back surface of the widow is propagated in opposite direction of the laser beam.

The LIBS spectra of the impurity layer recorded in back-ablation geometry in the laboratory is shown in Fig. 5. The spectrum exhibits the spectral signature of various elements such

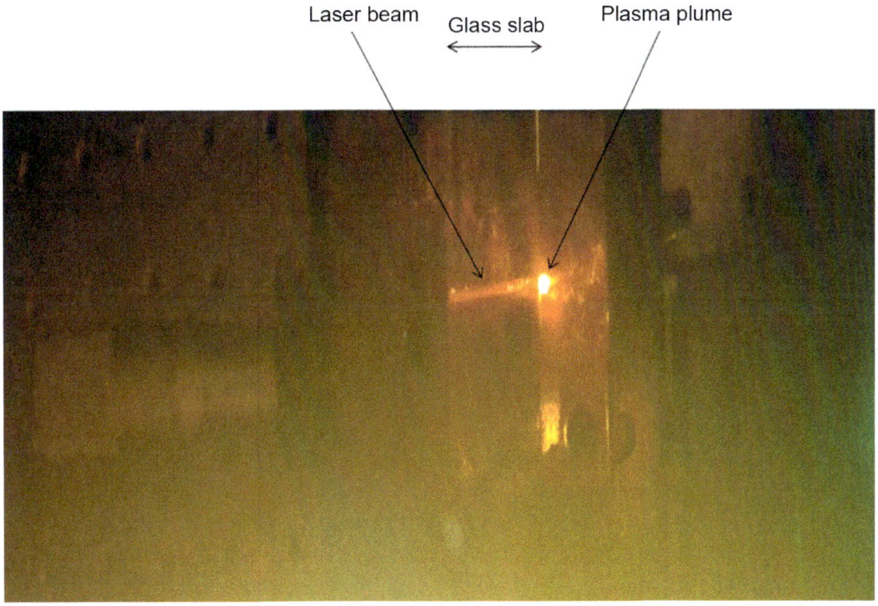

FIG. 4 The actual photograph of the plasma plume in the back-collection system.

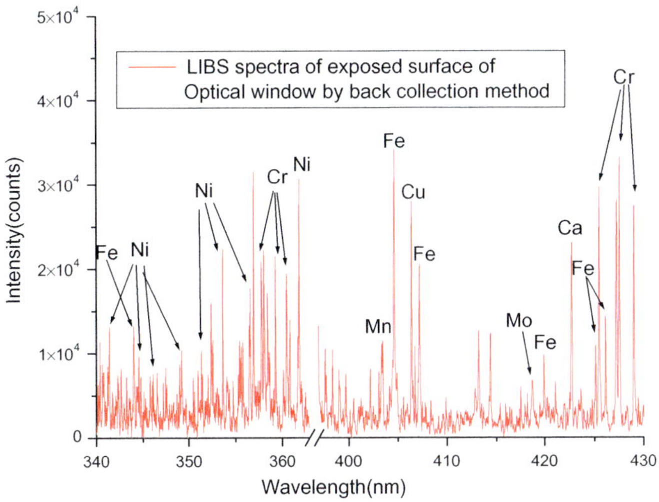

FIG. 5 Laser-induced breakdown spectroscopy spectra of the impurity layer deposited on the optical window recorded using the back-collection method. *Reproduced from G.S. Maurya, R. Kumar, A. Kumar, A.K. Rai, Rev. Sci. Instrum. 86 (2015) 123112, https://doi.org/10.1063/1.4938176, with the permission of AIP Publishing.*

as Fe, Ni, Cr, Mo, Mn, Cu, C, and Ca, similar to the LIBS spectra recorded using the conventional LIBS method [3–5, 8].

The LIBS spectra of different PFCs such as optical window, flange, and limiter are recorded using the conventional method and they show the presence of Fe, Ni, Cr, Mo, Mn, Cu, C, Ca, and Mg in their impurity layers [3–5, 8, 10, 17]. These elements are deposited on the inner surfaces (exposed parts) due to plasma-wall interaction. The high temperature inside the tokamak corrodes the inner exposed surfaces of the walls and the material starts circulating with the plasma. These impurity materials are deposited on various PFCs of the tokamak. Thus the elements found as impurities on the exposed part of the window are same as the constituents of the tokamak walls. Since this deposition is the result of multiple on/off cycles of the plasma, the thickness of the impurity may differ and hence the concentration of elements in different layers may differ.

To analyze the different deposited layers on the optical window, many single laser shot LIBS spectra at a particular spot/point on the surface of the optical window have been recorded. In the first laser shot, the spectral intensities of impurities are higher than the spectral intensities of successive laser shots (Fig. 6). It is observed that the intensity of the spectral line decreases rapidly with the increase in the number of laser shots [5, 8]. It was found that the deposited impurity had been totally removed in the ninth laser shot. In the ninth laser shot the appearance of new spectral lines belonging to elements such as Si, Al, and B can be seen from Fig. 6. The absence of these spectral lines in spectra from previous laser shots indicates that these elements are related to the matrix material of the optical window (boro-aluminum silicate) [5, 8].

To investigate the relative presence of Ca and Mg, shot-wise spectral analysis was performed and the variation of spectral intensity is shown in Fig. 7. It is found that the spectral intensity of Mg in the first laser shot is less than that of Ca, but from the second to the fifth shot the spectral intensity of Mg remains higher than that of Ca. The spectral intensities of the two elements become comparable and are extremely small for the sixth shot onward [5, 8].

Since the spectral intensity of Ca continuously decreases from the first shot onward, it can be assumed that there are two thin layers of coating: the first layer contains Ca and the second layer contains Mg. To ensure that the Mg and Ca are not generated from impurities,

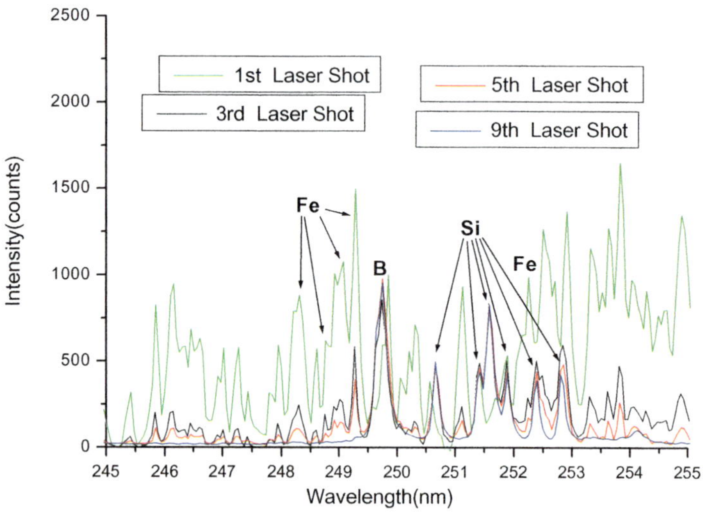

FIG. 6 Intensity variation of spectral lines with number of laser shot. *Reproduced with permission from G.S. Maurya, A. Jyotsana, R. Kumar, A. Kumar, A.K. Rai, Analysis of deposited impurity material on the surface of the optical window of the Tokamak using LIBS, Phys. Scr. 89 (2014) 075601 (12 p), https://doi.org/10.1088/0031-8949/89/7/075601, © IOP Publishing. All rights reserved.*

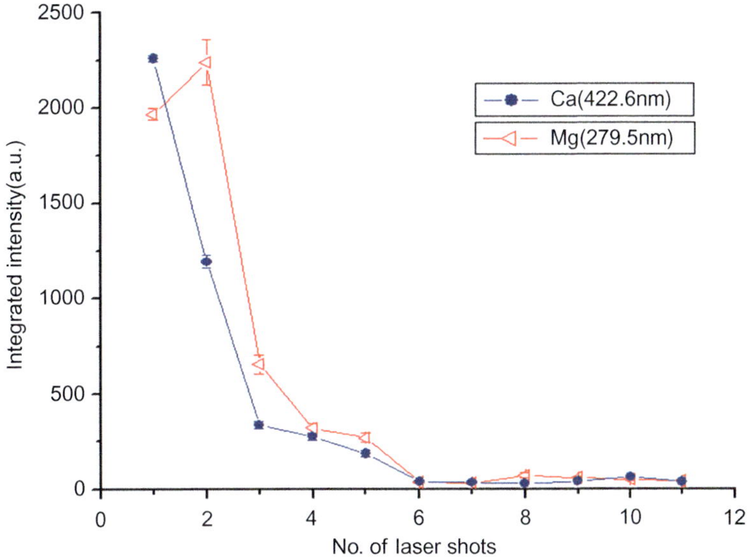

FIG. 7 Intensity variation of spectral lines of Mg (279.5 nm) and Ca (422.6 nm) with successive laser shots focused at the same point of the exposed surface of the optical window. *Reproduced with permission from G.S. Maurya, A. Jyotsana, R. Kumar, A. Kumar, A.K. Rai, Analysis of deposited impurity material on the surface of the optical window of the Tokamak using LIBS, Phys. Scr. 89 (2014) 075601 (12 p), https://doi.org/10.1088/0031-8949/89/7/075601,* *© IOP Publishing. All rights reserved.*

the LIBS spectra of the unexposed surface has also been recorded, and the spectral signatures of Ca, Mg, B, Al, and Si are observed [5, 8].

3.1 Quantitative analysis of the impurities deposited on PFCs

Quantification of constituents in the sample using the LIBS technique is basically performed by two methods: one is the calibration curve method and the other is the calibration-free method. Normally in the calibration curve method, a calibration curve is plotted between known concentrations of a particular element against its measured spectral intensities. The concentration of that element in the unknown sample can be estimated from its intensity on the calibration curve. One major issue with the calibration curve method is availability of matrix-matched reference material having known concentrations of different elements with the unknown sample. If the matrix of the unknown sample and reference/standard material is different, then an error may occur in the analysis. To overcome this issue, Cuicci et al. [18] proposed a new method known as calibration-free LIBS (CF-LIBS) in which there is no need for matrix-matched reference/standard materials. The method is based on the intensities of spectral lines of the constituents in the LIBS spectra of the samples.

In the present work for the quantification of impurities, the CF-LIBS method is preferred because it is not possible to have matrix-matched certified reference materials similar to PFCs. Spectral intensity of the element depends on its concentration as given in Eq. (1). In the

CF-LIBS method, the concentration of constituents is determined by measuring the intensity of spectral lines of elements present in the LIBS spectra of the sample, which is given by the Boltzmann equation:

$$I_\lambda^{ki} = C_s F A_{ki} g_k \frac{e^{-(E_k/k_B T)}}{U(T)} \quad (1)$$

where I_λ^{ki} is the intensity of a spectral line of wavelength λ, k is the Boltzmann constant, C_s is the concentration of species, $U(T)$ is the partition function, E_k is the upper energy level, A_{ki} is the spontaneous emission probability, and F is an experimental parameter.

It is clear from Eq. (1) that if the physical parameters are constant, then the concentration of species is directly proportional to the intensity of the spectral line. Thus if the physical parameters are known, then the concentration of elements may be calculated using Eq. (1). However, before measuring the intensity of the spectral line of the element present in the LIBS spectra of the sample, laser-induced plasma must satisfy certain conditions: (1) laser ablation must be stoichiometric in nature, (2) plasma must be optically thin, and (3) plasma must be in local thermodynamic equilibrium.

3.1.1 *Stoichiometric nature of laser ablation*

For concentration determination of constituent elements in a sample, laser ablation must be stoichiometric in nature, so that all the constituents of the samples should be ablated during plasma formation. Thus the plasma will be a true representative of the target sample in terms of its constituent compositions. If the laser irradiance at the focus spot is $>10^9$ W cm^{-2} to avoid vaporization, laser ablation of the sample occurs and the plasma is stoichiometric in nature.

The irradiance (power density) at the focused spot of the sample surface is calculated by measuring the diameter of the focused spot using Eq. (2) and the irradiance using Eq. (3):

$$D = 4\lambda f / \pi d \quad (2)$$

$$\text{Irradiance} = \frac{\text{Power (W)}}{\text{Area (cm}^2)} \quad (3)$$

where λ is the wavelength of the laser beam, f is the focal length of the lens, and d is the beam diameter.

For our experiment the calculated irradiance is 9.37×10^{11} W cm^{-2}, which is $>10^9$ W cm^{-2} and hence the ablation is stoichiometric in nature.

3.1.2 *Optical thinness of the plasma*

Laser-induced plasma must be thin so that reabsorption of the photon within the plasma does not take place. If the plasma is thick there is a chance of reabsorption of the photons emitted from the inner core. Optical thinness of the plasma can be checked by measuring the intensity ratios of two emission lines with similar upper energy levels and ratio of product of their transition probability, degeneracy, and inverse of wavelength, i.e.,

$$\frac{I}{I'} = \frac{A_{ki} g_k \lambda'}{A'_{ki} g'_k \lambda} \quad (4)$$

where I and I' are the intensities of spectral lines of the same species of the sample element having similar upper state and A_{ki}, g_k, λ and A'_{ki}, g'_k, λ' are corresponding transition probability, degeneracy of upper state, and wavelength of the spectral lines.

3.1.3 *Local thermodynamic equilibrium*

Plasma temperature in a local region must be the same and this is true when the McWhirter condition is fulfilled. For McWhirter criteria [19], electron density measured in laser-induced plasma must be greater than the lower limit given by Eq. (5):

$$N_e\,(\text{cm}^{-3}) > 1.6 \times 10^{12}\,[T\,(\text{K})]^{1/2}\,[\Delta E\,(\text{eV})]^3 \quad (5)$$

where N_e is the electron density, T is the temperature in kelvin, and ΔE is the maximum energy difference between two states of an atom.

To measure the electron density, FWHM of the spectral line of any element can be used, which is related to N_e as given in Eq. (6):

$$N_e = 10^{16}\,\Delta\lambda/2w \tag{6}$$

where w is the electron impact parameter [20] and $\Delta\lambda$ is the FWHM of the Stark-broadened line.

To calculate the lower limit of electron density (Eq. 5), plasma temperature is required. The plasma temperature can be estimated using the Boltzmann equation (Eq. 1). Rearranging Eq. (1) we get:

$$\ln\frac{I_\lambda^{ki}}{A_{ki}g_k} = -\frac{E_k}{kT} + \ln\frac{FC_s}{U(T)} \tag{7}$$

To calculate the experimental parameter F, we assume that the sum of concentrations of all species is one, i.e.,

$$\sum C_i = 1. \tag{8}$$

A graph between $\ln\frac{I_\lambda^{ki}}{A_{ki}g_k}$ and E_k is known as the Boltzmann plot, whose slope $(-1/kT)$ is related to the plasma temperature, and intercept $(\ln\frac{FC_s}{U(T)})$ is related to the species concentration C_s and the experimental parameter. Plasma temperature can be calculated by measuring the slope of the plot between the upper energy level and $\ln(\frac{I_\lambda^{ki}}{A_{ki}g_k})$. This has been plotted for different elements and we get a similar slope for each element, which shows that the plasma is in thermal equilibrium. A McWhirter criterion is a necessary but not sufficient condition for plasma to be in local thermodynamic equilibrium (LTE). Under sufficient conditions the ionization temperature should be equal (within 10% limit) to the excitation temperature. To calculate the ionization temperature, the Saha-Boltzmann equation (Eq. 9) for ionic/atomic emission intensity ratio can be used:

$$\ln\frac{I_{ij}^{II}A_{mn}^{I}g_m^{I}}{I_{mn}^{I}A_{ij}^{II}g_i^{II}} = -\frac{\left(E_{ion}-dE+E_i^{II}-E_m^{I}\right)}{kT} + \ln\frac{2(2\pi m_e kT)^{3/2}}{N_e h^3} \tag{9}$$

where m_e is the electron mass, h is Planck's constant, E_{ion} is the first ionization potential of the element, dE is the lowering correction parameter, which is the correction term in the first ionization potential arising due to high pressure in the plasma plume, and E_i^{II} and E_m^{I} are the upper energy levels of ionic and atomic species of the elements having transition probabilities A_{ij}^{II} and A_{mn}^{I} and statistical weights g_i^{II} and g_m^{I}. If we plot $E_{ion} - \mathrm{d}E + E_i^{II} - E_m^{I}$ and $\ln\frac{I_{ij}^{II}A_{mn}^{I}g_m^{I}}{I_{mn}^{I}A_{ij}^{II}g_i^{II}}$, then ionization temperature can be calculated from its slope. The ionization temperature thus calculated by the Saha-Boltzmann plot and excitation temperature calculated from the Boltzmann plot are very close to each other and hence a sufficient condition for LTE is also satisfied.

After verifying all the required criteria for the laser-induced plasma, the concentration of the impurities deposited on different PFCs is determined using Eq. (7) and the Boltzmann plot. Table 1 represents the concentration of Fe, Cr, Ni, Mo, and Mn on the exposed surface (impurity-deposited surface) of the optical window measured by the CF-LIBS method and the corresponding concentration present in stainless steel (SS316L). It is clear from Table 1 that the concentration of impurities determined by the present method are very similar to the composition of the constituents of the material (SS316L) of the wall of the tokamak [5, 8]. Thus it clearly demonstrates that the impurity depositions on different PFCs are due to erosion of the material of the wall by the very hot plasma in the tokamak and deposition of these eroded materials on different PFCs during plasma wall and plasma PFC interactions.

3.2 Multivariate analysis of the spectral signature observed in LIBS

LIBS spectra contain a large set of data in the form of wavelength versus intensity, because in the present case 5851 pixels collect emissions from the plasma to create a spectrum. To extract information from this huge dataset without losing any valuable information is a challenging task.

TABLE 1 Comparison of concentration of impurity elements to that present in SS316L material.

Concentration of impurities determined in the present work		Concentration of elements present in SS316L (%)	
Element	**%**	**Min**	**Max**
Fe	63.76	61	69
Cr	13.83	16	18
Ni	12.30	10	14
Mo	3.45	2	3
Mn	1.72	2	2

Reproduced with permission from G.S. Maurya, A. Jyotsana, R. Kumar, A. Kumar, A.K. Rai, Analysis of deposited impurity material on the surface of the optical window of the Tokamak using LIBS, Phys. Scr. 89 (2014) 075601 (12p), https://doi.org/10.1088/0031-8949/89/7/075601, © IOP Publishing. All rights reserved.

Multivariate analysis is a method of analyzing the large dataset for classification as well as quantification. For classification, principal component analysis (PCA) is used, and for quantification, partial least square regression (PLSR), principal component regression (PCR), etc. are used. PCA is a very useful tool, and is a linear technique for projecting multidimensional data to lower-dimensional subspace with minimal loss of information. For the classification of different types of patterns in the deposited impurities on the PFCs, PCA is used. It is seen from Fig. 8 that the exposed and unexposed surfaces of the optical window are clearly differentiated by PC1 because they are forming two different clusters [5]. Furthermore, it is clear from Fig. 8 that LIBS data obtained from the first laser shot on the unexposed surface of the window typically contains information of anti-reflection coating elements (Ca and Mg), which are uniformly coated/distributed and hence form one closely packed cluster (U1). In the case of LIBS spectra obtained from the first laser shot on the exposed part of the optical window, the ablated material typically contains information of the impurity layer elements (Fe, Cr, Ni, Mo, Mn, Cu, C, Ca, and Mg) and is relatively less uniformly deposited/distributed, resulting in another cluster but one that is not closely packed (E1). We have also tried to differentiate the LIBS spectra of different laser shots of the exposed part. For this we have recorded the single laser shot spectra at different spots of the exposed surface of the optical window. In contrast to the unexposed surface, the LIBS data obtained from different laser shots on the exposed surface of the optical window are scattered because different amounts of materials are ablated in each laser shot, and hence the LIBS spectra of different laser shots have different spectral line intensities of impurity elements. It is clear from Fig. 9 that the LIBS data obtained from the first and second laser shots have formed two different clusters (S1 and S2) because the material compositions ablated in these two laser shots are different. The LIBS data obtained after six laser shots are clustered together (S6–S11) because the impurity layer is almost removed after the sixth laser shot and the material ablated is the matrix of the window, which is uniform and gives a similar spectral line intensity of matrix elements. It is also clear from Fig. 9 that the clusters of the LIBS data of the third (S3), fourth (S4), and fifth (S5) laser shot are very close to the clusters of the sixth, seventh, etc. because a large amount of impurity layer is removed after the first and second laser shots. The results of the present study clearly reveal that the impurity on the PFCs is very thin and it is almost removed after the sixth laser shot.

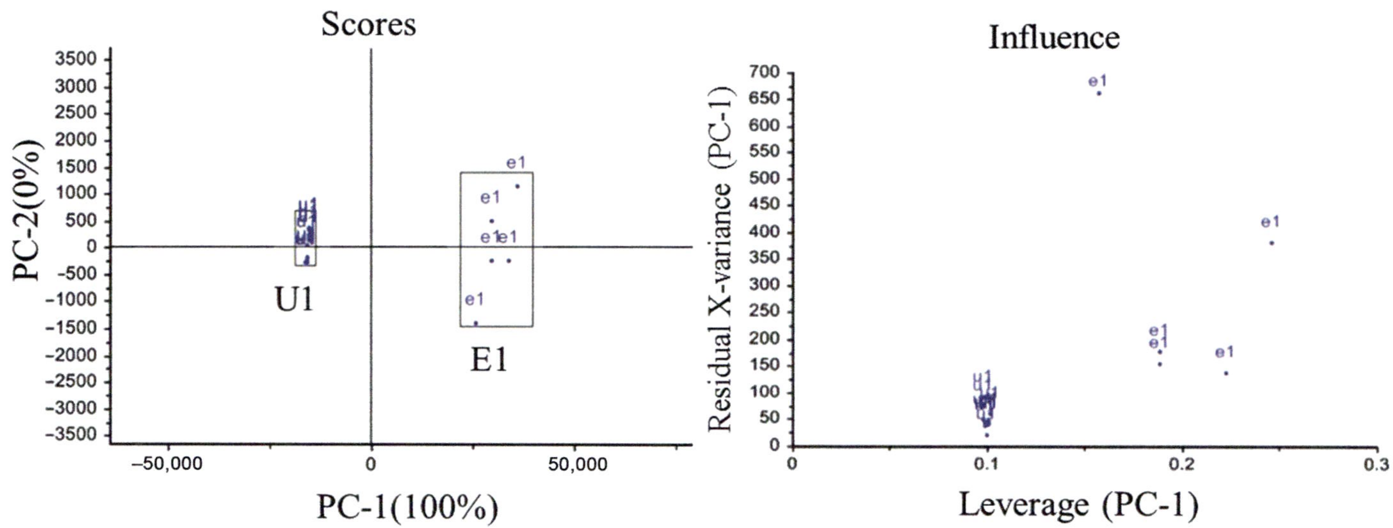

FIG. 8 Principal component analysis plot for exposed and unexposed surfaces of the plasma-facing component. *Reproduced with permission from G.S. Maurya, A. Jyotsana, R. Kumar, A. Kumar, A.K. Rai, Analysis of deposited impurity material on the surface of the optical window of the Tokamak using LIBS, Phys. Scr. 89 (2014) 075601 (12 p), https://doi.org/10.1088/0031-8949/89/7/075601,*

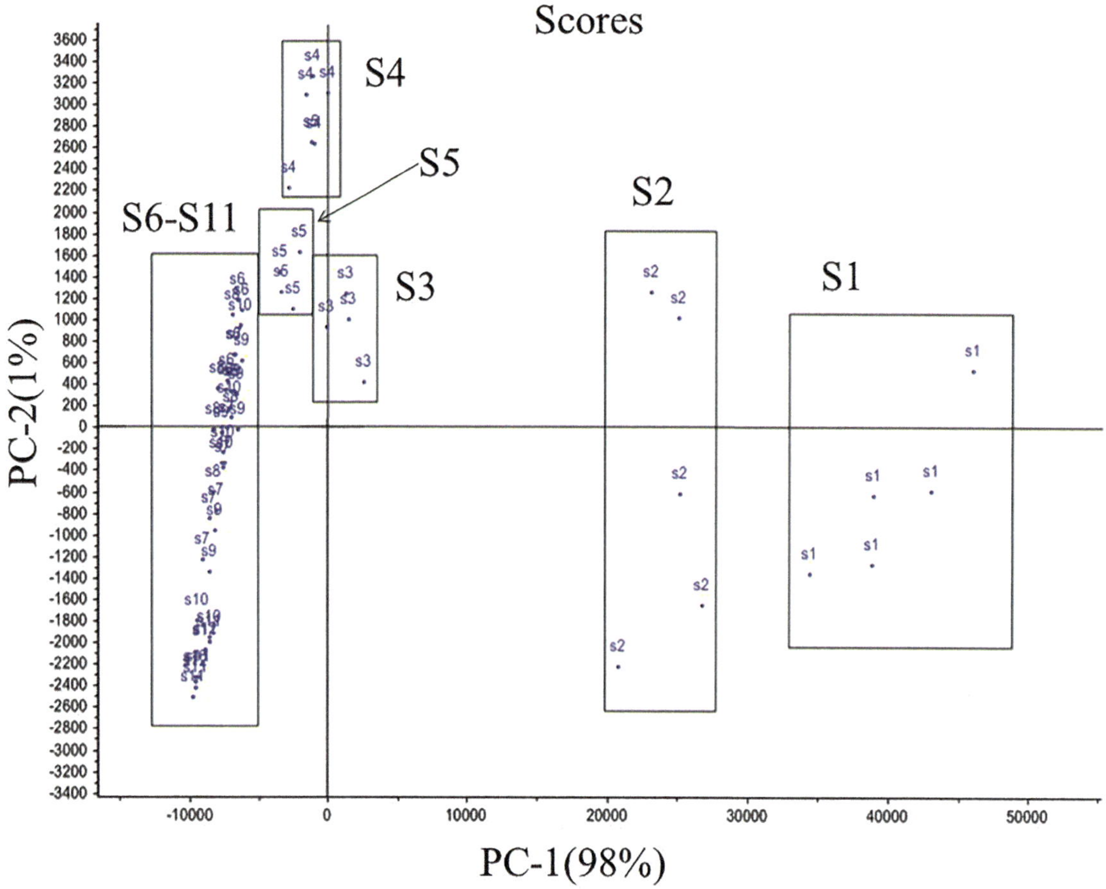

FIG. 9 Principal component analysis plot for shot-wise variation of the exposed part of the optical window. *Reproduced with permission from G.S. Maurya, A. Jyotsana, R. Kumar, A. Kumar, A.K. Rai, Analysis of deposited impurity material on the surface of the optical window of the Tokamak using LIBS, Phys. Scr. 89 (2014) 075601 (12 p), https://doi.org/10.1088/0031-8949/89/7/075601, © IOP Publishing. All rights reserved.*

3.3 Experimental setup and analysis at IPR

3.3.1 *Optimization of experimental setup*

After analyzing the LIBS spectra of different PFCs using the experimental setup shown in Fig. 1A and 1B in a laboratory at atmospheric pressure, our aim is to demonstrate the ability of the LIBS technique for online analysis of the impurities deposited on different PFCs of the Aditya tokamak. Therefore we have developed the experimental setup shown in Fig. 2 at IPR, Gandhinagar, and have recorded the LIBS spectra of the optical window and limiter in a vacuum chamber. To record the LIBS spectra, different experimental parameters such as ambient pressure, gate delay, gate width, laser energy, lens-to-sample distance, collection angle, etc. have been optimized. For the *online* LIBS study, the optimization of LIBS spectra has been performed using an aluminum (Al) sample at 45 mJ of laser energy. Using this

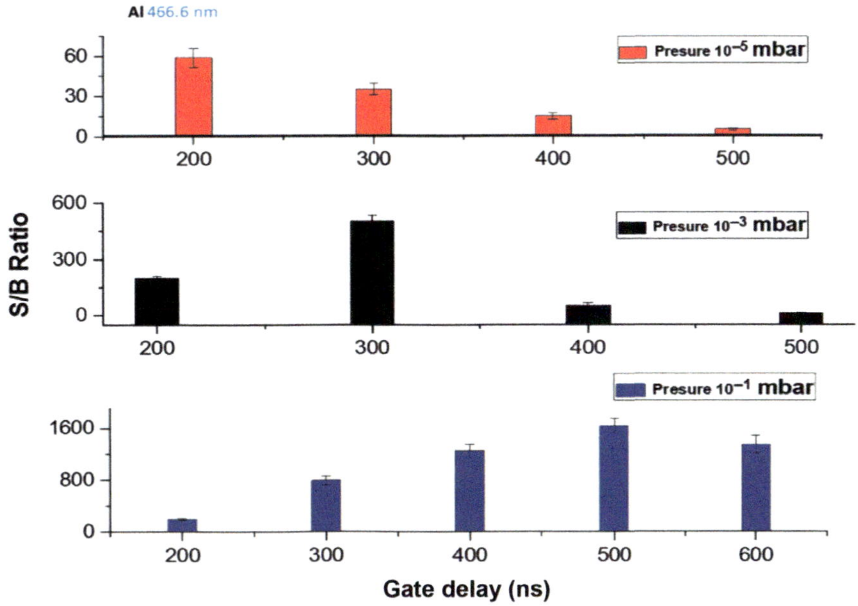

FIG. 10 Variation of signal-to-background ratio for Al (466.6 nm) with gate delay (ns) at difference pressures in a vacuum chamber.

sample the signal-to-background ratios have been optimized at different gate delays and atmospheric pressures, which are shown in Fig. 10. Fig. 10 indicates that for 45 mJ per pulse laser energy, best signal-to-background ratios are observed at gate delay of 200, 300, and 500 ns at atmospheric pressure of 10^{-5}, 10^{-3}, and 10^{-1}, mbar respectively.

After optimizing all experimental parameters, analysis of impurities deposited on the first wall of the Aditya tokamak, e.g., limiter and window, has been performed and different compositions of major and minor elements have been observed [16].

3.3.2 Spatial analysis of the limiter surface in a vacuum

For the identification of impurity-deposited elements and composition of the limiter, LIBS spectra of the limiter have been recorded and analyzed in a vacuum (10^{-5} mbar pressure) at a 200-ns gate delay. We observed impurity elements Fe, Cr, Ni, Mo, Mn, Cu, C, Ca, and Mg at the surface of the tokamak limiter. Variations of the spectral intensity of impurity elements with successive locations on contaminated limiter surfaces have been studied and it was observed that the impurity deposition of the elements is not homogeneous on the limiter surface (Fig. 11). The relative variation of Fe I (414.3 nm)/Fe I (438.3 nm) is shown in Fig. 11.

Thus we recorded the LIBS spectra at various pressures without breaking the vacuum and measured the intensity of the spectral line of different elements deposited on the surface of the PFCs of the Aditya tokamak. The present results show that the LIBS technique can be an efficient tool for the in situ and *real-time* analysis of impurities deposited on the first wall of the Aditya tokamak.

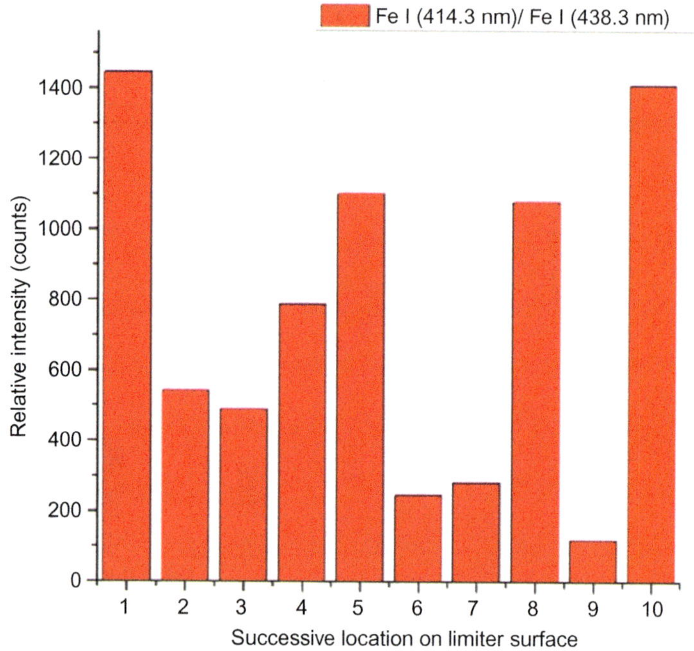

FIG. 11 Relative intensity variation of Iron (Fe I (414.3 nm)/Fe I (438.3 nm)) at different locations on the limiter surface.

4 Conclusion

In the present work, we developed an experimental setup to record *online* LIBS spectra of PFCs of the Aditya tokamak. To optimize the new setup, initial experiments were performed in the laboratory using the optical window of the tokamak. The present results demonstrate the level of performance of the developed system and its use in impurity analysis of PFCs such as mirror, flange, limiter, diverter, or gaseous impurity material within the vacuum vessel. This study also revealed that a thin layer of impurities such as Fe, Cr, Ni, Mo, and Mn along with C is deposited on the surface of the optical window. Ca and Mg are the constituents of the antireflection coating and Al, Si, and B are the matrix elements of the optical window. The impurity layer is very thin because the spectral intensity of impurity elements goes to zero after five or six successive laser shots. Furthermore, based on a practical situation, a back-collection method has been developed for *online* analysis of the rear contaminated surface of the optical window without breaking the vacuum of the system. It has been observed that CF-LIBS is highly suitable for quantitative analysis of PFCs like heavy samples. Also, for the classification of a large dataset, multivariates such as PCA will be helpful in the LIBS-based analysis of PFCs of the tokamak.

Acknowledgments

The Board of Research in Fusion Science and Technology and the Board of Research in Nuclear Sciences, IPR, Gandhinagar, Gujarat, India, are gratefully acknowledged for financial assistance. Dr. G. S. Maurya is grateful to IIT Kanpur for providing financial assistance as an institute postdoctoral fellow (PDF-98). Dr. P. K. Tiwari is also thankful to the Board of Research in Nuclear Sciences, India, for providing senior research fellowship support for the work for project Sanction No. 39/14/30/2016-BRNS/34434, 24.01.17. We are thankful to retired Prof. Ajai Kumar, IPR, for his valuable suggestions and support.

References

[1] Culham Centre for Fusion Energy. http://www.ccfe.ac.uk/Why_fusion.aspx.

[2] S.H. Hong, W.C. Kim, Developing in-situ ellipsometry for tokamak discharges in KSTAR, J. Phys. Conf. Ser. 227 (2010) 1–5. 012047.

[3] G.S. Maurya, A. Jyotsana, R. Kumar, A. Kumar, A. K. Rai, In situ analysis of impurities deposited on the Tokamak flange using laser induced breakdown spectroscopy, J. Nucl. Mater. 444 (2014) 23–29.

[4] G.S. Maurya, A. Jyotsana, A.K. Pathak, A. Kumar, A. K. Rai, Spatial analysis of impurities on the surface of flange and optical window of the tokamak using laser induced breakdown spectroscopy, Opt. Lasers Eng. 56 (2014) 13–18.

[5] G.S. Maurya, A. Jyotsana, R. Kumar, A. Kumar, A. K. Rai, Analysis of deposited impurity material on the surface of optical window of the tokamak using LIBS, Phys. Scr. 89 (2014) 075601.

[6] Q. Xiao, R. Hai, H. Ding, A. Huber, V. Philipps, N. Gierse, G. Sergienko, In-situ analysis of the first wall by laser-induced breakdown spectroscopy in the TEXTOR tokamak: dependence on the magnetic field strength, J. Nucl. Mater. 463 (2015) 911–914.

[7] S. Almaviva, L. Caneve, F. Colao, R. Fantoni, G. Maddaluno, Remote-LIBS characterization of ITER-like plasma facing materials, J. Nucl. Mater. 421 (2012) 73–79.

[8] G.S. Maurya, Study of Impurities Deposited on the Surface of Different Parts of the Aditya Tokamak Using Laser Induced Breakdown Spectroscopy (Ph.D. Thesis), Department of Physics, University of Allahabad, Prayagraj, 2015.

[9] E. Mukhin, K. Vukolov, V. Semenov, S. Tolstyakov, M. Kochergin, G. Kurskiev, K. Podushnikova, A. Razdobarin, A. Gorodetsky, R. Zalavutdinov, V. Bukhovets, A. Zakharov, S. Bulovich, V. Veiko, E. Shakshno, Progress in the development of deposition prevention and cleaning techniques of in-vessel optics in ITER, Nucl. Fusion 49 (2009) 085032.

[10] G.S. Maurya, R. Kumar, A. Kumar, A.K. Rai, Proof-of-concept experiment for on-line LIBS analysis of impurity layer deposited on optical window and other plasma facing components of Aditya Tokamak, Rev. Sci. Instrum. 86 (2015) 123112.

[11] M.C. Madamba, W.M. Mullett, S. Debnath, E. Kwong, Characterization of tablet film coatings using a laser-induced breakdown spectroscopic technique, AAPS PharmSciTech. 8 (4) (2007) 184–190.

[12] A.K. Pathak, R. Kumar, V.K. Singh, R. Agrawal, S. Rai, A.K. Rai, Assessment of LIBS for spectrochemical analysis: a review, Appl. Spectrosc. Rev. 47 (2012) 14–40.

[13] A.K. Pathak, V.K. Singh, N.K. Rai, A.K. Rai, P.K. Rai, P. K. Rai, S. Rai, G.D. Baruah, Study of different concentric rings inside gallstones with LIBS, Lasers Med. Sci. 26 (4) (2011) 531–537.

[14] S. Pandhija, A.K. Rai, Laser-induced breakdown spectroscopy: a versatile tool for monitoring traces in materials, Pramana J. Phys. 70 (3) (2008) 553–563.

[15] C. Grisolia, A. Semerok, J.M. Weulersse, F. Le Guern, S. Fomichev, F. Brygo, P. Fichet, P.Y. Thro, P. Coad, N. Bekris, M. Stamp, S. Rosanvallon, G. Piazza, In-situ tokamak laser applications for detritiation and co-deposited layers studies, J. Nucl. Mater. 363–365 (2007) 1138–1147.

[16] Final progress report of the project entitled "On-line analysis of impurities coated on optical windows and first wall of Aditya Tokamak using LIBS" funded by BRNS, Sanction Number: 39/14/30/2016-BRNS.

[17] GS. Maurya, R. Kumar, A. Kumar, A.K. Rai, Analysis of impurities on the contaminated surface of the tokamak limiter using laser induced breakdown spectroscopy, Spectrochim. Acta B At. Spectrosc. 126 (2016) 17–22.

[18] A. Ciucci, M. Corsi, V. Palleschi, V. Rastelli, A. Salvetti, E. Tognoni, A new procedure for quantitative elemental analyses by laser-induced plasma spectroscopy, Appl. Spectrosc. 53 (8) (1999) 960–964.

[19] J.P. Singh, S.N. Thakur (Eds.), Laser Induced Breakdown Spectroscopy, Elsevier, Amsterdam, 2007, pp. 23–48.

[20] H.R. Griem, Plasma Spectroscopy, McGraw-Hill, New York, 1964.

CHAPTER

18

High-energy materials application

Manoj Kumar Gundawar

Advanced Centre of Research in High Energy Materials (ACRHEM), University of Hyderabad, Gachibowli, Hyderabad, Telangana, India

1 Introduction

A high-energy material (HEM) is a substance that can undergo a rapid chemical reaction when stimulated. The stimulus can be in the form of a spark, friction, flame, impact, shock, heat, or light energy. The reaction is very rapid and can result in very high pressures and the production of lighter gaseous products. While the production of gaseous products is of vital importance for space applications where thrust is needed, the high pressures are useful for military applications. HEMs are also known by another name—explosives. Jai Prakash Agrawal said "...In order to camouflage research on explosives, propellants and pyrotechnics, a new term 'high energy materials (HEMs)' was coined by the explosives community for them. Thus all explosives, propellants and pyrotechnics can be referred to as high energy materials or energetic materials (EMs). In other words, the other name of HEMs/EMs is explosives, propellants and pyrotechnics depending on their formulations and intended use. Nowadays, the term HEMs/EMs is generally used for any material that can attain a highly energetic state mostly by chemical reactions" [1, 2].

These materials are divided into two groups on the basis of expansion after explosion: low explosives and high explosives. Low explosives include black powder, smokeless gunpowder, and low vulnerability ammunition. High explosives are further divided into three subgroups of primary, secondary, and tertiary explosives. Primary explosives include compounds like lead azide and hexamethylene triperoxide diamine, and secondary explosives comprise triacetone triperoxide, royal demolition explosive, C4, high melting explosive, dynamite, and trinitrotoluene (TNT). Tertiary explosives include ammonium nitrate and aluminum powder and ammonium nitrate fuel oil. Given their extensive use in civilian applications, HEMs are also categorized based on their use as military and civilian explosives.

Driven by homeland security and antiterrorism applications, explosive detection has gained momentum as a subject of research at the international level. There are two broad and important applications: trace and bulk detection. Trace detection involves the sensing of vapors or explosive particles adhered to surfaces, whereas in bulk detection more quantity of material is available. The ability to sense these materials from a

Laser-Induced Breakdown Spectroscopy
https://doi.org/10.1016/B978-0-12-818829-3.00018-6

distance is an important requirement. Given the nature of the material and needs of the application, remote and standoff detection is an important criterion for choosing an appropriate method of detection [3].

The proposals [4] made by the NATO committee for standoff detection can mostly be achieved through laser-based techniques, which are very well suited for environmental and security applications where there is a need for remote and standoff capability. In particular, Raman spectroscopy [5, 6], laser-induced fluorescence [7], infrared absorption [8], and laser-induced breakdown spectroscopy (LIBS) [9] have received considerable attention in this direction. Raman and infrared spectroscopies provide molecular information, and LIBS yields elemental information. Also, because of intense emission, it offers higher signal sensitivity. Minimal requirements on sample preparation, ability to detect all the elements in the periodic table, and simpler experimental geometry together with the possibility to perform real-time assessment in remote/standoff mode make it an attractive choice for the detection of HEMs. The capability of LIBS can be gauged from extensive studies in diverse fields ranging from archeology [10], planetary exploration [11, 12], HEM screening [13–17], pharmaceutical formulation identification [18–20], as well as in biomedical [21–23], soil analysis [24, 25], and polymer applications [26–28]. The marriage of multivariate analysis and LIBS has further enhanced the attractiveness of this technique. The studies reported by Miziolek and coworkers [29] and standoff LIBS at Los Alamos National Laboratory [30] are noteworthy in this direction. In this chapter, we present the LIBS-based detection of HEMs. Being a laser-based spectroscopy method, it is a point detection technique. Its implementation for scanning a wide area is a challenging task. There are other challenges even when it can be used as a potential point detection technique. Most of the secondary explosives have carbon, nitrogen, oxygen, and hydrogen as the elemental building blocks. Any organic matter is a possible interferent as they have similar constituents. The presence of oxygen and nitrogen in the atmosphere is another challenge as detection is supposed to be achieved in the open atmosphere. Variation in humidity can also pose serious concerns as its constituents are hydrogen and oxygen. Another issue of concern is the "substrate" on which the material can be found. The substrate can be a metal, soil, plastic, etc. and this can greatly influence the formation of plasma and the emitted LIBS signal.

There are two important research goals specific to HEM detection: first is to identify a material as an HEM and second is to classify it as a specific HEM. Broadly, the application scenarios can be categorized as point detection and wide area detection. When the amount of material present is of the order of ppm or below, it can be described as trace detection. When the amount present is high it can be categorized as bulk detection. The following situations/scenarios are some of the instances where there is a need for detection of explosives:

1. Airport baggage check
2. Land mines
3. Antipersonnel mines
4. Post bomb blast
5. Suspect powder found in a sensitive place or in postal parcels
6. At border checkpoints
7. Sanitizing an area before the arrival of VIPs
8. Movement of armed personnel in a conflict zone
9. Mass gathering of people
10. Improvised explosive devices

It is worth emphasizing that no single technique can possibly be employed in all the scenarios mentioned and for all the explosives of interest. The choice will be guided by many considerations: application scenario, trace or bulk detection, kind of explosive involved, and, most importantly, use of hybrid techniques where multiple detection methods are employed. Given the conditions under which the explosives can be present in real applications, the spectral interference emanating from various materials such as oils [31], grease, and biomaterials [32], oxygen and nitrogen present in the atmosphere, and hydrogen and oxygen in atmospheric moisture must be accounted to achieve reliable detection. The substrate can also further complicate the detection challenges because of considerable changes to the light emission, which have been detailed by Gottfried et al. [33]. Hence, nonanalyte-specific peaks can also be observed in the spectra because of the atmospheric air as well as the substrate [34]. Hence, the basic approach of utilizing intensity ratios may not be reliable.

In the last decade, the application of multivariate analysis methods to LIBS data has enabled the extraction of useful insights that are not possible from simple human examination. In the context of spectroscopy, the application of multivariate analysis methods is also referred to as chemometrics. The combination of LIBS data and chemometrics has been shown to have great potential in the identification of various materials [17–19, 25, 27, 35, 36]. The spectra of several samples can show significant complexity, but a comparison of hundreds of such spectra reveal reproducible patterns that can be used for objective sample identification. Different multivariate analysis methods, including partial least squares-discriminant analysis (PLS-DA) [29, 34, 35], principal component analysis (PCA) [29, 37], soft independent modeling of class analogy (SIMCA) [34], support vector machine (SVM) [38], artificial neural network (ANN) [39], etc., have been exploited to analyze the rich LIBS datasets. The work in this chapter focuses on the application of PCA for exploratory data analysis and ANN for the identification of HEMs/non-HEMs. The data were collected from different HEMs and non-HEMs. Also, feature selection, a method to reduce the number of data points required for numerical analysis, is employed by selectively choosing the wavelength regions of interest.

2 Experimental details

The HEM detection LIBS experiment can be divided into four steps:

1. Sample selection
2. Sample preparation
3. Spectrum recording
4. Analysis of the data

The question of identification of explosives becomes very complex given the fact that it is never present as a pure sample but is always mixed or combined with other similar substances and interferents. Hence, any attempt to apply LIBS to the identification of HEMs needs to be conducted taking into account the interferents and any other substances on which it may be found. The experimental method must include a comprehensive list of materials that include:

1. Explosives
2. Mixtures
3. Matrices
4. Interferents

The ultimate goal of applying LIBS to HEM detection is to identify these materials in outdoor field conditions where the sample can be present as a scattered powder. However, the first step toward achieving this goal is to understand these materials and their LIBS signal in laboratory conditions. In the literature, different studies have been reported where pellets were prepared for recording the LIBS signal [40]. Some groups have also reported experiments with trace amounts of samples [14]. Sufficient care must be taken in the preparation of the pellets. Also, the following precautions are recommended while conducting the LIBS experiment:

1. Remote operation of the hydraulic machine. This means the machine should be located in a separate room and operation performed from another room at a distance.
2. While recording the LIBS signal, the sample holder should be fortified with stainless-steel sheets on all sides.
3. The pellets should never be left unattended and should be stored properly.
4. While grinding, it is suggested first to mix the sample with acetone and also be aware of the friction sensitivity of the sample.
5. Most importantly, before undertaking the study it is highly recommended to talk to a chemistry expert to understand the sample being analyzed.

Most of the reports available on HEMs studied secondary explosives, which are safer to work with as their friction and impact sensitivities are low. In the literature, there are no reports of any accident when LIBS experiments with HEM materials were conducted. However, extra caution is always a good practice while working with these samples. The spectra for all the materials and mixtures of interest are first recorded. Then, a model is established with the available data. Analysis of the data requires the application of the multivariate methods to achieve the task of identification.

Early studies on explosives with LIBS have used ratios of peak intensities [13, 31, 41, 42]. In the last decade and a half, several multivariate approaches, such as PCA [13], SIMCA [34], SVM [38], PLS-DA [17, 32, 43], and ANN [44], have been implemented successfully for LIBS data on different materials, including HEMs.

In this study, a total of 10 organic materials, which includes a set of 5 secondary explosives and 5 nonexplosives, are considered. Information about the samples is shown in Table 1. The HEM samples were acquired from the High Energy Materials Research Laboratory (HEMRL), Pune, India. The choice of non-HEMs was deliberate as they have a CHNO type of molecular structure similar to that of HEMs. First, the samples were mixed with acetone and ground to make a fine powder. Then, 1 cm diameter pellets were made by pressing the powder in a die-hydraulic pressing machine with a pressure of two tons. The light from the second harmonic of an Nd:YAG laser was focused on the sample, which was placed on an X-Y-Z translation stage. Laser pulses of 7 ns duration operated in 1 Hz mode with an energy of 20 mJ/pulse were used to irradiate the sample after passing through a convex lens of 100 mm focal length. The spectral emissions from the plasma were collected by collection optics (ME-OPT-007) and fed to a spectrometer (Mechelle ME5000) via an optical fiber of core diameter 600 μm. The spectrometer was coupled with an intensified charged coupled device (ICCD, Andor-istar, DH734). The Pockels pulse from the laser triggers the delay generator (SRS DG535), which in turn provides a transistor-transistor logic pulse (4–5 V) to trigger the ICCD.

3 Spectral description

A total of 350 spectra were acquired from the HEM and non-HEM compounds (each 175 spectra). Fig. 1 shows representative spectra of two HEM materials and two non-HEM materials. The spectra are normalized to the peak of emission line from oxygen at 777.72 nm. All the spectra showed emission lines corresponding to elemental as well as two molecular entities. Carbon, hydrogen, nitrogen, and oxygen lines correspond to the major elemental atomic constituents of the material. Also, peaks corresponding to sodium, calcium, potassium, and magnesium have been observed. In addition, bands corresponding to CN and C_2 were also present in the spectra. The presence of these bands can be considered to be indicative of organic molecules. The CN violet bands are observed at 357–360, 384–389, and 414–423 nm.

TABLE 1 Details of samples, including name, formula, and structure.

S. No.	Name	Formula	Structure
1	HMX	$C_4H_8N_8O_8$	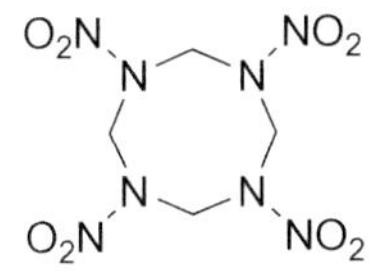
2	NTO	$C_2H_2N_4O_3$	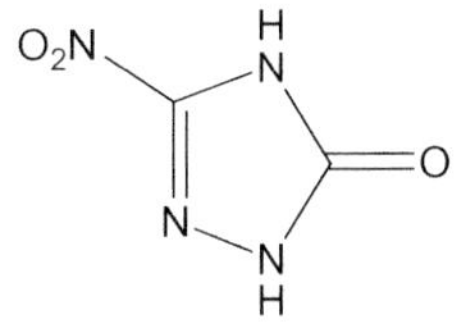
3	PETN	$C_5H_8N_4O_{12}$	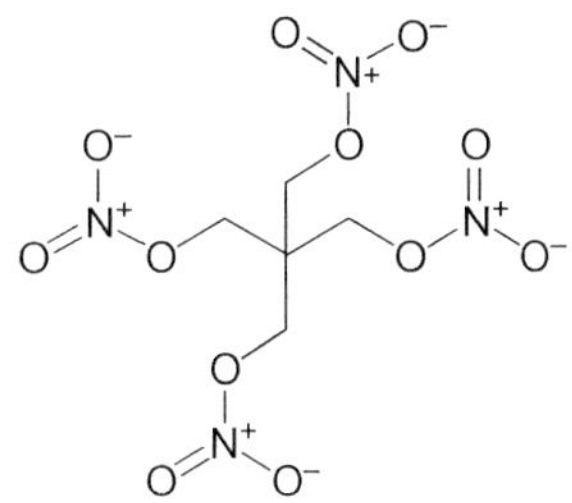
4	RDX	$C_3H_6N_6O_6$	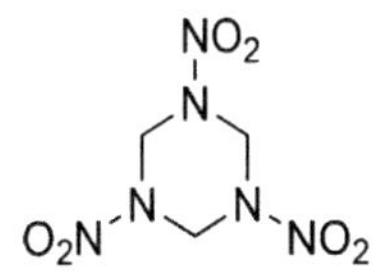
5	TNT	$C_7H_5N_3O_6$	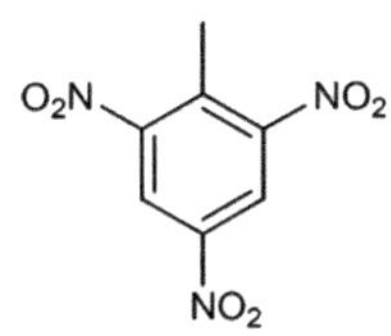
6	Glycine	$C_2H_5NO_2$	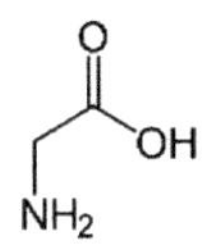
7	Methionine	$C_5H_{11}NO_2S$	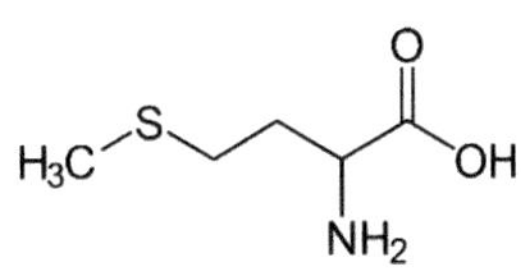

Continued

TABLE 1 Details of samples, including name, formula, and structure—cont'd

S. No.	Name	Formula	Structure
8	Neomycin	$C_{23}H_{46}N_6O_{13}$	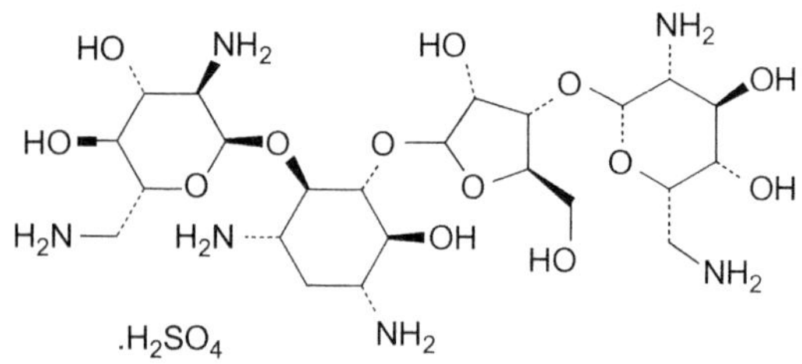
9	Ethylenediaminetetraacetic acid	$C_{10}H_{16}N_2O_8$	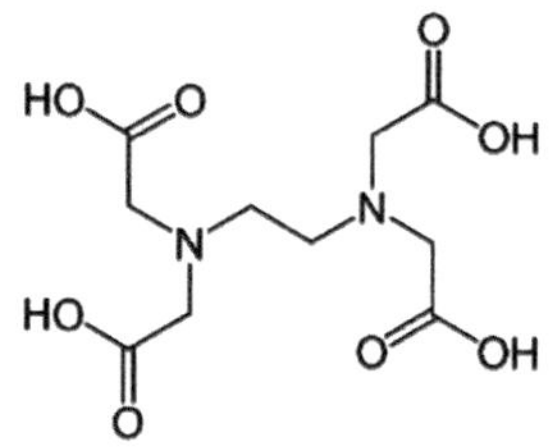
10	L-Tryptophan	$C_{11}H_{12}N_2O_2$	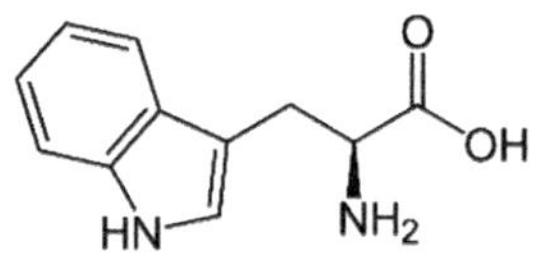

HMX, high melting explosive; *NTO*, 3-nitro-1,2,4-triazol-5-one; *PETN*, pentaerythritol tetranitrate; *RDX*, royal demolition explosive; *TNT*, trinitrotoluene.

They correspond to $\Delta \nu = -1$, 0, and 1 for the transition $B^2\Sigma^+ \rightarrow X^2\Sigma^+$. C_2 swan bands are also observed. Different mechanisms could be responsible for the presence of these molecular fragments in the plume [27, 37, 45, 46]: (1) Direct release of radicals because of sample fragmentation, (2) reaction of atomic entities in the plasma, and (3) recombination of atomic constituents of the sample with constituents from the surrounding atmosphere. Extensive work has been reported in the literature, which deals with the aspect of formation of the molecular entities [27, 45, 47–49].

LIBS spectra are a rich source of information individually; however, a gross examination shows near-identical peaks for all the samples. This is expected as LIBS is an atomic spectroscopic technique and all the samples included in the study contain carbon, hydrogen, oxygen, and nitrogen. Hence, identification of the sample cannot be accomplished just by considering the peak positions alone.

In Fig. 2, variation of the observed peak intensities for major emission lines of nitrogen, oxygen, hydrogen, and carbon for all the samples is shown. Here, the intensity refers to the area under the spectral line obtained from the Lorentzian fit. It is evident from the figure that there is no correlation between the number of atomic constituents in a molecule and their peak intensities. While oxygen and nitrogen show almost the same intensities irrespective of the sample, carbon and hydrogen show some variations. For carbon peak intensity, except for TNT, all

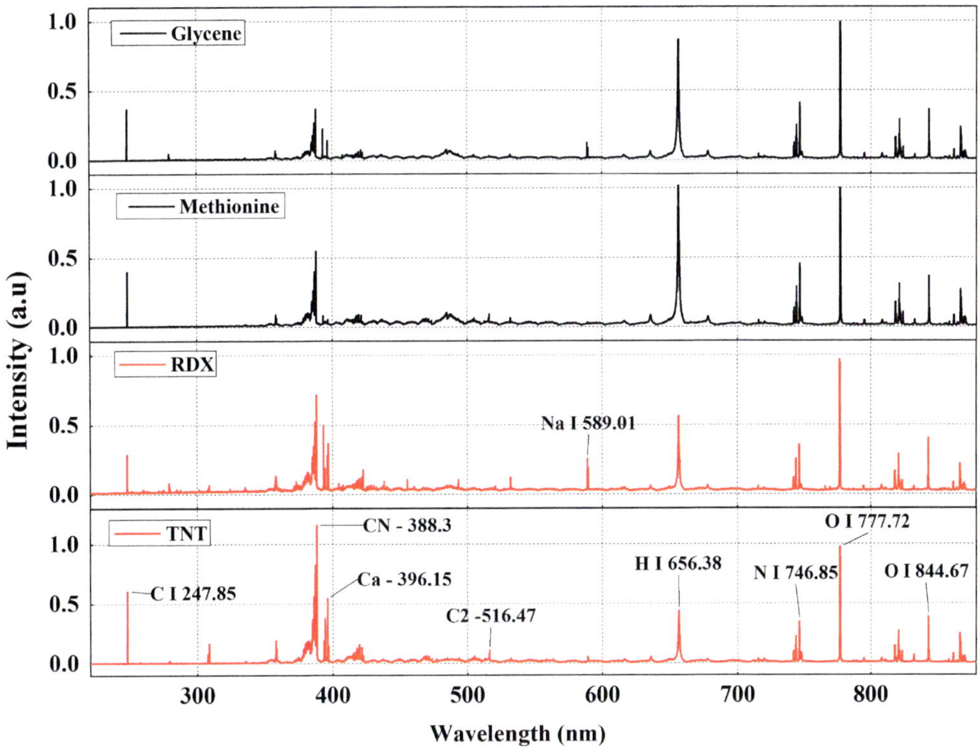

FIG. 1 Shows representative spectra of two HEMs and two non-HEMs. The spectra are normalized to the intensity corresponding to the oxygen emission peak at 777.72 nm.

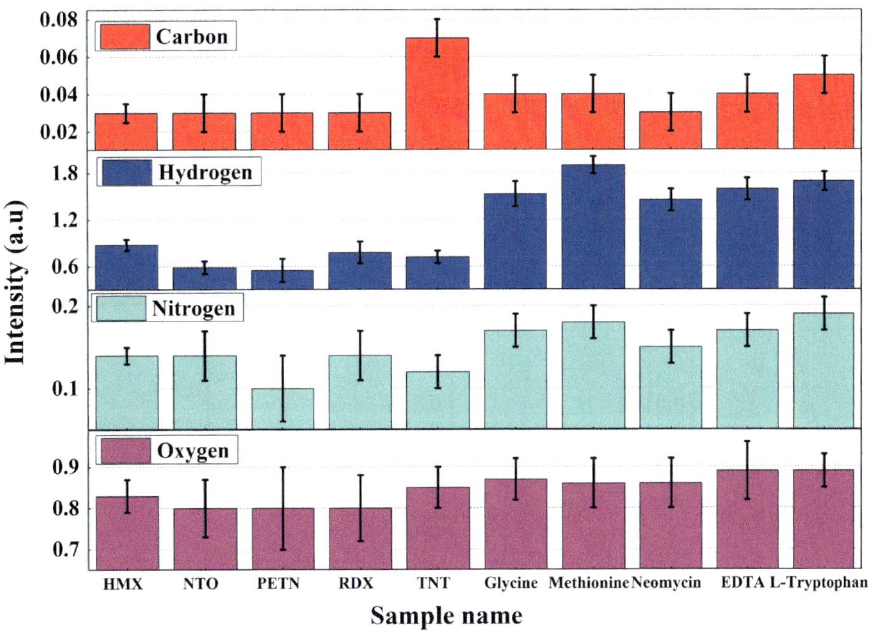

FIG. 2 The variation of the peak intensities at carbon, hydrogen, oxygen, and nitrogen peak maximum for all the samples.

the other samples show similar intensities. This is true for hydrogen also, though the intensities do not show any correlation but seem to suggest a binary division with HEMs having less intensity compared to non-HEMs. These observations put together indicate a highly stochastic nature of the laser-generated plasma. The intensities do not show any statistically significant variation, even though the number of atomic entities varies from sample to sample. But, a careful examination reveals finer but reproducible differences in the peak intensity/shape of all the samples. The individual peaks of carbon, nitrogen, oxygen, and hydrogen are depicted in Fig. 3. The peak shapes and heights vary considerably from sample to sample. The differences in the peak shape of hydrogen and that of oxygen can be observed very clearly.

4 Results and discussion

Each spectrum consisted of intensities at 25,699 wavelengths in the range of 200–850 nm. Though at a coarse level of observation it may appear to be similar, considerable variations in the spectra exist when observed carefully. These variations across the spectrum can be harvested to classify and identify a given sample. PCA has been employed as an exploratory data analysis tool [26]. ANN has been employed as a tool to perform the task of identification. The data has been treated in two different modes. First, the full spectrum is used for modeling the data. In the second approach, data from different spectral windows is considered. PCA performs a transformation, where the new variables, known as the principal components (PCs), are a

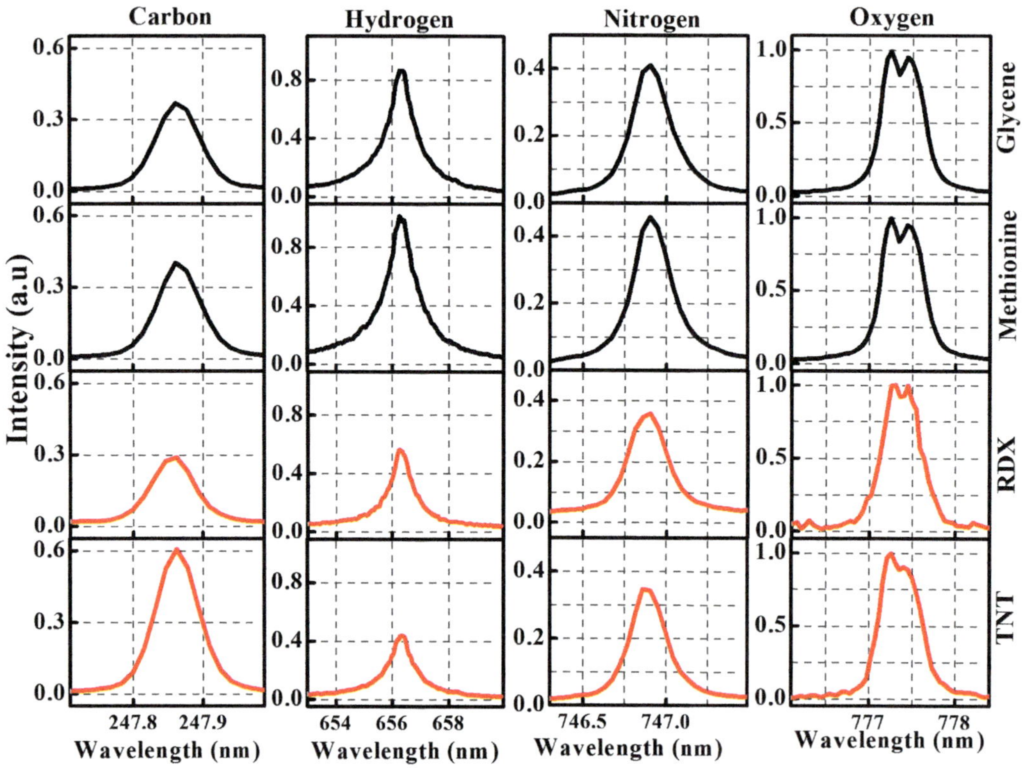

FIG. 3 The emissions at four peaks of carbon, hydrogen, nitrogen, and oxygen for two samples of HEMs and non-HEMs.

superposition of the old variables. The PCs thus form a multidimensional orthogonal space. The spectra are represented in the PC space by scores and the PCs themselves are referred to as loadings. The data projected on PCs will be referred to as score plot and visualization of the PCs as loadings plot. Most of the variance can be explained by the first few PCs, typically 3–5, and most of the remaining PCs are dominated by noise. Thus PCA can also be considered as a data compression/dimensionality reduction technique in addition to transformation to an orthonormal coordinate system. The score plot obtained as a result of the application of PCA for the full spectra model is shown in Fig. 4.

The area under the graph in each spectrum was used to normalize each spectrum. Data normalization can lead to reduction in the fluctuations caused in the score and also leads to enhanced discrimination among the localized clusters. Each point in the plot represents a spectrum by three ordinates in the PC space. Clearly, the spectra corresponding to HEMs and non-HEMs form distinct clusters. Though all the spectra showed the presence of peaks at the same emission wavelengths, this result indicates that there are subtle differences in the spectra that can be attributed to the chemical differences specific to HEMs and non-HEMs. Fig. 3 showed the emission profiles for important elements where we could observe considerable differences among them. Fig. 5 shows the scree plot corresponding to the PCs shown in Fig. 6. The scree plot shows the variance explained by individual PCs as a bar chart and cumulative variance is superimposed on it as a dot and line plot.

The first five PCs together account for about 77.2% of the variance in the data. From the fifth PC onward the variance explained per PC is very small. Hence, in the PC hyperplane, a spectrum can be represented by a smaller number of coordinates instead of 25,699 variables. It is worth looking at the prominent features in the spectrum, which aid in the process of scattering HEMs and non-HEMs as separate clusters. This can be visualized by looking at the PCs, which are linear combinations of the original variables. The loadings plot is shown in Fig. 6.

Most of the coefficients of the first PC have a positive correlation with carbon, oxygen, hydrogen, nitrogen, and CN. The second PC shows a positive correlation with carbon, CN, and sodium, whereas hydrogen shows a

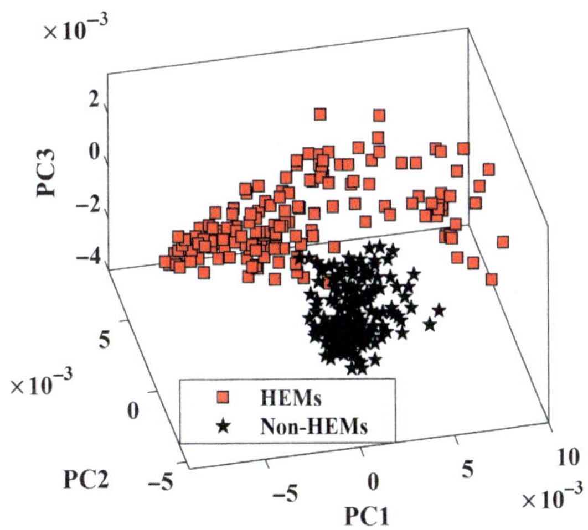

FIG. 4 Principal component analysis results showing the scores of the first three principal components.

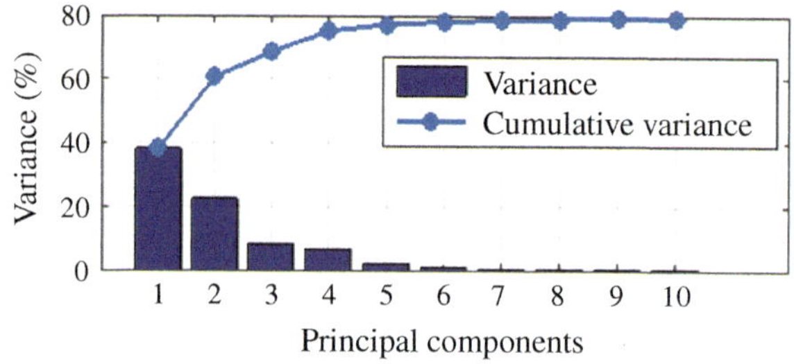

FIG. 5 The scree plot of the first 10 PCs. The *bar graph* represents the individual variance of the PCs and the *solid line* corresponds to the cumulative variance.

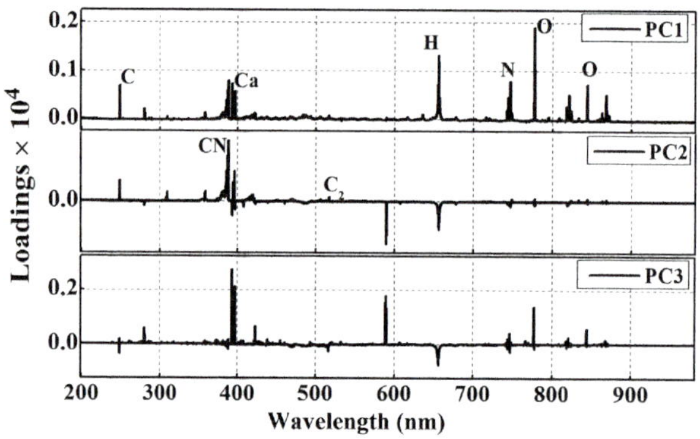

FIG. 6 PC loadings plot of the first three PCs total region as input to the PCA.

negative correlation. The third PC has contributions from calcium, oxygen, and sodium. Taking together all the information from Figs. 4–6, PCA has mostly picked up the features associated with carbon, hydrogen, nitrogen, and oxygen along with inorganic peaks of calcium and sodium. Peaks corresponding to inorganic elements are due to impurities present in the sample as minor constituents, which cannot be filtered before testing the sample. This is a challenge as the applications related to HEMs are envisaged in real time and the user cannot selectively avoid them while recording the data.

In terms of the total number of wavelengths, a substantial portion of a spectrum is composed of a flat noisy baseline. However, the full LIBS spectrum, which contains noise [50] and/or continuum [51], is still a gold standard and is regularly used in classification. This may not significantly affect the prediction performance; feature selection [52, 53], in general, is critical and can eliminate spurious correlations especially in applications where the interclass differences are subtle. Feature selection refers to choosing a subspectrum for the analysis. When working with an ICCD-based spectrograph, the dataset is largely asymmetrical as there are about ~25,700 variables and 350 observations. An effort has been made to explore and understand the variables that are responsible for contributing useful information toward classification. The variable selection was chosen using the manual approach, i.e., knowledge based, in other words, using a priori knowledge of the material to select the regions of interest.

Four different spectral regions have been considered based on the important emission features of CN, C_2, nitrogen, carbon, oxygen, and hydrogen as detailed in Table 2. First and foremost, region 1, referred to as R1, is constructed using carbon and hydrogen peaks as they are dominant constituents and less influenced by the surrounding ambient atmosphere. Region R2 spans the entire wavelength region from carbon to CN. As discussed earlier, CN indicates the sample being organic and hence it is included in R2. The formation of CN depends on several factors, including the content of carbon in the sample, besides the possible nitrogen contribution also from the atmosphere [27, 45, 47].

The presence of aromatic double bonds (C=C) [54] in a sample can be inferred from the presence of C_2 emission in the spectrum. Region R2 is extended by inclusion of C_2 and is referred to as R3. Finally, the entire spectral region after the hydrogen peak (656 nm) is designated as R4. As this part of the spectrum predominantly contains oxygen and nitrogen peaks, it may contain cumulative information of the sample as well as the atmosphere. These four regions and the entire spectrum are used for the analysis separately. The data is normalized for each region separately prior to the application of PCA. The score plot obtained when R1 is considered is shown in Fig. 7. The spectra corresponding to HEMs and non-HEMs are seen to be classified. The cluster for HEM is larger than that of non-HEMs. There are three outliers for HEMs and about five for non-HEMs.

TABLE 2 Details of selected spectral data regions based on sample compositional information.

S. No.	Region name	Description	Wavelengths (nm)	% of full spectrum
1	R1	C and H peaks	246.98–248.05; 640–670	3
2	R2	C to CN region	246.98–424.04	34
3	R3	C to H region	246.98–670.04	62
4	R4	Post-H region	670–875.99	17
5	Full	Entire spectrum	199–981	100

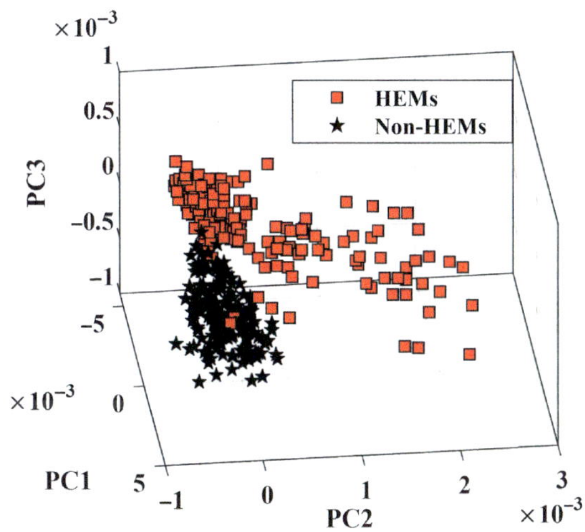

FIG. 7 The score plot obtained from the first three PCs for the R1 region.

The loadings show a positive correlation inevitably with carbon and hydrogen peaks the separation between classes in score plot is due to these peaks. The loadings plot is shown in Fig. 8. While the first PC has both carbon and hydrogen, the second is dominated by carbon and the third by hydrogen.

The first PC can explain 85.49% of the variance present in the data. The subsequent four PCs explain 4.16%, 1.27%, 0.68%, and 0.4% of the variance present in the data, respectively. The clear separation between the clusters implicates the use of this region for a quick check for HEM identification before proceeding for supervised statistical methods.

The subspectrum R2 contains CN peaks, in addition to carbon and other inorganic peaks such as aluminum and calcium. The score plot using R2 is shown in Fig. 9. Two localized clusters of both classes can be found in the score plot with two outliers for non-HEMs.

The scatter corresponds more to HEMs; however, there is again a clear classification of HEMs and non-HEMs. From the loadings plot shown in Fig. 10, it is evident that carbon, CN, and singly ionized calcium have a positive correlation with the first PC. The second PC has a dominant contribution from calcium and the third from carbon.

The first PC can explain a variance of 57.52%, the second PC 16.96%, the third PC 2.5%, the fourth PC 1.34%, and the fifth PC 1.14%. The first five PCs cumulatively explain a variance of 79.46%.

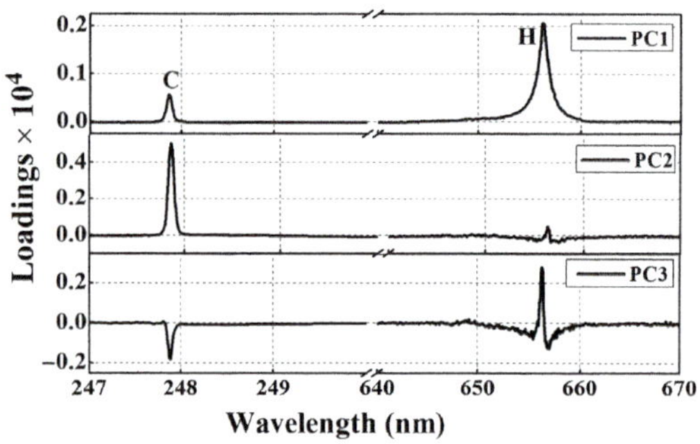

FIG. 8 Loadings plot obtained for the R1 region.

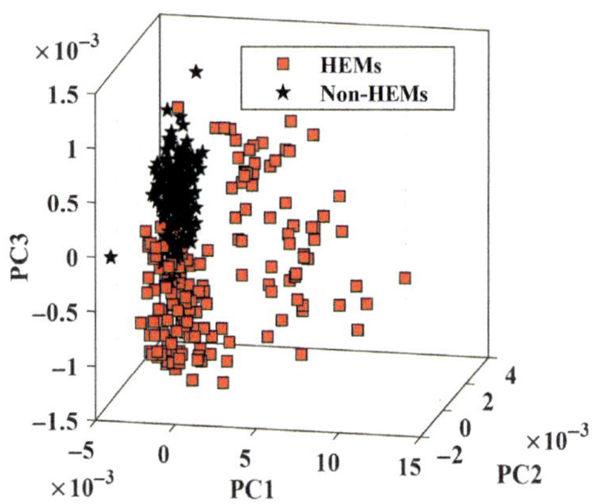

FIG. 9 3D score plot of the PCA for the spectral region of R2.

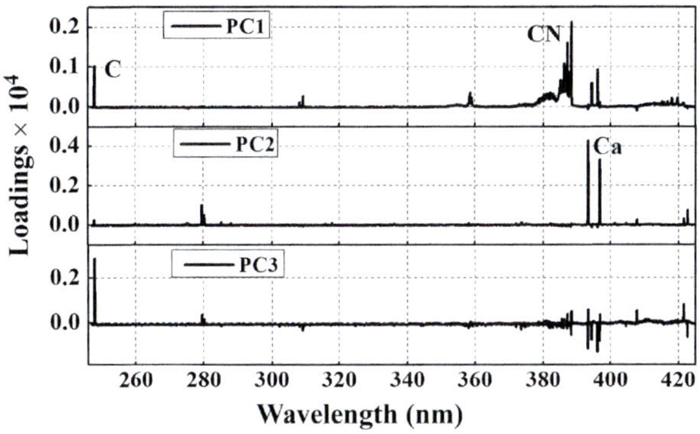

FIG. 10 The loadings of the first three PCs achieved from the PCA for the R2 region.

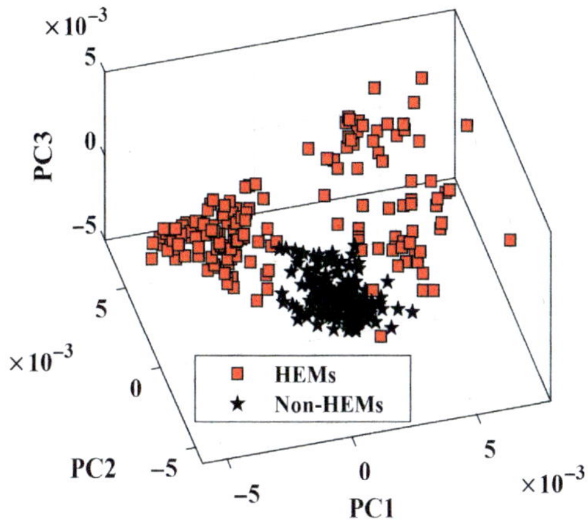

FIG. 11 The score plot achieved by the application of PCA to R3.

The subspectrum R3 is an extension of R2 with additional peaks such as C_2 and hydrogen. The score plot depicted in Fig. 11 shows a good separation among HEMs and non-HEMs. However, there are two slightly separated clusters among HEMs. This could be a result of the inclusion of hydrogen. Though the full spectrum and R1 also had hydrogen, the presence of other peaks could have helped in merging these two regions. While two outliers can be seen for HEMs, there is a larger region of confusion compared to clusters obtained for other regions.

The loadings plot depicted in Fig. 12 shows a positive correlation with carbon, CN, and hydrogen. The second PC has a dominance of hydrogen and calcium. Cumulatively, the first five PCs explain a variance of 77.61%.

The region R4 mainly contains oxygen and nitrogen peaks, which are also the constituents of the surrounding atmosphere. The score plot and corresponding loadings plot are shown in Figs. 13 and 14. Here also, HEMs show a slight tendency for two different clusters. As this spectral region consists of oxygen and nitrogen, which can also have substantial contributions from the atmosphere, the resulting changes in the spectrum could be a possible reason behind this observation. The first five PCs cumulatively explain a variance of 83.25%.

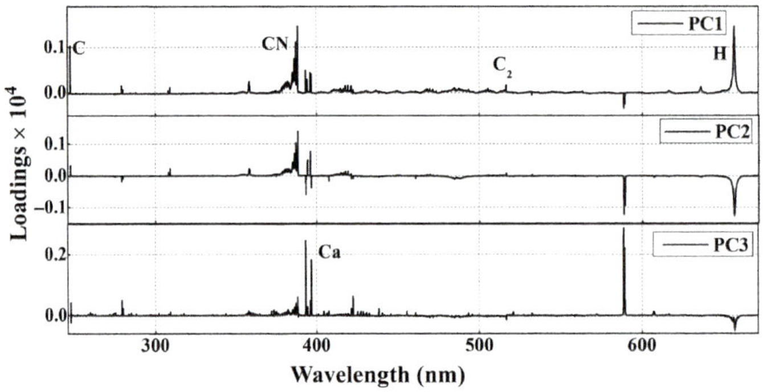

FIG. 12 The loadings of the first three PCs achieved from the PCA for the R3 region.

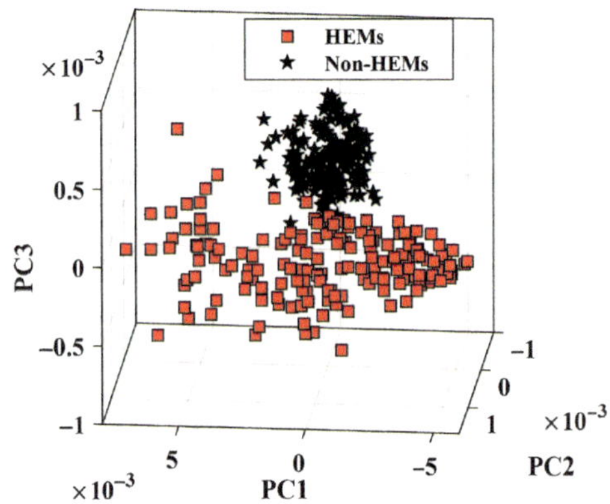

FIG. 13 Score plot construction of the first three consecutive PCs of region R4.

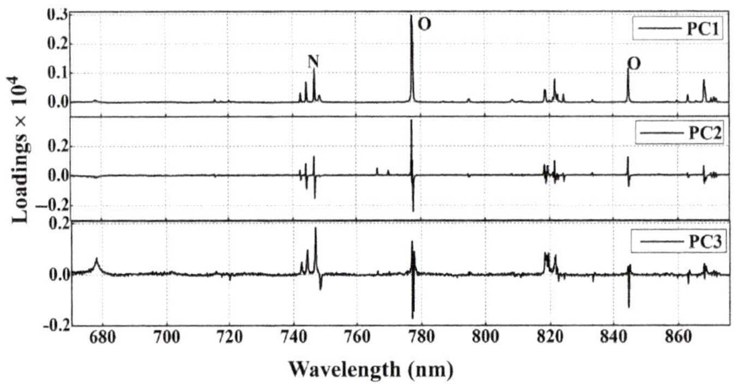

FIG. 14 The loadings of the first three PCs obtained from the R4 region.

The results imply that the full spectrum may not be necessary to obtain a classification. Although the four regions contain fewer intensity points, they can give similar classification results as compared to the full spectrum. Particularly, it is worth noting that region 1 with only 3% of the full data is also sufficient enough to achieve classification. Though commercially available off-the-shelf spectrometers have full spectrum sensitivity, and are routinely used, these results suggest that it is possible to tailor the spectrometers with specified spectral sensitivity for specific LIBS application.

The application of PCA resulted in good clusters corresponding to HEMs and non-HEMs, indicating sufficient information being present in the data that is distinct to these two classes. While PCA does not require any class information, it transforms the given data and using a score plot the differences and similarities in the data can be visualized. As indicated in the introduction, the main application would require the identification of a given sample as either HEM or non-HEM. The application of supervised classification techniques can be used to assign class membership [55]. The ANN method is employed here as a supervised technique.

Lately, ANN has received attention in LIBS analysis. It belongs to a class of nonparametric nonlinear, classification, and predictive methods. A three-layer feed-forward network model was used for classification consisting of one input layer, one hidden layer, and one output layer. The input layer comprised intensity values at various wavelengths. Sigmoid functions were used as neurons for the input and hidden layers. The final output consisted of binary encoding used as the target, where 1 was for correct assignment and 0 otherwise. It was trained with a classical scaled conjugate gradient back-propagation algorithm. Seventy percent of the data were used for training the network, which was adjusted according to its error. Fifteen percent of the spectra were used to measure network generalization and to halt training, indicating an increase in the mean square error of the validation samples. The remaining 15% spectra were used for testing and the validation process to provide an independent measure of network performance.

The results are depicted using a confusion matrix after performing ANN and are shown in Table 3. The diagonal entries show percent correctly classified rates, while off-diagonal elements are percent misclassified rates. More specifically, the first entry on the diagonal shows the capability of detecting an HEM as an HEM and is referred to as sensitivity. The second diagonal element shows the capability of detecting a non-HEM as a non-HEM and is referred to as specificity. For ideal detection, both sensitivity and specificity should be high.

TABLE 3 Confusion matrix computed using a neural network using HEMs and non-HEMs data. Diagonal elements in the matrix represent correctly identified, while off-diagonal elements are called misclassified.

		Predicted labels	
		HEM	Non-HEM
Targeted labels	HEM	98.29	1.71
	Non-HEM	1.03	98.97

HEM, high-energy material.

When the full-length spectrum was used, the classifier was able to predict 98.3% as HEMs, where misclassification was about 1.7%. In the case of non-HEMs, 99% was correctly classified and misclassification was 1%. Misclassification in the case of both classes may be due to the presence of outliers.

In the earlier section, PCA was applied on a subspectrum to observe the classification using variables associated with the loadings. ANN is also applied on subspectra to understand the influence of different spectral features on the classification model. The correct classification performance given by all subspectra was more than 93% as shown in Table 4. In the case subspectrum R1, whose variables are carbon and hydrogen, the average correct classification is 97.59%. But with the subspectrum R2, the misclassification increased to 6.87%, where inorganic peaks may produce undesirable effects in the classification pattern. The correct classification performance of R3 and R4 are 96.69% and 98.82%, respectively. These results suggest that irrespective of any variable, namely, either full spectrum or subspectrum, the neural network

TABLE 4 ANN results obtained for the different subspectra.

Subspectra	Correct classification (%)	Misclassification (%)
R1	97.59	2.41
R2	93.13	6.87
R3	96.69	3.31
R4	98.82	1.18

ANN, artificial neural network.

model can perform the task of classification. However, the number of spectral points used differs for each subspectrum region. R1 with ~3% of the spectral data is sufficient to achieve performance as good as a full spectrum.

5 Conclusion

A LIBS experiment was performed on a set of 10 materials containing 5 HEMs and 5 non-HEMs, both of which bear the molecular formula $C_aH_bN_cO_d$. The ability of combining the data with multivariate analysis approaches to identify the HEMs from the non-HEMs has been demonstrated. PCA is used as an exploratory data analysis tool and ANN classifier to perform the task of identification. Also, the power of wavelength selection is demonstrated by performing the analysis by choosing a subspectrum of data. The PCA score plot showed well-separated clusters for HEMs and non-HEMs for the full spectrum as well as for the subspectrum. When the full spectrum was used, a correct classification rate of ~98% was obtained. The correct classification rates obtained with subspectrum analysis are comparable to that of the full spectrum. Notably, regions R1 and R4 with data less than 3% and 17% of the full spectrum, respectively, yielded correct classification rates the same as that of the full spectrum. The results indicate that for the detection of HEMs from non-HEMs it may not be necessary to record the full spectrum.

Acknowledgments

The author acknowledges the funding support from DRDO, India, Dr. Ashwin Kumar Mykalwar for helping in the data collection and Mr. Rajendhar Junjuri for helping in the preparation of this chapter.

References

[1] J.P. Agrawal, High Energy Materials: Propellants, Explosives and Pyrotechnics, Wiley, 2010.

[2] J.P. Agrawal, R. Hodgson, Organic Chemistry of Explosives, Wiley, 2007.

[3] S. Wallin, A. Pettersson, H. Ostmark, A. Hobro, Laser-based standoff detection of explosives: a critical review, Anal. Bioanal. Chem. 395 (2009) 259–274, https://doi.org/10.1007/s00216-009-2844-3.

[4] M. Krausa, Vapour and Trace Detection of Explosives for Anti-terrorism Purposes, vol. 167, Springer Science & Business Media, 2012.

[5] D.S. Moore, R.J. Scharff, Portable Raman explosives detection, Anal. Bioanal. Chem. 393 (2009) 1571–1578, https://doi.org/10.1007/s00216-008-2499-5.

[6] J.C. Carter, S.M. Angel, M. Lawrence-Snyder, J. Scaffidi, R.E. Whipple, J.G. Reynolds, Standoff detection of high explosive materials at 50 meters in ambient light conditions using a small Raman instrument, Appl. Spectrosc. 59 (2005) 769–775, https://doi.org/10.1366/0003702054280612.

[7] C.M. Wynn, S. Palmacci, R.R. Kunz, M. Rothschild, Noncontact detection of homemade explosive constituents via photodissociation followed by laser-induced fluorescence, Opt. Express 18 (2010) 5399–5406, https://doi.org/10.1364/OE.18.005399.

[8] M.Á.F. de la Ossa, J.M. Amigo, C. García-Ruiz, Detection of residues from explosive manipulation by near infrared hyperspectral imaging: a promising forensic tool, Forensic Sci. Int. 242 (2014) 228–235, https://doi.org/10.1016/j.forsciint.2014.06.023.

[9] F.C. De Lucia Jr., R.S. Harmon, K.L. McNesby, R.J. Winkel Jr., A.W. Miziolek, Laser-induced breakdown spectroscopy analysis of energetic materials, Appl. Opt. 42 (2003) 6148–6152, https://doi.org/10.1364/ao. 42.006148.

[10] A. Botto, B. Campanella, S. Legnaioli, M. Lezzerini, G. Lorenzetti, S. Pagnotta, F. Poggialini, V. Palleschi, Applications of laser-induced breakdown spectroscopy in cultural heritage and archaeology: a critical review, J. Anal. At. Spectrom. 34 (2019) 81–103, https://doi.org/10.1039/c8ja00319j.

[11] A.S. Laxmiprasad, V.L.N.S. Raja, S. Menon, A. Goswami, M.V.H. Rao, K.A. Lohar, An in situ laser induced breakdown spectroscope (LIBS) for Chandrayaan-2 rover: ablation kinetics and emissivity estimations, Adv. Space Res. 52 (2013) 332–341, https://doi.org/10.1016/j.asr.2013.03.021.

[12] F. Rivera-Hernandez, D.Y. Sumner, N. Mangold, K.M. Stack, O. Forni, H. Newsom, A. Williams, M. Nachon, J. L'Harido, O. Gasnault, R. Wiens, S. Maurice, Using ChemCam LIBS data to constrain grain size in rocks on Mars: proof of concept and application to rocks at Yellowknife Bay and Pahrump Hills, Gale crater, Icarus 321 (2019) 82–98, https://doi.org/10.1016/j.icarus.2018.10.023.

[13] M.K. Gundawar, R. Junjuri, A.K. Myakalwar, Standoff detection of explosives at 1 m using laser induced breakdown spectroscopy, Def. Sci. J. 67 (2017) 623, https://doi.org/10.14429/dsj.67.11498.

[14] I. Gaona, J. Serrano, J. Moros, J.J. Laserna, Range-adaptive standoff recognition of explosive fingerprints on solid surfaces using a supervised learning method and laser-induced breakdown spectroscopy, Anal. Chem. 86 (2014) 5045–5052, https://doi.org/10.1021/ac500694j.

[15] A.K. Shaik, V.R. Soma, Standoff discrimination and trace detection of explosive molecules using femtosecond filament induced breakdown spectroscopy combined with silver nanoparticles, OSA Continuum 2 (2019) 554–562, https://doi.org/10.1364/OSAC.2.000554.

[16] Q.Q. Wang, G. Teng, C.Y. Li, Y. Zhao, Z. Peng, Identification and classification of explosives using semi-supervised learning and laser-induced breakdown spectroscopy, J. Hazard. Mater. 369 (2019) 423–429, https://doi.org/10.1016/j.jhazmat.2019.02.015.

[17] A.K. Myakalwar, N. Spegazzini, C. Zhang, S.K. Anubham, R.R. Dasari, I. Barman, M.K. Gundawar, Less is more: avoiding the LIBS dimensionality curse through judicious feature selection for explosive detection, Sci. Rep. 5 (2015) https://doi.org/10.1038/srep13169.

[18] A.K. Myakalwar, S. Sreedhar, I. Barman, N.C. Dingari, S.V. Rao, P.P. Kiran, S.P. Tewari, G.M. Kumar, Laser-induced breakdown spectroscopy-based investigation and classification of pharmaceutical tablets using multivariate chemometric analysis, Talanta 87 (2011) 53–59, https://doi.org/10.1016/j.talanta.2011.09.040.

[19] P.K. Tiwari, S. Awasthi, R. Kumar, R.K. Anand, P.K. Rai, A.K. Rai, Rapid analysis of pharmaceutical drugs using LIBS coupled with multivariate analysis, Lasers Med. Sci. 33 (2018) 263–270, https://doi.org/10.1007/s10103-017-2358-8.

[20] L.F. Zou, B. Kassim, J.P. Smith, J.D. Ormes, Y. Liu, Q. Tu, X.D. Bu, In situ analytical characterization and chemical imaging of tablet coatings using laser induced breakdown spectroscopy (LIBS), Analyst 143 (2018) 5000–5007, https://doi.org/10.1039/c8an01262h.

[21] V.K. Singh, A.K. Rai, Prospects for laser-induced breakdown spectroscopy for biomedical applications: a review, Lasers Med. Sci. 26 (2011) 673–687, https://doi.org/10.1007/s10103-011-0921-2.

[22] R. Kanawade, F. Mahari, F. Klampfl, M. Rohde, C. Knipfer, K. Tangermann-Gerk, W. Adler, M. Schmidt, F. Stelzle, Qualitative tissue differentiation by analysing the intensity ratios of atomic emission lines using laser induced breakdown spectroscopy (LIBS): prospects for a feedback mechanism for surgical laser systems, J. Biophotonics 8 (2015) 153–161, https://doi.org/10.1002/jbio.201300159.

[23] S.J. Rehse, H. Salimnia, A.W. Miziolek, Laser-induced breakdown spectroscopy (LIBS): an overview of recent progress and future potential for biomedical applications, J. Med. Eng. Technol. 36 (2012) 77–89, https://doi.org/10.3109/03091902.2011.645946.

[24] S. Chatterjee, M. Singh, B.P. Biswal, U.K. Sinha, S. Patbhaje, A. Sarkar, Application of laser-induced breakdown spectroscopy (LIBS) coupled with PCA for rapid classification of soil samples in geothermal areas, Anal. Bioanal. Chem. 411 (2019) 2855–2866, https://doi.org/10.1007/s00216-019-01731-3.

[25] S.C. Jantzi, J.R. Almirall, Characterization and forensic analysis of soil samples using laser-induced breakdown spectroscopy (LIBS), Anal. Bioanal. Chem. 400 (2011) 3341–3351, https://doi.org/10.1007/s00216-011-4869-7.

[26] V.K. Unnikrishnan, K.S. Choudhari, S.D. Kulkarni, R. Nayak, V.B. Kartha, C. Santhosh, Analytical predictive capabilities of laser induced breakdown spectroscopy (LIBS) with principal component analysis (PCA) for plastic classification, RSC Adv. 3 (2013) 25872–25880, https://doi.org/10.1039/c3ra44946g.

[27] R. Junjuri, M.K. Gundawar, Femtosecond laser-induced breakdown spectroscopy studies for the identification of plastics, J. Anal. At. Spectrom. 34 (2019) 1683–1692, https://doi.org/10.1039/c9ja00102f.

[28] J. Anzano, B. Bonilla, B. Montull-Ibor, J. Casas-Gonzalez, Plastic identification and comparison by multivariate techniques with laser-induced breakdown spectroscopy, J. Appl. Polym. Sci. 121 (2011) 2710–2716, https://doi.org/10.1002/app.33801.

[29] F.C. De Lucia, Jr, J.L. Gottfried, C.A. Munson, A.W. Miziolek, Multivariate analysis of standoff laser-induced breakdown spectroscopy spectra for classification of explosive-containing residues, Appl. Opt. 47 (2008) G112–G121, https://doi.org/10.1364/ao.47.00g112.

[30] B. Salle, J.L. Lacour, E. Vors, P. Fichet, S. Maurice, D.A. Cremers, R.C. Wiens, Laser-induced breakdown spectroscopy for Mars surface analysis: capabilities at stand-off distances and detection of chlorine and sulfur elements, Spectrochim. Acta B 59 (2004) 1413–1422, https://doi.org/10.1016/j.sab.2004.06.006.

[31] C. Lopez-Moreno, S. Palanco, J.J. Laserna, F. DeLucia Jr., A.W. Miziolek, J. Rose, R.A. Walters, A.I. Whitehouse, Test of a stand-off laser-induced breakdown spectroscopy sensor for the detection of explosive residues on solid surfaces, J. Anal. At. Spectrom. 21 (2006) 55–60, https://doi.org/10.1039/B508055J.

[32] J.L. Gottfried, F.C. De Lucia, C.A. Munson, A.W. Miziolek, Double-pulse standoff laser-induced breakdown spectroscopy for versatile hazardous materials detection, Spectrochim. Acta B 62 (2007) 1405–1411.

[33] J.L. Gottfried, F.C. De Lucia, A.W. Miziolek, Discrimination of explosive residues on organic and inorganic substrates using laser-induced breakdown spectroscopy, J. Anal. At. Spectrom. 24 (2009) 288–296, https://doi.org/10.1039/b818481j.

[34] J.L. Gottfried, F.C. De Lucia, C.A. Munson, A.W. Miziolek, Strategies for residue explosives detection using laser-induced breakdown spectroscopy, J. Anal. At. Spectrom. 23 (2008) 205–216, https://doi.org/10.1039/b703891g.

[35] J.B. Sirven, B. Salle, P. Mauchien, J.L. Lacour, S. Maurice, G. Manhes, Feasibility study of rock identification at the surface of Mars by remote laser-induced breakdown spectroscopy and three chemometric methods, J. Anal. At. Spectrom. 22 (2007) 1471–1480, https://doi.org/10.1039/b704868h.

[36] F.Y. Yueh, H.B. Zheng, J.P. Singh, S. Burgess, Preliminary evaluation of laser-induced breakdown spectroscopy for tissue classification, Spectrochim. Acta B 64 (2009) 1059–1067, https://doi.org/10.1016/j.sab.2009.07.025.

[37] C. Brown, M. Baudelet, C. Bridge, M. Fisher, M. Sigman, P. Dagdigian, M. Richardson, SPIE Defense, Security, and Sensing; International Society for Optics and Photonics, 2009. pp. 73041D-73041D-73012.

[38] N.C. Dingari, I. Barman, A.K. Myakalwar, S.P. Tewari, M. Kumar Gundawar, Incorporation of support vector machines in the LIBS toolbox for sensitive and robust classification amidst unexpected sample and system variability, Anal. Chem. 84 (2012) 2686–2694, https://doi.org/10.1021/ac202755e.

[39] A.K. Myakalwar, N.C. Dingari, R.R. Dasari, I Barman, M. K. Gundawar. Non-gated laser induced breakdown spectroscopy provides a powerful segmentation tool on concomitant treatment of characteristic and continuum emission. PLoS One. 9 (8) (2014) 103546, https://doi.org/10.1371/journal.pone.0103546.

[40] A.K. Myakalwar, S.K. Anubham, S.K. Paidi, I. Barman, M.K. Gundawar, Real-time fingerprinting of structural isomers using laser induced breakdown spectroscopy, Analyst 141 (2016) 3077–3083, https://doi.org/10.1039/C6AN00408C.

[41] S. Sreedhar, M.K. Gundawar, S.V. Rao, Laser induced breakdown spectroscopy for classification of high energy materials using elemental intensity ratios, Def. Sci. J. 64 (2014) 332, https://doi.org/10.14429/dsj.64.4741.

[42] R.D. Waterbury, A. Pal, D.K. Killinger, J. Rose, E.L. Dottery, G. Ontai, Chemical, Biological, Radiological, Nuclear, and Explosives (CBRNE) Sensing IX, International Society for Optics and Photonics, 2008, 695409.

[43] F.C. De Lucia, J.L. Gottfried, Classification of explosive residues on organic substrates using laser induced breakdown spectroscopy, Appl. Opt. 51 (2012) B83–B92, https://doi.org/10.1364/AO.51.000B83.

[44] A.H. Farhadian, M.K. Tehrani, M.H. Keshavarz, S.M.R. Darbani, Energetic materials identification by laser-induced breakdown spectroscopy combined with artificial neural network, Appl. Opt. 56 (2017) 3372–3377, https://doi.org/10.1364/AO.56.003372.

[45] S.J. Mousavi, M.H. Farsani, S.M.R. Darbani, A. Mousaviazar, M. Soltanolkotabi, A.E. Majd, CN and C-2 vibrational spectra analysis in molecular LIBS of organic materials, Appl. Phys. B Lasers Opt. 122 (2016) 106, https://doi.org/10.1007/S00340-016-6371-6.

[46] V.I. Babushok, F.C. DeLucia, P.J. Dagdigian, J.L. Gottfried, C.A. Munson, M.J. Nusea, A.W. Miziolek, Kinetic modeling study of the laser-induced plasma plume of cyclotrimethylenetrinitramine (RDX), Spectrochim. Acta B 62 (2007) 1321–1328, https://doi.org/10.1016/j.sab.2007.10.029.

[47] S.A. Kalam, N.L. Murthy, P. Mathi, N. Kommu, A.K. Singh, S.V. Rao, Correlation of molecular, atomic emissions with detonation parameters in femtosecond and nanosecond LIBS plasma of high energy materials, J. Anal. At. Spectrom. 32 (2017) 1535–1546, https://doi.org/10.1039/c7ja00136c.

[48] E.N. Rao, P. Mathi, S.A. Kalam, S. Sreedhar, A.K. Singh, B.N. Jagatap, S.V. Rao, Femtosecond and nanosecond LIBS studies of nitroimidazoles: correlation between molecular structure and LIBS data, J. Anal. At.

Spectrom. 31 (2016) 737–750, https://doi.org/10.1039/c5ja00445d.

[49] J. Serrano, J. Moros, J.J. Laserna, Sensing signatures mediated by chemical structure of molecular solids in laser-induced plasmas, Anal. Chem. 87 (2015) 2794–2801, https://doi.org/10.1021/acs.analchem.5b00212.

[50] J.B. Sirven, B. Bousquet, L. Canioni, L. Sarger, Laser-induced breakdown spectroscopy of composite samples: comparison of advanced chemometrics methods, Anal. Chem. 78 (2006) 1462–1469, https://doi.org/10.1021/ac051721p.

[51] A.K. Myakalwar, N.C. Dingari, R.R. Dasari, I. Barman, M.K. Gundawar, Non-gated laser induced breakdown spectroscopy provides a powerful segmentation tool on concomitant treatment of characteristic and continuum emission, PLoS One 9 (2014) e103546, https://doi.org/10.1371/journal.pone.0103546.

[52] K. Liu, D. Tian, H.X. Wang, G. Yang, Rapid classification of plastics by laser-induced breakdown spectroscopy (LIBS) coupled with partial least squares discrimination analysis based on variable importance (VI-PLS-DA), Anal. Methods 11 (2019) 1174–1179, https://doi.org/10.1039/c8ay02755b.

[53] M.J.C. Pontes, J. Cortez, R.K.H. Galvao, C. Pasquini, M.C.U. Araujo, R.M. Coelho, M.K. Chiba, M.F. de Abreu, B.E. Madari, Classification of Brazilian soils by using LIBS and variable selection in the wavelet domain, Anal. Chim. Acta 642 (2009) 12–18, https://doi.org/10.1016/j.aca.2009.03.001.

[54] P. Lucena, A. Dona, L.M. Tobaria, J.J. Laserna, New challenges and insights in the detection and spectral identification of organic explosives by laser induced breakdown spectroscopy, Spectrochim. Acta B 66 (2011) 12–20, https://doi.org/10.1016/j.sab.2010.11.012.

[55] R. Junjuri, C. Zhang, I. Barman, M.K. Gundawar, Identification of post-consumer plastics using laser-induced breakdown spectroscopy, Polym. Test. 76 (2019) 101–108, https://doi.org/10.1016/j.polymertesting.2019.03.012.

CHAPTER

19

Industrial applications

Reinhard Noll[a,b]

[a]Fraunhofer-Institut für Lasertechnik (ILT), Aachen, Germany [b]RWTH Aachen University, Lehrstuhl für Lasertechnik (LLT), Aachen, Germany

1 Introduction

There is no other analytical method offering such a wide range of measuring distances for chemical analyses than laser-induced breakdown spectroscopy (LIBS). These distances range from a few centimeters to up to hundreds of meters, spanning four orders of magnitude. This enables an inimitably broad variety of applications. For industrial applications, Fig. 1 illustrates this range of measuring distances for LIBS (the term "measuring distance" or "working distance" [WD] is used here to describe the free distance between the last optics of the LIBS system and the measuring object; this distance is primarily given by the focal length of the optics focusing the laser beam or the receiving optics).

Short WDs of a few centimeters enable a fine focusing of the laser beam and thus a high spatial resolution for chemical analysis such as, e.g., applied in industry for multielement mapping of steel samples [1]. At distances of typically 10–30 cm, LIBS systems can identify steel grades in rolling mills to perform a mix-up inspection. Here, laser pulse parameters are tailored to the task of removing locally, in a first step non-representative surface layers with thicknesses of several 100 μm prior to bulk analysis [2]. Even greater working distances in the range of up to 4 m are required for inline analysis tasks of hot objects such as liquid metals or liquid slag [3–5], or up to 1 m for objects moving fast on transport lines [6].

During 2014–19 two main development directions for industrial LIBS applications became obvious: on the one hand, stationary automated LIBS systems for inline measuring tasks embedded in a process line and, on the other hand, handheld LIBS units mainly for fast positive material identification [7]. Both branches reflect the unique feature of the LIBS method with a pure optical excitation of the measuring object over a distance being predestined for the fast analysis of metals and oxidic substances. In that sense, LIBS has the potential to become an important tool for Industry 4.0 scenarios.

At the same time period a series of publications appeared focusing on or giving at least first perspectives for industrial LIBS applications. A query in the Web of Science database from July 2019 with the search terms "Laser-induced breakdown spectroscopy OR LIBS) AND TOPIC: (application) AND TOPIC: (industr*) NOT TOPIC: (lithium-ion battery)" yielded 115

Laser-Induced Breakdown Spectroscopy
https://doi.org/10.1016/B978-0-12-818829-3.00019-8

hits for publications in the timespan from 2014 to 2019 (the search term "lithium-ion battery" is excluded, otherwise too many nonadequate papers were found, since they make use of the abbreviation "LIBs" to denote the plural form of lithium-ion batteries).

From these 115 hits, 30 papers have dominant Chinese authorships corresponding to a share of ca. 26%. Seventeen papers of the 115 were found not fit the query chosen, deal with completely different topics, are anonymous, or offer only little scientific/technological content. Most of the remaining 98 papers can be grouped in terms of the studied substances as follows (some of the publications can be allocated to more than one category; some cannot be allocated since they treat, e.g., a dominantly instrumental aspect): (1) analysis/identification of metals or alloys—28 publications; (2) analysis of raw materials, coal, oxides, minerals, recycling materials—16 publications; and (3) others, e.g., nuclear materials, water, rubber, paper, biological substances—41 publications. In the following we have omitted papers that are only available in the Chinese language.

Looking at those publications with a clear focus on industrial applications or a clear perspective for such applications or having more than five citations, and omitting those that present a more fundamental character and objects of research in the microparticle or geological field, then these publications in group (a) comprise 17 papers [5, 8–23].

Noharet et al. described a portable LIBS prototype for online analysis of steel scrap pieces from a production process [8]. The LIBS unit is equipped with a low resolving spectrometer and is triggered by a distance measurement based on the triangulation principle; WD is in the range of 0.5–1 m. Field tests have been performed with production scrap in a steel plant and at scrap yards. Li et al. studied the quantitative analysis of phosphorus in steel at atmospheric conditions and at a WD of 15 cm [9]. With the P I line at 214.91 nm detected by a high-resolving Czerny-Turner spectrometer the

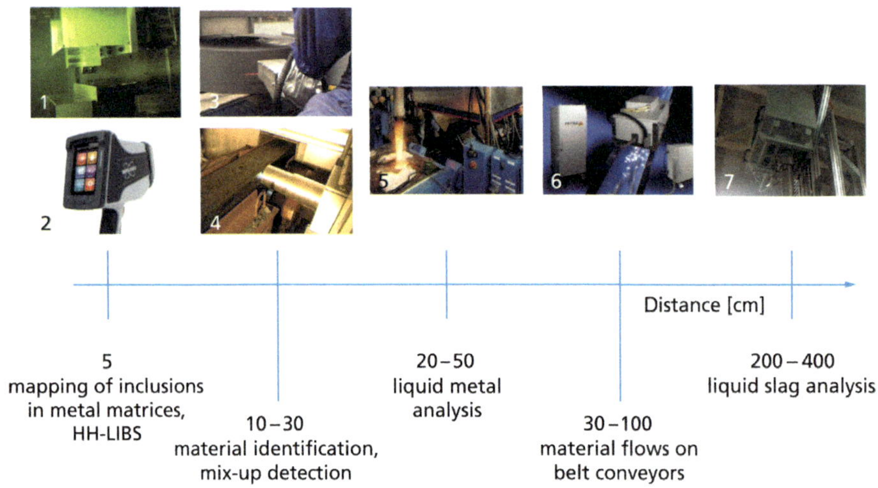

FIG. 1 LIBS systems for industrial applications with working distances from a few centimeters to a meter and more. *Photos from left to right and top to bottom*: High-speed inclusion mapping of steel samples (1), handheld LIBS system (2), identification of coils in a rolling mill (3), identification of blooms on a roller table (4), direct analysis of liquid metal melts (5), high-speed identification and sorting of scrap pieces (6), and direct analysis of liquid slag (7). *HH-LIBS*, handheld LIBS system. *Photos by courtesy of (1) and (3)—Laser Analytical Systems and Automation GmbH, Aachen, Germany; (2) Bruker Nanoanalytics Division, Kennewick, USA; and (4) to (7)—Fraunhofer ILT, and its industrial partners.*

phosphorus content in pig iron and low-alloy steel was determined yielding limits of detection (LODs) in the range of 9–12 μg/g. Lorenzetti et al. studied LIBS to monitor the transition layer of steel billets from a casting machine using double pulses and artificial neural networks [10]. The WD was presumably 10 cm. A set of samples was taken from such a billet and changes of nickel and chromium concentration from a level of about 1.5 wt% prior to the transition zone down to ≈0.9 wt% after that transition were detected. Merk et al. studied the identification of scrap metal by a LIBS setup with an echelle spectrometer at a WD of 80 cm [11]. Measuring objects were shredded pieces of metals with a single main element such as Ag, Al, Cr, Cu, Fe, Ni, Pb, Sn, Zn, and the secondary alloy brass. Sun et al. studied the in situ analysis of steel melts with LIBS and a Cassegrain telescope with WDs of 2.8 and 4.5 m [5]. They used an immersion probe, which dipped into the steel melt. The Cassegrain telescope collected the plasma light to fiber optics. Tests in a steel plant were conducted to quantitatively determine carbon, silicon, and manganese within a measuring time of 1 min. The authors claimed to determine carbon without resolving the 247.86-nm carbon line just by using the partial least square (PLS) method. Zeng et al. studied LIBS via optical fibers to analyze Mn and Ti in pig iron at a WD of 15 cm [12]. A laser pulse at 532 nm, duration at 7 ns, and energy at 24 mJ are transmitted by a 3-m long optical fiber having a core diameter of 1 mm to the focusing system. LODs shown for Mn and Ti are about a factor of 10 higher than those achieved with conventional LIBS. Noharet et al. [13] studied a LIBS prototype with a compact diode-pumped Nd:YAG laser generating 150 μJ at 7 kHz. In a WD of 50 cm, tests were performed to measure the analytes Si, Mg, and Cu in aluminum scrap samples. Wang et al. gave an overview of applications of LIBS for elemental monitoring of iron- and steelmaking processes [14]. Afgan et al. investigated a handheld LIBS instrument (from B&Wtek) to quantitatively determine eight elements—among these carbon—in steel [15]. With a dominant factor-based PLS the average standard deviation for the elements Si, Mn, Ni, and Cu in steel were better for LIBS than those achieved with a commercial handheld X-ray fluorescence (XRF) instrument. Bengtson presented a review comparing LIBS with conventional optical emission techniques (OES) for the analysis of metals [16]. He concluded that the extremely simple design of the electric spark source was well suited to automation, which has had a huge impact on routine process control analysis in the metallurgical industry. Whereas for spark discharge OES (SD-OES) and glow discharge OES there exist international standards, there is still a lack of such standards for LIBS with implications on a mid-term timescale.

Tortschanoff et al. presented application tests of a compact diode-pumped solid-state laser (DPSSL) for the quantitative analysis of steel [17]. The laser source running at 1064 nm delivered 2.5-ns pulses with an energy of 30 mJ. They were focused at WDs of up to 50 mm. Zhao et al. analyzed melts of nickel-base alloys at a WD of 1.5–2.5 m and ambient pressures of about 5 Pa [18]. Coaxial convex and concave mirrors focus the laser beam and collect the plasma radiation. The fiber-coupled echelle spectrometer was temperature stabilized to 30 ± 1°C. Calibration curves of Cr and Al were gained with relative standard deviations in the range from 1.1% to 2.8%. Zeng et al. studied the corrosion behavior of structure materials in direct contact with molten metals [19]. The compositional changes of molten aluminum in crucibles made of 304 stainless steel were obtained online at 1000°C with a WD of 300 mm. They detected the rise of line emissions of iron, manganese, and chromium as a function of the soaking time with different stabilization times.

Sun and coworkers studied the monitoring of molten aluminum and electrolyte components by LIBS to determine the molecular ratio of NaF to AlF_3 representing the acidity of the electrolyte [20]. The WD was about 750 mm. For the measurements in aluminum reduction cells the measuring beams were guided within a ceramic lance composed of SiC flushed with argon. Experiments were carried out in a Chinese aluminum electrolysis plant. The lance was dipped to different levels inside the melt. However, quantitative analyses could not be completed due to strong magnetic interferences caused by the large induction current of the cells, which caused changes of the laser pulse energy. Luo et al. used a DPSSL with pulse energies up to 200 mJ, whose beam was focused onto low-alloy steel samples placed on an argon-flushed sample stand [21]. The WD amounted to 150 mm. The spectrum was detected via a Paschen-Runge spectrometer equipped with 30 photomultipliers. They compared the analytical results with SD-OES and concluded that the analytical performance of LIBS was slightly worse than that of SD-OES, which they attributed to the small size of the focused laser beam and the small sample mass vaporized. Yang et al. proposed a method to transfer information on spectra gained at room temperature to spectra at high temperature to assist in building a better regression model [22]. They carried out experiments on 15 certified alloy steel standard samples to analyze Cr concentrations (c(Cr) ranges from 1.46 to 24.5 m%) at room temperature and on four standard samples at a temperature of ca. 800°C. The WD used was 250–300 mm. Zeng et al. developed a portable fiber-optic LIBS system and analyzed minor elements in pig iron and carbon steel [23]. A pulse of about 30 mJ at 532 nm was delivered via a 1-mm fiber and was focused by a $f = 50$-mm lens. Calibration curves for the analytes Mn, Ti, V, and Cr were determined. In the studied concentration ranges the average relative errors were comparable with or even slightly better than those of benchtop LIBS systems. LODs were inferior to those of traditional LIBS, which was ascribed to the low energy of the laser pulses delivered via the fiber.

From the 16 publications of group (b), 7 fulfill the foregoing criteria [24–30]. Six of these seven publications in group (b) treat raw materials and oxides [24–29]. Gaft et al. presented a LIBS online analyzer mounted in an overhead position above a belt conveyor to measure raw materials passing by at a WD of 35 cm [24]. As an example, phosphorus-bearing apatite and magnesium-bearing dolomite rocks were identified and pebble portions with elevated Mg were moved aside. Further applications discussed are iron sintering mix control and composition stabilization, magnesite crushed ore composition control and sorting, potash quality control, and copper and nickel ores online quality control. Hou et al. [25] developed a method to analyze pressed pellet coal samples at a WD of 10 cm. A dominant factor PLS model combined with spectrum standardization is applied and by this the sample-to-sample relative standard deviation of carbon measurements was reduced to 0.3% from 2.9% for the univariate model. The average measurement error of carbon content was reduced to 0.42%—for a carbon concentration range of 40% to 80%—which met the national Chinese standard (1%). Ljutic et al. detected phosphorus in industrial oxide materials, such as iron oxide ceramics and slag from industrial steel production, in the UV range using the line P I at 214.91 nm [26]. An LOD of 50 ppm was demonstrated for P_2O_5 in Fe_3O_4 ceramics and 70 ppm for homogeneous sintered slag samples. A WD was not specified.

Trichard and coauthors studied a quantitative elemental imaging of palladium-based porous alumina catalysts [27]. Samples were moved on a platform in two directions to perform a raster scan for LIBS measurements. The laser source was running at 10 Hz and a top hat pulse of <1 mJ was irradiated. Spectral images of palladium, aluminum, iron, and CN (cyanic radical) of a catalyst section were shown. LIBS measurements were calibrated via electron probe microanalysis due to a lack of reference samples. By this, quantitative elemental images over three orders of magnitude in concentration were gained (Pd concentration ranged from 10 to 10^4 μg/g). In [28] a compact vacuum ultraviolet spectrometer is presented for elemental imaging applied to mine core characterization. An argon-flushed light transmission channel is designed to couple the radiation of the laser-induced plasma to the entrance slit of the spectrometer (spectral range 150–250 nm), which is also argon flushed. Spectra of the main mineral phases of mine core samples are shown in the spectral range from 170 to 250 nm. Elements detected are as follows: C, Fe, S, Cu, Al, B, Si, P, Ca, Mo, As, Zn, and Ti in mining core. A single-pulse detection limit for S of 0.2 m% is stated. The combination of LIBS spectral information with space-resolved analysis allows the identification of mineral phases, such as pyrite, tourmaline, chalcopyrite, molybdenite, or renierite.

Legnaioli et al. studied the determination of the ash content of coal samples (size range 3–20 mm) in a laboratory environment focusing double pulses (30 + 60 mJ, $\Delta t = 1$ μs) at a WD of 70 cm [29]. The plasma emission was collected via concave mirrors and coupled into fiber optics connected to two spectrometers for the UV and the visible/infrared range. From a total of 120 spectra per sample, those were selected where the intensity of the carbon 247.9-nm line exceeded 90% of the maximum value observed for that sample. An artificial neural network was applied trained with nine coal samples and validated with eight further samples all covering an ash content range of 4–17 m%. The root mean square deviation found was 0.5 m% with a maximum deviation of about 4 m%.

One publication in group (b) examined recycling materials with LIBS [30]. Aquino et al. applied a commercial LIBS instrument (J200 LIBS system of Applied Spectra, USA) to measure scrap pieces of computer and mobile phones (MPHs) and evaluated the spectra with principal component analysis to determine Au and Ag. No information on the WD was given.

In group (c), Gong et al. studied the remote quantitative analysis of cerium by LIBS to monitor a pyrochemical process using as surrogate sample a mixture of cerium oxide and potassium chloride [31]. Excitation by the laser beam and observation of the plasma emission were performed through a plastic shielding window with a thickness of 1.4 cm. The WD amounted to 1.45 m. The plasma emission was collected by a Schmidt-Cassegrain telescope. The authors studied the analytical performance of LIBS with this shielding window, which was used in a real demonstration nuclear facility. Finally, Karpate et al. studied LIBS as a potential tool for industrial/agricultural wastewater analysis [32]. A pulse of a 532-nm laser was focused at a WD of 10 cm onto a vertically oriented water jet. Water samples consisting of Millipore water doped with copper and silver were studied. For a 1000-ppm solution of these elements, emission lines of copper and silver were identified as well as the hydrogen line at 656 nm and the oxygen triplet lines at 777 nm.

In the following sections, recent R&D activities performed at the Fraunhofer Institute for Laser Technology, Aachen, and RWTH Aachen University, referring to automated LIBS systems for inline analyses tasks in industry, are highlighted such as: combined use of inline measured 3D geometry information and LIBS analyses for high-speed sorting tasks of piece goods, sorting of refractories and alloyed metal scrap, identification of steel blooms in a rolling mill, and inverse production scenario for the recovery of valuable materials from end-of-life electronic equipment. At-line or offline LIBS measurements are not considered here.

2 Inline measured 3D geometry information and LIBS analyses for high-speed sorting tasks of piece goods

Short measurement times of LIBS in the order of microseconds allow the measurement of moving objects. Two classes of moving objects have to be considered: bulk material and singularized objects. Examples of bulk material are granular coal and minerals [24, 33–36]. In this case the laser beam direction is chosen to be stationary, directed from an overhead position downward to the moving material. Varying height levels are compensated for by an autofocusing optics or a triggered measuring mode, i.e., the laser pulse is emitted while the height level of the bulk material reaches a defined height interval [13].

Singularized objects are, e.g., scrap pieces or bricks that are sorted piece by piece in different fractions depending on their chemical composition. In this case the object moving on a belt conveyor is tracked with a galvanometer scanner guiding the laser beam to the right position on the object. The 3D geometry of the singularized object is measured upstream of the LIBS measuring volume and evaluated in real time to control the tracking procedure via the scanner. Fig. 2 illustrates this approach [6]. The dashed box indicates the measuring volume in which a measuring location for LIBS can be positioned.

Features of the first-generation LIBS machines based on this concept are given in Table 1, second column [6]. Up to 40 objects can be tracked per second while passing through the measuring volume (typical dimensions $400 \times 400 \times 120\ mm^3$). Since 2015, Fraunhofer ILT started the development of second-generation LIBS machines, called i*Sort*, for singularized object analysis and sorting, which now allow for higher speeds of the belt conveyor of up to 5 m/s for measuring up to 100 pieces per second. A significantly wider range of object sizes can be handled starting from small pieces of a few millimeters to >200-mm large objects. A multi-CCD Paschen-Runge spectrometer covers a wide spectral range with a high spectral resolution for comprehensive detection of line emissions even in line-rich LIBS spectra [37].

Fig. 3 illustrates schematically different industrial application cases of first- and second-generation LIBS machines in the coordinate's WD versus particle size range. The diameters of the circles qualitatively indicate the measuring frequency, and the thickness of the boundary circle marks the complexity of the shape (a simpler shape is presented as a thin line; a more complex one as a bold line).

For 3D measurement the laser light section method is applied where a laser beam is formed as a line and is projected from the

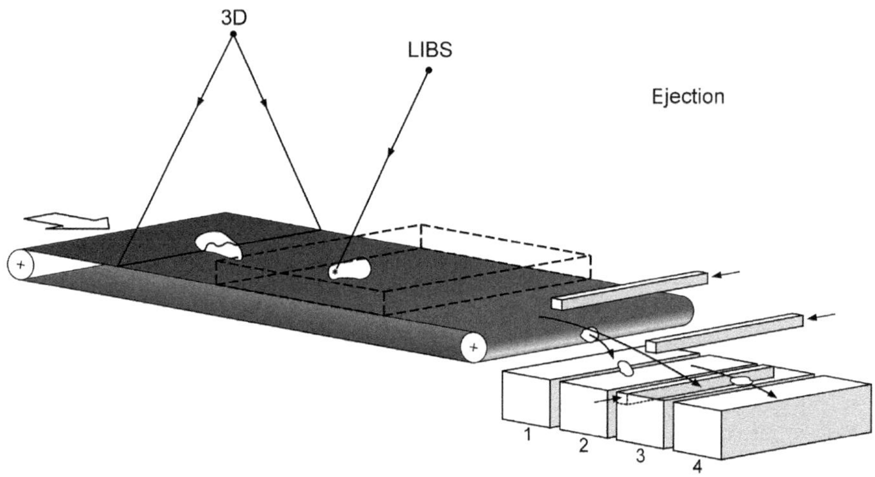

FIG. 2 Schematic for high-speed sorting of piece goods. The process flow runs from left to right with the following sequence: inline 3D measurement of single objects transported on a belt conveyor, laser-induced breakdown spectroscopy analysis, and sorting to different fractions 1–4 by pneumatic ejection.

TABLE 1 Data of first- and second-generation machines for automated inline 3D measurements of piece goods moving on a belt conveyor, piece-by-piece LIBS analyses, and sorting. Working distance refers to the vertical distance between the last optics of the LIBS unit to the surface of the belt conveyor.

	First-generation machine	Second-generation machine
Speed of belt conveyor (m/s)	≤3	≤5
Working distance (cm)	60–90	40–150
Object rate (s^{-1})	≤40	≤100
Object sizes (mm)	25–200	3–250
Simultaneously detected elements	16	>50

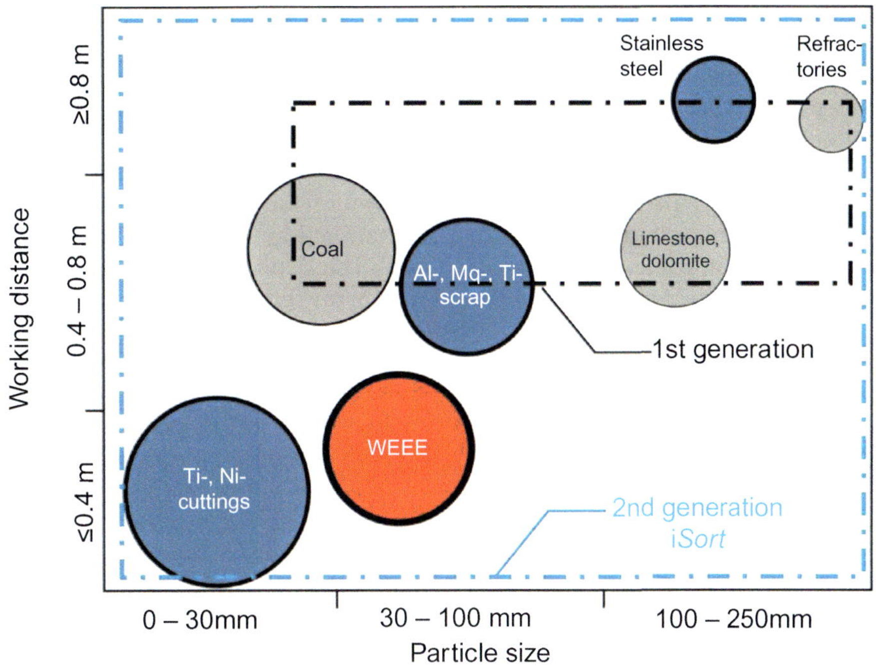

FIG. 3 Application fields of inline monitoring and identification tasks by LIBS on moving objects illustrated in the coordinate's working distance versus particle size. The *black dashed rectangle* indicates the working range of first-generation LIBS machines, the *blue rectangle* refers to second-generation LIBS systems, cf. Table 1. Filling-in colors are assigned as follows: *blue* = metals, *gray* = nonconductive materials, *red* = mixture of metals and nonmetals. *WEEE*, Waste electric and electronic equipment.

top onto the object passing by. This laser line is observed by a camera positioned at an angle to the plane of incident laser radiation. An example for the 3D measurement of small particles such as cuttings is shown in Fig. 4. The sequence of laser light sections is evaluated to generate a 3D image, see Fig. 4, right. Within this 3D image a set of spots is selected having a surface normal within predefined angular limits. The enlarged rectangle with

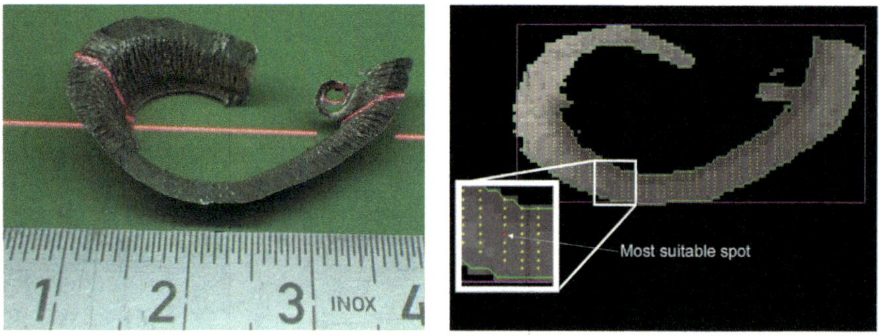

FIG. 4 *Left*: 3D detection of object topology with the laser light section method exemplarily shown for a cutting; a *red* laser line is projected onto the object. *Right*: Processed 3D image of that cutting with determined bounding box and selected spot suited for laser-induced breakdown spectroscopy analysis based on geometric criteria, see *red* spot within enlarged white rectangle (most suitable spot). *Photos: Fraunhofer ILT.*

the marked red spot shows the preferred position for LIBS measurement. In contrast to other approaches reported so far [8, 11, 36] where the laser beam direction is kept stationary, the configuration illustrated with Figs. 2 and 4 offers a significantly higher quality of measuring results by considering the local geometric features of the measuring objects.

3 Sorting of refractories

The goal of the European project REFRA-SORT is the development and validation of separation technologies for the high-grade recycling of refractory waste [38]. A dominant user of refractories is the steel industry where different types of refractory bricks are deployed as lining material for the various furnaces and ladles. At the end of a furnace campaign the lining is dismantled. The generated refractory waste consisting of different types of refractories is often used for low-grade applications such as, e.g., road beds or deposited in landfilling. To enable high-grade applications the refractories have to be sorted and separated piece by piece to gain high-purity waste streams as valuable resources and refeed for refractory production.

Eight different types of refractory materials are to be distinguished: MgO–C without antioxidant, MgO–C with antioxidant, fired MgO, fired doloma, carbon-bonded doloma, fired bauxite, fired analusite, and fired chamotte. The material has an intrinsic inhomogeneity with different grain sizes. The pieces to be handled are in the size range of 60–300 mm and weigh up to 20 kg. For a throughput of 10 t/h, one brick has to be identified and sorted per second on average.

Due to their interaction with the processed substances in the furnaces and ladles the surface of the refractory bricks has undergone a chemical and physical modification, i.e., the surface layer is not representative of the bulk composition. Fig. 5 shows a refractory brick with such a modified surface (left; cf., e.g., the broken surface at the bottom left having a different visual appearance to the material that was not exposed to the processed hot substances). LIBS-measuring spots penetrating this nonrepresentative layer are shown in the middle of Fig. 5. At the right a detailed view of a single interaction region is shown with a crater diameter in the order of 300 μm. To cope with the inhomogeneity of these measuring objects, several LIBS measurements are performed per piece. The experimental parameters are as follows: WD 730 mm, tailored pulse train consisting of cleaning pulse and double pulse, Paschen-Runge

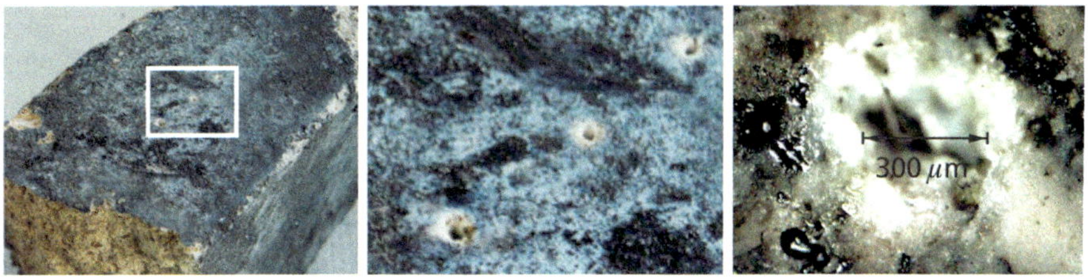

FIG. 5 LIBS-measuring spots on an end-of-life refractory brick. *Left*: Refractory brick with three LIBS measuring spots, see white rectangle. *Bottom left*: Broken surface of this material showing a different color of the internal material not being affected by the hot substances during use of this brick. *Middle*: Enlarged detail corresponding to the white rectangle shown in the left image. *Right*: Detailed view of generated crater with a diameter of about 300 μm. *Photos: Fraunhofer ILT.*

spectrometer optics with a Rowland circle of 500 mm and eight CCD detectors, and an evaluated spectral range of 240–404 nm.

The developed prototype for the sorting of refractory bricks based on inline LIBS measurements is shown in Fig. 6. The pieces are fed to a belt conveyor from the left, singularized, pass through the 3D and LIBS measurement (cf. Section 2), and are then pushed to different sorting fractions based on the measured and evaluated data.

The results of a 30-*t*-test with mixed used refractories are summarized in Table 2. In the majority the analyte concentrations found in the sorted fractions of magnesite and dolomite fulfill the target specifications, the only exception is a slightly higher content of SiO_2 in the dolomite fraction.

4 Sensor-based sorting of alloyed metal scrap

The economics of recycling scrap alloys depends, on the one hand, on the alloy, the element concentrations involved, and their value, and, on the other hand, on the recycling methods used and their costs. Scraps are traded on the market like products. They are differentiated into different qualities. For a CrNi-stainless steel, the content of chromium and nickel is crucial.

FIG. 6 Prototype sorting machine for refractories with an integrated LIBS unit (material flow from left to right). 1: Feeding, 2: singularization, 3: 3D camera, 4: LIBS unit, 5: sorting. *Photo: Fraunhofer ILT at the facilities of Orbix, Genk, Belgium.*

TABLE 2 Composition of sorted fractions of a 30-*t*-test and comparison with target composition. The chemical composition of the sorted fractions is verified with conventional analytical methods (performed by the project partner Magnesita).

Sorting results	(wt%)	CaO	MgO	SiO_2	Fe_2O_2	Al_2O_3	LOI
Magnesite	Achieved	2.94	93.21	1.2	1.17	0.95	1.34
	Target	<3.00	>88	<3.50	<2	<4	<5
Dolomite	Achieved	37.26	56.92	3.03	1.55	0.79	0.9
	Target	<60	>35	<2.5	<2	<2	–

LOI, loss on ignition.

To be able to set and bill the corresponding prices for the purchased alloys, the alloy composition must be known. Hence, the core of the inspection of incoming goods in the trade with alloyed scrap is the identification of the delivered alloy, the determination of the concentrations of the contained elements, and, if necessary, the sorting.

Common tools in the identification of alloys are portable spectrometers, which are, e.g., based on XRF. In the mass markets of scrap processing, an incomplete but representative review of the incoming goods takes place.

For certain material flows, the smelters require a sorted preparation. This may include a complete inspection of all scrap pieces. Examples of such niche markets are: nickel-base superalloys, titanium alloys, high-speed steels (HSSs), and hard metals (HMs). The state of the art in terms of sorting is hand picking and visual inspection supported by spectrometric methods (XRF, SD-OES). For metal scrap with low individual weights per piece and low-grade alloys, this sorting process is often not economical. This creates material flows in which not all elements are specifically recycled and recovered leading to downcycling. To prevent this, the project PLUS was launched to develop a pilot plant that identifies individual scrap pieces of material classes HSS, HM, and titanium and discharges them into several target fractions [39, 40].

The methodical approach for object recognition and LIBS measurement follows the setup shown in Fig. 2. Fig. 7, left, shows exemplarily photos of the processed pieces, in this case thread cutters. The 3D data gained by laser light section are processed to recognize the object contour, to define circumscribing rectangles, and last but not least to identify potential measuring locations for subsequent LIBS measurements, see Fig. 7, right.

A tailored laser pulse train is irradiated to penetrate nonrepresentative surface layers prior to LIBS measurement. The laser beam tracks the moving piece so that cleaning and measuring pulses are focused on the same spot. The speed of the belt conveyor can be varied between 0.15 and 5 m/s. Depending on the requirements of the analyses, either up to 60 measurements can be carried out on one sample or a high sample throughput can be generated with one measurement per piece. Fig. 8 shows photos of LIBS measurements at a twist drill moving at 0.3 m/s.

The classification logic in PLUS follows the conditions of recycling alloyed metal scrap. The three main classes of HSS, HM, and titanium are basically separated when recycling the scrap, but may have low impurities.

The aim of the classification in PLUS is the distinction of subclasses, which are related to the alloying elements. The specified limit values result from the composition of the alloys. Identifiers for main and subclasses are introduced. They are composed of the main class (1xx, 2xx, or 3xx) and the respective subclass (xx = 01–11). The matrix element for the main class 100/HSS is iron, for 200/titanium is titanium, and for

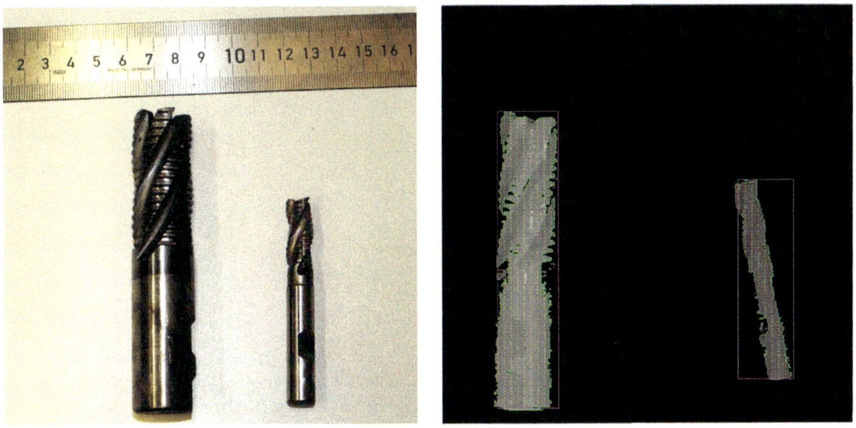

FIG. 7 *Left*: Photo of two different thread cutters. *Right*: Evaluated 3D data of these objects: contour *(green)*, circumscribing rectangles *(purple)*, height information *(gray scale)*, and indication of potential measuring spots for subsequent LIBS measurements *(yellow)*.

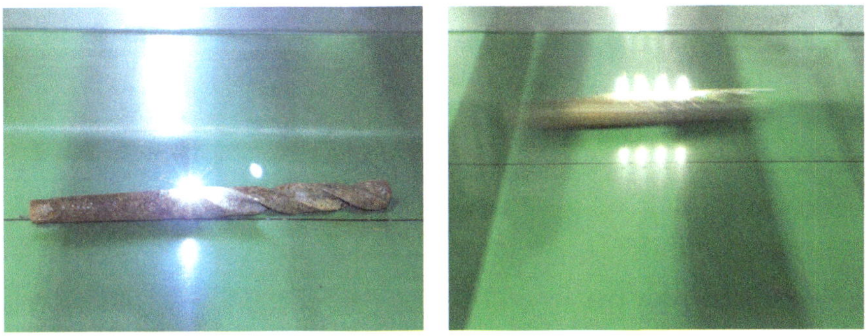

FIG. 8 *Left*: Photo of a single LIBS measurement at a twist drill transported on a belt conveyor. *Right*: Photo with an extended exposure time to visualize a sequence of LIBS measurements at a single measuring object.

300/HM is tungsten. To differentiate the individual alloy classes, the contents of the elements Fe, Ti, W, Co, Al, Nb, and Ni are measured with LIBS. Furthermore, the elements Mo, V, Cr, Mn, Sn, Zr, C, Cu, and Pd are examined.

For LIBS measurements, signals from >60 spectral lines were selected, integrated in each spectrum and averaged over 5 measurements each. The obtained measurement vectors are standardized and evaluated by linear discriminant analysis according to their class membership. Fig. 9 shows scatter diagrams for the various subclasses belonging to the main class 100/HSS. With a set of four vectors (see axes Vec2, Vec3, and Vec1, Vec2 in Fig. 9) all eight subclasses can be discriminated clearly.

After the metal scrap has been identified by the LIBS module of the pilot plant, the material on the belt conveyor reaches the automated discharge station. A delta robot hanging above the belt conveyor is used to offer high handling capacity, flexibility, process reliability, and economic efficiency.

One challenge is the occurrence of variability of object geometries. The first test gripper designed is shown in Fig. 10. It is a pneumatic gripper that is particularly suited for rod- or

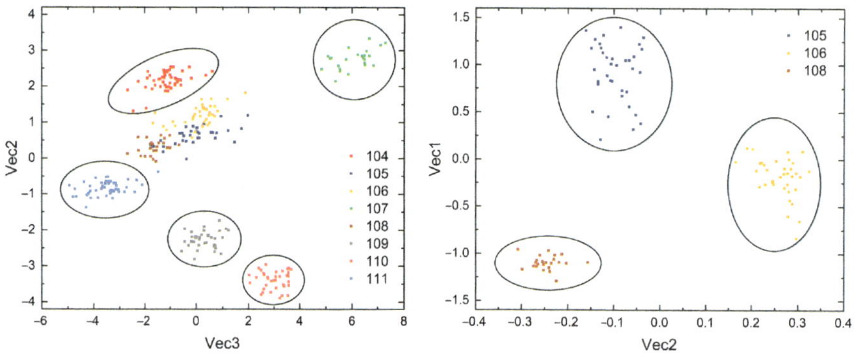

FIG. 9 *Left*: Scatterplot of high-speed steel subclasses 104–111. In the shown vector space five subclasses are well separated, whereas the subclasses 105, 106, and 108 are partially overlapping. *Right*: Using further vectors (Vec1, Vec2) the classes 105, 106, and 108 can be discriminated as well.

FIG. 10 Gripper of the delta robot with a piece of high-speed steel scrap.

pin-shaped objects. After grabbing the scrap parts from the belt these are dropped on up to six different discard slides guiding the pieces to the specified sorting fractions.

5 Identification of steel blooms in a rolling mill

An inline LIBS measurement at metallic piece goods in a rolling mill is shown in Fig. 11. The fed blooms have scale layers with nonrepresentative surface composition [2, 41]. The task is a mix-up detection to assure that the correct steel grade is always transported to the subsequent furnace and first roll stand. In Fig. 11, left, the bloom is just arriving on the roller table coming from the left; at the bottom the tip of the laser measuring lance is visible. The optical WD is 300 mm. The bloom is stopped for about 1 min, while the measuring lance approaches the side face and the laser ablation and LIBS analysis is activated, cf. Fig. 11, right. The approach of the lance is executed to improve the gas flow conditions in the interaction zone. The stop period of 1 min lies within the regular feeding rate of the subsequent

FIG. 11 Inline measurement of steel blooms. *Left*: The bloom (BM) is transported on a roller table and arrives from the left. *Right*: The measuring lance (ML) approaches the side face of the bloom, and laser irradiation is activated to locally ablate the scale layer and to analyze the bulk composition. Diameter of ML: 170 mm [41].

continuous furnace and thus does not imply any retarding in the production flow.

A process was developed consisting of two phases as follows: (1) local removal of scale layer with cleaning laser pulses, and (2) LIBS analysis using laser double pulses [2, 41]. In the first phase a scale layer thickness of 200–600 μm has to be penetrated. Within 20 s an effective crater depth of >2 mm in the scale material is achieved.

Fig. 12 shows results of laboratory tests at ground (i.e., nonscaled) and primary-scaled surfaces for steel grades with chromium contents ranging from 0.1 wt% to 17 wt%. The analysis curve (straight line in Fig. 12) was set up on the basis of 64 ground samples. To characterize the prediction error of Cr concentrations of scaled blooms, 53 different scaled steel samples were measured. The root mean square error amounts to 0.32 wt% for this set of scaled samples. The developed two-phase measuring process yields about the same results for ground and scaled samples and is thus capable of providing bulk analysis of blooms in their "natural" state with primary-scaled surfaces.

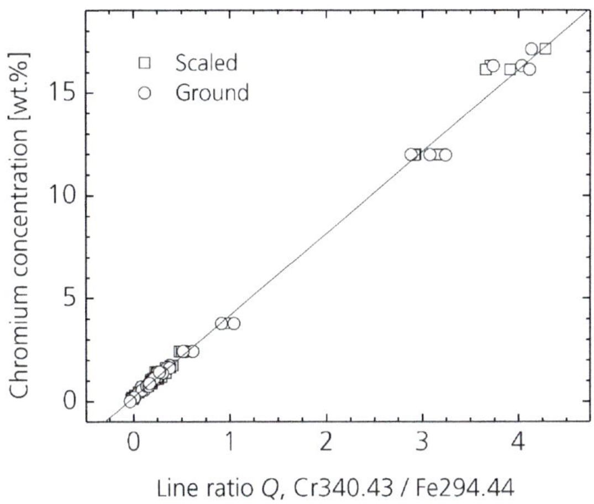

FIG. 12 Analysis curve of chromium for scaled and ground surfaces of steel grades deploying a two-phase measuring sequence of laser ablation and LIBS analysis [2].

6 Inverse production scenario for the recovery of valuable materials from electronic equipment

Electronic equipment, such as mobile phones (MPH) and electronic boards from computers, servers, and switching electronics, comprises a lot of valuable materials such as technology metals. On the other hand, their life-cycle is decreasing. Thus the amount of MPHs no longer used increases continuously, reaching an accumulated level of >190 million units in 2014 in the European Union [42]. The estimated total number of end-of-life MPHs worldwide is 1.75 billion units per year.

This mass flux of valuable material is often lost for the European economy by exports to Africa or Asia [43]. If recycled in the European Union, the conventional approach is focused on mass stream concepts such as shredder processes, partially followed by magnetic or eddy-current separators and pyrometallurgical procedures to extract the high-value metallic constituents, i.e., gold, silver, palladium, and copper from MPHs and printed circuit boards (PCBs). However, a series of critical elements cannot be recovered efficiently or is even lost completely in dust or in residual fractions with a high degree of dilution.

How can this mass flow be tackled in an efficient way? That is the point on which the European project "Next generation urban mining—automated disassembly, separation and recovery of valuable materials from electronic equipment," short title ADIR, is focused [44]. The goal of ADIR is to demonstrate the feasibility of a key technology for next-generation urban mining by selective treatment of the piece goods MPHs and PCBs, e.g., from servers and personal computers, for efficient recovery of all valuable materials instead of using only mechanical, such as crushing or shredding, and metallurgical methods designed mainly for the recovery of the aforementioned metals only.

The overall objective of ADIR is to open the door for next-generation urban mining by exploring and demonstrating innovative technologies for automated disassembly, separation, and recovery of valuable materials from electronic equipment. The process chain addressed by ADIR ranges from the collection of end-of-life MPHs and PCBs from computers/servers, their preparation by a recycler, and piece-by-piece processing in the ADIR demonstrator up to the recovery of valuable materials [45].

The technological concept of the ADIR demonstrator is shown schematically in Fig. 13. A robot and a manipulator pick single objects from the input stream and sequentially feed machines I, II, and III for selective disassembly of MPHs and PCBs yielding the following sorting fractions: S0 batteries, S1–S3 enriched fractions of valuable materials, RF residual fraction.

The collected end-of-life MPHs are piled in a container and fed to a conveyor (not shown in Fig. 13). A robot picks a single object and positions this piece in the first machine I, where the case of the MPH is opened and the rechargeable battery is removed and put into bin S0 (sorting fraction zero). These actions are based on robotic processes and milling actions. The PCB is taken out from machine I and a manipulator transfers this PCB to the next machine II. Machine II performs various measurements on the PCB. It takes 2D and 3D (by laser light section) images, and processes these to identify interesting components potentially containing high amounts of valuable material. LIBS ("material analysis" in Fig. 13) analyzes inline the constituents of these components, such as, e.g., tantalum in capacitors, as the basis for extraction and sorting decisions. These data are used to feed a database and to extract coordinates for laser beam control to unsolder components from the PCB or to cut out parts of that PCB in machine III.

The disassembled components are discharged to sorting fractions S1 to S3. After removal of the selected components the processed PCB goes to a residual fraction (RF). This fraction undergoes postprocessing by strong shockwaves generated by a pulsed-power machine (not shown in Fig. 13) to remove all electronic components

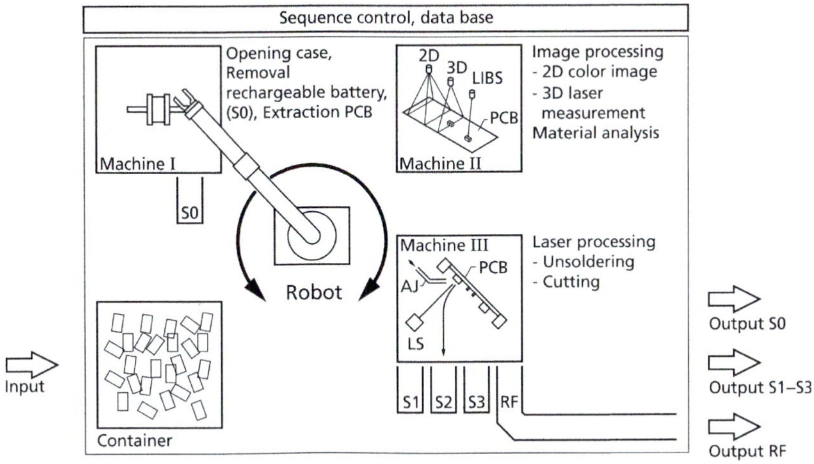

FIG. 13 Concept of the ADIR demonstrator. 2D/3D = two-/three-dimensional measurement of the geometry of PCBs, LIBS = laser-induced breakdown spectroscopy for material analysis, LS = laser source and scanner for unsoldering and cutting, AJ = air jet.

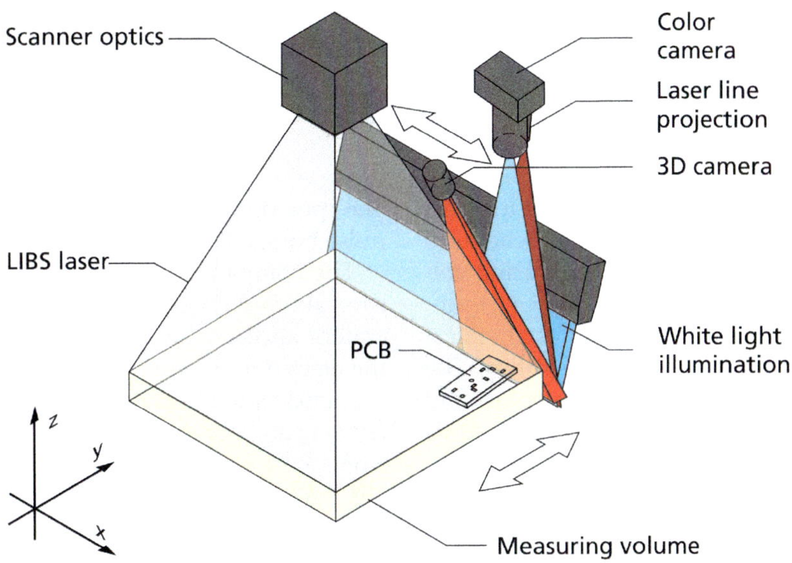

FIG. 14 Principal setup of machine II of the ADIR demonstrator with 2D (color) and 3D cameras and a LIBS unit with scanner optics. The measuring object—a PCB—is placed on a translational stage moving in the y-direction during image acquisition.

from the PCB. The generated two fractions—electronic components and bare PCB—are the input for specific hydrometallurgical and pyrometallurgical treatments to recover the valuable materials contained therein.

The principal setup of machine II is shown in Fig. 14. The PCBs to be measured are positioned in a measuring volume. By relative movement of a 2D- and 3D-camera, high-resolution images of the PCBs are taken. In the next step, using

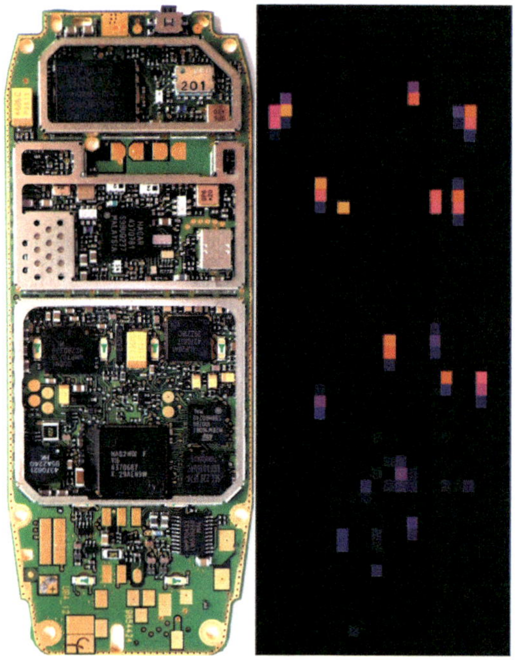

FIG. 15 *Left*: Photograph of the front side of a PCB from a MPH. *Right*: Raster scan by laser-induced breakdown spectroscopy showing in false colors (*orange/red* color presents high intensities) a chemical image of spectroscopic signatures of tantalum [46].

scanner optics, a laser beam is scanned to select electronic components to measure chemical information with LIBS [44]. The measuring volume has the dimensions $500 \times 500 \times 60\,mm^3$. By this approach, physical as well as chemical information is gained in the same setup. The LIBS WD between the last scanner optics and the bottom of the measuring volume amounts to 650 mm.

For test purposes a raster scan of a complete PCB from an MPH was performed with a step size of 2 mm for LIBS measurements to determine spectroscopic signals from, e.g., tantalum as a function of the *xy*-position on the PCB. The laser source runs at 1064 nm, 15 Hz with a tailored pulse train consisting of a cleaning pulse (here used for local ablation/drilling) and a double pulse. The laser spectroscopic method enables a multielement analysis. The measurement time per spot (including a series of irradiated laser pulse bursts) is in the order of 67 ms.

The complete PCB of an MPH was rastered several times. By this approach the formed laser craters successively penetrate the housings of the electronic components on the PCB to access the subjacent bulk material. Fig. 15 shows on the left side a photograph of the studied PCB (taken from a Nokia 3310 cell phone, size of PCB $108 \times 42\,mm^2$) and on the right the result of the 14th LIBS raster scan is shown as a false color plot to indicate tantalum. "14th raster scan" means that at each raster spot a sequence of 13 pulse bursts was irradiated plus the cleaning pulse of the 14th burst, and the LIBS signal generated by the last double pulse is evaluated. An overlay of the geometric 2D image and the chemical 2D image yields the identified valuable components.

Based on these data fed to the database a component list is generated comprising a set of coordinates to be used for the subsequent processes of selective laser unsoldering and cutting out of electronic components from PCBs (cf. Fig. 13, machine III).

Since the end of 2018 the ADIR demonstrator consisting of seven interlinked machines was set up at a recycling plant in Germany. These machines implement processes for feeding singularized MPHs, automated disassembly of MPHs, extraction of batteries and PCBs from MPHs, 2D/3D measurement of PCBs (extracted from MPHs or PCBs from computers/servers), laser spectroscopic analysis (LIBS), laser unsoldering and cutting, pulsed-power fragmentation (to release vibration alerts and loudspeakers from residues of disassembled MPHs), image recognition of fragmented pieces, and finally sorting of the extracted components to defined sorting fractions.

Hydrometallurgical tests were performed demonstrating that the recovery of elements from sorted fractions such as tantalum, tungsten, and neodymium is possible, achieving recovery rates of >90%.

During 2019 the ADIR demonstrator was tested in extended field tests to show the capability of automatic disassembly, separation, and recovery of valuable materials from end-of-life electronic equipment [47].

7 Conclusion

During the period 2014–19 two main development directions for industrial LIBS applications became obvious. On the one hand, stationary automated LIBS systems were developed and some of these were already put into routine use at production lines. They gained inline chemical information for monitoring and control purposes on material-dependent processes. Examples were shown for high-speed sorting tasks, sorting of refractories and alloyed metal scrap, identification of steel blooms, and the characterization of PCBs for the selective recovery of valuable materials. WDs in the range of 10 cm to about 1 m are predestinated for inline measuring tasks by automated LIBS systems in production and process lines since this allows for a simplified integration in the industrial environment considering the size and physical state of the measuring object.

For short WDs the reduced sizes of laser sources have stimulated the evolution of handheld LIBS systems in a competitive market, which will lead to lot sizes of LIBS units much higher than those achieved so far for stationary LIBS machines.

Acknowledgments

A part of the presented results was gained within the frame of a series of national and international R&D projects. The author wishes to express his thanks for the cofinancing of these projects by industrial partners: the Federal Ministry of Education and Research of Germany (BMBF), the Federal Ministry of Economics and Technology of Germany (BMWi), the European Commission, and the Fraunhofer Gesellschaft für Angewandte Forschung e.V., Germany. The project PLUS is funded by the BMBF within the program "r +Impuls—Innovative Technologien für Ressourceneffizienz, grant no. 033R181. The project concerning the identification of steel blooms, short title "Laser-ID," was cofinanced by the European Union and the Federal State of North-Rhine Westphalia, Germany: "Europäischer Fonds für regionale Entwicklung (EFRE) für NRW," grant agreement no. 603809. The project ADIR has received funding from the European Union's Horizon 2020 research and innovation program under grant agreement no. 680449.

References

[1] F. Boué-Bigne, Laser-induced breakdown spectroscopy and multivariate statistics for the rapid identification of oxide inclusions in steel products, Spectrochim. Acta B 119 (2016) 25–35.

[2] C. Meinhardt, V. Sturm, R. Fleige, C. Fricke-Begemann, R. Noll, Laser-induced breakdown spectroscopy of scaled steel samples taken from continuous casting blooms, Spectrochim. Acta B 123 (2016) 171–178.

[3] V. Sturm, R. Fleige, M. de Kanter, R. Leitner, K. Pilz, D. Fischer, G. Hubmer, R. Noll, Laser-induced breakdown spectroscopy for 24/7 automatic liquid slag analysis at a steel works, Anal. Chem. 86 (2014) 9687–9692.
[4] S. Hudson, J. Craparo, R. De Saro, D. Apelian, Laser-induced breakdown spectroscopy: a new tool for real time melt cognition, Metall. Ital. 6 (2016) 5–8.
[5] L. Sun, H. Yua, Z. Cong, Y. Xina, Y. Li, L. Qi, In situ analysis of steel melt by double-pulse laser-induced breakdown spectroscopy with a Cassegrain telescope, Spectrochim. Acta B 112 (2015) 40–48.
[6] P. Werheit, C. Fricke-Begemann, M. Gesing, R. Noll, Fast single piece identification with a 3D scanning LIBS for aluminium cast and wrought alloys recycling, J. Anal. At. Spectrom. 26 (2011) 2166–2174.
[7] R. Noll, C. Fricke-Begemann, S. Connemann, C. Meinhardt, V. Sturm, LIBS analyses for industrial applications—an overview of developments from 2014 to 2018, J. Anal. At. Spectrom. 33 (2018) 945–956.
[8] B. Noharet, E. Zetterlund, O. Tarasenko, M. Lindblom, J. Gurell, A. Bengtson, P. Lundi, Spectroscopy-based photonic instrumentation for the manufacturing industry: contactless measurements of distances, temperatures, and chemical composition, Proc. SPIE 8992 (2014). 89920R-1, -8.
[9] C. Li, Z. Zou, X. Yang, Z. Hao, L. Guo, X. Li, Y. Lu, X. Zeng, Quantitative analysis of phosphorus in steel using laser-induced breakdown spectroscopy in air atmosphere, J. Anal. At. Spectrom. 29 (2014) 1432–1437.
[10] G. Lorenzetti, S. Legnaioli, E. Grifoni, S. Pagnotta, V. Palleschi, Laser-based continuous monitoring and resolution of steel grades in sequence casting machines, Spectrochim. Acta B 112 (2015) 1–5.
[11] S. Merk, C. Scholz, S. Florek, D. Mory, Increased identification rate of scrap metal using laser-induced breakdown spectroscopy echelle spectra, Spectrochim. Acta B 112 (2015) 10–15.
[12] Q. Zeng, L. Guo, X. Li, C. He, M. Shen, K. Li, J. Duan, X. Zeng, Y. Lu, Laser-induced breakdown spectroscopy using laser pulses delivered by optical fibers for analyzing Mn and Ti elements in pig iron, J. Anal. At. Spectrom. 30 (2015) 403–409.
[13] B. Noharet, C. Sterner, T. Irebo, J. Gurell, A. Bengtson, R. Vainik, H. Karlsson, E. Illy, A compact LIBS system for industrial applications, Proc. SPIE 9369 (2015). 936904-1, -7.
[14] Z. Wang, Y. Deguchi, F. Shiou, J. Yan, J. Liu, Application of laser-induced breakdown spectroscopy to real-time elemental monitoring of iron and steel making processes, ISIJ Int. 56 (2016) 723–735.
[15] M. Afgan, Z. Hou, Z. Wang, Quantitative analysis of common elements in steel using a handheld μ-LIBS instrument, J. Anal. At. Spectrom. 32 (2017) 1905–1915.
[16] A. Bengtson, Laser-induced breakdown spectroscopy compared with conventional plasma optical emission techniques for the analysis of metals—a review of applications and analytical performance, Spectrochim. Acta B 134 (2017) 123–132.
[17] A. Tortschanoff, M. Baumgart, G. Kroupa, Application of a compact diode pumped solid-state laser source for quantitative laser-induced breakdown spectroscopy analysis of steel, Optim. Eng. 56 (2017). 124104-1, -8.
[18] T. Zhao, Z. Fan, F. Lian, Y. Liu, W. Lin, Z. Mo, S. Nie, P. Wang, H. Xiao, X. Li, Q. Zhong, H. Zhang, Using laser-induced breakdown spectroscopy on vacuum alloys-production process for elements concentration analysis, Spectrochim. Acta B 137 (2017) 64–69.
[19] Q. Zeng, C. Pan, C. Li, T. Fei, X. Ding, X. Du, Q. Wang, Online monitoring of corrosion behavior in molten metal using laser-induced breakdown spectroscopy, Spectrochim. Acta B 142 (2018) 68–73.
[20] L. Sun, H. Yu, Z. Cong, H. Lu, B. Cao, P. Zeng, W. Dong, Y. Li, Applications of laser-induced breakdown spectroscopy in the aluminum electrolysis industry, Spectrochim. Acta B 142 (2018) 29–36.
[21] D. Luo, Y. Liu, X. Li, Z. Zhao, S. Wang, Y. Zhang, Quantitative analysis of C, Si, Mn, Ni, Cr and Cu in low-alloy steel under ambient conditions via laser-induced breakdown spectroscopy, Plasma Sci. Technol. 20 (2018) 075504.
[22] J. Yang, X. Li, H. Lu, J. Xu, H. Li, A LIBS quantitative analysis method for alloy steel at high temperature based on transfer learning, J. Anal. At. Spectrom. 33 (2018) 1184–1195.
[23] Q. Zeng, F. Deng, Z. Zhu, Y. Tang, B. Wang, Y. Xiao, L. Xiong, H. Yu, L. Guo, X. Li, Portable fiber-optic laer-induced breakdown spectroscopy system for the quantitative analysis of minor elements in steel, Plasma Sci. Technol. 21 (2019) 034006.
[24] M. Gaft, L. Nagli, Y. Groisman, A. Barishnikov, Industrial online raw materials analyzer based on laser-induced breakdown spectroscopy, Appl. Spectrosc. 68 (2014) 1004–1015.
[25] Z. Hou, Z. Wang, T. Yuan, J. Liu, Z. Li, W. Ni, A hybrid quantification model and its application for coal analysis using laser-induced breakdown spectroscopy, J. Anal. At. Spectrom. 31 (2016) 722–736.
[26] E. Ljutic, P. Kolmhofer, H. Heilbrunner, B. Praher, N. Huber, R. Rössler, H. Wolfmeir, E. Arenholz, J. Heitz, J. Pedarnig, Detection of phosphorous in industrial oxide materials by laser-induced breakdown spectroscopy in the UV spectral range, Roman. Rep. Phys. 66 (2014) 1155–1164.
[27] F. Trichard, L. Sorbier, S. Moncayo, Y. Blouet, C. Lienemann, V. Motto-Ros, Quantitative elemental imaging of heterogeneous catalysts using laser-induced breakdown spectroscopy, Spectrochim. Acta B 133 (2017) 45–51.

[28] F. Trichard, S. Moncayo, D. Devismes, F. Peascini, J. Maurelli, A. Feugier, C. Sasseville, F. Suma, V. Motto-Ros, Evaluation of a compact VUV spectrometer for elemental imaging by laser-induced breakdown spectroscopy: application to mine core characterization, J. Anal. At. Spectrom. 32 (2017) 1527–1534.

[29] S. Legnaioli, B. Campanella, S. Pagnotta, F. Poggialini, V. Palleschi, Determination of ash content of coal by laser-induced breakdown spectroscopy, Spectrochim. Acta B 155 (2019) 123–126.

[30] F. Aquino, J. Santos, R. Carvalho, J. Coelho, E. Pereira-Filho, Obtaining information about valuable metals in computer and mobile phone scrap using laser-induced breakdown spectroscopy (LIBS), RSC Adv. 5 (2015) 67001–67010.

[31] Y. Gong, D. Choi, B. Han, J. Yoo, S. Han, Y. Lee, Remote quantitative analysis of cerium through a shielding window by stand-off laser-induced breakdown spectroscopy, J. Nucl. Mater. 453 (2014) 8–15.

[32] T. Karpate, M. Shameem, R. Nayak, V. Unnikrishnan, C. Santhosh, LIBS: a potential tool for industrial/agricultural waste water analysis, Proc. SPIE 9893 (2016). 989317-1, -7.

[33] R. Noll, C. Fricke-Begemann, M. Brunk, S. Connemann, C. Meinhardt, M. Scharun, V. Sturm, J. Makowe, C. Gehlen, Laser-induced breakdown spectroscopy expands into industrial applications, Spectrochim. Acta B 93 (2014) 41–51.

[34] M. Gaft, L. Nagli, I. Fasaki, M. Kompitsas, G. Wilsch, Laser-induced breakdown spectroscopy for on-line sulfur analyses of minerals in ambient conditions, Spectrochim. Acta B 64 (2009) 1098–1104.

[35] H. Xia, M. Bakker, Reliable classification of moving waste materials with LIBS in concrete recycling, Talanta 120 (2014) 239–247.

[36] Product MopaLIBS of the company Secopta, Berlin, http://www.secopta.de/produkte/mopalibs.

[37] R. Noll, Laser-Induced Breakdown Spectroscopy—Fundamentals and Applications, Springer-Verlag, Heidelberg, ISBN: 978-3-642-20667-2, 2012.

[38] REFRASORT, Innovative Separation Technologies for High Grade Recycling of Refractory Waste Using Non Destructive technologies, Final report summary, Grant Agreement no. 603809, http://cordis.europa.eu/result/rcn/194560_en.html, 2016.

[39] M. Geisler, S. Connemann, C. Fricke-Begemann, Sensorgestützte Sortierung von legierten Metallschrotten, in: S. Thiel, O.H.E. Thomé-Kozmiensky, D. Goldmann, B. Friedrich (Eds.), Recycling und Rohstoffe, Thomé-Kozmiensky Verlag GmbH, Neuruppin, Germany, ISBN: 978-3-944310-46-6, 2019, pp. 567–579.

[40] funded by BMBF, r+Impuls—Innovative Technologien für Ressourceneffizienz, PLUS—Pilotanlage zur laserge stützten Sortierung von Sonderlegierungen, 033R181, 1.1.2017–31.12.2019.

[41] V. Sturm, C. Meinhardt, R. Fleige, C. Fricke-Begemann, J. Eisbach, Fast identification of steel bloom composition at a rolling mill by laser-induced breakdown spectroscopy, Spectrochim. Acta B 136 (2017) 66–72.

[42] https://de.statista.com/statistik/daten/studie/192704/umfrage/absatz-von-mobiltelefonen-weltweit-seit-2005/.

[43] M. Kaya, Recovery of metals and nonmetals from electronic waste by physical and chemoca recycling processes, Waste Manag. 57 (2016) 64–90.

[44] R. Noll, G. Benke, M. Bergamos, H. Brumm, M. Eschen, C. Fricke-Begemann, K. Leszczynska-Sejda, J. Makowe, F. Schreckenberg, A. Tori, F. Veglia, Interlinked machinery for the automated disassembly, separation and recovery of valuable materials from electronic equipment—progress of the ADIR project, in: European Metallurgical Conference, EMC 2019, Vol. 3—Sustainable Production, ISBN: 978-3-940276-89-6, 2019, pp. 1209–1225.

[45] www.ADIR.eu.

[46] R. Noll, R. Ambrosch, K. Bergmann, S. Britten, H. Brumm, A. Chmielarz, S. Connemann, M. Eschen, A. Frank, C. Fricke-Begemann, C. Gehlen, T. Gorewoda, M. Guolo, W. Kurylak, J. Makowe, G. Sellin, M. Siguier, A. Tori, F. Veglia, Next generation urban mining—automated disassembly, separation and recovery of valuable materials from electronic equipment: overview of R&D approaches and first results of the European project ADIR, in: European Metallurgical Conference, EMC 2017, Vol. 2—Strategies for Resource Savings, ISBN: 978-3-940276-73-5, 2017, pp. 799–816.

[47] R. Noll, K. Bergmann, C. Fricke-Begemann, Schreckenberg, Inverse produktion für nachhaltige wertstoffkreisläufe – aktuelle entwicklungen zur automatisierten demontage und entstückung von elektronikplatinen, Chem. Ing. Technol 92 (2020). https://doi.org/10.1002/cite.201900123.

CHAPTER

20

Laser-induced breakdown spectroscopy in planetary science

Roger C. Wiens[a], *Xiong Wan*[b], *Jeremie Lasue*[c], *Sylvestre Maurice*[c]

[a]Los Alamos National Laboratory, Los Alamos, NM, United States

[b]Shanghai Institute of Technical Physics (SITP), Chinese Academy of Sciences (CAS), Shanghai, China

[c]IRAP, Univ. Paul Sabatier –CNRS –Obs. Midi-Pyrenees, Toulouse, France

1 Introduction

Laser-induced breakdown spectroscopy (LIBS) is known as a rugged technique that can be used in many different environments. As such, the use of LIBS in planetary science is natural. Perhaps more than anywhere else, it is critical to be able to analyze a sample with minimal preparation and handling, and to use compact and rugged instrumentation. Compact lasers and spectrometers can be made to fit into a package as small as a few kilograms. A number of other factors also make LIBS especially appealing to planetary science. These include the ability to remove dust from the analytical surface thanks to the shockwave from the plasma (Fig. 1B), and to perform depth profiles, observing surface weathering [1] and probing for layers in the targets [2, 3] (Fig. 1). The ability to perform rapid analysis is also important to space experiments, and a particular benefit is the complete spectrum of elements that can be observed, including the light elements (atomic mass <20) that are lacking in analyses by X-ray fluorescence instruments that have traditionally been used for surface composition measurements for planetary science. These light elements are potentially of particular interest, as hydrogen is important to the detection of clay minerals and other products of the weathering of rocks. On Mars, these characteristics are important, as they help to characterize the past climatic conditions (e.g., Refs. [4–7]). In other situations, the presence of organic materials could be determined by the observation and characterization of carbon, nitrogen, and oxygen emission lines. And in locations where ice is prevalent, LIBS analyses can provide unique compositional measurements of ice [8, 9], distinguishing, for example, trace elements [10] or excesses of oxygen or the presence of organic materials within the ice. The idea of using LIBS in planetary science is several decades old, and was first presented in 1992 [11]. At that time, NASA had not landed any planetary missions since the early 1970s. With the establishment of NASA's Mars program with the successful *Sojourner* rover in 1997 and more missions in the planning phase, development began at Los Alamos and eventually other places on application of LIBS to Mars. This included work in the

Laser-Induced Breakdown Spectroscopy
https://doi.org/10.1016/B978-0-12-818829-3.00020-4

FIG. 1 ChemCam, the first LIBS instrument on another planet (A), peers through a window in the *Curiosity* rover's mast enclosure. An inverted image of the background is seen in the telescope's primary mirror. LIBS pits (B; some indicated with *black arrows*) in Mars target Stephen (MSL mission day 627), with dust moved by the plasma shockwaves to the upper and left-most portions of the rock surface. Prior to LIBS, the rock had the same rust-red color as the surroundings. Laser removal of the dust revealed the dark surface. The LIBS pits *(arrows)* show light-colored interiors contrasting with the dark, manganese-rich rock coating. The spectra of corresponding laser shots show that the laser beam profiled through the thin coating, revealing the interior composition of the rock. The rock is the size of a penny, shot from about 2.4 m distance by the ChemCam instrument. The image was taken by MAHLI at a distance of 3 cm. The shadow of part of the arm is seen across the target. *Image: NASA/JPL-Caltech/MSSS; public domain.*

laboratory [12], development of a commercial, off-the-shelf-based prototype for demonstration on a NASA Mars analog rover [13], and eventually plans for flight instrumentation (e.g., Ref. [14]). The initial work showed several very important details: (1) that LIBS could work and be calibrated in different atmospheres, including that of Mars, and lower pressures [12] such as for airless bodies like the Moon [15] or Europa [9], and for higher-pressure atmospheres such as Venus [16, 17]; (2) that compact lasers (e.g., Ref. [13]) and spectrometers [14] could be used for this work; and (3) that long-distance remote LIBS could be performed with reasonable elemental abundance accuracy and precision [12].

After the 2004 selection of the ChemCam LIBS instrument [18, 19] (Fig. 1A) for the Mars Science Laboratory (MSL) rover there was a surge of interest in developing LIBS for planetary science. Initially, the ExoMars rover was planned to have a joint LIBS-Raman instrument; however, the LIBS capability was removed. In 2009, a joint LIBS-Raman instrument was selected by NASA for Phase-A development of the Surface and Atmosphere Geochemical Explorer (SAGE), a Venus lander mission [17]. Unfortunately, this mission was not ultimately selected for flight development. Development of a Venus LIBS and Raman spectroscopy instrument is currently (in 2020) taking place at Los Alamos under NASA New Frontiers Mission Office funding.

The Indian Space Research Organization (ISRO) launched a lunar rover equipped with a LIBS instrument in 2019, though it failed to land successfully. Two other rovers with LIBS instruments are currently being readied for launch. NASA's Mars 2020 rover is equipped with a remote LIBS, Raman, imaging, and VISIR instrument called SuperCam [20]. The Chinese Space Agency is preparing to launch a Mars rover in 2020 that is also equipped with a remote LIBS instrument. This brings to three the number of planetary LIBS experiments that may be operational by 2021.

This chapter describes what we know about these instruments, their calibration, and results so far, and additional possibilities for future planetary exploration.

2 ChemCam

2.1 Instrument description

ChemCam is a combined remote LIBS and imaging instrument designed for the MSL rover named *Curiosity*. It was built in two separate units. The laser and telescope required mounting on the rover's mast so that it could be aimed at targets using the azimuth and elevation gimbals of the rover's remote-sensing mast (RSM) structure. However, the RSM was designed to hold a minimal amount of mass, and so the spectrometers and the larger share of the electronics of the instrument were designed to fit in the body of the rover. *Curiosity*'s predecessors, *Spirit* and *Opportunity*, had somewhat similar architecture with the Mini-TES instrument [21], which used periscope optics to direct light acquired by the mast down through the tube comprising the mast to the main part of the instrument in the body of the rover. However, the alignment of the periscope was a challenge. In the case of ChemCam, light is passed to the rover body by a fiber optic cable, which is easier. ChemCam was developed and is operated as a partnership between Los Alamos National Laboratory (LANL), funded by NASA, and l'Institut de Recherche en Astrophysique et Planetologie (IRAP), funded by CNES. LANL built the body unit and leads the project, while IRAP built the mast unit and shares a 50% role in ChemCam's operation. The mast and body units are shown in Fig. 2A and B, and a schematic diagram is shown in Fig. 3A. A complete description of the instrument is provided in [18, 19]; here we provide a brief description.

The mast unit telescope, a 110-mm diameter aperture Schmidt-Cassegrain design, is used both for projection of the laser light and to collect the return light from the plasma, eliminating parallax and providing simultaneous focusing of the laser beam and the returning light. Fig. 4 shows in red the path of the outgoing laser light. The beam is expanded through a small Galilean telescope and is then dumped into the primary telescope by reflecting off a dichroic mirror at the center of the primary mirror. The laser beam is then expanded to the diameter of the telescope, which focuses it on the target up to 7 m away. The returning light collected by the telescope passes through the laser dichroic at the center of the primary mirror and is directed into an optical fiber behind the telescope. The telescope is focused by a linear stage that moves the secondary mirror. The instrument can be focused from 1.2 m to infinity [19].

Thales Optronique SA provided the LIBS laser for ChemCam. It is passively cooled, providing >24 mJ from the laser across the range of −20°C to +20°C. To achieve this, a neodymium-doped potassium gadolinium tungstate (Nd:KGW) rod is used in the oscillator section, pumped by a 700-W diode stack. The oscillator is Q-switched with a rubidium titanyl phosphate Pockels cell to provide nanosecond pulses at 1067 nm. Two diode-pumped amplifiers enhance the output energy to meet the requirement. Custom diode stacks were used to optimize the power output near −5°C [19]. In operation, the laser is warmed to −10°C prior to operation; self-heating rarely exceeds −5°C in the cold Martian environment. The laser was designed for 1–10 Hz but has been operated only at 3 Hz on Mars. The laser is rated for 3 million shots; so far, on Mars it has surpassed 700,000 shots, each with a corresponding LIBS spectrum sent back to Earth.

The LIBS beam spot size on the targets is in the range of 350–550 μm depending on the distance [19] (Fig. 1). ChemCam's stated distance

FIG. 2 ChemCam and SuperCam instruments for the NASA *Curiosity* and Mars 2020 rovers, respectively. ChemCam body unit (A) resides in the rover body and contains an optical demultiplexer, three reflection spectrometers, and electronics. ChemCam's mast unit (B) is mounted in the rover's mast and contains the laser, telescope, Remote Micro-Imager, and electronics. SuperCam's body unit (C) contains ChemCam's elements, but with a transmission spectrometer substituting for one of ChemCam's spectrometers. Different thermoelectric cooling systems can be seen in the lower portions of each panel in (A) and (C). A bracket for easier mounting of cables extending to the rover is included in SuperCam's body unit, seen on the right side of (C). SuperCam's mast unit (D) includes a second laser path for a green Raman beam (including the angled mirror at the center of the SuperCam telescope), and an infrared spectrometer and microphone (not shown). Both SuperCam units are shown instrumented for vibration testing with accelerometers and blue wires that are not part of the instrument.

limit is 7 m; however, most observations are between 2 m (straight down to the ground from the mast) and 4 m [22]. The main reason for this shorter distance for most targets is the limited resolution of the rover's Navcams (navigation cameras) that are usually used to select targets. Small features cannot be seen on Navcam images beyond about this distance. However, the ChemCam distance limit is somewhat fluid, as the team has shot an iron meteorite at a distance of >9 m, clearly identifying the lower-abundance nickel peaks in the spectrum.

The mast unit optical system contains two other subsystems. The Remote Micro-Imager (RMI) is a 1-megapixel Atmel panchromatic charge-coupled device (CCD) camera that records images through the telescope [19, 22]. A beam splitter in the telescope objective diverts a small amount (~10%) of the light to the imager. The images have a field of view of 20 mrad and an effective resolution to about 40 μrad at best focus. The RMI provides the highest resolution remote images from the rover. Its purpose is generally to provide context images for the LIBS targets. Given that the LIBS

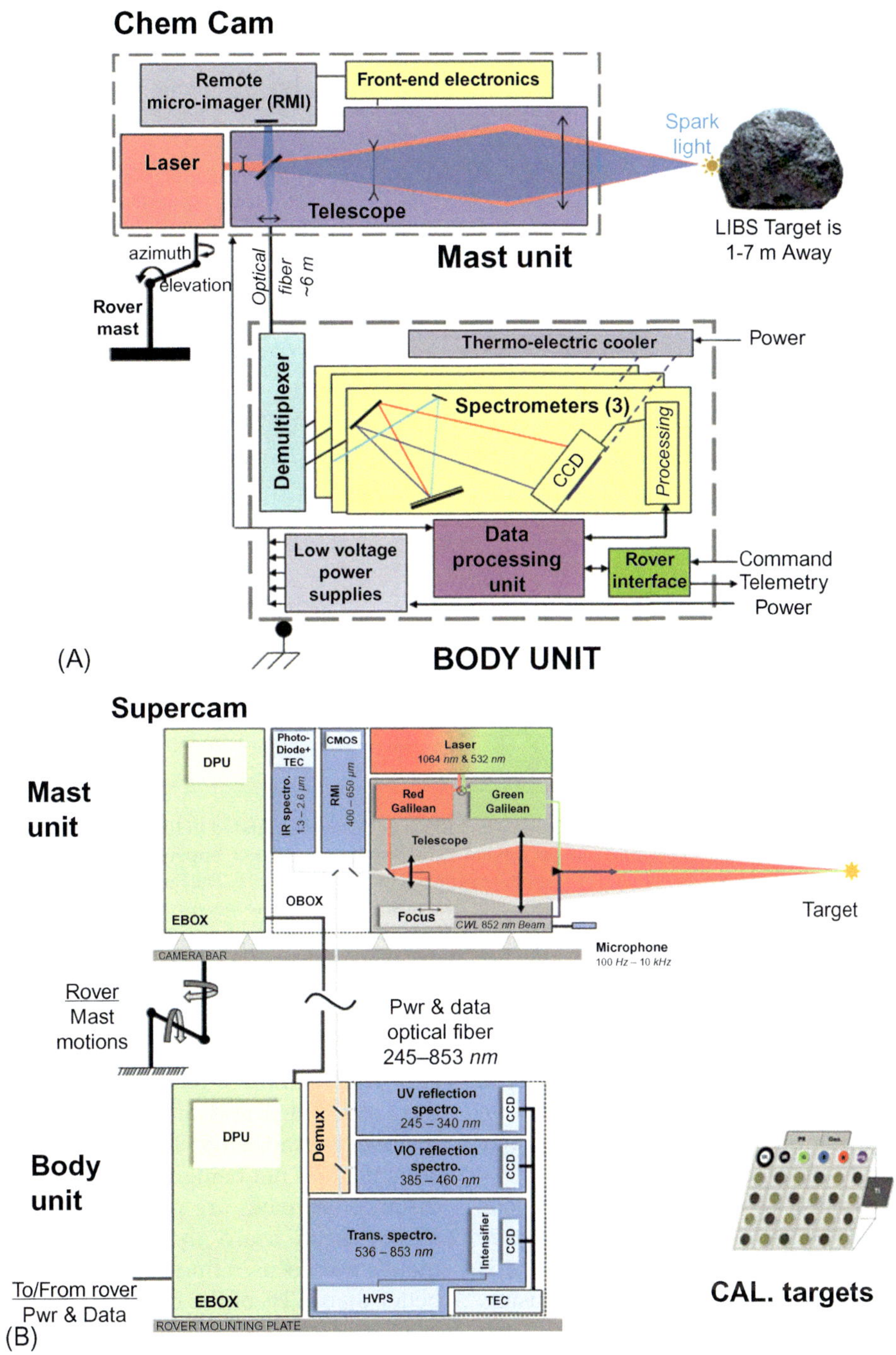

FIG. 3 Block diagrams of ChemCam (A) and SuperCam (B). The overall architecture is the same, with added features in both the mast and body units of SuperCam. Both ChemCam and SuperCam incorporated on-board calibration targets. However, SuperCam's calibration target assembly is far more extensive. *CCD*, charge-coupled device; *Demux*, demultiplexer; *DPU*, data processing unit; *EBOX*, electronics box; *HVPS*, high-voltage power supply; *IR*, infrared; *TEC*, thermoelectric cooler; *UV*, ultraviolet; *VIO*, violet.

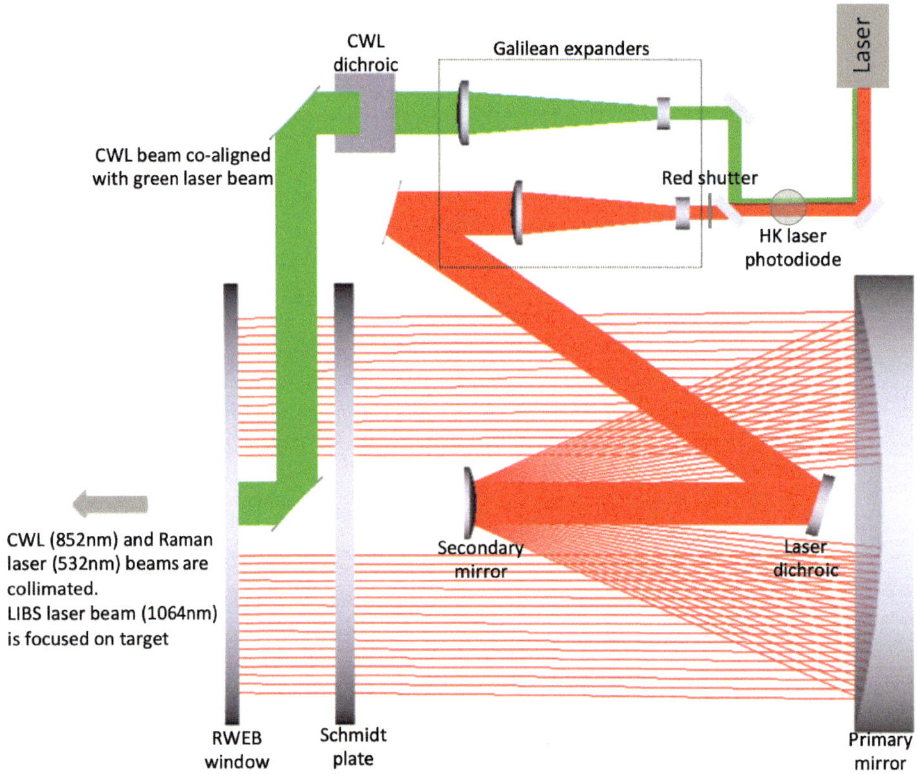

FIG. 4 Laser beam path for ChemCam is shown in *red*. The LIBS laser beam is expanded to 100 mm, nearly the entire diameter of the Schmidt-Cassegrain telescope, and then focused to a small spot on the target. SuperCam uses the same design for the LIBS beam path. Also shown is the path for the SuperCam green Raman beam *(green)*. The SuperCam Raman system uses a frequency-doubled parallel beam projected from the center of the telescope via a periscope. A continuous-wave laser (CWL) is built into the center of the telescope on ChemCam, and is coaligned with the Raman periscope on the SuperCam design, as shown in this diagram. Wavelengths indicated are for SuperCam. ChemCam's wavelengths are 1067 and 785 nm for the LIBS laser and CWL, respectively.

observation points are submillimeter in size, such a high-resolution context imager is a necessity to place the chemical compositions from LIBS in the context of the morphology and mineralogy of the rock and soil targets.

The other optical system is for autonomous focusing. At the center of the telescope aperture a small continuous-wave diode laser (CWL) is mounted (Fig. 2B) to provide a signal for focusing. It emits at 785 nm with up to 25 mW, modulated at 10 kHz. A photodiode at the back of the telescope is tuned to this wavelength and modulation frequency. With the CWL on, the photodiode collects signal while the focus stage is scanned. The resulting curve is used to find the optimum focus [19]. This method of focusing was not used after the CWL failed and an image-based autofocus method was adopted. The new method has the advantage of not being limited in distance. Seed distance estimates are provided from the stereo Navcams on board the rover, and the instrument's autofocus refines these to an accuracy of better than 0.2% of the distance.

ChemCam's body unit (Figs. 2A and 3A) contains an optical demultiplexer that receives the light from the mast-to-body fiber, splits the light into three different bands using dichroic mirrors, and passes the light to the three LIBS spectrometers. Light travels to the body unit via a 300-μm core ultraviolet (UV)-optimized, low-OH optical fiber. Between the demultiplexer and the spectrometers, the light travels through fiber bundles consisting of 12–19 fibers with 50- to 70-μm cores that are circularly arranged at the receiving end on the demultiplexer and linearly arranged at the spectrometer end, where the aperture slits are attached directly onto the ends of the fibers. The spectrometers use a crossed Czerny-Turner design from the Ocean Optics HR2000 commercial spectrometers. The spectral ranges are 240–342 nm (UV), 382–469 nm (violet), and 474–907 nm (visible and near infrared; VNIR). The three spectrometers are identical except for the optical coatings and gratings. The coatings of the mirrors and CCDs have been separately optimized for each spectral band. The detectors are Grade-0 E2V42-10 CCDs operated in advanced inverted mode. The active areas consist of 2048 × 515 pixels, each 13.5 μm on a side. The image pixels are summed in the serial register before being converted to digital 14-bit data [18]. During a burst of laser pulses for LIBS, the spectrum produced by each laser pulse is recorded separately and downlinked to the ground, so that a record of changes with depth within the laser pit is recorded. There is no shutter, so the active CCD region is exposed to light until the last pixel is read out, a process that takes ~13 ms. Because this results in a nonnegligible amount of ambient light, nonlaser background spectra are recorded under identical conditions for subtraction from the LIBS spectra. These background data are used for mineralogical information as passive reflectance spectra.

A detector cooling system was added late in the development phase. The body of the rover is heated by a fluid loop that rejects heat from the radiothermal isotope generator. The mounting plate for the body unit cycles over a differential range of 20–25°C for each 24-h 38-min day/night cycle on Mars, always staying within 0–35°C, slightly warmer in summer and cooler in winter. The thermoelectric cooler (TEC) system is turned on about a half hour before ChemCam is used, cooling each of the CCDs by about 25°C relative to the surroundings, reducing the detector noise.

The body unit houses three electronic boards: a low-voltage power board, a command and data handling (C&DH) board, and a spectrometer electronics board. The latter board operates the CCDs and retrieves data from them. The low-voltage power board receives the 28-V power from the rover, passes the mast unit power directly through, and filters and provides a number of analog and digital voltages for the CCDs, housekeeping circuits, heaters, and the boards themselves. The C&DH board handles communications with the rover, the spectrometer board, and the mast unit. It coordinates the timing of the laser firing and the CCD exposures. It also controls heater circuits and works with the mast unit to focus the telescope. For the latter, data are passed from the mast to the C&DH for processing and determination of the best focus [18]. Early in the mission, this consisted of focus photodiode data. After May 2015, this has been replaced by code to determine the best focus via a modified Laplacian that is implemented on pixels within a region of interest on successive images at different focus positions.

To save weight, the demultiplexer and electronics box were built of magnesium, and the spectrometers were built of beryllium. Mirrors and gratings were bonded to metal mounts, matching the coefficient of thermal expansion between the materials. Essentially, the entire mast unit was fabricated of aluminum. The mass of the body unit is 2445 g; the TEC is 2344 g; and the mast unit is 5778 g. The mast unit outside dimensions are 384 × 219 × 166 mm and the body unit with TEC is 197 × 236 × 154 mm [18].

2.2 Operational description

2.2.1 *Operations*

The sequence of events associated with a typical ChemCam activity are shown in Fig. 5. Before operation, mast unit parts must be

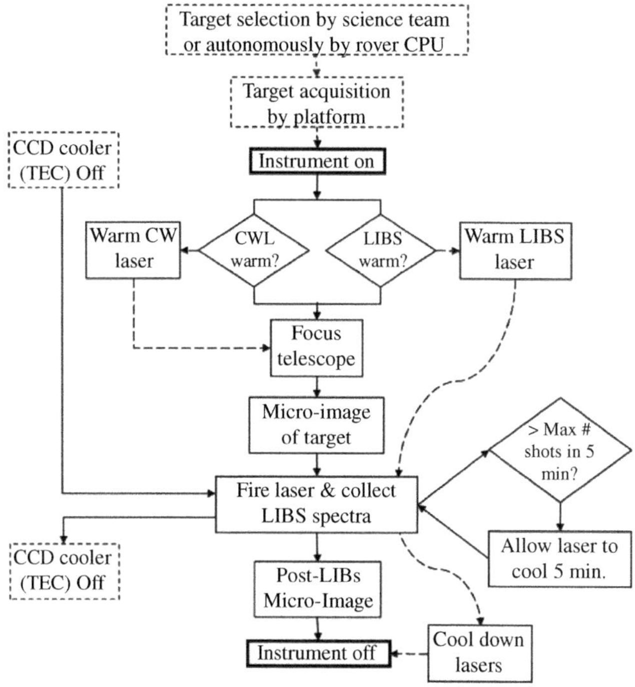

FIG. 5 Typical operation sequence for ChemCam. Cooling the CCDs using the TEC takes 30 min, commandable independently of the instrument; the laser warm-up stage takes about 10 min; and focusing the telescope takes about 1 min. The CWL is used for focusing. Once a single observation point is completed, the rover's mast points to the next location and the laser is fired again. The telescope is refocused and a new image is taken as often as is needed during a raster sequence. *CPU*, central processing unit; *CWL*, continuous-wave laser.

warmed. The mast unit is not heated by the rover's fluid loop; rather, it has a survival heater that maintains $\geq -40°C$; the laser is warmed to $-10°C$ before operating. Likewise, the TEC is turned on to cool the CCDs in the body unit. Once the systems are thermally ready, the rover points the instrument at the appropriate location and ChemCam proceeds with an autofocus, taking between 1 and 2 min. An RMI image is taken, and then LIBS analysis is done. Thirty laser shots and corresponding spectra are typically used. More than one LIBS observation is needed per target to check heterogeneity and improve statistics. So, after completing one LIBS observation, the rover will then point the instrument to a new location on the same target to develop a raster pattern (Fig. 1). These patterns usually consist of line scans of 5, 10, or 20 observation points (each with its own laser burst), or a grid pattern such as 2×2, 3×3, or 4×4. The raster points are typically between 0.5 and 2 mrad apart, corresponding to 1.25–5 mm spacing for a typical observation at 2.5-m distance. As more points are added to the raster, additional RMI images may be needed to cover the observational space. A new RMI image is taken after the instrument pointing has been moved 5 mrad or more from the last image. Likewise, depending on the target topography, additional autofocus routines may be required to keep the observations in focus during the raster. After the last raster point is shot, a final RMI image is taken, and then the instrument is either vectored to observe another target or a shutdown routine is performed. The latter consists of moving the focus table back to a sun-safe position (so as not to damage the instrument if the sun goes through the field of view), turning off the TEC, cooling the laser at a safe rate, and finally powering off.

Besides the typical LIBS raster patterns, several other types of LIBS analyses can be carried out. These include depth profiles, in which up to ~500 laser pulses are used to profile deeper into the target, and observations of calibration targets on the rover (described later). Relatively rarely ChemCam performs alignment tests and LIBS z-stacks in which the focus is moved between several laser bursts at the same location to check the effects of focus on the spectra. Passive (nonlaser) observations (e.g., Ref. [23]) and RMI-only sequences (e.g., Ref. [24]) are also performed, which will not be described here.

ChemCam accomplishes these activities by a series of nested commands. When a target is selected in a Navcam image, the user selects a command sequence that is designed to handle all details for a given observation, including autofocus, RMI context images, and a raster of LIBS observation points. The sequence is associated with a target by providing a target name and coordinates. The sequence consists of a series of spacecraft commands that include commands to the rover (turn the instrument power on, command the instrument, wait for command responses, receive data, point the RSM, and eventually turn the instrument off) and to the instrument (i-commands). The i-commands are received by ChemCam and are parsed into its own commands to the mast unit and to different parts of the body unit.

Mars is too distant to operate the rover in real time, as communications take between 3 and 20 min one way, depending on the relative positions of Earth and Mars. In this way, a Mars rover differs substantially from, e.g., a lunar rover, where responses can be received within seconds, and real-time commanding is feasible. Because of this, rover operations are preprogrammed for the entire day (or weekend or longer). ChemCam commands are delivered from Los Alamos or Toulouse to the Jet Propulsion Laboratory where they are combined with other rover commands to produce 1 day's worth (or weekend's worth) of rover activities. The rover receives its command set at the beginning of its day, and it operates autonomously for the entire day (or weekend) until it receives its next set of commands. Downlinks are programmed into the daily activities. These occur by streaming data, typically via a UHF antenna to a satellite (Mars Reconnaissance Orbiter, Mars Odyssey, MAVEN, Mars Express, or Trace Gas Orbiter) as it passes overhead. The duration of the pass is the limiting factor in the data volume. After the pass is finished, the satellite is commanded to transmit its data to NASA's Deep Space Network on Earth. Over the course of the mission, the rover has averaged ~60 MB per day.

Resources: A typical LIBS raster takes between about 20 and 40 min to complete. This consists of acquiring the target with the mast, autofocus (~2 min each), RMI images (~2 min each, mostly data transfer time to the body unit and then to the rover), LIBS observations (~60 s each, mostly data transfer time), and LIBS background observations (~60 s each). Each LIBS spectrum is 11 kB of data; one observation point of 30 spectra plus a background is ~380 kB, which is losslessly compressed for downlink. RMI images are 1.4 MB, but they undergo lossy compression for a factor of ~10× reduction in volume. The power required for the instrument is dominated by heating and cooling. The instrument itself uses < 10 Wh per target, but the thermal management uses ~40–50 Wh.

2.2.2 Data processing and calibration

Processing of the LIBS data is shown in Fig. 6. The raw data are denoised and background subtracted, calibrated for wavelength, and corrected for instrument optical response and a zeroth-order correction for distance. A resulting "clean

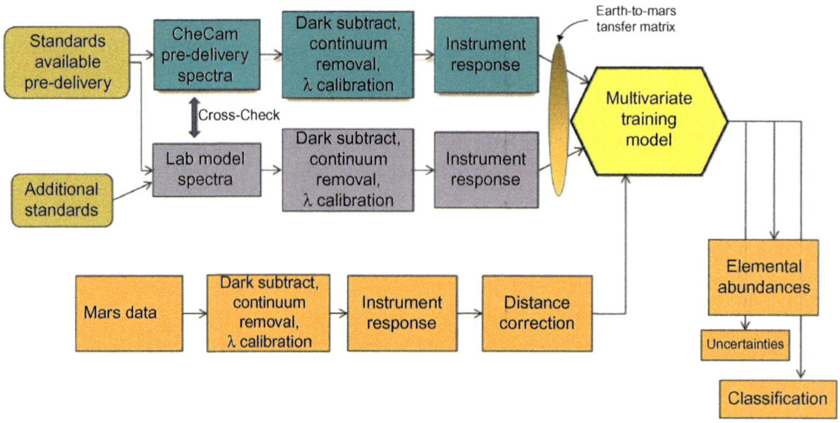

FIG. 6 Processing of LIBS data to obtain quantitative major element abundances on Mars with ChemCam, after Ref. [25]. Dark subtraction refers to taking a spectrum of the ambient light with the laser off and subtracting it. Other aspects are explained in the text.

calibrated spectrum" is shown in Fig. 7, and is provided in units of photons emitted by the plasma.

Calibration for ChemCam has several aspects. These include wavelength calibration, characterization of the instrument optical response (intensity), and calibration with laboratory standards to yield quantitative elemental abundances. We will describe these in the order mentioned.

Wavelength calibration: Laboratory instruments are typically operated under temperature-controlled conditions, such that wavelength calibration is good for the life of the instrument. The temperature of ChemCam's spectrometers varies by 20–25°C over the course of a day/night

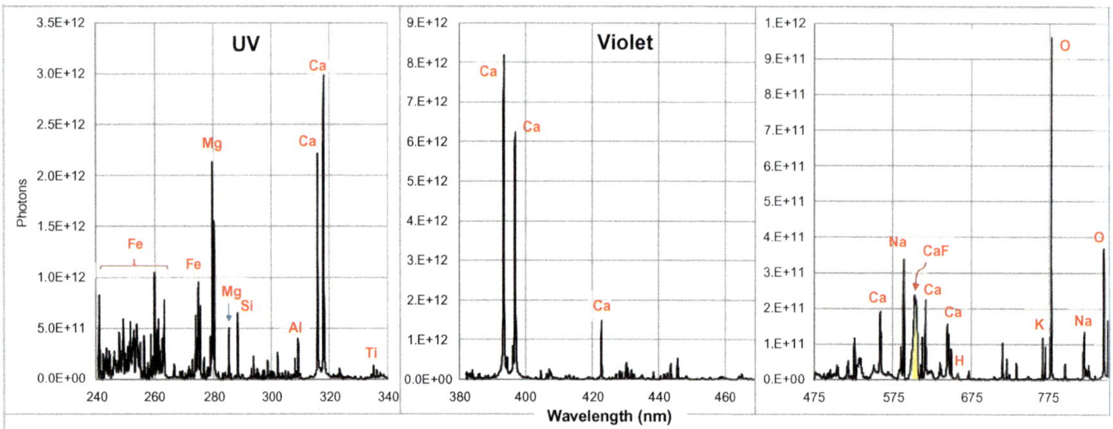

FIG. 7 Processed ChemCam LIBS spectrum of raster point 5 of the Epworth target on Sol 72 of the mission. Major emission peaks are labeled. The CaF molecular emission line is highlighted in *yellow*. The rest of the spectrum consists of atomic emission lines.

cycle and also by seasons inside the rover body, resulting in slight changes in the optical path of up to 2–3 pixels. The relationship with temperature is highly reproducible, and so the wavelength-to-pixel relationship is adjusted as a function of the spectrometer temperature, which is recorded during the LIBS observation. Tests with the algorithm that provides elemental compositions indicate that calibration within 0.3–0.5 pixels is needed, and the temperature calibration provides this level of accuracy [25].

Instrument optical response: This was measured after ChemCam was installed in the rover. The measurement included the flight mast-to-body fiber and the window in front of the instrument [18]. After the rover landed on Mars, there was no way of determining precisely if the instrument optical response had changed over time. The Mars Hand Lens Instrument regularly images the front of the telescope, showing that the optics still look pristine and are clear of any dust. LIBS observations of the calibration targets are compared early in the mission, but as the individual emission lines can vary somewhat from spectrum to spectrum, it is difficult to determine a small change in optical response.

Calibration targets: The rover carries 10 LIBS calibration targets that are located directly behind the rover's mast at a distance of 1.56 m [18]. These targets consist mostly of (1) glasses with compositions that were expected to be relevant for Mars basaltic rocks [26], and (2) ceramics intended to mimic the composition of sedimentary rocks using a lithium tetraborate binder [27]. The targets have a number of drawbacks in that the matrices (glass and lithium-dominated ceramic) are relatively different from the rock matrices on Mars. Additionally, the ceramics are somewhat heterogeneous at the scale of the laser beam diameter. Nevertheless, the rover calibration targets play an important role in ChemCam's calibration, as described next.

For major elements (Si, Ti, Al, Fe, Mg, Ca, Na, K), compositions were calibrated using multivariate analysis algorithms. For minor and trace elements (Li, B, F, Cr, Mn, Ni, Zn, Cu, Rb, Sr, Ba), univariate calibration curves are employed [28–34]. Both cases utilize a large library of spectra taken from >400 standards using a copy of the ChemCam instrument in the laboratory, with samples analyzed in a chamber containing CO_2 at Mars pressure. The univariate calibrations are relatively straightforward and do not need further description. The multivariate calibration is described briefly next.

The ChemCam team started the mission with a partial least squares (PLS) multivariate calibration model that used ~60 pressed rock powder standards that were observed with the ChemCam flight instrument prior to delivery to the rover [25]. The algorithm is known as PLS2 [35]. It is easier to use for multiple elements, as it regresses all elements simultaneously. However, the results were not satisfactory, in large part because they did not sufficiently cover the range of compositions observed on Mars for all elements. Subsequently, the team developed a library of ~400 standards that were observed by the ChemCam instrument copy in the laboratory. A calibration-transfer matrix correction was made to correct for differences between the two instruments and between the Earth and Mars environments using replicate standards on the rover and on Earth [18, 26, 27]. Two different multivariate algorithms were applied separately to each element: independent component analysis (ICA, e.g., Ref. [36]) and PLS1 [37]. For the latter, submodels were used to increase the accuracy in specific regions in composition space, such as using separate models for low, medium, and high abundances of a given element, and blending the results at cross-over regions [38]. The results of the ICA and PLS algorithms were then evaluated with current Mars data against known mineral compositions and were blended in a way that optimized the accuracy [37].

ChemCam's quantitative results can be described in terms of precision and accuracy.

The latter quantity is evaluated for the major elements by employing a test set of standards (a reasonable fraction of the ~400 standards, spanning the composition range for all major elements) that are held out of the calibration model and are used to evaluate its accuracy as a function of abundance [37]. The accuracy is expected to be valid across various rock types and can be used to compare mineral fractions and to make comparisons with other instruments. Its accuracy is generally in family with contact techniques such as the alpha particle X-ray spectrometer (APXS), which is also subject to chemical matrix effects, especially for elements like Al and Mg [39]. ChemCam's precision is checked by making multiple measurements of a homogeneous material, either its onboard calibration targets [40] or a homogeneous, fine-grained bedrock such as the Sheepbed Mudstone [41]. Based on 25 observations of a shergottite synthetic glass calibration target on the rover, the major elements (as oxides) have the following precisions in weight percent: SiO_2: ±1.53, TiO_2: ±0.14, Al_2O_3: ±0.57, FeO_T: ±1.83, MgO: ±0.49, CaO: ±0.42, Na_2O: ±0.49, K_2O: ±0.14 [40]. The precision observed from 480 observations of the Sheepbed Mudstone were comparable [41], indicating that distance to the target is not a significant factor in the precision, at least to the distances used (to 4.55 m) in that study. Both the rover targets and the mudstone can have heterogeneities that can artificially increase the reported precision, so one could argue for using the lowest values from the two studies for each element.

In any case, standard deviations within a given raster tend to be significantly smaller than for groups of targets shot over many different days. This has allowed the use of ChemCam point-to-point standard deviations as a proxy for grain sizes in the sediments [42]. The idea is that if grains of differing composition are much smaller than the size of the laser beam, a low standard deviation will be obtained, while if the size of the grains becomes larger, and more similar to the size of the laser beam, standard deviations increase significantly.

2.3 Science findings from ChemCam LIBS

Given the analytical and physical advantages listed in the Introduction, LIBS has made numerous new discoveries on the red planet, in many cases in applications that were previously unforeseen. A few of these discoveries are highlighted here.

ChemCam landed on Mars in August 2012. On Sol (Mars day) 19 of the mission, ChemCam made its first soil observation. Among the 250 spectra that were returned from five observation points, a variety of compositions was observed. As more soil observations were made and were correlated with the corresponding images, a trend was observed in which the finest-grained soils were found to be rich in hydrogen and lower in alkalis and silicon than coarse grains. The composition of the coarse grains was found to generally match the composition of the float rocks, indicating that physical comminution of the local rocks was being observed [43, 44]. The composition of the fine-grained soils generally matched that observed by previous missions. However, ChemCam was the first ground-based instrument to observe hydrogen, showing that the hydration of soils was ubiquitous and extended to airborne dust that settled on rocks and on the rover [45]. ChemCam was also able to determine abundances of other trace elements [28], helping to better constrain the origin of soils on Mars [44]. The abundance of water on Mars, and especially the fact that it remains bound in the soil in spite of the very low atmospheric water content (also quantified by ChemCam [46]), addresses the important topic of the habitability of Mars over time.

On Sol 72 of the mission, a LIBS spectrum of soil appeared with a large, broad peak spanning from 600 to 607 nm (Fig. 7). It was clearly not an atomic emission, and was eventually identified as a CaF molecular peak. Fluorine had also not been observed previously on Mars, as previous X-ray fluorescence instruments were not able to detect elements of this low atomic mass. ChemCam was also not expected to be sensitive to fluorine, as the atomic emission peaks in its spectra range are extremely weak. However, observation of the CaF peak opened up a new window into understanding Mars. The CaF peak is unique in that ChemCam has not clearly observed any other molecular emission peaks in Mars materials. The Ca and F do not need to be chemically bound in the rock matrix, as they are combined in the plasma for this observation.

As of early 2017, >600 fluorine detections had been made on Mars between a detection limit of ~0.2 wt% and >10 wt% [31, 47, 48]. In these observations, fluorine correlates with calcium and phosphorus in some cases, indicating the presence of the mineral fluorapatite. In other cases, fluorine trends with Al and K, suggesting the layered silicate muscovite. Finally, in other cases, fluorine appears to trend just with Ca, suggesting fluorite, as in the Garden City vein complex [48]. The presence of significant amounts of apatite, muscovite, and potentially topaz suggests that the igneous source rocks are relatively evolved and could represent late-stage magmas [31], something that was unexpected on Mars, which was thought to have relatively primitive basaltic compositions (e.g., Ref. [49]).

However, ChemCam revealed other clues suggesting that the Martian crust is far more diverse than previously thought. Although Gale crater's bedrock appears to be exclusively sedimentary, numerous igneous float rocks and pebbles were found on Bradbury Rise, the gravelly plain on which the rover landed. Indeed, Bradbury Rise lies at the distal end of an alluvial fan located just below the Peace Vallis channel that cuts down through the crater rim. The igneous rocks thus appear to have originated at the crater rim or within the Peace Vallis watershed. ChemCam observed ~60 rocks that were clearly igneous based on appearance in images and on compositions [50–52]. These include primitive compositions but extend to evolved (silica and alkali-rich) materials such as porphyritic trachyandesites and leucocratic trachytes [52]. These evolved basalts are lower density, reminiscent of continental crust on Earth, leaving open the question of whether significant portions of the Southern Highlands, which register somewhat lower density from satellite data, are replete with these evolved basalts [51].

Generally, ChemCam's remaining observations fall into two categories: bedrock and diagenetic (later-stage alteration) features. ChemCam provides regular surveys of bedrock compositions to understand changes in the sediments. ChemCam can provide these observations with much greater frequency than other instruments due to the ease of pointing and shooting compared to deploying the arm instruments. Fig. 8 shows the "chemostratigraphy" of the Murray lacustrine formation through which the *Curiosity* rover climbed for over 300 m of vertical elevation [53]. The details will not be described here.

ChemCam has excelled at observations of diagenetic features. Typically, these are groundwater precipitates found in the fractures in the rocks, and are thus small in size, typically <1 cm wide, meaning that ChemCam is the only probe small enough to sample these features. Two particular discoveries will be highlighted here: manganese oxide fracture fills and boron.

On Sol 620, ChemCam returned observations from Stephen, a strangely flat clast approximately 30 mm across (Fig. 1B). The spectra showed a surprisingly high manganese abundance on the first few laser shots, to >30 wt% (calculated as MnO), and tapering off to the end of the burst. After being cleared of dust around the observation points, the surface appeared dark but highly reflective. Reflectance spectra of the surface were consistent with manganese oxides. Several other clasts in the area showed the same features even though the

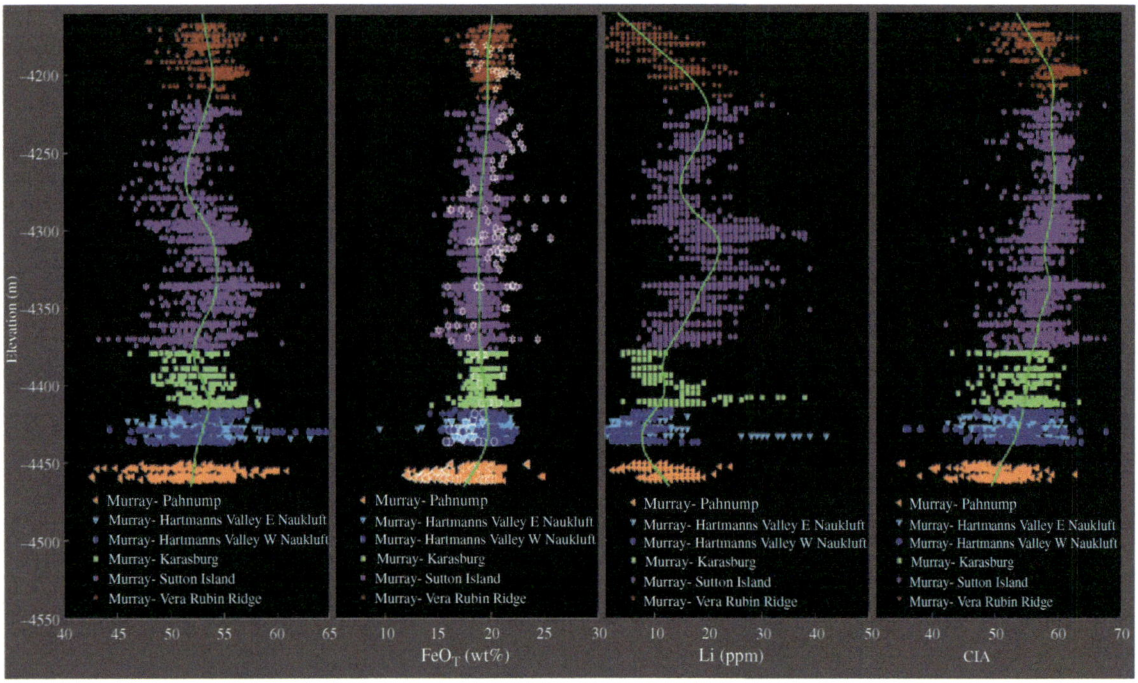

FIG. 8 The record of ChemCam measurements of selected elements and parameters as the *Curiosity* rover drove up the lower part of Mt. Sharp. Elevations are negative relative to the mean elevation of Mars due to the low elevation of Gale crater. The record shown here covers a large fraction of the fine-layered flat-lying lakebed sediments of the Murray formation spanning several geological members of that formation starting at Pahrump Hills and ending on Vera Rubin Ridge some 10 km south and 300 m vertically higher. Each data point is a ChemCam observation. Spline fit curves show the general trends. Only a few of the many recorded elements are shown here: silica and iron representing major elements, lithium representing trace elements, and the chemical index of alteration (CIA) indicating trends in the clay mineral content of the sediments, calculated from a ratio of aluminum abundance relative to more leachable elements.

surrounding bedrock was not enriched in Mn. These planar Mn oxide thin films were interpreted to be fracture-filling precipitates in which groundwater had infiltrated the bedding planes (hence, the very flat nature) and, with a source of oxidation, deposited the Mn oxide. LIBS was instrumental in (1) finding these features, which were not discovered by the arm instruments, and in fact were not very notable until the dust was removed by the laser shockwave; (2) discovering the thin nature of the Mn oxide films; and (3) observing the true Mn abundance. Observations by APXS showed only a few wt% MnO because the interaction depth of that instrument was significantly deeper than the thickness of the Mn oxide layer. The observation of Mn oxides here [3] and elsewhere [32, 54, 55] indicates that the Mars environment had a high oxidation potential at one time, suggestive of a more oxygen-rich atmosphere than the Earth had before the great oxygenation event [3]. On Earth, microbes apparently enriched the atmosphere in oxygen. On Mars, there are other potential mechanisms, although this finding of oxidized manganese has interesting implications for habitability.

One final example of LIBS discoveries on Mars to date is that of boron. Like fluorine, boron could not be detected on Mars prior to the arrival of ChemCam. Its LIBS instrumentation has a detection limit of ~100 ppm of B, although this varies depending on the presence of other elements, as there are several weak iron emission lines that interfere [30]. Boron was discovered in the Ca sulfate fracture fills that are ubiquitous in the bedrock of Gale crater [56], indicating that boron precipitated from the groundwater. Boron was observed at higher frequency and in higher abundances at higher elevations within the lacustrine Murray formation, suggesting that the source of the boron was somewhere above these layers, either as an evaporite formation or adsorbed in clay minerals [30]. Boron has interesting implications for the origin of life, as boron is required to stabilize ribose (a part of RNA) in water. As RNA is thought to have preceded DNA as a building block of life, boron's presence in groundwater would be needed for life-forming reactions to proceed (Ref. [30] and references therein).

ChemCam has discovered many other details of Mars geochemistry, and its success as the first fine-scale chemical probe on Mars [57] was noted in the Mars 2020 Science Definition Team Report [58]. As of this writing, ChemCam has returned >700,000 LIBS spectra from Mars. The spectra as well as the images and chemical compositions obtained from them are available to the public, archived in the NASA Planetary Data System at http://pds-geosciences.wustl.edu/missions/msl/chemcam.htm.

3 SuperCam LIBS

3.1 Instrument description

SuperCam is a second-generation ChemCam instrument, with accommodations to enable remote Raman spectroscopy, infrared spectroscopy, color microimaging, and LIBS acoustic signals [20]. Like ChemCam, it is a partnership between LANL and IRAP, but with Spanish and other international contributions to the more extensive onboard calibration target. The goal of SuperCam is thus to provide both remote chemistry and mineralogy, the latter of which was recommended by the Mars 2020 Science Definition Team [58]. On the outside, SuperCam looks very similar to ChemCam (Fig. 2C and D) because of a common architecture with separate mast and body units, a 104-mm diameter telescope, RMI, a laser in the mast unit, and spectrometers in the body unit, also shown in the block diagram (Fig. 3B). The latter shows that the infrared spectrometer was added to the mast unit. Overall, SuperCam is just slightly lower in mass than ChemCam.

Some of the major differences of SuperCam relative to ChemCam from the LIBS perspective are a different laser design; the blocking of light in the spectral region between 465 and 535 nm, necessitated in part by the need to block the green 532 nm beam used for Raman spectroscopy; and replacement of the VNIR spectrometer with a transmission spectrometer with a time-gated image intensifier. These are described next.

The SuperCam laser was supplied by the same company as ChemCam's laser, Thales Optronique SA, but a slightly different strategy was pursued for thermal accommodation. Rather than using the exotic KGW crystal, it was decided to use the more typical YAG. Normally, YAG is less thermally accommodating; however, the use of diode bars of slightly different peak wavelengths increases the thermal range over which a strong beam can be provided. The laser is frequency doubled to provide a green (532 nm) beam for the remote Raman spectroscopy experiments. Fig. 4 shows the path of the green beam. Some photons are produced at the fundamental frequency of the laser, and so a beam block is employed in the Galilean expander used by the LIBS beam.

The spectrometers used for LIBS on SuperCam consist of two reflection spectrometers that are nearly identical to those of ChemCam for the UV and violet spectral regions. To facilitate the Raman spectral technique, the third reflection spectrometer was replaced with a transmission spectrometer that contains a gated optical intensifier (Figs. 2C and 3B). A pair of optical filters designed to remove the 532-nm laser light for the Raman experiment limits the LIBS spectral range to >536 nm in this spectrometer. SuperCam has higher spectral resolution just above 536 nm (~0.35 nm full width at half maximum compared to ~0.6 nm on ChemCam), again made to optimize the Raman experiment. The intensifier gain can be increased to amplify low LIBS signal as needed. The exposure duration is expected to be 10 μs for LIBS, significantly reducing the ambient light background, as compared to the ungated CCDs of ChemCam. The duration can be adjusted as short as 100 ns and the delay of the exposure can be adjusted by 10-ns intervals, providing wide flexibility for other LIBS settings. The intensifier will result in somewhat higher signal noise than with the unintensified CCDs.

After selection by NASA of the SuperCam instrument, the team decided to add a microphone to the experiment package. The primary motivation for the microphone was to determine physical properties of the rock from the LIBS acoustic signal. It has been known for some time that the LIBS acoustic signal intensity can provide information about the properties of the target and its ablation as well as other details such as the focus quality (e.g., Ref. [59]). Understanding the physical properties of the rock are important both to the science and for practical aspects. For example, on the *Curiosity* rover, a lot of time has been expended trying to drill into rocks that ended up being too hard to drill. Knowing the hardness of the rock in advance would have saved weeks of work. And so we decided that it would be worth including a microphone.

The SuperCam microphone uses a commercially available Knowles model, mounted on the rover's mast just outside of the remote warm electronics box (Fig. 3B, not shown in Fig. 2). It has several gain settings to record sounds between 100 Hz and 100 kHz [60]. Expected sounds include wind-induced sounds, rover-induced sounds, and the sound produced by LIBS on Mars targets. The sound spectrum and intensity of laser pulses on various targets has been studied since at least the 1980s (e.g., Ref. [59]). Recent studies show that the evolution of the acoustic signal as a function of laser shot correlates with the depth of penetration and the hardness of the material [61]. By using this relationship, SuperCam can effectively determine the hardness of rocks without the rover ever having to drive up and touch them. In spite of the poor propagation of sound through the thin Mars atmosphere, SuperCam should be able to record the LIBS acoustic emission to a distance of 4 m based on experiments at Mars pressure [60].

SuperCam's other techniques—the infrared spectrometer, the Raman spectroscopy, and the color RMI—will be described elsewhere.

3.2 Operations and calibration

SuperCam is planned to be operated in much the same way as ChemCam, but with expanded capabilities as already described. Because it will be employing several different techniques, each raster will take longer to complete. The result may be somewhat fewer LIBS analyses than have been accomplished with ChemCam even though efforts have been made to increase the efficiency of the operation in terms of power and duration. The advantage, of course, is the ability to correlate the elemental composition with important mineralogical information and rock physical properties.

For its calibration, SuperCam benefits from the same library of ~400 standards as ChemCam. Although an extended calibration of ~330 standards was completed with the flight body unit and the qualification model mast unit, there was no opportunity to obtain such a spectral library with both flight body and mast unit prior to launch. Because of this, SuperCam will need to rely on cross-comparisons of its onboard standards and replicates on the ground to cross-calibrate laboratory, partial flight, and flight models (Fig. 6). SuperCam has a significant advantage in its onboard calibration targets: instead of the 10 on ChemCam [18, 26, 27], it employs 22 LIBS targets that were much better homogenized, with most of them using a spark sintering technique [62]. So, while a calibration model has not yet been completed for SuperCam, the team is optimistic that it will perform as well as and likely better than ChemCam in terms of accuracy and precision.

NASA's Mars 2020 rover is planned for launch in July 2020 with a landing on February 18, 2021. It is designed to collect drill core samples for return to Earth on a future mission. The planned mission lifetime is at least 1.5 Mars years (about 2.5 Earth years). The selected landing site is Jezero crater, an ~45-km diameter impact feature at 18.9°N, 77.5°E on the edge of the Isidis Basin. It was selected for its combination of suspected igneous terrain, lacustrine sediments, and a large deltaic formation. Orbital spectra show evidence of magnesium carbonates, serpentine, olivine, and extensive clay minerals (e.g., Ref. [63]). Jezero crater is thus one of the few sites on Mars observed to have carbonates, in spite of the CO_2 atmosphere of Mars. As such, SuperCam should encounter a much different range of chemistry relative to ChemCam in Gale crater. Study of the carbonates may reveal why Mars' crust contains relatively few carbonate bodies in spite of its atmosphere, and in sharp contrast to the sedimentary cycle on Earth.

4 MarsCode LIBS on the HX-1 Rover

4.1 Instrument description

MarsCode is China's first LIBS-based instrument for deep-space material detection. It will be mounted on the Mars rover of China's first Mars exploration project HX-1, as shown in Fig. 9. MarsCode consists of two parts: one

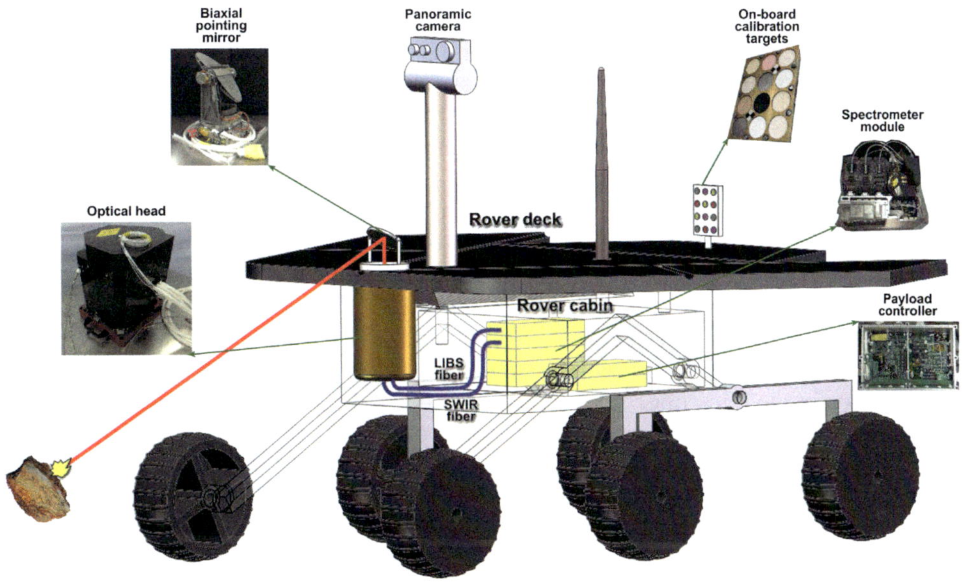

FIG. 9 Schematic diagram of MarsCode. An optical head, mounted outside of the rover's cabin, emits laser pulses, which are reflected by a biaxial pointing mirror and excite LIBS signals from targets. The signals are collected by the optical head, and then transmitted to a spectrometer module in the rover cabin via a LIBS fiber. A payload controller is mainly used to set LIBS working parameters.

inside the rover's body, or "cabin," and one outside the cabin. The internal unit includes a payload controller and a spectrometer module, which mainly contains two spectrometers: one for LIBS and the other for short-wave infrared (SWIR) detection. The external unit consists of an optical head, a biaxial pointing mirror (BPM), and a LIBS and SWIR calibration target assembly.

The optical head consists primarily of a Ritchey-Chretien telescope, a LIBS laser, an RMI, and a self-focusing module (SFM). The Ritchey-Chretien telescope consists of two hyperbolic primary and secondary mirrors, and a Schmidt window, which has no chromatic aberration, spherical aberration, or coma aberration. The telescope has a 106-mm diameter aperture, and can be used to focus the LIBS laser beam to the targets. Collection of the LIBS plasma light, RMI imaging, and SFM ranging are also accomplished through the telescope. Fig. 10 shows the optical design of MarsCode. The LIBS laser beam is collimated through a collimating lens, and is then sent to the secondary mirror by reflecting off a reflector and a central dichroic in turn. The secondary mirror reflects and expands the laser beam to the primary mirror. The beam is focused by the primary mirror and directed by the BPM to targets at 2–5 m distance. Light from the LIBS plasma is collected by the telescope, and passes through the central dichroic. The passed LIBS light is focused through central lenses, reflected by an RMI dichroic, and then coupled into a LIBS fiber through a LIBS dichroic. The secondary mirror is installed on a linear stage that can be moved to focus the telescope.

The MarsCode LIBS laser is a passively Q-switched semiconductor-pumped solid-state laser, which has a more compact structure and smaller volume compared with actively Q-switched lasers. However, passively Q-switched lasers have the problem of unstable time intervals between the trigger signals and the emission. To mitigate this, the laser has a trigger-out port that sends a signal when the laser fires, allowing proper timing of the LIBS spectrometer detectors. laser beam has wavelength of 1064 nm and a repetition rate adjustable from 1 to 3 Hz. The laser energy of a single shot is 21 mJ and its pulse width is 4.5 ns.

The RMI is used to obtain images of ablation holes' positions after the laser fires. It uses a CMOSIS CMV4000 sensor, which is a high-sensitivity, pipelined global-shutter CMOS image sensor with 2048 × 2048 pixels capable of HD format. The spectral range of RMI is 900–1000 nm. The emitted light from the micro-LIBS detection area travels along the receiving light path (Fig. 10), passes through the central dichroic, and is focused by the central lenses; it is then sensed by RMI via the RMI dichroic and two reflectors. The angular resolution of RMI is better than 100 μrad, corresponding to 0.2 mm at a 2-m distance. Due to the limitation of data transmission, the original 4-MB image will be compressed into 64 × 64 pixels for downlink.

The SFM contains a 1550-nm laser diode (LD) and a self-focusing receiver. The laser beam of the LD is collimated first and then passes through an LD dichroic. Afterward it is reflected successively by a central reflector located in front of the secondary mirror and the BPM before it reaches the target. The reflected light from the target passes through the telescope's central dichroic and is focused by the central lenses; it is then sensed by the self-focusing receiver, a PIN photodiode, after reflection by the LIBS dichroic and a reflector (Fig. 10). As the secondary mirror is moved along the linear stage, the output of the PIN photodiode is monitored in real time. The movement of the secondary mirror is stopped when the output reaches a peak, which accomplishes the focusing. The ranging accuracy of SFM is better than 1 cm.

The BPM adopts a worm gear drive structure and cryogenic-rated stepping motors to achieve two-axis rotation. The azimuth and pitch rotation ranges are 210 and 45 degrees, respectively. Two azimuth and pitch angle sensors are used to monitor the rotation position of the BPM.

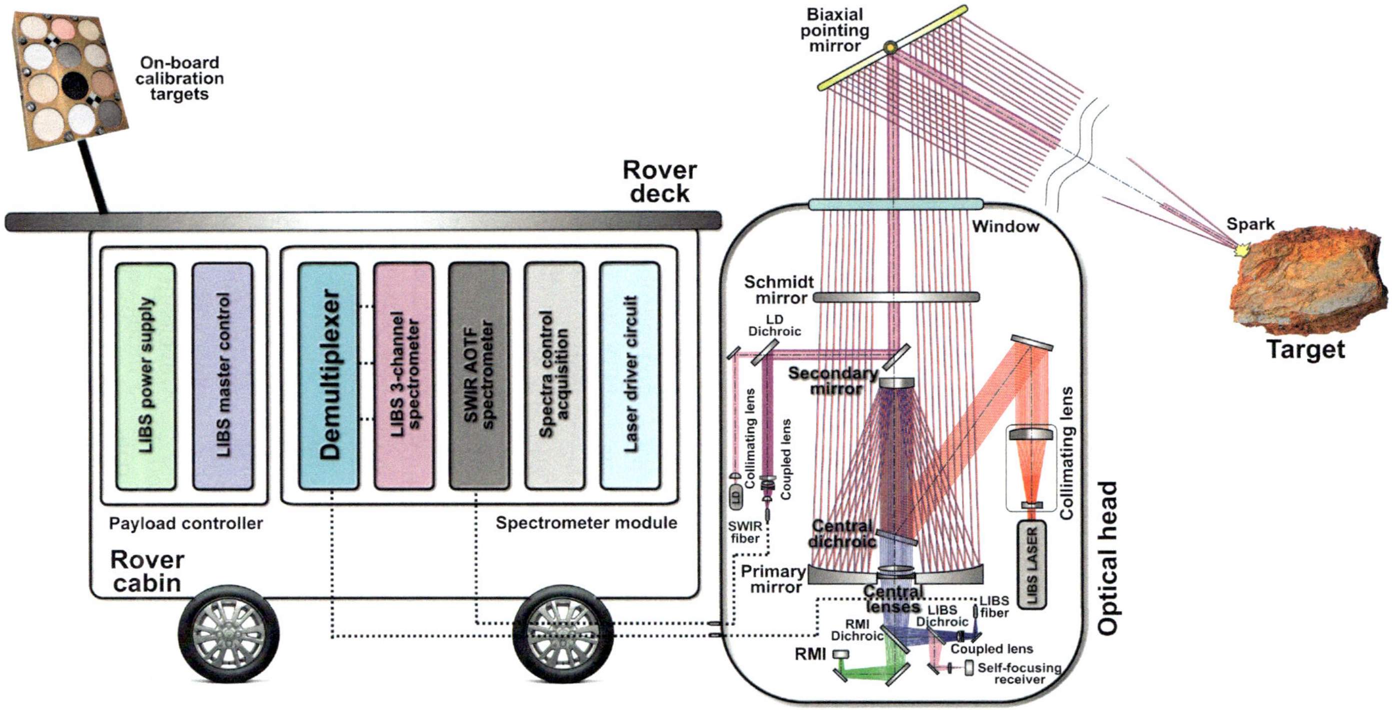

FIG. 10 Optical design of MarsCode. The LIBS laser beam is collimated first and then focused through a 106-mm diameter telescope to a distant target. A laser diode (LD), whose emitting beam is coaligned with a passive SWIR light path, is used for self-focusing. A biaxial pointing mirror is used to point to targets in front of the rover and calibration targets on the back of the rover.

The pointing accuracy is 0.1 degree. The material of the BPM is beryllium to save weight.

MarsCode also includes a passive infrared spectrometer. The infrared light from the target is reflected successively by the biaxial pointing mirror, the central reflector, and the LD dichroic, and then coupled into an SWIR fiber (Figs. 9 and 10).

The inner part of MarsCode consists of the payload controller and the spectrometer module. The payload controller mainly contains the LIBS power supply and the LIBS master controller. The latter is used to adjust the working parameters of LIBS. The spectrometer module includes a demultiplexer, a three-channel LIBS spectrometer, the SWIR spectrometer, a spectral control acquisition, and a laser driver circuit (Fig. 10).

The LIBS fiber is connected to the optical head with the demultiplexer, which uses two dichroic mirrors to split the light into three spectral bands, and then couple the light into the three LIBS spectral channels via individual fibers. The spectral ranges of the three channels are 240–340, 340–540, and 540–850 nm, respectively, whose corresponding resolutions are 0.1, 0.2, and 0.3 nm. Hamamatsu UV-enhanced CCDs are used as the spectral sensors, which are identical for the three channels. Among the 2048 effective pixels, the first 128 pixels and the last 120 pixels are discarded, and the middle 1800 pixels (129–1928) are taken as the effective data for each channel. The framework of the LIBS spectrometer is made of aluminum-based silicon carbide to save weight.

The SWIR fiber in the optical head guides the infrared emission from the target into the SWIR spectrometer, which uses an acousto-optic tunable filter to spectrally disperse the infrared light. The SWIR spectrometer covers 850–2400 nm with optional 64 or 330 spectral bands, and a spectral resolution better than 12 nm. Its signal-to-noise ratio is >40 dB with a solar illumination angle of 30 degrees and target albedo of 0.3. The SWIR is used to classify the mineralogy of Mars targets.

4.2 Operations and calibration

The LIBS operational scheme of MarsCode includes the following steps. First, five calibration targets are observed; 30 LIBS spectra are averaged for each calibration target. Next, for each Mars surface target, 50 laser pulses are emitted, and the 50 individual spectra are averaged to reduce the error. For each observation, the first five LIBS spectra are discarded to reduce the interference effects of other substances on the surface. In depth-profile mode, 1500 laser shots are used to form a ~1-mm deep hole. One LIBS spectrum is 84 kbits. Before and after each LIBS observation, a 64 × 64-pixel image is obtained by RMI.

The LIBS spectrometers of MarsCode undergo wavelength and radiometric calibration before launch. Radiometric calibration is implemented by an Avantes deuterium halogen-balanced light source, Avalight-DH-S-BAL, which provides stable continuous spectra in the UV, visible, and near-infrared bands from 215 to 2500 nm. For laboratory wavelength calibration, four standard lamps, including mercury, argon, neon, zinc, and cadmium, are utilized to provide 68 spectral lines for wavelength calibration. After fitting with a quadratic polynomial, an initial wavelength calibration result at 20°C is shown in Fig. 11.

The LIBS spectrometer is installed inside the rover's cabin, where the possible temperature range on Mars is from −40°C to +50°C, resulting in a maximal pixel drift of 100+ compared to 20°C. Due to the limitation of power consumption, it is impossible to narrow the wide working temperature range. Hence, an on-Mars real-time wavelength calibration is needed. Among the 12 rover standard targets, Ti has 100+ stable observed spectral lines within the 240–850 nm wavelength range, which can be used to obtain a fitting equation with comparable accuracy to that obtained by standard lamp fitting.

The Mars LIBS data are processed to calibrate the wavelength in the following steps, as shown in Fig. 12. First, wavelength calibration functions

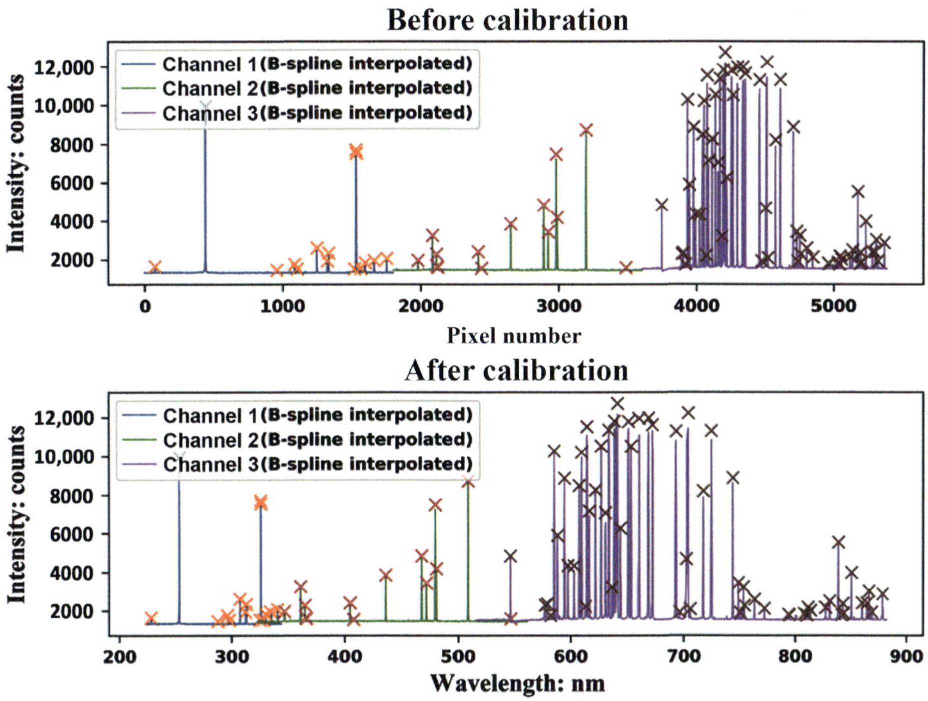

FIG. 11 Initial MarsCode LIBS wavelength calibration at 20°C. Quadratic polynomial fitting is after peak seeking of 68 spectral lines of standard lamps. B-spline interpolation is adopted to improve the peak-seeking precision. The average deviation between the calibrated wavelengths and the true wavelengths is 0.043 nm.

of different temperatures (CFDT) are obtained using standard lamps in the laboratory. Based on the LIBS on-Mars detection scheme of MarsCode, raw data of the Ti standard target and the Mars surface target will be downloaded sequentially. Raw detection data of MarsCode contains the LIBS and environmental data, from which the ambient temperature combined with CFDT can be used to generate an estimated calibration function. This is used to calibrate the raw LIBS data from the Ti standard target, obtaining a calibrated Ti spectrum (CTS). Referring to the standard wavelength values of all spectral lines of Ti, the peaks are readily searched in the CTS. Then, the corresponding pixel numbers of these peaks are listed. Hence, an accurate calibration function (ACF) is obtained from the Ti LIBS spectrum. Finally, combining the ACF and the raw LIBS data of the detected target, wavelength-calibrated LIBS data are obtained.

5 Indian lunar LIBS on Chandrayaan-2

In 2019 ISRO launched its second lunar exploration mission, Chandrayaan-2. After the successful Chandrayaan orbiter of 2008–09, this new mission was much more complex, with an orbiter and a lander (named *Vikram*) containing a rover (named *Pragyan*) (Fig. 13). The mission was targeted to land close to the South Pole (further poleward than 70°S). The lander and rover were designed to operate for a lunar day, equivalent to 14 Earth days. There were 13 scientific instruments onboard: 8 on the orbiter, 3 on the lander, and 2 on the rover [64]. Unfortunately,

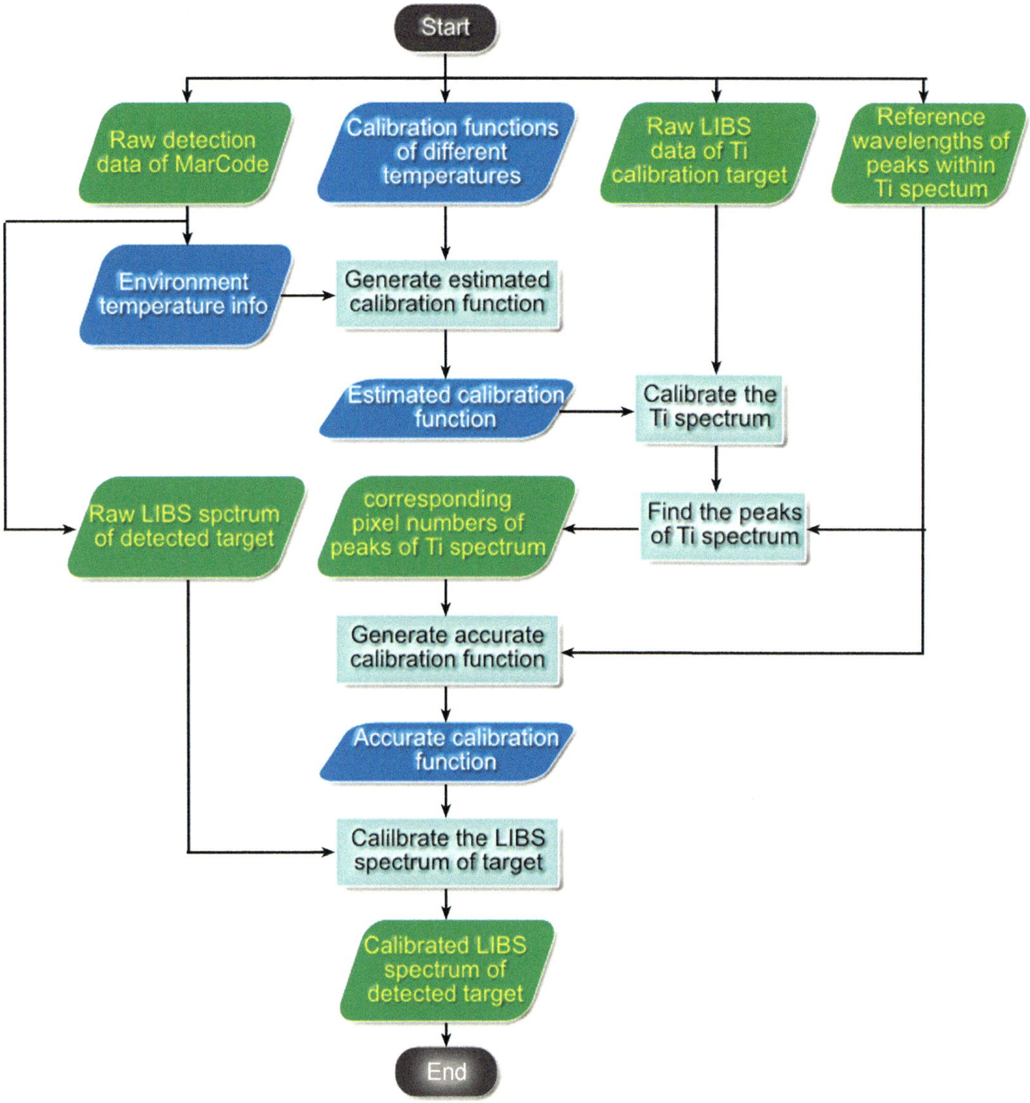

FIG. 12 Processing steps of MarsCode LIBS wavelength data on Mars. A Ti calibration target is used as a wavelength calibration standard.

the Chandrayaan-2 lander malfunctioned when it was approximately 1 km from the surface of the Moon, and the mission was lost.

The rover had a mass of ~20 kg, was to be powered by vertical solar panels, and was expected to travel 500 m. The vehicle had six wheels, similar in structure to the NASA pathfinder Mars rover (Fig. 14). In addition to the two cameras located on the rover's mast, it carried two scientific experiments dedicated to measuring the chemical composition of the lunar surface: an APXS (Physical Research Laboratory,

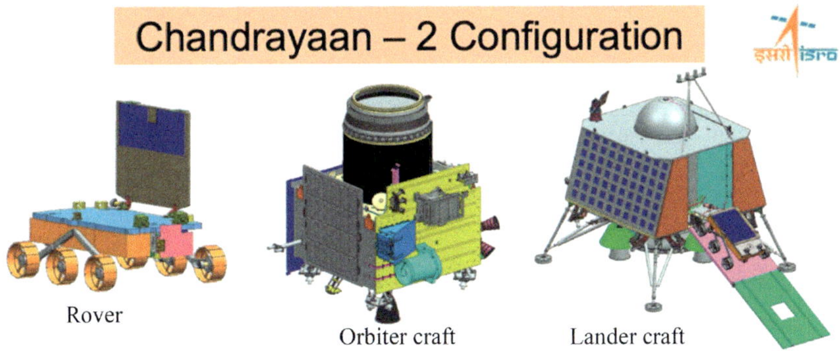

FIG. 13 Chandrayaan-2 mission configuration. *Credit: ISRO.*

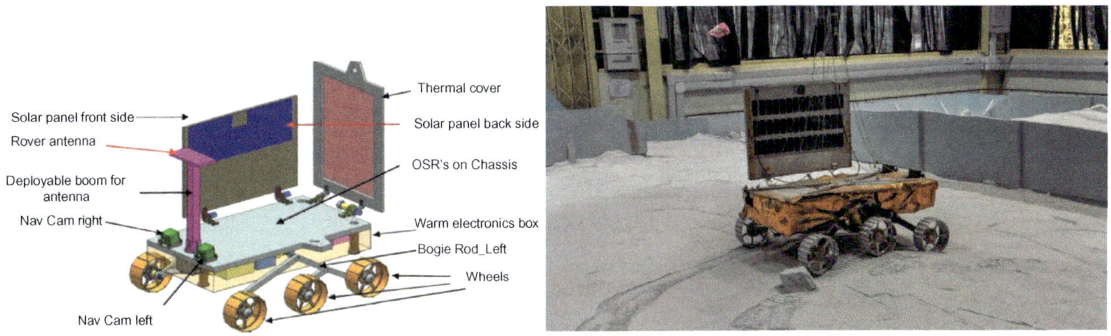

FIG. 14 Architecture of the *Pragyan* rover and test of its engineering model at ISRO facilities [64]. *Credit: ISRO.*

India) and a miniaturized LIBS instrument (Laboratory for Electro-Optic Systems, India).

5.1 Instrument description

The LIBS instrument was miniaturized and without moving parts. It weighed <1.2 kg and consumed <5 W of power (see specifications in Table 1). The instrument was developed at the Laboratory for Electro-Optics Systems (LEOS) in Bangalore, India. Fig. 15 shows an early computer-aided design of the instrument with the Focusing Optics Unit, the Collecting Optics Unit, and the space taken by the spectrometer and electronics. The expected operating temperatures ranged from −20°C to +55°C. The LIBS laser source was a Yb:Er:phosphate glass laser operating at a 1534 nm wavelength with a pulse energy of 2–3 mJ and pulse duration of 7 ns [66]. Its repetition rate was adjustable from 1 to 5 Hz. The peak power reached <1 MW, and a three-lens focusing optics allowed a power density of 120 MW/mm^2 on an ~50-μm spot. The laser was focused at a constant distance of 200 mm (±20%) corresponding to the distance between the lower deck of the rover where the instrument was located and the position of the lunar soil located between the wheels of the rover.

The collecting optics used a single lens system, followed by twofold mirror geometry with an F number of 2.2 and a field of view of 14 degrees [65]. The light from the plasma was injected in a flat-field reflective (concave) aberration-corrected holographic-grating single spectrograph with a linear 2048-pixel CCD with a pixel size of

TABLE 1 List of the laser-induced breakdown spectroscopy parameters for the Chandrayaan-2 rover [65].

Specification	Value
Volume	180 × 150 × 80 mm
Power	<5 W
Mass	<1.2 kg
Laser diode	1.534 μm, 2–3 mJ, 7 ns, 1–5 Hz
Coupling optics	Lens-coupled mechanism
Focusing optics	Three-lens system
1. Spot size	<50 μm
Collection optics	
2. F number	2.2
3. Field of view	±14 degrees
Grating	Flat-field holographic
Spectral characteristics	220–800 nm (resolution <1 nm)

14 × 200 μm. The wavelength range for detection was 220–800 nm (resolution <1 nm). The signal could be integrated over 8 μs up to 1 ms as needed. While the spectra did not have high wavelength resolution, the wavelength range included emission lines of 17 major elements constituting the lunar rock-forming minerals and volatiles of interest (Si, Na, Mg, Al, K, Ca, Fe, Cr, Mn, Ti, H, He, C, N, O, P, and S). There were no direct imaging capabilities of the analyzed area, but it was possible to image the region after analysis by moving the rover so the LIBS pits were in the imagers' field of view. No calibration targets were available to calibrate the instrument during the mission. The duration of the mission was expected to be relatively short (14 Earth days); this may not have been an issue with respect to temporal changes in the instrument. However, a good calibration database prior to launch was mandatory for data analysis.

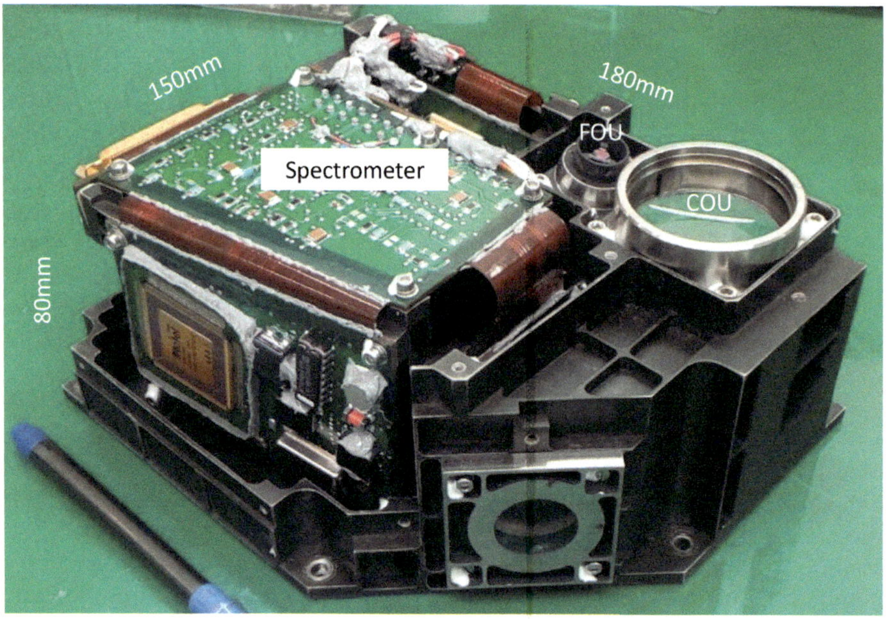

FIG. 15 Early CAD design of the lunar Chandrayaan-2 LIBS instrument with dimensions. *COU*, collecting optics unit; *FOU*, focusing optics unit. *Credit: LEOS; ISRO; and A.S. Laxmiprasad, V.L.N. Sridhar Raja, S. Menon, A. Goswami, M.V.H. Rao, K.A. Lohar, An in situ laser induced breakdown spectroscope (LIBS) for Chandrayaan-2 rover: ablation kinetics and emissivity estimations. Adv. Space Res. 52(2) (2013) 332–341.*

5.2 Operations and calibration

Laboratory demonstrations of the feasibility of the experiment were carried out in the early stages of the project using commercial components for the laser, optics, and detection system. Samples included the lunar soil simulant JSC-1A, which was used to understand the variation of the LIBS spectral intensity as a function of the pressure levels ranging from 760 to 6×10^{-5} Torr. The intensity of the spectral lines decreased with pressure to 10^{-3} Torr, below which no further decrease in emission was observed [65]. This is consistent with the asymptotic behavior determined by [12] for planetary applications of LIBS.

Further experiments were developed by LEOS using a commercial LIBS setup consisting of a Q- switched Nd:YAG laser (DIVA, Thales) operating at 1064 nm and 10 Hz, with a pulse duration of 10 ns and energy of 3 mJ. The samples were analyzed under close-to-vacuum conditions of 6×10^{-5} Torr. The plasma emission was collected with a telescope (TOP 100, Instrument Systems) and directed to the spectrometer (CAS 140B, Instrument Systems) equipped with a back-illuminated CCD detector system via a 2-m optical fiber. Different types of samples were analyzed, including terrestrial soils (marine soil and fly-ash class C and class F) and lunar regolith simulant JSC-1A. A study of the signal-to-noise ratio of the instrument using this configuration was made, showing the signal-to-noise ratio to be >100 for most of the major element lines necessary for the analysis. Qualitative analysis was successfully performed [67] and an illustration of the LIBS spectra obtained with the setup for the JSC-1 lunar analog at varying pressure conditions is shown in Fig. 16.

To achieve quantitative analysis of samples in situ, the team developed a geochemical

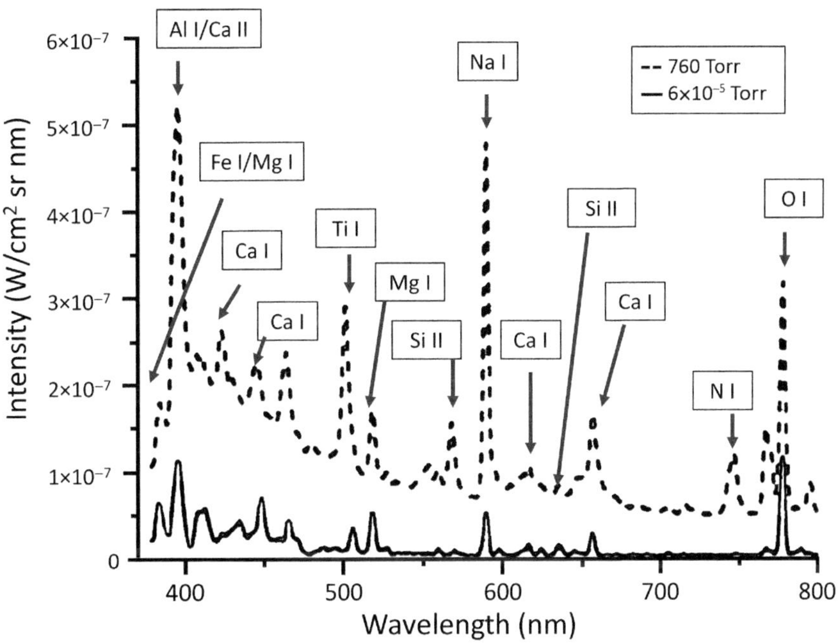

FIG. 16 Example LIBS spectra obtained with the Chandrayaan-2 LIBS experimental setup with a JSC-1 lunar analog sample under 760 Torr and vacuum conditions. Most major element emission lines are detectable [67].

standard/certified sample database of 60 different types of standards, such as SRM 688/691/699, GBW 07104, GBW 07015, GBW 07114, VS 811-89, JF-1, VS 2119-81, VS R28, SARM 4, JR-1, VS 2887-84, plagioclase, ilmenite, and anorthite. The first model was to use calibration curves for the most abundant elements of lunar soil like Mg, Al, Si, Ca, Fe, Ti, and Mn. Some trace elements like C, S, and P were also expected to be analyzed this way [68] though they are known to be more challenging than major elements [15]. Further analysis techniques for the spectra were to include the calibration-free LIBS methodology. Elemental abundance calibrations were in progress on the acquired spectra [66]. Hopefully, there will be a follow-up mission to capitalize on the development that has been done.

6 Future predictions

Overall, planetary surface analysis appears to be an excellent application of LIBS, which will likely grow, given the success so far. One new opportunity that has been under consideration over the last decade is on a Venus lander. The SAGE mission that was selected for Phase-A study by NASA in late 2009 contained a combination of LIBS and Raman spectroscopy instruments [17]. The Venus environment presents a severe challenge for surface analyses due to the high temperature (740 K) and pressure (>92 atm), requiring any probe to be housed in a protective shell. As such it can be expected to operate up to several hours on the surface before succumbing to the environment. In this environment, rapid analyses and the ability to perform measurements on multiple targets through a window in the protective shell are significant advantages of the LIBS technique. However, the high pressure of the atmosphere results in smaller plasmas and atomic emission lines that are much more susceptible to self-absorption [16]. In spite of this, using a gated intensifier to minimize self-adsorption, Clegg et al. [17] were able to produce LIBS elemental abundance calibration curves from spectra obtained at Venus surface temperature and pressure.

Another application that is likely in the future is the study of icy moon surfaces in the outer solar system. LIBS possesses the capability to detect and quantify volatile elements such as H, C, N, and O that likely exist in abundance in these environments, and also to quantify silicates, sulfates, and other species that exist within the icy matrices. By contrast, an X-ray-based analytical tool would lack the capability to observe the elements lighter in atomic mass than sodium, missing the bulk of the materials that are present in these materials. Given the LIBS advantages, Pavlov et al. [9] describe an instrument concept and measurements relevant to the study of an icy moon with LIBS. Similarly, LIBS is a highly relevant technique for probing cometary materials. The use of LIBS to study ice may even be relevant to studies of the lunar polar regions, as mentioned in [15].

If astronauts travel to the Moon, asteroids, and/or Mars, LIBS can provide a rapid and convenient hand-held tool for compositional analyses. In this case the visor of the helmet can provide eye protection, rendering LIBS to be completely safe for astronauts to use.

In summary, we may be seeing the beginning of a revolution in planetary science in which the LIBS technique becomes a dominant analytical tool in many different types of planetary environments. Cross-calibration between LIBS instruments, likely first on Mars and then on other planetary bodies, will be very important to improve accuracy and reproducibility. International dialog between planetary scientists using the LIBS technique is encouraged to refine calibration techniques to obtain the most accurate results.

Acknowledgments

Support for this work in the United States was provided by NASA Mars Exploration Program grant NNH15A35I. This work also benefited significantly from a meeting of an International Team Meeting on Planetary LIBS, organized and hosted by the International Space Science Institute in Bern, Switzerland.

References

[1] M. Salvatore, K. Truitt, K. Roszell, N. Lanza, E. Rampe, N. Mangold, E. Dehouck, R.C. Wiens, S. Clegg, Surface alteration of sedimentary rocks throughout the Transantarctic Mountains: implications for "modern" weathering processes on Mars, Icarus 319 (2018) 669–684, https://doi.org/10.1016/j.icarus.2018.10.007.

[2] N. Lanza, A. Ollila, A. Cousin, R.C. Wiens, S. Clegg, N. Mangold, N. Bridges, D. Cooper, M. Schmidt, J. Berger, R. Arvidson, N. Melikechi, H. Newsom, R. Tokar, C. Hardgrove, A. Mezzacappa, R. Jackson, B. Clark, O. Forni, S. Maurice, M. Nachon, R. B. Anderson, J. Blank, M. Deans, D. Delapp, R. Leveillé, R. McInroy, R. Martinez, P.-Y. Meslin, The MSL Team, Understanding the signature of rock coatings in laser-induced breakdown spectroscopy data, Icarus 249 (2015) 43–61, https://doi.org/10.1016/j.icarus.2014.05.038.

[3] N.L. Lanza, R.C. Wiens, R.E. Arvidson, B.C. Clark, W. W. Fischer, R. Gellert, J.P. Grotzinger, J.A. Hurowitz, S.M. McLennan, R.V. Morris, M.S. Rice, J.F. Bell III, J. A. Berger, D.L. Blaney, N.T. Bridges, F. Calef III, J. L. Campbell, S.M. Clegg, A. Cousin, K.S. Edgett, C. Fabre, M.R. Fisk, O. Forni, J. Frydenvang, K. R. Hardy, C. Hardgrove, J.R. Johnson, J. Lasue, S. Le Mouelic, M.C. Malin, N. Mangold, J. Martin-Torres, S. Maurice, M.J. McBride, D.W. Ming, H.E. Newsom, S. Schroeder, L.M. Thompson, A.H. Treiman, S. VanBommel, D.T. Vaniman, M.-P. Zorzano, Oxidation of manganese in an ancient aquifer, Kimberley formation, Gale crater, Mars, Geophys. Res. Lett. 43 (2016) 7398–7407, https://doi.org/10.1002/2016GL069109.

[4] W. Rapin, P.-Y. Meslin, S. Maurice, D. Vaniman, M. Nachon, N. Mangold, S. Schroeder, O. Gasnault, O. Forni, R.C. Wiens, G.M. Martinez, A. Cousin, V. Sautter, J. Lasue, D. Archer, The MSL science team, Hydration state of calcium sulfates in Gale crater: identification of bassanite veins, Earth Planet. Sci. Lett. 452 (2016) 197–205, https://doi.org/10.1016/j.epsl.2016.07.045.

[5] W. Rapin, P.-Y. Meslin, S. Maurice, R.C. Wiens, D. Laporte, B. Chauviré, O. Gasnault, S. Schroeder, P. Beck, S. Bender, O. Beyssac, A. Cousin, E. Dehouck, C. Drouet, O. Forni, M. Nachon, N. Melikechi, B. Rondeau, N. Mangold, N. Thomas, Quantification of water content by laser induced breakdown spectroscopy on Mars, Spectrochim. Acta B 130 (2017) 82–100, https://doi.org/10.1016/j.sab.2017.02.007.

[6] W. Rapin, B. Bousquet, J. Lasue, P.-Y. Meslin, J.-L. Lacour, C. Fabre, R.C. Wiens, J. Frydenvang, E. Dehouck, S. Maurice, O. Gasnault, O. Forni, A. Cousin, Roughness effects on the hydrogen signal in laser-induced breakdown spectroscopy, Spectrochim. Acta B 137 (2017) 13–22, https://doi.org/10.1016/j.sab.2017.09.003.

[7] W. Rapin, B. Chauviré, T. Gabriel, A. McAdams, P.-Y. Meslin, B.L. Ehlmann, B. Rondeau, E. Dehouck, N. Mangold, C. Hardgrove, S. Maurice, R.C. Wiens, O. Gasnault, A. Cousin, O. Forni, J. Frydenvang, S. Schroeder, In situ analysis of opal in Gale crater, Mars, J. Geophys. Res. Planets (2018) https://doi.org/10.1029/2017JE005483.

[8] Z.A. Arp, D.A. Cremers, R.C. Wiens, D.M. Wayne, B. Sallé, S. Maurice, Analysis of water ice and water ice/soil mixtures using laser-induced breakdown spectroscopy: application to Mars polar exploration, Appl. Spectrosc. 58 (2004) 897–909.

[9] S.G. Pavlov, E.K. Jessberger, H.-W. Huebers, S. Schroeder, I. Rauschenbach, S. Florek, J. Neumann, H. Henkel, S. Klinkner, Miniature laser-induced plasma spectroscopy for planetary in situ analysis—the case for Jupiter's moon Europa, Adv. Space Res. 48 (2011) 764–778.

[10] J.O. Caceres, J.T. Lopez, H.H. Telle, A.G. Urena, Quantitative analysis of trace metal ions in ice using laser-induced breakdown spectroscopy, Spectrochim. Acta B 56 (2001) 831–838.

[11] J.D. Blacic, D.R. Petit, D.A. Cremers, N. Roessler, Laser-induced breakdown spectroscopy for remote elemental analysis of planetary surfaces, in: Proceeding of the International Symposium on Spectral Sensing Research, Maui, HI, 1992, , pp. 302–312.

[12] A.K. Knight, N.L. Scherbarth, D.A. Cremers, M.J. Ferris, Characterization of laser-induced breakdown spectroscopy (LIBS) for application to space exploration, Appl. Spectrosc. 54 (2000) 331–340.

[13] R.C. Wiens, F.P. Seelos IV, M.J. Ferris, R.E. Arvidson, D. A. Cremers, J.D. Blacic, K. Deal, Combined remote mineralogical and elemental identification from rovers: field and laboratory tests using reflectance and laser-induced breakdown spectroscopy, J. Geophys. Res. Planets (2002) https://doi.org/10.1029/2000JE001439.

[14] B. Sallé, D.A. Cremers, S. Maurice, R.C. Wiens, P. Fichet, Evaluation of a compact spectrograph for in-situ and

stand-off laser-induced breakdown spectroscopy analyses of geological samples on Mars missions, Spectrochim. Acta B 60 (2005) 805–815.

[15] J. Lasue, R.C. Wiens, S.M. Clegg, D.T. Vaniman, K. H. Joy, S. Humphries, A. Mezzacappa, N. Melikechi, R.E. McInroy, S. Bender, Laser induced breakdown spectroscopy (LIBS) for lunar exploration, J. Geophys. Res. Planets 117 (2012) E01002, https://doi.org/10.1029/2011JE003898.

[16] Z.A. Arp, D.A. Cremers, R.D. Harris, D.M. Oschwald, G.R. Parker Jr., D.M. Wayne, Feasibility of generating a useful laser-induced breakdown spectroscopy plasma on rocks at high pressure: preliminary study for a Venus mission, Spectrochim. Acta B 59 (2004) 987–999.

[17] S.M. Clegg, R.C. Wiens, A.K. Misra, S.K. Sharma, J. Lambert, S. Bender, R. Newell, K. Nowak-Lovato, S. Smrekar, M.D. Dyar, S. Maurice, Planetary geochemical investigations by Raman-LIBS spectroscopy (RLS), Appl. Spectrosc. 68 (2014) 925–936.

[18] R.C. Wiens, S. Maurice, B. Barraclough, M. Saccoccio, W.C. Barkley, J.F. Bell III, S. Bender, J. Bernardin, D. Blaney, J. Blank, M. Bouye, N. Bridges, P. Cais, R. C. Clanton, B. Clark, S. Clegg, A. Cousin, D. Cremers, A. Cros, L. DeFlores, D. Delapp, R. Dingler, C. D'Uston, M.D. Dyar, T. Elliott, D. Enemark, C. Fabre, M. Flores, O. Forni, O. Gasnault, T. Hale, C. Hays, K. Herkenhoff, E. Kan, L. Kirkland, D. Kouach, D. Landis, Y. Langevin, N. Lanza, F. LaRocca, J. Lasue, J. Latino, D. Limonadi, C. Lindensmith, C. Little, N. Mangold, G. Manhes, P. Mauchien, C. McKay, E. Miller, J. Mooney, R. V. Morris, L. Morrison, T. Nelson, H. Newsom, A. Ollila, M. Ott, L. Pares, R. Perez, F. Poitrasson, C. Provost, J.W. Reiter, T. Roberts, F. Romero, V. Sautter, S. Salazar, J.J. Simmonds, R. Stiglich, S. Storms, N. Streibig, J.-J. Thocaven, T. Trujillo, M. Ulibarri, D. Vaniman, N. Warner, R. Waterbury, R. Whitaker, J. Witt, B. Wong-Swanson, The ChemCam instruments on the Mars Science Laboratory (MSL) rover: body unit and combined system performance, Space Sci. Rev. 170 (2012) 167–227, https://doi.org/10.1007/S11214-012-9902-4.

[19] S. Maurice, R.C. Wiens, M. Saccoccio, B. Barraclough, O. Gasnault, O. Forni, N. Mangold, D. Baratoux, S. Bender, G. Berger, J. Bernardin, M. Berthé, N. Bridges, D. Blaney, M. Bouyé, P. Cais, B. Clark, S. Clegg, A. Cousin, D. Cremers, A. Cros, L. DeFlores, C. Derycke, B. Dingler, G. Dromart, B. Dubois, M. Dupieux, E. Durand, L. d'Uston, C. Fabre, B. Faure, A. Gaboriaud, T. Gharsa, K. Herkenhoff, E. Kan, L. Kirkland, D. Kouach, J.-L. Lacour, Y. Langevin, J. Lasue, S. Le Mouélic, M. Lescure, E. Lewin, D. Limonadi, G. Manhes, P. Mauchien, C. McKay, P.-Y. Meslin, Y. Michel, E. Miller, H. E. Newsom, G. Orttner, A. Paillet, L. Pares, Y. Parot, R. Perez, P. Pinet, F. Poitrasson, B. Quertier, B. Sallé, C. Sotin, V. Sautter, H. Seran, J.J. Simmonds, J.-. B. Sirven, R. Stiglich, N. Streibig, J.-J. Thocaven, M. Toplis, D. Vaniman, The ChemCam instruments on the Mars Science Laboratory (MSL) rover: science objectives and mast unit, Space Sci. Rev. 170 (2012) 95–166, https://doi.org/10.1007/s11214-012-9912-2.

[20] R.C. Wiens, S. Maurice, F. Rull Perez, The SuperCam remote sensing instrument suite for the Mars 2020 rover mission: a preview, Spectroscopy 32 (5) (2017) 50–55. http://www.spectroscopyonline.com/supercam-remote-sensing-instrument-suite-mars-2020-rover-preview.

[21] S.W. Squyres, R.E. Arvidson, E.T. Baumgartner, J. F. Bell III, P.R. Christensen, S. Gorevan, K. E. Herkenhoff, G. Klingelhoefer, M.B. Madsen, R. V. Morris, R. Rieder, P.A. Romero, Athena Mars rover science investigation, J. Geophys. Res. 108 (2003) 8062, https://doi.org/10.1029/2003JE002121.

[22] S. Maurice, S. Clegg, R.C. Wiens, O. Gasnault, W. Rapin, O. Forni, A. Cousin, V. Sautter, N. Mangold, L. Le Deit, M. Nachon, R. Anderson, N. Lanza, C. Fabre, V. Payré, J. Lasue, P.-Y. Meslin, J.-B. Sirven, N. Melikechi, S. Le Mouelic, J. Frydenvang, A. Vasavada, J.C. Bridges, S. C. Bender, S. Schroeder, R. Francis, P. Pinet, H. Newsom, A. Ollila, K. Herkenhoff, M.B. Madsen, G. Dromart, P. Beck, E. Lewin, J.-L. Lacour, Y. Langevin, B. Gondet, L. d'Uston, G. Berger, M. Toplis, J.R. Johnson, M.D. Dyar, N. Bridges, D. Vaniman, B. Barraclough, ChemCam activities and discoveries during the Mars Science Laboratory nominal mission in Gale crater, Mars, J. Anal. At. Spectrom. (2016) https://doi.org/10.1039/c5ja00417a.

[23] J.R. Johnson, J.F. Bell III, S. Bender, D. Blaney, E. Cloutis, L. DeFlores, B. Helmann, O. Gasnault, B. Gondet, K. Kinch, M. Lemmon, S. Le Mouelic, S. Maurice, M. Rice, R. Wiens, The MSL Science Team, ChemCam passive reflectance spectroscopy of surface materials at the curiosity landing site, Mars, Icarus 249 (2015) 74–92, https://doi.org/10.1016/j.icarus.2014.02.028.

[24] S. Le Mouelic, O. Gasnault, K.E. Herkenhoff, N. T. Bridges, Y. Langevin, N. Mangold, S. Maurice, R. C. Wiens, P. Pinet, H.E. Newsom, R.G. Deen, J. F. Bell III, J.R. Johnson, W. Rapin, B. Barraclough, D. L. Blaney, L. Deflores, J. Maki, M.C. Malin, R. Perez, M. Saccoccio, The MSL Science Team, The ChemCam remote micro-imager at Gale crater: review of the first year of operations on Mars, Icarus 249 (2015) 93–107, https://doi.org/10.1016/j.icarus.2014.05.030.

[25] R.C. Wiens, S. Maurice, J. Lasue, O. Forni, R. B. Anderson, S. Clegg, S. Bender, B.L. Barraclough, L. DeFlores, D. Blaney, R. Perez, N. Lanza, A. Ollila, A. Cousin, O. Gasnault, D. Vaniman, M.D. Dyar, C. Fabre, V. Sautter, D. Delapp, H. Newsom, N. Melikechi, The ChemCam team, Pre-flight

calibration and initial data processing for the ChemCam laser-induced breakdown spectroscopy instrument on the Mars Science Laboratory rover, Spectrochim. Acta B 82 (2013) 1–27, https://doi.org/10.1016/j.sab.2013.02.003.

[26] C. Fabre, S. Maurice, A. Cousin, R.C. Wiens, O. Forni, V. Sautter, D. Guillaume, Onboard calibration igneous targets for the Mars Science Laboratory curiosity rover and the chemistry camera laser induced breakdown spectroscopy instrument, Spectrochim. Acta B 66 (2011) 280–289, https://doi.org/10.1016/j.sab.2011.03.012.

[27] D. Vaniman, M.D. Dyar, R.C. Wiens, A. Ollila, N. Lanza, J. Lasue, M. Rhodes, S.M. Clegg, Ceramic ChemCam calibration targets on Mars Science Laboratory, Space Sci. Rev. 170 (2012) 229–255, https://doi.org/10.1007/s11214-012-9886-0.

[28] A.M. Ollila, H.E. Newsom, B. Clark III, R.C. Wiens, A. Cousin, J.G. Blank, N. Mangold, V. Sautter, S. Maurice, S.M. Clegg, O. Gasnault, O. Forni, R. Tokar, E. Lewin, M.D. Dyar, J. Lasue, R. Anderson, S.M. McLennan, J. Bridges, D. Vaniman, N. Lanza, C. Fabre, N. Melikechi, G.M. Perrett, J.L. Campbell, P. L. King, B. Barraclough, D. Delapp, S. Johnstone, P.-Y. Meslin, A. Rosen-Gooding, J. Williams, The MSL Science Team, Trace element geochemistry (Li, Ba, Sr, and Rb) using curiosity's ChemCam: early results for Gale crater from Bradbury Landing Site to Rocknest, J. Geophys. Res. 119 (2014) 255–285, https://doi.org/10.1002/2013JE004517.

[29] V. Payré, C. Fabre, A. Cousin, V. Sautter, R.C. Wiens, O. Forni, O. Gasnault, J. Lasue, A. Ollila, W. Rapin, P.-Y. Meslin, S. Maurice, M. Nachon, L. Le Deit, N. Lanza, S.M. Clegg, Alkali trace elements with ChemCam: calibration update and geological implications of the occurrence of alkaline rocks in Gale crater, Mars, J. Geophys. Res. 122 (2017) https://doi.org/10.1002/2016JE005201.

[30] P.J. Gasda, E.B. Haldeman, R.C. Wiens, W. Rapin, T. F. Bristow, J.C. Bridges, S.P. Schwenzer, B. Clark, K. Herkenhoff, J. Frydenvang, N.L. Lanza, S. Maurice, S. Clegg, D.M. Delapp, V.L. Sanford, M.R. Bodine, R. McInroy, In situ detection of boron by ChemCam on Mars, Geophys. Rev. Lett. 44 (2017) https://doi.org/10.1002/2017GL074480.

[31] O. Forni, M. Gaft, M. Toplis, S.M. Clegg, V. Sautter, S. Maurice, R.C. Wiens, N. Mangold, O. Gasnault, V. Sautter, S. Le Mouelic, P.-Y. Meslin, M. Nachon, R. E. McInroy, A.M. Ollila, A. Cousin, J.C. Bridges, N. L. Lanza, M.D. Dyar, First detection of fluorine on mars: implications on Gale crater's geochemistry, Geophys. Res. Lett. 42 (2015) https://doi.org/10.1002/2014GL062742.

[32] N.L. Lanza, W.W. Fischer, R.C. Wiens, J.P. Grotzinger, A. Ollila, R.B. Anderson, B.C. Clark, N. Mangold, S. Maurice, S. Le Mouelic, M. Nachon, M. Schmidt, J. Berger, S.M. Clegg, M. Fisk, O. Forni, C. Hardgrove, N. Melikechi, H.E. Newsom, V. Sautter, et al., High manganese concentrations in rocks at Gale crater, Mars, Geophys. Res. Lett. 41 (2014) 5755–5763, https://doi.org/10.1002/2014GL060329.

[33] J. Lasue, S.M. Clegg, O. Forni, A. Cousin, R.C. Wiens, N. Lanza, N. Mangold, L. Le Deit, C. Fabre, O. Gasnault, S. Maurice, D. Blaney, J.R. Johnson, S. Le Mouelic, J. Berger, V. Payré, M. Nachon, W. Goetz, Observation of >5 wt. % zinc at the Kimberley outcrop, Gale crater, Mars, J. Geophys. Res. 121 (2016) 338–352, https://doi.org/10.1002/2015JE004946.

[34] V. Payré, C. Fabre, V. Sautter, A. Cousin, N. Mangold, L. Le Deit, O. Forni, W. Goetz, R.C. Wiens, O. Gasnault, P.-Y. Meslin, J. Lasue, W. Rapin, B. Clark, M. Nachon, N. Lanza, S. Maurice, Copper enrichments in Kimberley formation, Gale crater, Mars: evidence for a Cu deposit at the source, Icarus 321 (2019) 736–751, https://doi.org/10.1016/j.icarus.2018.12.015.

[35] S.M. Clegg, R.C. Wiens, J.E. Barefield, E. Sklute, M. D. Dyar, Quantitative remote laser induced breakdown spectroscopy by multivariate analysis, Spectrochim. Acta B 64 (2009) 79–88.

[36] O. Forni, S. Maurice, O. Gasnault, R.C. Wiens, A. Cousin, S.M. Clegg, J.-B. Sirven, J. Lasue, Independent component analysis classification of laser induced breakdown spectroscopy spectra, Spectrochim. Acta B 86 (2013) 31–41.

[37] S.M. Clegg, R.C. Wiens, R.B. Anderson, O. Forni, J. Frydenvang, J. Lasue, A. Pilleri, V. Payre, T. Boucher, M.D. Dyar, S.M. McLennan, R.V. Morris, T.G. Graff, S.A. Mertzman, B.L. Ehlmann, S.C. Bender, R.L. Tokar, I. Belgacem, H. Newsom, R.E. McInroy, R. Martinez, P. Gasda, O. Gasnault, S. Maurice, Recalibration of the Mars Science Laboratory ChemCam instrument with an expanded geochemical database, Spectrochim. Acta B 129 (2017) 64–85.

[38] R.B. Anderson, S.M. Clegg, J. Frydenvang, R.C. Wiens, S. McLennan, R.V. Morris, B. Ehlmann, M.D. Dyar, Improved accuracy in quantitative laser-induced breakdown spectroscopy using sub-model partial least squares, Spectrochim. Acta B 129 (2017) 49–57, https://doi.org/10.1016/j.sab.2016.12.002.

[39] J.L. Campbell, G.M. Perrett, R. Gellert, S. M. Andrushenko, N.I. Boyd, J.A. Maxwell, P.L. King, C.D.M. Schofield, Calibration of the Mass Science Laboratory alpha particle X-ray spectrometer, Space Sci. Rev. 170 (2012) 319–340.

[40] D. Blaney, R.C. Wiens, S. Maurice, S.M. Clegg, R. B. Anderson, L.C. Kah, S. Le Mouelic, A. Ollila, N. Bridges, R. Tokar, G. Berger, J. Bridges, A. Cousin, B. Clark, M.D. Dyar, B. Ehlmann, P.L. King, N. Lanza, N. Mangold, P.-Y. Meslin, H. Newsom, S. Schroeder, S. Rowland, J. Johnson, L. Edgar, O. Gasnault,

O. Forni, The MSL Science eam, Chemistry and texture of the rocks at "Rocknest", Gale crater: evidence for sedimentary origin and diagenetic alteration, J. Geophys. Res. 119 (2014) 2109–2131, https://doi.org/10.1002/2013JE004590.

[41] N. Mangold, O. Forni, G. Dromart, O. Gasnault, M. Nachon, R.C. Wiens, R.B. Anderson, B. Barraclough, J.F. Bell III, G. Berger, D.L. Blaney, J. C. Bridges, F. Calef, B. Clark, S.M. Clegg, A. Cousin, L. Edgar, K. Edgett, B. Ehlmann, C. Fabre, M. Fisk, J. Grotzinger, S. Gupta, K.E. Herkenhoff, J. Hurowitz, J.R. Johnson, L.C. Kah, N. Lanza, J. Lasue, S. Le Mouelic, R. Leveille, E. Lewin, M. Malin, S. McLennan, S. Maurice, P.-Y. Meslin, R. Milliken, H. Newsom, A. Ollila, S. Rowland, V. Sautter, M. Schmidt, S. Schroeder, K. Stack, D.Y. Sumner, C. D'Uston, D. Vaniman, R. Williams, Chemical variations in Yellowknife Bay formation sediments analyzed by the Curiosity rover on Mars, J. Geophys. Res. 120 (2015) 452–482, https://doi.org/10.1002/2014JE004681.

[42] F. Rivera-Hernandez, D.Y. Sumner, N. Mangold, K. M. Stack, O. Forni, H. Newsom, A. Williams, M. Nachon, J. L'Haridon, O. Gasnault, R.C. Wiens, S. Maurice, Using ChemCam LIBS data to constrain grain size in rocks on Mars: proof of concept and application to rocks at Yellowknife Bay and Pahrump Hills, Gale crater, Icarus 321 (2019) 82–98, https://doi.org/10.1016/j.icarus.2018.10.023.

[43] P.-Y. Meslin, O. Gasnault, O. Forni, S. Schröder, S. Clegg, G. Berger, J. Lasue, A. Cousin, S. Le Mouelic, S. Maurice, N. Mangold, C. Fabre, R. Wiens, B. Ehlmann, N. Lanza, P. Pinet, R. Anderson, D. Archer, D. Bish, D. Blake, D. Blaney, N. Bridges, B. Clark, G. Dromart, M. D. Dyar, M. Fisk, W. Goetz, K. Herkenhoff, J.-L. Lacour, Y. Langevin, L. Leshin, R. Leveille, E. Lewin, M.B. Madsen, T.H. McConnochie, J. E. Moores, H. Newsom, A. Ollila, R. Perez, E. Rampe, N. Renno, V. Sautter, J.-B. Sirven, M. de la Torre, L. d'Uston, D. Vaniman, The MSL Science Team, Soil diversity and hydration as observed by ChemCam at Gale crater, Mar. Sci. 341 (2013) https://doi.org/10.1126/science.1238670.

[44] A. Cousin, P.-Y. Meslin, R.C. Wiens, W. Rapin, N. Mangold, C. Fabre, O. Gasnault, O. Forni, R. Tokar, J. Lasue, D. Vaniman, A. Ollila, S. Schroeder, V. Sautter, D. Blaney, S. Le Mouelic, M. Nachon, G. Dromart, H. Newsom, S. Maurice, M.D. Dyar, N. Lanza, B. Clark, S. Clegg, W. Goetz, The MSL Science Team, Compositions of sub-millimeter-size clasts and fine particles in the Martian soils at Gale: a window into the production of soils, Icarus 249 (2015) 22–42, https://doi.org/10.1016/j.icarus.2014.04.052.

[45] J. Lasue, A. Cousin, P.-Y. Meslin, R.C. Wiens, O. Gasnault, W. Rapin, S. Schroeder, A. Ollila, C. Fabre, N. Mangold, G. Berger, S. Le Mouelic, E. Dehouck, O. Forni, S. Maurice, R. Anderson, N. Bridges, B. Clark, S.M. Clegg, C. D'Uston, W. Goetz, J. Johnson, N. Lanza, M.B. Madsen, N. Melikechi, A. Mezzacappa, H. Newsom, V. Sautter, J. Martin-Torres, M.P. Zorzano, The MSL Science Team, Martian eolian dust probed by ChemCam, Geophys. Res. Lett. 45 (2018) https://doi.org/10.1029/2018GL079210.

[46] T.H. McConnochie, M.D. Smith, M.J. Wolff, S. Bender, M. Lemmon, R.C. Wiens, S. Maurice, O. Gasnault, J. Lasue, P.-Y. Meslin, A.-M. Harri, M. Genzer, O. Kemppinen, G.M. Martinez, L. DeFlores, D. Blaney, J.R. Johnson, J.F. Bell III, Retrieval of water vapor column abundance and aerosol properties from ChemCam passive sky spectroscopy, Icarus (2017) https://doi.org/10.1016/j.icarus.2017.10.043.

[47] O. Forni, P.-Y. Meslin, J. L'Haridon, W. Rapin, M. Nachon, H. Newsom, N. Mangold, D. Anderson, R.B. Anderson, D.L. Blaney, S.M. Clegg, A. Cousin, E. Dehouck, O. Gasnault, J.R. Johnson, N.L. Lanza, S. Maurice, R.C. Wiens, Detection of fluorine-rich phases, phosphates, and halite in the Stimson-Murray units, Gale crater, Mars, in: Lunar Planet. Sci. XLVIII, The Lunar and Planetary Institute, Houston, TX, 2017, p. 1838.

[48] M. Nachon, N. Mangold, O. Forni, L. Kah, A. Cousin, W. Rapin, M. Fisk, R. Kronyak, D. Sumner, K. Stack, S. M. Clegg, D.L. Blaney, S. Schroeder, P.-Y. Meslin, S. Maurice, R.C. Wiens, R.B. Anderson, C. Fabre, J. P. Grotzinger, O. Gasnault, N.L. Lanza, J. Lasue, L. Le Deit, J.G. Blank, F. Calef, S. Le Mouelic, D.Z. Oehler, V. Payre, Chemistry of diagenetic features analyzed by ChemCam at Pahrump Hills, Gale crater, Mars, Icarus (2016) https://doi.org/10.1016/j.icarus.2016.08.026.

[49] H.Y. McSween Jr., G.J. Taylor, M.B. Wyatt, Elemental composition of the martian crust, Science 324 (2009) 736–739.

[50] V. Sautter, C. Fabre, O. Forni, M. Toplis, A. Cousin, A. M. Ollila, P.-Y. Meslin, S. Maurice, R.C. Wiens, N. Mangold, S. Le Mouelic, O. Gasnault, J. Lasue, G. Berger, E. Lewin, M. Schmidt, P. Pinet, D. Baratoux, B.L. Ehlmann, J. Bridges, M.D. Dyar, B. Clark, The MSL Science Team, Igneous mineralogy at Bradbury rise: the first ChemCamcampaign, J. Geophys. Res. 119 (2014) 30–46, https://doi.org/10.1002/2013JE004472.

[51] V. Sautter, M. Toplis, A. Cousin, O. Forni, R.C. Wiens, C. Fabre, O. Gasnault, M. Fisk, S. Maurice, A. Ollila, N. Mangold, L. Le Deit, P. Beck, W. Rapin, P. Pinet, J. Blank, S. Clegg, P.-Y. Meslin, M.D. Dyar, H. Newsom, N. Bridges, N. Lanza, S. Le Mouelic, D. Vaniman, In-situ evidence for continental crust on early Mars, Nat. Geosci. 8 (2015) 605–609, https://doi.org/10.1038/NGEO2474.

[52] A. Cousin, V. Sautter, V. Payré, O. Forni, N. Mangold, O. Gasnault, L. Le Deit, J. Johnson, S. Maurice, M. Salvatore, R.C. Wiens, P. Gasda, W. Rapin, Classification of igneous rocks analyzed by ChemCam at Gale crater, Mars, Icarus 288 (2017) 265–283.

[53] J. Frydenvang, N. Mangold, R.C. Wiens, A.A. Fraeman, L. A. Edgar, C. Fedo, J. L'Haridon, C.C. Bedford, S. Gupta, J. P. Grotzinger, J.C. Bridges, B.C. Clark, E.B. Rampe, O. Gasnault, S. Maurice, P.J. Gasda, N.L. Lanza, A. M. Ollila, P.-Y. Meslin, V. Payré, F. Calef, M. Salvatore, C.H. House, T.S.J. Gabriel, The role of diagenesis at Vera Rubin Ridge in Gale crater, Mars, and the chemostratigraphy of the Murray formation as observed by the ChemCam instrument, in: Proceeding of the 51st Lunar and Planetary Science Conference, 2020, p. 1479.

[54] R.E. Arvidson, S.W. Squyres, R.V. Morris, A.H. Knoll, R. Gellert, B.C. Clark, J.C. Catalano, B.L. Jolliff, S. M. McLennan, K.E. Herkenhoff, S. van Bommel, D. W. Mittlefehldt, J.P. Grotzinger, E.A. Guinness, J. R. Johnson, J.F.I.I.I. Bell, W.H. Farrand, N. Stein, V. K. Fox, M.P. Golombek, A.G. Hinkle, W.M. Calvin, P. A. de Souza Jr., High concentrations of manganese and sulfur in deposits on Murray Ridge, Endeavour Crater, Mars, Am. Minerol. 101 (2016) 1389–1485.

[55] S.N. Lamm, N.L. Lanza, P.J. Gasda, R.C. Wiens, P.-Y. Meslin, M.F. Kirk, Manganese observations from ChemCam during sols 1650-1750: implications for a changing redox environment, in: Lunar Planet. Sci. XXXXVIIII, The Lunar and Planetary Institute, Houston, TX, 2018, p. 2903.

[56] M. Nachon, S.M. Clegg, N. Mangold, S. Schroeder, L. C. Kah, G. Dromart, A. Ollila, J. Johnson, D. Oehler, J. Bridges, S. Le Mouelic, R.C. Wiens, R. Anderson, D. Blaney, J.F. Bell, B. Clark, A. Cousin, M.D. Dyar, B. Ehlmann, C. Fabre, O. Forni, O. Gasnault, J. Grotzinger, J. Lasue, E. Lewin, R. Leveillé, S. McLennan, S. Maurice, P.-Y. Meslin, M. Rice, K. Stack, D. Vaniman, D. Wellington, The MSL Science Team, Calcium sulfate veins characterized by the ChemCam instrument at Gale crater, Mars, J. Geophys. Res. 119 (2014) 1991–2016, https://doi.org/10.1002/2013JE004588.

[57] R.C. Wiens, S. Maurice, The ChemCam and MSL Science Teams, ChemCam: chemostratigraphy by the first Mars microprobe, Elements 11 (2015) 33–38.

[58] J.F. Mustard, M. Adler, A. Allwood, D.S. Bass, D. W. Beaty, J.F. Bell III, W.B. Brinckerhoff, M. Carr, D. J. Des Marais, B. Drake, K.S. Edgett, J. Eigenbrode, L. T. Elkins-Tanton, J.A. Grant, S.M. Milkovich, D. Ming, C. Moore, S. Murchie, T.C. Onstott, S.W. Ruff, M. A. Sephton, A. Steele, A. Treiman, Report of the Mars 2020 Science Definition Team, Mars Exploration Program Analysis Group (MEPAG), 2013, 154. http://mepag.jpl.nasa.gov/reports/MEP/Mars_2020_SDT_Report_Final.pdf.

[59] G. Chen, E.S. Yeung, Acoustic signal as an internal standard for quantitation in laser-generated plumes, Anal. Chem. 60 (1988) 2258–2263.

[60] N. Murdoch, B. Chide, J. Lasue, A. Cadu, A. Sournac, M. Bassas-Ports, J. Merrison, J.J. Iversen, C. Moretto, C. Valsco, L. Pares, A. Hynes, V. Godiver, R. D. Lorenz, P. Cais, P. Bernardi, S. Maurice, R. C. Wiens, D. Mimoun, Laser-induced breakdown spectroscopy acoustic testing of the Mars 2020 Microphone, Planet. Space Sci. (2018) https://doi.org/10.1016/j.pss.2018.09.009.

[61] B. Chide, S. Maurice, N. Murdoch, J. Lasue, B. Bousquet, X. Jacob, A. Cousin, P.-Y. Meslin, D. Mimoun, R. C. Wiens, Laser-induced breakdown spectroscopy coupled with acoustic measurements and crater morphology, Spectrochim. Acta B 153 (2019) 150–160, https://doi.org/10.1016/j.sab.2019.01.008.

[62] G. Montagnac, G. Dromart, P. Beck, F. Mercier, B. Reynard, A. Cousin, S. Maurice, R.C. Wiens, Spark plasma sintering preparation of reference targets for field spectroscopy on Mars, J. Raman Spectrosc. (2018) https://doi.org/10.1002/jrs.5406.

[63] B.L. Ehlmann, J.F. Mustard, S.L. Murchie, F. Poulet, J. L. Bishop, A.J. Brown, W.M. Calvin, R.N. Clark, D. J. DesMarais, R.E. Milliken, L.H. Roach, T.L. Roush, G. A. Swayze, J.J. Wray, Orbital identification of carbonate-bearing rocks on Mars, Science 322 (2008) 1828–1832.

[64] V. Sundararajan, Overview and technical architecture of India's Chandrayaan-2 Mission to the Moon, in: AIAA Aerospace Sciences Meeting, 2018, p. 2178.

[65] A.S. Laxmiprasad, V.S. Raja, A. Goswami, S. Menon, M. V.H. Rao, K.A. Lohar, Laser induced breakdown spectrometer (LIBS) for Chandrayaan-2 rover, Curr. Sci. 1 (3) (2011) 14–16.

[66] A.S. Laxmiprasad, A. Goswami, V.S. Raja, K.A. Lohar, M.V.H. Rao, M. Mahajan, B. Raha, The LIBS instrument for Chandrayaan-2 rover: engineering model development aspects, in: 67th International Astronautical Congress, 2016 #34004.

[67] A.S. Laxmiprasad, V.L.N. Sridhar Raja, S. Menon, A. Goswami, M.V.H. Rao, K.A. Lohar, An in situ laser induced breakdown spectroscope (LIBS) for Chandrayaan-2 rover: ablation kinetics and emissivity estimations, Adv. Space Res. 52 (2) (2013) 332–341.

[68] A.S. Laxmiprasad, V.L.N. Sridhar Raja, A. Goswami, M. V.H. Rao, K.A. Lohar, S. Menon, Realization of empirical database and generation of calibration curves for precise quantitative analysis employing a mini-LIBS for Chandrayaan-2 rover, in: 39th CoSPAr Scientific Assembly, vol. 39, 2012, p. 1.

CHAPTER

21

Coal analysis

Zhe Wang[a], *Zongyu Hou*[a], *Lei Zhang*[b], *Shunchun Yao*[c], *Sahar Sheta*[a], *Muhammad Sher Afgan*[a]

[a]State Key Lab of Power Systems, International Joint Laboratory on Low Carbon Clean Energy Innovation, Department of Energy and Power Engineering, Tsinghua University, Beijing, China

[b]State Key Laboratory of Quantum Optics and Quantum Optics Devices, Institute of Laser Spectroscopy, Shanxi University, Taiyuan, China

[c]School of Electric Power, South China University of Technology, Guangzhou, China

1 Introduction

Coal analysis mainly includes ultimate analysis, proximate analysis, and heat value analysis. Ultimate analysis includes the measurement of major elements (C, H, O, N, and S) and other elements, while proximate analysis includes moisture content, ash content, volatile matter, and fixed carbon measurement. There is a high demand to develop a safe and economic real-time coal property analysis technology for coal production and utilization, such as for coal washing, blending, combustion optimization and pollution reduction, and pricing. A chemical processing-based coal analysis method was established as the standard method and had been accepted widely for off-line coal analysis as well as for coal price determination. The disadvantage of the standard method is that it needs more than 10 h to finish a complete set of coal analyses in addition to the fact that only a few grams of coal were analyzed, which, in turn, means less representativeness for tons of coal every hour in production and utilization in real industry. For the coal-related industry, a real-time coal analysis technology, either online or in situ, is more desirable to improve utilization efficiency and reduce environmental pollution.

A safer real-time technology with low cost and satisfactory analytical performance is still greatly needed to realize better coal production/utilization control and a huge amount of coal trading. Currently, there are several technologies available for real-time coal analysis: X-ray fluorescence (XRF) [1], prompt gamma neutron activation analysis (PGNAA) [2], near-infrared spectrometry (NIRS) [3], and microwave analysis [4]. Among them, XRF and PGNAA are most preferred since they are able to provide most of the information required for coal analysis. However, both of them suffer from low accuracy and radiation risk. Besides, XRF is not able to measure light elements,

Laser-Induced Breakdown Spectroscopy
https://doi.org/10.1016/B978-0-12-818829-3.00021-6

including the major elements C and H in coal; while PGNAA is too expensive to install and maintain, and is governed by strict nuclear radiation source control requirements. NIRS is susceptible to environmental variation and the measurement is not precise enough. Microwave analysis is able to provide only moisture content.

Laser-induced breakdown spectroscopy (LIBS) has been regarded as the most promising technology for coal analysis due to its instant, multielement measurement and standoff analysis capability. Compared with XRF, LIBS is more sensitive to light elements and no X-ray radiation is involved, and is available at almost the same price. Relative to other coal analysis techniques, LIBS can discriminate itself with high sensitivity to low atomic number elements (unlike XRF), trace elements detection and safe operation (unlike PGNAA), and is comparatively cheap with portable analyzers available [5].

Research on LIBS applied to coal analysis/utilization was reported in the late 1980s by Sandia National Laboratories [6]. Since then, coal analysis research using LIBS has been progressively carried out in both laboratory and industrial sites with reported literature from Australia [7–9], Bangladesh [10, 11], Belarus [12], China [13–59], Colombia [60–62], France [63], Hungary [64], Israel [65–67], Italy [68, 69], Indonesia [70, 71], Japan [72–75], Spain [76–79], and the USA [80–89].

At present, the application of LIBS to coal analysis is facing a critical time to achieve wide commercialization in the power industry, coal chemical processing industry, and coal mine industry [90]. The purpose of this chapter is to offer up-to-date progress on LIBS in coal analysis with experimental and instrumental challenges as the focus. With continuous research efforts, LIBS will be technically mature for coal analysis at various industrial sites.

2 Main impact factors on coal analysis

Coal consists of complex organic and inorganic molecules with different sizes and structures. Coal contains nearly all the elements in the earth's crust (minor and trace elements) in different concentrations and distributions with C, H, O, and N as the major elements. Due to the complicated composition of coal, LIBS coal analysis suffers from strong matrix effects and interferences. For example, the volatile content, ash content, and water content can change the plasma characteristics and lead to strong effects, and the metallic elemental lines and molecular lines (such as CN and C_2) can lead to severe interference with different lines. In addition, coal is a heterogeneous material; thus the representativeness of sampling should be considered for LIBS analysis. Optimizing the experimental parameters of LIBS is undoubtedly helpful to perform accurate and repeatable quantitative measurements. The following influencing factors of LIBS for coal analysis can be classified into four parts: (1) experimental configuration, including the wavelength, pulse energy or repetition rate of excitation laser, and the optical configuration; (2) environmental factors, including the type or pressure of ambient gas and the spatial confinement; (3) coal matrix, including both the physical and chemical matrices of coal, such as the particular form of coal (e.g., particle size and surface smoothness), and the sample composition (e.g., adhesives, moisture, volatile matter, and ash content); and (4) data processing method, including data pretreatment and quantification models.

2.1 Experimental configuration

The measurement accuracy of coal is very sensitive to many experimental factors. To improve measurement repeatability, it is necessary to optimize the experimental parameters, such as excitation, optical collection, etc.

Excitation parameter. A high-power pulsed laser is a common excitation source in LIBS experiments. Its characteristics, including wavelength, pulse energy, and repetition rate, have great influence on measurement accuracy. Mateo et al. [76] compared the quantitative results of inorganic species in coal by LIBS using laser radiation at 355 and1064 nm, and revealed that the ultraviolet (UV) wavelength provided better results than infrared (IR) radiation in terms of accuracy. Li et al. [35] also found that the calibration curve of carbon in coal created by 266-nm wavelength laser ablation had a much higher linearity than that by 1064-nm wavelength, and explained that this is because the UV wavelength can provide better crater morphology, better laser sample coupling, and less plasma shielding. Qian et al. [91] analyzed LIBS spectra of coal using laser wavelengths at 1064, 532, and 355 nm, and demonstrated that the 532-nm wavelength showed the highest line intensity and maintained a high coal quality measurement. Wang et al. [39] optimized the laser energy and repetition rate to 100 mJ and 10 Hz, respectively, by comparing the SNRs and RSDs of C (I) 247.9-nm and Si (I) 288.2-nm line intensities at different laser energies and repetition rates.

Optical configuration. The geometrical configuration of a LIBS measurement system, such as the distance from the focusing lens to the target surface, the method of plasma fluorescence collection, etc., will greatly affect the power density, ablation mass, and even analysis results. For an excitation configuration in the case of industrial applications, Gaft et al. [66] employed signals from an ultrasonic height sensor above the belt to maintain a fixed optical focus with changes in coal height. Wang et al. [39] expanded the laser beam prior to focusing to reduce the spot size and investigated the laser focal position on spectral signals. They found that at the vicinity of 7.5 mm above the surface, they could achieve the best analysis results. For their optical collection configuration, Redoglio et al. [69] designed a detection system comprising two spherical mirrors and a flat mirror to collect emissions from the plasma plume, and a 178-mm depth of field was obtained when the distance between the holed spherical mirror and the plasma was 1000 mm. Chen et al. [29] compared the plasma emissions collected through backside and side paths in a two-phase pulverized coal stream, and the backside path was found to provide a higher repeatability but weaker line intensities.

2.2 Environmental factors

Ambient gas. Ambient gas can significantly impact the emission properties and evolution process of plasmas ignited from coal samples. Dong et al. [84] studied time-resolved laser-induced coal plasmas in different ambient gases, and the plasma in ambient argon showed both higher temperature and electron density than those in air or helium. Similarly, Wang et al. [24] also observed significant line enhancement of C and H in bituminous coal under an argon environment. Pan et al. [92] studied the formation routes of a CN band in air, argon, and nitrogen, and confirmed that CN mainly comes from the reaction between carbon from coal ash and nitrogen from air. Li et al. [16] demonstrated that several lines of coal plasma in argon were enhanced by three to four times compared to those in air or nitrogen.

Ambient pressure. Ambient pressure can affect the evolution and morphology of plasmas. Low pressure indicates low collision probability of excitation and ionization, while high pressure indicates short average free path and high collision velocity. Noda et al. [72] found that with an increase in gas pressure, the carbon signal from coal plasma decreased drastically due to severe plasma quenching, and the optimal delay time became shorter and narrower. Yang et al. [19]

found that with an increase in gas pressure in the coal combustion cell, the minimum ignition energy decreased sharply due to the increase in molecular number density in the focusing zone and the enhancement of laser absorption efficiency.

Spatial confinement. The use of spatial confinement can effectively enhance the LIBS signal by shockwave compression, plasma radiation reflection, etc. Wang et al. [23] verified that the spatial confinement of plasma by using a cylindrical cavity could enhance the line intensity and reduce the shot-to-shot fluctuation. Li et al. [93] studied the dependence of spectral signal enhancement on relative permittivity of the confinement ring, and concluded that the confinement ring with higher relative permittivity can induce the stronger enhancement. Moreover, multicavity and spark discharge-coupled confinement were proved to further stabilize the plasma morphology and enhance the signal as well as SNR [55]. Li et al. [56] combined a partial least square (PLS) model with spatial confinement for analysis of carbon content in coal, and achieved an estimated average absolute error (AAE) of 1.0%.

2.3 Coal matrix

The physical and chemical matrix effects that exist in coal, which are brought about by its complex chemical composition and changeable physical properties, make it a challenge to quantitatively analyze.

2.3.1 Physical matrices

Particle size. Yu et al. [13] studied the effects of coal particle size on plasma characteristics; the decrease in the powder particle size was shown to lead to a remarkable increase in the plasma temperature. Li et al. [16] investigated the effect of particle size on plasma stability, and found that the plasmas induced from coal samples with particle sizes of 0–0.1 or 0.3–0.6 mm tended to be more stable.

Pressing force. Wang et al. [39] optimized the pressing force on coal pellets by varying the machine pressure in the range of 5–30 MPa, and an optimal pressure of 25 MPa was selected to reach the maximum SNR and the minimum RSD. Li et al. [16] found that the RSD of line intensity decreased with an increase in pressing force and reached its maximum at 10 t, which was attributed to the improvement of relative stability of pulse-to-pulse ablated mass.

2.3.2 Chemical matrices

Moisture. In LIBS measurements, the moisture content in coal can consume part of the laser energy and result in serious reduction in most analytical lines, thus increasing the uncertainty of quantitative results. For example, Chen et al. [36] showed that with an increase in moisture content, most of the spectral lines decreased in intensity and the RSDs increased. Gaft et al. [66] reported that variation in moisture content could lead to significant errors in evaluation of the coal ash content for an online LIBS system. Yuan et al. [28] found it difficult to employ the PLS model to associate the line intensities of O and H and the moisture content, and claimed that LIBS measurements should be carried out on an air-dried basis.

Volatile matter. Volatile matter is the vapors driven off during pyrolysis; these vapors mainly consist of hydrocarbons, aromatic hydrocarbons, etc. Yao et al. [52] experimentally verified that the volatile matter in coal could consume laser energy in the process of evaporation, dissociation, and ionization. With an increase in volatile content, more vaporized volatiles were pushed to the upper layers of plasma plumes, thus inhibiting the spectral intensities of mineral elements.

Addition of binder. In the quantitative analysis of coal by LIBS, it is essential to add binder in pulverized coal to enhance the mechanical strength of pellets, thus improving the spectral intensity as well as the measurement repeatability. However, factors such as chemical composition and mixing proportion of binder should be considered prior to addition. Yuan et al. [23] compared the measurement uncertainties of anthracite pellets by mixing with three binders; $Na_2SiO_3 \cdot 9H_2O$ was found to be an applicable binder for anthracite samples with an optimal weight percentage at 50%. In another study, Yao et al. [47] reported that the optimal percentage of KBr binder for the quantitative analysis of carbon content in coal was 60%. Ctvrtnickova et al. [77, 78] compared the spectra of KBr, wax, polyvinyl alcohol, and carbon tape, and concluded that KBr should be selected as the most suitable binder for carbon calibration because it contains no carbon and thus introduces little interference.

2.4 Spectra processing

The spectra of coal are extremely complex and full of redundant and perplexing information. The transformation from coal spectra to quantitative results can be achieved by using proper spectra pretreatment and calibration models.

2.4.1 *Spectra pretreatment*

Generally, various spectra pretreatment methods are combined together to improve the repeatability and SNR, or reduce matrix effects. Accumulation and averaging of spectra are the simplest methods to improve measurement repeatability and SNRs [94]. However, with an increase in accumulation or averaging number, the improvement of signal repeatability will become weaker and weaker. Spectral filtering (including smoothing, mean filtering, Kalman filtering, Wiener filtering, etc.) is used to eliminate noise from the spectra. Li et al. [51] combined 11-point smoothing with second-order derivation, and the root mean squared error cross validation (RMSECV) and root mean squared error of prediction (RMSEP) were optimized compared with the traditional PLS model by 75.1% and 88.3%, respectively. Normalization is usually utilized to reduce the signal uncertainty. "Bode rule/DC level" normalization [18], internal normalization [66], pulse-to-pulse intensity normalization [95], etc. were validated to be efficient normalization methods for achieving improvement in signal-elemental concentration correlation. Spectral standardization is also used to reduce the measurement uncertainty. It uses plasma parameters to compensate for signal fluctuations and converts each measured spectrum into a standard state spectrum [96]. Standardization is a more direct method with physical and theoretical background, usually results in better repeatability, and has a wider application scope as compared to normalization. Moreover, the spectrum standardization method works effectively to reduce sample-to-sample measurement uncertainty for repeated measured samples, though the method is designed for reducing pulse-to-pulse RSD [40].

2.4.2 *Quantification models*

A quantification model, such as univariate or multivariate calibration, is used to establish the relationship between elemental concentration and the related line intensity. However, the univariate quantification model usually suffers from matrix effects and nonlinearity at high concentrations due to self-absorption [13]. Therefore several chemometric methods involved in the multivariate quantification model (such as multiple linear regression [22], principal component analysis [PCA] [97], PLS regression [98], dominant factor-based PLS [28, 40, 56], artificial neural network [ANN] [48], linear discriminant analysis [99], and kernel extreme learning machine [K-ELM] [100, 101]) were employed to yield the best classification or measurement accuracy of coal. It should be noted that the commonly used multivariate model usually ignores the physical background of LIBS measurement and is mostly dependent on statistical correlation or curve fitting, which may lead to noise overfitting and eventually ruin the measurement accuracy for samples outside the matrix of the calibration sample set. The dominant factor-based PLS method combines the advantages of the univariate model and the traditional PLS method. The conventional univariate model with physical background was applied to model the major concentration information and was referred to as the dominant factor; PLS was further applied to compensate for the residual errors of the dominant factor. Therefore the dominant factor-based PLS method can achieve better performance and has a wider applicable range.

In this section, we summarized the previous research contributions of influencing factors on coal analysis, including experimental configuration, environmental factor, coal matrix, and spectra processing. The following conclusions are drawn to improve both quantitative analysis and reproducibility performances of LIBS:

1. Experimental configuration. The use of commercial UV radiation sources in the coal industry produces better analytical LIBS performance. The laser energy, laser-to-sample distance, delay time, etc. should be optimized based on specific experimental conditions.
2. Environmental factors. Compared with air, He, and N_2, Ar is the best choice for coal analysis due to high intensity and good repeatability. Spatial confinement or other enhancement technology is also an option for coal analysis. In real industry applications, the environmental temperature and humidity should be considered for equipment stability.
3. Coal matrices. The addition of binders and the compressive force (about 10 t) applied to coal powder with optimized particle size (as small as possible) are proven ways to decrease matrix effects and enhance results. In addition, LIBS experiments are preferred to be performed on an air-dried basis to reduce obstruction by the moisture content to analytical lines.
4. Data processing method. Multivariate calibration models using multiple lines information greatly improve the measurement accuracy. Establishing a hybrid model by combining data pretreatment (such as accumulation and averaging, data filtering, normalization, and spectrum standardization) and a dominant factor-based model is the best possible choice at present.

3 The application of LIBS to the offline analysis of coal

Offline analysis of coal usually deals with manually prepared pressed pellets. At present, coal analysis using pressed pellets is considered to be the best solution to obtain precise LIBS results. By grinding and mixing the coal powder, the representativeness of LIBS analysis can be improved. Moreover, pressed pellets provide a smoother surface for LIBS to achieve better analytical performance due to more stable plasma and repeatable signals compared to direct analysis on coal blocks or coal powder flow.

3.1 Ultimate analysis

Ultimate analysis is the quantitative analysis of elements in coal, including major elements (C, H, O, N, and S) and other elements. Table 1 shows the requirements of repeatability and reproducibility for coal ultimate analysis by Chinese national standards for chemical methods (GB/T 476-2001 and GB/T 214-2007).

Carbon content. Carbon is the major element of coal and the main source of heat. The requirement for chemical methods set by Chinese national standards is an absolute error of C measurement of $<1.0\%$. This standard is also the goal for C-content measurement for other technologies.

TABLE 1 Requirements for coal ultimate analysis by Chinese national standards [102].

Element		Repeatability requirement (%)	Reproducibility requirement (%)
C		0.5	1
H		0.15	0.25
S	<1.5%	0.05	0.1
	1.5%–4%	0.1	0.2
	>4%	0.2	0.3
N		0.08	0.15

Redoglio et al. [68, 69] developed a coal rotation unit to simulate an industrial conveyor belt. Carbon calibration curves showed a good linear relation between molar concentrations and the spectral lines intensities (C I 193.0- and 247.9-nm lines) within seven different coal samples.

Li et al. [56] utilized a cylindrical cavity drilled in polytetrafluoroethylene to improve the performance of C-content measurement. A dominant factor-based PLS model was built with 24 bituminous coal pellet samples. Results showed that R^2, RMSEP, average relative error (ARE), and AAE were 0.99, 1.35%, 1.66%, and 1.08% with cavity, and 0.99, 1.63%, 1.82%, and 1.27% without cavity, respectively. The error of 1.08% is pretty close to the value required by the Chinese national standard for chemical methods (1.0%). Li et al. [45] applied nitrile-butadiene rubber as a confinement ring to enhance the C-content measurement performance in 58 coal pellet samples. Using a combined PCA/PLS model, results showed that R^2, RMSEP, and ARE were 0.99, 0.68%, and 1.17%, respectively.

Yuan et al. [23] prepared 22 anthracite coal pellets using three different binders: $Na_2SiO_3 \cdot 9H_2O$, H_3BO_3, and KBr. Best results for C-content measurement were obtained using $Na_3SiO_3 \cdot 9H_2O$, and the Si I 390.55-nm line was used for internal calibration. The R^2, RMSEP, max relative error, and ARE were 0.76, 6.25%, 18.99%, and 10.14%, respectively, for the optimized binder condition. Using bituminous and lignite coal samples, Yao et al. [47] achieved best results for C-content measurement using 60 wt% KBr as a binder and an Si I 288.16-nm line as the internal standard ($R^2 = 0.983$).

Using different laser wavelengths, Li et al. [34, 35] prepared 18 standard bituminous coal pellets to compare 266-nm and 1064-nm wavelength performance for C-concentration measurements. The integrated intensity of C I 247.856 nm showed better RSDs of 5.62% by 266 nm. Using a PLS model, the calibration curves showed R^2 and RMSEP values of 0.99 and 1.64% for 266 nm, while results were 0.99 and 1.55% for 1064 nm. Results were further improved to 0.99 and 1.51%, respectively, for 266-nm wavelength source after applying a dominant factor-based PLS model.

In different ambient gases, Wang et al. [24] plotted the C-calibration curve using 24 bituminous coal samples. The PLS model had the optimal performance using Ar as ambient gas, with RSD for C I 193.09 nm of 8.50%, RMSEP of 3.49% (17 samples for calibration and 7 for validation), root mean squared error (RMSE) of 2.69%, ARE of 2.98%, and R^2 of 0.95.

Yuan et al. [32] proposed a PLS-wavelet transform hybrid model to determine the C-content in 24 bituminous coal samples. The wavelet coefficients were used as the input of the PLS model. Calibration and prediction results yielded R^2 of 0.98, RMSEP of 1.94% (17 samples for calibration and 7 for validation), and ARE of 1.67% using the hybrid model. Yan et al. [43] compared K-ELM, support-vector machine (SVM), least-squares support-vector machine (LS-SVM), and back propagation artificial neural network (BP-ANN) for C-content measurements using 26 standard coal samples. The best results were achieved using the K-ELM model to yield R^2 and RMSE of 0.9994 and 0.3762 wt%, respectively. To improve the sample-to-sample reproducibility, Hou et al. [40] proposed a hybrid quantification model. The model applied the spectrum standardization method to reduce the uncertainty of all the spectral lines, and then built a dominant factor-based PLS model, and finally integrated an identification model based on a self-adaptive database to further improve the sample-to-sample reproducibility. The sample-to-sample RSD of carbon content measurements was reduced to 0.3% from 6.47% for the traditional PLS model and 2.9% for the univariate model. Results showed average measurement error of 0.42% for C-content measurement using the proposed hybrid model.

Oxygen content. Oxygen in dry coal samples exists in both organic and inorganic forms. Zhang et al. [15] built a calibration curve for organic oxygen using eight anthracite coal samples by subtracting the inorganic oxygen from the total oxygen content. A linear relationship was obtained with a nonzero intercept due to atmospheric oxygen excitation. Results showed R^2 >0.98, measurement accuracy of 1.15%–1.37%, and ARE of 19.39%. Yu et al. [13] plotted a calibration curve for O using 777.19-nm spectral line intensities for six pulverized coal samples (<100 μm). The univariate calibration curve showed nonlinearity for higher O concentrations (1.834%–4.931%).

Sulfur content. Accurate quantitative measurement of sulfur is important to reduce the SO_x emissions of coal utilization. However, the observed sulfur emission lines mainly lie in the vacuum ultraviolet (VUV) (125–180 nm) and NIR (>900 nm) spectral ranges, and the intensity is very weak in air.

Gaft et al. [67] performed a detailed study on the S emission spectra in the VUV-to-IR range on coal samples mixed with different S concentrations (1%–50%) and real coal samples with S content between 0.4% and 0.7%. The best correlation was obtained in the VUV emission range (180.8-, 182.1-, and 182.6-nm S lines) using an N_2-filled spectrometer and a sample spectrograph distance of 20 cm or less. The S calibration curve showed R^2 of 0.88, standard deviation (SD) of 0.11, and ARE of 7.5% for two sets of coal samples. Burakov et al. [12] applied a double-pulse LIBS system for S analysis. Five pressed coal samples with S concentrations in the range of 0.5–16 wt% were tested. Using an S II 545.38-nm line, the minimal detectable S concentration was 2 wt% (higher than the required detectable values of 0.1–0.5 wt%). The calibration curve showed a good linearity with $R^2 = 0.989$. The authors suggested the importance of high-resolution spectrometers for sulfur line detection in the UV and NIR ranges. Yan et al. [43] built a K-ELM model for S-content measurement with S concentrations in the range of 0.20–6.45 wt %. Results showed that the K-ELM model had better performance compared to SVM, LS-SVM, and BP-ANN models with R^2 of 0.9832 and RMSE of 0.7704 wt% (eight samples as a prediction set). Zhang et al. [26] reported that S I lines (921.3, 922.8, and 923.8 nm) were strongly interfered with O I 926.1 nm, so high laser energy was required (160 mJ/pulse). The calibration curve showed R^2 of 0.97 and SD of 23.1%.

To achieve satisfactory results for S-detection and measurement, using a vacuum or N_2/He-filled spectrometer and shortening the sample-detector distances can enhance the detection intensity. Moreover, since S usually exists as Fe_2S_3 or $Fe_2(SO_4)_3$, using chemometric methods could achieve good results [103, 104].

Other elements. Wallis et al. [7] plotted calibration curves for Ca, Al, Na, Fe, Mg, and Si elements in low-ash lignite coals. Results showed that limits of detection were 60 ppm for Ca and Al, 70 ppm for Na, 90 ppm for Fe, and 200 ppm for Mg and Si. Mateo et al. [76] constructed calibration curves for Fe, Mg, Si, Al, and Ca and found that 355-nm laser irradiation showed better results than 1064-nm laser irradiation. Table 2 summarizes the results of ultimate analysis.

TABLE 2 Summary of coal ultimate analysis.

Element	Experimental	Data processing	Results	References
C	1064 nm, 4 ns, 100 mJ/pulse, gate width 1.1 ms; coal rocks	Univariate model (7 samples)	Linear relation	[68, 69]
	532 nm, 5 ns, 120 mJ/pulse, delay time 2 μs, gate width 1 ms, with/without 1.5H × 3D-mm polytetrafluoroethylene cylindrical cavity; pellets	Spectrum standardization and dominant factor-based PLS model (16 samples for calibration and 8 for validation)	*Without cavity*: $R^2 = 0.99$, RMSEP = 1.63%, ARE = 1.82%, AAE = 1.27% *With cavity*: $R^2 = 0.99$, RMSEP = 1.35%, ARE = 1.66%, AAE = 1.08%	[56]
	1064 nm, 5 ns, 30 mJ/pulse, with nitrile-butadiene rubber cylindrical cavity; pellets	PCA/PLS combined model (40 samples for calibration and 18 for validation)	$R^2 = 0.99$, RMSEP = 0.68, ARE = 1.17%	[45]
	532 nm, 5 ns, 120 mJ/pulse, delay time 2 μs, gate width 1 ms; pellets with $Na_2SiO_3 \cdot 9H_2O$ binder	Univariate model (15 samples for calibration and 7 for validation)	$R^2 = 0.76$, RMSEP = 6.25%, ARE = 10.14%	[23]
	266 nm, 4 ns, 16.2 mJ/pulse, delay time 0.7 μs, gate width 2 ms; pellets with KBr binder	Multivariate model (9 samples)	$R^2 = 0.983$	[47]
	1064 and 266 nm, 7 ns, 40 mJ/pulse, delay time 1 μs, gate width 1 ms; pellets	PLS model (12 samples for calibration and 6 for validation)	*For 1064 nm*: $R^2 = 0.99$, RMSEP = 1.55% *For 266 nm*: $R^2 = 0.99$, RMSEP = 1.64%	[34, 35]
	532 nm, 7 ns, 120 mJ/pulse, delay time 2 μs, gate width 1 ms, ambient conditions (air, He, and Ar); pellets	PLS model (17 samples for calibration and 7 for validation)	*In air*: RSD = 9.76%, RMSEP = 4.25%, RMSE = 3.91%, $R^2 = 0.87$, ARE = 4.96% *In He*: RSD = 13.25%, RMSEP = 3.32%, RMSE = 2.43%, $R^2 = 0.96$, ARE = 2.81% *In Ar*: RSD = 8.50%, RMSEP = 3.49%, RMSE = 2.69%, $R^2 = 0.95$, ARE = 2.98%	[24]

Continued

TABLE 2 Summary of coal ultimate analysis—cont'd

Element	Experimental	Data processing	Results	References
	532 nm, 5 ns, 120 mJ/pulse, delay time 2 μs, gate width 1 ms, He ambient condition; pellets	PLS/wavelet transform combined model (17 samples for calibration and 7 for validation)	$R^2 = 0.98$, RMSEP = 1.94%, ARE = 1.67%	[32]
	1064 nm, 8 ns, 90 mJ/pulse, delay time 3 μs, gate width 100 μs; pellets	K-ELM model (18 samples for calibration and 8 for validation)	*K-ELM*: $R^2 = 0.9994$, RMSE = 0.3762 wt% *SVM*: $R^2 = 0.9485$, RMSE = 5.5791 wt% *LS-SVM*: $R^2 = 0.9356$, RMSE = 5.0273 wt% *BP-ANN*: $R^2 = 0.9317$, RMSE = 14.6857 wt%	[43]
	1064 nm, 8 ns, 90 mJ/pulse, delay time 0.5 μs, gate width 1 ms; pellets	Hybrid model (spectrum standardization, identification, and a dominant factor PLS; 69 samples for calibration and 8 for validation)	Calibration set error = 0, validation set error = 2.32, all samples error = 0.42, mean RSD = 0.3%	[40]
H	532 nm, 5 ns, 50–120 mJ/pulse, delay time 0.4 μs, gate width 0.4 μs; pulverized coal (<100 μm)	Univariate model (6 samples)	Deviation from linearity with high concentrations	[13]
	532 nm, 7 ns, 120 mJ/pulse, delay time 2 μs, gate width 1 ms, ambient conditions (air, He, and Ar); pellets	PLS model (17 samples for calibration and 7 for validation)	*In air*: RSD = 6.75%, RMSEP = 0.48%, RMSE = 0.34%, $R^2 = 0.82$, ARE = 6.69% *In He*: RSD = 9.84%, RMSEP = 0.27%, RMSE = 0.21%, $R^2 = 0.91$, ARE = 4.56% *In Ar*: RSD = 6.19%, RMSEP = 0.24%, RMSE = 0.15%, $R^2 = 0.98$, ARE = 2.78%	[24]
	1064 nm, 8 ns, 90 mJ/pulse, delay time 0.5 μs, gate width 1 ms; pellets	Hybrid model (spectrum standardization, identification, and a dominant factor PLS; 69 samples for calibration and 8 for validation)	Calibration set error = 0, validation set error = 0.08, all samples error = 0.049, mean RSD = 0.546%	[40]

TABLE 2 Summary of coal ultimate analysis—cont'd

Element	Experimental	Data processing	Results	References
	1064 nm, 4 ns, 100 mJ/pulse, gate width 1.1 ms; coal rocks	Univariate model (7 samples)	Linear relation	[68, 69]
O	1064 nm, 120 mJ/pulse, delay time 200 ns, gate width 10 ms; pulverized coal	Univariate model (8 samples)	$R^2 > 0.95$, measurement accuracy = 1.15%–1.37%, ARE = 19.39%	[15]
N	532 nm, 5 ns, 50–120 mJ/pulse, delay time 0.4 μs, gate width 0.4 μs; pulverized coal (<100 μm)	Univariate model (6 samples)	Deviation from linearity with high concentrations	[13]
S	1064 nm, 8–10 ns, 50 mJ/pulse, delay time 1 μs, gate width 5 μs; pellets	Univariate model (14 samples)	$R^2 = 0.88$, SD = 0.11, ARE = 7.5%	[67]
	1064 nm, 60 mJ/pulse, delay time 0.5 μs, gate width 2 ms, double-pulse mode; pellets	Univariate model (5 samples)	$R^2 = 0.989$	[12]
	1064 nm, 8 ns, 90 mJ/pulse, delay time 3 μs, gate width 100 μs; pellets	K-ELM model (18 samples for calibration and 8 for validation)	*K-ELM*: $R^2 = 0.9832$, RMSE = 0.7704 wt% *SVM*: $R^2 = 0.9572$, RMSE = 1.6179 wt% *LS-SVM*: $R^2 = 0.9686$, RMSE = 1.5944 wt% *BP-ANN*: $R^2 = 0.9413$, RMSE = 2.0655 wt%	[43]

AAE, average absolute error; *ARE*, average relative error; *BP-ANN*, back propagation artificial neural network; *K-ELM*, kernel extreme learning machine; *LS-SVM*, least-squares support-vector machine; *PCA*, principal component analysis; *PLS*, partial least square; *RMSE*, root mean squared error; *RMSEP*, root mean squared error of prediction; *RSD*, relative standard deviation; *SD*, standard deviation.

3.2 Proximate analysis

Proximate analysis of coal includes moisture content, ash content, volatile matter, and fixed carbon. Proximate analysis reveals the coal quality, and hence impacts the coal pricing and the determination of the most efficient use for each coal shipment. Table 3 specifies repeatability and reproducibility requirements for proximate analysis of coal, allocated by the Chinese national standards using traditional chemical analysis.

Moisture content. There are mainly two forms of water existing in coal: free water and bound water. Moisture content in coal refers to the free water. Usually, the coal samples used in the research are air dried, which means samples had a relatively low moisture content that could be ignored. However, in on-site measurements, the moisture content of the coal samples may vary drastically in a wide range and, indeed, affect the quantitative measurement of different elements in coal. Literature work proved observable fluctuations and high uncertainty in LIBS

TABLE 3 Requirements of proximate analysis by Chinese national standards with traditional methods [102].

Property	Content range (%)	Repeatability requirement (%)	Reproducibility requirement (%)
Moisture	<5	0.2	–
	5~10	0.3	–
	>10	0.6	–
Ash	<15	0.2	0.3
	15~30	0.3	0.5
	>30	0.5	0.7
Volatile	<20	0.3	0.5
	20~40	0.5	1.0
	>40	0.8	1.5

spectra with moisture content variations. Gaft et al. [66] found that strong moisture variations lead to substantial alternations in the spectral results, showing errors in ash evaluation. Chen et al. [36] studied the effects of moisture content variations on the spectra and plasma morphology. Plasma images, spectra evolution, and spatially resolved spectra illustrated that moisture evaporates at earlier stages of laser radiation compared with coal constituents. Yuan et al. [28] measured the moisture content in 53 air-dried coal samples, where the moisture was regarded as internal free water. Results showed R^2, RMSEP, RMSE, and ARE of 0.97, 0.87%, 0.72%, and 26.2%, respectively.

Volatile matter content. Volatile matter is the coal components that are released at high temperature (900 ± 5°C) in the absence of air. It is a mixture of short- and long-chain hydrocarbons, aromatic hydrocarbons, and sulfur, with main elements of C, H, O, and N. Generally, laser-induced coal plasma is formed in air ambient conditions. Therefore O and N originating from the surrounding air have to be considered in data processing. In addition, sulfur could be volatilized as sulfur dioxide, with hardly detected lines in many experiments.

Dong et al. [20] employed a PCA and multiple regression analysis method to yield $R^2 = 0.991$ for volatile measurement. Zhang et al. [38] used PCA and SVM models with 500 samples for calibration and 50 samples for prediction. Results showed that an SVM/PCA combination had better performance than the SVM model, yielding R^2, RMSEP, and ARE of 0.95, 1.22%, and 4.42%, respectively. Moreover, AAE was reduced to 1.09% for samples with volatile matter <20%. Li et al. [45] measured the volatile matter content of 58 coal slices with a confinement ring and a PCA/PLS combined model (40 samples for calibration and 18 samples for validation). Results showed R^2, RMSEP, and ARE were 0.99, 0.66%, and 1.96%, respectively. Yuan et al. [28] used a dominant factor PLS model to predict the volatile matter. The dominant factor was built by line intensities of the organic elements (C, H, O, and N). Results showed R^2, RMSEP, RMSE, and ARE of 0.97, 1.41%, 1.42%, and 5.47%, respectively (40 samples for calibration and 13 for validation). Best results for volatile matter measurement were achieved by Hou et al. [40] with a hybrid model (including spectrum standardization, a large spectral identification database, and a dominant factor PLS). Results showed that errors for samples with volatile matter content <20% and samples with content ranging from 20% to 40% were 0.03% and 0.11%, which met the requirement of Chinese national standards (<0.5% and <1%, respectively), while samples with volatile matter contents >40% have still not been reported yet. Table 4 summarizes the results of volatile matter measurement.

Ash content. Ash is the noncombustible residue left after coal combustion and it is primarily composed of oxides and sulfates. Ash content is important for slagging and the viscosity of coal ash in furnaces. Ash composition of raw coal depends on the mining location and natural variations within the deposit. For LIBS, ash content measurement is a challenge, similar to volatile content measurement. The line intensities of

TABLE 4 Summary of volatile matter content analytical results.

Experimental	Data processing	Results	References
532 nm, 8 ns, 100 mJ/pulse, 1 Hz, delay time 1417 ns, gate width 2 ms; pellets	Partial correlation and PCA (17 samples for calibration and 3 for validation)	$R^2 = 0.991$, ARE < 10%, AAE < 3%	[20]
1064 nm, 8 ns, 60 mJ/pulse, 10 Hz, delay time 2 μs, gate width 10 ms; pellets	SVM/PCA combined model (500 samples for calibration and 50 for validation)	*SVM model*: $R^2 = 0.91$, RMSEP = 1.69%, ARE = 5.83%, AAE = 1.11% (volatile content <20%), and 1.52% (volatile content >20%) *SVM/PCA combined model*: $R^2 = 0.95$, RMSEP = 1.22%, ARE = 4.42%, AAE = 1.09% (volatile content <20%), and 1.02% (volatile content >20%)	[38]
532 nm, 5 ns, 70 mJ/pulse, 1 Hz, delay time 1.5 μs, gate width 1 ms; pellets	Dominant factor PLS model (40 samples for calibration and 13 for validation)	*Conventional PLS model*: $R^2 = 0.96$, RMSEP = 1.69%, RMSE = 1.54%, ARE = 5.95% *Dominant factor PLS model*: $R^2 = 0.97$, RMSEP = 1.41%, RMSE = 1.42%, ARE = 5.47%	[28]
1064 nm, 8 ns, 90 mJ/pulse, delay time 0.5 μs, gate width 1 ms; pellets	Hybrid model (spectrum standardization, identification, and a dominant factor PLS; 69 samples for calibration and 8 for validation)	Measurement errors = 0.03% (volatile content <20%) and 0.11% (volatile content in the range of 20%–40%)	[40]
1064 nm, 5 ns, 30 mJ/pulse, 5 Hz, nitrile-butadiene rubber confinement ring; coal slices; pellets	PCA/PLS combined model (40 samples for calibration and 18 for validation)	$R^2 = 0.99$, RMSEP = 0.66%, ARE = 1.96%	[45]

AAE, average absolute error; *ARE*, average relative error; *PCA*, principal component analysis; *PLS*, partial least square; *RMSE*, root mean squared error; *RMSEP*, root mean squared error of prediction; *SVM*, support-vector machine.

various oxides (such as Al_2O_3, SiO_2, Na_2O, K_2O, CaO, MgO, Fe_2O_3) are able to contribute to ash content measurements. The line interference and matrix effects should be considered when constructing a calibration model.

Gaft et al. [66] measured the ash content by the sum of Mg, Al, Si, and Fe line intensities with C as internal standardization. Results showed $R^2 = 0.99$ and SD = 0.21. Yuan et al. [28] measured the ash content using a dominant factor-based PLS model and achieved a better result than conventional PLS. The line intensities of minerals were selected to construct the dominant factor. The R^2, RMSEP, RMSE, and ARE were 0.93, 3.49%, 3.18%, and 12%, respectively, for the dominant factor PLS model. Zhang et al. [38] applied a PCA/SVM combined model to measure the ash content in 550 air-dried coal samples. The combined model showed better performance than that of the SVM model. Results showed R^2, RMSEP, and ARE of 0.96, 1.82%, and 5.48%, respectively (500 samples for calibration and 50 for validation). The AAE was 1.37% for ash content <30% and 1.77% for the ash content >30%. The hybrid model was successfully applied for the ash content by Hou et al. [40] using 77 standard coal samples. For ash content values <15%, 15%–30%, >30%, the hybrid model revealed measurement errors of 0.07%, 0.17%, and 0.23%, respectively (Chinese national standard requirements are <0.3%, <0.5%, and <0.7%, respectively). Yin et al. [18] utilized a fully software controlled unit to measure the ash content in pulverized coal and the error was in the range of 2.29%–13.47%. Li et al. [45] measured the ash content in 58 coal slices with a cylindrical cavity confinement. Using a PCA/PLS combined model, R^2, RMSEP, and ARE were 0.98, 0.77%, and 4.029%, respectively (40 samples for calibration and 18 for validation). Yao et al. [22] built a PLS model to extract the ash content from LIBS spectra. For four samples used for the model validation, the correlation coefficient was 0.97 and the RSD of measurements was in the range of 6.53%–7.71%. Wang et al. [39] added a beam expander to improve the signals' quality. Using 30 samples for calibration and 5 for validation in a PCA/SVM combined model, the results showed R^2 (linear regression coefficient) of 0.9976, measurement accuracy of 0.31%, and AAE of 2.36%. Wei et al. [42] predicted the concentrations of ash oxides using wavelet neural network (WNN) and ANN models. WNN showed better performance for seven ash oxides (SiO_2, Al_2O_3, Fe_2O_3, CaO, MgO, MnO_2, and TiO_2) with $R^2 > 0.9725$ and RMSE < 0.2591. Table 5, summarizes the analytical results of the ash content measurement.

Fixed carbon. Fixed carbon is the material excluded from moisture, volatile, and ash

TABLE 5 Summary of ash content analytical results.

Experimental	Data processing	Results	References
1064 nm, 120 mJ/pulse, 10 Hz, delay time 200 ns, gate width 10 ms; pulverized coal	Univariate calibration (8 samples)	ARE = 2.29%–13.47%	[11]
532 nm, 5 ns, 70 mJ/pulse, 1 Hz, delay time 1.5 μs, gate width 1 ms; pellets	Dominant factor PLS model (40 samples for calibration and 13 for validation)	*Conventional PLS model*: $R^2 = 0.91$, RMSEP = 4.13%, RMSE = 3.73%, ARE = 13.2% *Dominant factor PLS model*: $R^2 = 0.93$, RMSEP = 3.49%, RMSE = 3.18%, ARE = 12%	[28]
1064 nm, 8 ns, 60 mJ/pulse, 10 Hz, delay time 2 μs, gate width 10 ms; pellets	SVM/PCA combined model (500 samples for calibration and 50 for validation)	*SVM model*: $R^2 = 0.92$, RMSEP = 2.74%, ARE = 8.30%, AAE = 2.08 (ash content <30%), and 2.56% (ash content >30%) *SVM/PCA combined model*: $R^2 = 0.96$, RMSEP = 1.82%, ARE = 5.48%, AAE = 1.37 (ash content <30%), and 1.77% (ash content >30%)	[38]
1064 nm, 8 ns, 90 mJ/pulse, delay time 0.5 μs, gate width 1 ms; pellets	Hybrid model (spectrum standardization, identification, and a dominant factor PLS; 69 samples for calibration and 8 for validation)	Measurement errors of 0.07%, 0.17%, and 0.23%, for ash content values <15%, 15%–30%, >30%	[40]

TABLE 5 Summary of ash content analytical results—cont'd

Experimental	Data processing	Results	References
532 nm, 8 ns, 70 mJ/pulse, 1 Hz, delay time 1.4 μs, gate width 2 ms; pellets	PLS model (13 samples for calibration and 4 for validation)	$R^2 = 0.97$	[22]
1064 nm, 8 ns, 80 mJ/pulse, 10 Hz, delay time 2 μs, gate width 10 ms, beam expander; pellets	SVM/PCA combined model (30 samples for calibration and 5 for validation)	$R^2 = 0.9976$, measurement accuracy = 0.31%, AAE = 2.36%	[39]
1064 nm, 10 ns, 100 mJ/pulse, 5 Hz, delay time 1.5 μs, gate width 100 μs, KBr binder; pellets	WNN model	$R^2 > 0.925$ and RMSE <0.2591 for 7 oxides	[42]
1064 nm, 5 ns, 30 mJ/pulse, 5 Hz, nitrile-butadiene rubber confinement ring; slices	PCA/PLS combined model (40 samples for calibration and 18 for validation)	$R^2 = 0.98$, RMSEP = 0.77%, ARE = 4.029%	[45]
Single pulse: 1064 nm, 6 ns, 100 mJ/pulse, delay time 1 μs, gate width 10 μs; H_3BO_3 binder for pellets *Double pulse*: 1064 nm, 6 ns, 300 mJ/pulse, interpulse delay 1 μs; H_3BO_3 binder for pellets	Univariate calibration (9 samples)	Good agreement with standard results by ICP-OES for Al_2O_3, SiO_2, TiO_2, Fe_2O_3, K_2O, MgO, CaO, and BaO	[77, 78]

AAE, average absolute error; *ARE*, average relative error; *ICP-OES*, inductively copled plasma-optical emission spectrometry; *PCA*, principal component analysis; *PLS*, partial least square; *RMSE*, root mean squared error; *RMSEP*, root mean squared error of prediction; *SVM*, support-vector machine; *WNN*, wavelet neural network.

contents. It differs from carbon content in ultimate analysis because some carbon has already been driven off as hydrocarbons with the volatile matter. Fixed carbon is significant to the combustion characteristic of coal and can impact the efficiency of coal combustion equipment. For traditional chemical methods, fixed carbon is indirectly measured by subtracting the moisture, volatile, and ash contents from the whole coal mass. For LIBS, Zheng et al. [33] measured the fixed carbon in a particle coal flow (particles diameter <200 μm) by discarding the weak spectra. The correlation coefficient, RMSECV, and RMSEP were improved from 0.921 to 0.930, 2.86% to 2.92%, and 7.30% to 2.40%, respectively, after discarding.

3.3 Calorific value

Calorific value (also called heat value or heating value) is the total heat released from a complete combustion of coal [105]. It is an important indicator for coal combustion efficiency and used to control the combustion temperature of coal utilization devices. Moreover, heat value is used to determine coal ranks and prices.

Noda et al. [72] measured the caloric heat values using four different coals under high-pressure/temperature conditions for a gasification power plant. A gas mixture (N_2 98% and O_2 2%) was pumped at a temperature of 720 K and a pressure of 3.0 MPa into the analysis furnace. Results were compared with conventional methods (JIS M8814) and showed good agreement. Zhang et al. [38] determined the caloric heat value using the support vector regression model combined with PCA (500 samples for calibration and 50 for validation). The R^2, RMSEP, ARE, and AAE were 0.91, 0.85 MJ/kg, 3.68%, and 0.65 MJ/kg, respectively. Yuan et al. [28] determined the caloric heat value using the dominant factor PLS model and achieved a remarkable improvement over the conventional PLS model. The R^2 improved from 0.94 to 0.97, while RMSEP, RMSE, and ARE decreased from 1.63 to 1.33 MJ/kg, 1.05 to 0.82 MJ/kg, and 3.55% to 2.71%, respectively, with 40 samples for calibration and 13 for validation. Li et al. [51] built a PLS model with different preprocessing methods for calorific heat value measurement. Better results were obtained using the PLS model with 11-point smoothing combined with the second-order derivation. The R_c (correlation coefficient of calibration set), R_v (correlation coefficient of validation set), RMSECV, and RMSEP were 0.9909, 0.9972, 0.47 MJ/kg, and 0.276 MJ/kg, respectively, using 33 samples for calibration and 11 for validation. Hou et al. [40] used the hybrid model (spectral line standardization, spectral identification, and dominant factor-based PLS model) to measure the heat value in 77 standard bituminous coal samples. Results showed AAE of 0.067 MJ/kg, fully meeting the requirements of the national standards (<0.3 MJ/kg). Table 6 summarizes the calorific

TABLE 6 Summary of heat value analytical results.

Experimental	Data processing	Results	References
1064 nm, 8 ns, 70 mJ/pulse, 10 Hz, 2 μs, 10 ms; pellets	SVM/PCA combined model (500 samples for calibration and 50 for validation)	*SVM model*: $R^2 = 0.90$, RMSEP = 1.23 MJ/kg, ARE = 5.4%, AAE = 0.91% *SVM combined with PCA model*: $R^2 = 0.91$, RMSEP = 0.85 MJ/kg, ARE = 3.68%, AAE = 0.65%	[38]
532 nm, 5 ns, 70 mJ/pulse, 1 Hz, 1.5 μs, 1 ms; pellets	Dominant factor PLS model (40 samples for calibration and 13 for validation)	*Conventional PLS model*: $R^2 = 0.94$, RMSEP = 1.63 MJ/kg, RMSE = 1.05 MJ/kg, ARE = 3.55% *Dominant factor PLS model*: $R^2 = 0.97$, RMSEP = 1.33 MJ/kg, RMSE = 0.82 MJ/kg, ARE = 2.71%	[28]
1064 nm, 6 ns, 50.75 mJ/pulse, 1 Hz, 1.664 μs, 1.1 ms; pellets	PLS model with 11-point smoothing and second derivative (33 samples for calibration and 11 for validation)	*PLS model*: $R_c = 0.855$, $R_v = 0.724$, RMSECV = 1.89 MJ/kg, RMSEP = 2.36 MJ/kg *PLS model with 11-point smoothing*: $R_c = 0.9906$, $R_v = 0.9961$, RMSECV = 0.481 MJ/kg, RMSEP = 0.305 MJ/kg *PLS model with 11-point smoothing and second derivative*: $R_c = 0.9909$, $R_v = 0.9972$, RMSECV = 0.47 MJ/kg, RMSEP = 0.276 MJ/kg	[51]
1064 nm, 8 ns, 90 mJ/pulse, 1 Hz, 0.5 μs, 1 ms; pellets	Hybrid model (spectrum standardization, identification, and a dominant factor PLS; 69 samples for calibration and 8 for validation)	Calibration set error = 0.0 MJ/kg, validation set error = 0.856 MJ/kg, all error = 0.067 MJ/kg, mean RSD = 0.317%	[40]

ARE, average relative error; *PCA*, principal component analysis; *PLS*, partial least square; *RMSECV*, root mean squared error cross validation; *RMSEP*, root mean squared error of prediction; *RSD*, relative standard deviation; *SVM*, support-vector machine.

heat value results using LIBS experiments through the literature.

4 The application of LIBS to the online analysis of coal

With the main advantages of rapid analysis, no (or little) sample preparation, and multielement analysis, LIBS is extremely promising for the online analysis of coal. According to whether the sample is prepared during the analysis process, there are two different situations for LIBS online analysis of coal: sample preparation analysis and direct analysis. The online analyzer based on sample preparation analysis usually has an automated sample preparation system coupled to the LIBS analysis system. The coal is sampled from the coal steam and then grinded, mixed, pressed into pellets, analyzed, and returned to coal steam. Direct analysis is without the sample preparation process and the laser pulses are directly focused on a coal block or coal powder. Since sample preparation analysis can provide reliable results and direct analysis can realize real-time online analysis, numerous LIBS online analysis systems based on the two methods were developed, ranging from measurement of elements to monitoring of coal properties.

The application of LIBS to the online analysis of coal begins with detection of the elemental composition of coal. In the 1990s, Sandia National Laboratories (Livermore, California) applied LIBS to in situ quantitative analysis of inorganic elements (such as Si, Al, Mg, Ca, Fe, and Ti) in single coal particles [106–108]. Pulverized coal flow with a particle size of 105–125 μm was introduced from a premixed flat-flame burner by a syringe-type feeder and continuously analyzed by LIBS. The results indicated that average compositions from several hundred particles and that from standard bulk analyses were in good agreement for Si, Al, Mg, and Ti. However, the accuracy of the results was limited by the heterogeneity of the particles.

One approach to achieve improved accuracy is to press powdered coal into pellets prior to analysis. Chadwick et al. [7–9, 95] designed a LIBS analyzer to detect the inorganic components of coal (Al, Si, Mg, Ca, Fe, Na, K) and the principal coal components (C, H, O, N) in the Yallourn mine, Victoria, Australia. Coal and mineral samples are crushed to <10 μm and mixed prior to analysis. Then, the crushed samples are pressed into pellets and transferred into the analyzer. It takes a few minutes (about 2 min of sample preparation and 30 s for the LIBS analysis) from the sampling to obtain an accurate analysis result of the sample. Results show detection limits are on the order of 0.01% and vary with the element. Measurement repeatability and accuracy are within ±10% absolute and meet the requirements of standard analysis procedures for heterogeneous materials.

Another important application of LIBS for detection of element composition is determining mineral elements [80, 81] of fly ash in a combustion environment (Figs. 1 and 2). The detection suffers from high temperature and high particle loadings in this situation. To reduce the impact of high temperature, Blevins et al. [81] designed a longer probe and an external, annular, steel

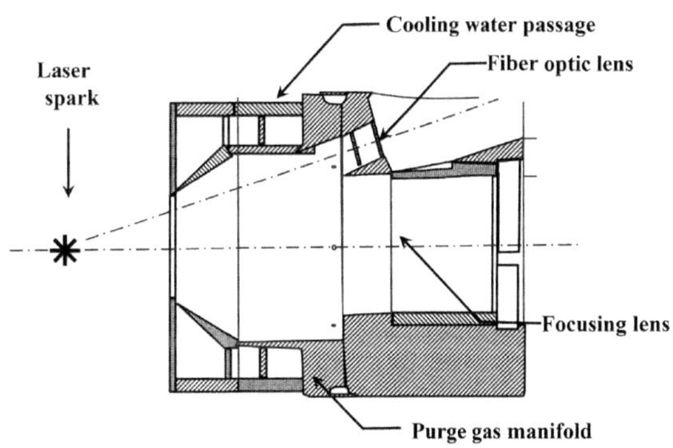

FIG. 1 The long-extension laser-induced breakdown spectroscopy probe [81].

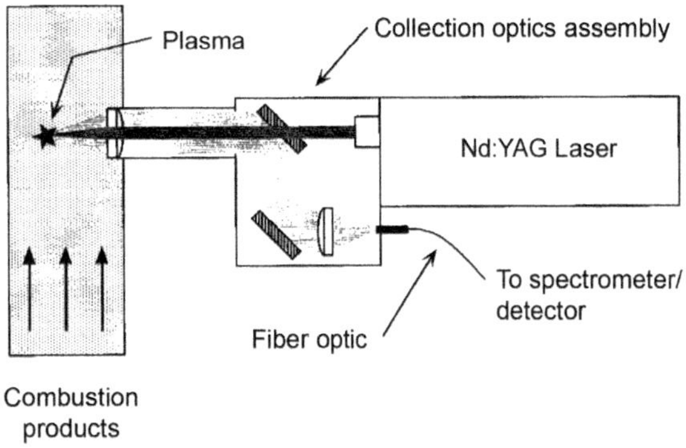

FIG. 2 Schematic of the laser-induced breakdown spectroscopy system used for the power generation boiler and the glass furnace [81].

water jacket (Fig. 1) to slip over the LIBS probe and keep it cool. The LIBS probe was applied to measure elements Mg, Si, Ti, Fe, Ca, Al, Na, and K in a power generation boiler and a glass furnace (temperature of 1170 K and 18 L/min flow of N_2 purge). The measured results were in good agreement with the chemical composition of deposits collected upstream of the LIBS measurement location.

The same situation exists in the measurement of unburned carbon in fly ash with LIBS, and related research was reported by Deguchi et al. [72, 73]. Fig. 3 shows a LIBS system applied to measure the carbon content in fly ash, char, and pulverized coal under high-temperature and high-pressure conditions. The system is composed of a furnace and an automated LIBS unit. The furnace allows the system to operate in an environment with a temperature range of 300–1000 K and a pressure range of 0.1–3 MPa. The automated LIBS unit gives emission intensities of Si, Al, Fe, Ca, and carbon, which are used to calculate the carbon content by a relational equation. This system can complete the detection within 1 min, as compared to over 30 min required by conventional methods.

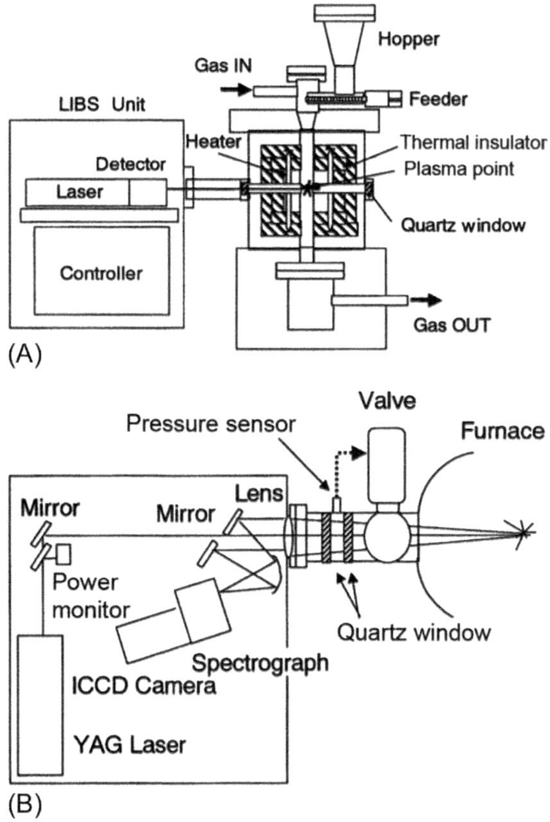

FIG. 3 Laser-induced breakdown spectroscopy system: (A) high-pressure and high-temperature furnace, and (B) optical setup [72].

Fig. 4 shows a LIBS machine installed on the Wemco Drum product conveyor at Optimum Colliery in South Africa to directly detect coal ash content [65, 66]. The LIBS machine with 150 × 80 × 130 cm dimensions and 250-kg weight contains a 1064-nm Q-switched Nd:YAG laser (Big Sky Ultra CFR) and a charge-couple device spectrometer (Avantes, AvaSpec). The nominal focal length is 35 cm and the depth of field is ±2 cm. To fix the focal distance, an ultrasonic height sensor is employed to detect changes in coal block height on the moving belt and the focusing optical system moves according to the detected height change. The calibration model in the system was built using coal samples with ash content varying from 12.5% to 53.9%. Moreover, the effect of moisture

FIG. 4 The laser-induced breakdown spectroscopy machine installed on the Wemco Drum product conveyor at Optimum Colliery in South Africa [65].

content on ash content was corrected in the model. Finally, the best-fit linear regression between laboratory ash analysis and system analysis is characterized by $R^2 = 0.99$ and an SD of 0.21. Compared with the result of the PGNAA analyzer installed on a conveyer belt in line, the LIBS machine gives an average standard error of 0.32% for ash content.

Similarly, a fully software-controlled LIBS system (Fig. 5) consisting of a LIBS apparatus

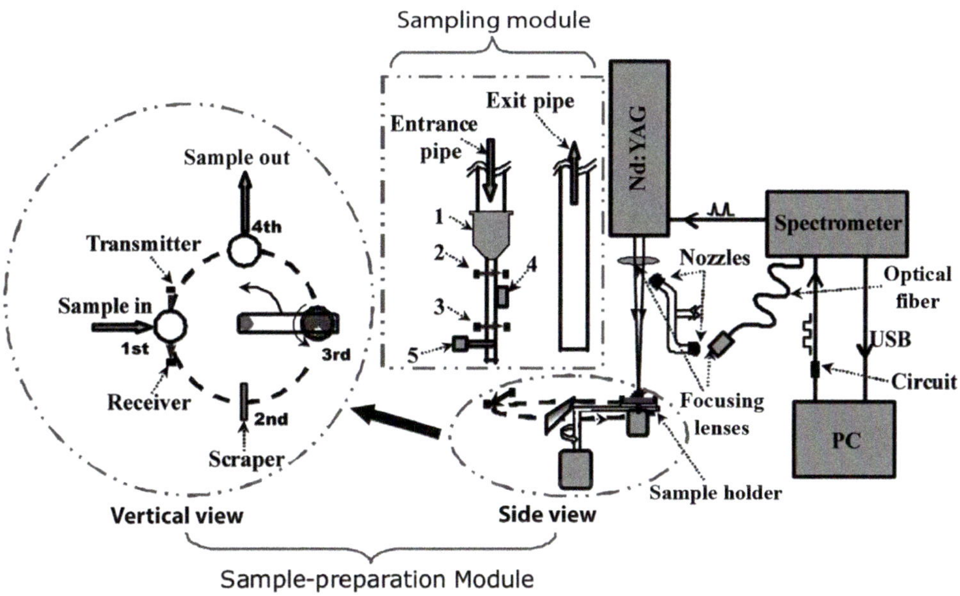

FIG. 5 Fully software-controlled laser-induced breakdown spectroscopy system [18].

and sampling equipment has been developed for online analysis of pulverized coal in power plants in China [18, 109]. The fine coal particles are brought to the bottom of the entrance pipe by airflow generated by a negative pressure generator, and then the coal is loaded into the sample holder, which is fixed at the end of a 0.25-m long rotation axis and rotates at a speed of 2 rev/min. The holder moves to the laser focus position and the coal is analyzed by LIBS. The optical components are cleaned by a high-pressure gas-cleaning device. This LIBS system can achieve the best measurements for C, H, Si, Na, Mg, Fe, Al, and Ti with measurement errors less than 10%. The measurement error of ash (air dried) was in the range of 2.29%–13.47%.

Fig. 6 displays a LIBS system developed by Energy Research Company (ERCo) for real-time monitoring of coal properties and ash fusion temperature [110]. The LIBS system was installed on a conveyor belt at PPL Generation's Montour Power Station, Pennsylvania, USA. The software embedded in the LIBS system was developed based on ANN (standard back-propagation and Kohonen unsupervised algorithms). The system runs continuously for 73 h and the results showed good accuracy and repeatability for the elements measurement. Moreover, the root mean square deviations of iron, sulfur, fusion temperature, and coal heating value were 1.07%, 0.14%, 33.16°F, and 75.86 Btu/lb., respectively.

LIBS has a poor ability to measure moisture content in coal. One approach to solve this problem is combining LIBS with other techniques. A progression Titan CCA unit (Fig. 7) consisting of a magnetic resonance (MR) module and a LIBS module was installed in Powder River Basin mine (USA) to analyze sulfur, ash, moisture, and Na_2O in coal [111]. A coal sample was taken from the conveyor and delivered to the unit for analysis through a two-stage sweep-arm sampling system. The MR module determined the moisture content, while the LIBS module analyzed sulfur, six major ash constituents (Si, Al, Fe, Ca, K, and Ti), and Na and Mg, which cannot be measured by PGNAA. A 3-week period of operation showed that the unit performed better than other elemental analyzers, especially in the measurement of calorific value, ash, moisture, and Na_2O.

FIG. 6 Real-time monitoring of coal properties and ash fusion temperature system [110].

FIG. 7 Progression Titan CCA unit [111].

5 Summary and future perspectives

Though renewable energy, like wind, biomass, solar, and geothermal energies, is growing fast in the world, coal is still an important resource for cost-effective power generation and industrial chemicals production. Fast/real-time coal analysis is urgently needed for efficient and environment-friendly utilization of coal. LIBS has the capability of providing fast qualitative and quantitative information for coal pricing, coal washing, and coal blending, as well as coal combustion optimization. From this point of view, China, due to its huge market of coal production and consumption, provides LIBS with great potential to realize wide applications.

The feasibility of quantitative analysis of coal using LIBS has been verified, and industrial demonstration projects are also under deployment. However, there are still two main challenges in the way of full commercialization and industrialization of LIBS coal analysis: (1) quantitative analysis performance, particularly long-term repeatability and reproducibility under a harsh industrial environment, needs to be improved, and (2) the adaptability of a LIBS coal analyzer for different kinds of coal needs to be promoted, which may need much effort to study the impact of coal matrixes on LIBS signals. Thus much research work on fundamental data processing and applications is still necessary to realize wide commercialization of LIBS coal analysis. For fundamental research, the mechanisms of interaction between laser, coal, plasma, and ambient gas need to be deeply investigated. For data processing, to improve the adaptability of the calibration model, the physical background should be considered and integrated into the model. Selecting a set of calibration samples with good representativeness also helps to improve adaptability. Improving the quantitative performance of sulfur and trace elements is also an important topic. LIBS combined with other technologies (microwave, XRF, Raman, etc.) may be the final solution.

LIBS must maximize its unique advantages based on its ability to perform inline and at-line analysis. To open the inline and at-line market, first, LIBS has to solve its own problems, such as calibration establishment for inline/online systems, long-term stability, sampling representativeness, and adaptability to different coal types. Considering these problems, at-line mode may be a preferable choice. Second, LIBS coal analysis has to be integrated with automatic control systems to bring benefits to the end user (increase efficiency of coal utilization, save coal consumption, reduce pollutant emission, increase safety of operation, etc.). The application of LIBS coal analyzers to coal pricing is still at an early and complicated stage due to economic interests among different stakeholders in the coal production and utilization chain. Before different stakeholders cooperate to establish a national/industry standard for LIBS coal analysis, it will be difficult for LIBS to enter the coal-pricing market, although the market is huge. In summary, coal analysis provides a huge application field and market for LIBS technology. A successful application of LIBS for coal analysis will boost LIBS technology to other fields and help LIBS become a real "superstar."

References

[1] Y. Zhang, X. Zhang, W. Jia, Q. Shan, Y. Ling, D. Hei, D. Chen, Online X-ray fluorescence (XRF) analysis of heavy metals in pulverized coal on a conveyor Belt, Appl. Spectrosc. 70 (2016) 272–278.

[2] J.H. Marshall, J.F. Zumberge, Online measurements of bulk coal using prompt gamma neutron-activation analysis, Nucl. Geophys. 3 (1989) 445–459.

[3] M. Kaihara, T. Takahashi, T. Akazawa, T. Sato, S. Takahashi, Application of near infrared spectroscopy to rapid analysis of coals, Spectrosc. Lett. 35 (2002) 369–376.

[4] E. Binner, E. Lester, S.W. Kingman, C. Dodds, J.P. Robinson, T. Wu, P. Wardle, J.P. Mathews, A review of microwave coal processing, J. Microw. Power Electromagn. Energy. 48 (2014) 35–60.

[5] S. Sheta, M.S. Afgan, Z. Hou, S. Yao, L. Zhang, Z. Li, Z. Wang, Coal analysis by laser-induced breakdown spectroscopy: a tutorial review, J. Anal. At. Spectrom. 34 (2019) 1047–1082.

[6] D.K. Ottesen, In situ real-time analysis of particulates using laser spark spectroscopy, in: H.A. G. Chermin (Ed.), 1987 International Conference on Coal Science: Proceedings, Elsevier Science Ltd, New York, 1987, p. 984.

[7] F.J. Wallis, B.L. Chadwick, R.J.S. Morrison, Analysis of lignite using laser-induced breakdown spectroscopy, Appl. Spectrosc. 54 (2000) 1231–1235.

[8] D. Body, B.L. Chadwick, Simultaneous elemental analysis system using laser induced breakdown spectroscopy, Rev. Sci. Instrum. 72 (2001) 1625–1629.

[9] B.L. Chadwick, D. Body, Development and commercial evaluation of laser-induced breakdown spectroscopy chemical analysis technology in the coal power generation industry, Appl. Spectrosc. 56 (2002) 70–74.

[10] A.F.M.Y. Haider, M.A. Rony, R.S. Lubna, K. M. Abedin, Detection of multiple elements in coal samples from Bangladesh by laser-induced breakdown spectroscopy, Opt. Laser Technol. 43 (2011) 1405–1410.

[11] A.F.M.Y. Haider, M.A. Rony, K.M. Abedin, Determination of the ash content of coal without ashing: a simple technique using laser-induced breakdown spectroscopy, Energy Fuel 27 (2013) 3725–3729.

[12] V.S. Burakov, N.V. Tarasenko, M.I. Nedelko, V.A. Kononov, N.N. Vasilev, S.N. Isakov, Analysis of lead and sulfur in environmental samples by double pulse laser induced breakdown spectroscopy, Spectrochim. Acta B 64 (2009) 141–146.

[13] L.Y. Yu, J.D. Lu, W. Chen, G. Wu, K. Shen, W. Feng, Analysis of pulverized coal by laser-induced breakdown spectroscopy, Plasma Sci. Technol. 7 (2005) 3041–3044.

[14] K.L. Deng, J.T. Wu, Z. Wang, B. Lee, R. Guida, On-line compositional analysis in coal gasification environment using laser-induced plasma technology, Proc. SPIE 6314 (2006).

[15] L. Zhang, L. Dong, H.P. Dou, W.B. Yin, S.T. Jia, Laser-induced breakdown spectroscopy for determination of the organic oxygen content in anthracite coal under atmospheric conditions, Appl. Spectrosc. 62 (2008) 458–463.

[16] J. Li, J. Lu, Z. Lin, S. Gong, C. Xie, L. Chang, L. Yang, P. Li, Effects of experimental parameters on elemental analysis of coal by laser-induced breakdown spectroscopy, Opt. Laser Technol. 41 (2009) 907–913.

[17] S. Yao, J. Lu, Z. Lu, C. Xie, J. Li, S. Pan, M. Jiang, P. Li, Influence of sample morphology on laser ablation properties of coal, Acta Opt. Sin. 29 (2009) 1126–1130.

[18] W. Yin, L. Zhang, L. Dong, W. Ma, S. Jia, Design of a laser-induced breakdown spectroscopy system for on-line quality analysis of pulverized coal in power plants, Appl. Spectrosc. 63 (2009) 865–872.

[19] Q. Yang, Z. Peng, Characteristics of plasma induced by interaction of a free-oscillated laser pulse with a coal target in air and combustible gas, Int. J. Hydrog. Energy 35 (2010) 4715–4722.

[20] M. Dong, J. Lu, S. Yao, J. Li, J. Li, Z. Zhong, W. Lu, Application of LIBS for direct determination of volatile matter content in coal, J. Anal. Atom. Spectrom. 26 (2011) 2183–2188.

[21] J. Feng, Z. Wang, L. West, Z. Li, W. Ni, A PLS model based on dominant factor for coal analysis using laser-induced breakdown spectroscopy, Anal. Bioanal. Chem. 400 (2011) 3261–3271.

[22] S. Yao, J. Lu, M. Dong, K. Chen, J. Li, J. Li, Extracting coal ash content from laser-induced breakdown spectroscopy (LIBS) spectra by multivariate analysis, Appl. Spectrosc. 65 (2011) 1197–1201.

[23] T. Yuan, Z. Wang, L. Li, Z. Hou, Z. Li, W. Ni, Quantitative carbon measurement in anthracite using laser-induced breakdown spectroscopy with binder, Appl. Opt. 51 (2012) B22–B29.

[24] Z. Wang, T.B. Yuan, S.L. Lui, Z.Y. Hou, X.W. Li, Z. Li, W.D. Ni, Major elements analysis in bituminous coals under different ambient gases by laser-induced breakdown spectroscopy with PLS modeling, Front. Phys. 7 (2012) 708–713.

[25] W. Yin, L. Zhang, L. Wang, L. Dong, W. Ma, S. Jia, Research on accurate measurement of oxygen content in coal using laser-induced breakdown spectroscopy in air environment, Spectrosc. Spectr. Anal. 32 (2012) 200–203.

[26] L. Zhang, Z.Y. Hu, W.B. Yin, D. Huang, W.G. Ma, L. Dong, H.P. Wu, Z.X. Li, L.T. Xiao, S.T. Jia, Recent progress on laser-induced breakdown spectroscopy for the monitoring of coal quality and unburned carbon in fly ash, Front. Phys. 7 (2012) 690–700.

[27] J. Feng, Z. Wang, L. Li, Z. Li, W. Ni, A nonlinearized multivariate dominant factor-based partial least squares (PLS) model for coal analysis by using laser-induced breakdown spectroscopy, Appl. Spectrosc. 67 (2013) 291–300.

[28] T. Yuan, Z. Wang, S.-L. Lui, Y. Fu, Z. Li, J. Liu, W. Ni, Coal property analysis using laser-induced breakdown spectroscopy, J. Anal. At. Spectrom. 28 (2013) 1045–1053.

[29] S. Chen, J. Lu, Z. Zhong, F. Pan, G. Pan, X. Zhang, S. Yao, J. Luo, J. Li, Investigation on collection performances of plasma signal for pulverized particle coal flow in different optical collection, Acta Photon. Sin. 42 (2013) 326–330.

[30] Z. Hu, L. Zhang, W. Yin, W. Ma, L. Dong, S. Jia, Application of laser-induced breakdown spectroscopy to coal-fired power plants and soil contaminants on-line monitoring, J. Atmos. Environ. Opt. 8 (2013) 26–35.

[31] X. Li, Z. Wang, Y. Fu, Z. Li, W. Ni, A model combining spectrum standardization and dominant factor based partial least square method for carbon analysis in coal using laser-induced breakdown spectroscopy, Spectrochim. Acta B 99 (2014) 82–86.

[32] T.B. Yuan, Z. Wang, Z. Li, W.D. Ni, J.M. Liu, A partial least squares and wavelet-transform hybrid model to analyze carbon content in coal using laser-induced breakdown spectroscopy, Anal. Chim. Acta 807 (2014) 29–35.

[33] J. Zheng, J. Lu, B. Zhang, M. Dong, S. Yao, W. Lu, X. Dong, Experimental study of laser-induced breakdown spectroscopy (LIBS) for direct analysis of coal particle flow, Appl. Spectrosc. 68 (2014) 672–679.

[34] X. Li, X. Mao, Z. Wang, R.E. Russo, Quantitative analysis of carbon content in bituminous coal by laser-induced breakdown spectroscopy using UV laser radiation, Plasma Sci. Technol. 17 (2015) 928–932.

[35] X. Li, Z. Wang, Y. Fu, Z. Li, W. Ni, Wavelength dependence in the analysis of carbon content in coal by nanosecond 266 nm and 1064 nm laser induced breakdown spectroscopy, Plasma Sci. Technol. 17 (2015) 621–624.

[36] M. Chen, T. Yuan, Z. Hou, Z. Wang, Y. Wang, Effects of moisture content on coal analysis using laser-induced breakdown spectroscopy, Spectrochim. Acta B 112 (2015) 23–33.

[37] S. Yao, J. Xu, X. Dong, B. Zhang, J. Zheng, J. Lu, Optimization of laser-induced breakdown spectroscopy for coal powder analysis with different particle flow diameters, Spectrochim. Acta B 110 (2015) 146–150.

[38] L. Zhang, Y. Gong, Y. Li, X. Wang, J. Fan, L. Dong, W. Ma, W. Yin, S. Jia, Development of a coal quality analyzer for application to power plants based on laser-induced breakdown spectroscopy, Spectrochim. Acta B 113 (2015) 167–173.

[39] X. Wang, L. Zhang, J. Fan, Y. Li, Y. Gong, L. Dong, W. Ma, W. Yin, S. Jia, Parameters optimization of laser-induced breakdown spectroscopy experimental setup for the case with beam expander, Plasma Sci. Technol. 17 (2015) 914–918.

[40] Z. Hou, Z. Wang, T. Yuan, J. Liu, Z. Li, W. Ni, A hybrid quantification model and its application for coal analysis using laser induced breakdown spectroscopy, J. Anal. At. Spectrom. 31 (2016) 722–736.

[41] D. Wang, J.D. Lu, M.R. Dong, S.C. Yao, J. Fan, Z. H. Tian, L. Wang, S.S. Li, A new calibrated model of coal calorific value detection with LIBS, Spectrosc. Spectr. Anal. 36 (2016) 2607–2612.

[42] J. Wei, J. Dong, T. Zhang, Z. Wang, H. Li, Quantitative analysis of the major components of coal ash using laser induced breakdown spectroscopy coupled with a wavelet neural network (WNN), Anal. Methods 8 (2016) 1674–1680.

[43] C.H. Yan, J. Qi, J.X. Ma, H.S. Tang, T.L. Zhang, H. Li, Determination of carbon and sulfur content in coal by laser induced breakdown spectroscopy combined with

kernel-based extreme learning machine, Chemom. Intell. Lab. Syst. 167 (2017) 226–231.

[44] T.L. Zhang, C.H. Yan, J. Qi, H.S. Tang, H. Li, Classification and discrimination of coal ash by laser-induced breakdown spectroscopy (LIBS) coupled with advanced chemometric methods, J. Anal. Atom. Spectrom. 32 (2017) 1960–1965.

[45] A. Li, S. Guo, N. Wazir, K. Chai, L. Liang, M. Zhang, Y. Hao, P.F. Nan, R.B. Liu, Accuracy enhancement of laser induced breakdown spectra using permittivity and size optimized plasma confinement rings, Opt. Express 25 (2017) 27559–27569.

[46] S. Yao, J. Xu, J. Zhao, K. Bai, J. Lu, Z. Lu, Characterization of Fly ash laser-induced plasma for improving the on-line measurement of unburned carbon in gas-solid flow, Energy Fuel 31 (2017) 4681–4686.

[47] S. Yao, J. Zhao, J. Xu, Z. Lu, J. Lu, Optimizing the binder percentage to reduce matrix effects for the LIBS analysis of carbon in coal, J. Anal. Atom. Spectrom. 32 (2017) 766–772.

[48] Z.M. Lu, J.H. Mo, S.C. Yao, J.B. Zhao, J.D. Lu, Rapid determination of the gross calorific value of coal using laser-induced breakdown spectroscopy coupled with artificial neural networks and genetic algorithm, Energy Fuel 31 (2017) 3849–3855.

[49] Y.S. Li, W.Y. Lu, J.B. Zhao, G.X. Feng, D.M. Wei, J. D. Lu, S.C. Yao, Z.M. Lu, Detection of caloric value of coal using laser-induced breakdown spectroscopy combined with BP neural networks, Spectrosc. Spectr. Anal. 37 (2017) 2575–2579.

[50] G. Pan, M.R. Dong, J.H. Yu, J.D. Lu, Accuracy improvement of quantitative analysis of unburned carbon content in fly ash using laser induced breakdown spectroscopy, Spectrochim. Acta B 131 (2017) 26–31.

[51] W. Li, J. Lu, M. Dong, S. Lu, J. Yu, S. Li, J. Huang, J. Liu, Quantitative analysis of calorific value of coal based on spectral preprocessing by laser-induced breakdown spectroscopy (LIBS), Energy Fuel 32 (2017) 24–32.

[52] S. Yao, J. Zhao, Z. Wang, Y. Deguchi, Z. Lu, J. Lu, Analysis of spectral properties for coal with different volatile contents by laser-induced breakdown spectroscopy, Spectrochim. Acta B At. Spectrosc. 149 (2018) 249–255.

[53] K. Bai, S. Yao, J. Lu, J. Zhao, J. Xu, Z. Lu, Correction of C–Fe line interference for the measurement of unburned carbon in fly ash by LIBS, J. Anal. Atom. Spectrom. 31 (2016) 2418–2426.

[54] Z. Hou, Z. Wang, J. Liu, W. Ni, Z. Li, Signal quality improvement using cylindrical confinement for laser induced breakdown spectroscopy, Opt. Express 21 (2013) 15974–15979.

[55] Z.Y. Hou, Z. Wang, J.M. Liu, W.D. Ni, Z. Li, Combination of cylindrical confinement and spark discharge for signal improvement using laser induced breakdown spectroscopy, Opt. Express 22 (2014) 12909–12914.

[56] X. Li, H. Yin, Z. Wang, Y. Fu, Z. Li, W. Ni, Quantitative carbon analysis in coal by combining data processing and spatial confinement in laser-induced breakdown spectroscopy, Spectrochim. Acta B 111 (2015) 102–107.

[57] C. Yan, J. Qi, J. Liang, T. Zhang, H. Li, Determination of coal properties using laser induced breakdown spectroscopy combined with kernel extreme learning machine and variable selection, J. Anal. Atom. Spectrom. 33 (2018) 2089–2097.

[58] S. Yao, L. Zhang, K. Yin, K. Bai, J. Xu, Z. Lu, J. Lu, Identifying laser-induced plasma emission spectra of particles in a gas–solid flow based on the standard deviation of intensity across an emission line, J. Anal. At. Spectrom. 33 (2018) 1676–1682.

[59] P. Haobin, C. Guohua, C. Xiaoxuan, L.U. Zhimin, Y.A.O. Shunchun, Hybrid classification of coal and biomass by laser-induced breakdown spectroscopy combined with K-means and SVM, Plasma Sci. Technol. 21 (2019) 034008.

[60] P. Rebolledo, P. Pacheco, R. Sarmiento, R. Cabanzo, E. Mejia-Ospino, Temporary spectral analysis of a laser plasma of mineral coal, in: 2nd International Meeting for Researchers in Materials and Plasma Technology, vol. 466, 2013.

[61] L.P. Vera, J.A. Perez, H. Riascos, Spectroscopic study of emission coal mineral plasma produced by laser ablation, in: 15th International Congress on Plasma Physics (Icpp2010) & 13th Latin American Workshop on Plasma Physics (Lawpp2010), vol. 511, 2014.

[62] L.P. Vera-Londono, J.A. Perez-Taborda, H. Riascos-Landazuri, Spectroscopic analysis of coal plasma emission produced by laser ablation, Rev. Fac. Ing.-Univ. Ant. (2016) 69–72.

[63] A. Stankova, N. Gilon, L. Dutruch, V. Kanicky, A simple LIBS method for fast quantitative analysis of fly ashes, Fuel 89 (2010) 3468–3474.

[64] A. Metzinger, D.J. Palasti, E. Kovacs-Szeles, T. Ajtai, Z. Bozoki, Z. Konya, G. Galbacs, Qualitative discrimination analysis of coals based on their laser induced breakdown spectra, Energy Fuel 30 (2016) 10306–10313.

[65] M. Gaft, I. Sapir-Sofer, H. Modiano, R. Stana, Laser induced breakdown spectroscopy for bulk minerals online analyses, Spectrochim. Acta B At. Spectrosc. 62 (2007) 1496–1503.

[66] M. Gaft, E. Dvir, H. Modiano, U. Schone, Laser induced breakdown spectroscopy machine for online ash analyses in coal, Spectrochim. Acta B 63 (2008) 1177–1182.

[67] M. Gaft, L. Nagli, I. Fasaki, M. Kompitsas, G. Wilsch, Laser-induced breakdown spectroscopy for on-line sulfur analyses of minerals in ambient conditions, Spectrochim. Acta B 64 (2009) 1098–1104.

[68] D. Redoglio, E. Golinelli, S. Musazzi, U. Perini, F. Barberis, A large depth of field LIBS measuring system for elemental analysis of moving samples of raw coal, Spectrochim. Acta B 116 (2016) 46–50.

[69] D.A. Redoglio, E. Golinelli, S. Musazzi, U. Perini, F. Barberis, Development of a large depth of field collection optics for on-line laser-induced breakdown spectroscopy applications, Spectrochim. Acta B 123 (2016) 179–183.

[70] N. Idris, M. Ramli, R. Mahidin, Z.S. Hedwig, K.H.K. Lie, Examination of the capability of the laser-induced breakdown spectroscopy (LIBS) technique as the emerging laser-based analytical tool for analyzing trace elements in coal, AIP Conf. Proc. 1617 (2014) 175–178.

[71] K. Lahna, R. Idroes, N. Idris, S.N. Abdulmadjid, K.H. Kurniawan, M.O. Tjia, M. Pardede, K. Kagawa, Formation and emission characteristics of CN molecules in laser induced low pressure He plasma and its applications to N analysis in coal and fossilization study, Appl. Opt. 55 (2016) 1731–1737.

[72] M. Noda, Y. Deguchi, S. Iwasaki, N. Yoshikawa, Detection of carbon content in a high-temperature and high-pressure environment using laser-induced breakdown spectroscopy, Spectrochim. Acta B 57 (2002) 701–709.

[73] M. Kurihara, K. Ikeda, Y. Izawa, Y. Deguchi, H. Tarui, Optimal boiler control through real-time monitoring of unburned carbon in fly ash by laser-induced breakdown spectroscopy, Appl. Opt. 42 (2003) 6159–6165.

[74] Y. Deguchi, T. Kamimoto, Z.Z. Wang, J.J. Yan, J.P. Liu, H. Watanabe, R. Kurose, Applications of laser diagnostics to thermal power plants and engines, Appl. Therm. Eng. 73 (2014) 1453–1464.

[75] Z.Z. Wang, Y. Deguchi, M. Kuwahara, T. Taira, X.B. Zhang, J.J. Yan, J.P. Liu, H. Watanabe, R. Kurose, Quantitative elemental detection of size-segregated particles using laser-induced breakdown spectroscopy, Spectrochim. Acta B 87 (2013) 130–138.

[76] M.P. Mateo, G. Nicolas, A. Yanez, Characterization of inorganic species in coal by laser-induced breakdown spectroscopy using UV and IR radiations, Appl. Surf. Sci. 254 (2007) 868–872.

[77] T. Ctvrtnickova, M.P. Mateo, A. Yanez, G. Nicolas, Characterization of coal fly ash components by laser-induced breakdown spectroscopy, Spectrochim. Acta B At. Spectrosc. 64 (2009) 1093–1097.

[78] T. Ctvrtnickova, M.P. Mateo, A. Yanez, G. Nicolas, Laser induced breakdown spectroscopy application for ash characterisation for a coal fired power plant, Spectrochim. Acta B 65 (2010) 734–737.

[79] T. Ctvrtnickova, M.P. Mateo, A. Yanez, G. Nicolas, Application of LIBS and TMA for the determination of combustion predictive indices of coals and coal blends, Appl. Surf. Sci. 257 (2011) 5447–5451.

[80] H. Zhang, J.P. Singh, F.-Y. Yueh, R.L. Cook, Laser-induced breakdown spectra in a coal-fired MHD facility, Appl. Spectrosc. 49 (1995) 1617–1623.

[81] L.G. Blevins, C.R. Shaddix, S.M. Sickafoose, P. M. Walsh, Laser-induced breakdown spectroscopy at high temperatures in industrial boilers and furnaces, Appl. Opt. 42 (2003) 6107–6118.

[82] C.E. Romero, R. De Saro, J. Craparo, A. Weisberg, R. Moreno, Z. Yao, Laser-induced breakdown spectroscopy for coal characterization and assessing slagging propensity, Energy Fuel 24 (2010) 510–517.

[83] Z.Y. Carlos, E. Romero, R. De Saro, J. Craparo, S. Lam, R. Silfies, R. Plangemann, F. Lyter, Development and demonstration of laser-induced breakdown soectroscopy for in-situ, on-line coal analysis, in: 2011 International Pittsburgh Coal Conference, Pittsburgh, Pennsylvania, USA, 2011, pp. 1–12.

[84] M. Dong, X. Mao, J. Gonzalez, J. Lu, R. Russo, Time-resolved LIBS of atomic and molecular carbon from coal in air, argon and helium, J. Anal. At. Spectrom. 27 (2012) 2066–2075.

[85] C.B. Stipe, A.L. Miller, J. Brown, E. Guevara, E. Cauda, Evaluation of laser-induced breakdown spectroscopy (LIBS) for measurement of silica on filter samples of coal dust, Appl. Spectrosc. 66 (2012) 1286–1293.

[86] C.E. Romero, R. De Saro, LIBS analysis for coal, in: Laser-Induced Breakdown Spectroscopy, Springer Series in Optical Sciences, vol. 182, Springer, 2014, pp. 511–529.

[87] T.X. Phuoc, P. Wang, D. McIntyre, Detection of rare earth elements in Powder River Basin sub-bituminous coal ash using laser-induced breakdown spectroscopy (LIBS), Fuel 163 (2016) 129–132.

[88] M. Dong, D. Oropeza, J. Chirinos, J.J. González, J. Lu, X. Mao, R.E. Russo, Elemental analysis of coal by tandem laser induced breakdown spectroscopy and laser ablation inductively coupled plasma time of flight mass spectrometry, Spectrochim. Acta B At. Spectrosc. 109 (2015) 44–50.

[89] A.A. Bol'shakov, X.L. Mao, J. Jain, D.L. McIntyre, R.E. Russo, Laser ablation molecular isotopic spectrometry of carbon isotopes, Spectrochim. Acta B 113 (2015) 106–112.

[90] P.M. Lemieux, J.V. Ryan, N.B. French, W.J. Haas, S. Priebe, D.B. Burns, Results of the September 1997 DOE/EPA demonstration of multimetal continuous emission monitoring technologies, Waste Manag. 18 (1998) 385–391.

[91] Y. Li, D. Tian, Y. Ding, G. Yang, K. Liu, C. Wang, X. Han, A review of laser-induced breakdown spectroscopy signal enhancement, Appl. Spectrosc. Rev. (2017) 1–35.

[92] P. Gang, L. Jidong, Y. DongMeirong, X. Shunchun, F.J. Zixin, A study on the characteristics of

carbon-related spectral lines from a laser-induced fly ash plasma, Plasma Sci. Technol. 17 (2015) 625–631.

[93] C.H. Wang, L. Huang, T.B. Chen, M.H. Liu, H. Yang, H.U. Hui-Qin, M.Y. Yao, Feasibility of analyzing Cr in rice husk and coarse rice with LIBS, Spectrosc. Spectr. Anal. 37 (2017) 3590–3593.

[94] B.C. Castle, K. Talabardon, B.W. Smith, J.D. Winefordner, Variables influencing the precision of laser-induced breakdown spectroscopy measurements, Appl. Spectrosc. 52 (1998) 649–657.

[95] D. Body, B.L. Chadwick, Optimization of the spectral data processing in a LIBS simultaneous elemental analysis system, Spectrochim. Acta B At. Spectrosc. 56 (2001) 725–736.

[96] A. Safi, B. Campanella, E. Grifoni, S. Legnaioli, G. Lorenzetti, S. Pagnotta, F. Poggialini, L. Ripoll-Seguer, M. Hidalgo, V. Palleschi, Multivariate calibration in laser-induced breakdown spectroscopy quantitative analysis: the dangers of a 'black box' approach and how to avoid them, Spectrochim. Acta B At. Spectrosc. 144 (2018) 46–54.

[97] I. Jolliffe, Principal Component Analysis, vol. 1986, Spring-verlag, New York, 1986.

[98] H. Abdi, Partial least square regression (PLS regression), in: Encyclopedia of Social Sciences Research Methods, vol. 6, 2003, , pp. 792–795.

[99] L. Breiman, J.H. Friedman, R.A. Olshen, C.J. Stone, Classification and Regression Trees, Wadsworth International Group, Belmont, CA, 1984, p. 432.

[100] S. Ding, L. Guo, Y. Hou, Extreme learning machine with kernel model based on deep learning, Neural Comput. Applic. 28 (2017) 1975–1984.

[101] G. Huang, Q. Zhu, C. Siew, Extreme learning machine: theory and applications, Neurocomputing 70 (2006) 489–501.

[102] C.N. Standards, http://en.cnis.gov.cn.

[103] B. Paris, Direct Determination of Organic Sulfur in Raw Coals, vol. 174, American Chemical Society, 1977, p. 1.

[104] W.E. Straszheim, R.T. Greer, R. Markuszewski, Direct determination of organic Sulfur in raw and chemically desulfurized coals, Fuel 62 (1983) 1070–1075.

[105] D.M. Mason, K.N. Gandhi, Formulas for calculating the calorific value of coal and coal chars: development, tests, and uses, Fuel Process. Technol. 7 (1983) 11–22.

[106] D.K. Ottesen, J.C.F. Wang, L.J. Radziemski, Real-time laser spark spectroscopy of particulates in combustion environments, Appl. Spectrosc. 43 (1989) 967–976.

[107] D.K. Ottesen, L.L. Baxter, L.J. Radziemski, J.F. Burrows, Laser spark emission spectroscopy for in-situ, real-time monitoring of pulverized coal particle composition, Energy Fuel 5 (1991) 304–312.

[108] D.K. Ottesen, Laser spark emission spectroscopy of individual coal particles, in: H.L.C. Meuzelaar (Ed.), Advances in Coal Spectroscopy, Springer US, Boston, MA, 1992, , pp. 91–118.

[109] L. Zhang, W. Ma, L. Dong, X. Yan, Z. Hu, Z. Li, Y. Zhang, W. Le, W. Yin, S. Jia, Development of an apparatus for on-line analysis of unburned carbon in fly ash using laser-induced breakdown spectroscopy (LIBS), Appl. Spectrosc. 65 (2011) 790–796.

[110] C. Sanchez-Ake, M. Bolanos, C.Z. Ramirez, Emission enhancement using two orthogonal targets in double pulse laser-induced breakdown spectroscopy, Spectrochim. Acta B 64 (2009) 857–862.

[111] J.Y. Ye, L. Balogh, T.B. Norris, Enhancement of laser-induced optical breakdown using metal/dendrimer nanocomposites, Appl. Phys. Lett. 80 (2002) 1713–1715.

CHAPTER

22

LIBS for aerosol analysis

Daniel Diaz[a,b], *David W. Hahn*[a,b], *Ulrich Panne*[c,d]

[a]Mechanical and Aerospace Engineering Department, University of Florida, Gainesville, FL, United States [b]College of Engineering, University of Arizona, Tucson, AZ, United States [c]Federal Institute for Materials Research and Testing (BAM), Berlin, Germany [d]Department of Chemistry, Humboldt University of Berlin, Berlin, Germany

1 Introduction

The analysis of aerosol samples presents a challenging analytical problem due to the wide variation in both aerosol particles and measurement needs. For example, ambient air aerosol particles are a complex mixture originating from both natural and anthropogenic sources, and are generated either via direct emission processes (primary particles) or via gas-to-particle conversion processes (secondary particles). Aerosol particles in the ambient atmosphere originate from several sources: wind-raised dust, agricultural activities and open fields, sea spray, industrial activity, traffic, volcanoes, wildfires and combustion processes, as well as photochemical conversion of gases to particles. Chemical components include sulfates, ammonium, nitrates, chlorides, trace metals, carbonaceous materials, crustal elements, and water. Because laser-induced breakdown spectroscopy (LIBS) is essentially an elemental analysis technique, attention must be given to aerosol particles containing elements that stand in contrast to the background gaseous matrix. For example, nitrogen (e.g., as nitrates), oxygen (e.g., as metal oxides), and carbon (e.g., as soot or organic) may all be present in particulates, but such elements also arise from the purely gas phase, thereby making the quantification of the gas-phase and particulate-phase partitioning difficult. Given these difficulties, much of the LIBS analysis of aerosols has focused on elements not found in the gas phase, such as trace metals.

Trace metals are ubiquitous in various raw materials, such as fossil fuels and metal ores, as well as in industrial products, hence, many trace metals evaporate entirely or partially from raw materials during the high-temperature production of industrial goods, combustion of fuels, and incineration of municipal and industrial wastes, thereby entering the ambient air with exhaust gases. Most of the time, aerosols from such sources have very distinct particle distributions with a low geometric standard deviation and a mean particle size on the order of 0.1 μm, yielding relatively low quantities on a particle-by-particle basis. For example, a solid spherical particle of 100 nm diameter with a density of 2 g cm^{-3} has a total mass of about 1 femtogram (fg).

Laser-Induced Breakdown Spectroscopy
https://doi.org/10.1016/B978-0-12-818829-3.00022-8

Natural sources are related primarily to the geological presence of trace metals in the crustal material and are transformed into aerosol particles during various natural physical, chemical, biological, and meteorological processes. For example, soil-derived dust accounts for over 50% of the total Cr, Mn, and V emissions, as well as for 20%–30% of the Cu, Mo, Ni, Pb, Sb, and Zn released annually to the atmosphere. Furthermore, volcanic emissions appear as a significant source, which account for 40%–50% of the total natural Cd and Hg, and 20%–40% of the total natural As, Cr, Cu, Ni, Pb, and Sb emitted yearly. Sea-salt aerosols seem to account for <10% of atmospheric trace metals from natural sources. In certain parts of the world, forest fires are the major emission source and >10% of atmospheric Cu, Pb, and Zn from natural sources can originate from fires.

Combustion of fossil fuels to produce electricity and heat is the main source of anthropogenic emissions of atmospheric Be, Co, Hg, Mo, Ni, Sb, Se, Sn, and V [1] and an important source of As, Cr, Cu, Mn, and Zn. In general, the amount of emissions from a conventional thermal power plant depends on the content of trace metals in the fuels, the physical and chemical properties of trace metals during combustion, technological conditions of a burner, and the type and efficiency of emission control equipment. The most commonly inventoried heavy metals are As, Be, Cd, Cr, Hg, Pb, and Zn. This is due to their effects on environmental and human health, as well as their ubiquitous appearance in the environment.

Given the broad range of aerosol size and composition, as already briefly summarized, knowledge of aerosol size distributions is essential because the particle size significantly affects ambient transport and deposition processes as well as the described uptake in the respiratory system. Moreover, elemental size distributions can give an indication of the source of the element [2, 3]. The typical characteristics of size distributions vary from element to element and from sampling location to sampling location, but generally they display a trimodal or bimodal distribution [4]. Not surprisingly, many tasks in aerosol analysis stem ultimately from the potential health effects of aerosols. The applications range from aerosol analysis in combustion to process analysis and control in various industrial production processes. For example, the major issue in occupational hygiene is the question of exposure, which may be considered as the time-averaged concentration of the agent under study at the relevant interface between the environment and the biological system (i.e., the worker). Not only the concentration, but also other characteristics such as composition, morphology, and particles size distribution are presently coming into the focus of exposure threshold definitions.

Due to the wide range of aerosol sampling needs in conjunction with the great variation of particles presented to any analytical instrument, it is not surprising that one might envision a wide number of desirable characteristics in such a tool. The "ideal" aerosol analyzer was described by Friedlander, who classified in his seminal 1970/71 work different aerosol measurement techniques on the type of information given by the instrument [5, 6]. Although today's aerosol analysis by mass spectrometry approaches the ideal of a perfect single particle counter, which should provide the complete size-resolved chemical composition of an aerosol in real time, most current methods are far from this ideal. The application of LIBS for aerosol analysis offers inherent advantages that overcome many of the limitations inherent in current aerosol analysis tools. The LIBS technique has been developed as a novel means for the quantitative, direct measurement of particle size and composition of individual particulates, although many fundamental issues remain regarding the interactions of the laser-induced plasma with aerosol particles. As a starting point for LIBS-based aerosol analysis, the following sections provide a detailed review of the laser-induced breakdown process, as focused on the gas-phase processes associated with plasma initiation and propagation.

2 Laser-induced breakdown of gases

With the advent of the first laser more than 50 years ago, nearly simultaneously the first laser-induced plasmas in gases were reported by Maker et al. [7] and Meyerand et al. [8]. Since then, a considerable body of literature aimed at understanding the various aspects of gas- and aerosol-induced breakdown has been written. For detailed information the reader is referred to the reviews [9–11] and corresponding chapters in the monographs [12–15] and references therein.

In contrast to the breakdown on solids or liquids, the irradiance for breakdown, that is, plasma formation, in gases is in excess of 10^9 W cm^{-2} due to the involved mechanism of multiphoton ionization (MPI) and/or cascade ionization [16]. While high-energy photons can ionize gases in single-photon interactions, it is not obvious in which way a laser photon of low energy (1–2 eV), compared to the ionization potential of common gases, can generate a breakdown. The formation of a laser plasma in a neutral gas follows three distinct but overlapping stages: (i) plasma ignition, (ii) plasma growth and interaction with the laser pulse (in the case of nanosecond laser pulses), and (iii) plasma development accompanied by shock waves generation and propagation in the surrounding gas.

Plasma ignition comprises the growth of the free electron and ion concentrations with the arrival of the first laser photons. The growth stage is characterized by a fast amplification of free electrons and ions. The term breakdown is rather arbitrarily and loosely defined in the literature, often the luminous plasma emission or the acoustical detection of the shock wave is taken as the only criterion. In the following, breakdown will imply an electron density $(N_e) > 10^{13}$ cm^{-3} or a degree of ionization about 10^{-3}, which permits a significant absorption and scattering of the incident laser radiation and leads to a fully developed plasma very quickly (N_e typically 10^{17}–10^{19} cm^{-3}). With the onset of breakdown a highly ionized plasma develops in which, in the case of nanosecond pulses, further absorption and photoionization occurs. With the end of the laser pulse, the plasma dies gradually away as a result of radiation and conduction of thermal energy, diffusion, attachment, and recombination of ions and electrons, until local thermodynamic equilibrium with the surrounding gas is restored.

For nanosecond and microsecond pulses at pressures above 10^4 Pa (100 mbar), the breakdown at long wavelengths (>1 μm) proceeds via an electron avalanche or cascade by inverse bremsstrahlung (IB), i.e., free–free absorption. In this way, electrons absorb photons in the presence of a neutral gas atom for momentum transfer and acquire ultimately sufficient energy for collisional ionization of gas atoms and generation of further electrons. At visible or ultraviolet (UV) wavelengths, highly excited states of the gas molecules or atoms can be readily photoionized over times much shorter than the typical nanosecond pulse widths. This reduces the observed breakdown thresholds considerably, and reduces the losses in electron density due to diffusion.

Under low-pressure conditions, the breakdown is initialized via electrons from multiphoton absorption and ionization, while at latter times the cascade process often overtakes the electron generation. MPI also comes back into play when considering the source of the "first" electron for the cascade process. In contrast to MPI, cascade ionization is not self-sufficient, but requires at least one electron in the focal volume. Due to natural local radioactivity (e.g., cosmic rays or UV radiation) ions occur naturally in the atmosphere at a concentration of 10^2–10^3 cm^{-3}, although free electrons are immediately attached to O_2 yielding O_2^-. The mean lifetime of O_2^-, which can be treated as a free electron due to electron tunneling, is in the order of 10^{-7} s, therefore the probability of encountering an electron in an interaction volume of 10^{-6} m^3 during a nanosecond laser pulse is rather negligible. The breakdown threshold depends weakly on the gas pressure through $p^{-1/m}$, where m is the necessary number of electrons for MPI to occur. The breakdown starts when a fraction of atoms (on the order of 10^{-3}) present in the interaction volume is ionized. In air, a MPI-initiated breakdown at the fundamental wavelength of a Nd:YAG (1064 nm) is an 8- or 10-photon process for O_2 and N_2 [17–22]. At visible and UV wavelengths, highly excited states can be readily photoionized over times much shorter than the laser pulse width, with thresholds significantly lower (5–10 GW cm^{-2}). Stark shift and broadening of intermediate levels can additionally bring the levels into resonance with the laser wavelength. Also, if a sufficient N_e is generated by MPI early enough in the pulse and affects the diffusion of electrons out of the focal volume, the diffusion becomes affected by the space charge of the ions remaining in the focal region. This ambipolar nature reduces the overall diffusion, and the breakdown threshold will be lowered especially in experiments with small focal spots where diffusion losses can be important [23, 24]. Finally, it must be remembered that laser modes can lead to local fields that significantly exceed the average value over the focal volume. Hence, the effective irradiance of reported experimental values are probably somewhat larger than the required irradiance. At longer wavelengths the problem of generating the first electron in the absence of gaseous or aerosol impurities becomes more serious. Breakdown at 1064 nm can then increase to an irradiance level of 10^{12} W cm^{-2}, where the electric field induces tunneling of an electron through its potential barrier [25–28].

While production of the first electron is crucial for the start of the breakdown, the irradiance threshold is governed by the cascade or avalanche growth of ionization, which is fed by absorption of the laser light. Electrons in a laser field will gain energy to ionize and increase in number through electron-neutral IB. Microwave breakdown theory [29–33] has been successful in describing breakdown thresholds at wavelengths beyond 1 μm, where IB is the dominant mechanism, while at visible and near-UV wavelengths data are not so well understood [24, 34–37]. If electron-impact ionization was the dominant mechanism leading to gas breakdown, the breakdown threshold for a given gas at a given pressure would scale inversely with the IB absorption coefficient (approximately with λ^{-2}). Experimental observations [37–43] revealed that the breakdown thresholds peaked in the middle of the visible spectrum due to competition between IB and MPI of ground and excited states [44–49].

The cascade process will lead to an exponential growth of electron density. Assuming a fully developed plasma at $N_e \approx 10^{17}$ cm^{-3}, about 40 generations are required to grow from an assumed initial value of $N_{e,0} \approx 1$–10 cm^{-3} in the focal volume, i.e., 99% of the ionization is produced in the last 7 generations. When the electron concentration exceeds the onset of the breakdown ($\approx 10^{13}$ cm^{-3}) electron-electron collisions will tend to populate the tail of the electron distribution function and this has a dramatic effect on the cascade rate. Quantities such as the growth and losses from the cascade and the time to breakdown are determined by conditions at times when the electron concentration is small. Losses can be through several inelastic processes such as vibrational and rotational excitation (for polyatomic gases), excitation of electronic levels of atoms and molecules, elastic collisions, attachment, recombination, and diffusion.

Although the theoretical modeling of the breakdown in gases has developed to quite a sophisticated level [24, 34], to elucidate the exact mechanism of the plasma formation in gases, strict control of the experimental parameters such as focusing, laser modes, and contamination of the gas and cell is still needed, accompanied by suitable plasma diagnostics [50, 51].

The study of interactions between laser light and aerosols began in the 1960s with the availability of the first lasers. Haught et al. reported in 1966 the first observations of particle-induced breakdown in gases due to the breakdown on an electrostatically levitated single 20-μm particle of LiH [52]. It became immediately clear that the propagation of lasers through the atmosphere and hence the operative efficiency of lasers for applications from nuclear fusion to military weapons research were intimately connected to aerosols (see [53–55] and references therein).

Once irradiated with the laser beam, an aerosol particle starts to absorb, which leads to heating and further melting, boiling, and gradual evaporation (sublimation) of particle material. See Ref. [56] for an visualization through a molecular dynamics simulation. The process may not be uniform because the distribution of the electromagnetic field inside the particle is not necessarily uniform, especially with internally mixed particles. Hot spots in the nodes of the field can virtually explode the particle before the thermal conductivity smooths out the temperature distribution. For particles small in comparison to the laser wavelength, the absorptivity decreases with particle diameter, d_p. The particle heating depends strongly on the heat losses caused by contact with the carrier gas and evaporation. The breakdown is promoted through aerosol particles by heating the surrounding gas and providing the breakdown zone with additional electrons caused by a number of mechanisms: heat explosion, shock waves in the surrounding carrier gas, breakdown in evaporated matter, and electron emission (thermo-, photo-, and tribo-electrons). The increase in the initial electron density can be up to $10^7\ cm^{-3}$. Naturally, this decreases the threshold and reduces the time for full plasma development. The presence of aerosol particles in the focal volume will lower the threshold for gas breakdown by several orders of magnitude (typically around $10^7\ W\ cm^{-2}$ compared to breakdown in pure air at $10^{11}\ W\ cm^{-2}$). Aerosol breakdown depends upon the pulse width, wavelength, focusing of the laser, and particle size, and the threshold usually scales with the particle diameter (d_p^{-1}–d_p^{-2}). Aerosol-induced breakdown has been described by several authors with different wavelengths between 1064 and 248 nm [57–68]. For microsecond pulses a wavelength dependence of λ^{-1} was observed, whereas with the usually employed nanosecond pulses a λ^{-2} scaling was found [65]. The observed increase of the threshold with a decrease of the focal beam diameter is explained through inclusion of larger particles in the interaction volume, which produce a higher initial increase in electron density [69]. The breakdown of droplets is modified through the curved liquid-gas interface [55, 70–74]. The droplet can be envisioned as a lens that concentrates the incident light wave to a localized region just within the droplet shadow face and focuses it to a localized region just outside the droplet shadow face. For large transparent droplets and moderate irradiance, the breakdown occurs just outside the shadow face at which the irradiance is highest. Whether the breakdown is initiated in the gas outside the droplet or within the droplet depends on parameters such as the breakdown threshold of the gas and liquid, as well as the droplet morphology [75–77].

3 Analysis of aerosol particles by LIBS

The analysis of aerosol samples presents a unique application for LIBS, in that the possibility exists to bring together the point-to-point sampling nature of laser-induced plasmas with the discrete nature of aerosol particles. The following sections elaborate on different approaches to optimize the spectroscopic information from aerosol-derived analytes.

A fundamental issue regarding to the analysis of aerosol particles with the LIBS technique concerns the relationship between the discrete aerosol particles and the finite-sized laser-induced plasma volume. With inductively coupled plasma (ICP) atomic emission spectroscopy, the plasma source is continuous, and analytes in solution are fed into the plasma at a constant rate. The resulting analyte signal represents an average analyte concentration, and this mode of operation is characteristic of many analytical methods. However, with the LIBS technique, the coupling of the repetitive, finite-sized plasma with the spatial distribution of aerosol particles must be considered. For example, if the spatial distribution of solid-phase aerosol is of such a value that the laser-induced plasma samples particles with a high probability, then average analyte signals are reflective of the particle-phase contribution as well as the gas-phase species contributions. In contrast, if the aerosol particle concentration is small such that the probability of sampling an aerosol particle in a given laser-induced plasma is very small, then the contribution of particle-derived analyte to the total signal may become a negligible fraction. With this latter case, the average LIBS signal may only reflect the gaseous components, with the result being a nondetect condition with respect to the aerosol particle constituent species. The LIBS-based analysis of aerosol particles may be naturally partitioned into these two statistical sampling regimes, namely one condition where the aerosol particle sampling rate is limited and another condition where the probability of aerosol particle sampling is sufficiently high. The latter condition is well suited to traditional ensemble averaging techniques, where many LIBS spectra are averaged together in an effort to eliminate random, single-pulse spectral noise. With the former condition, advanced spectral processing schemes are required to compensate for the large amount of spectral data that contain no information regarding the aerosol-derived analytes of interest. The methodologies and implementation of LIBS-based aerosol analysis schemes are presented and discussed for each of these operating approaches.

3.1 Spectral ensemble-averaging

Most laser-induced plasma spectral data are collected using intensified charge-coupled device (ICCD) array detectors, although increasingly nonintensified CCD detectors are being examined [78, 79]. The primary advantage of ICCD systems is the ability to temporally gate the detector, which provides for optimization of the analyte atomic emission signals with respect to the plasma continuum emission signals due to the differing decay rates of these two emission processes. An example of the optimal temporal gating for several toxic metals species is discussed in several publications [80–82], showing that considerable time differences (some tens of microseconds) can exist between optimal detection windows. The use of ICCD detectors, however, can result in a significant amount of spectral noise generally originating from the intensifier. Furthermore, the absolute intensity of the LIBS signal (both continuum emission and atomic emission) can vary significantly on a pulse-to-pulse basis. Hence, spectral data generated with the LIBS technique is inherently noisy due to the combination of natural plasma fluctuations and intensifier/detector noise. In some of the early LIBS research toward analysis of gaseous/aerosol samples, Radziemski et al. noted that self-normalization of the atomic emission signal by the laser-induced plasma continuum emission provided a more robust analyte signal [83–85]. In several papers, Hahn and coworkers examined the pulse-to-pulse fluctuations in detail for a number of gas-phase analyte species, including carbon, nitrogen, and hydrogen emission lines [86, 87]. For example, it was found that absolute values of the continuum emission intensity and atomic emission intensity of the 247.8-nm carbon line varied by a factor of 4 and 6, respectively, between the minimum and maximum values for a sequence of 100 consecutive laser pulses [86]. The corresponding relative standard deviations (RSD) of the continuum and atomic emission signals were 9.5% and 11%, respectively. Alternatively, the carbon atomic emission peak intensity normalized by the nearby continuum emission intensity (peak-to-base or P/B ratio) varied by no more than a factor of two for the same 100 spectra, with a corresponding RSD of 6%. Clearly, the benefit of spectral normalization was demonstrated by the nearly 50% improvement in precision, as realized in the reduction of RSD. A review of several normalization techniques in analytical atomic spectrometry with laser sampling can be found in the manuscript published by Zorov et al. [88]. The spectral noise characteristics of LIBS are readily quantified through the examination of representative single-pulse and ensemble-averaged spectra.

Carranza et al. performed a detailed analysis of the noise associated with ICCD detectors as part of a comparison between ICCD and nonintensified CCD detectors [79]. In that study, both single-pulse and ensemble-averaged data were examined. Fig. 1 is adapted from that study, and shows the resulting spectra from a single-pulse measurement with the ICCD and a 200-pulse ensemble-averaged spectrum corresponding to the same characteristics. For these two spectra, the root-mean-square (RMS) noise of the continuum emission intensity was calculated, as measured over relatively smooth spectral regions on each side of the calcium peaks. The average RMS noise was 2.37 and 14.0 for the ensemble-averaged and single-pulse spectra, respectively. This sixfold reduction in spectral noise translates into enhanced detection limits, as reported by Carranza and coworkers. For example, over a range of temporal delays between the incident laser pulse and initiation of the detector gate, the signal-to-noise ratio (SNR) of the 397.4-nm Ca emission line was increased by an average of 6.5 when comparing the ensemble-averaged data to the single-pulse data. Similarly, when comparing the single-pulse versus ensemble-averaged data for the nonintensified CCD, the SNR was also improved by a factor of 7.

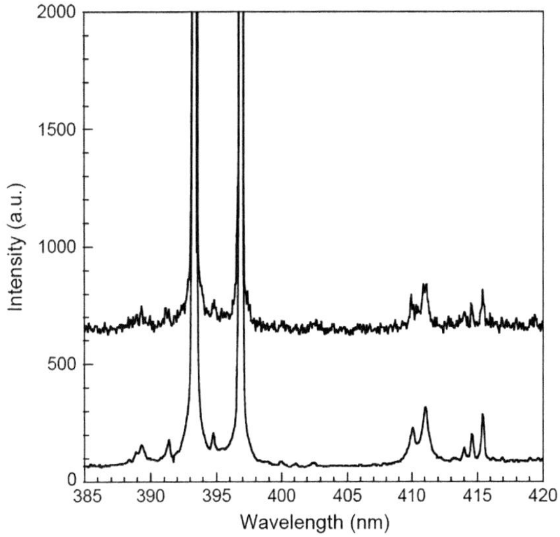

FIG. 1 LIBS spectra corresponding to a calcium-rich aerosol system for a 200-shot ensemble average and for a single shot. The ICCD detector delay was fixed at 5 μs following plasma initiation for all measurements. Both spectra are presented at the same intensity scale, and the single-shot spectrum has been shifted up by 600 counts.

A significant benefit (i.e., improvements in analytical figures of merit) is realized using spectral ensemble averaging with the LIBS technique. Historically, ensemble-averaging of hundreds or thousands of spectra is by far the most widely implemented approach for LIBS-based analysis of gaseous samples. With such an approach, the resulting average spectrum is characterized by relatively low signal noise, and provides an analyte signal representative of the average constitutive species. The constitutive species contribute to the analyte signal, which may contain contributions from both gas-phase and particulate-phase compounds as already discussed. Successful implementation of ensemble-averaging for LIBS-based analysis of aerosol samples is performed similarly to the analysis of gaseous species; however, the following issues must be given attention: (a) the aerosol particle concentration (particles/volume) must be sufficiently large such that a sufficient number of laser-induced plasmas (i.e., laser pulses) actually sample aerosol particles; (b) the size distribution of the targeted aerosol particles should lie within a size regime such that the largest particles can be totally vaporized within the laser-induced plasma, thereby presenting a linear analyte mass response with regard to the atomic emission signal; and (c) the analyte calibration source stream must be designed to ensure that the true LIBS analyte response is realized, including possible differences in analyte response due to gas-phase and solid-phase analyte sources.

First, consider condition (a), namely the exact percentage of required laser pulses to ensure adequate sampling of an aerosol source, as measured by a significant analyte signal in the ensemble-averaged spectrum with respect to the method detection limit. The exact sampling rate is difficult to specify in general, but must consider both the overall fraction of laser-induced plasmas that sample a particle and the relative emission strength of the targeted analyte. For hundreds of laser pulses, the baseline noise generally reaches a limiting value beyond which additional signal averaging provides no added benefit. Since this is the region desirable for ensemble-averaging, the following comments will assume that the spectral noise has reached the limiting value. Under such conditions, the analyte signal intensity is reduced by a factor equal to the inverse of the aerosol particle sampling rate. For example, if the aerosol particle sampling fraction were 0.10 (i.e., 10% hit rate), then the analyte signal stemming from the particulate-phase species would be degraded by a factor of 1/0.1, or a factor of 10. In other words, for every analyte-containing spectrum recorded, 10 additional nonanalyte-containing spectra are recorded, which when ensemble-averaged together, produces the 10-fold decrease in particulate-phase-derived emission signal. To estimate the necessary sampling rate for this mode of operation, divide the limit of detection by the expected analyte response of a single particle. However, with many signal responses unknown for different aerosol particle sizes and composition, it is often difficult to make such a calculation. In practice, one can simply examine the nature of the analyte signal produced via ensemble-averaging. If the analyte signal produces a significant SNR, then ensemble-averaging is a successful strategy that enjoys the simplicity of the overall LIBS technique. Conversely, if no strong analyte response is observed in the generated average spectrum, while analyte-containing aerosol particles are known to be present, then additional spectral processing schemes, as related later, are required.

The second criteria (b) regarding the quantitative analysis of aerosol samples concerns the upper particle size limit for complete particle dissociation, and the resulting independence of analyte signal on particle size (assuming an equal analyte mass). Cremers and Radziemski [89] explored LIBS for the detection of beryllium-rich particles deposited on filters using a range of conditions, including (i) particle diameters of about 50 nm, (ii) an ensemble collection of particles ranging from 0.5 to 5 μm, and (iii) for nominally 15-μm sized particles. They used a cylindrical lens to focus a laser beam directly on the surface of the filters, thereby producing a laser-induced plasma that engulfed the deposited beryllium (Be) particles. Plasma emission was collected and analyzed using the Be atomic emission line, which is one of the stronger emitting elements on a mass basis with LIBS. Cremers and Radziemski reported a different analyte response, as realized with different calibration curve slopes, for these three different particle size classes, and concluded based on their experimental observations that incomplete particle vaporization occurred for particles with diameters greater than about 15 μm. In other studies [90, 91], this group also explored direct LIBS-based analysis of Be aerosols using particles <10 μm in diameter, noting in their latter study that such a particle size is consistent with complete particle vaporization. Over the ensuing decades, many LIBS researchers have cited an upper size limit for complete particle vaporization on the order of 10 μm. While this value is consistent with the original research of Radziemski et al. [90], one must consider the context of this early work. No detailed studies of the limiting range for linear analyte response as a function of particle diameter were performed. However, the premise of a linear analyte response for quantitative LIBS analysis of aerosol particles is predicated on the complete vaporization; hence, more recent investigations have specifically addressed this issue.

Carranza and Hahn [92] investigated the laser-induced plasma vaporization of individual silica microspheres in an aerosolized air stream by examining the analyte response (silicon P/B) for progressively larger, monodisperse aerosol streams. A linear mass response (noting mass is proportional to the diameter cubed) was observed to an upper diameter limit of 2.1 μm for a laser pulse energy of about 320 mJ. The results are summarized in Fig. 2, which shows the silicon analyte response (P/B ratio) as a function of particle mass per plasma volume for single micron-sized microspheres. The most significant result is the linear relation between the silicon P/B ratio as a function of silicon particle mass (plotted as diameter cubed) for the smallest three silica particle diameters investigated (1.0, 1.5, and 2.1 μm) and the abrupt deviation from this linear trend for the particle diameters larger than 2.1 μm. For complete silica particle dissociation and vaporization, the resulting analyte signal should scale as the analyte mass contained in the particle, hence as the diameter cubed. With this in mind, Carranza and Hahn [92] interpreted the threshold in the Fig. 2 data as the upper size limit in which an aerosolized silica particle is completely vaporized in the laser-induced plasma for the experimental parameters reported in their study. To examine if effects such as self-absorption were the reason for the departure from linearity, additional measurements were made by nebulizing solutions of dissolved silicon. As reported in their paper, such a configuration results in a high number density of submicron-sized silicon-rich aerosol particles. The response to the silicon-rich nanoparticles was linear to an equivalent silicon mass of an approximately 8-μm sized silicate particle. The authors speculated that the different results for different size particles are due perhaps to a rate-limiting step of particle vaporization and dissociation. Therefore, the larger particles may simply not have had time to completely vaporize by the time the measurements were recorded, some tens of microseconds following plasma initiation. In addition, more recent imaging measurements, as discussed later, show that the plasma–particle interaction is limited to a spatial region about the particle, hence local plasma conditions may be affected by the presence of large (i.e., micron-sized) particles. Two more recent studies examined the complete vaporization of aerosol particles. Vors and Salmon examined the complete vaporization of carbon-rich particles (specifically glucose particles and sodium hydrogen

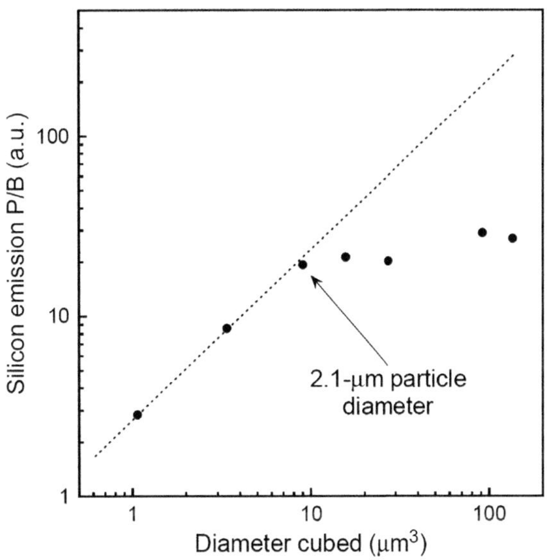

FIG. 2 Peak-to-base (P/B) ratio of the 288.16-nm silicon atomic emission line for ensemble-averaged spectra of individually detected monodisperse silica microspheres as a function of the cube of the silica particle diameter [92]. The dashed line is a linear fit of the first three data points.

carbonate particles) in a laser-induced plasma, and reported an upper size limit of 5 μm for complete vaporization [93]; and Gallou et al. reported an upper size limit of 7 μm for complete vaporization of Cu particles sampled in air [94]. Such a size difference most likely reflects the marked difference in melting points and volatility of the analytes as well as the different experimental conditions used. Additional experimental work and plasma modeling are needed to further determine the exact processes that govern particle vaporization, as well as to determine particle size limits for different laser pulse energies, wavelengths, focusing optics, and particle types. It is noted that these reported limiting particle sizes of 2–7 μm are below the frequently used 10-μm particle diameter limit, and should be taken into consideration for LIBS-based analysis of aerosol samples. Nonetheless, the upper size limits discussed here are consistent with the general needs of PM2.5 monitoring, where assumption of complete vaporization for particles smaller than about 2.5 μm is reasonable given the typical uncertainties associated with measurement of atmospheric aerosols.

To gain additional insight into the role of plasma–particle interactions, it is useful to consider the relative mass fraction due to an aerosol particle within a laser-induced plasma. Complete particle dissociation and subsequent heating of the particulate mass to the plasma state requires an amount of energy equal to the sum of the latent heat (i.e., heat of melting and heat of vaporization), the sensible heat (i.e., specific heat of gas-phase species), and the dissociation energies of the constituent species. Assuming comparable specific heats for various dissociated and gaseous species, the dissociation energy will scale as the particle mass fraction with respect to total energy deposited into the plasma. Using reasonable values for both aerosol particles and plasma properties, about 0.1% of the plasma energy will be consumed to heat the dissociated mass of a nominal 10-μm sized particle. However, heats of vaporization are typically two to three orders of magnitude greater than specific heat, hence, the energy required to initially vaporize the particle is expected to be significant. Assuming a conservative value of the heat of vaporization a factor of 100 greater than the specific heat, the initial vaporization of the 10-μm diameter particle could consume a quantity of energy on the order of 10% of the energy required to heat the particle-free gaseous plasma. In contrast, a 2-μm particle would only consume approximately 0.1% of this energy budget. A significant assumption of quantitative LIBS is that the presence of an aerosol particle does not exert a significant influence on the overall plasma evolution and plasma characteristics. Such effects are reviewed in more detail later, but clearly the energy budget for particle vaporization of an ≈2–2.5-μm aerosol particle is consistent with this primary tenet of LIBS-based aerosol analysis: the additive nature of the overall plasma spectral emission and the aerosol-derived analyte atomic emission. An important caveat, however, concerns the degree of homogeneity of the particle-derived analyte with respect to the overall plasma volume. If the atoms derived from the aerosol particle remain locally confined, as demonstrated and discussed in Section 6 (Future directions) and then perhaps the direct comparison of analyte mass to total plasma is too conservative of an approach.

The final issue to address with LIBS-based ensemble-averaging is the need for calibration schemes, which must produce an accurate analyte response for a known mass concentration. While it is traditionally accepted that the resulting atomic emission from a laser-induced plasma is independent of the actual analyte source (i.e., atomic, molecular, or aerosol particle), recent research suggests significant departures in emission response for gas-phase and particulate-phase analyte sources. Specifically, Hohreiter and Hahn reported marked differences in the atomic emission signal from carbon when comparing calibration streams of gas-phase and submicron-sized solid-phase carbon species [95]. The resulting calibration curve slopes varied by a factor of eight over a comparable range of atomic carbon concentrations for five different analyte sources, while the plasma electron density and temperature remained essentially constant. As an example, Fig. 3 presents the background-corrected LIBS spectra for two carbon sources (particulate and gaseous) present under equal overall mass loadings. Such findings challenge the widely held assumption that complete dissociation of constituent species within the highly energetic laser-induced plasma results in independence of the analyte atomic emission signal on the analyte source. A physical model of the plasma-analyte interaction was proposed that provides a framework to account for the observed dependence on the physical state of the analyte. The framework put forth must account for the enhancement of analyte signal with particulates less than the ≈2 μm limit for complete vaporization; hence, incomplete vaporization is not an issue. The authors suggested that the relative inertial differences between molecular and particle species play a role as the laser-induced plasma and subsequent plasma wave force species outward from the plasma kernel. The heavier particulates resist this outward push, resulting in a "concentration" of particulate-phase analyte species within the hotter plasma central region, resulting in an enhanced particulate-phase analyte response.

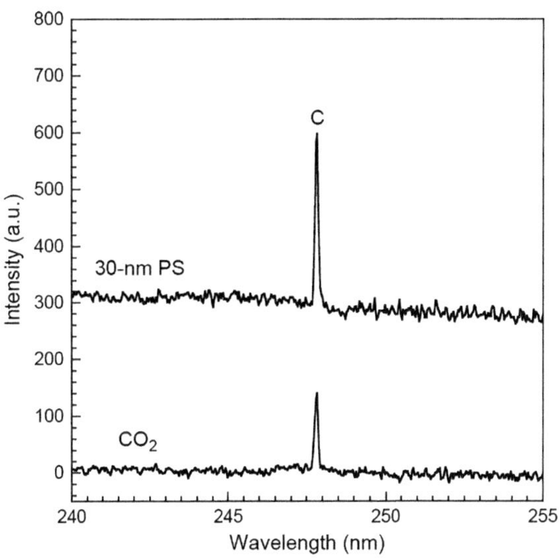

FIG. 3 LIBS spectra in the vicinity of the 247.9-nm carbon atomic emission line for two different sources of carbon: CO_2 and 30 nm polystyrene particles. Both spectra are the ensemble average of 1000 laser shots, and the concentration of atomic carbon was equal for both measurements [95]. Both spectra have the same scale and baseline intensity, and have been shifted vertically for clarity.

Evidently, one must pay attention to the calibration scheme in the context of the actual targeted analyte species. An ideal calibration approach would be the ensemble-averaging of thousands of laser pulses in an aerosol stream comprised of a high aerosol particle number density of particles that are matched to the particle size of interest. These conditions may be realized by nebulizing aqueous solutions or suspensions of the targeted analyte. For example, the nebulization of aqueous solutions and subsequently drying the droplets in a gaseous flow stream can produce a high-number density of nominally 100-nm aerosols, as documented in the past for a range of analytes [96]. The highly linear analyte response that is often characteristic of the LIBS technique is certainly achievable, but care must be given to provide as accurate a match as possible between the targeted analyte and the calibration source stream.

3.2 Statistical aerosol sampling with LIBS

As outlined in the foregoing sections, the analysis of aerosol particles with LIBS differs significantly from other common analytical techniques due to the discrete nature of the laser-induced plasma, which forms the sample volume. Consider once again the necessary conditions previously enumerated for ensemble-averaging, namely the first condition (a). If the aerosol number density is not sufficient to provide neither a suitable sampling rate nor a suitable analyte signal in the ensemble-averaged spectrum, then additional steps must be taken. Perhaps the most unique aspect of LIBS for analysis of aerosols is that the individual spectrum may contain additional information about the discrete analyte nature of the sample stream. For aerosol analysis, the LIBS technique is well suited to utilize the discrete sampling nature of laser-induced plasmas, thereby enabling optimal sampling strategies and single-particle analysis schemes. An appropriate starting point for further development of LIBS aerosol analysis is the consideration of aerosol sampling rates, which are presented here following the treatment reported by Hahn and coworkers [97, 98].

The overall aerosol particle mass concentration (i.e., particulate mass per unit volume), Y_A, may be given by the expression:

$$Y_A = \rho \frac{\pi}{6} N \left\{ \int_{r=0}^{\infty} r^3 p(r) dr \right\}^{1/3} \tag{1}$$

where ρ is the bulk particle mass density (mass/volume), N is the aerosol number density (particles/volume of gas), and the integral represents the volume mean diameter of the aerosol particle size distribution (volume), which is defined in terms of the normalized aerosol particle size distribution function $p(r)$. Typically, $p(r)$ is parameterized using a log-normal size distribution, which best reflects a wide range of actual aerosols. Note that the assumption of spherical particles is inherent in Eq. (1), although the distribution function may in fact represent the distribution of mass-based *equivalent* spherical diameters.

An additional parameter is needed to define the aerosol sampling problem, namely a measure of the effective sampling volume of the laser-induced plasma. The plasma volume is a complex function of the laser beam geometry, focusing optics, irradiance, and gas stream conditions. Plasma data based on plasma imaging studies, transmission measurements, and sampling considerations have been reported in the literature (see, e.g., Refs. [90, 99–101]). As reported in Carranza and Hahn [101], the actual volume of the laser-induced plasma can be defined in a number of manners. In their work, a physical plasma volume was mapped out using a secondary low-energy probe laser to probe the plasma boundary based on transmissivity (i.e., plasma absorption). The plasma is nearly opaque to incident laser energy in the first few tens of nanoseconds following plasma breakdown [102]. Using this approach, an elliptical plasma shape was defined, enabling calculation of the plasma volume. Note, however, that such an approach is dependent on the definition of the plasma edges, which was defined by a transmission value of 90% in the Carranza study, resulting in a plasma volume of 1.4 mm^3. Alternatively, the plasma volume was calculated in the same study using statistical methods. With that approach, the effective plasma sampling volume was reported as 1.2 mm^3. In general, the plasma volume may be considered on the order of 10^{-3}–10^{-4} cm^3, with variations depending on the laser pulse energy, the configuration of the focusing optics, as well as the sample matrix. Knowing the plasma volume and aerosol size distribution parameters, one can then calculate the average number of aerosol particles expected within a single plasma volume, as well as the overall aerosol particle sampling rates. The product of the plasma volume (V_{plasma}) and the aerosol number density (N) yields the average number of aerosol particles per laser-induced plasma, μ:

$$\mu = N V_{plasma} \tag{2}$$

Eqs. (1), (2) may be combined to define the average number of particles per plasma volume in terms of the aerosol mass concentration, particle size distribution, and the effective plasma volume:

$$\mu = \frac{6 Y_A V_{\text{plasma}}}{\pi \rho \left\{ \int_{r=0}^{\infty} r^3 p(r) dr \right\}^{1/3}} \tag{3}$$

Note that the particle size distribution and resulting integral may be replaced with the cube of the volume mean diameter. For discrete aerosol particles, the probability distribution of the expected number of aerosol particles sampled per plasma volume may be expressed using the Poisson distribution:

$$P_n = \frac{\mu^n}{n!} \exp(-\mu) \tag{4}$$

where P_n defines the probability of finding n discrete aerosol particles within a given plasma volume. The aerosol sampling rate (R_A) may be defined as the percentage of laser-induced plasmas (i.e., laser pulses) expected to sample one or more aerosol particles. The sampling rate is directly calculated as the sum of all probabilities for sampling a finite number of aerosol particles, namely:

$$R_A = 100 \times \sum_{n=1}^{\infty} P_n \tag{5}$$

Noting that the sum of all probabilities must be unity, it is observed that one minus the probability of sampling zero particles (P_o) also yields the sampling rate:

$$R_A = 100 \times (1 - P_o) = 100 \times (1 - e^{-\mu}) \tag{6}$$

To demonstrate these LIBS-based sampling statistics, a set of data recorded in ambient air is analyzed. LIBS was used to monitor ambient air in a manner that was previously reported [103, 104]. Spectral analysis of each LIBS spectrum was performed based on the pronounced 393.4-nm calcium (Ca) atomic emission peak. Ca is a common element in ambient air particulates, originating from many minerals or calcium hydroxide. For one session, LIBS data were collected in 1000-pulse laser sequences for a total of 20,000 laser pulses. During this period, 37 spectra were identified as containing a pronounced Ca atomic emission peak; hence, they were considered to represent the sampling of a calcium-rich aerosol particle. Assuming a plasma sampling volume of 1.2 mm^3, the corresponding number of sampled particles and total volume of air sampled by the plasma (1.2 mm^3/pulse $\times$ 20,000 pulses) yields the measured number density of calcium-rich aerosol particles. This value is calculated as 1541 calcium-rich particles/L of air. Using Eq. (2), the Poisson parameter is readily calculated as $\mu = 0.00185$. This value of μ may then be multiplied by the number of laser pulses per sampling interval, namely 1000 to yield the expected value of 1.85 particle hits per 1000 pulses. This value of 1.85 may then be used with Eq. (4) to predict the sampling distribution of calcium-based particle hits per 1000 pulses. This distribution is plotted in Fig. 4, along with the experimentally measured sampling distribution over the 20, 1000-pulse data collection intervals. The ideal Poisson distribution and the experimental sampling rates are in excellent agreement. Specifically, the probability of recording zero particle hits is equal to 15.7% and 10% for the predicted and experimental data, respectively, while the most probable sampling rate of two particles per sequence corresponds to 29.1% and 35% for the predicted and experimental data, respectively. Overall, the Fig. 4 data demonstrate the statistical nature of LIBS-based aerosol sampling, and provide corroboration of Poisson-based models to describe the sampling problem.

The aerosol particle sampling rate enables an examination of the LIBS-based aerosol analysis problem in the context of discrete aerosol particles and a finite number of discrete plasma sample volumes, thereby elucidating key regimes suited to ensemble averaging or perhaps better suited to more sophisticated data analysis approaches due to aerosol sampling limitations. Low aerosol particle sampling rates for LIBS-based analysis brings with it a potential decrease in method

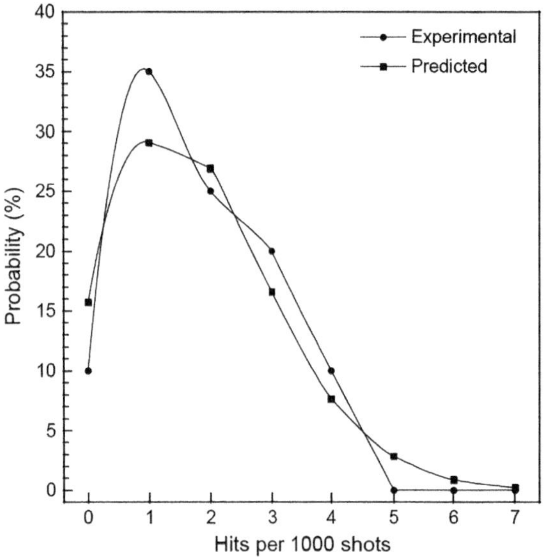

FIG. 4 Particle sampling probability as a function of the expected number of particle hits per 1000 laser shots. Experimental data are presented for sampling of calcium-based aerosol particles in ambient air for 20 consecutive 1000-shot trials, and the corresponding distribution is predicted using Poisson statistics as per Eq. (4).

sensitivity (i.e., atomic emission intensity) for a given analyte species when ensemble averaging all collected spectra. For example, for large numbers of laser pulses (e.g., ≈1000 pulses), the reduction in analyte signal scales directly as the sample rate, as ensemble-averaging reaches an upper limit in the reduction of spectral noise as already discussed. In the following section, LIBS-based sampling strategies and data analysis schemes are formulated in the context of aerosol sampling rates with a goal of maximizing the analyte signal and corresponding detection limits.

3.3 Conditional analysis for spectral processing

The above dataset demonstrates the concept of a limiting particle sampling rate in the context of aerosol analysis. Specifically, for this data, the 37 calcium-rich particle hits were recorded from a total of 20,000 laser pulses, which results in a particle sampling rate of 0.185%, or about one particle hit per 550 laser pulses. Clearly, with a sampling rate much <1%, the large fraction of spectral data with no analyte information (i.e., no Ca atomic emission lines present) will result in a greatly diminished analyte response with an ensemble-averaging approach. This point is demonstrated in Fig. 5, where the ensemble-average of the 20,000 laser pulses is presented, along with the ensemble-average of the identified 37 pulses containing significant Ca emission (i.e., calcium-rich particle hits). In the ensemble-averaged spectrum, the 393.4- and 396.7-nm Ca atomic emission lines are undetectable, while in contrast, these two lines are very pronounced in the spectrum corresponding to the 37 individual calcium-based particle hits. By averaging together so many blank spectra with respect to the Ca emission line, the effect of null data on the

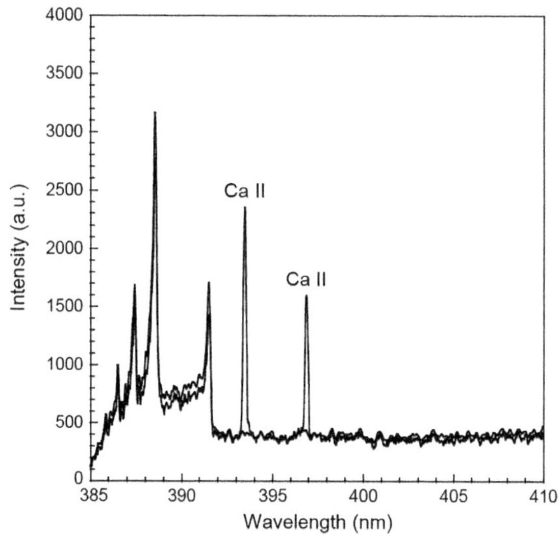

FIG. 5 Ensemble-averaged spectrum of 20,000 laser shots recorded in ambient air, along with the ensemble-average of a subset of 37 spectra corresponding to calcium-rich particle hits, as based on the presence of the 393.4 and 396.7-nm Ca atomic emission lines. The corresponding particles sampling rate is 0.185% [104]. Both spectra have the same scale.

actual analyte signals is readily apparent. Data such as those reported in Fig. 5 suggest the use of a suitable conditional data analysis approach to identify the subset of analyte-rich spectral data corresponding to the presence of particles within given plasma volumes as a means to greatly enhance the overall LIBS signal response. Such an approach was reported specifically for the analysis of aerosol particles based on the conditional analysis of plasma emission spectra on a pulse-to-pulse basis [98], while other researchers have used conditional schemes to increase the analyte response by rejecting both irregular spectra as well as spectra with poor analyte response [105]. Prior to implementation of a conditional data analysis routine for LIBS-based aerosol analysis, several fundamental issues must be addressed: (a) a threshold criteria must be established for processing individual LIBS spectra to determine the presence of an aerosol particle-derived analyte signal, which is used to identify and confirm the sampling of an aerosol particle in a given plasma; (b) the total number of actual particles to be sampled must be established such that a valid statistical sample of the aerosol distribution is achieved; and (c) a means for processing the identified spectrum corresponding to aerosol particle sampling must be identified for use with traditional calibration approaches.

For conditional processing of LIBS spectra, the most fundamental issue is the determination of a suitable means to detect the presence of a targeted analyte-rich aerosol particle in a given laser-induced plasma; that is, identification of a particle hit for a given laser pulse. A suitable threshold criterion should be selected taking into consideration the relatively large spectral noise associated with typical single-pulse LIBS spectra, notably with intensified CCD detectors, as demonstrated in Fig. 1. In addition to noise, the recorded intensity of the plasma emission is generally characterized by considerable variance on a pulse-to-pulse basis, as already discussed, hence, it is desirable to normalize the analyte signal. To obtain optimal performance for any thresholding technique, the most useful metrics are the ratio of the analyte atomic emission peak (either integrated peak or peak intensity) to the intensity of the adjacent continuum emission (P/B ratio), or the ratio of the analyte atomic emission to the noise of the adjacent continuum emission (SNR). With a suitable metric selected, all that remains is to select a threshold above which the metric is considered to correspond to an actual aerosol particle-derived emission signal, and below which the metric is considered to correspond to random spectral noise.

Hahn et al. [97, 98] outlined a basic approach for the selection of thresholds from an overall strategy point of view, while in a more recent work, Carranza et al. [106] explored in detail the various trade-offs between relatively high and low threshold values. The treatment here will follow the approaches put forth in these studies. The most straightforward method of selecting a threshold is to force the spectral processing to be in effect a binary process, which involves the selection of a threshold such that no spectra exceed the value as a result of spectral noise alone. With this scheme, the threshold value is selected such that it exceeds the maximum value of the spectral metric (P/B or SNR) that is obtained with the absence of any aerosol-derived analyte species (i.e., no aerosol particles present). Hence, the threshold value is selected to exclude any fluctuations in the normal spectral noise. This is readily accomplished by increasing the threshold value, with no aerosol particles present, until no spectrum exceeds the threshold for many thousands of laser pulses. The approach dictates that the extreme fluctuations of the spectral metric are insufficient to trigger an analyte hit. For monitoring of many sample streams, it is possible to pass the gaseous stream through a high-efficiency particulate air filter prior to the LIBS sample volume, thereby eliminating the aerosol source for calculating the appropriate threshold. However, if the aerosol source cannot be readily removed from the source stream for this procedure, the threshold value may be obtained as based on an adjacent, surrogate spectral region of comparable intensity that is free from any analyte atomic emission lines. Generally, the continuum emission signal is sufficiently smooth such that a suitable surrogate region is available within a few nanometers of the targeted analyte atomic emission line.

Carranza et al. provide a more detailed discussion of threshold selection in the context of the SNR of the resulting ensemble-average of spectra corresponding to selected particle hits [106]. Clearly, one must consider that there exists a region where the resulting atomic emission peak from aerosol-derived analytes will fall below the noise-free threshold related earlier, but above the method detection limit. This is because the extreme spectral noise, although infrequent at a specific spectral location, results in a necessarily large threshold value if all such noise events are to be rejected. To avoid such a strict spectral filter, the threshold value may be relaxed to yield a relatively low false hit rate, on the order of 0.05%–0.1% of total pulses, but not exactly zero. This method enhances the sensitivity of the conditional analysis method and extends detection to smaller analyte-containing particles; however, one must carefully optimize this procedure, as false analyte signals are being introduced into the spectral dataset. Toward this end, Fig. 6 presents analysis of the analyte P/B ratio and the expected contribution of noise to the analyte peak signal (i.e., false hits) as a function of the statistical threshold based on the 288.1-nm silicon emission line resulting from the detection of 2-μm sized silica spheres, as adapted from reference [106]. The figure demonstrates that an optimal value exists for a false hit rate of about 0.05%–0.1%, as reflected in the threshold of the P/B and expected noise contribution curves. A few false hits (≈1 per 1000 pulses) are tolerable, in that contribution to the genuine particle-hit data is negligible,

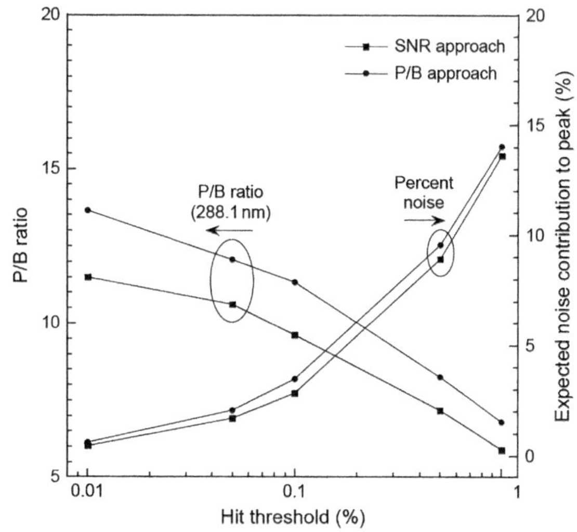

FIG. 6 Influence of the conditional processing threshold (as expected false hit rate) on the average P/B intensity of the silica emission line for identified hits, along with the expected noise as a weighted percentage of the on-peak signal [106]. The conditional processing rates are performed using the Si emission line as a signal-to-noise ratio (SNR approach) and as a peak-to-base ratio (P/B approach).

while the extension of the detection limits into the extreme envelope of statistical noise is beneficial.

While the above comments provide additional insight into the selection of appropriate thresholds for conditional spectral analysis, attention must always be paid to the issue of false hits when only a single analyte emission line is utilized. The difficulties of spectral thresholding are significantly reduced if additional analyte atomic emission lines are available in a given spectral window. If two or more emission lines are present on the recorded spectra, the additional atomic emission lines should also be used for quantitative analysis. Specifically, because spectral noise is considered white noise, it is improbable that the extreme noise fluctuations that trigger false hits will appear simultaneously on multiple atomic emission lines. Therefore a useful approach for conditional processing is to relax the threshold criteria to something on the order of 1%–2% false rates, which further lowers the overall detection limit with respect to identifying individual aerosol particles. Then, all candidate spectra (i.e., those that exceed the threshold based on a single atomic emission line) are checked for atomic emission above the noise threshold on additional emission lines. This approach was first detailed by Hahn and Lunden [97], and then utilized in a number of aerosol studies [103, 107]. An example of such an approach is presented in Fig. 7, which shows two single-pulse spectra recorded in a sample stream containing a dilute concentration of *Bacillus* spores. Conditional-based sampling was performed using the 396.9-nm Ca II atomic emission line, as reported by Dixon and Hahn [107]. In Fig. 7, both spectra exceeded the sampling threshold based on the single 369.9-nm Ca emission line; however, only the upper spectrum would be confirmed to correspond to an actual *Bacillus* spore hit, based on the correctly proportioned presence of the 393.4-nm Ca II emission line. This additional emission line is absent in the lower spectrum, hence this spectrum would generally be dismissed as a noise event, rather than an actual sampled spore. Together, the two spectra illustrate the issues associated with conditional spectral processing of LIBS data, including the corresponding single-pulse spectral noise and statistical fluctuations, as well as the excellent analyte sensitivity that may be realized with proper detection algorithms.

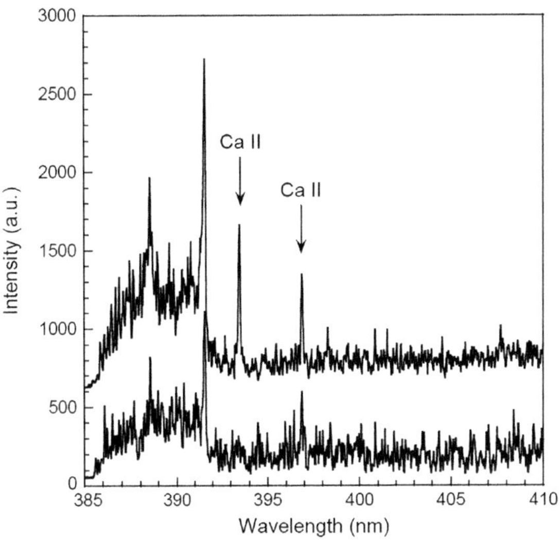

FIG. 7 Representative single-shot spectra corresponding to laser-induced breakdown spectroscopy-based measurements in an aerosol stream seeded with *Bacillus atrophaeous* spores. The upper spectrum corresponds to an individual *B. atrophaeous* spore, while the lower spectrum corresponds to spectral noise only, as triggered on the 396.9-nm Ca emission line using a conditional analysis routine. Both spectra have the same intensity scale, with the upper spectrum shifted vertically for clarity.

The above treatment of conditional analysis processing schemes provides insight into segregating LIBS spectral data into those spectra that contain significant analyte emission signals corresponding to a targeted analyte (presumably derived from analyte-rich particles), and those spectra that contain no detectable analyte emission. The identified analyte-containing spectra, which may be referred to as particle hits, may then be processed together or treated individually. The spectra corresponding to identified particle hits may then be ensemble-averaged, yielding an analyte emission-rich spectrum that enjoys the benefits of preconcentration with respect to the resulting analyte atomic emission lines in combination with the reduction in signal noise always realized with ensemble-averaging. An example of this outcome is presented in Fig. 5, as discussed earlier, in which the corresponding particle sampling rate (i.e., hit rate) was 0.185%, or about two particles per 1000 pulses. The rejection of 99.8% of the spectral data, namely the null data with respect to the targeted analyte Ca, results in a nearly 500-fold increase in the analyte emission peak intensity as compared to ensemble-averaging of all collected spectra. The RMS of the continuum emission intensity, a direct measure of signal noise, in the spectral regions adjacent to the Ca peaks increased by only 15% when comparing the

37-pulse ensemble-average to the average of all 20,000 spectra. Taking all factors into account, the Ca emission signal, as measured by the SNR, was enhanced by a factor of 470 with the conditional analysis routine, enabling excellent analyte sensitivity. In contrast, the corresponding 20,000-pulse ensemble-averaged spectrum is characterized by essentially a nondetect with respect to the two Ca emission lines.

Using conditional analysis, the actual concentration (mol/volume) of a given analyte species is determined using a linear combination of the analyte signal as based on the subset of the conditionally analyzed spectra and the corresponding aerosol sampling rate (i.e., hit rate). The ensemble-averaged spectrum corresponding to the selected analyte hits should be used with a traditional LIBS calibration curve relating the LIBS emission signal (e.g., P/B) to a known analyte concentration range. As related earlier, care must be taken to carefully match the analyte source in the calibration stream to the actual analyte source in the stream of interest, with specific attention given to the issue of gaseous-phase and particulate-phase analyte. Once the equivalent concentration of the hits spectrum is known, X_{hits}, this value is then adjusted by the sampling rate to yield the actual analyte concentration, X_{true}, as reflected by the relationship:

$$X_{true} = (X_{hits}) \times (\text{total hits}/\text{total pulses}), \quad (7)$$

where the latter term corresponds to the particle sampling rate. It is noted that as the sampling rate approaches 100%, the conditional analysis scheme converges to a classic LIBS methodology, where the ensemble-average of all pulses is directly related to the corresponding calibration curve. If two analyte emission peaks are used as discussed earlier (one for triggering and one for analysis), the effect of any false analyte hits is minimal because a false hit reduces the intensity of the *analysis* emission peak, but simultaneously increases the sampling rate, thereby producing the correct actual analyte concentration based on the foregoing relation.

A final issue concerns the statistical nature of the particle sampling rate, namely the number of hits divided by the total number of laser pulses. Because the sample rate can range considerably over limited sample periods, care must be taken to accurately quantify this parameter. The total number of particle hits necessary to adequately assess the overall mass concentration of aerosol particles follows the same guidelines as discussed earlier regarding ensemble-averaging, namely that on the order of 20 aerosol particles are sufficient to characterize the average aerosol mass for rather broad distributions of submicrometer to micron-sized aerosol distributions [98]. For a given average aerosol particle sampling rate, the total number of laser pulses should then be selected to ensure at least 20–30 aerosol particle hits are recorded. Hahn et al. explored this topic in detail, and noted that better accuracy is obtained if the conditional processing algorithm is terminated after a preselected total number of pulses rather than a preselected number of particle hits [98].

Although the conditional analysis methodology has been mostly reported for aerosol analysis, Diaz et al. reported on the use of this technique for the analysis of gold-bearing ores [108]. On the assumption of a low analyte concentration and discrete analyte distribution, a linear calibration curve for Au was obtained when analyzing ores prepared as pressed pellets. The effect of the number of laser pulses and beam irradiance was also addressed, demonstrating that conditional analysis may also work for solid samples containing micro- and nanoparticles [108, 109].

3.4 Analysis of individual aerosol particles

The foregoing treatment of LIBS spectra via conditional processing schemes is all based on the concept of discrete particles interacting with a specific plasma volume in a way such that characteristic spectral features (i.e., atomic emission lines) make their identification possible. Once individual spectra are identified as corresponding to an aerosol particle sampling event, they may be grouped together via ensemble-averaging to enhance the sensitivity for detection of aerosol-derived species, as already related. Alternatively, each individual spectrum may be analyzed on its own, thereby revealing information about individual aerosol particles on a laser pulse-to-pulse basis. Such an approach takes advantage of the single-point sampling nature of LIBS, which is unavailable with continuous analysis systems such as ICP.

The LIBS-based quantitative analysis of individual particles was first reported by Hahn [106], where a calibration scheme was proposed making use of microspheres with a predetermined concentration of analyte, in that case iron. As implemented, the methodology enabled determination of the absolute mass of the targeted analyte per particle, as determined from a single-pulse LIBS spectrum under dilute aerosol sampling conditions. In a follow-up paper, Hahn and Lunden [97] presented a formal treatment of the statistical sampling problem combined with the concept of an analyte signal and the associated method calibration curve response. Following their analysis, in combination with the concepts related earlier on the use of conditional analysis for determination of total analyte concentrations, the following relation may be developed:

$$X_{\text{hits}} R_A = f \rho \frac{\pi}{6} r^3 N \tag{8}$$

In this equation, the left side is an expression for the true analyte concentration as presented earlier in Eq. (7), where R_A is the sampling frequency equal to the ratio of the number of particle hits to the total number of laser pulses, and X_{hits} is the equivalent analyte mass concentration based on the spectrum of hits and the calibration curve. The right side of Eq. (8) is an alternative expression for the calculated analyte mass concentration (see Eq. 1) given by the bulk particle density ρ, the volume mean particle size r, the analyte mass fraction of the aerosol f, and the aerosol number density N. As related by Hahn and Lunden, when the aerosol sampling rate R_A is small (e.g., $\approx$1%–10% or less), the sampling rate may be approximated based on Poisson statistics as the product of the aerosol number density N and the effective plasma sampling volume V_{plasma} [97]. Using the above relations, the following expression is derived:

$$X_{\text{hit}} = \text{analyte mass/plasma volume} \tag{9}$$

which presents a direct relation between the equivalent mass concentration based on a calibrated LIBS analyte response, the discrete plasma volume, and the *absolute* analyte mass. If the plasma volume is known, Eq. (9) may be used to calculate the absolute analyte mass corresponding to a single aerosol particle as the simple product of the characteristic plasma volume and the equivalent mass concentration (i.e., calibrated response). The latter quantity is readily calculated from the corresponding single-pulse spectrum and the corresponding LIBS analyte calibration curve. One may readily calculate an equivalent spherical diameter of the individual particle present in the plasma volume based on the calculated analyte mass via the relation:

$$r_{eq} = \left(\frac{6\, X_{\text{hit}} V_{\text{plasma}}}{\pi \rho f} \right)^{1/3} \tag{10}$$

where ρ is the particle bulk density (mass/volume) and f is the mass fraction of the analyte with respect to the overall bulk particle mass. For a pure, homogeneous particle, f equals unity and the density is the actual density of the analyte.

Single-pulse LIBS-based spectral analysis can yield the absolute analyte mass and the corresponding equivalent spherical diameter using the relations developed earlier based on two parameters, the equivalent concentration (i.e., atomic emission response) and the characteristic plasma volume. A calibration approach was outlined by Hahn [110] using microspheres of known mass composition and size. To gain further insight into the issues of plasma volume, Carranza and Hahn investigated plasma size under laser-induced plasma conditions frequently used for aerosol- and gas-phase analysis [101]. In their study, three distinct plasma volume measurements were made, including a physical volume based on transmission measurements using a spatially resolved probe laser as already described, an emission-based diameter similar to the approach of Hahn [110] in which the emission response and a known mass are used to solve directly for plasma volume via Eq. (9), and a statistical plasma volume that was recorded based on aerosol sampling rates modeled with Poisson sampling statistics. The resulting plasma volumes varied from 1.2 to 2.4 mm^3, with the former corresponding to the statistical sampling volume and the latter corresponding to the emission-based volume.

The second parameter necessary for single-pulse mass measurements is the equivalent concentration, which as discussed above, is readily calculated directly from the spectral data (e.g., P/B ratio) and the corresponding calibration curve. The issue for concern arises from the nature of the calibration source stream. Recall the above discussion regarding the effects of analyte phase on the analyte atomic emission response, with particulate species yielding as much as an eightfold increase in analyte signal as compared to identical mole fractions of analyte in the gas phase [95]. Hence, to achieve the most accuracy in the calibration methodology, one should make use of a source stream that closely matches the range of aerosol data expected in the unknown sample stream.

Fig. 7 provides an illustrative spectrum generated from conditional-based processing of LIBS for the identification of single-pulse spectra corresponding to individual aerosol particles. The absolute calcium mass corresponding to the Fig. 7 data is about 3 fg based on the Eq. (9) analysis, noting that the calibration curves used were generated from a particulate-phase calcium calibration source. For the nominally 1-μm sized spores used to generate the spectral data, ≈ 3 fg calcium mass corresponds to a nominal calcium mass concentration of about 0.5% in the original spores. Such data demonstrate the overall sensitivity of the LIBS technique for single-particle analysis, as observed by the relative strength of the atomic emission line corresponding to only a few fg of analyte.

Finally, a few additional comments are offered on the overall sensitivity and precision of single-pulse LIBS-based aerosol analysis, including consideration of key parameters such as laser energy, laser beam stability, and the presence of aerosol particles themselves. Carranza and Hahn explored the role of laser pulse energy on the resulting precision of gas-phase analyte signals, where precision was quantified by the RSD of the pulse-to-pulse measurements of absolute analyte atomic emission peaks and normalized P/B ratios [86]. The RSD of the carbon atomic emission signal, the gas-phase analyte used in their study, was found to decrease with increased laser pulse energy up to a certain level, and then plateau for additional pulse energies. Interestingly, the threshold in the RSD versus laser pulse energy plot corresponded exactly to the saturation value for laser pulse energy absorption by the plasma. A significant conclusion was that single-pulse LIBS-based measurements should be made with sufficient laser pulse energy to achieve saturation with respect to absorbed laser pulse energy, as well as made with suitable collection geometry (e.g., backscatter mode) to minimize spatial variability. Another study examined the role of laser beam stability and the presence of aerosol particles on the initiation of laser-induced breakdown, and on the overall analyte sensitivity as measured for three different gas-phase species (hydrogen, nitrogen, and carbon) [87]. Improvements in the temporal stability of the breakdown ignition volume were observed with laser cavity seeding. However, laser cavity seeding produced no significant improvement in analyte precision for the range of gas-phase atomic emission lines. In contrast, greater pulse-to-pulse analyte precision, as measured by a nearly 60% reduction in RSD, was realized with the elimination of concomitant particles from the analyte sample stream. Clearly, one must live with the presence of particles for LIBS-based analysis of aerosols; however, this study provides insight into the expected precision and accuracy to be expected for the interactions of ambient aerosol particles and the laser-induced breakdown process.

4 Alternative methodologies for aerosol analysis

Much of the previous discussion is focused on the statistical nature of discrete particles and how aerosol sampling couples to the finite laser-induced plasma volume. While successful strategies for direct aerosol analysis were enumerated, including single-particle analysis, an alternative approach is the collection (i.e., concentration) of particles on an appropriate filter medium prior to LIBS analysis. Perhaps the greatest advantage of filter collection and subsequent LIBS analysis is the potential to enhance detection limits. With automated filter handling, including continuous spools of filter material, this methodology can be implemented online and in near real time. An additional benefit stems from the minimally destructive nature of LIBS, which enables the filter media to be archived for future analysis, including conventional benchtop chemistry such as digestion and ICP-MS analysis. The primary limitation of filter collection is linked to the very nature of particle concentration, namely the loss of information regarding individual particles. By collection of particles on a filter substrate, individual laser-induced plasmas may sample tens to hundreds of particles, with the result being a spectral signature representative of the entire particle ensemble. As such, this approach is naturally suited to emission monitoring, notably for toxic metals, where regulations are typically set for total elemental concentrations, which is consistent with this LIBS-based approach of filter collection and analysis. It is not surprising therefore that many of the published studies have been applied to environmental monitoring.

Beryllium is an excellent analyte for this methodology, as its high toxicity makes it of interest in emissions monitoring and industrial hygiene. Furthermore, Be enjoys excellent detection limits in atomic emission schemes, including LIBS, while it is undetectable with alternative schemes such as X-ray fluorescence spectroscopy. Cremers and coworkers at Los Alamos National Laboratory successfully used the filter collection approach for analysis of Be and Ta aerosols [89, 111]. Detection limits were below 1 ng cm^{-2} for Be and increased to 40 ng cm^{-2} for Ta. As already discussed, one issue with laser-induced plasma analysis is the complete vaporization of large diameter particles, namely those in the micron size range. In their studies, Cremers et al. did report some signal saturation at high particle loadings, which was attributed to particle size effects. Due to the large particle diameters (micrometer size range) investigated and the corresponding high filter loadings, a particle size dependence of the signal was observed due to saturation. Panne et al. used the filter collection approach for emission monitoring and ambient aerosol analysis, successfully extending the method to ultrafine aerosol particles [112–114]. For the waste incineration facilities, analyte volume concentrations between 0.1 and 5 μg m^{-3} were detected with this LIBS scheme [114]; however, the authors noted that problems may exist with independent reference analysis at such low concentrations. Rather surprisingly, because the laser-induced plasma samples only a certain depth of the filter (i.e., the surface layer that is enriched with deposited aerosols), matrix effects may actually be much more pronounced with traditional reference analysis that digests the entire filter substrate. As with any filtration method, filter selection must take into consideration temperature resistance for emissions monitoring applications, sampling efficiency (notably size dependency), and blank values for a range of analyte species. More recent studies have evaluated the use of filter collection for analysis of airborne silica dust simulating coal mine environments [115], Cu particles [94] that simulate air pollution from industrial sources, Cu aerosols from a pilot-scale melting furnace [116], and in-field detection of C, Ca, K, Mg, and Na from Arctic ambient aerosols [117]. Gallou et al. reported on a comparison between direct and indirect (i.e., filter collection) analysis of micrometer size Cu particles [94]. The authors reported Cu limits of detection of 15 and 60 $\mu g\ m^{-3}$ for direct and indirect sampling, respectively. This difference can be partially attributed to the relatively large filter area (sampling diameter of 36 mm), low sampling flow rate (5 lpm) in the indirect LIBS analysis, as well as the increased sensitivity produced by the use of a conditional analysis routine in the case of direct particle sampling. An upper particle size limit for complete vaporization of 7 μm for Cu particles was also reported [94]. In a posterior paper by the same French group, Dutouquet et al. performed on-site direct and indirect LIBS measurements of Cu aerosols in an exhaust duct from a pilot-scale furnace [116]. Due to high matrix effects produced by the collection filters, two calibration curves were necessary to fit the Cu concentrations range. However, calibration curves produced in the laboratory through direct LIBS analysis superimposed to each other quite well, and no matrix effects were reported.

5 Applications of LIBS-based aerosol analysis

A number of reviews have been published regarding the LIBS technique, including applications to aerosol analysis [118–124]. Belaev and coworkers are perhaps the first recognized group to suggest the use of laser-induced gas breakdown for chemical analysis of aerosols [125]. However, the modern foundation of LIBS as an analytical technique for aerosol analysis was established in a series of papers by Radziemski and Cremers in the early 1980s [83–85, 90, 91]. These papers systematically explored the issues of detection, including temporal gating (i.e., time-resolved analysis) to optimize the analyte atomic emission signals with respect to continuum emission, plasma formation and growth, and the issues of local thermodynamic equilibrium. In their landmark 1983 paper, they presented detection limits for Cd-, Pb-, and Zn-based aerosols on the order of 200 $\mu g\ m^{-3}$ for particles in the micron size range [90], while also exploring the issues of analyte source and plasma temperature. In additional communications, Cremers and Radziemski reported detection limits for Be aerosols below 1 $\mu g\ m^{-3}$. In all of this early work, ensemble-averaging schemes were used to reduce spectral noise, while aerosol source streams were generally of sufficient number density such that sampling limitations were of no concern. Following the pioneering work of the Los Alamos group, a wide range of studies and applications has been reported in the literature regarding LIBS-based aerosol analysis. What follows is not a thorough review of the published body of related papers, but rather an overview of select papers that focus on either unique applications or elucidate fundamental issues associated with this topic.

As a starting place for aerosol analysis, one may first consider breakdown of the gas-phase matrix. Research on the wavelength dependence of gas breakdown and the ensuing plasma characteristics [39, 126–131], the effects of He addition to the gaseous analyte sample stream as a means to improve the LIBS signal response [132], and temporal evolution of plasma formation at the earliest times of inception [133] has been carried out. For calibration accuracy, it is desirable to have as much independence of the analyte emission on the overall gas composition. Yalcin et al. [99, 134] investigated the temporally and spatially resolved plasma temperature and electron density for several gaseous species at ambient pressure, including N_2, He, and SF_6. Consistent values were found for electron densities and plasma temperatures with changing gas species, laser energy, particulate levels, and humidity levels. This important work has been widely cited in support of overall plasma invariance with respect to changing ambient conditions. While the effect of background matrices is obviously of importance for aerosol analysis, the issue of calibration (i.e., emission response of analyte) is the key step for any practical application. Essien and coworkers produced a submicron-sized aerosol for LIBS calibration [91]. Generally, linear calibration curves were generated for Cd, Pb, and Zn, although various degrees of saturation were observed at higher concentrations, which was attributed to incomplete vaporization of particles, as already discussed. An important finding was their general agreement (within 10%) of Pb atomic emission signals when nebulizing either lead acetate, lead chloride, or lead nitrate, thereby showing independence of the analyte source. However, Cd revealed a 27% difference in analyte response when comparing nebulized solutions of cadmium nitrate and cadmium chloride, although detailed size measurements of the calibration aerosol were not reported. In analogous experiments, the relative independence of analyte signals on molecular source for purely gas-phase species was reported in several studies [135, 136]. Specifically, Dudragne et al. demonstrated that analyte signals for C, Cl, F, and S scaled with the number of analyte atoms in the constituent molecules for a wide range of compounds, concluding that the parent molecules were fully dissociated in the laser-induced plasma of their study [135]. Similarly, Tran et al. verified that SF_6 and HF yielded identical fluorine atomic emission signals when the gas composition of different mixtures was adjusted to the same atomic fluorine mole fraction [136].

The previous studies lead to two general conclusions, namely that the laser-induced plasma, as measured by temperature and electron density, is remarkably robust with respect to changes in gas-phase composition, and that the resulting analyte emission is generally independent of the analyte source. However, a very important point must be made with regard to the latter comment, namely analyte independence has been observed when the various analyte sources are all in a similar state (i.e., all gas phase or all particulate phase with a similar particle size distribution). As already discussed, the effects of analyte phase on the calibration response for a typical LIBS system were investigated for a range of carbon species [95]. Significant differences in the atomic emission signals from carbon were observed when comparing calibrations of gas-phase and submicron-sized solid-phase carbon species. Consistent with the above comments, the plasma electron density and temperature remained essentially constant. Such findings challenge a widely held assumption that complete dissociation of constituent species within a highly energetic laser-induced plasma results in independence of the analyte atomic emission signal on the analyte source. The authors proposed a physical model of the plasma-analyte interaction to account for the observed dependence on the physical state of the analyte by considering the rapidly expanding plasma as a shock wave-like phenomenon, in which the pressure and electron wave rapidly expand from the original plasma kernel. As the plasma wave expands, molecular and particulate species are pushed toward the edge of the plasma volume; however, due to the many orders of magnitude difference in the mass of gas-phase species (molecules) and solid-phase species (particulates), it was proposed that the efficiency at which a given species is carried by the plasma wave will scale inversely with particle mass. The authors referred to this behavior as an effective analyte slip factor as the plasma wave expands. An important point from such studies is the need to produce calibration schemes that reflect as much as possible the physical state of the analyte species of interest. Overall, more research is necessary to fully understand the exact nature of the plasma-particle interactions associated with LIBS-based aerosol analysis. Nonetheless, a number of practical applications have been reported.

The analysis of toxic metals, such as As, Be, Cd, Cr, Pb, and Hg, has received considerable attention over the years, most notably in the context of a LIBS-based real-time emissions and process monitor. In the early 1990s, a LIBS-based monitoring effort was initiated at Sandia National Laboratories, with early trials reported for the detection of a chromium-rich aerosol [137, 138]. In a contemporary effort similar to the Sandia work, Singh et al. from the DIAL (now ICET) laboratory at Mississippi State University initiated a LIBS program geared toward process analysis and emissions monitoring [139–141]. They successfully studied different optical arrangements at a coal-fired flow facility and achieved detection limits for different heavy metals between 1 and 600 $\mu g\ m^{-3}$. In fact, both the DIAL and Sandia LIBS teams participated in a jointly sponsored Department of Energy/Environmental Protection Agency (DOE/EPA) demonstration test of continuous emission monitoring technologies conducted in September 1997 [140, 142]. The demonstrations were conducted using a pilot-scale rotary kiln incinerator that was seeded with the toxic metals As, Be, Cd, Cr, Pb, Hg, and Sb at concentrations between 15 and 100 $\mu g\ m^{-3}$, which were consistent with the US maximum achievable control technology rules. Although the two employed LIBS systems demonstrated a fast response time, their overall precision and accuracy (between 50% and 150%) did not meet the required 20% relative accuracy mandated by US regulatory agencies. Much of the Sandia LIBS-based emissions monitoring was summarized in Buckley et al.'s paper, where conditional analysis as described earlier was found to provide considerably better results in terms of detection limits and accuracy, including excellent agreement with independent extractive sampling [143]. LIBS remains a viable technology for real-time emissions and process monitoring, with several research groups still pursuing associated issues. However, to date, the widespread implementation of real-time monitoring technologies has not been widely mandated for toxic metal species, hence this potential LIBS application remains a work in progress.

Online LIBS air monitoring schemes have continued to improve. For example, Neuhauser et al. reported the successful detection of Cr-rich aerosol using LIBS for fast monitoring of particulate emissions in an electroplating facility [144]. Their prototype LIBS system was tested in cooperation with an independent laboratory that provided corroborating data as to the total chromium content. The LIBS system provided both the necessary time-resolution and detection limits (14 $\mu g\ m^{-3}$) for emission monitoring. As mentioned previously, Gallou et al. [94] and Dutouquet et al. [116] compared direct and indirect LIBS analysis of Cu aerosols. On-site measurements in a copper-melting furnace produced calibration curves that matched quite well those produced in laboratory setups. The better performance of direct analysis was attributed to the lack of observable matrix effects. Kim et al. studied Arctic ambient aerosols through filter collection during a sampling period of 8 days (September 2012). The detection of Ca, K, Mg, and Na suggested that minerals and sea salts are the major constituents of Arctic aerosols at the sampling location. A comparison with LIBS analysis from soils in nearby areas suggested that the effect of local soils on the elemental distribution in aerosols is small [117].

Due to detection limits on the order of $\mu g\ m^{-3}$, lower with conditional analysis, not many studies were focused on atmospheric aerosols [145]. Carranza et al. measured ambient concentrations of Al, Ca, Mg, and Na as low as 5 ng m^{-3} using conditional signal processing, and were able to detect transient changes in magnesium and aluminum-based aerosols due to the discharge of fireworks [103]. Nunez et al. studied the detection of sulfuric acid, which is relevant for several atmospheric processes [146]. A direct measurement of sulfur via the S I 182.034 nm emission line yielded a detection limit of 165 ppbv after a 15-min integration time. However, laser photofragmentation (LPF) after interaction of NaCl and H_2SO_4 gave improved detection limits of 46.5 ppbV in only 10 s. Lithgow et al. reported on a LIBS-based ambient air measurement campaign as part of the Pittsburgh Aerosol Supersite study [147]. They used single-pulse conditional analysis similar to the methodology developed by Hahn et al. [98], and reported aerosol measurements that contained the elements Al, Ca, Cr, Cu, Mg, Mn, and Na. A significant difference between their measurements and the Carranza et al. [103] study was the overall particle hit rates, with Lithgow et al. reporting hit rates that were generally one to two orders of magnitudes less. Differences can be attributed to actual differences in ambient air loadings of the targeted particles, or possibly to additional combinations of reduced pulse energy (40 mJ vs. 375 mJ) resulting in a smaller plasma sampling volume, orthogonal plasma emission collection versus back-collection, or differences in sampling transport. Hettinger et al. reported a direct comparison of LIBS-based aerosol particle sampling rates (calcium-based and sodium-based particles) with independent light scattering-based particle sampling rates [104]. A typical comparison is presented in Fig. 8, where the correlation between the LIBS-based sampling and the light-scattering sampling (particles between 500 nm and 2.5 μm) is very strong. These studies support the use of LIBS for real-time measurement of ambient air particulate matter, although outstanding issues include overall limits of detection for select species, calibration and particle size effects, and maximization of particle sampling rates. More recent research on the use of LIBS for aerosol analysis has been dedicated to the real-time detection of airborne interfering particles (i.e., $CaCl_2$, $MgCl_2$, NaCl, and KCl) in semiconductor manufacturing [148], Pb aerosols simulating atmospheric pollution [149], and monitoring of aqueous aerosols in benchtop [150, 151] and standoff LIBS configurations [152, 153].

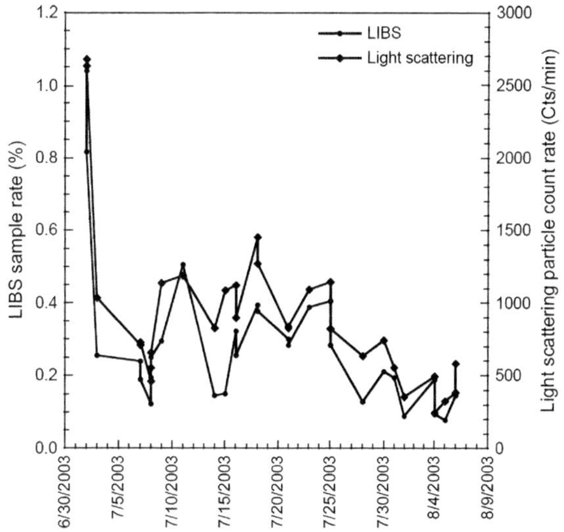

FIG. 8 LIBS calcium-based particle sampling rate (calcium-containing spectra/total spectra) and the light scattering-based total aerosol sampling rate for the cumulative size range between 500 nm and 2.5 μm [104]. The data are presented for thirty, 1-h sampling intervals as measured for ambient air sampled through a PM2.5 inlet.

In a complementary scheme, LPF and subsequent optical detection of atomic emission can be in some cases a valuable alternative to LIBS, especially for alkali species, which are actively involved in the breakdown, corrosion, and erosion of power plant materials [154–156]. Similar observations were reported for the detection of Hg by Tong et al. [157, 158]. The metalloids as well as the halides are problematic elements for current LIBS systems due to the low sensitivity of the detection systems and the low emission strength in the near infrared. Consequently, a buffer gas (e.g., He) can be employed to increase the emissivity for the near-infrared lines. In this way, Tran et al. demonstrated the detection of gaseous and particulate fluorides [159]. For particulate fluorides in air, limits of detection were 9 mg m^{-3} and for detection in He 0.5 mg m^{-3}. In addition, significant improvements were reported for filter-sampled aerosols, which allowed for a 5-μg m^{-3} detection limit in He after a 10-min sampling time at 10 lpm.

Another extension of the direct analysis approach is the use of laser-induced plasma only as an atomization reservoir, with a subsequent single element analysis by excitation of atomic fluorescence with a second laser (laser-enhanced atomic fluorescence). Neuhauser et al. demonstrated for ultrafine lead aerosol particles produced with a differential mobility analyzer a detection limit in the ng m^{-3} range [160]. However, the detection limit increased from 55 ng m^{-3} for a particle diameter of 48 nm to 130 ng m^{-3} for a particle diameter of 300 nm. The increasing detection limit with increasing particle diameter was probably due to the incomplete atomization of larger particles in the colder periphery of the plasma. A suitable way for an improved LIBS analysis of some of the relevant heavy metal species (e.g., Sn, Hg, and As) is the formation of volatile hydrides [139, 161], respectively, and in the case of mercury the direct detection of mainly gaseous species [162]. In all cases, detection limits on the order of 50 μg m^{-3} were observed.

6 Future directions

Possibilities for improvement of the direct analysis approach are either technology driven or in the methodology. The advent of echelle spectrographs extends the possibilities to cope with spectral interferences and improve the approach to calibration. Especially, when calibration strategies without any a priori information on the sample are employed, the simultaneous detection of all major constituents is an essential prerequisite [163, 164]. The combination of a kHz-diode laser-pumped Nd:YAG and fast readout ICCD cameras could improve the sampling efficiency and hence the overall analytical performance of the direct approach to LIBS aerosol analysis and conditional sampling by orders of magnitude. The approach described by Noll et al. for LIBS microanalysis with a kHz sampling rate, that is, a multiplexed Rowland spectrometer with photomultiplier, might be a suitable alternative to improve the analytical figures of merit [165]. The use of high repetition rate lasers for solid analysis has also been described by Cristoforetti [166] and Pořízka [167]. The extension of spectrometer, especially echelle, technology to the visible UV range will permit the analysis of nonmetals such as S, P, Cl, or Br, which are of considerable relevance for both process analysis and atmospheric aerosols [168, 169]. The direct approach can also benefit from aerodynamic lens systems that not only allow the focusing of an aerosol stream for a short distance [170–178], and hence a considerable enrichment, but also a simultaneous sizing of ultrafine aerosols. New spectral data processing for direct aerosol analysis has also been proposed. Álvarez-Trujillo et al. [179] developed a spectra processing methodology for aerosol analysis based on the standard deviation of the collected spectra. Experiments and Monte Carlo simulations showed that this methodology can produce a monotonic analyte response with analyte concentrations up to a point at which the analyte signal starts to transition from a nonhomogeneous to a homogeneous system.

Another interesting option to reduce possible matrix effects is online condensation of a matrix on the aerosols. This could reduce the variability of the plasma ignition and subsequent elemental emissions [180, 181].

Necessary detection limits on the order of $1\ \mu g\ m^{-3}$ and lower make future applications to atmospheric sciences challenging. Process control is expected to be a significant LIBS application in the future. Due to the improved emission control measures for all types of industrial emission, the requirements for the detection limits will, however, be tightened in the future and the emphasis will shift to ultrafine particles. In that case, conditional analysis will be an indispensable condition to meet user requirements. The DOE/EPA test of 1997 showed that aerosol LIBS has similar problems as other LIBS applications in terms of accuracy and precision, so there is certainly room for considerable improvement in the general LIBS methodology. The filter-based approach offers some advantages for long-term monitoring of atmospheric aerosols, especially from remote or rural areas. Automatic filter band sampling could be easily combined with other aerosol characterization methods such as optical absorption for black carbon, Raman, and/or fluorescence spectroscopy. The latter could also be combined with the direct approach utilizing the fourth harmonic of a Nd:YAG laser for both plasma ignition and fluorescence spectroscopy, for example, for bioaerosols. In this context, echelle systems, which permit a combination of different techniques, that is, Raman, LIBS, or molecular fluorescence in a single instrument, could provide a decisive advantage.

Alternative methodologies for aerosol particle concentration have also been studied. Diwakar et al. reported on the preconcentration and posterior LIBS analysis of fg to ng particles directly from the sampled aerosol stream through electrostatic deposition of charged particles onto a flat-tip microneedle electrode [182, 183]. Absolute mass detection limits from 0.018 to 5 ng were obtained for analytes such as Cd, Cr, Cu, Mn, Na, and Ti. Whereas Laserna's group reported on the use of optical catapulting and trapping for the generation and analysis of aerosol particles [184–186].

As discussed in other chapters in this book, the use of double-pulse or dual-pulse LIBS has proved a successful configuration for enhancing the analyte response with analysis of solid- or liquid-phase systems, but only a few studies have been reported for aerosol analysis. Windom et al. reported on the effects of an orthogonal double-pulse laser configuration on the atomic emission response for purified air seeded with calcium-rich particles [187], which revealed a marked improvement in calcium atomic emission P/B ratio (≈twofold increase) and SNR (≈fourfold increase) with the double-pulse configuration. Representative spectra as recorded with a single laser and a double-pulse system are shown in Fig. 9, which shows the increased particle-phase analyte response, as measured from the calcium emission stemming from calcium-rich aerosol particles. In addition

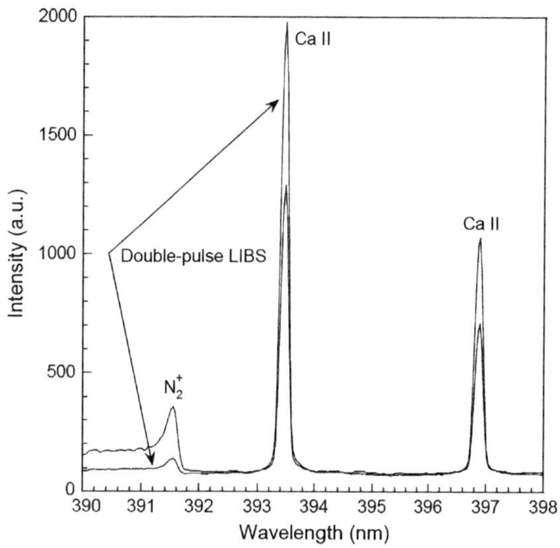

FIG. 9 Spectra corresponding to the single-pulse LIBS configuration and to the double-pulse LIBS configurations with a 750-ns delay. Both measurements were performed in an aerosol flow of submicron-sized calcium-rich particles. The Ca II atomic emission lines at 393.4 and 396.9 nm are enhanced with double-pulse LIBS, while the purely gaseous-phase emission lines given by N_2^+ are degraded with double-pulse LIBS [187]. Both spectra have the same intensity scale.

to increased analyte response, the double-pulse LIBS system yielded an enhanced single-particle sampling rate when compared to conventional LIBS. Additional plasma transmission measurements with respect to the plasma-creating laser pulse were recorded for both single- and double-pulse methods over a range of temporal delays, where it was found that an optimal temporal region existed which yielded moderate coupling of the second laser pulse into the plasma created by the first laser. Too short a laser–laser delay (<100 ns) resulted in near total absorption of the second laser pulse into the first plasma, but only modest improvements in analyte response. Too long a laser-laser delay (>20 μs) resulted in significantly reduced coupling of the second laser pulse into the first plasma due to rarefaction within the plasma volume, negating the benefits of double-pulse LIBS. The optimal region for enhanced analyte response was a pulse-to-pulse separation between about 750 ns and 5 μs. In consideration of the overall spectroscopic and transmission data, the plasma-analyte interactions realized with a double-pulse methodology were explained in terms of the interaction with the expanding plasma, which differs between gaseous- and particulate-phase analytes, as reported in a recent study [95]. Double-pulse LIBS has also been used to differentiate between solid-phase and gaseous-phase analytes in an aerosol stream [188].

The analysis of aerosols generated (i.e., ablated) from solid samples has also been studied and represents a novel methodology to minimize matrix effects, improve analyte response, and use nonmatrix-matched standards [189–191]. In this new LIBS scheme, referred as to laser ablation—laser-induced breakdown spectroscopy, two independent laser processes are used to analyze the sample. The first laser samples the material, and the ablation plume containing the analyte is transported using a carrier gas to a second plasma, which is formed directly in the carrier gas stream by a second laser pulse. The second plasma is equivalent to performing LIBS in an aerosol stream. By decoupling the sampling laser ablation process from the analytical LIBS plasma a single calibration curve was obtained from a range of disparate reference materials (from pure iron, to copper nickel, to aluminum alloys), showing the potential of this technique to overcome matrix effects.

Perhaps the final frontier in fully understanding the governing processes in LIBS-based aerosol analysis concerns the overall interactions of the laser-induced plasma and the individual aerosol particle. Such processes speak directly to the issues of complete particle vaporization, size dependency of analyte response, and issues of plasma inhomogeneity, method detection limits, and analyte signal precision and accuracy. Recent efforts toward this end include the study of matrix effects for aerosol particles [192, 193], numerical modeling [194], time-resolved analysis [195], and plasma imaging in the presence of aerosol particles [196–198]. Fig. 10 presents an image of a single laser-induced plasma that captured an individual borosilicate glass particle. The image was captured at a delay of 2 μs following plasma initiation using a narrow line filter centered at the calcium II line (396.2-nm, 3-nm full width at half maximum (FWHM)) and an ICCD detector.

The image shows that at this delay time, the material that has vaporized and subsequently dissociated to Ca atoms remains localized about the original aerosol particle. In fact, the high degree of spherical symmetry in the reported images suggests that the atoms diffuse away from the particle on a timescale of some tens of microseconds. Time-resolved measurements provided an estimate of the overall diffusion coefficient in the range of $0.04\ m^2\ s^{-1}$, the first direct measurements of atomic diffusion in a laser-induced plasma subsequent to particle vaporization/dissociation. In a posterior study by Hahn's group, single microdroplets were introduced into laser-induced plasmas to study

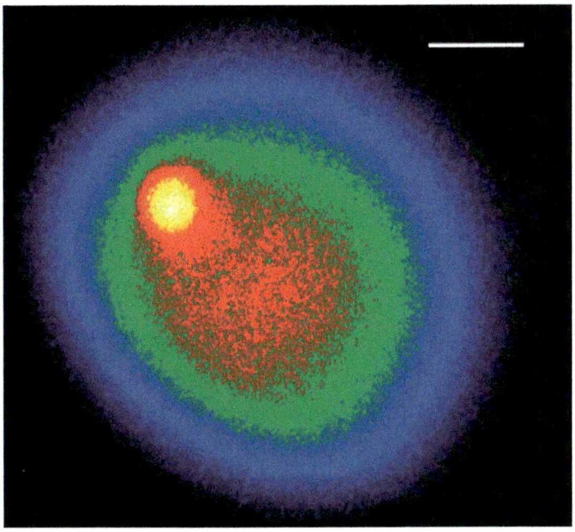

FIG. 10 Laser-induced plasma image recorded in the presence of a borosilicate glass particle corresponding to a delay time of 2 μs with respect to plasma initiation [196]. The image was recorded using a 396.2-nm narrow band (3-nm FWHM) interference filter. Scale bar = 600 μm.

desolvation, atomization, and atomic diffusion of Ca and H. The value of the H diffusion coefficient following analyte desolvation ranged from 0.022 to 0.26 $m^2 s^{-1}$ at 6000 and 20,000 K, respectively, whereas the Ca diffusion coefficient value ranged from 0.0005 to 0.0243 $m^2 s^{-1}$ at 3000 and 25,000 K, respectively [197]. Such measurements provide critical data to complement more advanced plasma modeling efforts, for example, as reported by Gornushkin et al. in a series of recent studies [199, 200]. Overall, as LIBS researchers start to discern a timescale for the dissociation of aerosol particles and subsequent atomic diffusion within laser-induced plasmas, a more complete picture of the analyte response begins to emerge. To further advance the analytical rigor of laser-induced plasma spectroscopy, rate-limited particle dissociation and mass transport should be further explored in the context of LIBS, as well as for plasmas from other systems (e.g., ICP) and plasma interactions with other forms of matter.

References

[1] J.O. Nriagu, J. Pacyna, Quantitative assessment of worldwide contamination of air, water and soil by trace metals, Nature 5886 (1988) 134–139.

[2] B. Veronesi, M. Oortgiesen, Neurogenic inflammation and particulate matter (PM) air pollutants, Neurotoxicology 22 (2001) 795–810, https://doi.org/10.1016/S0161-813X(01)00062-6.

[3] A.G. Allen, E. Nemitz, J.P. Shi, R.M. Harrison, J.C. Greenwood, Size distributions of trace metals in atmospheric aerosols in the United Kingdom, Atmos. Environ. 35 (2001) 4581–4591, https://doi.org/10.1016/S1352-2310(01)00190-X.

[4] A.E. Suarez, J.M. Ondov, Ambient aerosol concentrations of elements resolved by size and by source: contributions of some cytokine-active metals from coal- and oil-fired power plants, Energy Fuel 16 (2002) 562–568, https://doi.org/10.1021/ef010170c.

[5] S.K. Friedlander, The characterization of aerosols distributed with respect to size and chemical composition-II. Classification and design of aerosol measuring devices, J. Aerosol Sci. 2 (1971) 331–340, https://doi.org/10.1016/0021-8502(71)90058-9.

[6] S.K. Friedlander, The characterization of aerosols distributed with respect to size and chemical composition, J. Aerosol Sci. 1 (1970) 295–307.

[7] P.D. Maker, R.W. Ternune, S.M. Savage, Optical third harmonic generation in various solids and liquids, in: P. Grivet, N. Bloembergen (Eds.), Proc. 3rd Int. Conf. Quantum Electron., Paris, 1964, , pp. 1559–1565.

[8] R.G. Meyerand, A.F. Haught, Gas breakdown at optical frequencies, Phys. Rev. Lett. 11 (1963) 401–403, https://doi.org/10.1103/PhysRevLett.11.401.

[9] C. DeMichelis, Laser induced gas breakdown: a bibliographical review, IEEE J. Quantum Electron. 5 (1969) 188–202.

[10] C.G. Morgan, Laser-induced breakdown of gases, Rep. Prog. Phys. 38 (1975) 621–665, https://doi.org/10.1088/0034-4885/38/5/002.

[11] N. Bloembergen, A brief history of laser breakdown, J. Nonlinear Opt. Phys. Mater. 6 (1997) 377–385.

[12] T.P. Hughes, Plasmas and Laser Light, Wiley, New York, 1975.

[13] Y.B. Sel'dovich, Y.P. Raizer, Physics of Shock Waves and High-Temperature Hydrodynamic Phenomena, Dover Publications, New York, 2002.

[14] G. Bekefi, Principles of Laser Plasmas, New York, 1976.

[15] J.F. Ready, Effects of High-Power Laser Radiation, Academic Press, New York, 1971.

[16] J.R. Bettis, Correlation among the laser-induced breakdown thresholds in solids, liquids, and gases, Appl. Opt. 31 (1992) 3448–3452, https://doi.org/10.1364/AO.31.003448.

[17] G.S. Voronov, Dependence of the probability of multiphoton ionization of atoms on the photon flux intensity, J. Exp. Theor. Phys. 24 (1967) 1009–1011.

[18] G. Voronov, G. Delone, N. Delone, Multiphoton ionization of atoms. II. Ionization of krypton by ruby-laser radiation, Sov. J. Exp. Theor. Phys. 24 (1967) 1122.

[19] P. Agostini, G. Barjot, G. Mainfray, C. Manus, J. Thebault, Multiphoton ionization of rare gases at 1.06 μ and 0.53 μ, Phys. Lett. A 31 (1970) 367–368, https://doi.org/10.1016/0375-9601(70)90987-4.

[20] P. Agostini, G. Barjot, J.F. Bonnal, G. Mainfray, C. Manus, J. Morellec, U-5-multiphoton ionization of hydrogen and rare gases, IEEE J. Quantum Electron. 4 (1968) 667–669, https://doi.org/10.1109/JQE.1968.1074955.

[21] G. Baravian, G. Sultan, Determination of multiphoton ionization cross sections in focused laser experiments, Physica B+C 128 (1985) 343–352, https://doi.org/10.1016/0378-4363(85)90014-2.

[22] A. L'Huillier, Multiple ionization of atoms through multiphoton absorption, J. Phys. Colloq. 48 (1987) C9–415-C9-425, https://doi.org/10.1051/jphyscol:1987971.

[23] C.H. Chan, C.D. Moody, W.B. McKnight, Significant loss mechanisms in gas breakdown at 10.6 μ, J. Appl. Phys. 44 (1973) 1179–1188, https://doi.org/10.1063/1.1662325.

[24] G.M. Weyl, D. Rosen, Laser-induced breakdown in argon at 0.35 1/4 m: theory and experiments, Phys. Rev. A 31 (1985) 2300–2313, https://doi.org/10.1103/PhysRevA.31.2300.

[25] E. Yablonovitch, Similarity principles for laser-induced breakdown in gases, Appl. Phys. Lett. 23 (1973) 121–122, https://doi.org/10.1063/1.1654827.

[26] E. Yablonovitch, Self-phase modulation of light in a laser-breakdown plasma, Phys. Rev. Lett. 32 (1974) 1101–1104, https://doi.org/10.1103/PhysRevLett.32.1101.

[27] E. Yablonovitch, Self-phase modulation and short-pulse generation from laser-breakdown plasmas, Phys. Rev. A 10 (1974) 1888–1895, https://doi.org/10.1103/PhysRevA.10.1888.

[28] L.V. Keldysh, Ionization in the field of a strong electromagnetic wave, J. Exp. Theor. Phys. 20 (1965) 1307–1314.

[29] A.D. McDonald, Microwave Breakdown in Gases, Wiley, New York, 1966.

[30] A.V. Phelps, Theory of growth of ionization during laser breakdown, in: P.L. Kelley, B. Lax, P.E. Tannenwald (Eds.), Phys. Quantum Electron, McGraw-Hill, New York, 1966.

[31] P.F. Browne, Mechanism of gas breakdown by lasers, Proc. Phys. Soc. 86 (1965) 1323–1332, https://doi.org/10.1088/0370-1328/86/6/320.

[32] J.K. Wright, Theory of the electrical breakdown of gases by intense pulses of light, Proc. Phys. Soc. 84 (1964) 41–46, https://doi.org/10.1088/0370-1328/84/1/307.

[33] E.K. Damon, R.G. Tomlinson, Observation of ionization of gases by a ruby laser, Appl. Opt. 2 (1963) 546, https://doi.org/10.1364/ao.2.000546.

[34] N. Kroll, K.M. Watson, Theoretical study of ionization of air by intense laser pulses, Phys. Rev. A 5 (1972) 1883–1905, https://doi.org/10.1103/PhysRevA.5.1883.

[35] G.M. Weyl, D.I. Rosen, J. Wilson, W. Seka, Laser-induced breakdown of argon at 0.35 m, Phys. Rev. A 26 (1982) 1164–1167, https://doi.org/10.1103/PhysRevA.26.1164.

[36] D.I. Rosen, G. Weyl, Laser-induced breakdown in nitrogen and the rare gases at 0.53 and 0.357μm, J. Phys. D. Appl. Phys. 20 (1987) 1264–1276, https://doi.org/10.1088/0022-3727/20/10/009.

[37] H.T. Buscher, R.G. Tomlinson, E.K. Damon, Frequency dependence of optically induced gas breakdown, Phys. Rev. Lett. 15 (1965) 847–849, https://doi.org/10.1103/PhysRevLett.15.847.

[38] K.C. Byron, G.J. Pert, Measurement of the wavelength dependence of the threshold of laser-induced gas breakdown, J. Phys. D. Appl. Phys. 12 (1979) 409–416, https://doi.org/10.1088/0022-3727/12/3/010.

[39] J.B. Simeonsson, A.W. Miziolek, Spectroscopic studies of laser-produced plasmas formed in CO and CO_2 using 193, 266, 355, 532 and 1064 nm laser radiation, Appl. Phys. B Lasers Opt. 59 (1994) 1–9, https://doi.org/10.1007/BF01081722.

[40] R. Tambay, R.K. Thareja, Laser-induced breakdown studies of laboratory air at 0.266, 0.355, 0.532, and 1.064 μm, J. Appl. Phys. 70 (1991) 2890–2892.

[41] J.P. Davis, A.L. Smith, C. Giranda, M. Squicciarini, Laser-induced plasma formation in Xe, Ar, N_2, and O_2 at the first four Nd:YAG harmonics, Appl. Opt. 30 (1991) 4358, https://doi.org/10.1364/ao.30.004358.

[42] J. Stricker, J.G. Parker, Experimental investigation of electrical breakdown in nitrogen and oxygen induced by focused laser radiation at 1.064 μ, J. Appl. Phys. 53 (1982) 851–855, https://doi.org/10.1063/1.330592.

[43] R.A. Armstrong, R.A. Lucht, W.T. Rawlins, Spectroscopic investigation of laser-initiated low-pressure plasmas in atmospheric gases, Appl. Opt. 22 (1983) 1573, https://doi.org/10.1364/ao.22.001573.

[44] R.J. Dewhurst, Breakdown in the rare gases using single picosecond ruby laser pulses, J. Phys. D. Appl. Phys. 10 (1977) 283–289, https://doi.org/10.1088/0022-3727/10/3/008.

[45] R.J. Dewhurst, G.J. Pert, S.A. Ramsden, Laser-induced breakdown in the rare gases using picosecond Nd: Glass laser pulses, J. Phys. B: At. Mol. Phys. 7 (1974) 2281–2290, https://doi.org/10.1088/0022-3700/7/16/025.

[46] C.L.M. Ireland, C.G. Morgan, Gas breakdown by a short laser pulse, J. Phys. D. Appl. Phys. 6 (1973) 720–729, https://doi.org/10.1088/0022-3727/6/6/314.

[47] Y.E.E.D. Gamal, M.A. Harith, Molecular gas breakdown by short laser pulses, J. Phys. D. Appl. Phys. 14 (1981) 2209–2214, https://doi.org/10.1088/0022-3727/14/12/009.

[48] M. Hanafi, M.M. Omar, Y.E.E.D. Gamal, Study of laser-induced breakdown spectroscopy of gases, Radiat. Phys. Chem. 57 (2000) 11–20, https://doi.org/10.1016/S0969-806X(99)00344-8.

[49] L.M. Davis, L.Q. Li, D.R. Keefer, Picosecond resolved evolution of laser breakdown in gases, J. Phys. D. Appl. Phys. 26 (1993) 222–230, https://doi.org/10.1088/0022-3727/26/2/009.

[50] C.L.M. Ireland, A. Yi, J.M. Aaron, C.G. Morgan, Focal-length dependence of air breakdown by a 20-psec laser pulse, Appl. Phys. Lett. 24 (1974) 175–177, https://doi.org/10.1063/1.1655141.

[51] H. Sobral, M. Villagrán-Muniz, R. Navarro-González, A.C. Raga, Temporal evolution of the shock wave and hot core air in laser induced plasma, Appl. Phys. Lett. 77 (2000) 3158–3160, https://doi.org/10.1063/1.1324986.

[52] A.F. Haught, D.H. Polk, High-temperature plasmas produced by laser beam irradiation of single solid particles, Phys. Fluids 9 (1966) 2047–2056, https://doi.org/10.1063/1.1761564.

[53] V.E. Zuev, A.A. Zemlyanov, Y.D. Kopytin, A.V. Kuzikovskii, High-power laser radiation in atmospheric aerosols, D. Reidel Publishing Company, Dordrecht, 1985.

[54] A.A. Lushnikov, A.E. Negin, Aerosols in strong laser beams, J. Aerosol Sci. 24 (1993) 707–735, https://doi.org/10.1016/0021-8502(93)90042-8.

[55] R.L. Armstrong, Aerosol heating and vaporization by pulsed light beams, Appl. Opt. 23 (1984) 148, https://doi.org/10.1364/ao.23.000148.

[56] T.A. Schoolcraft, G.S. Constable, L.V. Zhigilei, B.J. Garrison, Molecular dynamics simulation of the laser disintegration of aerosol particles, Anal. Chem. 72 (2000) 5143–5150, https://doi.org/10.1021/ac0007635.

[57] S.V. Zakharchenko, L.P. Semenov, A.M. Skripkin, Low-threshold optical discharge in a disperse aerosol medium, Sov. J. Quantum Electron. 14 (1984) 1642–1645, https://doi.org/10.1070/qe1984v014n12abeh006244.

[58] D.E. Lencioni, L.C. Pettingill, The dynamics of air breakdown initiated by a particle in a laser beam, J. Appl. Phys. 48 (1977) 1848–1851, https://doi.org/10.1063/1.323937.

[59] D.E. Lencioni, Laser-induced air breakdown for 1.06-um radiation, Appl. Phys. Lett. 15 (1974) 15–17.

[60] D.E. Poulain, D.R. Alexander, J.P. Barton, S.A. Schaub, J. Zhang, Interactions of intense ultraviolet laser radiation with solid aerosols, J. Appl. Phys. 67 (1990) 2283–2288, https://doi.org/10.1063/1.345522.

[61] I.Y. Borets-Pervak, V.S. Vorob'ev, Threshold for gas breakdown initiated by an interaction of laser light with aerosol particles, Quantum Electron. 23 (1993) 224–230, https://doi.org/10.1070/qe1993v023n03abeh002980.

[62] D.C. Smith, R.T. Brown, D.C. Smith, R.T. Brown, Aerosol-Induced Air Breakdown with CO_2 Laser Radiation, 1146, 2012.

[63] D.C. Smith, Gas breakdown initiated by laser radiation interaction with aerosols and solid surfaces, J. Appl. Phys. 48 (1977) 2217–2225, https://doi.org/10.1063/1.324025.

[64] G.M. Weyl, Ionisation path formation in gases using a laser with retractable focus, J. Phys. D. Appl. Phys. 12 (1979) 33–49, https://doi.org/10.1088/0022-3727/12/1/006.

[65] C. David, C. Smith, Laser radiation-induced air breakdown and plasma shielding, Opt. Eng. 20 (1981) 962–969.

[66] S.T. Amimoto, J.S. Whittier, F.G. Ronkowski, P.R. Valenzuela, G.N. Harper, R. Hofland Jr., DF-laser pulse breakdown induced by maritime aerosols, AIAA J 22 (1984) 1108–1114.

[67] J.E. Lowder, H. Kleiman, Long-pulse breakdown iwth 10.6-μm laser radiation, J. Appl. Phys. 44 (1973) 5504–5505.

[68] A.A. Boni, D.A. Meskan, Effect of impurities on laser-induced air breakdown at 1.06 μ, Opt. Commun. 14 (1975) 115–118, https://doi.org/10.1016/0030-4018(75)90071-1.

[69] G.H. Canavan, P.E. Nielsen, Focal spot size dependence of gas breakdown induced by particulate ionization, Appl. Phys. Lett. 22 (1973) 409–410, https://doi.org/10.1063/1.1654693.

[70] R.G. Pinnick, A. Biswas, J.D. Pendleton, R.L. Armstrong, Aerosol-induced laser breakdown thresholds: effect of resonant particles, Appl. Opt. 31 (1992) 311, https://doi.org/10.1364/ao.31.000311.

[71] P. Chýlek, M.A. Jarzembski, N.Y. Chou, R.G. Pinnick, Effect of size and material of liquid spherical particles on laser-induced breakdown, Appl. Phys. Lett. 49 (1986) 1475–1477, https://doi.org/10.1063/1.97307.

[72] A. Biswas, H. Latifi, P. Shah, L.J. Radziemski, R.L. Armstrong, Time-resolved spectroscopy of plasmas initiated on single, levitated aerosol droplets, Opt. Lett. 12 (1987) 313, https://doi.org/10.1364/ol.12.000313.

[73] A. Biswas, H. Latifi, L.J. Radziemski, R.L. Armstrong, Irradiance and laser wavelength dependence of plasma spectra from single levitated aerosol droplets, Appl. Opt. 27 (1988) 2366–2391.

[74] R.G. Pinnick, P. Chylek, M. Jarzembski, E. Creegan, V. Srivastava, G. Fernandez, J.D. Pendleton, A Biswas, aerosol-induced laser breakdown thresholds: wavelength dependence, Appl. Opt. 27 (1988) 987–996. http://www.ncbi.nlm.nih.gov/pubmed/20523719.

[75] P. Chýlek, M.A. Jarzembski, V. Srivastava, R.G. Pinnick, J.D. Pendleton, J.P. Cruncleton, Effect of spherical particles on laser-indcued breakdown of gases, Appl. Opt. 26 (1987) 760–762.

[76] W.-F. Hsieh, J.H. Eickmans, R.K. Chang, Internal and external laser-induced avalanche breakdown of single droplets in an argon atmosphere, J. Opt. Soc. Am. B 4 (1987) 1816–1820, https://doi.org/10.1364/josab.4.001816.

[77] R.K. Chang, J.H. Eickmans, W.-F. Hsieh, C.F. Wood, J.-Z. Zhang, J. Zheng, Laser-induced breakdown in large transparent water droplets, Appl. Opt. 27 (1988) 2377, https://doi.org/10.1364/ao.27.002377.

[78] M. Sabsabi, R. Héon, V. Detalle, L. St-Onge, A. Hamel, Comparison between intensified CCD and non-intensified gated CCD detectors for LIPS analysis of solid samples, in: ThE16, 2014. https://doi.org/10.1364/libs.2002.the16.

[79] J.E. Carranza, E. Gibb, B.W. Smith, D.W. Hahn, J.D. Winefordner, Comparison of nonintensified and intensified CCD detectors for laser-induced breakdown spectroscopy, Appl. Opt. 42 (2003) 6016, https://doi.org/10.1364/ao.42.006016.

[80] E. Voigtman, Gated peak integration versus peak detection in white noise, Appl. Spectrosc. 45 (1991) 237–241, https://doi.org/10.1366/0003702914337416.

[81] U. Panne, C. Haisch, M. Clara, R. Niessner, Analysis of glass and glass melts during the vitrification process of fly and bottom ashes by laser-induced plasma spectroscopy. Part I: Normalization and plasma diagnostics, Spectrochim. Acta B At. Spectrosc. 53 (1998) 1957–1968, https://doi.org/10.1016/S0584-8547(98)00238-9.

[82] B.T. Fisher, H.A. Johnsen, S.G. Buckley, D.W. Hahn, Temporal gating for the optimization of laser-induced breakdown spectroscopy detection and analysis of toxic metals, Appl. Spectrosc. 55 (2001) 1312–1319, https://doi.org/10.1366/0003702011953667.

[83] L.J. Radziemski, D.A. Cremers, T.R. Loree, Detection of beryllium by laser-induced-breakdown spectroscopy, Spectrochim. Acta Part B At. Spectrosc. 38 (1983) 349–355, https://doi.org/10.1016/0584-8547(83)80133-5.

[84] T.R. Loree, L.J. Radziemski, Laser-induced breakdown spectroscopy: time-integrated applications, Plasma Chem. Plasma Process. 1 (1981) 271–279.

[85] L.J. Radziemski, T.R. Loree, Laser-induced breakdown spectroscopy: time-resolved spectrochemical applications, Plasma Chem. Plasma Process. 1 (1981) 281–293, https://doi.org/10.1007/BF00568836.

[86] J.E. Carranza, D.W. Hahn, Sampling statistics and considerations for single-shot analysis using laser-induced breakdown spectroscopy, Spectrochim. Acta B At. Spectrosc. 57 (2002) 779–790, https://doi.org/10.1016/S0584-8547(02)00007-1.

[87] V. Hohreiter, A.J. Ball, D.W. Hahn, Effects of aerosols and laser cavity seeding on spectral and temporal stability of laser-induced plasmas: applications to LIBS, J. Anal. At. Spectrom. 19 (2004) 1289, https://doi.org/10.1039/b402190h.

[88] N.B. Zorov, A.A. Gorbatenko, T.A. Labutin, A.M. Popov, A review of normalization techniques in analytical atomic spectrometry with laser sampling: from single to multivariate correction, Spectrochim. Acta B At. Spectrosc. 65 (2010) 642–657, https://doi.org/10.1016/j.sab.2010.04.009.

[89] D.A. Cremers, L.J. Radziemski, Direct detection of beryllium on filters using the laser spark, Appl. Spectrosc. 39 (1985) 57–63, https://doi.org/10.1366/0003702854249349.

[90] L.J. Radziemski, T.R. Loree, D.A. Cremers, N. M. Hoffman, Time-resolved laser-induced breakdown spectrometry of aerosols, Anal. Chem. 55 (1983) 1246–1252, https://doi.org/10.1021/ac00259a016.

[91] M. Essien, L.J. Radziemski, J. Sneddon, Detection of cadmium, lead and zinc in aerosols by laser-induced breakdown spectrometry, J. Anal. At. Spectrom. 3 (1988) 985–988, https://doi.org/10.1039/JA9880300985.

[92] J.E. Carranza, D.W. Hahn, Assessment of the upper particle size limit for quantitative analysis of aerosols using laser-induced breakdown spectroscopy, Anal. Chem. 74 (2002) 5450–5454. http://www.ncbi.nlm.nih.gov/pubmed/12433072.

[93] E. Vors, L. Salmon, Laser-induced breakdown spectroscopy (LIBS) for carbon single shot analysis of micrometer-sized particles, Anal. Bioanal. Chem. 385 (2006) 281–286, https://doi.org/10.1007/s00216-006-0320-x.

[94] G. Gallou, J.B. Sirven, G. Gallou, C. Dutouquet, O. Le Bihan, E. Frejafon, Aerosols analysis by libs for monitoring of air pollution by industrial sources, Aerosol Sci. Technol. 45 (2011) 918–926, https://doi.org/10.1080/02786826.2011.566899.

[95] V. Hohreiter, D.W. Hahn, Calibration effects for laser-induced breakdown spectroscopy of gaseous sample streams: analyte response of gas-phase species versus solid-phase species, Anal. Chem. 77 (2005) 1118–1124. http://www.ncbi.nlm.nih.gov/pubmed/15858994.

[96] D.W. Hahn, J.E. Carranza, G.R. Arsenault, H.A. Johnsen, K.R. Hencken, Aerosol generation system for development and calibration of laser-induced breakdown spectroscopy instrumentation, Rev. Sci. Instrum. 72 (2001) 3706–3713, https://doi.org/10.1063/1.1391440.

[97] D.W. Hahn, M.M. Lunden, Detection and analysis of aerosol particles by laser-induced breakdown spectroscopy, Aerosol Sci. Technol. 33 (2000) 30–48.

[98] D.W. Hahn, W.L. Flower, K.R. Hencken, Discrete particle detection and metal emissions monitoring using laser-induced breakdown spectroscopy, Appl. Spectrosc. 51 (1997) 1836–1844, https://doi.org/10.1366/0003702971939659.

[99] Ş. Yalçin, D.R. Crosley, G.P. Smith, G.W. Faris, Influence of ambient conditions on the laser air spark, Appl. Phys. B Lasers Opt. 68 (1999) 121–130, https://doi.org/10.1007/s003400050596.

[100] Y. Chen, J.W.L. Lewis, C. Parigger, Spatial and temporal profiles of pulsed laser-induced air plasma emissions, J. Quant. Spectrosc. Radiat. Transf. 67 (2000) 91–103, https://doi.org/10.1016/S0022-4073(99)00196-X.

[101] J.E. Carranza, D.W. Hahn, Plasma volume considerations for analysis of gaseous and aerosol samples using laser-induced breakdown spectroscopy, J. Anal. At. Spectrom. 17 (2002) 1534–1539, https://doi.org/10.1039/b206140f.

[102] V. Hohreiter, J.E. Carranza, D.W. Hahn, Temporal analysis of laser-induced plasma properties as related to laser-induced breakdown spectroscopy, Spectrochim. Acta B At. Spectrosc. 59 (2004) 327–333, https://doi.org/10.1016/j.sab.2003.12.015.

[103] J.E. Carranza, B.T. Fisher, G.D. Yoder, D.W. Hahn, On-line analysis of ambient air aerosols using laser-induced breakdown spectroscopy, Spectrochim. Acta B At. Spectrosc. 56 (2001) 851–864, https://doi.org/10.1016/S0584-8547(01)00183-5.

[104] B. Hettinger, V. Hohreiter, M. Swingle, D.W. Hahn, Laser-induced breakdown spectroscopy for ambient air particulate monitoring: correlation of total and speciated aerosol particle counts, Appl. Spectrosc. 60 (2006) 237–245, https://doi.org/10.1366/000370206776342544.

[105] I. Schechter, Direct aerosol analysis by time resolved laser plasma spectroscopy - improvement by single shot measurements, Anal. Sci. Technol. 8 (1995) 779–786.

[106] J.E. Carranza, K. Iida, D.W. Hahn, Conditional data processing for single-shot spectral analysis by use of laser-induced breakdown spectroscopy, Appl. Opt. 42 (2003) 6022–6028, https://doi.org/10.1364/AO.42.006022.

[107] P.B. Dixon, D.W. Hahn, Feasibility of detection and identification of individual bioaerosols using laser-induced breakdown spectroscopy, Anal. Chem. 77 (2005) 631–638, https://doi.org/10.1021/ac048838i.

[108] D. Díaz, D.W. Hahn, A. Molina, Quantification of gold and silver in minerals by laser-induced breakdown spectroscopy, Spectrochim. Acta B At. Spectrosc. 136 (2017) 106–115, https://doi.org/10.1016/j.sab.2017.08.008.

[109] D. Díaz, A. Molina, D. Hahn, Effect of laser irradiance and wavelength on the analysis of gold- and silver-bearing minerals with laser-induced breakdown spectroscopy, Spectrochim. Acta B At. Spectrosc. 145 (2018) 86–95, https://doi.org/10.1016/j.sab.2018.04.007.

[110] D.W. Hahn, Laser-induced breakdown spectroscopy for sizing and elemental analysis of discrete aerosol particles, Appl. Phys. Lett. 72 (1998) 2960–2962, https://doi.org/10.1063/1.121507.

[111] S. Arnold, D.A. Cremers, Rapid determination of metal particles on air sampling filters using laser-induced breakdown spectroscopy, Am. Ind. Hyg. Assoc. J. 56 (1995) 1180–1186.

[112] R. Neuhauser, U. Panne, R. Niessner, Laser-induced plasma spectroscopy (LIPS): a versatile tool for monitoring heavy metal aerosols, Anal. Chim. Acta 392 (1999) 47–54, https://doi.org/10.1016/S0003-2670(99)00053-7.

[113] R.E. Neuhauser, U. Panne, R. Niessner, Utilization of fiber optics for remote sensing by laser-induced plasma spectroscopy (LIPS), Appl. Spectrosc. 54 (2000) 923–927, https://doi.org/10.1366/0003702001950337.

[114] U. Panne, R. Neuhauser, M. Theisen, H. Fink, R. Niessner, Analysis of heavy metal aerosols on filters by laser-induced plasma spectroscopy, Spectrochim. Acta B At. Spectrosc. 56 (2001) 839–850, https://doi.org/10.1016/S0584-8547(01)00209-9.

[115] C.B. Stipe, A.L. Miller, J. Brown, E. Guevara, E. Cauda, Evaluation of laser-induced breakdown spectroscopy (LIBS) for measurement of silica on filter samples of coal dust, Appl. Spectrosc. 66 (2012) 1286–1293, https://doi.org/10.1366/12-06671.

[116] C. Dutouquet, G. Gallou, O. Le Bihan, J.B. Sirven, A. Dermigny, B. Torralba, E. Frejafon, Monitoring of heavy metal particle emission in the exhaust duct of a foundry using LIBS, Talanta 127 (2014) 75–81, https://doi.org/10.1016/j.talanta.2014.03.063.

[117] G. Kim, Y.J. Yoon, H.A. Kim, H. joo Cho, K. Park, Elemental composition of Arctic soils and aerosols in Ny-Ålesund measured using laser-induced breakdown spectroscopy, Spectrochim. Acta B At. Spectrosc. 134 (2017) 17–24, https://doi.org/10.1016/j.sab.2017.06.006.

[118] M.Z. Martin, M.D. Cheng, R.C. Martin, Aerosol measurement by laser-induced plasma technique: a review, Aerosol Sci. Technol. 31 (1999) 409–421, https://doi.org/10.1080/027868299303968.

[119] B.W. Smith, D.W. Hahn, E. Gibb, I. Gornushkin, J.D. Winefordner, Laser induced plasma spectroscopy for the characterization of aerosols and particulates, KONA Powder Part. J. 19 (2001) 25–33, https://doi.org/10.14356/kona.2001008.

[120] D. Cremers, L. Radziemski, Handbook of Laser-Induced Breakdown Spectroscopy, John Wiley & Sons, Ltd., Chichester, England, 2006.

[121] D.W. Hahn, N. Omenetto, Laser-induced breakdown spectroscopy (LIBS), part I: review of basic diagnostics and plasma-particle interactions: atill-challenging issues within the analytical plasma community, Appl. Spectrosc. 64 (2010) 335–366, https://doi.org/10.1366/000370210793561691.

[122] D.W. Hahn, N. Omenetto, Laser-induced breakdown spectroscopy (LIBS), part II: review of instrumental and methodological approaches to material analysis and applications to different fields, Appl. Spectrosc. 66 (2012) 347–419, https://doi.org/10.1366/11-06574.

[123] R. Noll, Laser-Induced Breakdown Spectroscopy-Fundamentals and Applications, Springer-Verlag, Berlin, Germany, 2012.

[124] D.A. Rusak, B.C. Castle, B.W. Smith, J.D. Winefordner, Fundamentals and applications of laser-induced breakdown spectroscopy, Crit. Rev. Anal. Chem. 27 (1997) 257–290, https://doi.org/10.1080/10408349708050587.

[125] E.B. Belyaev, A.P. Godlevskiĭ, Y.D. Kopytin, Laser spectrochemical analysis of aerosols, Kvantovaya Elektron. 5 (1978) 2594–2601.

[126] Y.L. Chen, J.W.L. Lewis, C. Parigger, Probability distribution of laser-induced breakdown and ignition of ammonia, J. Quant. Spectrosc. Radiat. Transf. 66 (2000) 41–53, https://doi.org/10.1016/S0022-4073(99)00166-1.

[127] T.X. Phuoc, F.P. White, Laser-induced spark ignition of CH_4/air mixtures, Combust. Flame 119 (1999) 203–216, https://doi.org/10.1016/S0010-2180(99)00051-6.

[128] T.X. Phuoc, C.M. White, Experimental studies of the absorption and emissions from laser-induced spark in combustible gases, Opt. Commun. 181 (2000) 353–359, https://doi.org/10.1016/S0030-4018(00)00770-7.

[129] B. McGann, C.D. Carter, T.M. Ombrello, S. Hammack, T. Lee, H. Do, Gas property measurements in a supersonic combustor using nanosecond gated laser-induced breakdown spectroscopy with direct spectrum matching, Proc. Combust. Inst. 36 (2017) 2857–2864, https://doi.org/10.1016/j.proci.2016.06.089.

[130] A. Sircar, R.K. Dwivedi, R.K. Thareja, Laser induced breakdown of Ar, N_2 and O_2 gases using 1.064, 0.532, 0.355 and 0.266 μm radiation, Appl. Phys. B Lasers Opt. 63 (1996) 623–627, https://doi.org/10.1007/BF01831003.

[131] R.J. Nordstrom, Study of laser-induced plasma emission spectra of N_2, O_2, and ambient air in the region 350 nm to 950 nm, Appl. Spectrosc. 49 (1995) 1490–1499, https://doi.org/10.1366/0003702953965416.

[132] C.A. Henry, P.K. Diwakar, D.W. Hahn, Investigation of helium addition for laser-induced plasma spectroscopy of pure gas phase systems: analyte interactions and signal enhancement, Spectrochim. Acta B At. Spectrosc. 62 (2007) 1390–1398, https://doi.org/10.1016/j.sab.2007.10.002.

[133] P.K. Diwakar, D.W. Hahn, Study of early laser-induced plasma dynamics: transient electron density gradients via Thomson scattering and stark broadening, and the implications on laser-induced breakdown spectroscopy measurements, Spectrochim. Acta B At. Spectrosc. 63 (2008) 1038–1046, https://doi.org/10.1016/j.sab.2008.07.003.

[134] Ş. Yalçin, D.R. Crosley, G.P. Smith, G.W. Faris, Spectroscopic characterization of laser-produced plasmas for in situ toxic metal monitoring, Hazard. Waste Hazard. Mater. 13 (1996) 51–61, https://doi.org/10.1089/hwm.1996.13.51.

[135] L. Dudragne, P. Adam, J. Amouroux, Time-resolved laser-induced breakdown spectroscopy: application for qualitative and quantitative detection of fluorine, chlorine, sulfur, and carbon in air, Appl. Spectrosc. 52 (1998) 1321–1327, https://doi.org/10.1366/0003702981942654.

[136] M. Tran, B.W. Smith, D.W. Hahn, J.D. Winefordner, Detection of gaseous and particulate fluorides by laser-induced breakdown spectroscopy, Appl. Spectrosc. 55 (2001) 1455–1461, https://doi.org/10.1366/0003702011953865.

[137] L.W. Peng, W.L. Flower, K.R. Hencken, H.A. Johnsen, R.F. Renzi, N.B. French, A laser-based technique for continuously monitoring metal emissions from thermal waste treatme, Process. Control. Qual. 7 (1955) 39–49.

[138] W.L. Flower, L.W. Peng, M.P. Bonin, N.B. French, H.A. Johnsen, D.K. Ottesen, R.F. Renzi, L.V. Westbrook, A laser-based technique to continuously monitor metal aerosol emissions, Fuel Process. Technol. 39 (1994) 277–284, https://doi.org/10.1016/0378-3820(94)90186-4.

[139] J.P. Singh, H. Zhang, F.Y. Yueh, K.P. Carney, Investigation of the effects of atmospheric conditions on the quantification of metal hydrides using laser-induced breakdown spectroscopy, Appl. Spectrosc. 50 (1996) 764–773, https://doi.org/10.1366/0003702963905619.

[140] H. Zhang, F.Y. Yueh, J.P. Singh, Laser-induced breakdown spectrometry as a multimetal continuous-emission monitor, Appl. Opt. 38 (1999) 1459–1466, https://doi.org/10.1364/AO.38.001459.

[141] J.P. Singh, F.-Y. Yueh, Z. Hansheng, R.L. Cook, Study of laser induced breakdown spectroscopy as a process monitor and control tool for hazardous waste remediation, Process. Control. Qual. 10 (1997) 247–258.

[142] P.M. Lemieux, J.V. Ryan, N.B. French, W. Haas, S.J. Priebe, D.B. Burns, Results of the September 1997 DOE/EPA demonstration of multimetal continuous emission monitoring technologies, Environ. Prot. 18 (1998) 385–391.

[143] S.G. Buckley, H.A. Johnsen, K.R. Hencken, A.D. W. Hahn, Implementation of laser-induced breakdown spectroscopy as a continuous emissions monitor for toxic metals, Waste Manag. 20 (2000) 455–462, https://doi.org/10.1016/S0956-053X(00)00011-8.

[144] R.E. Neuhauser, U. Panne, R. Niessner, P. Wilbring, On-line monitoring of chromium aerosols in industrial exhaust streams by laser-induced plasma spectroscopy (LIPS), Fresenius J. Anal. Chem. 364 (1999) 720–726, https://doi.org/10.1007/s002160051421.

[145] M. Casini, M.A. Harith, V. Palleschi, A. Salvetti, D.P. Singh, M. Vaselli, Time-resolved LIBS experiment for quantitative determination of pollutant concentrations in air, Laser Part. Beams 9 (1991) 633–639, https://doi.org/10.1017/S0263034600003633.

[146] M. Hidalgo Núñez, P. Cavalli, G. Petrucci, N. Omenetto, Analysis of sulfuric acid aerosols by laser-induced breakdown spectroscopy and laser-induced photofragmentation, Appl. Spectrosc. 54 (2000) 1805–1816, https://doi.org/10.1366/0003702001948907.

[147] G.A. Lithgow, A.L. Robinson, S.G. Buckley, Ambient measurements of metal-containing PM2.5 in an urban environment using laser-induced breakdown spectroscopy, Atmos. Environ. 38 (2004) 3319–3328, https://doi.org/10.1016/j.atmosenv.2004.03.017.

[148] H.A.L. Ee, H.Y.M. Aeng, K.Y.K. Im, G.I.K. Im, Application of laser-induced breakdown spectroscopy for realtime detection of contamination particles during the manufacturing process, Appl. Opt. 57 (2018) 3288–3292.

[149] D.A. Redoglio, N. Palazzo, F. Migliorini, S. De Iuliis, Laser-induced breakdown spectroscopy analysis of lead aerosol in nitrogen and air atmosphere, Appl. Spectrosc. 72 (2018) 584–590, https://doi.org/10.1177/0003702817742314.

[150] A.N. Williams, S. Phongikaroon, Elemental detection of cerium and gadolinium in aqueous aerosol using laser-induced breakdown spectroscopy, Appl. Spectrosc. 70 (2016) 1700–1708, https://doi.org/10.1177/0003702816648327.

[151] X. Yang, L. Guo, J. Li, R. Yi, Z. Hao, M. Shen, R. Zhou, K. Li, X. Li, Y. Lu, X. Zeng, Laser-induced breakdown spectroscopy of liquid solutions: a comparative study on the forms of liquid surface and liquid aerosol, Appl. Opt. 55 (2016) 7406, https://doi.org/10.1364/ao.55.007406.

[152] L.A. Álvarez-Trujillo, A. Ferrero, J. Javier Laserna, Preliminary studies on stand-off laser induced breakdown spectroscopy detection of aerosols, J. Anal. At. Spectrom. 23 (2008) 885–888, https://doi.org/10.1039/b716762h.

[153] L.A. Álvarez-Trujillo, V. Lazic, J. Moros, J.J. Laserna, Standoff monitoring of aqueous aerosols using nanosecond laser-induced breakdown spectroscopy: droplet size and matrix effects, Appl. Opt. 56 (2017) 3773–3782, https://doi.org/10.1364/AO.56.003773.

[154] S.G. Buckley, C.S. Mcenally, R.F. Sawyer, C.P. Koshland, D. Lucas, Metal emissions monitoring using excimer laser fragmentation-fluorescence spectroscopy, Combust. Sci. Technol. 118 (1996) 169–188, https://doi.org/10.1080/00102209608951978.

[155] S.G. Buckley, R.F. Sawyer, C.P. Koshland, D. Lucas, Measurements of lead vapor and particulate in flames and post-flame gases, Combust. Flame 128 (2002) 435–446, https://doi.org/10.1016/S0010-2180(01)00366-2.

[156] K.T. Hartinger, P.B. Monkhouse, J. Wolfrum, Determination of alkali traces in coal combustion by excimer laser induced fragmentation fluorescence, Berichte Der Bunsengesellschaft Für Phys. Chemie. 97 (1993) 1731–1733, https://doi.org/10.1002/bbpc.19930971239.

[157] X. Tong, R.B. Barat, A.T. Poulos, A charge-coupled device-based laser photofragment fluorescence spectrometer for detection of mercury compounds, Rev. Sci. Instrum. 70 (1999) 4180–4184, https://doi.org/10.1063/1.1150049.

[158] X. Tong, R.B. Barat, A.T. Poulos, Detection of mercuric bromide in a gas phase flow cell by laser photofragment fluorescence spectroscopy, Environ. Sci. Technol. 33 (1999) 3260–3263, https://doi.org/10.1021/es9813461.

[159] M. Tran, Q. Sun, B.W. Smith, J.D. Winefordner, Determination of F, Cl, and Br in solid organic compounds by laser-induced plasma spectroscopy, Appl. Spectrosc. 55 (2001) 739–744, https://doi.org/10.1366/0003702011952433.

[160] R.E. Neuhauser, U. Panne, R. Niessner, G.A. Petrucci, P. Cavalli, N. Omenetto, On-line and in-situ detection of lead aerosols by plasma-spectroscopy and laser-excited atomic fluorescence spectroscopy, Anal. Chim. Acta 346 (1997) 37–48, https://doi.org/10.1016/S0003-2670(97)00244-4.

[161] E.A.P. Cheng, R.D. Fraser, J.G. Eden, Detection of trace concentrations of column III and V hydrides by laser-induced breakdown spectroscopy, Appl. Spectrosc. 45 (1991) 949–952, https://doi.org/10.1366/0003702914336273.

[162] C. Lazzari, M. De Rosa, S. Rastelli, A. Ciucci, V. Palleschi, A. Salvetti, Detection of mercury in air by time-resolved laser-induced breakdown spectroscopy technique, Laser Part. Beams 12 (1994) 525–530, https://doi.org/10.1017/S0263034600008387.

[163] A. Ciucci, M. Corsi, V. Palleschi, S. Rastelli, A. Salvetti, New procedure for quantitative elemental analysis by laser-induced plasma spectroscopy, Appl. Spectrosc. 53 (1999) 960–964.

[164] E. Tognoni, V. Palleschi, M. Corsi, G. Cristoforetti, Quantitative micro-analysis by laser-induced breakdown spectroscopy: a review of the experimental approaches, Spectrochim. Acta B At. Spectrosc. 57 (2002) 1115–1130, https://doi.org/10.1016/S0584-8547(02)00053-8.

[165] H. Bette, R. Noll, High speed laser-induced breakdown spectrometry for scanning microanalysis, J. Phys. D. Appl. Phys. 37 (2004) 1281–1288, https://doi.org/10.1088/0022-3727/37/8/018.

[166] G. Cristoforetti, S. Legnaioli, V. Palleschi, A. Salvetti, E. Tognoni, P. Alberto Benedetti, F. Brioschi, F. Ferrario, Quantitative analysis of aluminium alloys by low-energy, high-repetition rate laser-induced breakdown spectroscopy, J. Anal. At. Spectrom. 21 (2006) 697–702, https://doi.org/10.1039/b604628b.

[167] P. Pořízka, B. Klessen, J. Kaiser, I. Gornushkin, U. Panne, J. Riedel, High repetition rate laser-induced breakdown spectroscopy using acousto-optically gated detection, Rev. Sci. Instrum. 85 (2014) 1–8, https://doi.org/10.1063/1.4890337.

[168] I. Radivojevic, R. Niessner, C. Haisch, S. Florek, H. Becker-Ross, U. Panne, Detection of bromine in thermoplasts from consumer electronics by laser-induced plasma spectroscopy, Spectrochim. Acta B At. Spectrosc. 59 (2004) 335–343, https://doi.org/10.1016/j.sab.2004.01.003.

[169] I. Radivojevic, C. Haisch, R. Niessner, S. Florek, H. Becker-Ross, U. Panne, Microanalysis by laser-induced plasma spectroscopy in the vacuum ultraviolet, Anal. Chem. 76 (2004) 1648–1656, https://doi.org/10.1021/ac0351365.

[170] G.A. Petrucci, P.B. Farnsworth, P. Cavalli, N. Omenetto, A differentially pumped particle inlet for sampling of atmospheric aerosols into a time-of-flight mass spectrometer: optical characterization of the particle beam, Aerosol Sci. Technol. 33 (2000) 105–121, https://doi.org/10.1080/027868200410877.

[171] B. Dahneke, Aerosol beam spectrometry, Nat. Phys. Sci. 244 (1973) 54–55, https://doi.org/10.1038/246421a0.

[172] J. Schreiner, C. Voigt, K. Mauersberger, P. McMurry, P. Ziemann, Aerodynamic lens system for producing particle beams at stratospheric pressures, Aerosol Sci. Technol. 29 (1998) 50–56, https://doi.org/10.1080/02786829808965550.

[173] P. Liu, P.J. Ziemann, D.B. Kittelson, P.H. McMurry, Generating particle beams of controlled dimensions and divergence: I. Theory of particle motion in aerodynamic lenses and nozzle expansions, Aerosol Sci. Technol. 22 (1995) 293–313, https://doi.org/10.1080/02786829408959748.

[174] P. Liu, P.J. Ziemann, D.B. Kittelson, P.H. McMurry, Generating particle beams of controlled dimensions and divergence: II. Experimental evaluation of particle motion in aerodynamic lenses and nozzle expansions, Aerosol Sci. Technol. 22 (1995) 314–324, https://doi.org/10.1080/02786829408959749.

[175] M.D. Cheng, L. Karr, J. Kornuc, D. Staat, T. Wainman, B. Harre, B. Sugiyama, Evaluation of aerosol lead measurement techniques using laboratory generated super-micrometer particles, Microchem. J. 72 (2002) 209–219, https://doi.org/10.1016/S0026-265X(02)00036-X.

[176] M.D. Cheng, Real-time measurement of trace metals on fine particles by laser-induced plasma techniques, Fuel Process. Technol. 65 (2000) 219–229, https://doi.org/10.1016/S0378-3820(99)00099-5.

[177] J.W. Lee, M.Y. Yi, S.M. Lee, Inertial focusing of particles with an aerodynamic lens in the atmospheric pressure range, J. Aerosol Sci. 34 (2003) 211–224, https://doi.org/10.1016/S0021-8502(02)00158-1.

[178] K. Park, G. Cho, J. Kwak, Development of an aerosol focusing-laser induced breakdown spectroscopy (aerosol focusing-LIBS) for determination of fine and ultrafine metal aerosols, Aerosol Sci. Technol. 43 (2009) 375–386, https://doi.org/10.1080/02786820802662947.

[179] L.A. Álvarez-Trujillo, A. Ferrero, J.J. Laserna, D.W. Hahn, Alternative statistical methods for spectral data processing: applications to laser-induced breakdown spectroscopy of gaseous and aerosol systems, Appl. Spectrosc. 62 (2008) 1144–1152, https://doi.org/10.1366/000370208786049178.

[180] S.N. Jackson, K.K. Murray, Matrix addition by condensation for matrix-assisted laser desorption/ionization of collected aerosol particles, Anal. Chem. 74 (2002) 4841–4844, https://doi.org/10.1021/ac020181i.

[181] D.B. Kane, M.V. Johnston, Enhancing the detection of sulfate particles for laser ablation aerosol mass spectrometry, Anal. Chem. 73 (2001) 5365–5369, https://doi.org/10.1021/ac010469s.

[182] P. Diwakar, P. Kulkarni, M.E. Birch, New approach for near-real-time measurement of elemental composition of aerosol using laser-induced breakdown spectroscopy, Aerosol Sci. Technol. 46 (2012) 316–332, https://doi.org/10.1080/02786826.2011.625059.

[183] P.K. Diwakar, K.H. Loper, A.-M. Matiaske, D.W. Hahn, Laser-induced breakdown spectroscopy for analysis of micro and nanoparticles, J. Anal. At. Spectrom. 27 (2012) 1110, https://doi.org/10.1039/c2ja30012e.

[184] F.J. Fortes, L.M. Cabalín, J.J. Laserna, Laser-induced breakdown spectroscopy of solid aerosols produced by optical catapulting, Spectrochim. Acta B At. Spectrosc. 64 (2009) 642–648, https://doi.org/10.1016/j.sab.2009.05.006.

[185] M. Abdelhamid, F.J. Fortes, A. Fernández-Bravo, M.A. Harith, J.J. Laserna, Production of aerosols by optical catapulting: imaging, performance parameters and laser-induced plasma sampling rate, Spectrochim. Acta B At. Spectrosc. 89 (2013) 1–6, https://doi.org/10.1016/j.sab.2013.08.003.

[186] F.J. Fortes, A. Fernández-Bravo, J. Javier Laserna, Chemical characterization of single micro- and nanoparticles by optical catapulting-optical trapping-laser-induced breakdown spectroscopy, Spectrochim. Acta B At. Spectrosc. 100 (2014) 78–85, https://doi.org/10.1016/j.sab.2014.08.023.

[187] B.C. Windom, P.K. Diwakar, D.W. Hahn, Dual-pulse laser induced breakdown spectroscopy for analysis of gaseous and aerosol systems: plasma-analyte interactions, Spectrochim. Acta B At. Spectrosc. 61 (2006) 788–796, https://doi.org/10.1016/j.sab.2006.06.003.

[188] M.E. Asgill, M.S. Brown, K. Frische, W.M. Roquemore, D. W. Hahn, Double-pulse and single-pulse laser-induced breakdown spectroscopy for distinguishing between gaseous and particulate phase analytes, Appl. Opt. 49 (2010) C110, https://doi.org/10.1364/AO.49.00C110.

[189] B.C. Windom, D.W. Hahn, Laser ablation—laser induced breakdown spectroscopy (LA-LIBS): a means for overcoming matrix effects leading to improved analyte response, J. Anal. At. Spectrom. 24 (2009) 1665, https://doi.org/10.1039/b913495f.

[190] R. Glaus, D.W. Hahn, Double-pulse laser ablation coupled to laser-induced breakdown spectroscopy, Spectrochim. Acta B At. Spectrosc. 98 (2014) 48–53, https://doi.org/10.1016/j.sab.2014.05.012.

[191] D. Diaz, D.W. Hahn, Laser ablation at high repetition rate coupled to laser-induced breakdown spectroscopy for analysis of non-matrix matched standards, Spectrochim. Acta B At. Spectrosc. 166 (2020) 105795, https://doi.org/10.1016/j.sab.2020.105795.

[192] P.K. Diwakar, P.B. Jackson, D.W. Hahn, The effect of multi-component aerosol particles on quantitative laser-induced breakdown spectroscopy: consideration of localized matrix effects, Spectrochim. Acta B At. Spectrosc. 62 (2007) 1466–1474, https://doi.org/10.1016/j.sab.2007.10.001.

[193] M.E. Asgill, S. Groh, K. Niemax, D.W. Hahn, The use of multi-element aerosol particles for determining temporal variations in temperature and electron density in laser-induced plasmas in support of quantitative laser-induced breakdown spectroscopy, Spectrochim. Acta B At. Spectrosc. 109 (2015) 1–7, https://doi.org/10.1016/j.sab.2015.04.005.

[194] P.S. Dalyander, I.B. Gornushkin, D.W. Hahn, Numerical simulation of laser-induced breakdown spectroscopy: modeling of aerosol analysis with finite diffusion and vaporization effects, Spectrochim. Acta B At. Spectrosc. 63 (2008) 293–304, https://doi.org/10.1016/j.sab.2007.11.023.

[195] M.E. Asgill, D.W. Hahn, Particle size limits for quantitative aerosol analysis using laser-induced breakdown spectroscopy: temporal considerations, Spectrochim. Acta B At. Spectrosc. 64 (2009) 1153–1158, https://doi.org/10.1016/j.sab.2009.07.026.

[196] V. Hohreiter, D.W. Hahn, Plasma-particle interactions in a laser-induced plasma: implications for laser-induced breakdown spectroscopy, Anal. Chem. 78 (2006) 1509–1514, https://doi.org/10.1021/ac051872s.

[197] P.K. Diwakar, S. Groh, K. Niemax, D.W. Hahn, Study of analyte dissociation and diffusion in laser-induced plasmas: implications for laser-induced breakdown spectroscopy, J. Anal. At. Spectrom. 25 (2010) 1921–1930, https://doi.org/10.1039/c0ja00063a.

[198] M.S. Afgan, S. Sheta, Z. Hou, Z. Li, Z. Wang, Understanding the laser-induced aerosol ablation of sub-micron liquid particles: Via size-resolved spectral and image analyses, J. Anal. At. Spectrom. 34 (2019) 2385–2393, https://doi.org/10.1039/c9ja00278b.

[199] I.B. Gornushkin, A.Y. Kazakov, N. Omenetto, B.W. Smith, J.D. Winefordner, Radiation dynamics of post-breakdown laser induced plasma, Spectrochim. Acta B At. Spectrosc. 59 (2004) 401–418, https://doi.org/10.1016/j.sab.2003.12.023.

[200] A.Y. Kazakov, I.B. Gornushkin, N. Omenetto, B.W. Smith, J.D. Winefordner, Radiative model of post-breakdown laser-induced plasma expanding into ambient gas, Appl. Opt. 45 (2006) 2810–2820, https://doi.org/10.1364/AO.45.002810.

CHAPTER

23

Field-portable and handheld LIBS

Giorgio S. Senesi[a], *Russell S. Harmon*[b], *Richard R. Hark*[c]

[a]CNR—Istituto per la Scienza e Tecnologia dei Plasmi (ISTP)—Sede di Bari, Bari, Italy [b]Department of Marine, Earth, and Atmospheric Sciences, North Carolina State University, Raleigh, NC, United States [c]Institute for the Preservation of Cultural Heritage, Yale University, West Haven, CT, United States

1 Introduction

Laser-induced breakdown spectroscopy (LIBS) is an established technique of analytical atomic spectrometry that today is a convenient, versatile, and reliable technique for localized material compositional analysis. LIBS has a persuasive set of advantages that include its ability to rapidly analyze material in any state—solid, liquid, or gas—with little to no sample preparation. LIBS performance under laboratory conditions has been studied for many years and is presently well understood. A multitude of LIBS applications have been described in the literature (see, e.g., [1–7] and references therein). LIBS is sensitive to all elements and elemental concentrations for many elements can be readily determined down to the tens of ppm level or below. In addition, LIBS affords high spatial resolution for lateral and depth compositional profiling. A particularly attractive feature of LIBS is its capability for rapid chemical analysis outside the laboratory, for example, in an industrial facility or in the natural environment under ambient conditions. Quantitative analysis by LIBS is readily achieved using calibration curves generated from matrix-matched standards or through calibration-free methodologies and, when used in conjunction with chemometric techniques and preestablished databases, LIBS spectral analysis is capable of identifying and discriminating unknown materials. These LIBS capabilities become particularly attractive when analysis cannot be done in the laboratory and is especially desirable for applications where real-time analysis is required, such as in-field geological prospecting or environmental surveying, high-speed material identification, manufacturing process control, and interrogation of potentially hazardous materials.

2 Instrumentation design, implementation, and attributes

The development of laboratory-based instrumental techniques for chemical analysis is often accompanied by a desire for material analysis outside the laboratory in two contexts: for on-site analysis situations, such as in an industrial

Laser-Induced Breakdown Spectroscopy
https://doi.org/10.1016/B978-0-12-818829-3.00023-X

facility or museum, where it is difficult or impossible to remove and transport a material of interest to the laboratory and for in situ analysis of natural materials in the field. This demand first drove the development of portable and transportable LIBS systems and then handheld LIBS analyzers within the research community [8]. Because commercial interests have driven the development of handheld LIBS analyzers, technical descriptions are rare in the current published literature, although detailed specifications can be found in patent declarations. However, several recent general reviews are available (e.g., [9–11]). The primary difference between laboratory LIBS systems and their handheld derivatives is that the latter tend to use diode-pumped solid-state (DPSS) lasers with low pulse energies (microjoule to millijoule) and high repetition rate (5–10 kHz), as compared to high pulse energies (> 400 mJ) and low repetition rate (< 10 Hz) lasers employed in laboratory and some mobile systems.

The key components of any portable or handheld LIBS device, which will determine its dimensions, weight, and power requirements are the laser, the optics, and the spectrograph-detector system. From the information published by most instrument companies, 1064-nm Class 3B lasers with repetition rates in the 50 Hz–10 kHz range appear to be the most widely used in handheld LIBS systems, although a variety of lasers have been used in portable systems. In particular, flash lamp-pumped solid-state lasers that use a wavelength of 1500 nm should be preferred because they are eye safe. However, these lasers present several disadvantages, such as low energy conversion rate and poor energy stability, and often need to be equipped with a water-cooling system resulting in reduced portability. However, new systems have been developed recently that consist of small DPSS, microchip, and fiber lasers. In particular, DPSS lasers feature unique advantages, such as compact volume, good optical quality, and small energy fluctuation between pulses. Additionally, microchip lasers (typically semiconductor solid-state lasers) are very appropriate excitation sources for handheld LIBS instruments thanks to their small volume, low energy consumption, high repetition rate, low cost, and small size.

The other two key components of any LIBS analyzer are the spectrograph and the detector, whose operating parameters strongly influence the overall figures of merit of the instrument. Comparatively, not much technological improvement has been achieved in this area compared to laser sources. In particular, the most important parameters of the spectrometer and the detection system are the spectral range, spectral resolution, integration time, and dynamic range. The Czerny-Turner (C-T) and Echelle spectrometers are commonly used in LIBS instruments. A general-purpose laboratory LIBS system commonly requires a large wavelength range from ultraviolet to infrared (i.e., 180–1000 nm) to be covered at a high spectral resolution. This can be achieved via an Echelle design or with a stack of small spectrometers that cover different portions of the overall spectral range. Although an Echelle design appears unsuitable for use in a handheld instrument, the spectrometer optics may be simplified, if a specific analysis is targeted for the instrument, by careful instrument design employing just one or two spectral ranges addressed via one or two miniature spectrometers. Low cost and small size would favor the C-T spectrometer for a handheld LIBS analyzer, but the spectral range and spectral resolution constraints of this device require the use of multichannel combination to achieve a full-spectrum detection. However, the Echelle spectrometer features higher resolution and exhibits significant advantages in the analysis of samples in complex matrices, but its luminous flux is small so that a number of spectral events need to be accumulated to improve the signal-to-noise ratio.

Charge-coupled device (CCD) and intensified charge-coupled device (ICCD) photomultipliers are commonly used as detectors in portable and handheld LIBS systems. As the LIBS spectrum has time-resolved characteristics, the detector depends on the utilization of time gating. Furthermore, the characteristics of time evolution for plasma can be analyzed only by ICCD.

Distinct from the low repetition rate (about 10 Hz) and high pulse energies (about 400 mJ) lasers used in laboratory and in some portable and mobile systems, handheld LIBS instruments often make use of a high repetition rate laser (5–10-kHz range) with low pulse energies (microjoule to millijoule) and narrow pulse width to deliver sufficient energy to the sample to generate the plasma signal. The laser is focused onto a spot of approximately a few tens of micrometers to produce a high-temperature plasma at the sample surface. Thus a microscopic amount of material (around 1 ng) is removed, leaving only a tiny ablation crater of tens of micrometers diameter on the surface of a solid sample. Although the power density at the sample surface can exceed 1 GW/cm^2 during the pulse, the average power density of a single analysis is <1 W, so almost no sample heating occurs around the sampling area. The focusing lens and collection optics are generally mounted on a 3-D stage under processor control, which allows the laser to be rastered over the sample. This permits collection of spectra from multiple locations, which are subsequently averaged by the analyzer to account for spot-to-spot sample nonhomogeneity and for automated focal length adjustment. The spectrometer generally spans a wavelength range of 200–800 nm with approximately 0.1 nm full-width-at-half-maximum resolution, and is gated to reject the large amount of background generated mainly due to Bremsstrahlung radiation during the early stage of plasma creation. Some handheld instruments are provided with an on-board He or Ar canister stored in the analyzer handle that is able to provide a gas purge of the region around the plasma. This gas purge, combined with the spectrometer gating, can improve the limit of detection (LOD) by a factor of 5–20 times depending on the specific element, due to the increased signal from the inert gas environment and/or to the greatly reduced background from the spectrometer gating.

Chemometric methods can be used in LIBS for both signal analysis and data processing. Multivariate statistical analysis has been widely utilized as a means of relating LIBS spectral information to the composition of a sample for identification and discrimination of different materials (see recent reviews in [12–14]). Databases of LIBS spectra have been developed that can be used in chemometric analysis for different types of materials of common interest (e.g., metals, alloys, plastics, concretes, and glasses).

Until quite recently, accurate elemental analysis of different natural and manufactured materials in the field was traditionally and routinely performed using handheld X-ray fluorescence (XRF) and mobile optical emission spectroscopy instruments. In particular, the main competitor of handheld LIBS in field applications is handheld XRF, although both techniques have particular advantages and disadvantages. In the discussion that follows and in the remainder of the chapter, "hXRF" and "hLIBS" are used to designate handheld XRF and LIBS instruments, respectively. The label "pXRF" is also used in many publications to designate an hXRF system. LIBS systems that are portable but can not be considered as handheld instruments are identified as "pLIBS."

The hXRF technique has intrinsic physical limitations for the measurement of elements of low atomic number, typically Si and lighter elements, due to the dramatic drop in fluorescence yield with decreasing atomic number, and is unable to detect Mg and Na reliably due to the absorption of fluorescent X-rays by air. If a portable vacuum unit or He gas purge (options often supplied as accessories by the instrument manufacturer) is used the limit of detection for these lighter elements is improved. The most recent hXRF units are now available with graphene windows, which are strong and much thinner than the Be-foil windows typically used, allowing for enhanced detection of Na, Mg, and Si. Although XRF can be considered a technique that is more mature and quantitative than LIBS, hXRF is burdened by licensing requirements, special operating precautions, and operator safety training because of its open beam X-ray source that can also result in restrictions for transportation. Furthermore, the key hXRF components, i.e., X-ray tubes and detectors, especially Si-drift detectors, are relatively expensive items. Although LIBS generally provides fewer quantitative data unless matrix-matched calibrations are developed for specific applications, it is dramatically faster than XRF for immediate elemental identification. In principle, LIBS can detect any element of the periodic table, but is ideally suited for identification of light elements H, Li, Be, B, C, N, O, and F, an analysis that handheld energy dispersive energy dispersive X-ray fluorescence (ED-XRF) will never be able to perform. An hLIBS instrument does not emit ionizing radiation and, therefore, has no special considerations for operation or transportation, other than the use of an eye-safe laser or a locking system that will prohibit an hLIBS analyzer with a noneye-safe laser from firing unless in contact with a sample, a condition met by the current generation of commercial hLIBS analyzers.

Other important factors to be considered in comparing the two techniques include sample preparation and the spatial size of the area being interrogated for analysis. Being a surface technique, some surface cleaning and/or preparation is sometimes required with hXRF, whereas hLIBS can use the first few laser shots to ablate and clean the surface of a sample. An hXRF instrument typically has a spot size for analysis of several millimeters in diameter, whereas an individual hLIBS shot may cover an area of only 50–300 μm in diameter, so the spatial resolution of hLIBS is much higher than that of hXRF. Although hLIBS typically uses a rastering capability to acquire representative information, this capability also provides opportunity for compositional mapping over a small spatial scale.

The relative complexity of a LIBS spectrum means that it can contain dozens of characteristic emission lines in the ultraviolet, visible, or near-infrared regions for each element present in the sample. This allows for the choice of a specific emission line for analysis that avoids or minimizes interference from other spectral lines. By contrast, the XRF spectrum is relatively simple containing just two to five characteristic lines per element. This lack of alternative lines means that it is not possible to avoid the potential interferences that commonly arise in certain matrices, necessitating the use of fitting algorithms for deconvolution of the spectral overlaps. By contrast, LIBS requires a means to disperse the emitted light to separate it into different wavelength bands, whereas XRF, more specifically ED-XRF, does not require any additional hardware between sample and detector.

There are multiple considerations in comparing LIBS and XRF analyses. In general, a single LIBS spectrum can be obtained in a fraction of a second, whereas typically tens of seconds to minutes are needed to obtain an XRF spectrum. Thus, a first piece of compositional information is obtained more quickly by hLIBS than by hXRF. However, XRF is a completely nondestructive technique, whereas LIBS is microdestructive, as each laser shot removes a fraction of a few nanograms of material from the sample surface forming a small surface crater. Thus a LIBS analysis of a particular spot on a sample cannot be repeated. This represents an additional drawback for LIBS in the analysis of valuable objects such as cultural heritage artifacts, where even a microdestructive analysis may be unacceptable, or of a sample that is too small or too thin to yield a meaningful spectrum without destroying it completely. However, a positive aspect of LIBS is that the removal of material in a shot-by-shot fashion can reveal information about an object's composition as a function of depth, as each laser shot excavates a similar depth of a few tens of nanometers into the sample. Penetration depth is not so clear in the case of XRF, as many factors directly related to the energy of the X-rays are involved (i.e., the higher the X-ray energy, the deeper the X-ray penetrates). In general, X-rays are assumed to penetrate down a few micrometers to several millimeters depending on the nature of the sample matrix. At best, fluorescent X-rays are detectable from a few millimeters within the sample, although for different sample types this could be reduced to a few micrometers or less. This means that bulk analysis of geological samples with a crust or a cultural heritage object with a corrosion layer or patina using hXRF is not straightforward.

Overall, LIBS is certainly complementary to XRF for material analysis and may be more broadly suitable than XRF for handheld analysis because of its ability to identify light elements and perform a faster analysis, although XRF currently has performance advantages for the analysis of base metals and rare earth elements due to its lower limits of detection.

3 Portable, mobile, and handheld LIBS

Initially, attempts to take LIBS from the laboratory to the field centered around what can be described as a "LIBS in a box" approach, i.e., the compression of a typical laboratory apparatus into a briefcase-size box or small mobile cabinet operated off an external power source that could be transported to the site of sample analysis [15–20]. Typically, such systems used a Q-switched Nd:YAG laser source operating at its fundamental wavelength to deliver a low energy pulse of short duration, a gated detector and spectrograph/spectrometer array, and a fiber optic cable to both deliver the laser light to the sample and then collect the subsequent plasma emission.

The first pLIBS instrument, developed at the US Department of Energy Los Alamos National Laboratory in 1996, was constructed in a suitcase measuring 46 × 33 × 24 cm and weighed 14.6 kg [15]. Initially, powered by 115 V AC, it was subsequently upgraded to battery operation [16]. The plasma light produced by the passively Q-switched laser providing 15–20 mJ power at an operational frequency of 1 Hz over a pulse width of 4–8 ns was recorded by a multichannel spectrometer coupled to a CCD detector. This two-part system, consisting of a laser probe tethered to a control and data acquisition apparatus, was used to measure Pb in paint; Ba, Be, Pb, and Sr in soil; and Be and Pb particles collected on filters. Results using this unit were similar to those obtained by a conventional LIBS apparatus and by XRF, with a dependent relationship noted between LOD and laser energy. The US Army Research Laboratory sponsored the development of a similar field-portable LIBS system in 2001 [17]. This LIBS analyzer consisted of two parts in a 23 × 51 × 38-cm aluminum case—a laser-bearing sample probe with an optical fiber for laser light transmission and plasma signal collection and a central detector-analyzer unit housing the spectrometer-detector, timing, power, and data acquisition and analysis equipment. A passively Q-switched single-pulse lamp provided a nominal 15-mJ energy per pulse at 1064 nm at a laser spot size of 60 μm. The system could be operated by both a standard 12-V snowmobile battery and 115-V AC current. Data were stored in a palmtop-type personal computer. For analysis of Pb in paint and soil using the Pb I emission line at 406 nm, this portable unit exhibited similar performance to a laboratory LIBS system with regards to detection over the concentration range of 0.05–1 wt% Pb.

Subsequent pLIBS devices followed the concept of having an optical probe and a hardware unit as two separate components. Thus, initial backpack LIBS designs [21, 22] were a modification of the "LIBS in a box" concept, only in this case the laser and fiber optic cable for light delivery and acquisition were contained in a hand-operated wand and the control apparatus, detector, spectrometer, and data processing hardware were integrated into the backpack unit. The next major step forward toward today's hLIBS analyzers was made by Rakovský et al. [23], who placed the laser and optical elements within a drill-shaped box that was connected by an umbilical containing the optical fiber and wires to power and control the laser. This umbilical was connected to a second box containing a small C-T Ocean Optics HR 2000+ spectrometer with a bandwidth from 200 to 650 nm in wavelength and resolution of 0.4 nm, the batteries, and system control and data processing computer. Total system weight was about 5 kg. Ruggedized pLIBS systems using the umbilical concept were also designed and utilized for remote compositional interrogation inside nuclear power reactors [24] and for undersea analysis of metals [25].

An alternative path followed in taking LIBS outside the laboratory was the development of high-performance mobile systems that could be readily transported offsite to acquire LIBS analyses of the same quality as could be obtained in the laboratory. The best example of this approach is *MODI*, the LIBS system in a mobile cabinet jointly created by the CNR Applied Laser Spectroscopy Laboratory in Pisa and Marwan Technology [26]. *MODI* utilizes a dual-pulse laser that emits two collinear pulses between 50 and 150 mJ at a repetition rate of 10 Hz with a pulse delay between 0 and 60 μs. Measurements can be made either by a probe on an articulated arm or inside a closed chamber that has a motorized table for appropriate positioning of the sample. In either case, the plasma light emission is collected by an optical fiber and transmitted to a compact Echelle spectrometer coupled to an ICCD camera for spectral acquisition and analysis using the calibration-free method [27, 28]. Operation of the system is controlled, and data analysis accomplished, by an integrated computer.

The portable and mobile analyzers just described and those constructed by others based on similar ideas, although somewhat unwieldy and time consuming to employ, performed adequately for simple in-field applications such as the analysis of soil and sediment (e.g., [29–31], cave speleothems [32, 33], oil spill residues [34], historical buildings [35], cultural object analysis on-site in museums [36, 37], and radiological monitoring and forensics [38, 39]).

Further development of small and compact LIBS analyzers was strongly driven during the first decade of the current millennium by the selection of LIBS as one of the technologies to be contained in the *ChemCam* system designed by NASA to analyze the elemental composition of soils and rocks on the surface of Mars by the *Curiosity* rover [40, 41]. Then, the first truly hLIBS analyzer was introduced to the commercial market in 2013 and described in the literature shortly thereafter [42, 43]. Designed to utilize recent advances in optics, compact laser sources, miniaturized high-resolution spectrometers, microelectronics, and computers, several manufacturers that include Bruker [44], B&W Tek [45], Hitachi [46], Rigaku [47], SciAps [48], Thermo-Fisher [49], and TSI [50] are presently producing lightweight battery-powered hLIBS analyzers in a "cordless drill" form factor that can be used by an individual in the field for several hours (Fig. 1). This current generation of handheld instruments offers real-time spectral display (Fig. 2) plus many features present in laboratory systems, including variable gating, gas purging, sample rastering, video targeting, and on-board chemometric analysis that substantially enhance the application of LIBS outside the laboratory.

Today, hLIBS analyzers have a significant presence in the commercial marketplace [10] that addresses the strong, latent demand for real-time compositional analysis outside the research laboratory. In particular, there has been a surge of use by nonspecialists, particularly for real-time chemical analysis in industrial facilities and in the ambient field environment. Applications are as diverse as

FIG. 1 Examples of the current generation of commercial hLIBS analyzers [44, 46–50].

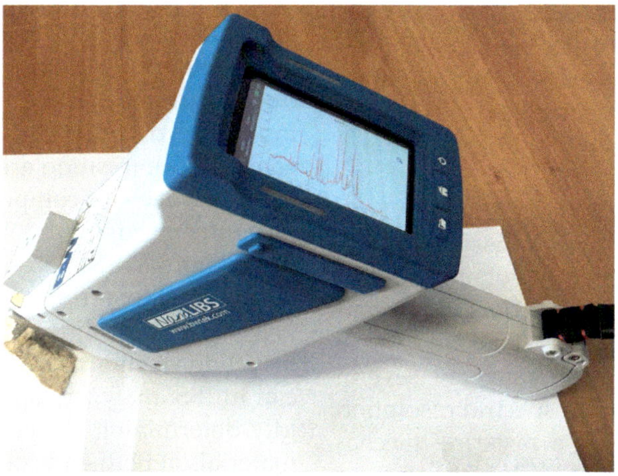

FIG. 2 The B&W Tek *NanoLIBS* handheld instrument showing a LIBS broadband spectrum for a limestone block from the Castello Svevo in Bari, Italy. *Reprinted from G.S. Senesi, D. Manzini, O. De Pascale, Application of a laser-induced breakdown spectroscopy handheld instrument to the diagnostic analysis of stone monuments, Appl. Geochem. 96 (2018) 87–91. Copyright 2018, with permission from Elsevier.*

natural and manufactured material compositional screening, manufacturing process quality control, identification of environmental contaminants, biomedical diagnostics, forensics, and hazardous material identification where speed of analysis is often crucial. Used in combination with an on-board global positioning system capability and with wireless data transmission, hLIBS analyzers are expected to find even more widespread application for real-time chemical analysis in the field.

4 Applications of handheld LIBS

Laboratory instruments will remain the best means for obtaining high-quality material compositional information because of the highly accurate and precise chemical analyses they produce. However, intrinsic disadvantages of laboratory analysis are the sample transportation, long analytical time, and high cost that this typically entails. Also, not all problems require the best possible analysis; often qualitative or survey data will suffice. Handheld instruments can provide rapid analysis in the field and therefore can fill two important needs. They can provide a rapid chemical analysis that is "fit for purpose," insofar as an appropriate answer to a question about composition of material at a site is provided. Alternatively, an expedient provisional analysis can be obtained that can then guide informed in-field collection of the most appropriate samples for later laboratory analysis.

The current generation of hLIBS analyzers, which are classified as either Class 1 or Class 3B laser devices, depending on whether or not they utilize an eye-safe laser, have a rugged design and weigh about 2 kg. Typically, these instruments cover a spectral range of several hundred nanometers, have a spectral resolution of 0.1–0.3 nm, and employ a rastering mechanism to provide either spatial averaging or highly spatially resolved analytical information. The sections that follow describe how hLIBS analysis has been utilized in the recent past and offer a guide to the future application of this exciting new technology. One configuration of a pLIBS analyzer is included, because of its small size and ability to be used in the same way as its handheld counterparts.

4.1 Industrial applications

Every analytical technology requires widespread utilization for success in the marketplace and for hLIBS the largest potential market undoubtedly is in industrial applications. This need, which is pervasive across a variety of different industries and likely to dominate the use of hLIBS into the foreseeable future, is for rapid material identification. This is a particular challenge in both the primary and secondary metals industries. For example, to be economically competitive, iron and steel of different types must be refined and produced at an economically competitive price. Knowledge of composition is critical since both bulk composition and elemental impurities affect behavioral characteristics such as strength, brittleness, and resistance to corrosion. Similarly, knowledge of composition is crucially important in the manufacture of metal alloys, glass, photovoltaic silicon, microelectronics, and many other such industrial materials. Thus accurate and reliable chemical analysis of feedstock, as well as elemental monitoring for quality control during the production process, is required, especially for the multitude of situations in which products of complex composition are produced or in which the grade of every component of a multimaterial assembly needs to be substantiated. Then, as such manufactured products are recycled, there is the ever-expanding challenge for the real-time identification of different metals and alloys in a recycling waste stream for separation. Commercial hLIBS analyzers are ideally suited for rapidly determining the composition of industrial materials at both ends of this industrial applications spectrum.

Positive material identification (PMI) is an approach to quality assurance across multiple industries in which nondestructive testing is utilized to distinguish metals and alloy grades and to guarantee compliance to specification to avoid selecting the wrong material for a specific use or prevent the incorrect mixing of materials.

Although hXRF is presently in routine use for PMI, hLIBS is another highly suitable tool [51], particularly as it can analyze the light elements that are out of reach for hXRF. As noted in [52], hLIBS uniquely can detect Be in bronze, Si and Mg in Al, Li in the Al alloys used in the automotive and aerospace industries, and Al and Mg in high-strength Ti alloys. The speed of analysis makes hLIBS preferable to hXRF for analysis of most steel and Al alloys, but hXRF has the advantage for Cu and Pb alloys, precious metal alloys, and solders because of its better sensitivity and accuracy.

To date, the most successful industrial application of hLIBS has been in steel analysis. The SciAps *Z-200C+* analyzer (Fig. 1) has found widespread use for the determination of C in steel [53]. This instrument employs a pulsed 1064-nm laser, operating at 5.5 mJ/pulse and 50-Hz repetition rate, has an onboard spectrometer that covers a spectral range from 190 to 620 nm, but features an additional dedicated spectrometer that observes the region of the 193-nm C peak at high resolution. The reliable determination of the C content of steel has been documented at 80-ppm LOD at high precision. Calibration is based on analysis of standards having C concentrations up to 0.15% and has proven satisfactory for distinguishing between L- and H-grade steel. It has also been used for C-content measurement in steel pipes to determine their weldability. With appropriate calibration, the SciAps *Z-200* can also be used to analyze other elements such as Si, Al, Ti, V, Cr, Mn, Co, Ni, Cu, Mo, and W in steel and iron-base alloys. For example, accurate Mn:C ratios can rapidly be determined at a precision of 0.03% using the *Z-200* [54].

Corrosion of steel is another concern, particularly in the hydrocarbon exploitation, geothermal, and nuclear power industries. Dry CO_2 in steel pipelines is not a problem, but when wet it generates carbonic acid and related hydrogen ions that result in electrochemical reactions with the metal surface to cause irreversible deterioration. The use of C-steel with elevated Cr content inhibits such corrosion. Thus, there is a need to rapidly determine the Cr content of C-steels. Although hXRF has been used effectively for this purpose, hLIBS offers several advantages as it requires less time for precision measurement at the 0.1% concentration range [50] and as a laser-based technique is not subject to the very restrictive regulations concerning the use of hXRF.

The use of hLIBS for steel analysis has been examined in detail by Afgan et al. [55], who used an hLIBS analyzer weighing just 1.8 kg that was specially designed and produced by B&W Tek (Fig. 2). This instrument employed a miniature DPSS laser operating at the high repetition rate of 2 kHz at a maximum energy output of 200 mW and pulse duration of <1 ns. LIBS spectra are recorded by a compact spectrometer covering a spectral window of 185–680 nm at an overall resolution of 0.4 nm operated in a nongated mode. Steel samples of different composition were analyzed and calibration curves developed for C, Cr, Mn, Mo, Si, V, and Cu (Fig. 3) using dominant factor-based partial least squares (PLS) regression with spectral standardization. Respective average absolute measurement errors of 0.019%, 0.039%, 0.013%, and 0.001% were achieved for Si, Cr, Mn, and Ni and overall average relative standard deviation for quantitative analysis of the eight elements was <5%.

During steel production, pig iron is an intermediate product derived from the mixing of iron ore, coke, and limestone in a blast furnace. A brittle metal with high C content in its solid state, pig iron is then melted and refined into steel in an oxygen furnace. The chemical composition of pig iron determines its class and fitness to meet the characteristics required for the production of a specific type of steel. Typically, the measurement of Si in the laboratory is undertaken to measure pig iron quality. Senesi et al. [56] used a B&W Tek *NanoLIBS* hLIBS instrument (Fig. 2) to analyze pig iron in open air. Emission lines for Si, Mn, and Fe were measured

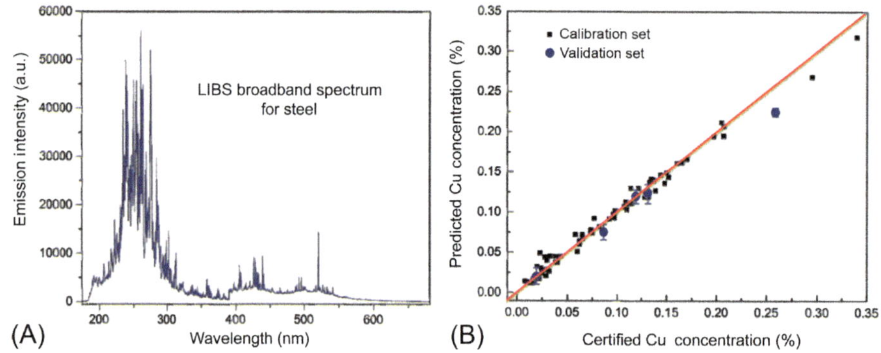

FIG. 3 (A) Broadband LIBS spectrum for steel acquired using a B&W Tek *NanoLIBS* hLIBS analyzer and (B) a calibration and validation data for Cu in different steels using a partial least squares model after spectrum normalization. *Reprinted and modified from M.S. Afgan, Z. Hou, Z. Wang, Quantitative analysis of common elements in steel using a handheld μ-LIBS instrument, J. Anal. Atom. Spectrom. 32(10) (2017) 1905–1915. Copyright 2017, with permission from the Royal Society of Chemistry.*

within a 280–315-nm spectral window, with results comparing favorably to those obtained using a typical LIBS benchtop system. The *NanoLIBS* instrument consists of a miniature diode-pumped, solid-state, short-pulsed laser emitting at 1064-nm wavelength and operating at a maximum pulse energy 150 μJ, a high repetition rate between 1 and 5 kHz, and a pulse duration of 500 ps. It utilizes a compact, low-resolution spectrometer covering the spectral range from 180 to 800 nm. This apparatus is contained in a lightweight handheld body. A rechargeable Li-ion battery provides up to 8 h of autonomous use.

The recycling of Al is important because of the reduced amount of energy required to produce Al from recycled stock compared to primary bauxite ore. However, impurities such Si, Mg, Fe, Cu, Cr, Mn Ni, Zn, and Pb in recycled Al material streams can be problematic and need to be identified. The Rigaku *Katana KT-100 μLIBS* analyzer (Fig. 1) was created with this application in mind [57], although it can also be used for metal-grade determination, alloy composition verification, and positive material identification. The laser utilized generates pulses of 120-μJ energy at a user-selected frequency between 100 and 1000 Hz and the C-T spectrometer-CCD detector covers a wavelength range from 200 to 480 nm at an average resolution of <0.2 nm. Calibrations were developed for Si and Mg, the two most important light elements in common Al alloys. Similarly, the Hitachi *Vulcan Expert* hLIBS analyzer (Fig. 1) was designed for the measurement of the minor constituents (Cu, Mg, Mn, Si, and Zn) in Al alloys [58]. The Rigaku *Katana KT-100 μLIBS* analyzer (Fig. 1) has been used to identify and quantify light atomic number elements in Cu alloys, such as Al-bronzes, Si-bronzes, and Be-coppers [59]. Similarly, the SciAps *Z-series* hLIBS analyzer (Fig. 1) was used to determine the amount of Li in Al alloys [84] employing the spectral line at 610 nm for quantifying Li down to a lower limit of about 0.01%.

Bennett et al. [60] presented calibration curves for a suite of commercial Al and Cu alloys obtained using a SciAps *Z-500* hLIBS analyzer (Fig. 1). This instrument uses a preset laser energy of 5 mJ/pulse and a pulse duration of 5 ns, with the plasma emission collected over a wavelength range of 188–675 nm. The laser variability in shot-to-shot laser energy per pulse is 6%–7%, producing an energy variation between 4.7 to 5.3 mJ/pulse. The LIBS investigation was carried out on the as-machined surface for four Al and Cu alloys without additional preparation. LIBS spectral data were collected for three different locations on each sample employing a 12-point grid pattern. The spectra for the three datasets were averaged and, after correction of emission line intensities by background subtraction, used to produce calibration curves for the matrix Cu and Al, as well as Mg, Mn, and Zn in the Al alloys and Fe, Ni, and Zn in the Cu alloys.

Lithium is an alkali metal that is used widely across the aerospace, battery, energy, glass and ceramic, lubricant, metal, pharmaceutical, and rubber industries, but information about its purity is required for most applications. For example, battery-grade Li typically contains traces of Na, K, and Ca. The catalyst-grade metallic Li used as a precursor in the synthesis of various pharmaceuticals and as a catalyst for polymerization can contain up to 1% Na. Metallic Li is commonly combined with Mg and Al to improve the strength of alloys and make them lighter, as in the Mg-Li alloy used for armor plating or the Al-Li alloys used in aircraft, bicycle frames, and high-speed trains. In all of these applications, compositional information about the trace element content of Li metal or the precise amount of Mg in Li-Mg alloy or Al in Li-Al alloy is necessary for production process control and product quality assurance. The B&W Tek *NanoLIBS* hLIBS analyzer (Fig. 2) has been utilized successfully for Li alloy analysis [61]. Calibration curves developed for Li-Mg and Li-Al alloys were used for accurate and precise quantitative analysis of Al and Mg in daily production quality control processes at an Li factory, eliminating the need to transport samples to an analytical laboratory.

The light element Be is used as a metal and in alloys in nuclear power reactors, aerospace applications, electrical equipment, and missile fuel. There is particular interest in the on-site analysis of Be during nuclear plant decommissioning. The SciAps *Z-series* hLIBS analyzers (Fig. 1) have been designed with this application in mind and can be used for in-field Be measurements as well as the quantification of Be in alloys, soils, and other materials at a 5-ppm LOD [62]. Alloys of Ti are also widely used in the aerospace industry for airframe, engine, helicopter, and space applications because of their favorable weight and resistance to corrosion. Rapid and reliable measurement of Ti contents of alloys at test stations during production or subsequently at scrapyards for identification and sorting is readily accomplished by hLIBS with customized software [63].

The production of hydrofluoric acid (HF) comprises the primary industrial use of F worldwide. Typically, the mineral fluorite (CaF_2) is used as the raw material for HF production, with polymetallic ores and skarns being a common natural source. Such material must be purified to the high grade required by a froth flotation process that requires accurate knowledge of feedstock F content, which is difficult and time consuming using conventional techniques. Foucaud et al. [64] used a SciAps *Z-300* hLIBS (Fig. 1) instrument to analyze mineral powders produced from rock samples containing fluorite and other F-bearing minerals derived from flotation tests on a tungsten-skarn ore. Samples were interrogated under a constant Ar gas flow to enhance the emission intensities and minimize the effect of air on the LIBS signal. The use of two CaF molecular bands located between 529 and 543 and 590–606 nm was investigated for F quantification over a wide range of F contents that ranged from 1.5 to 40.7 wt%. Particle size was shown to have a significant effect on LIBS signal intensity, with differences between the 10–150-μm and <20-μm size powders observed without a correlation between size range and signal magnitude. Normalized peak areas under the two molecular bands were processed by multivariate regression to correct for matrix effects. A formula was derived to predict F content at a mean average error of <2.2%, demonstrating that a precise and accurate F quantification by hLIBS during online processing of feedstock fluorite is possible.

4.2 Geological and environmental applications

Knowledge of the Earth's composition is necessary to understand its origin and continuing evolution. Therefore, geologists have been obtaining chemical analyses for minerals and rocks since the late 19th century and, more recently, chemical analysis has become a foundational tool of the environmental scientist. Because of the long-standing need for analytical technology for field use, there has been a rapid uptake of hLIBS for material analysis in the field within the geological and environmental science communities.

Rakovský et al. [23] were the first to demonstrate in 2012 that an hLIBS instrument could be used to analyze geological samples in the field. In this case, geochemical signatures were obtained to assess the potential of hLIBS as a tool for geologists in two circumstances: the recognition of tephra layers in lacustrine sediments and fossilization processes in ammonites using the bespoke hLIBS instrument described earlier. A clear difference in Al, Ca, Ti, Ba, and Na abundances was observed during depth profiling in a sequence of lacustrine chalk sediments of the Jura region and an intercalated Laacher See tephra layer. Al and Na were not detected by hXRF, whereas K, Cr, Mn, Co, Zn, Rb, Sr, and Zr were determined by hXRF, but not detected by hLIBS. Phosphatization is an important process during fossilization, but could not be examined because the spectral lines for P at 213.6, 214.9, 253.6, and 255.3 were not detected. Emission intensity depth variations for Fe, Ca, and Al as well as ratios of Fe/Ca and Al/Ca provided information about pyritization. The negative correlation of Ca and Fe is compatible with the replacement of Ca by Fe during the pyritization process.

Roux et al. 2015 [65] analyzed volcanic rocks in situ in the field in Iceland using a bespoke pLIBS instrument weighing 5 kg designed and assembled by the Laboratory Interdisciplinaire Carnot de Bourgogne (ICB) in a hand drill form factor. This analyzer incorporates a pulsed Nd:YAG flash lamp laser that delivers 40-mJ energy at a pulse duration of approximately 5 ns. A transparent silica window located between the focusing lens inside a hollow aluminum cone protects the laser from particles generated by ablation. Shot residues that could damage the silica window are evacuated by a membrane pump. The analyzer is made safe because laser shots are only possible when there is direct contact between the sample and aluminum cone. Shot location is displayed in a color camera pointed at the zone to be analyzed by two crossed red laser beams. A fiber optic in an umbilical cable transmits the plasma light to a C-T Ocean Optics HR 2000+ spectrometer with a spectral range of 200–650 nm in wavelength and an optical resolution of about 0.4 nm contained in a separate housing. Spectral data are analyzed by an embarked laptop using a software package developed at the ICB. An analysis of 19 atomic lines for Al, Ba, Ca, Cr, Cu, Fe, Mg, Mn, Na, Si, Sr, and Ti for 21 volcanic rocks was performed successfully in the field. This study demonstrated that hLIBS could be used to sort volcanic rocks belonging to three different magmatic series at a 90%–100% success rate.

A year later, Connors et al. [43] described use of a commercial hLIBS analyzer, the SciAps *Z-500* instrument, to analyze different types of geological samples. The *Z-500* uses the DPSS, short-pulsed, Q-switched proprietary PULSAR 1064-nm Nd:YAG laser with 50-μm focused beam size. This laser delivers 6 mJ of energy to the sample at 1-ns pulse duration with a user-selected laser firing rate between 1 and 10 Hz. The instrument is capable of gas purging that will deliver an inert gas (He or Ar) directly to the focusing area on the sample surface where the LIBS plasma formation occurs. The LIBS emission signal generated by the microplasma is collected by a fiber optic cable and delivered into four spectrometers with respective spectral ranges of 180–254, 254–314, 314–423, and

423–675 nm that use time-gated CCD detectors with resolution of 0.1 nm below 423 nm and 0.3 nm above 423 nm. Spectra are typically collected with a 1-μs delay time between the laser pulse and beginning of light collection, although variable gate delays from 250 ns up 100 μs as well as operation in a nongated mode are possible. The *Z-500* is equipped with a translational stage to enable 3-D rastering of the laser beam across a sample surface area up to 2 mm^2. All of the SciAps *Z-series* hLIBS instruments are equipped with a quartz window to protect the internal components from the particles generated via the ablation process. System control and data processing are accomplished by an internal computer with an Android operating system with graphic user interface. The *Z-500* analyzer, and its recently introduced *Z-300* companion instrument, have two data processing capabilities that are particularly useful for geochemical analysis. The *GeoChem* mode provides a quantitative analysis based upon either calibration curves loaded at the factory or bespoke calibration generated by the user using the SciAps proprietary *Profile Builder PC* software and allows conversion of elemental concentrations to oxides in user selectable units of measurement of wt% or ppm. The *GeoChem Pro* mode for qualitative analysis can identify the spectral peaks of specific elements, generates estimated elemental concentration maps based upon the relative intensities of these peaks across a raster pattern [66], and facilitates the collection of metadata. All data and maps can be exported for subsequent offline analysis using third-party geochemical software. Sensitivity has been documented for samples in a solid or powder form for a wide variety of elements that includes Al, B, C, Ca, Cu, Fe, Mg, Mn, Na, Ni, S, Ti, and Zn at percent levels; Cr, Ba, and Pb, at 1000 ppm levels; and Be and Li at <100-ppm levels. A set of application notes illustrates how the SciAps *Z-series* hLIBS analyzers can be used for the quantification of C in soil, which is of particular interest to the agricultural and climate research communities [67], and by the mineral exploration community for the detection, localization, and/or quantification of Au, Li, and base metals [66, 68, 69].

In a wide-ranging study of geological applications of hLIBS, Harmon et al. [70] used the SciAps *Z-500* analyzer to demonstrate how the concept of geochemical fingerprinting could be used to distinguish carbonate minerals and rocks, the oxide mineral pair columbite-tantalite, which is a common "conflict" mineral, the silicate mineral garnet, which is a "pathfinder" mineral in kimberlite exploration, and native gold of different provenances. The distinguishing chemical characteristics of the diverse types of geological material samples analyzed were identified through multivariate statistical analysis, demonstrating how hLIBS can be used to distinguish between similar samples based on minor and trace constituents. Fig. 4 illustrates the discrimination of the divalent metal carbonate minerals and the elemental loadings upon which the successful differentiation was based, whereas the successful discrimination of modern carbonate sediment accumulations across a wide range of depositional environments from palustrine and tidal flats to shallow marine and kilometer-deep basins is shown in Fig. 5. This study demonstrated how hLIBS could be used for real-time chemical analysis in the field for element or mineral identification and applications such as stratigraphic correlation, provenance determination, and natural resource exploration via spectral matching with a predetermined and assembled spectral library for materials of interest.

Fdez-Ortiz de Vallejuelo et al. [71] combined in situ measurement by an IVEA *EasyLIBS* analyzer and chemometric data processing using principal component analysis (PCA) of complete LIBS spectra after outlier identification and removal by a *k*-nearest neighbor (KNN) approach to assess water quality in the urban Deba River catchment in Spain during extreme rainfall events. The handheld instrument uses

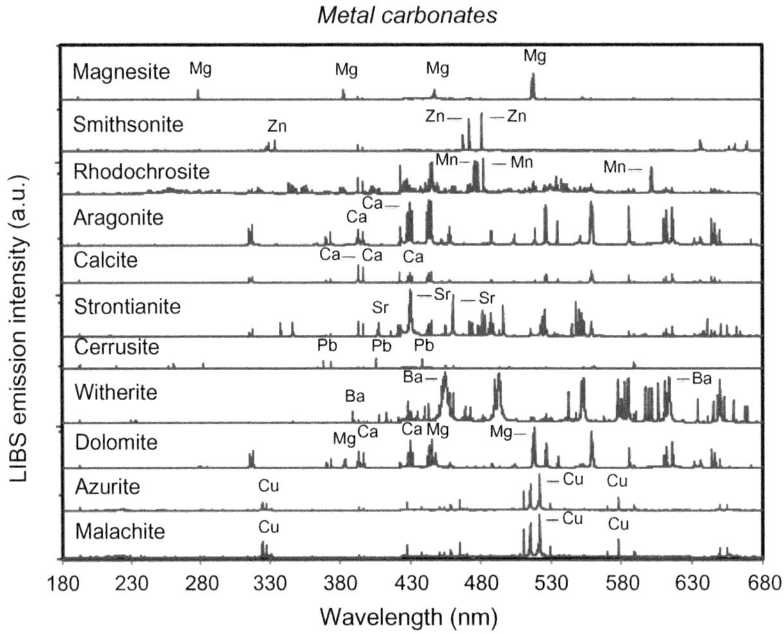

FIG. 4 Single-pulse LIBS spectra acquired with a SciAps *Z-500* handheld analyzer for a suite of divalent metal carbonate minerals. *From top to bottom*: magnesite—$MgCO_3$, smithsonite—$ZnCO_3$, rhodochrosite—$MnCO_3$, aragonite—orthorhombic $CaCO_3$, calcite—hexagonal $CaCO_3$, strontianite—$SrCO_3$, cerrusite—$PbCO_3$, witherite—$BaCO_3$, dolomite—$CaMg(CO_3)_2$, azurite—$Cu_3(CO_3)_2(OH)_2$, malachite—$Cu_2CO_3(OH)_2$. The chemical compositions of the different carbonate minerals are clearly differentiated in the LIBS spectra acquired across the spectral range of 180–680 nm on the basis of their primary cationic constituent—Mg, Zn, Mn, Ca, Sr, Pb, Ba, and Cu.

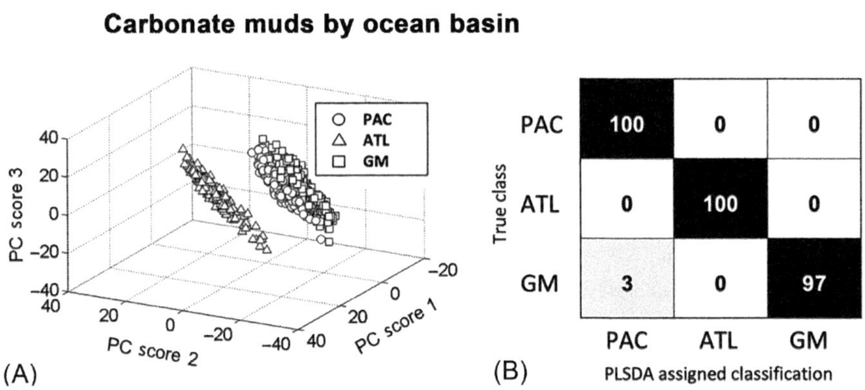

FIG. 5 Principal component analysis (PCA) score plot (A) and partial least squares-discriminant analysis classification matrix (B) for carbonate muds from the south-central Pacific Ocean (PAC: Maupiti Atoll and Nonouti Atoll), western Atlantic Ocean (ATL: Caicos Platform and Crooked-Acklins Platform), and Gulf of Mexico (GM: Alacran Reef). With 896 of 930 LIBS spectra acquired over the spectral range from 180 to 675-nm using a SciAps *Z-500* handheld analyzer correctly classified (B), these six localities of carbonate mud deposition were successfully discriminated at an overall success rate of >93%. The fact that the Atlantic and Pacific localities group together in the PCA plot (A) also implies that seawater chemistry likely influences the composition of carbonate mud deposits at the ocean basin spatial scale as well as at the local scale.

a pulsed Nd:YAG laser emitting at the fundamental wavelength of 1064 nm with a pulse energy in excess of 25 mJ at a pulse duration of 4–5 ns and repetition rate of 1 Hz. The laser is contained in a handheld probe attached via an umbilical to a larger unit that houses the spectrometers, batteries, and computer control components. Measurements were performed in the dual-pulse mode using an optimized delay time of 50 μs. Suspended particulate matter samples composed of colloids and particles of different sizes, including heterogeneous aggregates of mineral, organic, and biological phases, were collected every 2 h during five storm events at an automatic water sampler in three stations of an urban river in Spain and subsequently analyzed by the pLIBS system. After baseline correction and mean centering of the LIBS spectra, PCA analysis focused on the spectral range between 198 and 552 nm revealed the toxic elements Pb, Cr, Ni, and Cu in the suspended matter as important discriminant variables. Additional PCA analysis was able to differentiate between monitored localities and the timing of the storm event. The environmental risk of these riverine pollutants was assessed through correlation between variables representing the different sources (urban/municipal contamination, anthropogenic activities, etc.) of the contaminant elements.

Harmon et al. [72] compared broadband LIBS emission spectra for fresh and unaltered volcanic samples of varying age and composition from two tectonic settings utilizing both a bespoke laboratory system and the SciAps *Z-500* hLIBS analyzer. The different volcanic sources were differentiated through multivariate chemometric analysis after the chemical characteristics of the different volcanic regions and centers were distinguished (Fig. 6). Classification and discrimination were accomplished successfully using a pattern recognition approach. Novel use of similarity analysis, a technique not previously utilized to evaluate the quality of LIBS spectral data, indicated that it is necessary to capture the full spectral heterogeneity of different sample types to maximize performance of chemometric classifiers.

Senesi et al. [73] used a B&W Tek *NanoLIBS* hLIBS (Fig. 2) analyzer to compositionally interrogate and discriminate an actual iron meteorite from a suspected meteorite fragment labeled as a "meteor-wrong" and a pig iron product (Fig. 7). A calibration-free approach was used to quantify the major element contents of Fe, Ni, and Co plus the trace elements Ga and Ir in the iron meteorite and Fe, Mn, Si, and Ti in the other two fragments. As the LIBS quantitative data were only partially satisfactory for the identification and classification of metallic objects of extraterrestrial origin with respect to human artifacts, improvements of the hLIBS instrument were deemed necessary to increase analytical sensitivity.

In a companion paper, Senesi et al. [74] used the same hLIBS analyzer to develop an innovative strategy for meteorite classification based on the integration of automated feature selection and supervised learning in a fuzzy rule-based classifier that could be easily interpreted to provide information on the nature of meteorites and discriminate among iron, stone, stony-iron meteorites, and "meteor-wrongs." The statistical model, which was based on a set of fuzzy rules that could be easily interpreted by a nonspecialist operator, was able to classify correctly 25 of 26 samples, whereas only 18 samples were correctly discriminated by partial least squares-discriminant analysis (PLS-DA).

Gómez-Nubla et al. [75] presented quantitative analysis of Si, Al, Mg, Ca, Ba, Na, and Fe in weathered terrestrial analogs to meteorites using the *EasyLIBS* pLIBS analyzer. The predictive performance of a PLS regression model for discrimination was used to evaluate LIBS spectral data for glasses from Darwin, Australia; the Libyan Desert of western Egypt; and black steel slag residues from steelworks in the Basque country of Spain. Analytical results obtained by pLIBS were in good agreement with the range of

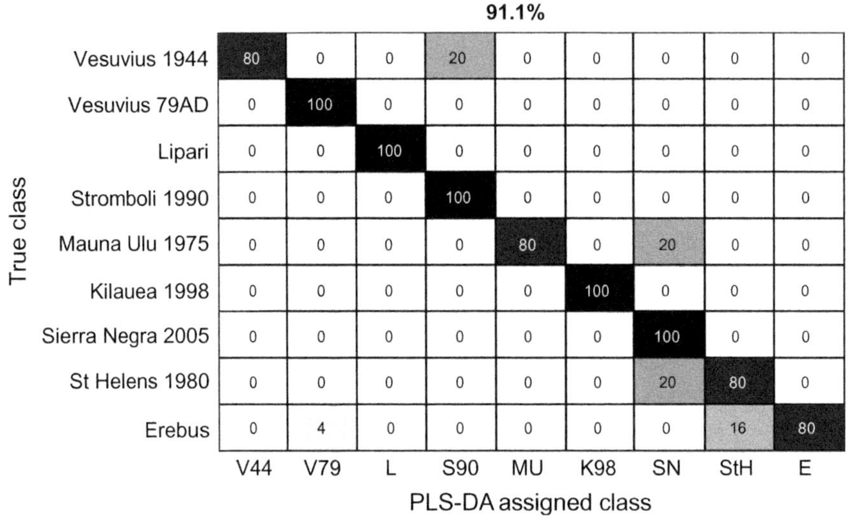

FIG. 6 PLS-DA classification matrix for nine geologically young volcanic centers in Ecuador (Sierra Negra in the Galapagos Islands), Italy (Vesuvius, Stromboli, and Lipari), Antarctica (Mount Erebus), and the USA (Mount St. Helens plus Mauna Ulu and Kilauea in the Hawaiian Islands) based upon analysis of five measurement sites on each sample by handheld LIBS over the spectral range from 180 to 675 nm using a SciAps *Z-500* handheld analyzer. Measurement cross-validation resulting in 45 folds (5 measurement sites × 9 samples) produced an overall classification success of >91%. *Reprinted from R.S. Harmon, C.S. Throckmorton, R.R. Hark, J.L. Gottfried, G. Wörner, K. Harpp, L. Collins, Discriminating volcanic centers with handheld laser-induced breakdown spectroscopy (LIBS), J. Archaeol. Sci. 98 (2018) 112–127. Copyright 2018, with permission from Elsevier.*

composition measured by XRF. The observation that the PLS regression approach provided better prediction than a univariate calibration for these types of silicate-based materials subjected to high temperatures during their formation is relevant for the development of calibration strategies for other types of similar materials, such as silicate meteorites and other extraterrestrial materials.

Yant et al. [76] and the Project Exploration Science Pathfinder Research for Enhancing Solar System Observations team utilized hLIBS in the context of human space exploration to answer specific geological questions at field analog sites. A SciAps *Z-300* LIBS instrument (Fig. 1) was used as part of a field campaign at the Potrillo Volcanic Field in southern New Mexico to measure the elemental composition of sequences of ash/pyroclastic material at the Kilbourne Hole maar and basaltic material/surface coatings in inflation domes. Three distinct color layers within the ~3.5-m sequence of a lava flow were measured with the default calibrations of the hLIBS analyzer. Then, the underlying portion of the lava flow was analyzed after removal of the superficial layer to determine if distinct compositions could be detected among surface coatings and the underlying rock with respect to the surface. The data acquired allowed the researchers to compare results from LIBS and Raman handheld devices and were helpful in evaluating the capacity of such field portable instruments to solve meaningful geochemical questions and assess the relationship of chemistry to both outcrop-scale compositional spatial variability and small-scale sedimentological heterogeneity (ash, lithics, xenoliths, diagenetic components, etc.). Both handheld instruments proved to be extremely attractive as candidate tools for geochemical mapping, resource prospecting, sample selection, and hazard identification by future astronauts.

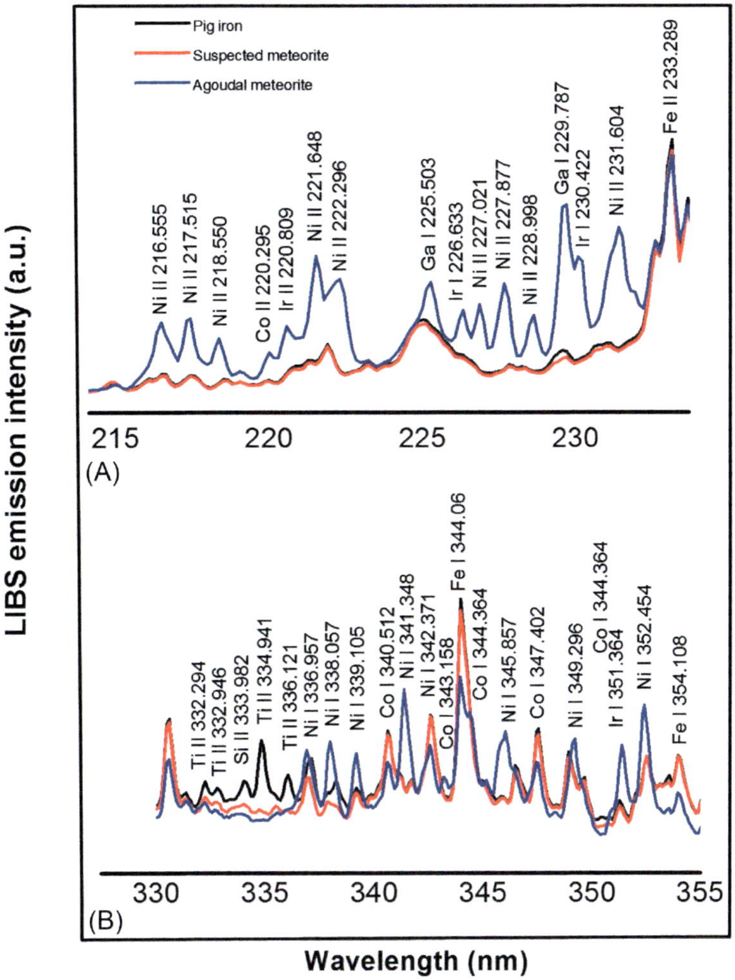

FIG. 7 LIBS spectra acquired on a B&W Tek *NanoLIBS* handheld LIBS analyzer in (A) the range 215–240 nm and (B) 330–355 nm of the Agoudal iron meteorite, a suspected meteorite, and a pig iron sample. *Reprinted from G.S. Senesi, P. Manzari, G. Tempesta, G. Agrosì, A.A. Touchnt, A. Ibhi, O. De Pascale, Handheld laser induced breakdown spectroscopy instrumentation applied to the rapid discrimination between iron meteorites and meteor-wrongs, Geostand. Geoanal. Res. 42 (4) (2018) 607–614. Copyright 2018, with permission from Wiley.*

Although hLIBS technology is very promising for use in geological field investigations, it has been noted by Senesi [77] that the technology still needs important technological advances, including improvements in detection limits and analytical precision. Importantly, matrix-matched calibration materials are not available for soil, mineral, and rock analysis. Also, the user interfaces of commercial instruments are not yet designed for specific use by geologists and environmental scientists. These deficiencies makes rapid interpretation of geochemical data in the field especially difficult. However, the manufacturers of commercial hLIBS instruments are continually refining and improving their products to expand the scientific user base.

In particular, more work is required to define the appropriate operational parameters for analysis of a wider range of lithologies and soil types, develop the best practice for measuring trace elements, and refine sampling protocols to be used with hLIBS instrumentation. Algorithms should be developed that allow data obtained by hLIBS to yield reliable geochemical fingerprints for a variety of geological materials in real time. Progress to date has demonstrated clearly that hLIBS can be an especially valuable tool and prospects for future applications in this area are particularly bright.

4.3 Recent applications in other areas

Traditional laboratory LIBS analysis has been demonstrated to be a valuable analytical tool across a broad spectrum of disciplines, particularly in areas such as biological and biomedical science, archeology and cultural resources management, forensic analysis, and security and defense, where rapid compositional analysis under ambient environmental conditions is required [1–7]. Thus, there is every reason to expect a rapid growth in the use of hLIBS in these and related areas. For example, compositional analysis of biological materials and as a diagnostic tool for health and biomedical use are beneficial in three areas: (i) the determination of elements in various media that are hazardous to human health, (ii) elemental detection and quantification in pathogenic organisms and biomedical specimens, and (iii) the use of unique elemental compositions for biomedical diagnostics. Some potential applications in the forensics and security domains include detection of gunshot residues and illicit drugs, material analysis at crime scenes, detection of explosive materials and toxic substances, and validation of currency and documents of uncertain origin. Art, cultural heritage objects, and artifacts at archeological sites consist of a wide variety of materials, sometimes in forms, structures, and dimensions that make analysis challenging. Since precious art and artifacts are unique, noninvasive or minimally invasive techniques are preferred for their analysis. Additionally, some objects, such as wall paintings and monument building stone, cannot be transferred to an analytical laboratory because of their fragility, their location outdoors, or as in situ parts of building structures. Even in situations where an object could be moved to the laboratory for analysis, transportation may be prohibited to avoid any risk of damage. At archeological sites, analysis of artifacts in real time during an excavation can provide in situ contextual information. Rapid analysis upon discovery can also be beneficial as compositional changes may occur to archeological objects once exposed to the ambient environment. Understanding the state of deterioration on the surface and along a depth profile of an object is necessary to make the most appropriate subsequent laboratory diagnostic measurements or to choose the most appropriate means for restoration or conservation intervention. Mobile or portable instruments for in situ chemical analysis are ideally suited for such applications. Indeed, hLIBS has been shown to be a valuable technique to perform microdestructive, in situ, diagnostic analysis and also in-depth elemental profiling of surficial contaminants, encrustations, and multilayer materials. The examples that follow illustrate the powerful potential of hLIBS in such areas.

Myers et al. [78] used a developmental pLIBS system comprised of a Kigre MK-88 Q-switched erbium glass laser operating at 1.54 μm with integrated optical beam launch, a detection system employing a Stellarnet EPP2000-UV2-14 fiber spectrometer covering the spectral range of 200–400 nm, and a laptop computer loaded with *SpectraWiz* and laser controller software for the noninvasive, in situ detection of malignant skin tissue. It was determined that relative changes in the intensities of Ca and Al emission lines in human skin provided a basis for discrimination between healthy and unhealthy tissue.

Senesi et al. [79] presented hLIBS analysis of a limestone fragment from a masonry block from the Castello Svevo in Bari, Italy, that is characterized by environmental degradation that has produced a black superficial crust. This study utilized the B&W Tek *Nano-LIBS* analyzer (Fig. 2) operating at a pulse energy of 150 μJ with a pulse duration of 500 ps and a repetition rate between 1 and 5 kHz. For this work, spectrometer integration time was set at 4 ms, which corresponds to 12 laser pulses. The rastering beam covered an area of about 300 μm diameter. Each measurement of 6 sec duration accumulated 1800 laser pulses and produced 150 spectra averaged for each analysis. Three such measurements at 10 different positions were averaged for each sample point on the surface crust and for the limestone core to obtain a single average geochemical fingerprint in the form of a LIBS broadband spectrum based on the presence of eight elements: Al, C, Ca, Fe, K, Mg, Na, and Si. As would be expected, Ca, Mg, and C are major constituents of the unaltered limestone, whereas the altered surface layer is characterized by the presence of Al, Fe, and Si attributed to solid particulates deposited from atmospheric pollution plus lesser amounts of Na and K derived from the marine aerosol that is pervasive across the coastal, urban area in which the monument is located.

Syvilay et al. [80] used the *EasyLIBS* pLIBS instrument to analyze and discriminate 228 obsidian samples from four source-islands in the western Mediterranean Sea. Obsidian is a volcanic glass widely used by prehistoric populations to make tools and weapons. Knowledge of the geological origin of obsidian artifacts is used by archeologists to reconstruct prehistoric population movements and interactions. LIBS spectra were analyzed by an advanced data processing supervised multivariate approach, called quadratic discriminant analysis (QDA), which enabled the successful discrimination and sorting of the obsidian samples on the basis of nine variables (the elements Al, Ba, Ca, Mg, Rb, Si, Fe, Zr, and Mn) whereas some of the elements detected (K, Na, Ti, and Sr) were not observed to contribute to discrimination. Thus, the combination of a pLIBS instrument with the QDA approach appears to be a viable, efficient, and promising option for on-site obsidian analysis and sourcing at archeological sites.

Illegal harvesting of trees listed as endangered by the Convention on International Trade in Endangered Species of Fauna and Flora, which includes 58 species of the genus *Dalbergia* (commonly known as rosewood), is a worldwide problem. Celani et al. [81] assessed the utility of hLIBS as a means of *Dalbergia* species identification. Broadband LIBS spectra were acquired from 10 different tree samples using a SciAps *Z-200C* hLIBS analyzer (Fig. 1) for each of the seven *Dalbergia* species from Madagascar, India, Malaysia, Tanzania, Brazil, Honduras, and Mexico and two other species of exotic hardwood from Central and South America. After data preprocessing to decrease baseline variability and Whittaker filtering, PLS-DA models based on a binary decision tree structure were applied to the spectral dataset using predictive KNN models determined from a 72-sample training set and then to an 18-sample test set that yielded a 94% success rate for classification. This success highlighted the potential for hLIBS as a tool for rapid customs screening of imported hardwoods at ports of entry to screen tree samples for a mismatch to the species reported on the cargo manifest.

Manard et al. [82] explored the use of hLIBS for its feasibility to perform rapid, semiquantitative analysis of Be on fibrous swipe materials for industrial hygiene applications. Quantification of Be on solid substrates having 0.63 and 4.98 mg Be was accomplished through swipe measurement by hLIBS. Spatial analysis of the distribution of Be on a surface was performed by rastering across a desired pattern that was translated into a chemical map.

5 Conclusions and perspectives

The development and construction of mobile and pLIBS systems for use outside the laboratory was facilitated by recent laser, spectrometer, and detector technological progress. The introduction of compact and user-friendly hLIBS by the commercial sector just a few years ago was followed by rapid uptake for applications in the industrial workplace and by researchers for real-time analysis in the field. Continuing technological advances are expected to drive down instrumentation cost, which presently can be prohibitive for small- and medium-size laboratories, thus further advancing analytical opportunities for geologists, archeologists, and conservators as well as for new biological, biomedical, forensic, security, and defense applications. For example, LIBS can be used to rapidly identify materials in the field and select only appropriate samples that may need further analysis by conventional laboratory approaches. The use of hLIBS appears to be particularly promising in the cultural heritage field for performing in situ analysis of outdoor monuments, at archeological sites, and of murals and cave paintings, and to guide restoration and conservation efforts.

The studies reviewed in this chapter provided robust evidence that analysis by hLIBS, when used in conjunction with chemometric data processing, is a powerful and convenient means for performing in situ geochemical analysis and for classifying and discriminating geological materials. The performance of hLIBS instruments to date has demonstrated that analyses performed in the field can yield results comparable to those obtained in the laboratory.

Although hLIBS is a very promising tool for geological field investigations, further technological progress is required, including the increase of elemental detection limits and improvements in analytical precision. Furthermore, the user interfaces and calibrated materials are not yet designed for specific use by geologists, environmental scientists, biologists, and other specialists, which renders difficult the rapid and direct interpretation of hLIBS data.

Commercial manufacturers are, however, refining their products by continuous development of new and emerging hLIBS analyzers with the aim of expanding the scientific user base. In particular, more work is required to define best operational parameters for analysis of a wider range of samples, to develop the best practices for measuring trace elements, and to refine appropriate sampling protocols to be used with hLIBS instrumentation. In this same context, appropriate data processing algorithms need to be developed for the real-time analysis of a wider range of materials, both natural and synthetic.

Although recent studies have demonstrated the potential of hLIBS, it will only be through continued investigation of its strengths and weaknesses, further operational improvements, and the development of best practice protocols for its use in the field that hLIBS will become an essential tool for environmental studies, archeology, terrestrial field geology, and even for future planetary field exploration.

References

[1] Y.I. Lee, K. Song, J. Sneddon, Laser-Induced Breakdown Spectrometry, Nova Publishers, Hauppauge, NY, 2000.

[2] A.W. Miziolek, V. Palleschi, I. Schechter (Eds.), Laser Induced Breakdown Spectroscopy, Cambridge University Press, Cambridge, 2006.

[3] J.P. Singh, S.N. Thakur (Eds.), Laser-Induced Breakdown Spectroscopy, Elsevier, Amsterdam, 2007.

[4] D.W. Hahn, N. Omenetto, Laser-induced breakdown spectroscopy (LIBS), part I: review of basic diagnostics and plasma–particle interactions: still-challenging issues within the analytical plasma community, Appl. Spectrosc. 64 (12) (2010) 335A–366A.

[5] D.W. Hahn, N. Omenetto, Laser-induced breakdown spectroscopy (LIBS), part II: review of instrumental and methodological approaches to material analysis and applications to different fields, Appl. Spectrosc. 66 (4) (2012) 347–419.

[6] D.A. Cremers, L.J. Radziemski, Handbook of Laser-Induced Breakdown Spectroscopy, Wiley, Hoboken, New Jersey, 2013.

[7] S. Musazzi, U. Perini (Eds.), Laser-Induced Breakdown Spectroscopy, Theory and Applications, Springer, New York, 2014.

[8] F.J. Fortes, J.J. Laserna, The development of fieldable laser-induced breakdown spectrometer: no limits on the horizon, Spectrochim. Acta B 65 (12) (2010) 975–990.

[9] J. Rakovský, P. Čermák, O. Musset, P. Veis, A review of the development of portable laser induced breakdown spectroscopy and its applications, Spectrochim. Acta B 101 (2014) 269–287.

[10] Q.-Y. Lin, Y.-X. Duan, Laser-induced breakdown spectroscopy: from experimental platform to field instrument, Chin. J. Anal. Chem. 45 (9) (2017) 1405–1414.

[11] R.A. Crocombe, Portable spectroscopy, Appl. Spectrosc. 72 (12) (2018) 1701–1751.

[12] T.L. Zhang, W.U. Shan, T.A.N.G. Hong-Sheng, W. Kang, D. Yi-Xiang, L.I. Hua, Progress of chemometrics in laser-induced breakdown spectroscopy analysis, Chin. J. Anal. Chem. 43 (6) (2015) 939–948.

[13] P. Pořízka, J. Klus, E. Képeš, D. Prochazka, D.W. Hahn, J. Kaiser, On the utilization of principal component analysis in laser-induced breakdown spectroscopy data analysis, a review, Spectrochim. Acta B 148 (2018) 65–82.

[14] T. Zhang, H. Tang, H. Li, Chemometrics in laser-induced breakdown spectroscopy, J. Chemometrics 32 (11) (2018) e2983.

[15] K.Y. Yamamoto, D.A. Cremers, M.J. Ferris, L.E. Foster, Detection of metals in the environment using a portable laser-induced breakdown spectroscopy instrument, Appl. Spectrosc. 50 (2) (1996) 222–233.

[16] B.C. Castle, A.K. Knight, K. Visser, J.D. Winefordner, Battery powered laser-induced plasma spectrometer for elemental determinations, J. Anal. At. Spectrom 13 (7) (1998) 589–595.

[17] R.T. Wainner, R.S. Harmon, A.W. Miziolek, K.L. McNesby, P.D. French, Analysis of environmental lead contamination: comparison of LIBS field and laboratory instruments, Spectrochim. Acta B 56 (6) (2001) 777–793.

[18] S. Palanco, A. Alises, J. Cunat, J. Baena, J.J. Laserna, Development of a portable laser-induced plasma spectrometer with fully-automated operation and quantitative analysis capabilities, J. Anal. At. Spectrom 18 (8) (2003) 933–938.

[19] G. Cristoforetti, S. Legnaioli, V. Palleschi, L. Pardini, A. Salvetti, E. Tognoni, New perspectives in LIBS analysis of polluted soils, in: International Conference on Lasers, Applications, and Technologies 2005: Laser Technologies for Environmental Monitoring and Ecological Applications, and Laser Technologies for Medicine, Proc. SPIE, vol. 6284, 2006, p. 628405.

[20] J. Goujon, A. Giakoumaki, V. Piñon, O. Musset, D. Anglos, E. Georgiou, J.P. Boquillon, A compact and portable laser-induced breakdown spectroscopy instrument for single and double pulse applications, Spectrochim. Acta B 63 (10) (2008) 1091–1096.

[21] F.C. DeLucia, A.C. Samuels, R.S. Harmon, R.A. Walters, K.L. McNesby, A. LaPointe, R.J. Winkel, A.W. Miziolek, Laser-induced breakdown spectroscopy (LIBS): a promising versatile chemical sensor technology for hazardous material detection, IEEE Sensors J. 5 (4) (2005) 681–689.

[22] J. Cuñat, F.J. Fortes, L.M. Cabalín, F. Carrasco, M. D. Simón, J.J. Laserna, Man-portable laser-induced breakdown spectroscopy system for in-situ characterization of karstic formations, Appl. Spectrosc. 62 (11) (2008) 1250–1255.

[23] J. Rakovský, O. Musset, J. Buoncristiani, V. Bichet, F. Monna, P. Neige, P. Veis, Testing a portable laser-induced breakdown spectroscopy system on geological samples, Spectrochim. Acta B 74 (2012) 57–65.

[24] A.I. Whitehouse, J. Young, I.M. Botheroyd, S. Lawson, C.P. Evans, J. Wright, Remote material analysis of nuclear power station steam generator tubes by laser-induced breakdown spectroscopy, Spectrochim. Acta B 56 (6) (2001) 821–830.

[25] S. Guirado, F.J. Fortes, V. Lazic, J.J. Laserna, Chemical analysis of archeological materials in submarine environments using laser-induced breakdown spectroscopy. On-site trials in the Mediterranean Sea, Spectrochim. Acta B 74 (2012) 137–143.

[26] A. Bertolini, G. Carelli, F. Francesconi, M. Francesconi, L. Marchesini, P. Marsili, F. Sorrentino, G. Cristoforetti, S. Legnaioli, V. Palleschi, L. Pardini, Modì: a new mobile instrument for in-situ double-pulse LIBS analysis, Anal. Bioanal. Chem. 385 (2) (2006) 240–247.

[27] A. Ciucci, M. Corsi, V. Palleschi, S. Rastelli, A. Salvetti, E. Tognoni, New procedure for quantitative elemental analysis by laser-induced plasma spectroscopy, Appl. Spectrosc. 53 (8) (1999) 960–964.

[28] E. Tognoni, G. Cristoforetti, S. Legnaioli, V. Palleschi, Calibration-free laser-induced breakdown spectroscopy: state of the art, Spectrochim. Acta B 65 (1) (2010) 1–14.

[29] R.S. Harmon, F.C. De Lucia, A.W. Miziolek, K. L. McNesby, R.A. Walters, P.D. French, Laser-induced breakdown spectroscopy (LIBS)–an emerging field-portable sensor technology for real-time, in-situ geochemical and environmental analysis, Geochem. Explor. Environ. Anal. 5 (1) (2005) 21–28.

[30] M. Corsi, G. Cristoforetti, M. Hidalgo, S. Legnaioli, V. Palleschi, A. Salvetti, E. Tognoni, C. Vallebona, Double pulse, calibration-free laser-induced breakdown spectroscopy: a new technique for in-situ standard-less analysis of polluted soils, J. Appl. Geochem. 21 (5) (2006) 748–755.

[31] J. Cuñat, F.J. Fortes, J.J. Laserna, Real time and in-situ determination of lead in road sediments using a man-portable laser-induced breakdown spectroscopy analyzer, Anal. Chim. Acta 633 (1) (2009) 38–42.

[32] J. Cuñat, S. Palanco, F. Carrasco, M.D. Simon, J. J. Laserna, Portable instrument and analytical method using laser-induced breakdown spectrometry for in-situ characterization of speleothems in karstic caves, J. Anal. At. Spectrom 20 (4) (2005) 295–300.

[33] F.J. Fortes, I. Vadillo, H. Stoll, M. Jiménez-Sánchez, A. Moreno, J.J. Laserna, Spatial distribution of paleoclimatic proxies in stalagmite slabs using laser-induced breakdown spectroscopy, J. Anal. At. Spectrom 27 (5) (2012) 868–873.

[34] F.J. Fortes, T. Ctvrtnícková, M.P. Mateo, L.M. Cabalín, G. Nicolas, J.J. Laserna, Spectrochemical study for the in-situ detection of oil spill residues using laser-induced breakdown spectroscopy, Anal. Chim. Acta 683 (1) (2010) 52–57.

[35] F.J. Fortes, J. Cuñat, L.M. Cabalín, J.J. Laserna, In-situ analytical assessment and chemical imaging of historical buildings using a man-portable laser system, Appl. Spectrosc. 61 (5) (2007) 558–564.

[36] M. Grozeva, P. Penkova, In-situ LIBS analysis of valuable museum objects: the Mogilanska tumulus ritual knemida, in: R. Radvan, S. Akyuz, M. Simileanu (Eds.), Proceedings of the Third Balkan Symposium on Archaeometry: The Unknown Face of the Artwork, İstanbul Kültür University, İstanbul, 2012, , pp. 103–107.

[37] K.L. Rasmussen, A.L. Tenorio, I. Bonaduce, M. P. Colombini, L. Birolo, E. Galano, A. Amoresano, G. Doudna, A.D. Bond, V. Palleschi, G. Lorenzetti, The constituents of the ink from a Qumran inkwell: new prospects for provenancing the ink on the Dead Sea Scrolls, J. Archaeol. Sci. 39 (9) (2012) 2956–2968.

[38] D.A. Cremers, A. Beddingfield, R. Smithwick, R. C. Chinni, C.R. Jones, B. Beardsley, L. Karch, Monitoring uranium, hydrogen, and lithium and their isotopes using a compact laser-induced breakdown spectroscopy (LIBS) probe and high-resolution spectrometer, Appl. Spectrosc. 66 (3) (2012) 250–261.

[39] E. Hondrogiannis, D. Andersen, A.W. Miziolek, The evaluation of a new technology for gunshot residue (GSR) analysis in the field, in: Next-Generation Spectroscopic Technologies VI, International Society for Optics and Photonics, SPIE Proc., vol. 8726, 2013, p. 87260P.

[40] R.C. Wiens, S. Maurice, B. Barraclough, M. Saccoccio, W.C. Barkley, J.F. Bell, S. Bender, J. Bernardin, D. Blaney, J. Blank, M. Bouyé, The ChemCam instrument suite on the Mars science laboratory (MSL) rover: body unit and combined system tests, Space Sci. Rev. 170 (1–4) (2012) 167–227.

[41] S. Maurice, R.C. Wiens, M. Saccoccio, B. Barraclough, O. Gasnault, O. Forni, N. Mangold, D. Baratoux, S. Bender, G. Berger, J. Bernardin, The ChemCam instrument suite on the Mars Science Laboratory (MSL) rover: science objectives and mast unit description, Space Sci. Rev. 170 (1–4) (2012) 95–166.

[42] D. Day, B. Connors, M. Jennings, J. Egan, K. Derman, P. Soucy, S. Moller, D. Sackett, A full featured handheld LIBS analyzer with early results for defense and security, in: international society for optics and photonics, next-generation spectroscopic technologies VIII, SPIE Proc. 9482 (2015) 948206.

[43] B. Connors, A. Somers, D. Day, Application of handheld laser-induced breakdown spectroscopy (LIBS) to geochemical analysis, Appl. Spectrosc. 70 (5) (2016) 810–815.

[44] Bruker, Inc., EOS 500—Handheld LIBS, https://www.bruker.com/products/x-ray-diffraction-and-elemental-analysis/libs/eos-500/overview.html, 2015. Accessed 22 November 2019.

[45] B&W Tek, Inc., NanoLIBS®, https://bwtek.com/products/nanolibs/, 2016. Accessed 22 November 2019.

[46] Hitachi, Inc., Handheld LIBS analyzers, Vulcan+ range, https://hha.hitachi-hightech.com/en/product-range/products/handheld-xrf-libs-analysers/handheld-libs-analysers, 2019. Accessed 22 November 2019.

[47] Rigaku, Inc., KT-100S—LIBS Analyzer, https://www.rigaku.com/en/products/libs/kt100s/overview, 2019. Accessed 22 November 2019.

[48] SciAps, Inc., Z-300 LIBS Analyzer, https://www.sciaps.com/libs-handheld-laser-analyzers/z-300/, 2016. Accessed 22 November 2019.

[49] ThermoFisher Scientific, Inc., Niton Apollo Handheld LIBS Analyzer, https://www.thermofisher.com/us/en/home/industrial/spectroscopy-elemental-isotope-analysis/portable-analysis-material-id/industrial-elemental-radiation-solutions/niton-apollo-handheld-libs-analyzer.html, 2019. Accessed 22 November 2019.

[50] TSI, Inc., ChemLite™ Plus Metal Analyzer 4240, https://tsi.com/products/metal-analyzers/chemlite-plus-laser-metals-analyzer-4240/, 2018. Accessed 22 November 2019.

[51] Rigaku, Inc., Positive Material Identification (PMI) of Metal Alloys Using Handheld Libs, Application Note. https://www.rigaku.com/en/products/libs/kt100s/app003, 2018. Accessed 22 November 2019.

[52] J.I. Patterson, HH-XRF and HH-LIBS for Alloy Analysis: Choosing the Right Tool for the Right Job, Recycling Today, (2015, October 2) 30. Available from. https://www.recyclingtoday.com/article/rt1015-xrf-libs-handheld-analyzers/.

[53] SciAps, Inc., Carbon Analysis in Stainless and Carbon Steels with Handheld LIBS, Application Note. https://www.sciaps.com/wp-content/uploads/2017/09/SciAps-ApNote_Carbon-Analysis-HH-LIBS.pdf?x24702, 2018. Accessed 22 November 2019.

[54] SciAps, Inc., In-field Measurement of Mn:C Ratios for Steel Embrittlement with Z LIBS Handheld Analyzer, Application Note. https://www.sciaps.com/wp-content/uploads/2016/07/SciAps-ApNote_Mn-C-ratios-steel.pdf?x24702, 2017. Accessed 22 November 2019.

[55] M.S. Afgan, Z. Hou, Z. Wang, Quantitative analysis of common elements in steel using a handheld μ-LIBS instrument, J. Anal. At. Spectrom. 32 (10) (2017) 1905–1915.

[56] G.S. Senesi, G. Dilecce, A. Bove, O. De Pascale, Laser-induced breakdown spectroscopy (LIBS) in open air applied to the elemental analysis, specifically Si, in: Pig Iron Samples. Comparison from a Handheld Instrument and a Bench-Top Apparatus, Thirteenth Frontiers in Low-Temperature Plasma Diagnostics & First Frontiers in Low-Temperature Plasma Simulations, Bad Honnef, Germany, May 13–16, 2019.

[57] S. Piorek, Rapid sorting of aluminum alloys with handheld μLIBS analyzer, Mater. Today-Proc. 10 (2019) 348–354.

[58] Hitachi, Inc., LIBS For Sorting Aluminium Alloys, Application Note. https://hha.hitachi-hightech.com/assets/uploads/downloads/handheld-analysers/vulcan/3626_HHA_Vulcan_Application%20Note_1_EN_DIGITAL.pdf, 2017. Accessed 22 November 2019.

[59] Rigaku, Inc., Accurate Identification of Light Elements in Copper Alloys Using Handheld LIBS, Application Note. https://www.rigaku.com/en/products/libs/kt100s/app001, 2016. Accessed 22 November 2019.

[60] B.N. Bennett, M.Z. Martin, D.N. Leonard, E. Garlea, Calibration curves for commercial copper and aluminum alloys using handheld laser-induced breakdown spectroscopy, Appl. Phys. B Lasers Opt. 124 (3) (2018) 42.

[61] B&W Tek, Inc., Lithium alloy measurement using B&W Tek's NanoLIBS-Q Handheld LIBS Elemental Analyzer, Application Note. http://bwtek.com/appnotes/lithium-alloy-measurement-using-bw-teks-nanolibs-q-handheld-libs-elemental-analyzer/, 2017. Accessed 13 September 2019.

[62] SciAps, Inc., In-Field Analysis of Beryllium in Alloys and Other Materials Using Hand Held Laser Induced Breakdown Spectroscopy [LIBS], Application Note. https://www.sciaps.com/wp-content/uploads/2016/08/SciAps_ApNote_Be-in-Alloys-and-Soil.pdf?x24702, 2017. Accessed 22 November 2019.

[63] TSI, Inc., Titanium Scrap Sorting and Alloy Identification, Application Note. https://www.tsi.com/solutions/solutions-by-application/scrap-metal-sorting-and-recycling/titanium-scrap-sorting-and-alloy-identification/, 2016. Accessed 22 November 2019.

[64] Y. Foucaud, C. Fabre, B. Demeusy, I.V. Filippova, L. O. Filippov, Optimisation of fast quantification of fluorine content using handheld laser induced breakdown spectroscopy, Spectrochim. Acta B 158 (2019) 105628–105639.

[65] C.P. Roux, J. Rakovský, O. Musset, F. Monna, J. F. Buoncristiani, P. Pellenard, C. Thomazo, In-situ laser induced breakdown spectroscopy as a tool to discriminate volcanic rocks and magmatic series, Iceland, Spectrochim. Acta B 103 (2015) 63–69.

[66] SciAps, Inc., Where's the Gold: Identification of Gold and its Mineralogical Associations Using SciAps Z-300 Handheld LIBS, Application Note. https://www.sciaps.com/wp-content/uploads/2018/12/SciAps_ApNote_wheresthegoldRGBfinal.pdf?x24702, 2018. Accessed 14 September 2019.

[67] SciAps, Inc., Rapid Measurement of Total Organic Carbon in Soil Using SciAps Z-300 Handheld LIBS, Application Note. https://www.sciaps.com/wp-content/uploads/2018/12/SciAps_ApNote_Carbon-in-SoilsFinalRGB.pdf?x24702, 2018. Accessed 14 September 2019.

[68] SciAps, Inc., Lithium Concentrations in Brines: Instant, On-Site Measurements Using Handheld LIBS, Application Note. https://www.sciaps.com/wp-content/uploads/2016/07/SciApsApNote_Lithium-Concentrations-in-Brines_Feb2017.pdf?x24702, 2017. Accessed 13 September 2019.

[69] SciAps, Inc., In-Field Measurements of Major Elements and Base Metals with Handheld LIBS, Application Note. https://www.sciaps.com/wp-content/uploads/2014/02/SciAps_Atomic_10.2015.pdf?x24702, 2015. Accessed 13 September 2019.

[70] R.S. Harmon, R.R. Hark, C.S. Throckmorton, E.C. Rankey, M.A. Wise, A.M. Somers, L.M. Collins, Geochemical fingerprinting by handheld laser-induced breakdown spectroscopy, Geostand. Geoanal. Res. 41 (4) (2017) 563–584.

[71] S. Fdez-Ortiz de Vallejuelo, A. Gredilla, L. Gomez-Nubla, E. Ruiz-Romera, A. Zabaleta, J.M. Madariaga, Portable laser induced breakdown spectrometry to characterize the environmental impact of potentially hazardous elements of suspended particulate matter transported during a storm event in an urban river catchment, Microchem. J. 135 (2017) 171–179.

[72] R.S. Harmon, C.S. Throckmorton, R.R. Hark, J.L. Gottfried, G. Wörner, K. Harpp, L. Collins, Discriminating volcanic centers with handheld laser-induced breakdown spectroscopy (LIBS), J. Archaeol. Sci. 98 (2018) 112–127.

[73] G.S. Senesi, P. Manzari, G. Tempesta, G. Agrosì, A. A. Touchnt, A. Ibhi, O. De Pascale, Handheld laser induced breakdown spectroscopy instrumentation applied to the rapid discrimination between iron meteorites and meteor-wrongs, Geostand. Geoanal. Res. 42 (4) (2018) 607–614.

[74] G.S. Senesi, P. Manzari, A. Consiglio, O. De Pascale, Identification and classification of meteorites using a handheld LIBS instrument coupled with a fuzzy logic-based method, J. Anal. At. Spectrom 33 (2018) 1664–1675.

[75] L. Gómez-Nubla, J. Aramendia, S.F.O. de Vallejuelo, J. M. Madariaga, Analytical methodology to elemental quantification of weathered terrestrial analogues to meteorites using a portable laser-induced breakdown spectroscopy (LIBS) instrument and partial least squares (PLS) as multivariate calibration technique, Microchem. J. 137 (2018) 392–401.

[76] M.H. Yant, K.W. Lewis, A.H. Parker, S.M. Hörst, A.C. McAdam, C.A. Knudson, The project ESPRESSO team, project ESPRESSO: exploration roles of handheld LIBS at the Potrillo volcanic field, in: Fiftieth Lunar and Planetary Science Conference 2019 (LPI Contrib. No. 2132), 2019.

[77] G.S. Senesi, Portable handheld laser-induced breakdown spectroscopy (LIBS) instrumentation for in-field elemental analysis of geological samples, Int. J. Earth Sci. 2 (2017) 146.

[78] M.J. Myers, J.D. Myers, B. Guo, C. Yang, C.R. Hardy, J.A. Myers, A.G. Myers, S.M. Christian, Non-invasive in-situ detection of malignant skin tissue and other abnormalities using portable LIBS system with fiber spectrometer and eye-safe erbium glass laser, International Society for Optics and Photonics, Optical Diagnostics and Sensing VIII, SPIE Proc. 6863 (2008) 68630W.

[79] G.S. Senesi, D. Manzini, O. De Pascale, Application of a laser-induced breakdown spectroscopy handheld instrument to the diagnostic analysis of stone monuments, Appl. Geochem. 96 (2018) 87–91.

[80] D. Syvilay, B. Bousquet, R. Chapoulie, M. Orange, F. X. Le Bourdonnec, Advanced statistical analysis of LIBS spectra for the sourcing of obsidian samples, J. Anal. At. Spectrom. 34 (5) (2019) 867–873.

[81] C.P. Celani, C.A. Lancaster, J.A. Jordan, E.O. Espinoza, K.S. Booksh, Assessing utility of handheld laser induced breakdown spectroscopy as a means of Dalbergia speciation, Analyst 144 (17) (2019) 5117–5126.

[82] B.T. Manard, M.F. Schappert, E.M. Wylie, G. E. McMath, Investigation of handheld laser induced breakdown spectroscopy (HH LIBS) for the analysis of beryllium on swipe surfaces, Anal. Methods 11 (6) (2019) 752–759.

Further reading

SciAps, Inc., Fast, Laser-based Cr Analysis for Flow Accelerated Corrosion (FAC) Applications, Application Note. https://www.sciaps.com/fast-laser-based-cr-analysis-for-flow-accelerated-corrosion-fac-applications/, 2016. Accessed 22 November 2019.

SciAps, Inc., The Z: Handheld LIBS Analysis for Lithium in Aluminum Alloys, Application Note. https://www.sciaps.com/wp-content/uploads/2014/02/LIBZ_Li-and-Al-alloys_MAR2016.pdf?x24702, 2016. Accessed 22 November 2019.

CHAPTER

24

Self-calibrated methods for LIBS quantitative analysis

Stefano Legnaioli, Beatrice Campanella, Stefano Pagnotta, Francesco Poggialini, Vincenzo Palleschi

Applied and Laser Spectroscopy Laboratory, Institute of Chemistry of Organometallic Compounds, Pisa, Italy

1 Introduction

The laser-induced breakdown spectroscopy (LIBS) technique offers unique advantages with respect to other traditional analytical methods. The use of the same tool—the laser—to sample and excite the materials under study greatly simplifies the experimental setup with respect to other methods that separate the two steps, for example, using laser ablation for sampling and inductively coupled plasma torch for excitation (laser ablation-inductively coupled plasma-optical emission spectrometry) or ionization (laser ablation-inductively coupled plasma-mass spectrometry).

On the other hand, what is gained in simplicity and flexibility of the experimental setup is lost in terms of interpretation of the LIBS spectra and increased difficulty in quantification of the results obtained. LIBS suffers from a very strong matrix effect; although a number of analytical workarounds have been developed in the last five decades to overcome these effects (for a critical discussion see, e.g., the works of Tognoni et al. [1, 2]) the problem is still perceived as important in all the LIBS analytical approaches involving the building of calibration curves. The quantitative analysis of minor elements in a given matrix is generally obtained by using matrix-matched standards for building calibration curves, which are then used for quantification of the analyte concentration in the unknown samples. However, in some situations it can be problematic obtaining enough matrix-matched standards for building a suitable calibration curve. It has been remarked in the past [3] that the standard IUPAC recommendation for quantitative analysis using spectrochemical methods [4] is often disattended in LIBS analysis, and the main reason for this is the difficulty of obtaining a sufficient number of matrix-matched standards for the calibration curve. The availability of matrix-matched standards is a potential limit to the development of important applications of the LIBS technique; studies of biological samples, nuclear industry materials, and forensic analysis are just a few examples.

Laser-Induced Breakdown Spectroscopy
https://doi.org/10.1016/B978-0-12-818829-3.00024-1

In a calibration-based strategy, the experimental parameters must be kept constant during the building of the calibration curve and, after that, during the measurements on the unknown samples; hence, the need for matrix-matched standards, since matrix composition variations may affect the measured LIBS line intensity of the analyte(s). However, the changes in concentration of the analyte during the process of construction of the calibration curve could be neglected only when the analyte is present at very low concentrations; on the contrary, LIBS is often used for the analysis of major components, whose change in concentration during the construction of the calibration curve might cause macroscopic changes in laser-sample interaction and evidence the sign of the matrix effect. In some cases, the LIBS signal is even found to decrease with the increase in analyte concentration (Fig. 1).

This apparently strange behavior has been discussed and explained by Borisov et al. [5] in the framework of an inductively coupled plasma mass spectrometric (ICP-MS) study of the laser ablation of binary Cu-Zn alloys. The authors found that the anomalous behavior of the Cu calibration curve is due to changes in the reflectivity of the sample surface associated with the changes in Zn concentration, which result in a decrease in the laser energy absorbed and, consequently, of the laser-ablated mass. Since the LIBS signal is proportional to the ablated mass, its reduction reflects in the corresponding reduction of the signal reported in Fig. 1.

In the case reported, a way for recovering a meaningful calibration curve is plotting the ratio between the LIBS intensities of two Zn and Cu lines versus the ratio of their concentrations (Fig. 2). Note that the intensity ratio is independent of the ablated mass and the corresponding calibration curve is acceptably linear, except at low Zn concentration, where the presence of other minor elements in the calibration sample alloy produces deviations from the ideal behavior of a binary alloy.

Even when the ablated mass does not vary appreciably during the building of the

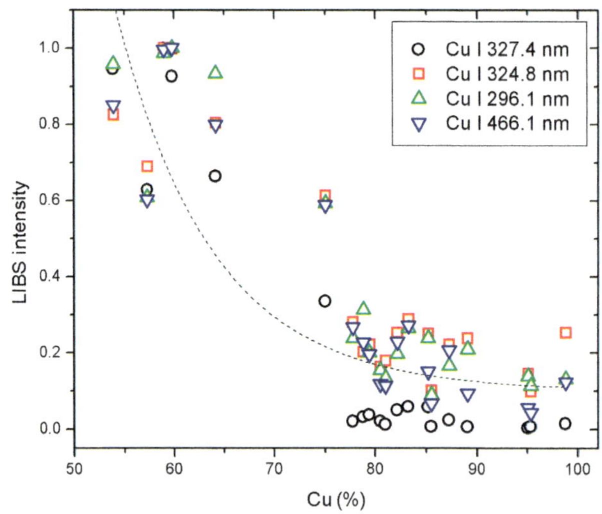

FIG. 1 Calibration curve of copper in brass (Cu-Zn) samples. Four Cu I lines are considered. The laser-induced breakdown spectroscopy line intensities were normalized at their maximum. The *dashed line* is a guide for the eye, showing the decrease in the Cu LIBS signal with the Cu concentration.

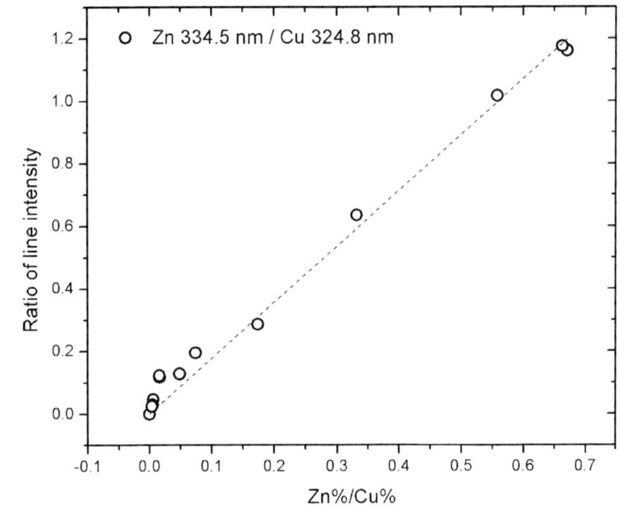

FIG. 2 Calibration curve for the concentration ratio Zn/Cu in brass samples. The *dashed line* is the best linear fit of the experimental data.

calibration curve, other effects can affect its linearity and precision. Panne et al. [6] in 1998 proposed to correct the experimental values of the LIBS intensities by measuring the electron temperature of the LIBS plasma and taking into account its variation from sample to sample during the construction of the calibration curve and during the analysis.

Under suitable approximations, the measured LIBS intensity I corresponding to a concentration C of a given analyte depends on the plasma electron temperature as:

$$I \propto C \frac{g_k e^{-\frac{E_k}{k_B T}}}{U(T)} A_{ki} \tag{1}$$

where g_k and E_k are degeneracy and energy of the upper level of the transition, respectively, k_B is the Boltzmann constant, $U(T)$ is the partition function, and A_{ki} is the transition probability. According to Panne, the line intensities can thus be normalized to the same effective plasma temperature T_0:

$$\bar{I} = I e^{-\frac{E_k}{k_B T_0} + \frac{E_k}{k_B T}} \tag{2}$$

where T_0 can be taken, for example, as the average electron temperature obtained for the calibration samples (the variation of the partition function U with temperature can usually be neglected).

The normalization of the LIBS intensities to the same temperature T_0 has been recently used by Castellon et al. [7] to improve the linearity of Hg calibration curves for LIBS measurements on dental amalgams (Fig. 3).

2 The calibration-free approach

At the end of the last century, the group led by Vincenzo Palleschi in Pisa arrived at the conclusion that the matrix effect in LIBS is produced by the changes in the physical and chemical properties of the laser-induced plasma produced by changes in the sample composition. In particular, changes in the concentration of a given element might influence the LIBS signal of all the other elements in the plasma, through changes of global parameters of the plasma, namely total particle density, electron number

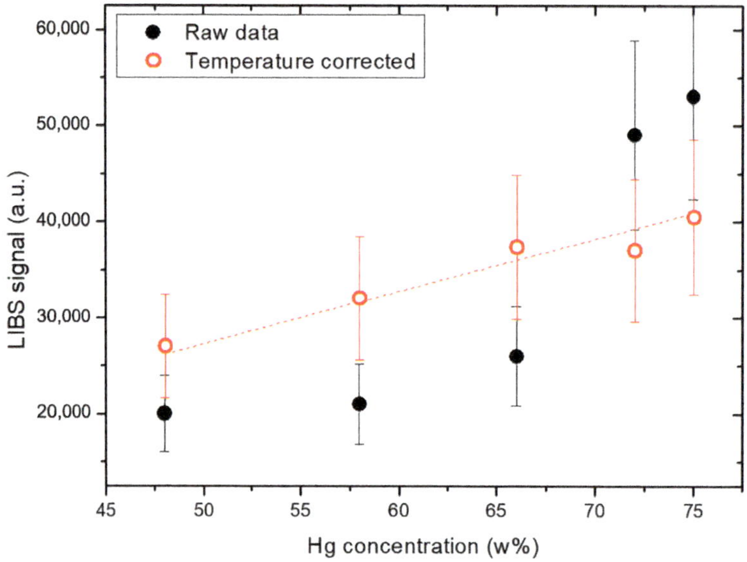

FIG. 3 Calibration curve for the concentration of Hg in dental amalgams. *Black points*: uncorrected laser-induced breakdown spectroscopy intensities. *Red points*: LIBS intensities corrected for plasma temperature (Eq. 1). The *dashed line* is the best linear fit of the corrected experimental data (zero intercept). *Data from reference E. Castellón, P. Pacheco Martínez, J. Álvarez, F. Bredice, F. Borges, M.-V. Muniz, C. Sánchez-Aké, V. Palleschi, R. Sarmiento, Elemental analysis of dental amalgams by laser-induced breakdown spectroscopy technique, Spectrochim. Acta B At. Spectrosc. 149 (2018) 229–235. https://doi.org/10.1016/j.sab.2018.05.025.*

density, and plasma temperature. Therefore by measuring these parameters from the LIBS spectra, one could in principle get rid of the matrix effect. Moreover, under suitable approximations, the composition of the sample could be derived from the plasma parameters (plasma temperature and electron number density), thus overcoming the need for using materials of known composition for the construction of calibration curves.

The calibration-free method (calibration-free [CF]-LIBS) was first introduced by Ciucci et al. [8] in 1999 starting from these considerations. The main hypotheses needed for the application of the CF-LIBS method are:

1. The plasma must be in local thermal equilibrium (LTE) [9] conditions, that is, all the processes in the plasma must be at equilibrium at the same temperature (excluding radiation).
2. The plasma composition must be representative of the sample composition (stoichiometric approximation).
3. The plasma must be homogeneous, that is, plasma composition, temperature, and electron number density must be the same all over the plasma plume.
4. The plasma must be optically thin (negligible self-absorption approximation).

These hypotheses and the consequences of their possible failure in real plasmas have been analyzed in detail in Ref. [10]. However, it is worth stressing that the fulfillment of the first hypothesis is the most important condition for the application of the method, because the same definition of a plasma temperature would not be possible out of the LTE approximation. De Giacomo et al. [11] proposed in 2007 a variation of the CF-LIBS method based on a relaxed approximation of partial LTE taking into account in the

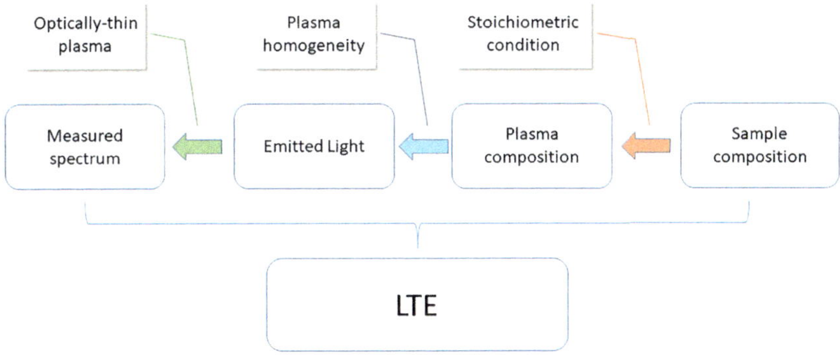

FIG. 4 Schematic view of the hypotheses needed for the application of the calibration-free laser-induced breakdown spectroscopy method.

calculation only the concentration of the neutral atoms. The method was called self-calibrated LIBS but it does not seem to have been further developed since its introduction.

The other three conditions derive logically from the need for univocally linking the LIBS spectrum to the composition of the sample; a graphical scheme explaining the importance of these hypotheses is shown in Fig. 4.

The CF-LIBS method determines the composition of the sample only indirectly through the analysis of the emission spectrum of the plasma. As shown in the figure, the sample composition can be derived from the LIBS spectrum, assuming LTE conditions (hypothesis 1), only if the plasma composition is representative of the sample's composition (hypothesis 2), the plasma emission is representative of the plasma composition (hypothesis 3), and, finally, the measured LIBS spectrum is representative of the plasma-emitted light (hypothesis 4).

In the last 20 years, the validity of CF-LIBS basic hypotheses has been extensively discussed in typical conditions of LIBS analysis; LTE conditions are usually fulfilled in a definite temporal window after the laser pulse, when the electron number density of the plasma is high enough to fulfill the McWhirter criterion [9] and, at the same time, the plasma expansion has slowed down to a condition where the spatial and temporal gradients are less important. The stoichiometric ablation conditions are usually fulfilled when the laser irradiance on the sample is high enough for the onset of phase explosion ablation [12, 13]. Methods have been proposed for correcting the effect of selective ablation (fractionation) in bronze [14, 15] and Al alloys [16]. Among the different hypotheses, the most difficult to control is the hypothesis of homogeneous plasma (several authors reported evidence of strong gradients in plasma electron temperature and electron number density at the boundary between plasma and ambient gas [17, 18]); on the contrary, the effect of self-absorption can be taken into account more easily [19–21] and even exploited for improving the analytical precision of CF-LIBS analysis [22].

3 The CF-LIBS algorithm

The two basic equations on which the CF-LIBS method is based are the Boltzmann equation, describing the equilibrium distribution of the electrons among the atomic energy levels:

$$B_s(E_k) = n_s \frac{g_k e^{-\frac{E_k}{k_B T}}}{U_s(T)} \tag{3}$$

where n_s is the total number of atoms of the species s, and the Saha equation, describing the equilibrium between the different species of the same element, which in the case of neutral and ionized species typically observed in LIBS spectra can be written as:

$$\frac{n_s^{II} n_e}{n_s^I} = S(T) \tag{4}$$

where $S(T)$ is the equilibrium constant of the processes of ionization and electron recombination, given by

$$S(T) = 2\frac{U_s^{II}(T)}{U_s^I(T)}\left(\frac{2\pi m_e k_B T}{h^2}\right)^{3/2} e^{-\frac{E_i}{k_B T}} \tag{5}$$

where E_i represents the ionization energy of the neutral species, m_e is the mass of the electron, and U_s^I and U_s^{II} are the partition functions of the neutral and ionized state, respectively.

The intensity of the emission (photons per second) at the wavelength $\lambda = c/\nu$, corresponding to the transition between two atomic energy levels E_k and E_i separated by the energy $hv = E_k - E_i$, can be written as:

$$I_\lambda = B_s(E_k) A_{ki} = n_s \frac{g_k e^{-\frac{E_k}{k_B T}}}{U_s(T)} A_{ki} \tag{6}$$

If self-absorption is negligible, the measured intensity corresponding to the transition can be written as:

$$\overline{I}_\lambda = FC_s \frac{g_k e^{-\frac{E_k}{k_B T}}}{U_s(T)} A_{ki} \tag{7}$$

where C_s is the concentration of the species and F is an experimental factor that may depend on λ but does not vary with the matrix of the samples. The dependence of F on the wavelength must be considered while determining the experimental line intensity, therefore in the following we will consider it as a numerical factor, which takes into account, for example, the total ablated mass and the collection efficiency of the optical apparatus.

Eq. (7) can be seen as a relation between the measured total intensity of the transition and the concentration of the emitting species; the temperature of the plasma and the F factor are not known a priori, while the other elements in the equation can be generally obtained from the available atomic databases (e.g., [23]).

Rearranging the terms of Eq. (7) we obtain:

$$\ln\left(\frac{\overline{I}_\lambda A_{ki}}{g_k}\right) = \ln\left(\frac{FC_s}{U_s(T)}\right) - \frac{1}{k_B T} \tag{8}$$

If at least two lines are available from the same species, Eq. (8) can be exploited for obtaining the plasma temperature T and the concentration of the species C_s (more precisely, the product FC_s) through a best fitting procedure. When the term at the left of the equation is plotted graphically versus the energy E_k of the upper level of the transition we have a so-called Boltzmann plot. In this representation, the plasma temperature is related to the inverse of the slope of the curve and the concentration of the species is related to the intercept of this curve with the y axis at $E_k = 0$ (Fig. 5).

Note that the plasma temperature is a global parameter of the plasma, according to the LTE approximation, therefore the concentration of the other species can be obtained using just a single emission line. This has been presented in the past as an "improvement" of the classical CF-LIBS called one-line CF-LIBS, but in fact this strategy is implicit with the CF-LIBS method and, moreover, it was already explicitly mentioned in 2006 in Ref. [2]. Therefore the use of the acronym to indicate something different from CF-LIBS must be discontinued. Similar "improvements" have been proposed from time to time; however, the use of different names for the same technique may be misleading and generate confusion; this habit should thus be discouraged.

Coming back to the CF-LIBS algorithm, the calculation of the intercepts in the Boltzmann plot gives all the information needed for the

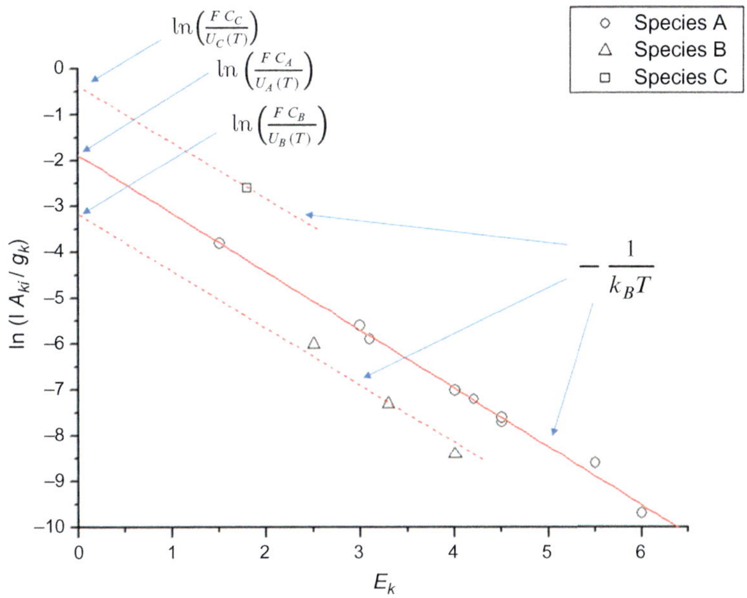

FIG. 5 Boltzmann plot for three species (A, B, and C). The temperature is evaluated using the data of species A, the intercepts are then calculated for the other species using the same temperature.

determination of the sample composition except for the yet unknown F factor. For calculating the F factor, the condition that the sum of all the species concentrations is equal to 100% could be imposed:

$$\sum_s C_s = 100\% \tag{9}$$

An alternative way is deriving F from the known concentration of an internal standard [24] or using the CF-LIBS method for calculating ratios of elemental concentrations [25]. In both the cases the F factor does not need to be calculated.

In discussing the possible normalization of the CF-LIBS results with internal standards or ratios, it is important to stress the difference between species and elements in LIBS. Knowing the concentration of a given element does not grant knowledge of the concentration of the neutral and ionized species. Since in typical LIBS experiments the concentrations of twice-ionized species are usually negligible, the elemental concentration is given by the sum of the concentrations of neutral and singly ionized species. In LTE conditions, the ratio between the two concentrations depends exponentially on the ratio between ionization energy and plasma temperature and, at a given temperature, linearly with the plasma electron number density (Eq. 4). The electron number density is a global parameter of the plasma that changes with the changes of the matrix of the sample. However, when needed, the electron number density can be obtained from the LIBS spectrum using either the ratio between an ionic and a neutral line of the same element:

$$n_e = \frac{I_s^I}{I_s^{II}} \frac{g_k A_{ki}}{g_l A_{lm}} \frac{U_s^I(T)}{U_s^{II}(T)} e^{-\frac{E_l - E_k}{k_B T}} \left(\frac{2\pi m_e}{h^2}\right)^{3/2} \frac{g_0^{II} (k_B T)^{3/2}}{g_0^I} e^{-\frac{E_i}{k_B T}} \tag{10}$$

or measuring the Stark broadening of a optically thin line of known Stark broadening coefficient w_s [26]:

$$\Delta\lambda_{FWHM} = w_s n_e \tag{11}$$

A good choice for the emission line to study to obtain the electron number density is the hydrogen Balmer alpha line at 656.28 nm [27]. As reported in several papers [28, 29], although the link between Stark broadening and electron number density is more complicated [26, 28] for hydrogen lines with respect to Eq. (11), the broadening of the hydrogen Balmer alpha line is very strong, and can be measured easily even using cheap low-resolution spectrometers. On the other hand, it must also be remarked that cheap (time-integrated) spectrometers might not be appropriate for CF-LIBS applications, unless the time-integrated LIBS spectra are treated to recover a sort of temporal resolution, as suggested in Ref. [30].

There are many good reasons for measuring the electron number density of the plasma in CF-LIBS applications: first, knowledge of the electron number density can be used for a quick check of the (necessary) condition for LTE given by the McWhirter criterion [9]; second, the determination of the ratio between neutral and ionized species of a given element can be essential for the analysis, when either the neutral species or the ionized species does not have useful emission lines in the spectral region available for the measurements.

In this case, Eq. (4) must be used for obtaining the concentration of the missing species and being able to apply the closure to 100% needed for eliminating the unknown F coefficient in Eq. (8).

Knowledge of the electron number density can be important for two other reasons:

- Obtaining a more precise determination of the plasma electron temperature by using the Saha-Boltzmann plot [31], a variation of the Boltzmann plot that puts on the same curve both neutral and ionized lines of an element;
- Calculating the effect of self-absorption and correcting it using, for example, the method of Ref. [32].

In summary, taking into account the fact that specific experimental needs would require tailored approaches, a sequence of actions is suggested for application of the CF-LIBS method:

1. Identification of the lines and measurement of their intensities in the LIBS spectrum. Integrated intensity should be preferred to peak intensity, because it is independent of the spectrometer resolution and it is less dependent on self-absorption [33]. The spectral response of the spectrometer must be known, since the measured line intensities must be corrected for the relative spectral response of the spectrometer [34].
2. Measurement of the electron number density (from hydrogen Balmer alpha line broadening, whenever possible). Since the Stark coefficient depends (weakly) on temperature, a reasonable starting value of the plasma temperature can be assumed for the calculation (typically $k_B T = 1$ eV).
3. Quick check of the fulfillment of the McWhirter criterion (necessary for validity of the LTE approximation).
4. Check and eventual correction of the effects of self-absorption, using one of the many methods proposed in the LIBS literature (Internal Reference Self-Absorption Correction [35], duplicating mirror [36], etc.). Use of the method reported in [32] is suggested, since it is quick and easy to implement, does not require changes in the experimental apparatus as in [36], and gives results more accurate than [35].
5. Measurement of the plasma temperature through the Saha-Boltzmann plot whenever possible. Depending on the situation, the plasma temperature can be calculated from a single element or as the average of several to obtain better precision.
6. Iteration of procedures 1–5 up to convergence (in general, convergence to the correct temperature is achieved in a few steps).

7. Calculation of the FC_s values for the species observed from the intercepts of the Boltzmann plot.
8. Calculation of the concentration of the species not observed (ionized from neutral or vice versa) using the Saha equation (Eqs. 4 and 5).
9. Determination of the F factor from Eq. (9).
10. Calculation of the elemental concentrations.
11. Check of the results for reliability and consistency with the hypothesis of CF-LIBS.

This procedure is schematically reported in Fig. 6.

Regarding Fig. 6, it must be remarked that errors in the identification of one or more LIBS lines to the right element and species might result in wrong results, therefore great care should be devoted to Step 1, which sometimes is erroneously taken for granted. In case of doubts in the identifications of the lines, they should not be inserted in the CF-LIBS analysis or, alternatively, an automated identification procedure should be used [37, 38]. As for the "optional" procedures, such as the correction for self-absorption effects, it should be remarked again that the recent habit of renaming the CF-LIBS procedure according to the spectral treatment used (self-absorption corrected CF-LIBS, for example) must be avoided, since the CF-LIBS procedure is always the same whatever correction to the LIBS lines intensity is applied (or not applied).

It is also important to remember that a careful check of the self-consistency of the results is of paramount importance when using algorithms such as CF-LIBS, which are based on quite strict assumptions. The hypothesis of plasma homogeneity, for example, could be questioned when self-reversed [39] spectral lines are detected in the LIBS spectrum. Most of all, one should always remember that the assumptions on which CF-LIBS is based are fulfilled; when this is the case, only in a relatively short temporal interval which may vary from sample to sample. This is the reason why in CF-LIBS applications the use of long gate times (larger than 1–2 μs) in the acquisition of the LIBS spectrum should be avoided when using time-resolved spectrometers, and more so the use of time-integrated spectrometers.

4 Other calibration-free approaches

4.1 Monte Carlo LIBS

One of the most interesting calibration-free approaches, not relying on the procedures outlined in the 1999 paper of Ciucci et al., was introduced in a series of papers published by Gornushkin et al. [40–42] and by Yaroshchyk et al. [43]. The method is based on the synthetic reconstruction of the experimental LIBS spectra using the theoretical emission intensities predicted in the framework of the LTE approximation and the consequent minimization of the differences between experimental and synthetic LIBS spectra. The optimization parameters are the plasma parameters (total density, electron number density, plasma temperature) and the concentrations of the elements in the sample.

Gornushkin and coworkers have realized a Monte Carlo-simulated annealing optimization code for CF-LIBS analysis called Monte Carlo LIBS (MC-LIBS). Besides LTE approximation, the condition of stoichiometric ablation is usually assumed in MC-LIBS, while the hypothesis of homogeneous, optically thin plasma can be relaxed, and other plasma nonidealities can be treated as soon as they can be theoretically modeled in the synthetic spectrum.

A comparison of the analytical capabilities of the two calibration-free approaches was done by Herrera et al. in 2009 [44], showing that conventional CF-LIBS performed better than MC-LIBS in the case under study (aluminum alloy standards under vacuum conditions).

However, the MC-LIBS has largely improved over time [45, 46] and represents a good

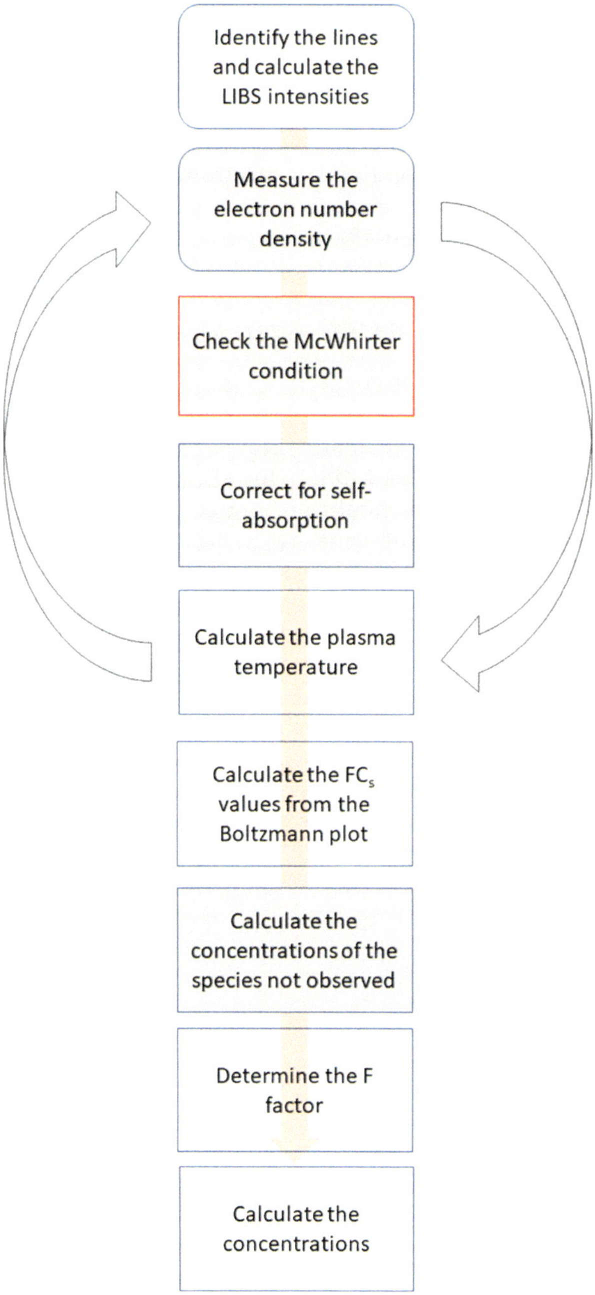

FIG. 6 Conceptual scheme of the calibration-free laser-induced breakdown spectroscopy procedure. The blocks in gray represent optional steps.

alternative to conventional CF-LIBS analysis; what is particularly interesting is the possibility of modeling nonideal plasmas and comparing the effects on the nonideality with the experimental LIBS spectra.

4.2 One-point calibration LIBS

Another alternative method to improve the analytical performances of CF-LIBS is one-point calibration-LIBS (OPC-LIBS), proposed by Cavalcanti et al. in 2013 [34]. As noted in [47], OPC-LIBS is not strictly a standardless technique; rather it belongs to the larger class of "fundamental parameter" techniques. In practice, the information acquired from a single standard of known composition (and the same matrix as the ones to be analyzed) is exploited for determining with precision the A_{ki} spectral coefficients of the lines used in a classical CF-LIBS calculation. Indetermination on the transition probability greatly influences the precision of CF-LIBS analysis, therefore the OPC approach may improve substantially the performances of standard CF-LIBS. If the response curve of the spectrometer is not precisely known, the OPC-LIBS method will also correct it. The OPC must be performed only once, since it would remain valid for all the following CF-LIBS analyses, provided that the matrix of the samples would not change. Note that, except for the need of a reference standard, the analysis does not differ from a classical CF-LIBS (plasma temperature and electron number density are determined independently in all the spectra acquired).

4.3 Inverse CF-LIBS

The inverse CF-LIBS method was proposed by Gaudiuso et al. [48] in 2012; also this method belongs to the category of fundamental parameter techniques and requires a standard of known composition for its implementation. The CF-LIBS analysis of this standard is then performed, and the plasma temperature is determined as the one that minimizes the difference between the predicted and the known composition of the standard. Finally, this temperature is fixed for the CF-LIBS analysis of the unknown samples (of the same matrix). It should be noted that the inverse CF-LIBS and the OPC approaches, although developed (independently) almost at the same time and apparently similar, are deeply different in the fact that the unavoidable variations in plasma temperature and electron number density from sample to sample can be measured and compensated using OPC, while the inverse CF-LIBS method in that respect is no different from a conventional calibration analysis, where the experimental parameters are supposed to remain constant during the building of the calibration curve and the measurement.

4.4 Artificial neural network CF-LIBS

CF-LIBS analysis using artificial neural networks (ANN-CF-LIBS) was proposed in 2015 by D'Andrea et al. [49] for the fast analysis of a large number of samples of similar matrix, without making any assumption of the fact that plasma temperature and electron number density would remain constant. The procedure involves a preliminary standard CF-LIBS analysis on a limited number of samples whose results (elemental concentrations, plasma temperature, and electron number density) are used for training a suitable ANN [50]. After that, the CF-LIBS analysis of the rest of the samples is performed in very short times (fractions of seconds) using the pretrained ANN. One of the major drawbacks of the classical CF-LIBS procedure (Fig. 6) is the fact that it is hardly automatable; moreover, the calculations can be relatively time consuming (several seconds per analysis). The ANN-CF-LIBS approach overcomes these problems, and appears to be particularly suited for the analysis of a very large number of spectra, such as those that can be acquired during the monitoring of industrial processes or by LIBS high-resolution elemental mapping [51].

4.5 Columnar density method

The columnar density method was introduced in 2013 by Cristoforetti and Tognoni [22], starting with the consideration that the use of a strongly self-absorbed line for quantitative analysis, commonly avoided, could offer instead a number of definite analytical advantages in a CF-LIBS approach. For the application of the method, the electron number density must be determined from the LIBS spectrum; from the knowledge of n_e a modified Saha-Boltzmann equation can be written for the ratio between the population of the ground states n_i^I and n_m^{II} of the transitions of neutral and ionized species, respectively.

Correspondingly, a modified Saha-Boltzmann plot can be built, having on the y axis a quantity linked to the logarithm of the columnar density $n_i^I l$ or $n_m^{II} l$ for neutral and ionized lines, where l is the length of the plasma column that the light must travel to reach the detector, and on the x axis the energy of the *lower level* of the transition. As in a conventional Saha-Boltzmann plot, the slope of the line that fits better the experimental data is $-\frac{1}{k_B T}$.

The main advantage of the Cristoforetti and Tognoni approach is that the lower-level population of strongly self-absorbed lines is, by definition, at thermal equilibrium, therefore their columnar density method does not suffer from possible deviations from the LTE conditions. In Ref. [22], the columnar density method was applied to the calibration-free determination of the composition of a binary Al-Mg alloy; the electron temperature was found to be practically identical for Al and Mg, as expected, and the ratio of Mg/Al concentration was in very good agreement with the nominal values.

4.6 Extended C-sigma method

In 2014/15, Aragón and Aguilera published two papers on the application of a generalized curve-of-growth method for quantitative LIBS analysis [52, 53]. In the original formulation, application of the method would require the use of a single matrix-matched standard for the characterization of the experimental apparatus; similarly to other fundamental parameter methods, also in the C-sigma method the LTE equations are exploited to derive the plasma parameters and obtain from them the composition of the samples to be analyzed. As for the columnar density method of Cristoforetti and Tognoni, the Aragón and Aguilera method exploits the presence of (in this case, moderately) self-absorbed lines for improving the analytical capability of the technique. The authors recently applied their method to the analysis of aluminum samples [54], obtaining results in very good agreement with the known composition.

In 2019, Safi et al. [55] extended the Aragón and Aguilera approach, demonstrating that the method can be easily modified to take into account arbitrary levels of self-absorption. In this way, the C-sigma treatment becomes completely equivalent to a classical CF-LIBS analysis (with self-absorption correction); the advantage of the method is the possibility of building separate, generalized C-sigma plots for neutral and ionized lines, thus potentially compensating for possible differences between atomic and ionic electron temperatures [56].

5 Applications

The classical CF-LIBS approach, as well as its variants already described, has been applied in many fields, ranging from industrial diagnostics, space research, biomedicine, cultural heritage, etc.

In the following we will mention a few of these works, focusing specifically on the CF-LIBS applications developed in fields where standard samples are not readily available for building conventional calibration curves or using multivariate approaches, such as partial least squares or ANN [3], which require the use of a suitable number of calibration standards for their use. We will focus, in particular, in the recent works that are not covered in the 2010 review of Tognoni et al. [10].

5.1 Food and health

In 2011, Lei et al. [57] studied the mineral elements in milk powder (Mg, K, and Ca) using CF-LIBS, obtaining very good agreement with ICP-AES measurements and nominal concentrations. Similar good agreement with atomic absorption spectroscopy was obtained by Agrawal et al. [58] in the analysis of food supplements. In 2017 Ahmed et al. [59] applied CF-LIBS to study and compare the composition of different cigarette brands. Beldjilali et al. [60] analyzed multivitamin tablets by CF-LIBS, assessing the possibility of using the technique for fast quality control of these materials. In 2018, Chen et al. [61] proposed a two-step procedure for the analysis of food, using different delays and gates for the CF-LIBS analysis of major and trace elements. Several papers have also been published on the CF-LIBS analysis of edible plants (see, e.g., Rehan et al. [62, 63] in 2018 and Jabbar et al. [64] in 2019).

5.2 Nuclear applications

In 2012, Almaviva et al. [65] demonstrated the feasibility of online CF-LIBS monitoring of the inner wall of a fusion vessel. CF-LIBS analysis was further developed by the same authors in 2013 [66] and 2015 [67]. Also, Karhunen et al. [68] in 2014 used CF-LIBS for the diagnostics of deposited layers on ITER tiles. ITER is a new-generation fusion device under construction in Cadarache (France). Maurya et al. [69] in 2016 analyzed by CF-LIBS the depth profile of impurities in the Aditya tokamak, while Pribula et al. [70] applied CF-LIBS to the analysis of W materials of interest for nuclear fusion research. Fantoni et al. [71] in 2017 and Almaviva et al. [72] in 2019 applied the CF-LIBS approach for determining the concentration of W-, Al-, and D-mixed layers on Mo samples for the analysis of plasma-facing components at ITER.

5.3 Industrial applications

Praher et al. [73] in 2011 analyzed the composition of industrial oxides using CF-LIBS, observing improved stability of the results on long runs of the system (better than 10% for the major elements), even in the presence of relatively high fluctuations in laser energy. In 2013, Cavalcanti et al. [74] applied the CF-LIBS method for a stratigraphic analysis of petroleum scales using a double-pulse LIBS setup. Ahamer et al. [75] in 2016 determined the concentration of CaO, Al_2O_3, MgO, SiO_2, FeO, MnO, and TiO_2 oxides in steel slag using CF-LIBS. Diaz Pace et al. [76] in 2017 studied with CF-LIBS the composition of foundry waste, detecting and quantifying the concentrations of Al, Ba, Fe, Li, Mg, Mn, Pb, Ti, Zr, and Zn. Igbal et al. [77] in 2019 used CF-LIBS for the analysis of coal.

5.4 Environment

Kwak et al. [78] in 2011 used CF-LIBS for the quantitative determination of the concentration of lead in soil, obtaining results in good agreement with ICP-MS for the analysis of historical mining sites. Kumar et al. [79] successfully used CF-LIBS for quantifying the concentration of toxic metals (Cr and Pb) in the sludges of industrial wastewaters. Boudhib et al. [80] in 2016 demonstrated the feasibility of the application of CF-LIBS analysis for the study of aerosols. The use of CF-LIBS has been demonstrated useful also for the analysis of plants used for phytoremediation in a paper by Jabbar et al. [81] published in 2019.

5.5 Cultural heritage and archeology

Gaudiuso et al. [82] in 2014 applied the inverse calibration-free method for the analysis of archeological samples, obtaining satisfactory results. They found that good results were also obtained using the temperature measured on standards with different matrixes with respect to the samples analyzed. In 2016 the same group [83] used the inverse CF-LIBS method for improving the quality of LIBS depth profile analysis.

In 2015, Takahashi et al. [84] applied CF-LIBS to the analysis of submersed brass objects, finding very good agreement between the analytical results and the nominal composition of the samples for the major components of the alloy. They also found that nonstoichiometric effects were negligible in their experimental conditions (250-ns laser pulses). Smaldone et al. [85] in 2016 used inverse calibration free for the analysis of brass using femtosecond LIBS. Pagnotta et al. [86] in 2018 used CF-LIBS for fast quantitative elemental mapping of highly inhomogeneous cultural heritage materials.

5.6 Space research

Dell'Aglio et al. [87] used CF-LIBS for the analysis of meteorites; similar results were also published by Horňáčková et al. [88] in the elemental analysis of meteorite fragments. The authors found good agreement between CF-LIBS analysis and ICP-MS results for elements such as Al, Ca, Cr, Fe, Mg, Mn, Na, Ni, and Si. Ozdin et al. [89] in 2015 applied CF-LIBS analysis to the study of a meteorite fallen in Košice (eastern Slovakia) in 2010. One year later, Senesi et al. [90] also used CF-LIBS and OPC-LIBS for the analysis of meteorites; the authors found good agreement of the CF-LIBS results with the ones obtained by traditional analytical techniques such as ICP-MS and energy-dispersive X-ray spectroscopy-scanning electron microscopy. Dell'Aglio et al. [91] in 2018 used CF-LIBS for standoff analysis of two meteorites (at a distance of 5 m). The same year, Tempesta et al. [92] successfully applied CF-LIBS to the μ-LIBS analysis of the Dronino meteorite, a 40-kg iron meteorite found in Russia in 2000.

5.7 Jewelry

Agrosì et al. [93] in 2014 applied CF-LIBS to the analysis of emeralds using a μ-LIBS double-pulse instrument [94]. Tempesta and Agrosì [95] used the same technique in 2016 for the analysis of red beryls. Ahmed et al. [96] in 2018 successfully used CF-LIBS for determining gold caratage in precious alloys, confirming the results of Corsi et al. [97] who, in 2001, first proposed this application.

6 Conclusion

In this chapter, we have presented the foundations of the CF-LIBS method as proposed by the Pisa group in 1999. After 20 years, the technique has evolved, although most of the current "improvements," often described with different names, are essentially different implementations of the traditional CF-LIBS approach, and as such they should be presented. Other calibration-free approaches, deeply different from the classical method, have also been discussed. Monte Carlo CF-LIBS is probably the most interesting alternative to classical CF-LIBS, for its (at least potential) characteristics of better dealing with plasma nonideality. The extended C-sigma method is also potentially very interesting for calibration-free analysis because of the way in which it naturally includes the self-absorbed lines in the analysis.

A short collection of the main applications of CF-LIBS papers presented in the scientific literature in the last decade has also been included. The number of papers published per year more than doubled in 2018 and 2019 with respect to the average of the previous 10 years; this clearly indicates that the CF-LIBS method is still receiving great attention as a powerful analytical technique, which exploits at best the advantages of LIBS in many fields where the availability of suitable matrix-matched standards is scarce.

References

[1] E. Tognoni, V. Palleschi, M. Corsi, G. Cristoforetti, Quantitative micro-analysis by laser-induced breakdown spectroscopy: a review of the experimental approaches, Spectrochim. Acta B At. Spectrosc. 57 (2002) 1115–1130, https://doi.org/10.1016/S0584-8547(02)00053-8.

[2] E. Tognoni, V. Palleschi, M. Corsi, G. Cristoforetti, N. Omenetto, I. Gornushkin, B.W. Smith, J.D. Winefordner, From Sample to Signal in Laser-Induced Breakdown Spectroscopy: A complex Route to Quantitative Analysis, Cambridge University Press, Cambridge, (2006), https://doi.org/10.1017/CBO9780511541261.004.

[3] A. Safi, B. Campanella, E. Grifoni, S. Legnaioli, G. Lorenzetti, S. Pagnotta, F. Poggialini, L. Ripoll-Seguer, M. Hidalgo, V. Palleschi, Multivariate calibration in Laser-Induced Breakdown Spectroscopy quantitative analysis: the dangers of a 'black box' approach and how to avoid them, Spectrochim. Acta B At. Spectrosc. 144 (2018), https://doi.org/10.1016/j.sab.2018.03.007.

[4] K. Danzer, L.A. Currie, Guidelines for calibration in analytical chemistry. Part I. Fundamentals and single component calibration (IUPAC recommendations 1998), Pure Appl. Chem. 70 (1998) 993–1014, https://doi.org/10.1351/pac199870040993.

[5] O.V. Borisov, X.L. Mao, A. Fernandez, M. Caetano, R.E. Russo, Inductively coupled plasma mass spectrometric study of non-linear calibration behavior during laser ablation of binary Cu–Zn alloys, Spectrochim. Acta B At. Spectrosc. 54 (1999) 1351–1365, https://doi.org/10.1016/S0584-8547(99)00083-X.

[6] U. Panne, C. Haisch, M. Clara, R. Niessner, Analysis of glass and glass melts during the vitrification process of fly and bottom ashes by laser-induced plasma spectroscopy. Part I: normalization and plasma diagnostics, Spectrochim. Acta B At. Spectrosc. 53 (1998) 1957–1968, https://doi.org/10.1016/S0584-8547(98)00238-9.

[7] E. Castellón, P. Pacheco Martínez, J. Álvarez, F. Bredice, F. Borges, M.-V. Muniz, C. Sánchez-Aké, V. Palleschi, R. Sarmiento, Elemental analysis of dental amalgams by laser-induced breakdown spectroscopy technique, Spectrochim. Acta B At. Spectrosc. 149 (2018) 229–235, https://doi.org/10.1016/j.sab.2018.05.025.

[8] A. Ciucci, M. Corsi, V. Palleschi, S. Rastelli, A. Salvetti, E. Tognoni, New procedure for quantitative elemental analysis by laser-induced plasma spectroscopy, Appl. Spectrosc. 53 (1999) 960–964, https://doi.org/10.1366/0003702991947612.

[9] G. Cristoforetti, A. De Giacomo, M. Dell'Aglio, S. Legnaioli, E. Tognoni, V. Palleschi, N. Omenetto, Local thermodynamic equilibrium in laser-induced breakdown spectroscopy: beyond the McWhirter criterion, Spectrochim. Acta B At. Spectrosc. 65 (2010) 86–95, https://doi.org/10.1016/j.sab.2009.11.005.

[10] E. Tognoni, G. Cristoforetti, S. Legnaioli, V. Palleschi, Calibration-free laser-induced breakdown spectroscopy: state of the art, Spectrochim. Acta B At. Spectrosc. 65 (2010) 1–14, https://doi.org/10.1016/j.sab.2009.11.006.

[11] A. De Giacomo, M. Dell'Aglio, O. De Pascale, R. Gaudiuso, R. Teghil, A. Santagata, G.P. Parisi, ns- and fs-LIBS of copper-based-alloys: a different approach, Appl. Surf. Sci. 253 (2007) 7677–7681, https://doi.org/10.1016/j.apsusc.2007.02.037.

[12] O.V. Borisov, X. Mao, R.E. Russo, Effects of crater development on fractionation and signal intensity during laser ablation inductively coupled plasma mass spectrometry, Spectrochim. Acta B At. Spectrosc. 55 (2000) 1693–1704, https://doi.org/10.1016/S0584-8547(00)00272-X.

[13] G. Cristoforetti, S. Legnaioli, V. Palleschi, E. Tognoni, P.A. Benedetti, Observation of different mass removal regimes during the laser ablation of an aluminium target in air, J. Anal. At. Spectrom 23 (2008), https://doi.org/10.1039/b800517f.

[14] L. Fornarini, R. Fantoni, F. Colao, A. Santagata, R. Teghil, A. Elhassan, M.A. Harith, Theoretical modeling of laser ablation of quaternary bronze alloys: case studies comparing femtosecond and nanosecond LIBS experimental data, J. Phys. Chem. A 113 (2009) 14364–14374, https://doi.org/10.1021/jp903279h.

[15] S.M. Pershin, F. Colao, V. Spizzichino, Quantitative analysis of bronze samples by laser-induced breakdown spectroscopy (LIBS): a new approach, model, and experiment, Laser Phys. 16 (2006) 455–467, https://doi.org/10.1134/S1054660X06030066.

[16] V.N. Lednev, S.M. Pershin, Plasma stoichiometry correction method in laser-induced breakdown

spectroscopy, Laser Phys. 18 (2008) 850–854, https://doi.org/10.1134/S1054660X08070086.

[17] J.A. Aguilera, C. Aragón, Characterization of a laser-induced plasma by spatially resolved spectroscopy of neutral atom and ion emissions. Comparison of local and spatially integrated measurements, Spectrochim. Acta B At. Spectrosc. 59 (2004) 1861–1876, https://doi.org/10.1016/j.sab.2004.08.003.

[18] C. Aragón, J.A. Aguilera, Characterization of laser induced plasmas by optical emission spectroscopy: a review of experiments and methods, Spectrochim. Acta B At. Spectrosc. 63 (2008) 893–916, https://doi.org/10.1016/j.sab.2008.05.010.

[19] D. Bulajic, M. Corsi, G. Cristoforetti, S. Legnaioli, V. Palleschi, A. Salvetti, E. Tognoni, A procedure for correcting self-absorption in calibration free-laser induced breakdown spectroscopy, Spectrochim. Acta B At. Spectrosc. 57 (2002), https://doi.org/10.1016/S0584-8547(01)00398-6.

[20] F.O. Bredice, H.O.D.I. Rocco, H.M. Sobral, M. Villagránmuniz, V. Palleschi, A new method for determination of self-absorption coefficients of emission lines in laser-induced breakdown spectroscopy experiments, Appl. Spectrosc. 64 (2010), https://doi.org/10.1366/000370210790918454.

[21] T. Li, Z. Hou, Y. Fu, J. Yu, W. Gu, Z. Wang, Correction of self-absorption effect in calibration-free laser-induced breakdown spectroscopy (CF-LIBS) with blackbody radiation reference, Anal. Chim. Acta 1058 (2019) 39–47, https://doi.org/10.1016/j.aca.2019.01.016.

[22] G. Cristoforetti, E. Tognoni, Calculation of elemental columnar density from self-absorbed lines in laser-induced breakdown spectroscopy: a resource for quantitative analysis, Spectrochim. Acta B At. Spectrosc. 79–80 (2013) 63–71, https://doi.org/10.1016/j.sab.2012.11.010.

[23] NIST: Atomic Spectra Database Lines Form, https://physics.nist.gov/PhysRefData/ASD/lines_form.html (n.d.). Accessed 17 March 2020.

[24] V.S. Burakov, S.N. Raikov, Quantitative analysis of alloys and glasses by a calibration-free method using laser-induced breakdown spectroscopy, Spectrochim. Acta B At. Spectrosc. 62 (2007) 217–223, https://doi.org/10.1016/j.sab.2007.03.021.

[25] M. Corsi, G. Cristoforetti, M. Hidalgo, S. Legnaioli, V. Palleschi, A. Salvetti, E. Tognoni, C. Vallebona, Application of laser-induced breakdown spectroscopy technique to hair tissue mineral analysis, Appl. Optics 42 (2003), https://doi.org/10.1364/AO.42.006133.

[26] H.R. Griem, Stark broadening, Adv. At. Mol. Phys. 11 (1976) 331–359, https://doi.org/10.1016/S0065-2199(08)60033-0.

[27] A.M. El Sherbini, T. El Sherbini, H. Hegazy, G. Cristoforetti, S. Legnaioli, L. Pardini, V. Palleschi, A. Salvetti, E. Tognoni, Measurement of the stark broadening of atomic emission lines in non-optically thin plasmas by laser-induced breakdown spectroscopy, Spectrosc. Lett. 40 (2007), https://doi.org/10.1080/00387010701300958.

[28] M.A. Gigosos, Stark broadening models for plasma diagnostics, J. Phys. D Appl. Phys. 47 (2014) 343001, https://doi.org/10.1088/0022-3727/47/34/343001.

[29] L. Pardini, S. Legnaioli, G. Lorenzetti, V. Palleschi, R. Gaudiuso, A. De Giacomo, D.M. Diaz Pace, F. Anabitarte Garcia, G. De Holanda Cavalcanti, C. Parigger, On the determination of plasma electron number density from Stark broadened hydrogen Balmer series lines in Laser-Induced Breakdown Spectroscopy experiments, Spectrochim. Acta B At. Spectrosc. 88 (2013) 98–103, https://doi.org/10.1016/j.sab.2013.05.030.

[30] E. Grifoni, S. Legnaioli, M. Lezzerini, G. Lorenzetti, S. Pagnotta, V. Palleschi, Extracting time-resolved information from time-integrated laser-induced breakdown spectra, J. Spectrosc. 2014 (2014), https://doi.org/10.1155/2014/849310.

[31] Ş. Yalçin, D.R. Crosley, G.P. Smith, G.W. Faris, Spectroscopic characterization of laser-produced plasmas for in situ toxic metal monitoring, Hazard. Waste Hazard. Mater. 13 (1996) 51–61, https://doi.org/10.1089/hwm.1996.13.51.

[32] A.M. El Sherbini, T.M. El Sherbini, H. Hegazy, G. Cristoforetti, S. Legnaioli, V. Palleschi, L. Pardini, A. Salvetti, E. Tognoni, Evaluation of self-absorption coefficients of aluminum emission lines in laser-induced breakdown spectroscopy measurements, Spectrochim. Acta B At. Spectrosc. 60 (2005) 1573–1579, https://doi.org/10.1016/j.sab.2005.10.011.

[33] B. Praher, V. Palleschi, R. Viskup, J. Heitz, J.D. Pedarnig, Calibration free laser-induced breakdown spectroscopy of oxide materials, Spectrochim. Acta B At. Spectrosc. 65 (2010) 671–679, https://doi.org/10.1016/j.sab.2010.03.010.

[34] G.H. Cavalcanti, D.V. Teixeira, S. Legnaioli, G. Lorenzetti, L. Pardini, V. Palleschi, One-point calibration for calibration-free laser-induced breakdown spectroscopy quantitative analysis, Spectrochim. Acta B At. Spectrosc. 87 (2013), https://doi.org/10.1016/j.sab.2013.05.016.

[35] L. Sun, H. Yu, Correction of self-absorption effect in calibration-free laser-induced breakdown spectroscopy by an internal reference method, Talanta 79 (2009) 388–395, https://doi.org/10.1016/j.talanta.2009.03.066.

[36] H.-Y. Moon, K.K. Herrera, N. Omenetto, B.W. Smith, J. D. Winefordner, On the usefulness of a duplicating

mirror to evaluate self-absorption effects in laser induced breakdown spectroscopy, Spectrochim. Acta B At. Spectrosc. 64 (2009) 702–713, https://doi.org/10.1016/j.sab.2009.06.011.

[37] G. Amato, G. Cristoforetti, S. Legnaioli, G. Lorenzetti, V. Palleschi, F. Sorrentino, E. Tognoni, Progress towards an unassisted element identification from Laser Induced Breakdown Spectra with automatic ranking techniques inspired by text retrieval, Spectrochim. Acta B At. Spectrosc. 65 (2010) 664–670, https://doi.org/10.1016/j.sab.2010.04.019.

[38] G. Amato, S. Legnaioli, G. Lorenzetti, V. Palleschi, L. Pardini, F. Rabitti, Element detection relying on information retrieval techniques applied to laser spectroscopy, in: Proceedings of 4th International Conference on Similarity Search and Applications, SISAP 2011, 2011, https://doi.org/10.1145/1995412.1995429.

[39] D. Karabourniotis, M. Ribiere, B.G. Cheron, Electron temperature and density determination in a nonequilibrium laser induced plasma by means of self-reversed-line spectroscopy, Appl. Phys. Lett. 93 (2008) 2012–2015, https://doi.org/10.1063/1.2963472.

[40] I.B. Gornushkin, A.Y. Kazakov, N. Omenetto, B. W. Smith, J.D. Winefordner, Radiation dynamics of post-breakdown laser induced plasma, Spectrochim. Acta B At. Spectrosc. 59 (2004) 401–418, https://doi.org/10.1016/j.sab.2003.12.023.

[41] A.Y. Kazakov, I.B. Gornushkin, N. Omenetto, B. W. Smith, J.D. Winefordner, Radiative model of post-breakdown laser-induced plasma expanding into ambient gas, Appl. Optics 45 (2006) 2810, https://doi.org/10.1364/AO.45.002810.

[42] I.B. Gornushkin, A.Y. Kazakov, N. Omenetto, B.W. Smith, J.D. Winefordner, Experimental verification of a radiative model of laser-induced plasma expanding into vacuum, Spectrochim. Acta B At. Spectrosc. 60 (2005) 215–230, https://doi.org/10.1016/j.sab.2004.11.009.

[43] P. Yaroshchyk, D. Body, R.J.S. Morrison, B.L. Chadwick, A semi-quantitative standard-less analysis method for laser-induced breakdown spectroscopy, Spectrochim. Acta B At. Spectrosc. 61 (2006) 200–209, https://doi.org/10.1016/j.sab.2006.01.004.

[44] K.K. Herrera, E. Tognoni, I.B. Gornushkin, N. Omenetto, B.W. Smith, J.D. Winefordner, Comparative study of two standard-free approaches in laser-induced breakdown spectroscopy as applied to the quantitative analysis of aluminum alloy standards under vacuum conditions, J. Anal. At. Spectrom 24 (2009) 426, https://doi.org/10.1039/b820494b.

[45] I.B. Gornushkin, T. Völker, A.Y. Kazakov, Extension and investigation by numerical simulations of algorithm for calibration-free laser induced breakdown spectroscopy, Spectrochim. Acta B At. Spectrosc. 147 (2018) 149–163, https://doi.org/10.1016/j.sab.2018.06.011.

[46] A. Demidov, S. Eschlböck-Fuchs, A.Y. Kazakov, I. B. Gornushkin, P.J. Kolmhofer, J.D. Pedarnig, N. Huber, J. Heitz, T. Schmid, R. Rössler, U. Panne, Monte Carlo standardless approach for laser induced breakdown spectroscopy based on massive parallel graphic processing unit computing, Spectrochim. Acta B At. Spectrosc. 125 (2016) 97–102, https://doi.org/10.1016/j.sab.2016.09.016.

[47] E. Grifoni, S. Legnaioli, G. Lorenzetti, S. Pagnotta, F. Poggialini, V. Palleschi, From calibration-free to fundamental parameters analysis: a comparison of three recently proposed approaches, Spectrochim. Acta B At. Spectrosc. 124 (2016), https://doi.org/10.1016/j.sab.2016.08.022.

[48] R. Gaudiuso, M. Dell'Aglio, O. De Pascale, A. Santagata, A. De Giacomo, Laser-induced plasma analysis of copper alloys based on Local Thermodynamic Equilibrium: an alternative approach to plasma temperature determination and archeometric applications, Spectrochim. Acta B At. Spectrosc. 74–75 (2012) 38–45, https://doi.org/10.1016/j.sab.2012.06.034.

[49] E. D'Andrea, S. Pagnotta, E. Grifoni, S. Legnaioli, G. Lorenzetti, V. Palleschi, B. Lazzerini, A hybrid calibration-free/artificial neural networks approach to the quantitative analysis of LIBS spectra, Appl. Phys. B Lasers Opt. 118 (2015) 353–360, https://doi.org/10.1007/s00340-014-5990-z.

[50] F.O. Borges, G.H. Cavalcanti, G.C. Gomes, V. Palleschi, A. Mello, A fast method for the calculation of electron number density and temperature in laser-induced breakdown spectroscopy plasmas using artificial neural networks, Appl. Phys. B Lasers Opt. 117 (2014), https://doi.org/10.1007/s00340-014-5852-8.

[51] S. Pagnotta, M. Lezzerini, B. Campanella, G. Gallello, E. Grifoni, S. Legnaioli, G. Lorenzetti, F. Poggialini, S. Raneri, A. Safi, V. Palleschi, Fast quantitative elemental mapping of highly inhomogeneous materials by micro-Laser-Induced Breakdown Spectroscopy, Spectrochim. Acta B At. Spectrosc. 146 (2018), https://doi.org/10.1016/j.sab.2018.04.018.

[52] C. Aragón, J.A. Aguilera, CSigma graphs: a new approach for plasma characterization in laser-induced breakdown spectroscopy, J. Quant. Spectrosc. Radiat. Transf. 149 (2014) 90–102, https://doi.org/10.1016/j.jqsrt.2014.07.026.

[53] C. Aragón, J.A. Aguilera, Quantitative analysis by laser-induced breakdown spectroscopy based on generalized curves of growth, Spectrochim. Acta B At. Spectrosc. 110 (2015) 124–133, https://doi.org/10.1016/j.sab.2015.06.010.

[54] C. Aragón, J.A. Aguilera, Direct analysis of aluminum alloys by CSigma laser-induced breakdown spectroscopy, Anal. Chim. Acta 1009 (2018) 12–19, https://doi.org/10.1016/j.aca.2018.01.019.

[55] A. Safi, S.H. Tavassoli, G. Cristoforetti, E. Tognoni, B. Campanella, S. Legnaioli, S. Pagnotta, F. Poggialini, V. Palleschi, Exploiting self-absorption for plasma characterization in laser-induced breakdown spectroscopy experiments: a comparison of two recent approaches, Anal. Chem. 91 (2019) 8595–8601, https://doi.org/10.1021/acs.analchem.9b01885.

[56] I.B. Gornushkin, S.V. Shabanov, S. Merk, E. Tognoni, U. Panne, Effects of non-uniformity of laser induced plasma on plasma temperature and concentrations determined by the Boltzmann plot method: implications from plasma modeling, J. Anal. At. Spectrom 25 (2010) 1643, https://doi.org/10.1039/c0ja00016g.

[57] W.Q. Lei, J. El Haddad, V. Motto-Ros, N. Gilon-Delepine, A. Stankova, Q.L. Ma, X.S. Bai, L.J. Zheng, H.P. Zeng, J. Yu, Comparative measurements of mineral elements in milk powders with laser-induced breakdown spectroscopy and inductively coupled plasma atomic emission spectroscopy, Anal. Bioanal. Chem. 400 (2011) 3303–3313, https://doi.org/10.1007/s00216-011-4813-x.

[58] R. Agrawal, R. Kumar, S. Rai, A.K. Pathak, A.K. Rai, G.K. Rai, LIBS: a quality control tool for food supplements, Food Biophys. 6 (2011) 527–533, https://doi.org/10.1007/s11483-011-9235-y.

[59] N. Ahmed, Z.A. Umar, R. Ahmed, M. Aslam Baig, On the elemental analysis of different cigarette brands using laser induced breakdown spectroscopy and laser-ablation time of flight mass spectrometry, Spectrochim. Acta B At. Spectrosc. 136 (2017) 39–44, https://doi.org/10.1016/j.sab.2017.08.006.

[60] S.A. Beldjilali, E. Axente, A. Belasri, T. Baba-Hamed, J. Hermann, Compositional analysis of drugs by laser-induced breakdown spectroscopy, J. Appl. Spectrosc. 84 (2017) 472–477, https://doi.org/10.1007/s10812-017-0494-3.

[61] C.-T. Chen, D. Banaru, T. Sarnet, J. Hermann, Two-step procedure for trace element analysis in food via calibration-free laser-induced breakdown spectroscopy, Spectrochim. Acta B At. Spectrosc. 150 (2018) 77–85, https://doi.org/10.1016/j.sab.2018.10.011.

[62] I. Rehan, K. Rehan, S. Sultana, M.Z. Khan, R. Muhammad, LIBS coupled with ICP/OES for the spectral analysis of betel leaves, Appl. Phys. B Lasers Opt. 124 (2018) 76, https://doi.org/10.1007/s00340-018-6947-4.

[63] I. Rehan, S. Sultana, K. Rehan, R. Muhammad, A. Mateen, M.A. Farooque, Application of laser induced breakdown in air in conjunction with atomic absorption spectroscopy for detection of trace elements in fennel seeds, Opt. Spectrosc. 125 (2018) 130–136, https://doi.org/10.1134/S0030400X18070238.

[64] A. Jabbar, M. Akhtar, A. Ali, S. Mehmood, S. Iftikhar, M. A. Baig, Elemental composition of rice using calibration free laser induced breakdown spectroscopy, Optoelectron. Lett. 15 (2019) 57–63, https://doi.org/10.1007/s11801-019-8099-0.

[65] S. Almaviva, L. Caneve, F. Colao, R. Fantoni, G. Maddaluno, Laboratory feasibility study of fusion vessel inner wall chemical analysis by Laser Induced Breakdown Spectroscopy, Chem. Phys. 398 (2012) 228–232, https://doi.org/10.1016/j.chemphys.2011.07.012.

[66] R. Fantoni, S. Almaviva, L. Caneve, F. Colao, A.M. Popov, G. Maddaluno, Development of Calibration-Free Laser-Induced-Breakdown-Spectroscopy based techniques for deposited layers diagnostics on ITER-like tiles, Spectrochim. Acta B At. Spectrosc. 87 (2013) 153–160, https://doi.org/10.1016/j.sab.2013.05.032.

[67] S. Almaviva, L. Caneve, F. Colao, P. Gasior, M. Kubkowska, M. Łepek, G. Maddaluno, Double pulse Laser Induced Breakdown Spectroscopy measurements on ITER-like samples, Fusion Eng. Des. 96–97 (2015) 848–851, https://doi.org/10.1016/j.fusengdes.2015.02.068.

[68] J. Karhunen, A. Hakola, J. Likonen, A. Lissovski, P. Paris, M. Laan, K. Piip, C. Porosnicu, C.P. Lungu, K. Sugiyama, Development of laser-induced breakdown spectroscopy for analyzing deposited layers in ITER, Phys. Scr. 2014 (2014), https://doi.org/10.1088/0031-8949/2014/T159/014067.

[69] G.S. Maurya, R. Kumar, A. Kumar, A.K. Rai, Analysis of impurities on contaminated surface of the tokamak limiter using laser induced breakdown spectroscopy, Spectrochim. Acta B At. Spectrosc. 126 (2016) 17–22, https://doi.org/10.1016/j.sab.2016.10.010.

[70] M. Pribula, J. Krištof, M. Suchoňová, M. Horňáčková, J. Plavčan, A. Hakola, P. Veis, Use of the near vacuum UV spectral range for the analysis of W-based materials for fusion applications using LIBS, Phys. Scr. T167 (2016) 014045, https://doi.org/10.1088/0031-8949/T167/1/014045.

[71] R. Fantoni, S. Almaviva, L. Caneve, F. Colao, G. Maddaluno, P. Gasior, M. Kubkowska, Hydrogen isotope detection in metal matrix using double-pulse laser-induced breakdown-spectroscopy, Spectrochim. Acta B At. Spectrosc. 129 (2017) 8–13, https://doi.org/10.1016/J.SAB.2016.12.008.

[72] S. Almaviva, L. Caneve, F. Colao, G. Maddaluno, Deuterium detection and quantification by laser-induced breakdown spectroscopy and calibration-free analysis in ITER relevant samples, Fusion Eng. Des. (2019), https://doi.org/10.1016/j.fusengdes.2019.03.109.

[73] B. Praher, R. Rössler, E. Arenholz, J. Heitz, J.D. Pedarnig, Quantitative determination of element concentrations in industrial oxide materials by laser-induced breakdown spectroscopy, Anal. Bioanal. Chem. 400 (2011) 3367–3375, https://doi.org/10.1007/s00216-011-5000-9.

[74] G.H. Cavalcanti, A.A. Rocha, R.N. Damasceno, S. Legnaioli, G. Lorenzetti, L. Pardini, V. Palleschi, Double-pulse laser-induced breakdown spectroscopy analysis of scales from petroleum pipelines, Spectrochim. Acta B At. Spectrosc. 87 (2013), https://doi.org/10.1016/j.sab.2013.05.018.

[75] C.M. Ahamer, S. Eschlböck-Fuchs, P.J. Kolmhofer, R. Rössler, N. Huber, J.D. Pedarnig, Laser-induced breakdown spectroscopy of major and minor oxides in steel slags: influence of detection geometry and signal normalization, Spectrochim. Acta B At. Spectrosc. 122 (2016) 157–164, https://doi.org/10.1016/j.sab.2016.06.008.

[76] D.M. Díaz Pace, R.E. Miguel, H.O. Di Rocco, F. Anabitarte García, L. Pardini, S. Legnaioli, G. Lorenzetti, V. Palleschi, Quantitative analysis of metals in waste foundry sands by calibration free-laser induced breakdown spectroscopy, Spectrochim. Acta B At. Spectrosc. 131 (2017), https://doi.org/10.1016/j.sab.2017.03.007.

[77] S.M.Z. Iqbal, Z. Uddin, N. Ahmed, Z.A. Umar, M. A. Baig, On the compositional analysis of coal using calibration free laser induced breakdown spectroscopy, Laser Phys. 29 (2019) 036101, https://doi.org/10.1088/1555-6611/aaff58.

[78] J. Kwak, K.-W. Kim, M. Park, J. Kim, K. Park, Determination of lead in soil at a historical mining and smelting site using laser-induced breakdown spectroscopy, Environ. Technol. 33 (2012) 2177–2184, https://doi.org/10.1080/09593330.2012.665485.

[79] R. Kumar, A.K. Rai, D. Alamelu, S.K. Aggarwal, Monitoring of toxic elements present in sludge of industrial waste using CF-LIBS, Environ. Monit. Assess. 185 (2013) 171–180, https://doi.org/10.1007/s10661-012-2541-0.

[80] M. Boudhib, J. Hermann, C. Dutouquet, Compositional analysis of aerosols using calibration-free laser-induced breakdown spectroscopy, Anal. Chem. 88 (2016) https://doi.org/10.1021/acs.analchem.6b00329.

[81] A. Jabbar, M. Akhtar, S. Mehmood, N. Ahmed, Z.A. Umar, R. Ahmed, M.A. Baig, On the detection of heavy elements in the *Euphorbia indica* plant using laser-induced breakdown spectroscopy and laser ablation time of flight mass spectrometry, J. Anal. At. Spectrom 34 (2019) 954–962, https://doi.org/10.1039/C9JA00053D.

[82] R. Gaudiuso, M. Dell'Aglio, O. De Pascale, S. Loperfido, A. Mangone, A. De Giacomo, Laser-induced breakdown spectroscopy of archaeological findings with calibration-free inverse method: comparison with classical laser-induced breakdown spectroscopy and conventional techniques, Anal. Chim. Acta 813 (2014) 15–24, https://doi.org/10.1016/j.aca.2014.01.020.

[83] R. Gaudiuso, Calibration-free inverse method for depth-profile analysis with laser-induced breakdown spectroscopy, Spectrochim. Acta B At. Spectrosc. 123 (2016) 105–113, https://doi.org/10.1016/j.sab.2016.08.002.

[84] T. Takahashi, B. Thornton, K. Ohki, T. Sakka, Calibration-free analysis of immersed brass alloys using long-ns-duration pulse laser-induced breakdown spectroscopy with and without correction for nonstoichiometric ablation, Spectrochim. Acta B At. Spectrosc. 111 (2015) 8–14, https://doi.org/10.1016/j.sab.2015.06.009.

[85] A. Smaldone, A. De Bonis, A. Galasso, A. Guarnaccio, A. Santagata, R. Teghil, Inverse calibration free fs-LIBS of copper-based alloys, Z. Phys. Chem. (2016), https://doi.org/10.1515/zpch-2015-0704.

[86] S. Pagnotta, M. Lezzerini, B. Campanella, G. Gallello, E. Grifoni, S. Legnaioli, G. Lorenzetti, F. Poggialini, S. Raneri, A. Safi, V. Palleschi, Fast quantitative elemental mapping of highly inhomogeneous materials by micro-Laser-Induced Breakdown Spectroscopy, Spectrochim. Acta B At. Spectrosc. 146 (2018) 9–15, https://doi.org/10.1016/j.sab.2018.04.018.

[87] M. Dell'Aglio, A. De Giacomo, R. Gaudiuso, O. De Pascale, S. Longo, Laser Induced Breakdown Spectroscopy of meteorites as a probe of the early solar system, Spectrochim. Acta B At. Spectrosc. 101 (2014) 68–75, https://doi.org/10.1016/j.sab.2014.07.011.

[88] M. Horňáčková, J. Plavčan, J. Rakovský, V. Porubčan, D. Ozdín, P. Veis, Calibration-free laser induced breakdown spectroscopy as an alternative method for found meteorite fragments analysis, Eur. Phys. J. Appl. Phys. 66 (2014) 10702, https://doi.org/10.1051/epjap/2014130465.

[89] D. Ozdín, J. Plavčan, M. Horňáčková, P. Uher, V. Porubčan, P. Veis, J. Rakovský, J. Tóth, P. Konečný, J. Svoreň, Mineralogy, petrography, geochemistry, and classification of the Košice meteorite, Meteorit. Planet. Sci. 50 (2015) 864–879, https://doi.org/10.1111/maps.12405.

[90] G.S. Senesi, G. Tempesta, P. Manzari, G. Agrosì, An innovative approach to meteorite analysis by laser-induced breakdown spectroscopy, Geostand. Geoanal. Res. 40 (2016), https://doi.org/10.1111/ggr.12126.

[91] M. Dell'Aglio, M. López-Claros, J.J. Laserna, S. Longo, A. De Giacomo, Stand-off laser induced breakdown spectroscopy on meteorites: calibration-free approach, Spectrochim. Acta B At. Spectrosc. 147 (2018) 87–92, https://doi.org/10.1016/j.sab.2018.05.024.
[92] G. Tempesta, G.S. Senesi, P. Manzari, G. Agrosì, New insights on the Dronino iron meteorite by double-pulse micro-Laser-Induced Breakdown Spectroscopy, Spectrochim. Acta B At. Spectrosc. 144 (2018) 75–81, https://doi.org/10.1016/j.sab.2018.03.014.
[93] G. Agrosì, G. Tempesta, E. Scandale, S. Legnaioli, G. Lorenzetti, S. Pagnotta, V. Palleschi, A. Mangone, M. Lezzerini, Application of Laser Induced Breakdown Spectroscopy to the identification of emeralds from different synthetic processes, Spectrochim. Acta B At. Spectrosc. 102 (2014) 48–51, https://doi.org/10.1016/j.sab.2014.10.012.
[94] R. Grassi, E. Grifoni, S. Gufoni, S. Legnaioli, G. Lorenzetti, N. Macro, L. Menichetti, S. Pagnotta, F. Poggialini, C. Schiavo, V. Palleschi, Three-dimensional compositional mapping using double-pulse micro-laser-induced breakdown spectroscopy technique, Spectrochim. Acta B At. Spectrosc. 127 (2017), https://doi.org/10.1016/j.sab.2016.11.004.
[95] G. Tempesta, G. Agrosì, Standardless, minimally destructive chemical analysis of red beryls by means of Laser Induced Breakdown Spectroscopy, Eur. J. Mineral. 28 (2016) 571–580, https://doi.org/10.1127/ejm/2016/0028-2529.
[96] N. Ahmed, R. Ahmed, M.A. Baig, Analytical analysis of different karats of gold using laser induced breakdown spectroscopy (LIBS) and laser ablation time of flight mass spectrometer (LA-TOF-MS), Plasma Chem. Plasma Process. 38 (2018) 207–222, https://doi.org/10.1007/s11090-017-9862-2.
[97] M. Corsi, G. Cristoforetti, V. Palleschi, A. Salvetti, E. Tognoni, A fast and accurate method for the determination of precious alloys caratage by Laser Induced Plasma Spectroscopy, Eur. Phys. J. D. 13 (2001), https://doi.org/10.1007/s100530170255.

CHAPTER

25

Scope of future development in LIBS

Jagdish P. Singh[a,b], *Fang Y. Yueh*[a,b], *Chet R. Bhatt*[c]

[a]Institute for Clean Energy Technology (ICET), Mississippi State University, Starkville, MS, United States [b]JPS Advanced Technology R & D LLC, Starkville, MS, United States [c]National Energy Technology Laboratory, U.S. Department of Energy, Morgantown, WV, United States

1 Introduction

The development of LIBS has progressed rapidly during the last couple of decades. Various groups have worked on improving LIBS measurements of different samples by using advanced lasers, detection systems, and data processing techniques. A number of recent publications are evidence of the versatility of LIBS as a promising analytical tool being used in a wide range of applications. The availability of high-resolution spectrometers, gated cameras with significant gain factor, and ready-to-go data analysis software buttress the possibility of the commercial use of LIBS for in situ and online elemental analysis. In this chapter, we discuss future LIBS developments both for its current applications as well as for possible near-term and far-term applications.

Many existing LIBS applications, especially for quantitative analysis, have not reached the state of practical use due to insufficient accuracy and precision. This is mainly because of issues associated with spectral data collection and analysis. The current level of precision and accuracy, reported in the literature, is enough for a process monitor and control unit, or for use as an on-site/online screening tool in an industrial environment. However, for LIBS to be accepted as an analytical technique for quality control, its precision and accuracy need to be further improved. Besides the general problems of analytical figure of merit, different sample types also have different technical challenges. The directions of research work needed for effective handling of a variety of sample types, for developing theoretical models to understand various physical and chemical processes in LIBS, and for developing the requisite hardware are briefly summarized in the following sections. Although it is not possible to indicate the exact course of development in a rapidly expanding field of LIBS applications, we have tried to enumerate some of the areas on the basis of current trends.

Laser-Induced Breakdown Spectroscopy
https://doi.org/10.1016/B978-0-12-818829-3.00025-3

2 Gas-phase LIBS

Early LIBS development in the gas phase concentrated on using it as a multimetal continuous emissions monitor (CEM). LIBS technology has been evaluated and compared with other techniques, such as inductively coupled plasma (ICP) developed by the US Department of Energy (DOE) and the US Environmental Protection Agency (DOE/EPA), in several field tests at an EPA site. During these tests, the limit of detection (LOD) for some of the RCRA metals (such as Be, Cr, Pb, Cd) satisfied the EPA requirements, but it was not so for Hg and As. To meet EPA requirements for a CEM, LIBS research has to focus on improving calibration methods, sensitivity, and accuracy. Also, a reliable calibration method is needed to provide real-time, on-line LIBS analysis for off-gas emission measurements (see Chapter 8). LIBS sensitivity for some toxic metals such as Hg, As, Sb, and Ce also needs to be further improved to reach EPA requirements. The sensitivity of LIBS measurement can be improved either by increasing the signal or reducing the background noise from the plasma. Research that can significantly improve the LIBS signal-to-noise ratio is needed. The pulse-to-pulse variation in laser energy and effects of particle size can also cause large LIBS signal variation. To improve measurement precision, work is needed to reduce pulse-to-pulse fluctuations of the laser-induced spark. Data processing techniques to select data over a certain threshold and to apply statistical averaging methods to the collected data might be very helpful for further improvement of the LIBS analytical figure of merit. Moreover, multivariate analysis would also improve accuracy compared to univariate calibration. Some toxic metals have their most sensitive analytical lines at wavelengths shorter than 200 nm. However, the sensitivity of the detectors decreases tremendously in this spectral region and work to improve measurements in this spectral region will greatly enhance the sensitivity of detection for some elements.

3 Liquid-phase LIBS

There are many types of samples in the liquid phase that have been studied by LIBS [1–9] (see also Chapter 9). The most common liquid samples comprise toxic metal-contaminated water. Others include solid metal or ceramic pieces in water, and slurry samples that are a mixture of solids and water. Deep-sea exploration using LIBS has also been reported [10, 11]. Since atmospheric parameters like pressure and temperature might also affect LIBS spectra significantly, studies are performed to investigate these parameters [7, 12–14]. LIBS has also demonstrated its potential for the direct and rapid analysis of pharmaceutical liquid formulations.

For liquid samples, direct LIBS measurements can be performed within the liquid (underwater LIBS) or on the surface of the liquid. Because of the quenching effect, plasma volume is small and lifetime is shorter in liquid, which ultimately contributes to poor sensitivity. For surface measurements, the splashing of liquid due to laser pulse-induced shockwaves is a major problem. It can alter the localized plasma and measurement locations and hence precision and accuracy are compromised. In addition, splashing can damage the optics. To clear the quenching effect and splashing problem, measurements from the jet of liquid samples could be useful [15–17]. During measurements from the jet of liquids, plasma expands in air, which produces strong LIBS signals. Recent studies have shown that multipulse LIBS can improve its sensitivity in liquid measurements [18–25]. Use of multiple pulses with an appropriate pulse gap helps to make larger plasma volumes and increase plasma lifetime. Significant improvement in the LOD of various elements is reported [25, 26]. Further work in this area needs to be carried out to obtain the best configuration for multipulse LIBS. Signal enhancement with multipulse LIBS is needed for all liquid phase samples to improve the LOD. The other main research areas of interest are studies of the signal-to-noise enhancement for the laser spark produced with different liquid surfaces and bulk liquids with pressure and temperature effects taken into consideration, and techniques to measure concentrations at various depths. Further research work in handling different types of liquid samples is also needed. Improved LIBS experimental geometry should be able to increase the signal and reduce the background noise from the plasma. Future work should also focus on direct application areas rather than just performing benchtop measurements.

4 Solid-phase LIBS

Solid samples such as soil, glass, alloy, concrete, paint, etc. have been explored by LIBS [27–32] (see also Chapter 10). The main issue for solid samples is to achieve a high degree of quantitative precision in measurements. Although at present measurements on solid samples have better precision compared to liquid- and gas-phase samples, further improvements in instrumentation are needed to compete with other analytical instruments in the market. Lately, the use of short-pulse lasers (picosecond and femtosecond [s]) for LIBS measurements have been demonstrated [33–36]. Fs lasers can ablate the sample more effectively and cause less damage to the target compared to that using nanosecond (ns-)LIBS. Fs-LIBS also produces less continuum background as compared to ns-LIBS leading to higher sensitivity. Further developments in fs-LIBS systems might be able to greatly improve their analytical capabilities. Hybrid techniques, combining LIBS with other discharge excitation sources, provide alternative ways to improve the analytical capabilities. Matrix effects have been known for years to affect composition measurements with LIBS and this area also needs further investigation. Generally, solid materials have been analyzed by surface ablation assuming or making the sample homogeneous. LIBS mechanisms can be developed to analyze the samples of varying elemental content at different layers, and the capability of elemental concentration measurements with depth will help in depth profiling.

5 LIBS of molten samples

Applications of LIBS as an online process monitor for molten materials such as alloys, steel, and glass have been demonstrated [37–40] (see also Chapter 10). The major focus in this field is improving precision and accuracy to establish LIBS as an online compositional analysis tool for process control. Improvements are also needed in the design of the LIBS probe for long-term operation under high-temperature environments. The design needs to be perfected so that the optics in the probe remain clean in the high-temperature furnace and the laser-produced spark is at the melt surface, not in the metal vapor above the surface. Moreover, with an aid of appropriate lasing material and optics for radiation environments, LIBS could be useful for online monitoring in nuclear reactors. Online calibration techniques for molten samples also need to be evaluated and tested for long-term operation.

6 LIBS for biomedical applications

LIBS is increasingly being used in the analysis of clinical samples like gall stones and kidney stones for trace elemental mapping and detection. The results of quantitative analysis of trace elements in gall stones has been found to be in agreement with inductively coupled plasma atomic emission spectroscopy measurements, and single-shot LIBS spectra of the cross-section reveal the variation of the constituents from the center to the surface [41, 42]. In the case of kidney stones, quantitative measurements of Mg, Cu, Zn, and Sr were carried out and it was concluded, on the basis of their distribution from the center to the surface, that Mg and Cu play important roles in nucleation [43]. A healthy tooth tissue has two components: the hardest material on the surface, enamel, and dentine, which lies under the enamel, and in the case of caries infection the hard enamel becomes more porous. LIBS can discriminate the healthy tooth material with high spatial resolution. Samek et al. [44] used chemometric analysis to distinguish between the caries and the healthy regions of tooth samples and have indicated the potential of LIBS as a useful tool for in vivo caries identification with a spatial resolution of 100–200 μm and a depth resolution of about 10 μm. Singh and Rai [45] observed a decrease in the concentration of Ti and an increase in the concentration of Cu in caries-affected tissues relative to healthy dental tissue. It is to be noted that Ti is a common element in dental implant material or it may come from TiO_2, which is a common additive to toothpaste; Cu may be absorbed during normal eating and drinking. Elemental analysis by LIBS of fingernail samples has led to classification of normal, alcoholic, and doping subjects [46] and to discriminate healthy subjects from those with an illness of the thyroid [47]. Singh and coworkers [48] used LIBS to distinguish normal and malignant cells in histological sections of tissue samples by observing higher concentrations of trace elements Na, Mg, Ca, and lower concentration of Cu in malignant cells relative to normal cells. In a later publication, they applied chemometric analysis to the LIBS data to identify biological tissue from different organs of chickens [49] and reached the conclusion that more work was needed to produce robust classification models. Rehse et al. [50] reviewed the applications of LIBS in the characterization and identification of various biological, biomedical, and clinical samples to explore the possibilities of their implementation for the benefit of humankind. Recently, Singh et al. [51] reviewed LIBS applications in the identification of microbes causing infectious diseases. It has been pointed out that chemometrics would play an increasingly important role in the future, and the convergence of LIBS, machine learning, and nanotechnology is likely to provide new diagnostic tools for disease detection and cure.

7 A theoretical model of laser-induced plasma

Laser-induced plasma is a product of the very complicated process of laser-matter interaction. When a short-pulsed, high-peak power laser beam is focused onto any target, the processes of absorption of laser energy, vaporization, ejection of atoms, ions, molecular species, and fragments take place in quick succession. Plasma initiation is followed by its expansion, production of shockwaves, and many other processes. Many theoretical models have been developed to describe these processes but they are usually applicable under idealized conditions [52–54]. A model to completely describe the processes of laser ablation and plasma formation does not exist at present and the development of such a model will be a great help in selecting the optimum experimental parameters and configurations for quantitative LIBS measurement in various environments. The power density of the focused laser beam and the thermo-optical properties of the material are two critical parameters that influence the laser ablation process, and a model that could predict the ablation processes for different materials will be very handy in determining the LIBS operation condition for different applications. Two (or more) pulse LIBS excitations have shown great promise for investigations of liquid and solid samples [19, 26], and theoretical work in this area is needed to understand the mechanisms involved in multipulse LIBS.

8 Commercialization of LIBS

LIBS has great potential for field applications for both the military and industry because of its in situ and online detection capabilities even in hazardous and hard-to-reach places.

Success of commercialization of LIBS is mainly determined by its advantages compared to other techniques and its overall cost. Its advantages include its analytical performance, applicability for different test materials in various circumstances, timing of results, and operational cost. In addition, its portability for field use is also a key factor.

Some commercial LIBS systems have found their way into the market for specific applications. J200 from Applied Spectra [55], ChemReveal and ChemLogix from TSI [56], and Z-Series from SciAps [57] are a few examples of analyzing units based on the LIBS technique available in the market.

The commercialization of a LIBS instrument for general composition analysis still needs a lot of research and development. Research efforts should focus not only on the analytical figure of merit, but also on methods to minimize the components for a compact, lightweight, portable system. A cheap LIBS detection system that could simultaneously measure all the elements of interest is yet to be developed. A telescope-type optical design is also needed for standoff detection. Developments in the field of micro-LIBS can produce future affordable and portable LIBS systems for industrial and military applications [58, 59]. Reported literature has shown the possibility of LIBS in different applications, and now the time has come to develop prototypes for different applications. Specific environmental conditions should be considered in these kinds of prototypes for their field use.

9 Future applications

LIBS has found applications in many areas of exploration related to scientific, industrial, medical, environmental, security, and social problems of current interest [60–68]. Possible directions of research and development that need further attention are briefly summarized here:

- Ambient air particle analysis

Fine particulate matter in air has posed serious risks to human health. LIBS has great potential in the real-time analysis of ambient air particles. Gallou et al. [69, 70] measured Cu particles in the size range 1–7 μm in the laboratory, which is in good agreement with online measurements to analyze stack emissions during Cu melting in a foundry. These results may be transposable to particles composed of other heavy metals like Cr, Ni, Pb, etc. (see also Chapter 22).

- Continuous emission monitor

Monitoring of the toxic metal emissions from coal-fired power plants/incinerators/industrial processes by LIBS needs further refinement.

- Detection of toxic elements in water

Groundwater contaminated by metals due to increased acid rain, livestock, and pesticides has become a health issue. LIBS can monitor the toxic elements in effluents from water treatment plants or groundwater. Zhao et al. [71] developed an automatic monitor based on LIBS technology for the detection of heavy metals in industrial wastewater, and LODs for Cr, Ni, Cu, Zn, Cd, and Pb were found to be several ppb. Ponce et al. [72] used a portable LIBS instrument for monitoring copper and lead in edible fish with LODs of 25 ppm for Pb and 100 ppm for Cu that are below the maximum tolerable levels of some international standards.

- Soil analysis

LIBS has been successfully used for detecting toxic metals in contaminated soil and in the measurement of total carbon content in soils. Guo et al. [73] used univariate and multivariate regression methods for analyzing LIBS data for the detection of Si, Al, Mg, Ca, Na, K, Mn, Ba, Ti, Cr, Cu, Sr, and P in different standard soils. They concluded that it would be meaningful to propose a quantitative reference among univariate linear regression, multivariate linear regression, and multivariate nonlinear regression methods for the multielemental analysis of soil samples under complex matrix conditions. He et al. [74] carried out a comparative study of the ability of single-pulse-LIBS and collinear double-pulse-LIBS for quantitative determination of soil nutrient elements. It was found that collinear double-pulse signals had a higher spectral intensity and better signal stability than single-pulse signals. Chemometrics models of collinear double-pulse signals showed better accuracy and reliability than those of SP signals (see also Chapter 16).

- Slurry sample analysis

Analysis of slurry samples in various industrial processes or waste processes can be very efficiently performed using LIBS. Cheng et al. [75] carried out the challenging on-stream LIBS analysis of iron ore slurry by developing a slurry circulation system. The effects of slurry flow rate on LIBS spectra were investigated to achieve an optimal detecting surface for better reproducibility and the results indicate that the method is effective for on-stream slurry analysis. Tantalum and niobium are two very important rare metals and a very rapid and sensitive LIBS method for quantitative detection of trace Ta and Nb elements in Ta-Nb ore has been developed by Liu and Hao [76] (see also Chapter 11).

- Powder sample composition

LIBS applications for the online analysis of powder samples (e.g., glass batch, fly ash, ash added to cement, etc.) in various industrial and waste management processes have been demonstrated.

- Radioactive analysis

The remote sensing capability of LIBS is very attractive for the detection and measurement of radioactive nuclei in different radioactive samples. It can be used to monitor radioactive glasses to store nuclear waste. Austenitic stainless-steel 304L is widely used in the nuclear waste storage and reprocessing operations. Polished surfaces of stainless-steel 304L contaminated with Sr and Cs have been analyzed by Lang et al. [77] in a very delicate experiment, in view of the fact that the stainless-steel matrix comprises a large number of alloying elements, thus yielding a complex LIBS spectrum. They have found that LIBS offers a reliable standoff analysis for Sr contamination in nuclear plants.

- Molten material

Online analysis of an alloy composition in the molten state containing Al, Fe, Mn, Cu, etc. has been demonstrated. This provides a process monitor for these molten metal industries.

- Military applications

Military applications include mine/unexploded ordnance detection, detection of chemical warfare agents, depleted uranium in aerosols/particulates, etc.

- Space exploration

The study of LIBS for future space exploration was initiated in 1992. Recent research is focused on developing standoff LIBS for application to Mars exploration. LIBS (ChemCam) was used in the Mars Curiosity rover in 2012 and will again be coupled in SuperCam in 2020.

- Phytoremediation analysis

Plant leaves can be analyzed on site to study the metals absorbed by the biomass. By combining LIBS data with other optical spectroscopy analytical techniques, one could obtain optimum control of the phytoremediation process. Euphorbia belongs to a large family of flowering plants having a poisonous, milky, white, latex-like sap. Jabbar et al. [78] used the LIBS technique to ascertain the presence of chromium in different parts of the euphorbia plant. The detection limit of the trace element was improved by applying an electric field across the laser-induced plasma plume to enhance the spectral intensity.

• Biomedical applications

LIBS has been successfully used for the analysis of trace elements in nail, hair, blood, urine, and dental samples. It has been used to distinguish between normal and cancerous tissue based on the difference in elemental concentration.

• Pharmaceutical applications

LIBS has been applied for the direct and rapid analysis of drugs and lubricants in tablets and saline solutions for the pharmaceutical industry.

• Food safety

LIBS can be applied to determine toxic metal levels in processed as well as unprocessed food. The analytical technique has many advantages as discussed earlier, but challenges remain in terms of matrix effects, instrument development, and model calibration.

• LIBS imaging

Spectral imaging or chemical imaging has emerged as a powerful tool to obtain the molecular mapping of material samples, including biological samples employing the techniques of Raman spectroscopy, fluorescence spectroscopy, and photoacoustic spectroscopy. The LIBS technique for elemental mapping of samples has evidenced a great upsurge in recent years in terms of the published literature. The annual number of research publications involving LIBS imaging were in single digits until 2015, reaching 20 in 2017, and have been rapidly increasing since then. Caceres et al. [79] reported for the first time the possibility of producing megapixel-sized images of large samples. In the future we hope there will be a variety of applications of LIBS imaging in the investigations of heterogeneous materials in the fields of biology, geology, and other industries [80–82].

• Nanoparticle-enhanced LIBS (NELIBS)

In recent times it has been found that the deposition of metallic nanoparticles on the sample surface before laser irradiation enhances the emission signal during LIBS. The mechanism of this large enhancement is related to a faster and more efficient production of electrons in the process of laser ablation. An optimum surface concentration of nanoparticles is needed for application of this method in the investigation of metals, transparent samples, and biological fluids [83, 84]. Wang et al. [85] demonstrated the effect of dielectric nanoparticles on fs-LIBS to illustrate the hydrodynamics of fs laser-induced plasma on a silica colloidal particle lens array-deposited substrate. Amyloid fibrils are systems for applications in water purification and medicine and are used to investigate the performance of laser-matter interaction in biological systems. Dell'Aglio et al. [86] employed LIBS for quantitative metal detection using amyloid fibrils coated with gold particles as enhancers to show the potential of NELIBS for quantification at the sub-ppm level of metallic elements (Cr, Pb, Tl, and Cd) even in the protein and biological environment. The characteristic features of NELIBS plasma have been found to be larger emission volume and longer persistence than LIBS plasma and the technique is likely to have many novel applications in the future.

The major difficulties common to most LIBS applications are:

1. Problems in achieving very high precision and accuracy in quantitative analysis.
2. It is too expensive for many applications due to the cost of the laser and the sensitive detection systems.

3. Matrix effects cause difficulty in obtaining suitable standard spectral lines for LIBS measurements.
Different elements have different optimal detection windows. It is hard to simultaneously obtain an optimum detection window for all the elements of interest.

To overcome the known problems and advance LIBS for future applications, we need to improve our knowledge of the technology through basic research. The fundamentals of LIBS plasma, improvement in LIBS quantification, increase in the understanding of pulse-to-pulse fluctuations, and study of laser-matter interactions are some of the broad areas for future research. Instrumentation development that can provide good broadband detector sensitivity and fast response, improved overall signal-to-noise ratio, and cheap rugged laser sources are needed for future LIBS systems. For specific applications (e.g., biomedical applications) the study of specific preparation procedures is needed to achieve the usefulness of LIBS data.

Since the first use of laser-induced plasma as an ablation source in 1962, LIBS has made significant progress toward a mature analytical tool. Various LIBS applications in solid, liquid, and gas samples have been explored by many research groups. Future applications for LIBS should focus on its real strengths of online, real-time, nonintrusive, and simultaneous multielement analysis. With rapid technology advances, more and more new LIBS applications will appear in the future. We should also see more commercial LIBS instruments for different applications in the market in the near future.

References

[1] D.A. Cremers, L.J. Radziemski, T.R. Loree, Appl. Spectrosc. 38 (1984) 721–729.

[2] O. Samek, D.C. Beddows, J. Kaiser, S. Kukhlevsky, M. Liška, H.H. Telle, J. Young, Opt. Eng. 39 (2000) 2248–2262.

[3] J.R. Wachter, D.A. Cremers, Appl. Spectrosc. 41 (1987) 1042–1048.

[4] L. Barrette, S. Turmel, Spectrochim. Acta B At. Spectrosc. 56 (2001) 715–723.

[5] L. St-Onge, E. Kwong, M. Sabsabi, E.B. Vadas, J. Pharm. Biomed. Anal. 36 (2004) 277–284.

[6] V. Lazic, S. Jovićević, Spectrochim. Acta B At. Spectrosc. 101 (2014) 288–311.

[7] C.R. Bhatt, J.C. Jain, H.M. Edenborn, D.L. McIntyre, Talanta 205 (2019) 120170.

[8] S.M. Angel, J. Bonvallet, M. Lawrence-Snyder, W.F. Pearman, J. Register, J. Anal. At. Spectrom 31 (2016) 328–336.

[9] C.R. Bhatt, J.C. Jain, C.L. Goueguel, D.L. McIntyre, J.P. Singh, Spectrochim. Acta B At. Spectrosc. 137 (2017) 8–12.

[10] A. De Giacomo, M. Dell'Aglio, F. Colao, R. Fantoni, Spectrochim. Acta B At. Spectrosc. 59 (2004) 1431–1438.

[11] A.P. Michel, M.J. Lawrence-Snyder, S.M. Angel, A.D. Chave, Oceanic applications of laser induced breakdown spectroscopy: laboratory validation, in: Proceedings of OCEANS 2005 MTS/IEEE, IEEE, 2005, pp. 741–747.

[12] C.R. Bhatt, J.C. Jain, D.L. McIntyre, Spectrochim. Acta B At. Spectrosc. 149 (2018) 42–47.

[13] C.L. Goueguel, J.C. Jain, D.L. McIntyre, C.G. Carson, H.M. Edenborn, J. Anal. At. Spectrom 31 (2016) 1374–1380.

[14] C. Hanson, S. Phongikaroon, J.R. Scott, Spectrochim. Acta B At. Spectrosc. 97 (2014) 79–85.

[15] D.-H. Lee, S.-C. Han, T.-H. Kim, J.-I. Yun, Anal. Chem. 83 (2011) 9456–9461.

[16] X. Fang, S.R. Ahmad, Appl. Spectrosc. 61 (2007) 1021–1024.

[17] A. Kumar, F.Y. Yueh, J.P. Singh, Appl. Optics 42 (2003) 6047–6051.

[18] J. Uebbing, J. Brust, W. Sdorra, F. Leis, K. Niemax, Appl. Spectrosc. 45 (1991) 1419–1423.

[19] R. Sattmann, V. Sturm, R. Noll, J. Phys. D Appl. Phys. 28 (1995) 2181.

[20] D.N. Stratis, K.L. Eland, S.M. Angel, Appl. Spectrosc. 55 (2001) 1297–1303.

[21] L. St-Onge, M. Sabsabi, P. Cielo, Spectrochim. Acta B At. Spectrosc. 53 (1998) 407–415.

[22] L. St-Onge, V. Detalle, M. Sabsabi, Spectrochim. Acta B At. Spectrosc. 57 (2002) 121–135.

[23] Y. Yu, W. Zhou, X. Su, Opt. Commun. 333 (2014) 62–66.

[24] G. Cristoforetti, M. Tiberi, A. Simonelli, P. Marsili, F. Giammanco, Appl. Optics 51 (2012) B30–B41.

[25] K. Rifai, S. Laville, F. Vidal, M. Sabsabi, M. Chaker, J. Anal. At. Spectrom 27 (2012) 276–283.

[26] A. De Giacomo, M. Dell'Aglio, F. Colao, R. Fantoni, V. Lazic, Appl. Surf. Sci. 247 (2005) 157–162.

[27] M. Hemmerlin, R. Meilland, H. Falk, P. Wintjens, L. Paulard, Spectrochim. Acta B At. Spectrosc. 56 (2001) 661–669.

[28] G.A. Theriault, S. Bodensteiner, S.H. Lieberman, Field Anal. Chem. Technol. 2 (1998) 117–125.

[29] T. Thiem, R. Salter, J. Gardner, Y. Lee, J. Sneddon, Appl. Spectrosc. 48 (1994) 58–64.

[30] S. Rosenwasser, G. Asimellis, B. Bromley, R. Hazlett, J. Martin, T. Pearce, A. Zigler, Spectrochim. Acta B At. Spectrosc. 56 (2001) 707–714.

[31] J.A. Millard, R.H. Dalling, L.J. Radziemski, Appl. Spectrosc. 40 (1986) 491–494.

[32] R. Wisbrun, I. Schechter, R. Niessner, H. Schroeder, K. L. Kompa, Anal. Chem. 66 (1994) 2964–2975.

[33] P. Rohwetter, J. Yu, G. Méjean, K. Stelmaszczyk, E. Salmon, J. Kasparian, J.-P. Wolf, L. Wöste, J. Anal. At. Spectrom 19 (2004) 437–444.

[34] B. Le Drogoff, F. Vidal, Y. Von Kaenel, M. Chaker, T. Johnston, S. Laville, M. Sabsabi, J. Margot, J. Appl. Phys. 89 (2001) 8247–8252.

[35] B. Le Drogoff, J. Margot, M. Chaker, M. Sabsabi, O. Barthelemy, T. Johnston, S. Laville, F. Vidal, Y. Von Kaenel, Spectrochim. Acta B At. Spectrosc. 56 (2001) 987–1002.

[36] K.L. Eland, D.N. Stratis, T. Lai, M.A. Berg, S.R. Goode, S. M. Angel, Appl. Spectrosc. 55 (2001) 279–285.

[37] L. Peter, V. Sturm, R. Noll, Appl. Optics 42 (2003) 6199–6204.

[38] C. Aragon, J. Aguilera, J. Campos, Appl. Spectrosc. 47 (1993) 606–608.

[39] A.K. Rai, F.-Y. Yueh, J.P. Singh, Appl. Optics 42 (2003) 2078–2084.

[40] J.-I. Yun, R. Klenze, J.-I. Kim, Appl. Spectrosc. 56 (2002) 852–858.

[41] V.K. Singh, V. Singh, A.K. Rai, S.N. Thakur, P.K. Rai, J.P. Singh, Appl. Optics 47 (2008) G38–G47.

[42] Z. Gazali, S. Thakur, A. Rai, Opt. Laser Technol. 111 (2019) 696–700.

[43] V.K. Singh, A. Rai, P. Rai, P. Jindal, Lasers Med. Sci. 24 (2009) 749–759.

[44] O. Samek, H.H. Telle, D.C. Beddows, BMC Oral Health, 1 (2001) 1.

[45] V.K. Singh, A.K. Rai, Lasers Med. Sci. 26 (2011) 307–315.

[46] Z. Hosseinimakarem, S.H. Tavassoli, J. Biomed. Opt. 16 (2011) 057002.

[47] M. Bahreini, S.H. Tavassoli, J. Lasers Med. Sci. 3 (2012) 127.

[48] A. Kumar, F.-Y. Yueh, J.P. Singh, S. Burgess, Appl. Optics 43 (2004) 5399–5403.

[49] F.-Y. Yueh, H. Zheng, J.P. Singh, S. Burgess, Spectrochim. Acta B At. Spectrosc. 64 (2009) 1059–1067.

[50] S. Rehse, H. Salimnia, A. Miziolek, J. Med. Eng. Technol. 36 (2012) 77–89.

[51] V.K. Singh, J. Sharma, A.K. Pathak, C.T. Ghany, M. Gondal, Biophys. Rev. 10 (2018) 1221–1239.

[52] R.E. Russo, X. Mao, S.S. Mao, Peer Reviewed: The Physics of Laser Ablation in Microchemical Analysis, ACS Publications, 2002.

[53] S.I. Anisimov, V.A. Khokhlov, Instabilities in Laser-Matter Interaction, CRC Press, 1995.

[54] R. Krasniker, V. Bulatov, I. Schechter, Spectrochim. Acta B At. Spectrosc. 56 (2001) 609–618.

[55] AppliedSpectra, J200 LIBS Instrument, n.d., https://appliedspectra.com/products/j200-libs-instrument.html.

[56] TSI, n.d. Elemental Analyzers LIBS, https://tsi.com/solutions-fr/solutions-par-type-d-applications/analyse-elementaire.

[57] SciAps, n.d. LIBS Analyzer, https://www.sciaps.com/libs-handheld-laser-analyzers/z-50/?gclid=CjwKCAjw95D0BRBFEiwAcO1KDOSupW22dqP90wF4Py5quz7P7zhQihlBCwR3MzS7UNLgLLSlHY9MhBoCqIwQAvD_BwE.

[58] D. Menut, P. Fichet, J.-L. Lacour, A. Rivoallan, P. Mauchien, Appl. Optics 42 (2003) 6063–6071.

[59] J.J. Zayhowski, J. Alloys Compd. 303 (2000) 393–400.

[60] J.-F. Archambault, A. Vintiloìu, E. Kwong, AAPS PharmSciTech 6 (2005) E253–E261.

[61] L. St-Onge, J.-F. Archambault, E. Kwong, M. Sabsabi, E.B. Vadas, J. Pharm. Pharm. Sci. 8 (2005) 272–288.

[62] A.K. Knight, N.L. Scherbarth, D.A. Cremers, M.J. Ferris, Appl. Spectrosc. 54 (2000) 331–340.

[63] B. Sallé, J.-L. Lacour, E. Vors, P. Fichet, S. Maurice, D.A. Cremers, R.C. Wiens, Spectrochim. Acta B At. Spectrosc. 59 (2004) 1413–1422.

[64] S. Morel, N. Leone, P. Adam, J. Amouroux, Appl. Optics 42 (2003) 6184–6191.

[65] A.C. Samuels, F.C. DeLucia, K.L. McNesby, A.W. Miziolek, Appl. Optics 42 (2003) 6205–6209.

[66] M. Baudelet, L. Guyon, J. Yu, J.-P. Wolf, T. Amodeo, E. Fréjafon, P. Laloi, J. Appl. Phys. 99 (2006) 084701.

[67] C. Bohling, D. Scheel, K. Hohmann, W. Schade, M. Reuter, G. Holl, Appl. Optics 45 (2006) 3817–3825.

[68] D.R. Alexander, Merrill Series on The Research Mission of Public Universities, (2002) pp. 57–60. https://journals.ku.edu/merrill/issue/view/1087.

[69] G. Gallou, J. Sirven, C. Dutouquet, O.L. Bihan, E. Frejafon, Aerosol Sci. Tech. 45 (2011) 918–926.

[70] C. Dutouquet, G. Gallou, O. Le Bihan, J. Sirven, A. Dermigny, B. Torralba, E. Frejafon, Talanta 127 (2014) 75–81.

[71] N. Zhao, D. Meng, Y. Jia, M. Ma, L. Fang, J. Liu, W. Liu, Opt. Express 27 (2019) A495–A506.

[72] L. Ponce, T. Flores, M. Sosa-Saldaña, F.C. Alvira, G. M. Bilmes, Appl. Optics 55 (2016) 254–258.

[73] G. Guo, G. Niu, Q. Shi, Q. Lin, D. Tian, Y. Duan, Anal. Methods 11 (2019) 3006–3013.

[74] Y. He, X. Liu, Y. Lv, F. Liu, J. Peng, T. Shen, Y. Zhao, Y. Tang, S. Luo, Sensors 18 (2018) 1526.
[75] X. Cheng, X. Yang, Z. Zhu, L. Guo, X. Li, Y. Lu, X. Zeng, Appl. Optics 56 (2017) 9144–9149.
[76] L. Liu, Z. Hao, Appl. Optics 58 (2019) 461–465.
[77] A. Lang, D. Engelberg, N.T. Smith, D. Trivedi, O. Horsfall, A. Banford, P.A. Martin, P. Coffey, W.R. Bower, C. Walther, J. Hazard. Mater. 345 (2018) 114–122.
[78] A. Jabbar, M. Akhtar, S. Mehmood, N. Ahmed, Z. A. Umar, R. Ahmed, M.A. Baig, J. Anal. At. Spectrom 34 (2019) 954–962.
[79] J. Cáceres, F. Pelascini, V. Motto-Ros, S. Moncayo, F. Trichard, G. Panczer, A. Marín-Roldán, J. Cruz, I. Coronado, J. Martín-Chivelet, Sci. Rep. 7 (2017) 5080.
[80] L. Sancey, V. Motto-Ros, B. Busser, S. Kotb, J. Benoit, A. Piednoir, F. Lux, O. Tillement, G. Panczer, J. Yu, Sci. Rep. 4 (2014) 6065.
[81] H. Bette, R. Noll, J. Phys. D Appl. Phys. 37 (2004) 1281.
[82] F. Boué-Bigne, Spectrochim. Acta B At. Spectrosc. 63 (2008) 1122–1129.
[83] M. Dell'Aglio, R. Alrifai, A. De Giacomo, Spectrochim. Acta B At. Spectrosc. 148 (2018) 105–112.
[84] A. De Giacomo, M. Dell'Aglio, R. Gaudiuso, C. Koral, G. Valenza, J. Anal. At. Spectrom 31 (2016) 1566–1573.
[85] M. Wang, L. Jiang, S. Wang, Q. Guo, F. Tian, Z. Chu, J. Zhang, X. Li, Y. Lu, Anal. Chem. 91 (2019) 9952–9961.
[86] M. Dell'Aglio, Z. Salajková, A. Mallardi, R. Mezzenga, L. Van't Hag, N. Cioffi, G. Palazzo, A. De Giacomo, Spectrochim. Acta B At. Spectrosc. 155 (2019) 115–122.

Index

Note: Page numbers followed by *f* indicate figures and *t* indicate tables.

D

N

O

9780128188293